Lecture Notes in Computer

Commenced Publication in 1973
Founding and Former Series Editors:
Gerhard Goos, Juris Hartmanis, and Jan van Leeuwen

Otfried Cheong
Kyung-Yong Chwa
Kunsoo Park (Eds.)

Algorithms
and Computation

21st International Symposium, ISAAC 2010
Jeju Island, Korea, December 15-17, 2010
Proceedings, Part II

 Springer

Volume Editors

Otfried Cheong
KAIST
Department of Computer Science
Daejeon 305-701, Korea
E-mail: otfried@kaist.edu

Kyung-Yong Chwa
KAIST
Department of Computer Science
Daejeon 305-701, Korea
E-mail: kychwa@jupiter.kaist.ac.kr

Kunsoo Park
Seoul National University
School of Computer Science and Engineering
Seoul 151-742, Korea
E-mail: kpark@snu.ac.kr

Library of Congress Control Number: 2010939852

CR Subject Classification (1998): F.2, I.3.5, E.1, C.2, G.2, F.1

LNCS Sublibrary: SL 1 – Theoretical Computer Science and General Issues

ISSN   0302-9743
ISBN-10   3-642-17513-9 Springer Berlin Heidelberg New York
ISBN-13   978-3-642-17513-8 Springer Berlin Heidelberg New York

springer.com

© Springer-Verlag Berlin Heidelberg 2010
Printed in Germany

Typesetting: Camera-ready by author, data conversion by Scientific Publishing Services, Chennai, India
Printed on acid-free paper   06/3180

Preface

This volume contains the proceedings of the 21st Annual International Symposium on Algorithms and Computations (ISAAC 2010), held in Jeju, Korea during December 15–17, 2010. Past editions have been held in Tokyo, Taipei, Nagoya, Hong Kong, Beijing, Cairns, Osaka, Singapore, Taejon, Chennai, Taipei, Christchurch, Vancouver, Kyoto, Hong Kong, Hainan, Kolkata, Sendai, Gold Coast, and Hawaii over the years 1990-2009.

ISAAC is an annual international symposium that covers the very wide range of topics in algorithms and computation. The main purpose of the symposium is to provide a forum for researchers working in algorithms and the theory of computation where they can exchange ideas in this active research community.

In response to the call for papers, ISAAC 2010 received 182 papers. Each submission was reviewed by at least three Program Committee members with the assistance of external referees. Since there were many high-quality papers, the Program Committee's task was extremely difficult. Through an extensive discussion, the Program Committee accepted 77 of the submissions to be presented at the conference. Two special issues, one of *Algorithmica* and one of the *International Journal of Computational Geometry and Applications*, were prepared with selected papers from ISAAC 2010.

The best paper award was given to "From Holant to #CSP and Back: Dichotomy for Holant^c Problems" by Jin-Yi Cai, Sangxia Huang and Pinyan Lu, and the best student paper award to "Satisfiability with Index Dependency" by Hongyu Liang and Jing He. Two eminent invited speakers, David Eppstein from University of California, Irvine, and Matt Franklin from University of California, Davis, also contributed to this volume.

We would like to thank all Program Committee members and external referees for their excellent work, especially given the demanding time constraints; they gave the conference its distinctive character. We thank all who submitted papers for consideration; they all contributed to the high quality of the conference. We also thank the Organizing Committee members for their dedicated contribution that made the conference possible and enjoyable. Finally, we thank our sponsor SIGTCS (Special Interest Group on the Theoretical Computer Science) of KIISE (The Korean Institute of Information Scientists and Engineers) for the assistance and support.

December 2010
Otfried Cheong
Kyung-Yong Chwa
Kunsoo Park

Organization

Program Chairs

Otfried Cheong	KAIST, Korea
Kyung-Yong Chwa	KAIST, Korea
Kunsoo Park	Seoul National University, Korea

Program Committee

Lars Arge	University of Aarhus, Denmark
Takao Asano	Chuo University, Japan
Danny Chen	University of Notre Dame, USA
Rudolf Fleischer	Fudan University, China
Satoshi Fujita	Hiroshima University, Japan
Mordecai Golin	Hong Kong UST, Hong Kong
Seok-Hee Hong	University of Sydney, Australia
Oscar Ibarra	University of California - Santa Barbara, USA
Giuseppe Italiano	University of Rome "Tor Vergata", Italy
Tao Jiang	UC Riverside, USA
Mihyun Kang	Humboldt University Berlin, Germany
Ming-Yang Kao	Northwestern University, USA
Tak-wah Lam	University of Hong Kong, Hong Kong
Gad Landau	University of Haifa, Israel
Peter Bro Miltersen	Aarhus University, Denmark
David Mount	University of Maryland, USA
Ian Munro	University of Waterloo, Canada
Yoshio Okamoto	Tokyo Institute of Technology, Japan
Frank Ruskey	University of Victoria, Canada
Kunihiko Sadakane	National Institute of Informatics, Japan
Steven Skiena	Stony Brook University, USA
Takeshi Tokuyama	Tohoku University, Japan
Ryuhei Uehara	Japan Advanced Institute of Science and Technology, Japan
Peter Widmayer	ETH Zurich, Switzerland
Chee Yap	Courant, NYU, USA
Hsu-Chun Yen	National Taiwan University, Taiwan
Afra Zomorodian	Dartmouth College, USA

Host Institute

KAIST (Korea Advanced Institute of Science and Technology)

Organizing Committee

Joon-Soo Choi	Kookmin University, Korea
Kyung-Yong Chwa	KAIST, Korea
Seung Bum Jo	KAIST, Korea
Hyunseob Lee	KAIST, Korea
Jung-Heum Park	The Catholic University of Korea, Korea

Referees

Mohammad Abam	Arash Farzan	Jiongxin Jin
Peyman Afshani	Sandor Fekete	Shinhaeng Jo
Hee-Kap Ahn	Andreas Feldmann	Daniel Johannsen
Laila El Aimani	Holger Flier	Naoyuki Kamiyama
Toru Araki	Mathew Francis	Tom Kamphans
Abdullah Arslan	Robert Fraser	Iyad Kanj
Ilia Averbouch	Sorelle Friedler	Bruce Kapron
Laszlo Babai	Hiroshi Fujiwara	Akinori Kawachi
Christian Bachmaier	Oliver Gableske	Daniel Keren
Jeremy Barbay	Joachim von zur Gathen	Shuji Kijima
Amir Barghi	Petr Golovach	Hyo-Sil Kim
Peter van Beek	Martin Golumbic	Jae-Hoon Kim
Renévan Bevern	Joachim Gudmundsson	Masashi Kiyomi
Binay Bhattacharya	Prosenjit Gupta	Jan Willem Klop
Danny Breslauer	Gregory Gutin	Koji Kobayashi
Gerth Stølting Brodal	Carsten Gutwenger	Darek Kowalski
Joshua Brody	Michel Habib	Kasper Dalgaard Larsen
Jonathan Buss	Kristoffer Arnsfelt	Hyunseob Lee
Ho-Leung Chan	Hansen	Lap-Kei Lee
Kun-Mao Chao	Sariel Har-Peled	Mira Lee
Ho-Lin Chen	Masud Hasan	Taehyung Lee
Ming-Yang Chen	Mohammad Khairul	Avivit Levy
Minkyoung Cho	Hasan	Chun-Cheng Lin
Marek Chrobak	Meng He	Maarten Loeffler
David Cohen	Pinar Heggernes	Daniel Lokshtanov
Amin Coja-Oghlan	Danny Hermelin	Alejandro Lopez-Ortiz
Maxime Crochemore	Tomio Hirata	Vadim Lozin
Pooya Davoodi	Christian Hoffmann	Jun Luo
Yann Disser	Ming-Deh Huang	Thomas Mølhave
Reza Dorrigiv	John Iacono	Khalegh Mamakani
David Doty	Keiko Imai	Maurice Margenstern
Scot Drysdale	Toshimasa Ishii	Dimitri Marinakis
Vida Dujmovic	Takehiro Ito	Tomomi Matsui
Robert Elsaesser	Kazuo Iwama	Yusuke Matsumoto
Guy Even	Li Jian	Yuichiro Miyamoto

Matthias Mnich
Morteza Monemizadeh
Petra Mutzel
Hiroshi Nagamochi
Shin-Ichi Nakano
C. Thach Nguyen
Patrick Nicholson
Takao Nishizeki
Martin Noellenburg
Katsuyuki Okeya
Hirotaka Ono
Yota Otachi
Konstantinos
    Panagiotou
Daniel Panario
Eunhui Park
Jeong-Hyeon Park
Jung-Heum Park
Anders Sune Pedersen
Benny Pinkas
Greg Plaxton
Sheung-Hung Poon
Sanguthevar
    Rajasekaran
Jörg Rambau

Bala Ravikumar
Iris Reinbacher
Daniel Roche
Juanjo Rué
Daniel Russel
Toshiki Saitoh
Jagan Sankaranarayanan
Ignasi Sau
Saket Saurabh
Joe Sawada
Dominik Scheder
Akiyoshi Shioura
Michiel Smid
Aaron Sterling
Mizuyo Takamatsu
Kenjiro Takazawa
Nicholas Tran
Xuehou Tan
Siamak Tazari
My Thai
Hing-Fung Ting
Alexander Tiskin
Etsuji Tomita
Lorenzo Traldi
Amitabh Trehan

Rahul Tripathi
Shi-Chun Tsai
Kostas Tsakalidas
Takeaki Uno
Yushi Uno
Leslie Valiant
Antoine Vigneron
Bow-Yaw Wang
Haitao Wang
Osamu Watanabe
Oren Weimann
Joel Wein
Renato Werneck
Aaron Williams
Peter Winkler
Alexander Wolff
Prudence Wong
Prudence W.H. Wong
Mutsunori Yagiura
Katsuhisa Yamanaka
Koichi Yamazaki
Deshi Ye
Xiao Zhou
Binhai Zhu
Anna Zych

Table of Contents – Part II

Session 7B. Graph Coloring II

Session 8A. Approximation Algorithm II

Session 8B. Online Algorithm

Session 10B. Computational Geometry III

Table of Contents – Part I

Session 4A. Computational Geometry I

Session 4B. Graph Coloring I

Session 5A. Fixed Parameter Tractability

Session 5B. Optimization

D²-Tree: A New Overlay with Deterministic Bounds

Gerth Stølting Brodal[1], Spyros Sioutas[2],
Kostas Tsichlas[3], and Christos Zaroliagis[4]

[1] MADALGO (Center for Massive Data Algorithmics, a Center of the Danish
National Research Foundation), Aarhus University
gerth@madalgo.au.dk
[2] Ionian University, Department of Informatics
sioutas@ionio.gr
[3] Aristotle University of Thessaloniki, Department of Informatics
tsichlas@csd.auth.gr
[4] CTI and Dept. of Computer Engineering & Informatics, University of Patras
zaro@ceid.upatras.gr

Abstract. We present a new overlay, called the *Deterministic Decentralized tree* (*D²*-tree). The D^2-tree compares favourably to other overlays for the following reasons: (a) it provides matching and better complexities, which are deterministic for the supported operations; (b) the management of nodes (peers) and elements are completely decoupled from each other; and (c) an efficient deterministic load-balancing mechanism is presented for the uniform distribution of elements into nodes, while at the same time probabilistic optimal bounds are provided for the congestion of operations at the nodes.

1 Introduction

Decentralized systems and in particular Peer-to-Peer (P2P) networks have become very popular of late and are widely used for sharing resources and store very large data sets. Data are stored at the nodes (or peers) and the most crucial operations are data search (identify the node that stores the requested information) and updates (insertions/deletions of data). Searching and updating is typically done by building a logical *overlay network* that facilitates the assignment and indexing of data at the nodes. Sometimes, we distinguish between the overlay structure per se and the indexing scheme used to access the data.

Following the typical modeling, a decentralized communication network is represented by a graph. Its nodes correspond to the network nodes, while its edges correspond to communication links. We assume constant size messages between nodes through links and asynchronous communication. It is assumed that the network provides an upper bound on the time needed for a node to send a message and receive an acknowledgment. The complexity of an operation is measured in terms of the number of messages issued during its execution. Throughout the paper, when we refer to cost we shall mean number of messages (internal computations at nodes are considered insignificant). The *overlay* is

O. Cheong, K.-Y. Chwa, and K. Park (Eds.): ISAAC 2010, Part II, LNCS 6507, pp. 1–12, 2010.

another graph defined over the communication network. The nodes of the overlay correspond to nodes of the original network, while its edges (links) may not correspond to existing communication links, but to communication paths.

With respect to its *structure*, the overlay supports the operations *Join* (of a new node v; v communicates with an existing node u in order to be inserted into the overlay), and *Departure*(of an existing node u; u leaves the overlay announcing its intent to other nodes of the overlay). The overlay is used to implement an *indexing scheme* for the stored data. Such a scheme supports the operations *search* for an element, *insert* a new element, *delete* an existing element, and *range query* for elements in a specific range.

In terms of efficiency, an overlay network should address the following issues:

- *Fast queries and updates:* updates and queries must be executed in a minimal number of communication rounds and using a minimal number of messages.
- *Ordered data:* keeping the data in order facilitates the implementation of various enumeration queries when compared to a simple dictionary that can only answer membership queries, including those arising in DNA databases, location-based services, and prefix searches for file names or data titles. Indeed, the ever-wider use of P2P infrastructures has found applications that require support for range queries (e.g., [6]).
- *Size of nodes (peers):* the size of a node is the routing information (links and related data) maintained by this node and it is not related to the number of data elements stored in it. Keeping the size of a node small allows for more efficient update operations, but in general reduces the efficiency of access operations while aggravating fault tolerance.
- *Fault Tolerance:* the structure should be able to discover and heal failures at nodes or links.
- *Congestion:* it refers to the distribution of the load of search (access) operations per node, aiming at distributing this load equally across all nodes. The congestion is an *expected* quantity defined as the maximum, among all nodes, of the fraction of the expected number of accesses of a node due to a random sequence of operations on the structure.
- *Load Balancing:* it refers to the distribution of data elements on the nodes. The goal of load balancing is to distribute equally the n elements stored in the N nodes of the network (typically $N \ll n$). That is, ideally each node should carry approximately k elements, where $\lfloor n/N \rfloor \le k \le \lfloor n/N \rfloor + 1$.

There has been considerable recent work in devising effective distributed search and update techniques. Existing structured P2P systems can be classified into two broad categories: distributed hash table (DHT)-based systems and tree-based systems. Examples of the former, which constitute the majority, include Chord [11], Pastry [14], Symphony [12], and Tapestry [17]. DHT-based systems support exact match queries well and use (successfully) probabilistic methods to distribute the workload among nodes equally. DHT-based systems work with little synchrony and high *churn* (the collective effect created by independent burstly arrivals and departures of nodes), a fundamental characteristic of the

Internet. Since hashing destroys the ordering on keys, DHT-based systems typically do not possess the functionality to support straightforwardly range queries, or more complex queries based on data ordering (e.g., nearest-neighbor and string prefix queries). The most recent effort towards range queries is reported in [16].

Tree-based systems are based on hierarchical structures. They support range queries more naturally and efficiently as well as a wider range of operations, since they maintain the ordering of data. On the other hand, they lack the simplicity of DHT-based systems, and they do not always guarantee data locality and load balancing in the whole system. Important examples of such systems include Skip Graphs (SG) [4,7], NoN SG [13], SkipNet (SN), Deterministic SN [9], Bucket SG [3], Family Trees [15], Skip Webs [1], BATON [10], Rainbow Skip Graphs (RSG) [8], and Strong RSG [8].

In this work, we focus on tree-based overlay networks that support directly range and more complex queries. Let N be the number of nodes present in the network and let n denote the size of data ($N \ll n$). Let M be the size of each node, $Q(n, N)$ be the cost of a single query, $U(n, N)$ be the cost of an update, $C(n, N)$ be the congestion per node (measuring the load) incurred by search operations, and let $L(n, N)$ be the cost for load balancing the overlay w.r.t. element updates. With respect to congestion, each node issues one operation, while the destination node of the operation is assumed to be selected uniformly at random among all nodes of the network. Congestion depends on the distribution of elements into nodes as well as on the topology of the overlay. It provides hints as to how well the structure avoids the existence of *hotspots* (i.e., nodes which are accessed multiple times during a sequence of operations – the root of a tree is usually a hotspot in decentralized tree structures).

A comparison of the aforementioned tree-based overlays is given in Table 1. We would like to emphasize that w.r.t. load balancing, there are solutions in the literature either as part of the overlay (e.g., [10]) or as a separate technique (e.g. [3,7]). These solutions are either heuristics, or provide expected bounds under

Table 1. A comparison between previous methods and the D^2-tree. By $\widehat{O}$ we represent expected bounds, by $\widetilde{O}$ we represent amortized bounds, and by $\overline{O}$ expected amortized bounds. All other bounds are worst-case. Typically, $N \ll n$.

Methods	N	M	$Q(n, N)$	$U(n, N)$	$C(n, N)$	$L(n, N)$
SG [4,7]	$\leq n$	$O(\log N)$	$\widehat{O}(\log N)$ w.h.p.	$\widehat{O}(\log N)$ w.h.p.	$\widehat{O}(\frac{\log N}{N})$	$\widetilde{O}(\log N)$
NoN SG [13]	n	$O(\log^2 n)$	$\widehat{O}(\frac{\log n}{\log \log n})$	$\widehat{O}(\log^2 n)$	$\widehat{O}(\frac{\log^2 n}{n})$	—
Determ. SN [9]	n	$O(\log n)$	$O(\log n)$	$O(\log^2 n)$	$O(\frac{n^{0,32}}{n})$	—
BATON [10]	$\leq n$	$O(\log N)$	$O(\log N)$	$O(\log N)$	—	$\overline{O}(\log n)$
Family Trees [15]	n	$O(1)$	$\widehat{O}(\log n)$	$\widehat{O}(\log n)$	$\widehat{O}(\frac{\log n}{n})$	—
Bucket SG [3]	$\leq n$	$O(\frac{n}{N} + \log N)$	$\widehat{O}(\log N)$	$\widehat{O}(\log N)$	$\widehat{O}(\frac{1}{N} + \frac{\log N}{n})$	No Bounds
Skip Webs [1]	n	$O(\log n)$	$\widehat{O}(\frac{\log n}{\log \log n})$	$\widehat{O}(\frac{\log n}{\log \log n})$	$\widehat{O}(\frac{\log n}{n})$	—
Rainbow SG [8]	n	$O(1)$	$\widehat{O}(\log n)$ w.h.p.	$\overline{O}(\log n)$ w.h.p.	$\widehat{O}(\frac{\log n}{n})$	—
Strong RSG [8]	n	$O(1)$	$O(\log n)$	$\widetilde{O}(\log n)$	$\widehat{O}(\frac{n^\epsilon}{n})$	—
D²-tree	$\leq n$	$O(1)$	$O(\log N)$	$\widetilde{O}(\log N)$	$\widehat{O}(\frac{\log N}{N})$	$\widetilde{O}(\log N)$

certain assumptions, or amortized bounds but at the expense of increasing the size per node (see [5] for a detailed discussion).

Our Contribution. In this paper we present a new tree-based overlay, called the *Deterministic Decentralized tree* or D^2-*tree*. The D^2-tree (see also Table 1) uses $O(1)$ space per node, achieves a deterministic $O(\log N)$ query bound and a deterministic (amortized) $O(\log N)$ update bound for elements as well as for node joins and departures, achieves *optimal* congestion, and exhibits a deterministic (amortized) $O(\log N)$ bound for load-balancing. Moreover, it supports ordered data queries optimally, and tolerates node failures.

The D^2-tree is an overlay consisting of two levels. The upper level is a perfect binary tree, while the lower level consists of buckets (sets of nodes), where each bucket is structured as a doubly linked list. Each bucket contains $O(\log N)$ nodes. Since N changes, the size of buckets is dynamically maintained by the overlay.

In the D^2-tree, we separate the index from the overlay structure using the load-balancing mechanism. The number of elements per node is dynamic w.r.t. node joins and departures and it is controlled by the load-balancing mechanism. Moreover, the number of nodes of the perfect binary tree is not connected by any means to the number of elements stored in the structure. The overlay structure supports the operations of node join and node departure, while at the same time it tackles failures of nodes whenever these are discovered.

Our load-balancing technique distributes almost equally the elements among nodes by making use of weights. Weights are used to define a metric of load-balance, which shows how uneven is the load between nodes. When the load is uneven, then a data migration process is initiated to equally distribute elements.

Our load-balancing technique is quite general and can be applied to any hierarchical decentralized overlay (e.g., BATON, Skip Graphs) with the following specifications: (i) The overlay structure must be a tree with height $O(\log N)$ with each node having $O(1)$ children. (ii) Nodes at level i having the same father have approximately (within constant factors) the same weight, which is $\Omega(i^4)$. (iii) Updates are performed at the leaves. Alternatively, if each node has access to a leaf in $O(1)$ messages then this is enough, since the update is simply forwarded to this leaf.

We discuss the load balancing technique in Section 2, and present the D^2-tree in Section 3. We conclude in Section 4. Due to space constraints, some details and proofs are deferred to the full version [5].

2   Deterministic Load Balancing

The load-balancing mechanism distributes almost equally the elements among nodes by making use of weights, which are used to define a metric showing how uneven is the load between nodes. When the load is uneven, then a data migration process is initiated to equally distribute elements.

A few definitions are in place. Assume that the overlay structure is a tree $\mathcal{T}$. Based on $\mathcal{T}$ ancestor-descendant relationships are defined. There is a node that has no ancestor (the *root*) and there are nodes with no descendants (the *leaves*).

All nodes which are not leaves are called *internal*. The subgraph induced by the descendants of node v (including v) in $\mathcal{T}$ is the *subtree* of v. The *weight* $w(v)$ of node v is equal to the number of elements stored in its subtree. The term weight will also be used to express other similar quantities at some parts of the paper, in which case we explicitly say so. The number of elements residing in node v is represented by $e(v)$. The *height* of node v is the length of the longest path from v to one of its leaves. The *depth* (or *level*) of node v is the length of the path from v to the root. Two nodes are called *brothers* when they have the same father and they are consecutive in his child list.

We describe the load-balancing mechanism in two steps. First, we provide a mechanism that allows for efficient and local update of weight information in a tree when elements are added or removed at the leaves. This is necessary to avoid hotspots. Then, we describe the load-balancing scheme in a tree overlay.

2.1   A Technique for Amortized Constant Weight Updating

We provide a technique that lazily updates the weights on the nodes of a tree. When an element is added or removed to/from a leaf u in $\mathcal{T}$ the weights on the path from u to the root must be updated. If the height of $\mathcal{T}$ is H, then the cost of the weight updating is $O(H)$. Assume that node v lies at height h and its children are $v_1, v_2, \ldots, v_s$ at height $h - 1$. We relax the weight of a node and its recomputation. We define the *virtual weight* $b(v)$ of v as the weight stored in node v. In particular, for node v the following invariants are maintained

Invariant 1. $b(v) > e(v) + (1 - \epsilon_h)\left(\sum_{i=1}^{s} b(v_i)\right)$

Invariant 2. $b(v) < e(v) + (1 + \epsilon'_h)\left(\sum_{i=1}^{s} b(v_i)\right)$

where ϵ_h and ϵ'_h are appropriate constants. These invariants imply that the weight information is approximate, at most by a multiplicative constant.

Assume that an update takes place at leaf u. Apparently, only the weight of its ancestors need to be updated by ± 1 and no other node is affected. We traverse the path from u to the root until we find a node z for which Invariants 1 and 2 hold. Let v be its child for which either Invariant 1 or 2 does not hold on this path. We recompute all weights on the path from u to v. In particular, for each node z on this path, we update its weight information by taking the sum of the weights written in its children plus the number of elements that z carries.

The following lemma states how frequently the weight information in each node changes. Its proof follows from the fact that the update of node v is a result of the violation of either of Invariants 1 or 2 and by taking into account that $\frac{1}{2} \cdot w(v) < b(v) < 2 \cdot w(v)$, if we choose $\epsilon_h = \epsilon'_h = \frac{1}{h^2}$ [5].

Lemma 1. *The minimum number of updates in the subtree of v, causing a weight update at v, is $\Theta(\epsilon_h w(v))$.*

The above lemma states that if we make $\epsilon_h w(v)$ update operations then the maximum number of weight changes at node v is 1, implying that the amortized cost per update operation at height h is $\frac{1}{\epsilon_h b(v)}$. Since a node on the path at height i has (by assumption) virtual weight $\Omega(i^4)$, it is not hard to see that the weight updating mechanism is efficient in an amortized sense.

Theorem 1. *The amortized cost of the weight update algorithm is $O(1)$.*

2.2   Updates and Load Balancing

We now investigate how load balancing is realized on the balanced tree structure $\mathcal{T}$. For clarity of exposition, we assume that $\mathcal{T}$ is a binary tree. The following discussion can be easily generalized for trees with $O(1)$ maximum degree, simply by looking between brother nodes.

First, bear in mind that this mechanism does not tamper with the structure of $\mathcal{T}$. An update operation (either insertion or deletion of an element) is initiated at node v. Node v issues a search for the involved element and the appropriate node u is returned. Then, the update request is forwarded from v to u. Node u executes the update operation and signals v for the status of the update. The load balancing mechanism redistributes the elements among nodes when the load between nodes is not distributed equally enough.

Assume that node v at height h has child p and its right brother q at height $h - 1$. Let $|v|$ denote the number of nodes of the subtree of v (including v) in the overlay structure. The *density* $d(v) = \frac{w(v)}{|v|}$ of node v represents the mean number of elements per node in the subtree of v. The *criticality* $c(p, q) = \frac{d(p)}{d(q)}$ of two brother nodes p and q represents their difference in densities. The following invariant guarantees that there will not be large differences between densities.

Invariant 3. *For two brothers p and q, it holds that $\frac{1}{c} \leq c(p, q) \leq c$, $1 < c \leq 2$.*

For example, choosing $c = 2$ we get that the density of any node can be at most twice or half of that of its brother. In the more general case where the number of children of node v is $O(1)$, we get that no child of v has more density than a constant factor w.r.t. the other children of v.

When an update takes place at leaf u, weights are updated by using the mechanism described in Section 2.1. In this way, we guarantee that no hotspot exists w.r.t. weight updating as implied by Lemma 1. Then, starting from u, the highest ancestor w is located that is unbalanced w.r.t. his brother z, meaning that Invariant 3 is violated. Finally, the elements in the subtree of their father v are redistributed uniformly so that the density of the brothers becomes equal; this procedure is henceforth called *redistribution* of node v. Assume that the redistribution phase has a cost of $O(f(w(v)))$, for some increasing function $f : \mathbb{N} \to \mathbb{N}$. The following theorem provides amortized bounds for the redistribution.

Theorem 2. *The load balancing has an amortized cost of $O\left(H\frac{f(n)}{n}\right)$.*

3   The D^2-Tree

In this section we design and analyze the D^2-*tree* overlay. We first describe the overlay structure, then move to the description of the index, and finally discuss efficiency issues regarding congestion and fault-tolerance.

3.1   The D^2-Tree Structure

The D^2-tree is a binary tree, where each node maintains an additional set of links to other nodes apart from the standard links which form the tree. Each node v in the tree maintains the following links:

1. Links to its father (if there is one) and its children.
2. Links to its adjacent nodes based on an inorder traversal of the tree.
3. Links to nodes at the same level as v. These links facilitate an exponential search on the nodes of the same level. Assume that node v lies at level ℓ. In a binary tree, the maximum number of nodes at level ℓ is equal to 2^ℓ. Node v maintains at most 2ℓ links: ℓ links to nodes to the right and ℓ links to nodes to the left. The links are distributed in exponential steps, that is the first link points to a node (if there is one) 2^0 positions to the left (right), the second 2^1 positions to the left (right) and the i-th link 2^{i-1} positions to the left (right). These links constitute the *routing table* of v.

The next lemma captures some important properties of the routing tables w.r.t. their construction. It follows immediately from the aforementioned link structure and the fixed distances between successive links in the routing tables.

Lemma 2. *(i) If a node v contains a link to node u in its routing table, then the parent of v also contains a link to the parent of u, unless u and v have the same father. (ii) If a node v contains a link to node u in its routing table, then the left (right) sibling of v also contains a link to the left (right) sibling of u, unless there are no such nodes. (iii) Every non-leaf node has two adjacent nodes in the inorder traversal, which are leaves.*

A Weight-Balanced Overlay. The overlay consists of two levels. The upper level of the overlay is a Perfect Binary Tree (PBT). The lower level of the overlay are the leaves of this tree, which are sets of nodes called *buckets* containing $O(\log N)$ nodes. Each bucket is structured as a doubly linked list. Each node of the bucket points to the node which is a leaf of the PBT and is called the *representative* of the bucket. Additionally, it maintains its routing table w.r.t. the nodes of all buckets.

When a node z makes a join request to v, then this node is forwarded to its adjacent leaf u w.r.t. the inorder traversal. Then, node z is added to the doubly linked list representing the bucket of u by manipulating a constant number of links. The routing table of z is updated by using Lemma 2(ii). When a node v leaves the network, then it is replaced by its right adjacent node u (if there is no right adjacent node then we choose the left one) which in turn is replaced by its first node z in its bucket. Link and data information are copied from v to u and from u to z. When a node v is discovered to be unreachable, its adjacent node u is first located. This is accomplished by traversing the path to the rightmost or leftmost leaf starting from the left or right child respectively. Node u fills the gap of v and the first child z in the bucket of u fills the gap left by u. The contents of u are not moved to another node except from the navigation data (routing tables and other links) which are moved to node z that take its place. Node u has its routing tables recomputed.

The join and departure of nodes may cause the size of the buckets to be uneven, which in the long run renders the structure unbalanced (imagine a bucket holding almost all nodes). To control the size of the buckets we employ a weight-based approach. Each node v of the PBT maintains its weight $|v|$, which is equal to the number of nodes in the buckets of its subtree. The size control is accomplished by using the method introduced in Section 2.1, in order to avoid the existence of hotspots.

The *node criticality* nc_v of a node v at level ℓ with left and right children w and z at level $\ell + 1$, respectively, is defined as $nc_v = \frac{|w|}{|v|}$. The following invariant bounds the criticality of nodes.

Invariant 4. *The node criticality of all nodes is in the range* $\left[\frac{1}{4}, \frac{3}{4}\right]$.

Invariant 4 implies that the number of nodes in buckets in the left subtree of a node v is at least $1/3$ and at most threefold the corresponding number of its right subtree (this definition can be easily generalized when v has a $O(1)$ number of children). When an update takes place at bucket x, then we locate the highest ancestor v of x whose node criticality is out of bounds, w.r.t. Invariant 4, and we redistribute the nodes in its subtree. The redistribution phase is described in [5]. The redistribution guarantees that if there are z nodes in total in the y buckets of the subtree of v, then after the redistribution each bucket maintains either $\lfloor z/y \rfloor$ or $\lfloor z/y \rfloor + 1$ nodes. However, the following discussion still holds (with minor changes) even if the redistribution phase guarantees that the minimum and maximum size of the buckets is within constant factors. The cost for the redistribution we propose for node v is $f(|v|) = O(|v|)$.

We guarantee that each bucket contains $O(\log N)$ nodes when subject to joins or departures of nodes by employing two operations on the PBT, the *contraction* and the *extension*. When a redistribution takes place at the root of the PBT, we also check whether any of these two operations can be applied to the PBT. The extension operation adds one more level of nodes at the PBT from existing nodes in the buckets, thus increasing its height by one. The contraction operation removes one level of nodes from the PBT and puts them into the buckets, thus decreasing its height by one. In order to decide whether the PBT needs extension or contraction we compare the size of the buckets B after the redistribution with the height of the PBT. Note that after redistribution, the sizes of all buckets may differ by at most 1. If the size is larger by at least 1 then an extension takes place. If the size of the bucket is smaller than the height of the PBT by at least 1 then a contraction takes place. The height of the PBT can be deduced by the size of the routing table in the nodes of the last level of the PBT. These two operations involve a reconstruction of the overlay which rarely happens as shown in the following lemma.

Lemma 3. *If a redistribution operation is performed at a node with weight s, then this node will be redistributed again after $\Omega(s)$ joins or departures have been performed in its subtree.*

Lemma 3 states that the expensive operations of extension and contraction take place when the number of nodes has at least doubled or halved. Assuming that the redistribution of v has $O(f(|v|))$ cost, it follows by Lemma 3 that the amortized cost for join/departure of a node v at height h is $O\left(\frac{f(|v|)}{|v|}\right)$. Since the PBT has height H, we establish the following.

Lemma 4. *The amortized cost of join/departure of a node v is $O\left(H\frac{f(N)}{N}\right)$.*

$O(1)$ **Space per Node.** The routing tables require $O(\log N)$ space for each node. To make the space consumption constant, one could apply on the overlay the schemes described in [8,15]. However, on the one hand the complexities will not be deterministic while on the other hand even in the case of the strong rainbow graphs [8] with deterministic bounds our congestion for searching is much better than theirs. To achieve constant space we distribute the routing tables to many nodes doing the same also for nodes in the buckets. A set of nodes with constant degree is grouped together and a routing table is distributed on all these nodes, such that each node uses constant space. Thus, a node can recreate approximately its routing table by accessing nodes inside the same group. We call each such group a *hypernode*.

A hypernode at level ℓ consists of at most ℓ nodes, numbered from left to right $1, 2, \ldots$. This number is the *rank* of the node within the hypernode. A node v with rank i maintains two links to the nodes that are approximately 2^i positions to the right and to the left. In particular, node v either points to a node z in the same hypernode whose distance is 2^i or to a node z' whose rank is i and lies in a different hypernode than that of v which contains a node whose distance is 2^i from v. The concatenation of all such links constitutes the routing table for the hypernode. Additionally, each node with rank i maintains two links to nodes with ranks $i-1$ and $i+1$, if there are such nodes. Finally, each node with rank i in the hypernode maintains a link to the node with the largest rank. The following lemma translates Lemma 2(ii) in the setting of hypernodes.

Lemma 5. *If node v contains a link to node u, then the left (right) sibling of v also contains a link to the left (right) sibling of u, unless $\nexists$ such nodes.*

Using Lemma 5 we can update the links of a node v by simply looking at the links of its siblings u and w and update the links of v by pointing to the adjacent nodes of the nodes pointed to by u and w. Hypernodes are static in the overlay and only in the case of contraction we destroy the hypernodes of the last level while in the case of extension we create new hypernodes for the new level. A faulty node inside a hypernode will not disconnect it since by accessing the parents we can find its siblings and reconstruct the missing routing information.

3.2 The Index Structure of the D^2-Tree

The overlay provides the infrastructure for the index to efficiently support various operations. The overlay is used as a node-oriented tree. The range of all values stored in the overlay is partitioned into subranges each one of which is assigned to a node of the overlay. An internal node v with range $[x_v, x_v']$ may have a left child u and a right child w with ranges $[x_u, x_u']$ and $[x_w, x_w']$ respectively such that $x_u < x_u' < x_v < x_v' < x_w < x_w'$. Thus, if an element $x \in [x_v, x_v']$ then it must be stored at node v. Ranges are dynamic in the sense that they depend on the values maintained by the node.

Search and Range Queries. The search for an element α in the overlay may be initiated from any node v at level ℓ. Let z be the node with range of values containing α. Assume $O(\log N)$ space per node and assume that w.l.o.g $x_v' < \alpha$. Then, by using the routing tables we search at level ℓ for a node u with right sibling w (if there is such sibling) such that $x_u' < \alpha$ and $x_w > \alpha$ unless α is in the range of u and the search terminates. This step has $O(\ell)$ cost, since we simulate a binary search. If the search continues, then node z will either be an ancestor of u or in the subtree rooted at u. If u is a leaf, then we move upwards (or in its corresponding bucket) until we find node z in $O(\log N)$ steps. If u is an internal node, by following the respective link we move to the left adjacent node y of u which is certainly a leaf (inorder traversal). If $x_y' > \alpha$ then an ordinary top down search from node u will suffice to find z in $O(\log N)$ steps (or in its bucket). Otherwise, node z is certainly an ancestor of u and thus we can move upwards from u until we find it in $O(\log N)$ steps. The case with $O(1)$ space per node, along with the proof of the following lemma, are given in [5].

Lemma 6. *The search for an element α in a D^2-tree of N nodes is carried out in $O(\log N)$ steps.*

A range query $[a, b]$ reports all elements x such that $x \in [a, b]$. A range query $[a, b]$ initiated at node v, invokes a search operation for element a. Node u that contains a returns to v all elements in this range. If all elements of u are reported then the range query is forwarded to the right adjacent node (inorder traversal) and continues until an element larger than b is reached for the first time.

Updates and Load Balancing. Assume that an update operation is initiated at node v involving element α. By invoking a search operation we locate node u with range containing element α. Finally, the update operation is performed on u. The main issue is how to balance the load to all nodes of the overlay as much equally as possible. To do that we employ the machinery developed in Section 2. Details can be found in [5].

The cost for the redistribution of a node v is $O(|v| \log N)$ for the case of $O(\log N)$ space per node or $O(|v|)$ for the case of $O(1)$ space per node. This is because, during the transfer of elements the routing tables must be reconstructed. The following lemma states that the load balancing is efficient in an amortized sense when the structure is subject to insertions and deletions of elements. It is a direct implication of Theorem 2 and the space used by the nodes.

Lemma 7. *The load rebalancing operation of the index has an amortized cost of $O(\log N)$.*

One final comment is that the redistribution of elements may be affected by the redistribution of nodes in the weight-balanced overlay. In order to avoid such a phenomenon, the redistribution of nodes in the subtree of node v in the overlay is preceded by a redistribution of elements.

3.3   Other Efficiency Issues and the Main Result

We are now ready to tackle the congestion and the fault-tolerance of the D^2-tree overlay, and to present the main results of this work.

Congestion. We assume that a sequence of searches $s_1, s_2, \ldots, s_N$ is initiated from each of the N nodes of the overlay. Assume that s_i is looking for an element residing in a node z_i (target node for s_i). The target nodes $z_1, z_2, \ldots, z_N$ are chosen independently and uniformly at random from all nodes of the overlay. There are two phases in the search. The first is the horizontal search, which makes use of the routing tables, and the second is the vertical search on a path from a node either towards the root or towards a leaf. The following theorem, whose proof can be found in [5], establishes the congestion bound.

Theorem 3. *The congestion due to the search operations is $O(\frac{\log N}{N})$ expected in a D^2-tree with N nodes, where each node uses $O(1)$ space.*

Fault Tolerance. If a node v discovers (during the execution of an operation) that node u is unreachable, then it contacts a sibling of u through the routing tables of the siblings of v (by making use of Lemma 2(ii)). This sibling of u is able by Lemma 2(ii) (or Lemma 5) to reconstruct all links of node u and a node departure for u is initiated, which resolves this failure. A more extensive discussion can be found in [5].

Main Result. We are now ready to establish the main results of this work stated in the Introduction and in Table 1. In particular, space usage is $O(1)$ by construction. The search cost follows from Lemma 6, which also dominates the cost for updating a data element. Node join and departures are $O(\log N)$ amortized by Lemma 4 and the fact that $f(n) = O(N)$. The congestion bound comes from Theorem 3. Finally, the load-balancing bound comes from Lemma 7.

4   Conclusions and Discussion

The load-balancing scheme can be applied straightforwardly to BATON [10]. BATON is a balanced tree-like overlay that satisfies the specifications set in the Introduction. The same goes also for Skip Graphs [4] with the exception that the specifications hold probabilistically and thus the bounds are also probabilistic. Additionally, it provides a mechanism to control the bucket size of [3].

We provide a technique that lazily updates the weights on the nodes of a tree. This technique is interesting by itself and can be straightforwardly applied to weighted balanced trees [2] in the Pointer Machine model of computation for single processor internal memory machines. In this manner, the update of balancing information is supported in $O(1)$ amortized time.

References

1. Arge, L., Eppstein, D., Goodrich, M.T.: Skip-Webs: Efficient Distributed Data Structures for Multidimensional Data Sets. In: Proc. of the 24th PODC, pp. 69–76 (2005)
2. Arge, L., Vitter, J.: Optimal External Memory Interval Management. SIAM Journal on Computing 32(6), 1488–1508 (2003)
3. Aspnes, J., Kirsch, J., Krishnamurthy, A.: Load-balancing and Locality in Range-Queriable Data Structures. In: Proc. of the 23rd PODC, pp. 115–124 (2004)
4. Aspnes, J., Shah, G.: Skip Graphs. In: Proc. of the 14th SODA, pp. 384–393 (2003)
5. Brodal, G.S., Sioutas, S., Tsichlas, K., Zaroliagis, C.: D^2-Tree: A New Overlay with Deterministic Bounds (September 2010), http://arxiv.org/abs/1009.3134
6. Li, D., Cao, J., Lu, X., Chan, K.C.C.: Efficient Range Query Processing in Peer-to-Peer Systems. IEEE Transactions on Knowledge and Data Engineering 21(1), 78–91 (2009)
7. Gasenan, P., Bawa, M., Garcia-Molina, H.: Online Balancing of range-Partitioned Data with Applications to Peer-to-Peer Systems. In: Proc. of the 13th VLDB, pp. 444–455 (2004)
8. Goodrich, M.T., Nelson, M.J., Sun, J.Z.: The Rainbow Skip Graph: A Fault-Tolerant Constant-Degree Distributed Data Structure. In: Proc. of the 17th SODA, pp. 384–393 (2006)
9. Harvey, N., Munro, J.I.: Deterministic SkipNet. In: Proc. of the 22nd PODC, pp. 152–153 (2003)
10. Jagadish, H.V., Ooi, B.C., Vu, Q.H.: BATON: a Balanced Tree Structure for Peer-to-Peer Networks. In: Proc. of the 31st VLDB, pp. 661–672 (2005)
11. Karger, D., Kaashoek, F., Stoica, I., Morris, R., Balakrishnan, H.: Chord: A Scalable Peer-to-Peer Lookup Service for Internet Applications. In: Proc. of the SIGCOMM, pp. 149–160 (2001)
12. Manku, G.S., Bawa, M., Raghavan, P.: Symphony: Distributed hashing in a small world. In: 4th USENIX Symp. on Internet Technologies and Systems (2003)
13. Manku, G.S., Naor, M., Wieder, U.: Know thy Neighbor's Neighbor: the Power of Lookahead in Randomized P2P Networks. In: Proc. of the 36th STOC, pp. 54–63 (2004)
14. Rowstron, A., Druschel, P.: Pastry: A Scalable, Decentralized Object Location, and routing for large-scale peer-to-peer systems. In: Liu, H. (ed.) Middleware 2001. LNCS, vol. 2218, pp. 329–350. Springer, Heidelberg (2001)
15. Zatloukal, K.C., Harvey, N.J.A.: Family trees: An Ordered Dictionary with Optimal Congestion, Locality, Degree and Search Time. In: Proc. of the 15th SODA, pp. 301–310 (2004)
16. Zhang, Y., Liu, L., Li, D., Liu, F., Lu, X.: DHT-Based Range Query Processing for Web Service Discovery. In: Proc. of the 2009 IEEE ICWS, pp. 477–484 (2009)
17. Zhao, B.Y., Huang, L., Stribling, J., Rhea, S.C., Joseph, A.D., Kubiatowicz, J.D.: Tapestry: A Resilient Global-scale Overlay for Service Deployment. IEEE Journal on Selected Areas in Communications 22(1), 41–53 (2004)

Efficient Indexes for the Positional Pattern Matching Problem and Two Related Problems over Small Alphabets*

Chih-Chiang Yu, Biing-Feng Wang, and Chung-Chin Kuo

Department of Computer Science, National Tsing Hua University
Hsinchu, Taiwan 30013, Republic of China
{littlejohn,bfwang,cckuo}@cs.nthu.edu.tw

Abstract. In this paper, we study the following three variants of the classical text indexing problem over small alphabets: the positional pattern matching problem, the position-restricted pattern matching problem, and the indexing version of the variable-length don't care pattern matching problem. Let n be the length of the text, p be the length of a query pattern, and Σ be the alphabet. Assume that $|\Sigma| = O(\text{polylog}(n))$. For the first and third problems, we present $O(n)$-word indexes with $O(p)$ query time. For the second problem, we show that each query can be answered in $O(n \log^\epsilon n)$ space and $O(p + occ)$ time, or in $O(n)$ space and $O(p + occ \log^\epsilon n)$ time, where occ is the number of outputs. When the alphabet size is $O(\text{polylog}(n))$, the indexes presented in this paper improve the results in [6, 10, 11, 22].

1 Introduction

In this paper, we first consider a variant of the classical text indexing problem, called the *positional pattern matching problem*, which is to construct an index for a text T so that the first occurrence of a pattern P in T at or after a given position s can be found efficiently. This problem was firstly considered by Keller *et al.* [11] as an application of the range successor problem [6, 11, 14, 22]. For the positional pattern matching problem, Keller *et al.* [11] had an $O(n \log n)$-word index with $O(p + \log \log n)$ query time; Crochemore *et al.* [6] had an $O(n^{1+\epsilon})$-word index with $O(p)$ query time; and Yu *et al.* [22] had an $O(n)$-word index with $O(p + \log n / \log \log n)$ query time, where n is the length of the text, p is the length of a query pattern, and $\epsilon > 0$ is an arbitrary small constant. For real world applications, the alphabet size of a string to be indexed is usually small so that there are many researches concerning string matching over small alphabets [19, 20]. For instance, nucleotide sequences are strings over an alphabet of size 4 [19]; amino acid sequences are strings over an alphabet of size 20 [20]; and the standard ASCII characters have values between 0 and 127 [20]. This inspires us to study the positional pattern matching problem over small alphabets. In

* This research is supported by the National Science Council of the Republic of China under grant NSC-98-2221-E-007-081.

O. Cheong, K.-Y. Chwa, and K. Park (Eds.): ISAAC 2010, Part II, LNCS 6507, pp. 13–24, 2010.

particular, for $|\Sigma| = O(\text{polylog}(n))$, we present an $O(n)$-word index with $O(p)$ query time. This is the first result which uses $O(n)$ space while achieving $O(p)$ query time. Table 1 summarizes the above results. Our model of computation is a unit-cost RAM with word size of $\log n$ bits.

Table 1. Indexes for positional pattern matching

	Space	Query time	Remarks
[11]	$O(n \log n)$	$O(p + \log \log n)$	
[6]	$O(n^{1+\varepsilon})$	$O(p)$	
[22]	$O(n)$	$O(p + \log n / \log \log n)$	
Ours	$O(n)$	$O(p)$	$|\Sigma| = O(\text{polylog}(n))$

A problem closely related to the positional pattern matching problem is called the *position-restricted pattern matching problem*, introduced by Mäkinen and Navarro [14]. The problem is to construct an index for a text T so that all occurrences of a pattern P in T within a given interval $[s,t]$ can be reported in sorted order efficiently. For this problem, Mäkinen and Navarro [14] had an $O(n)$-word index with $O(p + occ \log n)$ query time; Keller *et al.* [11] had an $O(n \log n)$-word index with $O(p + occ \log \log n)$ query time; Crochemore *et al.* [6] had an $O(n^{1+\varepsilon})$-word index with $O(p + occ)$ query time; and Yu *et al.* [22] had an $O(n)$-word index with $O(p + occ \log n / \log \log n)$ query time, where occ is the number of outputs. In addition, Mäkinen and Navarro [14] had an $O(n \log^{\varepsilon} n)$-word index with $O(p + \log \log n + occ)$ query time, yet the occurrences are not delivered in sorted order. In this paper, for $|\Sigma| = O(\text{polylog}(n))$, we show that a position-restricted query can be answered in $O(n \log^{\varepsilon} n)$ space and $O(p + occ)$ time, or in $O(n)$ space and $O(p + occ \log^{\varepsilon} n)$ time. Table 2 summarizes the results.

Table 2. Indexes for position-restricted pattern matching

	Space	Query time	Sorted	Remarks
[11]	$O(n \log n)$	$O(p + occ \log \log n)$	✓	
[6]	$O(n^{1+\varepsilon})$	$O(p + occ)$	✓	
[22]	$O(n)$	$O(p + occ \log n / \log \log n)$	✓	
[14]	$O(n \log^{\varepsilon} n)$	$O(p + \log \log n + occ)$		
Ours	$O(n)$	$O(p + occ \log^{\varepsilon} n)$	✓	$|\Sigma| = O(\text{polylog}(n))$
Ours	$O(n \log^{\varepsilon} n)$	$O(p + occ)$	✓	$|\Sigma| = O(\text{polylog}(n))$

In a Unix-like system, we may use ? or * in a directory listing command and the famous search engine, Google, allows keywords containing *, where ? denotes a *don't care symbol* that can match any single character and * denotes a *variable-length don't care symbol* that can match any number of characters. This feature not only facilitates the search of an approximate keyword in a text, but also benefits the search of DNA or amino acid patterns [12]. Therefore, pattern matching with don't care and variable-length don't care symbols has received much attention in the literature [2, 12, 13, 17]. Given a text T and a pattern P

that may contain variable-length don't care symbols, the *variable-length don't care pattern matching problem* is to determine whether the pattern P occurs in T. For this problem, efficient sequential and parallel algorithms had been proposed in [2, 12, 17]. For the indexing version of this problem, Inenaga *et al.* [10] had an index that supports each query in $O(p)$ time, using $O(n^2)$ space; and Crochemore *et al.*'s index in [6] can be used to answer each query in $O(p)$ time, using $O(n^{1+\epsilon})$ space. In this paper, we also consider the indexing version of the variable-length don't care pattern matching problem. For $|\Sigma| = O(\mathrm{polylog}(n))$, we give an $O(n)$-word index with $O(p)$ query time. Table 3 summarizes the above results.

Table 3. Indexes for pattern matching with variable-length don't care symbols

	Space	Query time	Remarks
[10]	$O(n^2)$	$O(p)$	
[6]	$O(n^{1+\varepsilon})$	$O(p)$	
Ours	$O(n)$	$O(p)$	$|\Sigma| = O(\mathrm{polylog}(n))$

2   Preliminaries

Let X be a string over an alphabet Σ. The length of X is denoted by $|X|$. The substring of X containing $X[i], X[i+1], \ldots, X[j]$, where $1 \le i \le j \le |X|$, is denoted by $X[i, j]$. For $1 \le i \le |X|$, the substring $X[1, i]$ is called a *prefix* of X, whereas the substring $X[i, |X|]$ is called a *suffix* of S. A string Y *occurs* in X at position i if Y is equal to $X[i, i + |Y| - 1]$. If a string Y occurs in X at a position i, we call i an *occurrence* of Y in X.

Let A be a sequence of numbers. The *successor* of a number x in A, denoted by $succ(A, x)$, is the smallest number in A whose value is greater than or equal to x. Let $succ^{-1}(A, x)$ be the position of the successor of x in A.

The improvement of our indexes stems from (1) a simple classification of patterns into types according to length, which in turn motivates the use of a different strategy for each type, and (2) a novel application of rank and select indexes [5, 16] on the suffix tree and the segment tree to facilitate the search. Suffix trees, segment trees, and rank and select indexes are introduced as follows.

Let T be a text of length n over an alphabet Σ. The *suffix tree* [15] of T, denoted by $\mathcal{ST}$, is a compact trie of all suffixes of T, where each edge is labeled by a non-empty substring of T. For each node v in the suffix tree, the concatenation of the edge labels along the path from the root to v is called the *path label* of v. The path label of each leaf is exactly a suffix of T, and we label a leaf by i if its path label is the suffix $T[i, n]$. For a pattern P, its *locus* $\mu(P)$ is the node v, nearest from the root, whose path label has the prefix P. Let L_v be the set of leaf labels in the subtree of $\mathcal{ST}$ rooted at v. We have the following.

Lemma 1 ([15]). *If $\mu(P)$ exists, all occurrences of P in T are those in $L_{\mu(P)}$.*

The set L_v is not explicitly stored. Rather, an array L is used to store the labels of the leaves of $\mathcal{ST}$, from left to right, and each node v maintains only two indices

b_v and e_v, so that L_v is equal to the set of numbers in $L[b_v, e_v]$. The maximum out-degree of each internal node v in $\mathcal{ST}$ is $|\Sigma|$. To facilitate the traversal of the tree, hash tables [9] can be used to achieve the following result.

Lemma 2 ([9, 11]). *Given a string T over an alphabet Σ, the suffix tree $\mathcal{ST}$ occupies $O(|T|)$ space. For any pattern P, $\mu(P)$ can be determined in $O(|P|)$ time, and all occurrences of P in T can be found in $O(|P| + occ)$ time, where occ is the number of occurrences.*

The *segment tree* $\mathcal{GT}$ of interval $[1, n]$ is an ordered balanced binary tree with n leaves, where the ith leaf represents the elementary interval $[i, i]$, and each internal node v represents the union of the intervals represented by its descendant leaves. All intervals represented by the nodes in the segment tree are *standard intervals*. We have the following.

Lemma 3 ([1]). *The segment tree $\mathcal{GT}$ of interval $[1, n]$ can be constructed in $O(n)$ time and $O(n)$ space. Any interval $[i, j]$, $1 \le i \le j \le n$, can be decomposed into a collection of $O(\log n)$ standard intervals in $O(\log n)$ time.*

Let $S[1, n]$ be a character string of length n over an alphabet Σ. For any $c \in \Sigma$ and any position i, a *rank query* $rank_c(S, i)$ reports the number of symbols c in $S[1, i]$. A *select query* $select_c(S, j)$ returns the position of the jth occurrence of c in S. For instance, if $T = 231131321$, then $rank_3(S, 6) = 2$ and $select_3(S, 2) = 5$. We have the following.

Lemma 4 ([5, 16]). *For a binary string S of length n, using $O(n)$ preprocessing time and $O(n)$ bits of space, rank and select queries can be answered in $O(1)$ time.*

Lemma 5 ([7]). *For a character string S of length n over an alphabet Σ of size $O(\mathrm{polylog}(n))$, we can construct an index for S using $O(n \log |\Sigma|)$ bits of space which supports rank and select queries in $O(1)$ time.*

3   The Positional Pattern Matching Problem

Let T be a text of length n and P be a pattern of length p. Our idea is to classify patterns into two types: *short patterns* and *long patterns*, and develop a different index for each of them. Let δ be a threshold to be specified later. A pattern P is *short* if $p \le \delta$, and is *long* otherwise. A positional pattern matching index is abbreviated as a *PPM index*, and a positional pattern matching query with a pattern P and position s is abbreviated as a *PPM query* (P, s).

   In Section 3.1, we first assume $|\Sigma| = O(1)$ and present a basic version of our PPM index, which uses $O(n)$ words and supports $O(p)$ query time. Then, in Section 3.2, we show how to extend the result to $|\Sigma| = O(\mathrm{polylog}(n))$.

3.1   An Index for Finite Alphabets

Throughout this subsection, we assume that $|\Sigma| = O(1)$ and set $\delta = \log n$.

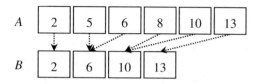

Fig. 1. An illustration of bridges

3.1.1   An Index for Short Patterns

Given a text T, we first construct a suffix tree $\mathcal{ST}$ of T. For ease of discussion, we assume that $\mathcal{ST}$ is a binary tree. In case it is not true, since the out-degree of each internal node is a constant, we simply transform $\mathcal{ST}$ into a binary tree. Recall that the set of all occurrences of a given pattern P in T is $L_{\mu(P)}$. For each node v, define A_v to be the sequence obtained by sorting the labels in L_v increasingly. Then, given a PPM query (P, s), the desired answer for the query is $succ(A_{\mu(P)}, s)$. The locus $\mu(P)$ in $\mathcal{ST}$ can be determined in $O(p)$ time. Therefore, to complete our index for short patterns, it suffices to show how to compute $succ(A_{\mu(P)}, s)$ in $O(p)$ time using $O(n)$ space. The main idea is based on the *bridging technique* [18]. Let A and B be two increasing sequences. The bridges from A to B are pointers from the elements of A to their respective successors in B. An example is given in Fig 1. As will be seen later, in our application, bridges are constructed only from a sequence to its subsequences. Assume that B is a subsequence of A. Then, it is easy to see that for any given number s, $succ(B, s)$ can be obtained in $O(1)$ time by using the bridge of $succ(A, s)$. Consider the example in Fig 1. Given that $succ(A, 7) = 8$, we can obtain $succ(B, 7) = 10$ immediately by following the bridge of the element 8.

Now, we are ready to describe our index for short patterns, which is obtained by augmenting the suffix tree $\mathcal{ST}$ with additional data structures. Let r be the root of $\mathcal{ST}$. We define the *depth* of a vertex v in $\mathcal{ST}$, denoted by $d(v)$, to be the number of edges on the path from the root r to v. For each node v with $d(v) < \log n$, bridges are constructed from A_v to A_u for each child u of v. As mentioned, a PPM query (P, s) can be answered by determining the value of $succ(A_{\mu(P)}, s)$. Let $(v_1 = r, v_2, \ldots, v_g = \mu(P))$ be the path from the root r of $\mathcal{ST}$ to $\mu(P)$. At the root vertex $v_1 = r$, since $L_r = \{1, 2, \ldots, n\}$, we have $succ(A_{v_1}, s) = s$ trivially. When going downwards from v_i to v_{i+1}, $1 \le i < g$, we compute $succ(A_{v_{i+1}}, s)$ in $O(1)$ time by using the bridge of $succ(A_{v_i}, s)$. In this way, when we reach the node $\mu(P)$, $succ(A_{\mu(P)}, s)$ is obtained. Thus, a PPM query can be answered in $O(p)$ time for a short pattern P.

A straightforward implementation of the above index requires $O(n \log n)$ space, as we need to store the sequences A_v and the bridges. In the following, we show that the bridges can be maintained with smaller space. More specifically, we show that rank indexes can be used to serve the function of bridges. Our scheme is motivated by the wavelet tree [8, 14] and thus has a similar flavor. However, as will be discussed in Remark 1 in Section 3.1.2, our index is essentially different from the wavelet tree.

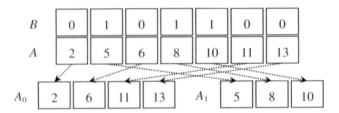

Fig. 2. An illustration of bit-vector B

Lemma 6. *Let A_0 and A_1 be two sorted sequences and let A be the sorted sequence obtained by merging A_0 and A_1. The bridges from A to A_0 and A_1 can be maintained using $O(|A|)$ bits of space.*

Proof. Rather than storing a pointer from each element of A to its successor in A_0, we employ binary rank query to serve the function of bridges. The details are as follows. We create a bit-vector B of size $|A|$ to indicate whether each element of A belongs to A_0 or A_1. Precisely, $B[i] = 0$ if $A[i]$ belongs to A_0 and $B[i] = 1$ otherwise. An example is illustrated in Fig 2. We preprocess B for binary rank queries. Consider a fixed element $A[i]$. If $B[i] = 0$, the successor of $A[i]$ in A_0 is itself and the successor of $A[i]$ in A_1 is the first number $A[i']$ with $B[i'] = 1$ in $A[i, |A|]$. Similarly, if $B[i] = 1$, the successor of $A[i]$ in A_1 is itself and the successor of $A[i]$ in A_0 is the first number $A[i']$ with $B[i'] = 0$ in $A[i, |A|]$. Thus, for each element $A[i]$, the position of its successor in A_j, $j \in \{0, 1\}$, can be obtained as follows: if $B[i] = 0$, we compute $succ^{-1}(A_0, A[i]) = rank_0(B, i)$ and $succ^{-1}(A_1, A[i]) = rank_1(B, i) + 1$; otherwise, we compute $succ^{-1}(A_0, A[i]) = rank_0(B, i) + 1$ and $succ^{-1}(A_1, A[i]) = rank_1(B, i)$. By Lemma 4, it takes $O(|A|)$ bits for supporting binary rank queries. Thus, the lemma holds. $\square$

Given a node v and its two children $left(v)$ and $right(v)$, by Lemma 6 we can construct bridges from A_v to $A_{left(v)}$ and $A_{right(v)}$ using $O(|A_v|)$ bits. The sorted sequences A_v, $A_{left(v)}$, and $A_{right(v)}$ are only needed for constructing the bridges. Thus, we carry out the construction level by level in a bottom-up fashion, keeping the sorted sequences only in two adjacent levels, so that the space remains $O(n)$ during the construction. For the bridge structures, we spend $O(n)$ bits at a fixed level $l < \log n$ of ST, resulting in $O(n \log n)$ bits of space in total.

For a PPM query (P, s), by following the bridges we can determine $k = succ^{-1}(A_{\mu(P)}, s)$ in $O(p)$ time. However, what we need is the value of $A_{\mu(P)}[k]$. In order to obtain $A_{\mu(P)}[k]$, our intent is not to store all the sequences A_v in advance but to follow the bridges back to the root. In other words, given a node v and a position k in A_v, our problem is to identify the number $A_v[k]$ without storing A_v. As shown in [8], this identification can be done efficiently by applying binary select indexes. To make this paper self contained, we give the following.

Lemma 7 ([8]). *Let A_0 and A_1 be two sorted sequences and let A be the sorted sequence obtained by merging A_0 and A_1. We can maintain a data structure of $O(|A|)$ bits, so that given a position k in A_0 (or A_1), the corresponding position of $A_0[k]$ (or $A_1[k]$) in A can be determined in $O(1)$ time.*

Proof. Similar to the proof of Lemma 6, we create a bit-vector B of size $|A|$ to indicate whether each number in A belongs to A_0 or A_1. We preprocess B for binary select queries. Then, for any position k in A_j, $j \in \{0, 1\}$, we have $A_j[k] = A[select_j(B, k)]$. Therefore, the lemma holds. $\square$

To identify $A_v[k]$ for some given node v and position k with $d(v) \le \log n$, we construct the structure of Lemma 7 for each node v of $\mathcal{ST}$ with $1 < d(v) \le \log n$ in a bottom-up fashion to facilitate the identification. It takes $O(n \log n)$ bits in total. Then, by Lemma 7, we can determine the position of $A_v[k]$ in A_r in $O(d(v))$ time by backtracking from v to the root r. Since $A_r[i] = i$ for $i = 1, 2, \ldots, n$, the value of $A_{\mu(P)}[k]$ can be found in $O(d(v))$ time.

In summary, given a PPM query (P, s), we first locate $\mu(P)$ in $\mathcal{ST}$. By following the bridges the value $k = succ^{-1}(A_{\mu(P)}, s)$ is also determined. Then, tracing from $\mu(P)$ back to the tree root r, we compute k' as the position of $A_{\mu(P)}[k]$ in A_r. Finally, we report k' as the answer. We have the following.

Theorem 1. *We can construct an $O(n)$-word index for a string T over a finite alphabet, so that a positional pattern matching query can be answered in $O(p)$ time for any short pattern P.*

3.1.2   An Index for Long Patterns

In this section, we present an $O(n)$-word PPM index with $O(p + \log n) = O(p)$ query time for a long pattern. Recall that in the suffix tree, we maintain an array L and two indices b_v, e_v at each internal node v such that L_v is the set of numbers in $L[b_v, e_v]$. Thus, given a PPM query (P, s), the desired answer is $succ(L[b_{\mu(P)}, e_{\mu(P)}], s)$. And therefore, our problem is to preprocess the array L, so that $succ(L[i, j], s)$ can be efficiently answered for any interval $[i, j]$ and number s. Before presenting our index, we remark that this problem can be solved in the desired time and space bounds by employing the range successor indexes in [14,22]. However, our scheme has the advantage that it can be extended to obtain more efficient indexes for the position-restricted pattern matching problem over an alphabet of size $O(\text{polylog}(n))$, as will be discussed in Section 4.

We build an index for the finding of $succ(L[i, j], s)$ as follows. First, construct the segment tree $\mathcal{GT}$ of $[1, n]$. Consider a fixed internal node v in $\mathcal{GT}$. Let $[b_v, e_v]$ be the interval represented by v, and define A_v to be the sequence obtained by sorting the elements in $L[b_v, e_v]$. Next, for each internal node v of $\mathcal{GT}$, construct the bridge structure of Lemma 6 from A_v to $A_{left(v)}$ and $A_{right(v)}$, where $left(v)$ and $right(v)$ are, respectively, the left and right children of v. We also maintain the structure of Lemma 7 so that the position in A_v of an element in $A_{left(v)}$ or $A_{right(v)}$ can be found in $O(1)$ time. Since $\mathcal{GT}$ is a binary tree and its height is $O(\log n)$, the structures of Lemmas 6 and 7 take $O(n \log n)$ bits of space.

With the above constructed data structure, $succ(L[i, j], s)$ is found as follows. By traversing downwards from the root r, we find a set C of $O(\log n)$ nodes which represent the $O(\log n)$ canonical pieces of the interval $[i, j]$. Then, we have $succ(L[i, j], s) = \min\{succ(L[b_v, e_v], s) \mid v \in C\} = \min\{succ(A_v, s) \mid v \in C\}$. Let $\mathcal{CT}$ be the subtree of $\mathcal{GT}$ that is induced by the nodes in C and their ancestors.

The size of $\mathcal{CT}$ is $O(\log n)$. For each node $v \in C$, the position $k_v = succ^{-1}(A_v, s)$ can be obtained in $O(\log n)$ time by using bridges. After the values of all k_v, where $v \in C$, are available, the task becomes to identify the minimum among all $A_v[k_v]$, without explicitly storing the sorted sequences A_v. From the proof of Lemma 7, using binary select queries, each $A_v[k_v]$ can be found in $O(\log n)$ time by tracing from v back to the root. However, finding all $A_v[k_v]$ requires $O(\log^2 n)$ time in total. Fortunately, the finding of all $A_v[k_v]$ is unnecessary, since only their minimum is required. Consequently, at each internal node u of $\mathcal{CT}$, we can abandon the finding of $A_v[k_v]$ for all of its descendant leaves v that are impossible to induce the answer. For example, let $v_1, v_2 \in C$ be two nodes and u be their parent. In addition, let s_1 and s_2 be the positions of $A_{v_1}[k_{v_1}]$ and $A_{v_2}[k_{v_2}]$ in A_u. If $s_1 > s_2$, then the finding of $A_{v_1}[k_{v_1}]$ can be ruled out, since $A_{v_1}[k_{v_1}]$ is not possible to be the final answer. Similarly, if $s_2 > s_1$, the finding of $A_{v_2}[k_{v_2}]$ can be ruled out. Therefore, with a bottom-up traversal on $\mathcal{CT}$, we can compute $\min\{succ(A_v, s) \mid v \in C\}$ in $O(\log n)$ time. Combining Theorem 1 and the above result, we obtain the following.

Theorem 2. *We can construct an $O(n)$-word index for a string T over a finite alphabet, so that a positional pattern matching query can be answered in $O(p)$ time for any pattern P.*

Remark 1. Similar to the wavelet tree [8], our segment tree structure for long patterns is a complete binary tree augmented with binary rank and select indexes. But, it is essentially different from the wavelet tree. Consider the array L. In our segment tree structure, the ith leaf represents $L[i]$ and each internal node v represents the sorted sequence of the numbers in $L[b_v, e_v]$. However, in the wavelet tree of L, the ith leaf represents i and each internal node v represents the subsequence of L containing only the numbers in Σ_v, where Σ_v is the set of numbers represented by the descendant leaves of v, which corresponds to a subinterval of $[1, n]$. For example, if $L = (3, 8, 2, 5, 4, 1, 7, 6)$, the root represents L and the node v with $\Sigma_v = \{5, 6, 7, 8\}$ represents the subsequence $(8, 5, 7, 6)$.

3.2   An Index for $|\Sigma| = O(\mathbf{polylog}(n))$

In this section, the threshold δ is set to be $\log_{|\Sigma|} n$. By definition, a long pattern P has length $p > \log_{|\Sigma|} n$ and thus $p = \Omega(\log n / \log \log n)$. Therefore, we can apply Yu et al.'s $O(n)$-word index [22] to achieve $O(p + \log n / \log \log n) = O(p)$ query time for long patterns. In the following, we sketch how to obtain an $O(n)$-word index for short patterns with $O(p)$ query time, which in turn completes our index for $|\Sigma| = O(\text{polylog}(n))$. (Due to the page limitation, the details are deferred to the full version of this paper.) Recall that our PPM index for $|\Sigma| = O(1)$ relies on binary rank and select indexes with $O(1)$ query time to facilitate the traversal on a binary tree. With some effort, we can apply Lemma 5 to obtain generalized results of Lemmas 6 and 7, which are beneficial for traversing a tree structure with branching factor $O(\text{polylog}(n))$. The generalized results can then be employed on each node v in $\mathcal{ST}$ with $d(v) \leq \log_{|\Sigma|} n$. Since the out-degree of each internal node in $\mathcal{ST}$ is $O(|\Sigma|) = O(\text{polylog}(n))$, the space usage is $O(n \log |\Sigma|)$ bits for a fixed level. In total, our index takes $O(n \log |\Sigma| \times \log_{|\Sigma|} n) = O(n \log n)$ bits. The query process is similar to that of the index in Section 3.1.1. In summary, we have the following.

Theorem 3. *We can construct an $O(n)$-word index for a string T over an alphabet of size $O(\mathrm{polylog}(n))$, so that a positional pattern matching query can be answered in $O(p)$ time for any pattern P.*

4   The Position-Restricted Pattern Matching Problem

For ease of exposition, we only present indexes for T that report all occurrences of P in $T[s, n]$ in sorted order for any pattern P and position s given later. It is easy to modify our solutions to report all occurrences of P in T in sorted order within a given query interval $[s, t]$.

The indexes are adapted from the index in Section 3. We classify patterns into three types: *short patterns*, *medium patterns*, and *long patterns*. A pattern P of length p is *short* if $p \leq \log_{|\Sigma|} n$ and is *long* if $p > \log n$. If $\log_{|\Sigma|} n < p \leq \log n$, then P is a *medium* pattern. Using $O(n \log n)$ space, we first show how to answer each query for each type of patterns in $O(p + occ)$ time. Next, the space requirement is reduced to the desired bounds.

An Index for Short Patterns. Our index for short patterns is the same as that in Section 3.2, except that we explicitly store A_v for each node v in $\mathcal{ST}$ with $d(v) \leq \log_{|\Sigma|} n$. Given a PPM query (P, s), it is easy to see that $A_{\mu(P)}(succ^{-1}(A_{\mu(P)}, s), |A_{\mu(P)}|)$ is the sorted occurrences of P in $T[s, n]$. Therefore, the query time is $O(p + occ)$ and the space usage is $O(n \log_{|\Sigma|} n)$.

An Index for Medium Patterns. Similar to the index for short patterns, we explicitly store A_v for each node v in $\mathcal{ST}$ with $\log_{|\Sigma|} n < d(v) \leq \log n$, so that all occurrences of P in $T[s, n]$ can be easily reported in sorted order once $succ^{-1}(A_{\mu(P)}, s)$ has been determined. To save space, we do not construct bridges for those nodes v with $\log_{|\Sigma|} n < d(v) \leq \log n$. Instead, for medium patterns we employ two *dominance queries* to determine $succ^{-1}(A_{\mu(P)}, s)$. Given an array A, a dominance query asks to return the number of elements in $A[1, i]$ that are less than or equal to s, where both i and s are specified in the query stage. We index the array L for dominance queries, so that we can obtain $succ^{-1}(A_{\mu(P)}, s)$ by determining the numbers of elements in $L[1, b_{\mu(P)} - 1]$ and in $L[1, e_{\mu(P)}]$, respectively, that are less than or equal to s. Since L consists of integers in $[1, n]$, we use the $O(n)$-word index in [3] to support each dominance query in $O(\log n / \log \log n)$ time. In summary, we answer the query of a medium pattern as follows. First, locate the locus $\mu(P)$. Then, invoke two dominance queries to compute $succ^{-1}(A_{\mu(P)}, s)$. Finally, output the subsequence $A_{\mu(P)}(succ^{-1}(A_{\mu(P)}, s), |A_{\mu(P)}|)$. The total space is $O(n \log n)$ and the query time is $O(p + \log n / \log \log n + occ) = O(p + occ)$.

An Index for Long Patterns. Our index for long patterns is the same as that in Section 3.1.2, except that we explicitly store A_v at each node v of the segment tree $\mathcal{GT}$. Let U be the sorted sequence of the occurrences of P in $T[s, n]$. Recall that to answer a PPM query, the interval $[b_{\mu(P)}, e_{\mu(P)}]$ is decomposed into $O(\log n)$ canonical pieces. Let C be the set of the $O(\log n)$ nodes that represent the pieces. Then, U can be obtained by merging the $|C|$ sequences $A_v(succ^{-1}(A_v, s), |A_v|)$, where $v \in C$. Here, we use Willard's Q*-heap [21] for the merge. The Q*-heap maintains a set S of $O(\text{polylog}(n))$ integers in $\{1, 2, \ldots, n\}$ using $O(|S|)$ space and supports each insertion, deletion, and find-min operation in $O(1)$ time, where a find-min operation returns the smallest element in the heap. With the Q*-heap, the numbers in U can be reported one by one with constant delay between two successive outputs. The total query time is $O(p + \log n + occ) = O(p + occ)$.

Reducing the Space Usage. So far, we have obtained an $O(n \log n)$-word index with $O(p + occ)$ query time. The $O(n \log n)$ space requirement comes from maintaining the sorted sequences A_v on the suffix tree $\mathcal{ST}$ and the segment tree $\mathcal{GT}$. To save the space, we utilize the following result by Chazelle [4].

Lemma 8 ([4]). *Let $\mathcal{BT}$ be a binary tree of height $O(\log n)$ with n leaves, where each leaf is associated with a number. For each internal node v, denote by S_v the sorted sequence of all numbers associated in its descendant leaves. Then, we can construct a data structure so that the retrieval of an arbitrary element in a sequence S_v can be accomplished in (1) $O(n \log^\epsilon n)$ space and $O(1)$ time; or (2) $O(n)$ space and $O(\log^\epsilon n)$ time.*

In the indexes for short patterns and medium patterns, we explicitly store the sorted sequence A_v at each node v in $\mathcal{ST}$ with $d(v) \leq \log n$, which requires $O(n \log n)$ space. Our intent is to apply the result in Lemma 8 to store the sequences A_v. More specifically, we intend to build a binary tree $\mathcal{BT}$ of height $O(\log n)$ with n leaves such that for each node v in $\mathcal{ST}$ with $d(v) \leq \log n$, there exists a corresponding node v' in $\mathcal{BT}$ with $L_v = L_{v'}$. The tree $\mathcal{BT}$ is obtained from $\mathcal{ST}$ as follows. First, we create $\mathcal{BT}$ as a copy of $\mathcal{ST}$. Then, for each internal node v with $d(v) = \log n$, we remove all other internal nodes in the subtree rooted at v and then connect v to each of its descendant leaves, making $\mathcal{BT}$ a tree of height $\log n + 1$. There is still a problem. The out-degree of each node in $\mathcal{BT}$ can be as large as $|\Sigma|$, whereas the result of Lemma 8 holds only for binary trees. To resolve this problem, we transform $\mathcal{BT}$ into a binary tree by using the technique in [18] as follows: for each interval node v with $k > 2$ children $v_1, v_2, \ldots, v_k$, we replace v and its children by a weight-balanced binary search tree with root v and leaves $v_1, v_2, \ldots, v_k$, in which node v_i has weight equal to its number of descendant leaves. The above transformation increases the tree height by $O(\log n)$ [18]. Thus, we can apply Chazelle's data structures on $\mathcal{BT}$ to store the sorted sequences A_v, where v is a node in $\mathcal{ST}$ with $d(v) \leq \log n$. For long patterns, Chazelle's data structures can be directly applied on $\mathcal{GT}$ to store the sorted sequences A_v. Consequently, we obtain the following.

Theorem 4. *We can construct an index for a text T over an alphabet of size $O(\text{polylog}(n))$, so that each position-restricted query is answered in (1) $O(n \log^\epsilon n)$ space and $O(p + occ)$ time; or (2) $O(n)$ space and $O(p + occ \log^\epsilon n)$ time.*

5   Variable-Length Don't Care Pattern Matching Problem

Let T be a text over an alphabet Σ and $P = P_1*P_2*\ldots*P_m$ be a pattern such that each P_i is a string over Σ. Pinter [17] had a linear-time algorithm for determining whether P occurs in T. His algorithm is based upon the following observation: if P_1 does not occur in T, then neither does P; otherwise, P occurs in T if and only if $P_2*\ldots*P_m$ occurs in $T[k_1 + |P_1|, n]$, where k_1 is the first occurrence of P_1 in T. Consider the example in Fig 3. The first occurrence of P_1 is at position 3. Thus, our problem reduces to determining whether P_2*P_3 occurs in $T[8, n]$. Similarly, since the first occurrence of P_2 in $T[8, n]$ is at position 11, the problem further reduces to determining whether P_3 occurs in $T[15, n]$. Finally, since P_3 occurs in $T[15, n]$, we conclude that P occurs in T.

$$
\begin{array}{ll}
T & \text{CACAATCAATCACAATCACACATGGCCTGCT} \\
P_1 & \quad\;\text{CAATC} \\
P_2 & \qquad\qquad\quad\text{CACA} \\
P_3 & \qquad\qquad\qquad\qquad\qquad\text{GC}
\end{array}
$$

Fig. 3. Finding a pattern $P_1*P_2*P_3$ in a text T

Inspired by Pinter's work, we construct the $O(n)$-word PPM index in Theorem 3 for T and then answer a query pattern $P = P_1*P_2*\ldots*P_m$ as follows. Initially, set $i = 1$ and $s = 1$. Then, iterate as follows. First, find the first occurrence k of P_i in $T[s, n]$ by a PPM query. Next, if k does not exist, report that P does not occur in T. Otherwise, do the following: if $i = m$, report that P occurs in T; and otherwise, increase i by one, set $s = k+|P_i|$, and then proceed to the next iteration. It is easy to see that the query time is $O(|P_1|+|P_2|+\ldots+|P_m|) = O(|P|)$. We obtain the following.

Theorem 5. *We can construct an $O(n)$-word index for a string T over an alphabet Σ of size $O(\mathrm{polylog}(n))$, so that we can determine in $O(p)$ time whether a pattern P over $\Sigma \cup \{*\}$ occurs in T.*

6   Concluding Remarks

Mäkinen and Navarro [14] had an $O(n)$-word index with $O(p+\log n)$ query time for the following problem: preprocess a text T, so that for any pattern P and integer k, the kth occurrence of P in T can be found efficiently. Using the result in [3], the query time can be improved to $O(p + \log n/ \log \log n)$. We remark that when $|\Sigma| = O(\mathrm{polylog}(n))$, using our data structure in Section 3.2, the query time can be further improved to $O(p)$. One direction for further study is to investigate more pattern matching problems which can be solved efficiently based on the idea in this paper.

References

1. Bentley, J.L.: Solutions to Klee's rectangle problems. Department of Computer Science, Carnegie Mellon University (1977) (manuscript)
2. Bertossi, A.A., Lodi, E.: Parallel string matching with variable length don't cares. J. Parallel Distrib. Comput. 22(2), 229–234 (1994)

3. Brodal, G.S., Jørgensen, A.G.: Data structures for range median queries. In: Dong, Y., Du, D.-Z., Ibarra, O.H. (eds.) ISAAC 2009. LNCS, vol. 5878, pp. 822–831. Springer, Heidelberg (2009)
4. Chazelle, B.: A functional approach to data structures and its use in multidimensional searching. SIAM J. Comput. 17(3), 427–462 (1988)
5. Clark, D.: Compact pat trees. PhD Thesis, Univ. Waterloo (1996)
6. Crochemore, M., Iliopoulos, C.S., Kubica, M., Rahman, M.S., Walen, T.: Improved algorithms for the range next value problem and applications. In: 25th Annual Symposium on Theoretical Aspects of Computer Science, pp. 205–216 (2008)
7. Ferragina, P., Manzini, G., Mäkinen, V., Navarro, G.: Compressed representations of sequences and full-text indexes. ACM Transactions on Algorithms 3(2) (2007)
8. Grossi, R., Gupta, A., Vitter, J.S.: High-order entropy-compressed text indexes. In: 14th Annual ACM-SIAM Symposium on Discrete Algorithms, pp. 841–850 (2003)
9. Hagerup, T., Miltersen, P.B., Pagh, R.: Deterministic dictionaries. J. Algorithms 41(1), 69–85 (2001)
10. Inenaga, S., Takeda, M., Shinohara, A., Hoshino, H., Arikawa, S.: The minimum DAWG for all suffixes of a string and its applications. In: Apostolico, A., Takeda, M. (eds.) CPM 2002. LNCS, vol. 2373, pp. 153–167. Springer, Heidelberg (2002)
11. Keller, O., Kopelowitz, T., Lewenstein, M.: Range non-overlapping indexing and successive list indexing. In: Dehne, F., Sack, J.-R., Zeh, N. (eds.) WADS 2007. LNCS, vol. 4619, pp. 625–636. Springer, Heidelberg (2007)
12. Kucherov, G., Rusinowitch, M.: Matching a set of strings with variable length don't cares. Theor. Comput. Sci. 178(1-2), 129–154 (1997)
13. Lam, T.-W., Sung, W.-K., Tam, S.-L., Yiu, S.-M.: Space efficient indexes for string matching with don't cares. In: Tokuyama, T. (ed.) ISAAC 2007. LNCS, vol. 4835, pp. 846–857. Springer, Heidelberg (2007)
14. Mäkinen, V., Navarro, G.: Rank and Select Revisited and Extended. Theor. Comput. Sci. 387(3), 332–347 (2007)
15. McCreight, E.M.: A space-economical suffix tree construction algorithm. J. ACM 23(2), 262–272 (1976)
16. Munro, J.I.: Tables. In: 16th Conference on Foundations of Software Technology and Theoretical Computer Science, pp. 37–42 (1996)
17. Pinter, R.Y.: Efficient string matching with don't-cares. Combinatorial Algorithms on Words 12, 11–29 (1985)
18. Preparata, F.P., Shamos, M.I.: Computational Geometry: An Introduction. Springer, Heidelberg (1985)
19. Sustik, M.A., Moore, J.S.: String searching over small alphabets. Technical Report TR-07-62, Department of Computer Sciences, University of Texas at Austin (2007)
20. Thathoo, R., Virmani, A., Lakshmi, S.S., Balakrishnan, N., Sekar, K.: TVSBS: A fast exact pattern matching algorithm for biological sequences. Current Sciences 91(1), 47–53 (2006)
21. Willard, D.E.: Examining computational geometry, van Emde Boas trees, and hashing from the perspective of the fusion tree. SIAM J. Comput. 29(3), 1030–1049 (2000)
22. Yu, C.-C., Hon, W.-K., Wang, B.-F.: Efficient data structures for the orthogonal range successor problem. In: Ngo, H.Q. (ed.) COCOON 2009. LNCS, vol. 5609, pp. 96–105. Springer, Heidelberg (2009)

Dynamic Range Reporting in External Memory

Yakov Nekrich

Department of Computer Science
University of Bonn
yasha@cs.uni-bonn.de

Abstract. In this paper we describe a dynamic external memory data structure that supports range reporting queries in three dimensions in $O(\log_B^2 N + \frac{k}{B})$ I/O operations, where k is the number of points in the answer and B is the block size. Our data structure uses $O(\frac{N}{B} \log_2^2 N \log_2^2 B)$ blocks of space and supports updates in $O(\log_B^3 N)$ amortized I/Os. This is the first dynamic data structure that answers three-dimensional range reporting queries in $\log_B^{O(1)} N + O(\frac{k}{B})$ I/Os.

1 Introduction

The orthogonal range reporting problem is to maintain a set of points S in a data structure so that for an arbitrary axis-aligned query rectangle Q all points in $Q \cap S$ can be reported. This is a fundamental problem with several important applications, such as geographic information systems, computer graphics, and databases. In this paper we present a dynamic external-memory data structure that supports three-dimensional range reporting queries in $O(\log_B^2 N + \frac{k}{B})$ I/O operations and updates in $O(\log_2^3 N)$ I/O operations, where k is the number of reported points and N is the number of points in the data structure.

In the external memory model the data is stored in *disk blocks* of size B, a block can be read into internal memory from disk (resp. written from internal memory into disk) with one *I/O operation (I/O)*, and computation can only be performed on data stored in the internal memory. The space usage is measured in the number of blocks, and the time complexity is measured in the number of I/O operations. A more detailed description of the external memory model can be found in e.g. [22] or [4]. Since we are interested in minimizing the number of I/O operations, an efficient data structure should support queries in $\log_B^{O(1)} N + O(\frac{k}{B})$ I/O operations.

In the RAM computation model, there are both static and dynamic data structures that use $N \log_2^{O(1)} N$ space and support d-dimensional orthogonal queries in $O(\log_2^{O(1)} N + k)$ time for any constant d; see e.g., [3] for a survey of previous results. In the external memory model, these results can be matched only in two dimensions (dynamic data structure) and three dimensions (static data structure). The dynamic data structure of Arge *et al.* [9] uses $O((N/B) \log_2 N / \log_2 \log_B N)$ blocks of space and supports two-dimensional range reporting queries and updates in $O(\log_B N + \frac{k}{B})$ and $O(\log_B N (\log_2 N / \log_2 \log_B N))$ I/O operations respectively. The static data structure of Vengroff and Vitter [23,22] supports

O. Cheong, K.-Y. Chwa, and K. Park (Eds.): ISAAC 2010, Part II, LNCS 6507, pp. 25–36, 2010.

Table 1. Upper bounds for orthogonal range reporting in RAM and external memory models in two and three dimensions. Only dynamic results in the RAM model are listed. For comparison, the space usage of data structures in the RAM model is specified in blocks of size B. We denote by ω and ε arbitrary constants such that $\varepsilon > 0$ and $\omega > 7/8$; our result is marked with an asterisk.

Model	Ref.	Query	Space	Update
$d = 2$:				
RAM	[13]	$O(\log_2 N/\log_2\log_2 N + k)$	$O((N/B)\log^\omega N)$	$O(\log^\omega N)$
IO	[9]	$O(\log_B N + \frac{k}{B})$	$O((N/B)\log_2 N/\log_2\log_B N)$	$O(\log_B N \log_2 N/\log_2\log_B N)$
$d = 3$				
RAM	[13]	$O((\log_2 N/\log_2\log_2 N)^2 + k)$	$O((N/B)\log^{\omega+1} N)$	$O(\log^{\omega+1} N)$
IO	[23]	$O(\log_B N + \frac{k}{B})$	$O((N/B)\log_2^4 N)$	-
IO	[1]	$O(\log_B N + \frac{k}{B})$	$O((N/B)\log_2^3 N)$	-
IO	[2]	$O(\log_B N + \frac{k}{B})$	$O((N/B)(\log_2 N/\log_2\log_B N)^3)$	-
IO	[2]	$O(\log_B N(\log_2 N/\log_2\log_B N) + \frac{k}{B})$	$O((N/B)(\log_2 N/\log_2\log_B N)^2)$	-
IO	[9]+[5]	$O(\log_B N(\log_2 N/\log_2\log_2 N) + \frac{k}{B})$	$O((N/B)\log_2^{2+\varepsilon} N)$	$O(\log_B N \log_2^{1+\varepsilon} N)$
IO	*	$O(\log_B^2 N + \frac{k}{B})$	$O((N/B)\log_2^2 N \log_2^2 B)$	$O(\log_2^3 N)$

three-dimensional range reporting queries in $O(\log_B N + \frac{k}{B})$ I/Os and uses ~~$O((N/B)\log_2^4 N)$~~ blocks of space. The space usage of a three-dimensional data structures was improved by Afshani [1] ~~and Afshani, Arge, and Larsen [2]~~ to $O((N/B)\log_2^3 N)$ and $O((N/B)(\log_2 N/\log_2\log_B N)^3)$ blocks respectively; see Table 1. The query cost can be improved if all point coordinates are positive integers bounded by a parameter U [15,16,1], and the space usage can be reduced for some special cases of orthogonal queries, such as dominance queries; we refer the reader to [1,2] for a more detailed description of special cases and to [7] for an extensive description of previous results.

Using range trees with fan-out $\log^\varepsilon N$ [5], we can transform a two-dimensional data structure into a data structure that supports d-dimensional orthogonal queries, so that the cost of queries increases by a $O(\log_2 N/\log_2\log_2 N)$ factor for each dimension, while the space usage and update cost increase by a factor of $O(\log_2^{1+\varepsilon} N)$ for each dimension. The recent (static) dimension reduction technique of [2] increases the cost of queries by a $O(\log_2 N/\log_2\log_B N)$ factor and the space usage also by a $O(\log_2 N/\log_2\log_B N)$ factor. These techniques can be used to obtain three-dimensional data structures that support queries with $O(\log_B N(\log_2 N/\log_2\log_2 N) + \frac{k}{B})$ and $O(\log_B N(\log_2 N/\log_2\log_B N) + \frac{k}{B})$ I/Os respectively; see Table 1. However, these data structures do not achieve $O(\log_B^c N)$ query bound for any B and a constant c. In the case when $B = \Omega((\log_2 N)^{f(N)})$ for some function $f(N) = \Omega(1)$, we need $\Omega(f(N)\log_B^2 N) + O(\frac{k}{B})$ I/O operations to answer queries using the combination of [9] and [5] or the result of [2]. We also don't know if there are efficient (static or dynamic) data structures for range reporting in $d \geq 4$ dimensions that report all points in $\log_B^{O(1)} N + O(\frac{k}{B})$ I/O operations.

In this paper we describe a data structure that uses $O(\frac{N}{B}\log_2^2 N \log_2^2 B)$ blocks of space, supports updates in $O(\log_2^3 N)$ amortized I/Os, and answers three-dimensional orthogonal range reporting queries in $O(\log_B^2 N + \frac{k}{B})$ I/Os. Thus our result "matches" the query complexity of the dynamic RAM data structure of [13]. Moreover, the space usage of our data structure differs by a

$O(\log_2^2 B(\log_2 \log_B N)^3 / \log_2 N)$ factor from the best previously known external memory static data structure [2]. Hence, when B is not very large, i.e., when $\log_2 B = o(\sqrt{\log_2 N / (\log_2 \log_B N)^3})$, our dynamic data structure uses less space than the static data structure of [2].

In section 2 we describe a dynamic data structure that supports dominance queries in $O(\frac{k}{B})$ I/Os when the set S contains $O(B^{4/3})$ points. Our data structure maintains $O(\log_2 B)$ t-approximate boundaries of [23], that will be defined in section 2. We show that each t-approximate boundary can be constructed with $O(B \log_2 B)$ I/O operations for $\geq B$ and a small set S. The cost of re-building the data structure is distributed among $O(B^{4/3})$ updates with the lazy updates approach: the newly inserted and deleted points are stored in two buffers for each t-approximate boundary, and each t-approximate boundary is re-built when one of its buffers contains the sufficient number of points. We further improve the update time by showing how to store only two buffers for all boundaries. The trick of storing inserted (deleted) points for different boundaries in the same buffer may be of independent interest. Using standard techniques, more general orthogonal range queries can be reduced to dominance queries as described in section 2.1.

In section 3 we describe a data structure that supports $(2, 1, 2)$-sided queries $Q = [a, b] \times [c, +\infty) \times [d, e]$ on a set of points S such that $p.z = O(B^f)$ for a small constant f and for any $p \in S$. Here and further we denote by $p.x$, $p.y$, and $p.z$ the x-, y-, and z-coordinates of a point p. The main idea of section 3 is to store points in a data structure $\mathcal{T}$ that is similar to the external memory priority search tree, but contains three-dimensional points. The data structure for small sets from section 2.1 is used to guide the search in each node of $\mathcal{T}$. The data structure that supports arbitrary $(2, 1, 2)$-sided queries is described in section 4. The data structure is based on a range tree with fan-out $\Theta(B^f)$ for a small constant f that is built on z-coordinates of points. The main idea of section 4 is to store the data structure F_v of section 3 in every node v of the range tree. The z-coordinate of each point p in F_v is replaced with an index bounded by $\Theta(B^f)$ that indicates which child of the node v contains p. We show how a general $(2, 1, 2)$-sided query can be reduced to $O(\log_B N)$ queries to data structures F_v. Finally, we can obtain a data structure for general three-dimensional queries from the data structure for $(2, 1, 2)$-sided queries using standard techniques.

Thus our approach is based on a combination of some previously known techniques with some novel ideas. In particular we believe that the data structures described in sections 3 and 4 and the general decomposition of the three-dimensional range reporting problem into subproblems are new.

2   Dominance Reporting for Small Sets

A point q dominates a point p if all coordinates of q are greater than or equal to the respective coordinates of p. The dominance reporting query is to report all points $p \in S$ that dominate a query point q. A three-dimensional dominance reporting query is equivalent to reporting all points in a product of three half-open intervals. In this section we describe a dynamic data structure that contains $O(B^{4/3})$

elements and supports dominance reporting queries and updates. The main idea of this data structure is that the t-approximate boundary [23] for a small set of elements can be efficiently maintained under insertions and deletions.

Overview. A three-dimensional t-approximate boundary was introduced by Vengroff and Vitter [23]. A t-approximate boundary for a three-dimensional set S is a surface $\mathcal{V}$ that satisfies the following properties: (1) $\mathcal{V}$ divides the space, i.e. every point either dominates a point on $\mathcal{V}$ or is dominated by a point of $\mathcal{V}$; (2) every point of $\mathcal{V}$ is dominated by at least t and at most $3t$ points of S. An example of a t-approximate boundary constructed with the algorithm of [23] is shown on Fig. 1. There are $O(|S|)$ points on $\mathcal{V}$ called *inward corners*, such that every point on $\mathcal{V}$ dominates an inward corner and an inward corner does not dominate any point on $\mathcal{V}$ (except of itself). There is a linear space data structure that finds an inward corner c of $\mathcal{V}$ that is dominated by a query point q, if such inward corner c exists, and reports all points of S that dominate c in $O(\log_B(|S|)+t/B)$ I/Os. We maintain $(\log_2 B)/6$ t-approximate boundaries $\mathcal{V}_1, \mathcal{V}_2, \ldots, \mathcal{V}_s$, where $\mathcal{V}_i$ is a $B \cdot 2^{2i}$-approximate boundary. Given a query point q, we examine $\mathcal{V}_1, \mathcal{V}_2, \ldots, \mathcal{V}_i$ and find the minimal index i, such that q dominates an inward corner c_j of $\mathcal{V}_i$ using the method described in [23]. We can test each $\mathcal{V}_i$ in $O(\log_B B^{4/3}) = O(1)$ I/Os and thus find the index i in $O(i)$ I/Os. If q dominates an inward corner c_j of $\mathcal{V}_i$ but does not dominate any point on $\mathcal{V}_{i-1}$, then q is dominated by $\Theta(2^{2i}B)$ points of S. Since q dominates c_j, all points that dominate q also dominate c_j. Hence, we can examine the list of points that dominate c_j and report all points that dominate q in $O(2^{2i}) = O(\frac{k}{B})$ I/O operations. Thus the total query cost is $O(\frac{k}{B})$. See [23] for a more detailed description.

We can construct a t-approximate boundary $\mathcal{V}_i$ with $O(B)$ I/O operations if S contains $O(B^{4/3})$ points and $t \geq B$; the algorithm is described in the full version of this paper [17]. In the next part of this section we show how the data structure for a small set of points can be dynamized by distributing the construction cost among $\Theta(B)$ updates. This is achieved by storing buffers with newly inserted and deleted points and periodically rebuilding the data structure. Then, we show that we can support updates in $O(1)$ amortized I/Os on the data structure that consists of $O(\log_2 B)$ boundaries by storing one buffer with recently inserted points and one buffer with recently deleted points for *all* t-approximate boundaries.

Deletion-only Data Structure. A t-approximate boundary $\mathcal{V}_i$ supports lazy deletions in $O(1)$ amortized I/O operations. When a point p is deleted, we simply add it to a list $\mathcal{D}$ of deleted elements that may contain up to $2^{2i-1}B$ points. Let T be the list of points that dominate a query point q; we can obtain T in $O(\frac{|T|}{B})$ I/Os as described in the beginning of this section. We can sort T in $O(\frac{|T|}{B} \log_B |T|) = O(2^{2i} \log_B(2^{2i}B)) = O(2^{2i})$ I/Os (we assume that each point in S has a unique integer identifier). We can also sort $\mathcal{D}$ in $O(2^{2i})$ I/Os. Then, we traverse T and $\mathcal{D}$ and remove from T all points that occur in $\mathcal{D}$ in $O(\frac{|T|+|\mathcal{D}|}{B}) = O(2^{2i})$ I/Os. Since we use $\mathcal{V}_i$ when $\frac{k}{B} = \Omega(2^{2i})$, the query cost remains unchanged. When the number of deleted points in $\mathcal{D}$ equals to $B \cdot 2^{2i}/2$,

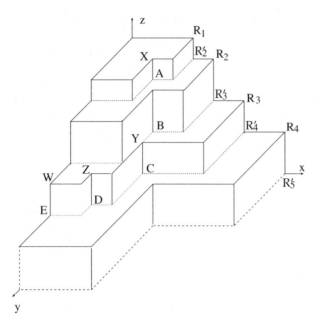

Fig. 1. An example of a t-approximate boundary. The points of the set S are not shown. Ridges R'_2, R'_3, R'_4, and R'_5 are drawn with dotted lines. Ridges R_1, R_2, R_3, and R_4 are drawn with solid lines. A, B, C, D, E are examples of inward corners. X, Y, Z, and W are examples of in-corners; X belongs to ridge R_1, and Y, Z, and W belong to ridge R_3.

we re-build the data structure for $\mathcal{V}_i$ without deleted points in $O(B)$ I/Os and empty the list $\mathcal{D}$.

Supporting Insertions. Insertions can be supported with a similar technique. Inserted points are stored in the list of new points $\mathcal{I}$ that may contain up to $2^{2i-1}B$ points. When a point p is deleted, we add it to a list $\mathcal{D}$ of deleted points as described above. If a point p stored in $\mathcal{I}$ is deleted, we simply remove p from $\mathcal{I}$. When $\mathcal{I}$ contains $2^{2i-1}B$ points, we re-build the data structure for $\mathcal{V}_i$. To answer a query, we examine all points from T that do not belong to $\mathcal{D}$ in $O(\frac{k}{B})$ I/Os as described in the previous paragraph. Then, we traverse the list $\mathcal{I}$ and report all point that dominate the query point in $O(2^{2i-1}) = O(\frac{k}{B})$ I/Os.

Updates with $O(1)$ Amortized Cost. Since our data structure consists of $O(\log_2 B)$ boundaries $\mathcal{V}_i$, the total cost of an update is $O(\log_2 B)$. We can reduce the amortized update cost to a constant by storing newly inserted points for all boundaries in one list $\mathcal{I}$ and newly deleted points for all boundaries in one list $\mathcal{D}$. An array D stores pointers to elements of $\mathcal{D}$, such that all elements between $D[i]$ and the end of $\mathcal{D}$ are removed from the data structure for $\mathcal{V}_i$. An array I stores pointers to elements of $\mathcal{I}$, such that all elements between $I[i]$ and the end of $\mathcal{I}$ are new elements that are not yet inserted into the data structure for $\mathcal{V}_i$. The pointer $\text{end}(\mathcal{D})$ ($\text{end}(\mathcal{I})$) points to the last (in chronological order) deleted (inserted) element stored in $\mathcal{D}$ ($\mathcal{I}$). Both $\mathcal{D}$ and $\mathcal{I}$ also contain one additional dummy element l_D (resp., l_I) that follows end($\mathcal{D}$) (resp., end($\mathcal{I}$)). When a new point p is inserted, we store p in the l_I, set the pointer end($\mathcal{I}$) so that it points to l_I, and append a new dummy element after end($\mathcal{I}$). A deleted element is appended at the end of $\mathcal{D}$ with the same procedure. After $2^{2i-1}B$ deletions we rebuild the data structure for $\mathcal{V}_i$ without deleted elements and change $D[i]$ so that it points to l_D. After $2^{2i-1}B$ insertions we rebuild the data structure for $\mathcal{V}_i$ with new elements and change $I[i]$ so that it points to l_I. After $\Theta(\log_2 B \cdot B)$ updates, we re-build the data structures for all $\mathcal{V}_i$ as well as the lists $\mathcal{I}$ and $\mathcal{D}$. This incurs an amortized cost $O(1)$. The total cost of re-building data structures and (pointers to) lists $\mathcal{D}$ and $\mathcal{I}$ in a sequence of $B^{4/3}$ update operations is $O(\sum_{j=0}^{r} 2^{r-j}B) = O(B^{4/3})$, where $r = \log_2 B/6 + O(1)$ is the index of the last t-approximate boundary $\mathcal{V}_s$. We can report all points that dominate an inward corner of $\mathcal{V}_i$ in $O(2^{2i})$ I/Os as described above. Hence, dominance queries can be supported in $O(\frac{k}{B})$ I/Os. This result is summarized in the following Lemma.

Lemma 1. *Elements of a set S such that $|S| = O(B^{4/3})$ can be stored in a data structure that uses $O(\frac{|S|}{B} \log_2 |S|)$ blocks of space and supports dominance queries in $O(\frac{k}{B})$ I/O operations and updates in $O(1)$ I/O operations amortized.*

2.1 $(1,1,2)$- and $(2,1,2)$-Sided Queries for Small Sets

Suppose that b_x, b_y, and b_z are natural constants such that $1 \leq b_x, b_y, b_z \leq 2$. We say that a query Q is a (b_x, b_y, b_z)-sided query if the projection of Q on the x-axis is bounded on b_x sides, the projection of Q on the y-axis is bounded on b_y sides and the projection of Q on the z-axis is bounded on b_z sides. Thus the projection of Q on the x-axis (resp., y- or z-axis) is a an infinite half-open interval if b_x (resp., b_y or b_z) equals 1, and the projection of Q on the x-axis (resp., y- or z-axis) is a finite closed interval if b_x (resp., b_y or b_z) equals 2. Dominance queries considered in section 2 are equivalent to $(1,1,1)$-sided queries. Using a standard reduction [11,21], we can transform a $O(s(N))$ space data structure that supports $(1,1,1)$-sided queries in $O(t(N) + k/B)$ time and updates in $O(u(N))$ time into a $O(s(N) \log_2^m N)$ space data structure that supports (b_x, b_y, b_z)-sided queries in $O(t(N) + \frac{k}{B})$ time and updates in $O(u(N) \log_2^m N)$ time; here $m = b_x + b_y + b_z - 3$. Applying this transformation to the external memory data structure of Lemma 1, we obtain the following result.

Lemma 2. *Let $1 \leq b_x, b_y, b_z \leq 2$ and $m = b_x + b_y + b_z - 3$. Elements of a set S such that $|S| = O(B^{4/3})$ can be stored in a data structure that uses $O(\frac{|S|}{B} \log_2^{m+1} |S|)$ blocks of space and supports (b_x, b_y, b_z)-sided queries in $O(\frac{k}{B})$ I/O operations and updates in $O(\log_2^m(|S|))$ I/O operations amortized.*

In particular, we can support $(2,1,2)$-sided queries in $O(\frac{k}{B})$ I/Os and updates in $O(\log_2^2 B)$ I/Os on a set S that contains $\Theta(B^{4/3})$ points using a data structure that needs $O(B^{1/3} \log_2^3 B)$ blocks of space.

3   Extended Three-Sided Queries

In this section we describe a data structure that supports $(2, 1, 2)$-sided reporting queries when z-coordinates of all points are positive integers bounded by $\Theta(B^f)$, $p.z = \Theta(B^f)$ for all points $p \in S$. Here f is a constant such that $f \le 1/6$.

Data Structure. Our data structure is a modification of the external memory priority search tree [9]. The (external) priority search tree is a tree built on x-coordinates of two-dimensional points. A point stored in a leaf is associated with an ancestor of l or with l itself, so that the following property is guaranteed: points associated with a node v have larger y-coordinates than points associated with descendants of v. The main idea of our modification is to maintain this property for every possible value of the z-coordinate. Thus, we maintain the data structure of section 2.1 in each tree node and use it to guide the search, i.e., to decide which descendants of a node must be visited.

We construct a tree $\mathcal{T}$ with fan-out $\Theta(B^f)$ on the set of x-coordinates of all points. We store $\Theta(B^{1+f})$ values, i.e., x-coordinates of $\Theta(B^{1+f})$ consecutive points of S, in each leaf node. The range of an internal node v is an interval $rng(v) = [a_v, b_v]$, where a_v and b_v are the smallest and the largest values stored in the leaf descendants of v.

We associate a set of points S_v with each node v of $\mathcal{T}$. Sets S_v can be constructed by visiting nodes of $\mathcal{T}$ in pre-order. For the root r of $\mathcal{T}$, let L_r be the set of all points in S sorted in increasing order by their y-coordinates, and let $L_r[j]$ be the set of all points $p \in S$, $p.z = j$, sorted in increasing order by their y-coordinates. The set $S_r[j]$ contains the last B points of $L_r[j]$, i.e., B points with largest y-coordinates. For each non-root node v of $\mathcal{T}$, the list L_v contains all points p such that $p.x$ belongs to the range of v and p does not belong to any S_w, where w is an ancestor of v; points in L_v are sorted in increasing order by their y-coordinates. The list $L_v[j]$ contains all points $p \in L_v$ such that $p.z = j$. If v is an internal node, the set $S_v[j]$ contains the last B points of $L_v[j]$. If v is a leaf, then $S_v[j]$ contains all points from $L_v[j]$. Note that $L_v[j]$ and $S_v[j]$ may contain less than B points or even be empty for some j. The set S_v is the union of all sets $S_v[j]$, $S_v = \cup_j S_v[j]$. For any node v, $|S_v| = O(B^{1+f})$ The set S'_v contains at most one point from each set $S_v[j]$. If $|S_v[j]| = B$, then S'_v contains the point $p \in S_v[j]$ with minimal y-coordinate; otherwise S'_v contains no points from $S_v[j]$.

We store data structures D_v and D'_v, implemented according to Lemma 2, in each internal node v of $\mathcal{T}$. The data structure D_v contains all points of S_{v_i} for every child v_i of v, and the data structure D'_v contains all points of S'_{v_i} for every child v_i of v. Thus D_v contains $O(B^{1+2f})$ points, and D'_v contains $O(B^{2f})$ points. By Lemma 2, D_v and D'_v can be stored in $O(B^{2f} \log_2^3 B)$ and $O(\log_2^3 B)$ blocks respectively and support $(2, 1, 2)$-sided queries in $O(1)$ I/O operations. In every node v of $\mathcal{T}$, we also store a data structure E_v that contains all points of S_v and supports $(2, 1, 2)$-sided queries. Note that lists L_v and $L_v[j]$ and sets $S_v[j]$ are not stored in the data structure; we only use them to simplify the description.

Search Procedure. Given a query $Q = [a, b] \times [c, +\infty) \times [d, e]$, we identify leaves l_a and l_b: l_a contains the smallest value that is greater than a and l_b contains the largest value that is smaller than b. Let π_a and π_b denote the paths from the root of $\mathcal{T}$ to l_a and l_b respectively. Let $\pi = \pi_a \cup \pi_b$ denote the set of all nodes of $\mathcal{T}$ that belong to π_a or π_b. Every point $p \in S$ such that $p.x \in [a, b]$ is stored in some set S_v such that either v belongs to π or v is a descendant of a node that belongs to π.

We can visit all nodes $v \in \pi$ and report all points in $S_v \cap Q$ in $O(\log_B N)$ I/Os using data structures E_v (we ignore the time needed to output points). Points in descendants of $v \in \pi$ can be found using the following property.

Fact 1. *Let v', $v' \notin \pi$, be a child of a node $v \in \pi$, and let w be a descendant of v'. If $S_w[j] \cap Q \neq \emptyset$, then $|S_{\mathrm{par}(w)}[j] \cap Q| = B$ where $\mathrm{par}(w)$ denotes the parent of a node w.*

Proof: Recall that $Q = [a, b] \times [c, +\infty) \times [d, e]$. For a child v' of v, such that $v' \notin \pi$, either $rng(v') \cap [a, b] = \emptyset$ or $rng(v') \subset [a, b]$. Hence, Fact 1 is non-trivial only in the case when $j \in [d, e]$ and $rng(v') \subset [a, b]$. In this case a point $p \in S_w[j]$ (resp., $p \in S_{\mathrm{par}(w)}[j]$) belongs to Q if and only if $p.y \geq c$. Suppose that some $p \in S_w[j]$ belongs to Q. Since $p.y \geq c$ and $p'.y > p.y$ for any point $p' \in S_{\mathrm{par}(w)}[j]$, all points $p' \in S_{\mathrm{par}(w)}[j]$ belong to Q. The set $S_{\mathrm{par}(w)}[j]$ contains B points because $S_w[j]$ is not empty. $\square$

Consider a node v, such that $v \in \pi_a$ and $v \notin \pi_b$. Suppose that the i-th child v_i of v belongs to π_a and $rng(v_{i+1}) = [a', b']$. We define the query $Q_v = [a', b] \times [c, +\infty) \times [d, e]$. For any point p stored in a descendant w of v, such that $w \notin \pi_a$, queries Q_v and Q are equivalent: p belongs to Q if and only if p belongs to Q_v. Points in $S_w \cap Q = S_w \cap Q_v$ for all descendants w of v, $w \notin \pi_a$, can be reported with the following recursive procedure. We report all points in $Q_v \cap S_{v_i}$ for all children v_i of v using the data structure D_v. All children v_i of v, such that $Q_v \cap S_{v_i}[j]$ contains at least B points for at least one j, can be identified using D'_v. We visit all such non-leaf nodes v_i and recursively call the same procedure.

Our procedure reports all points in $S_w \cap Q_v$: Suppose that $S_w[j] \cap Q_v \neq \emptyset$ for some w and j. Then $S_{\mathrm{par}(w)}[j] \cap Q_v$ contains B points by Fact 1. Let w' be ancestor of w that is a descendant of v, i.e., w' is situated on the path from w to v. By the same argument, $S_{w'}[j] \cap Q_v$ also contains B points. Hence, the parent of w will be visited and all points in $S_w \cap Q_v$ will be reported by querying the data structure $D_{\mathrm{par}(w)}$. If k_v is the total number of points in $S_w \cap Q_v$ for all w, then the search procedure takes $O(\frac{k_v}{B})$ I/O operations: Queries answered by D_w and D'_w in every visited node w take $O(1)$ I/O operations and a node w is visited only if $|S_w[j] \cap Q_v| = B$ for at least one value of j.

All points in $S_w \cap Q$ for all descendants w of a node v, such that $v \in \pi_b$ but $v \notin \pi_a$ or v is the lowest common ancestor of l_a and l_b, can be found with the same procedure. The only difference is that the query Q_v is defined differently: if $v \in \pi_b$, $v \notin \pi_a$, and the i-th child v_i of v belongs to π_b, then $Q_v = [a, b'] \times [c, +\infty) \times [d, e]$ where $rng(v_{i-1}) = [a', b']$. If v is the lowest common ancestor of l_a and l_b, then $v \in \pi_a$ and $v \in \pi_b$. Suppose that $v_i \in \pi_a$ and

$v_l \in \pi_b$ where v_i and v_l are the children of v. Then $Q_v = [a', b''] \times [c, +\infty) \times [d, e]$ where $rng(v_{i+1}) = [a', b']$ and $rng(v_{l-1}) = [a'', b'']$. Hence, a query Q can be answered with $O(\log_B N + \frac{k}{B})$ I/O operations.

Space Usage and Updates. Every data structure D_v contains $O(B^{1+2f})$ points and can be stored in $O(B^{2f} \log_2^3 B)$ blocks of space. Every D'_v contains $O(B^{2f})$ points and can be stored in $O(\log_2^3 B)$ blocks. There are $O(\frac{N}{B^{1+2f}})$ internal nodes in $\mathcal{T}$; hence, all D_v and D'_v use $O(\frac{N}{B} \log_2^3 B)$ blocks. Every data structure E_v contains $O(B^{1+f})$ points. Since the total number of nodes is $O(\frac{N}{B^{1+f}})$, all E_v can be stored in $O(\frac{N}{B} \log_2^3 B)$ blocks.

When a point p is inserted into S, we identify the leaf l_p in which $p.x$ must be stored and traverse the path π_p from l_p to the root until we find a node v such that $p.y < m_v.y$ and m_v is the point with maximal y-coordinate in $S_v[p.z]$. Then, we insert p into $S_v[p.z]$. Now $S_v[p.z]$ may contain $B+1$ points; if $|S_v[p.z]| = B+1$, the point s_v with the smallest y-coordinate must be removed from $S_v[p.z]$. We insert the point s_v into $S_{v_i}[p.z]$ where v_i is the child of v such that v_i belongs to π_p. If $S_{v_i}[p.z]$ contains $B + 1$ points, we move the point with the smallest y-coordinate from $S_{v_i}[p.z]$ to $S_u[p.z]$ where u is the child of v_i, $u \in \pi_p$. The procedure continues until $S_u[p.z]$ contains at most B points or the leaf l_p is reached. In every node u visited by the insertion procedure, one point is inserted into S_u and at most one point is deleted from S_u. Hence data structures E_u, D_w, and D'_w, where w denotes the parent of u, can be updated in $O(\log_2^2 B)$ I/Os. Since $O(\log_B N)$ nodes are visited, insertion takes $O(\log_2 N \log_2 B)$ I/O operations. Deletions can be supported with a similar procedure.

It remains to show how the tree $\mathcal{T}$ can be re-balanced after update operations, so that the height of $\mathcal{T}$ remains $O(\log_B N)$. We implement the base tree $\mathcal{T}$ as a WBB-tree [10] with leaf parameter $n_l = B^{1+f}$ and branching parameter $n_b = B^f$. In a WBB-tree with this choice of parameters the following invariants are maintained: each leaf contains between B^{1+f} and $2B^{1+f} - 1$ values and for each internal node v on level h (counting from the lowest level) there are between $B^{1+(h+1)f}/2$ and $2B^{1+(h+1)f} - 1$ values stored in leaf descendants of v. It is also shown in [10] that internal node has between $B^f/4$ and $4B^f$ children. Hence, the height of $\mathcal{T}$ is $O(\log_B N)$.

If the invariants of a WBB-tree are violated after an insertion, i.e., if a node v on level h contains $2B^{1+(h+1)f}$ values (resp., v contains $2B^{1+f}$ values if v is a leaf), then we split the node v into v' and v'' that contain $B^{1+(h+1)f}$ (B^{1+f}) values each. Splitting a node does not affect the children of v, i.e., every child of v becomes the child of v' or v'' after splitting. It can be shown [10] that a node v on level h is split at most once when a sequence of $B^{1+(h+1)f}/2$ values has been inserted into leaf descendants of v. See [10] for a complete description of the splitting procedure.

When a node v is split into v' and v'', data structures in nodes v', v'', and in their descendants may change. Since $S_v[j] = S_{v'} \cup S_{v''}$ for each j after the split procedure, at least one of $S_{v'}[j]$ and $S_{v''}[j]$ contains less than B points. Suppose that for some j, the set $S_{v'}[j]$ contains less than B points. If $S_{v_i}[j] \neq \emptyset$ for at least one child v_i of v', then some points must be moved from sets $S_{v_t}[j]$

into $S_v[j]$, where v_t is a child of v'. Let $d_v = \min(|\cup S_{v_t}[j]|, B - |S_v|)$. We can identify d_v points with largest y-coordinates in $\cup S_{v_t}$, using $D_{v'}$ and insert those points into $S_{v'}[j]$. Data structures $E_{v'}$, D_w, and D'_w where w is a parent of v' are updated accordingly. If $d_v > 0$, we recursively check sets S_{v_t} for all children v_t of v'. Data structures stored in the node v'' and the descendants of v'' are processed in the same way. Each point is moved only once and the total number of moved points does not exceed the total number of values stored in leaf descendants of v' and v''. When a point is moved, each affected data structure can be updated in $O(\log_2^2 B)$ I/Os. The number of values stored in a node v on level h and all its descendants is $\Theta(B^{1+(h+1)f})$. Since v is split at most once after $B^{1+(h+1)f}/2$ update operations, the amortized cost for splitting a node is $O(\log_2^2 B)$. Every leaf has $O(\log_B N)$ ancestors; hence, the total amortized costs of splits incurred by an inserted point is $O(\log_2 B \log_2 N)$. Thus the total cost of an insertion is $O(\log_2 N \log_2 B)$.

We implement deletions with the lazy deletions approach. Suppose that a point p such that $p.x$ is stored in a leaf l_p is deleted from S. Then we mark the value $p.x$ as deleted in l_p. When $N/2$ values stored in leaves of T are marked as deleted, we rebuild the tree T and all secondary data structures. This can be done in $O(N \log_2^2 B)$ I/O operations. Hence, rebuilding after deletions incurs an amortized cost of $O(\log_2^2 B)$.

The result of this section is summed up in the following Lemma.

Lemma 3. *Suppose that z-coordinates of all points are bounded by $O(B^f)$. There exists a $O(\frac{N}{B} \log_2^3 B)$ space data structure that supports $(2, 1, 2)$-sided queries in $O(\log_B N + \frac{k}{B})$ I/O operations and updates in $O(\log_2 N \log_2 B)$ amortized I/O operations.*

4   Range Reporting in Three Dimensions

Using range trees with fan-out $\Theta(B^f)$, we can transform the result of section 3 into a data structure for $(2, 1, 2)$-sided queries. For completeness, we sketch the data structure below.

We construct an external memory range tree on z-coordinates of the points in a set S: z-coordinates of all points are stored in leaves of the tree; each leaf contains $\Theta(B)$ values and each internal node has $\Theta(B^f)$ children. We denote by R_v the set of points whose z-coordinates are stored in leaf descendants of the node v. The data structure F_v contains one point for each point $p \in R_v$. If $p = (p.x, p.y, p.z)$, $p \in R_v$, is also stored in the i-th child v_i of v, then F_v contains the point $p' = (p.x, p.y, i)$. In other words, we replace the z-coordinate of each point $p \in R_v$ with an index $i \in [1, \Theta(B^f)]$, such that $p \in R_{v_i}$. F_v supports $(2, 1, 2)$-sided queries as described in Lemma 3.

For each internal node v, let $int(v, i, j)$ denote the interval $[\min_i, \max_j]$ where $\min_i$ denotes the minimal value stored in a leaf descendant of the i-th child of v, and $\max_j$ denotes the maximal value stored in a leaf descendant of the j-th child of v. For a query $Q = [a, b] \times [c, +\infty) \times [d, e]$, we can represent the interval

$[d, e]$ as a union of $O(\log_B N)$ intervals $int(v, i, j)$. Hence, Q can be answered by answering $O(\log_B N)$ queries of the form $[a, b] \times [c, +\infty) \times int(v, i, j)$. Every such query can be answered by the data structure F_v. Hence, a $(2, 1, 2)$-sided query can be answered with $O(\log_B^2 N + \frac{k}{B})$ I/O operations. Since each point is stored in $O(\log_B N)$ data structures F_v, the space usage and update cost increase by a factor $O(\log_B N)$ compared with the data structure of Lemma 3.

Lemma 4. *There exists a $O(\frac{N}{B} \log_2 N \log_2^2 B)$ space data structure that supports $(2, 1, 2)$-sided queries in $O(\log_B^2 N + \frac{k}{B})$ I/Os and updates in $O(\log_2^2 N)$ I/Os amortized.*

Finally, we apply the reduction described in section 2.1 and obtain the main result of this paper. The space usage and update cost increase by a factor $O(\log_2 N)$ in comparison with Lemma 4.

Theorem 1. *There exists a $O(\frac{N}{B} \log_2^2 N \log_2^2 B)$ space data structure that supports three-dimensional orthogonal range reporting queries in $O(\log_B^2 N + \frac{k}{B})$ I/O operations and updates in $O(\log_2^3 N)$ amortized I/O operations.*

5   Conclusion

In this paper we presented the first dynamic data structure that supports three-dimensional orthogonal range reporting queries in $O(\log_B^2 N + K/B)$ I/O operations. This query cost "matches" the query bound of the fastest internal memory data structure. The space usage of our data structure is quite comparable with the most space-efficient static external memory data structure [2]. It is an interesting open question, whether the $O(\log_2^3 N)$ update cost can be significantly improved.

Using our approach, we can also obtain data structures that support special cases of range reporting queries; these data structures answer queries in $O(\log_B^2 N)$ I/Os, but use less space and support faster update operations than the data structure of Theorem 1. In particular, we can obtain:

(i) The data structure for $(1, 1, 1)$-sided queries (three-dimensional dominance queries) that uses $O((N/B) \log_2 N)$ blocks of space and supports updates in $O(\log_B^2 N)$ I/Os.

(ii) The data structure for $(1, 1, 2)$-sided queries that uses $O((N/B) \log_2 N \log_2 B)$ blocks of space and supports updates in $O(\log_2 N \log_B N)$ I/Os.

(iii) The data structure for $(2, 1, 2)$-sided queries that uses $O((N/B) \log_2 N \log_2^2 B)$ blocks of space and supports updates in $O(\log_2^2 N)$ I/Os.

The data structure (iii) is the result of Lemma 4. We obtain the results (i) and (ii) by replacing the data structures D_v, D_v', and E_v in the proof of Lemma 3 with data structures that support $(1, 1, 1)$-sided queries (resp. $(1, 1, 2)$-sided queries) on a set with $O(B^{4/3})$ points. Details will be given in the full version of this paper.

References

1. Afshani, P.: On Dominance Reporting in 3D. In: Halperin, D., Mehlhorn, K. (eds.) ESA 2008. LNCS, vol. 5193, pp. 41–51. Springer, Heidelberg (2008)
2. Afshani, P., Arge, L., Larsen, K.D.: Orthogonal Range Reporting in Three and Higher Dimensions. In: Proc. FOCS 2009, pp. 149–158 (2009)
3. Agarwal, P.K., Erickson, J.: Geometric Range Searching and its Relatives. In: Chazelle, B., Goodman, J.E., Pollack, R. (eds.) Advances in Discrete and Computational Geometry, pp. 1–56. AMS Press, Providence (1999)
4. Aggarwal, A., Vitter, J.S.: The Input/Output Complexity of Sorting and Related Problems. Communications of the ACM 31(9), 1116–1127 (1988)
5. Alstrup, S., Brodal, G.S., Rauhe, T.: New Data Structures for Orthogonal Range Searching. In: Proc. FOCS 2000, pp. 198–207 (2000)
6. Alstrup, S., Husfeldt, T., Rauhe, T.: Marked Ancestor Problems. In: Proc. FOCS 1998, pp. 534–544 (1998)
7. Arge, L.: External Memory Data Structures. In: Meyer auf der Heide, F. (ed.) ESA 2001. LNCS, vol. 2161, pp. 1–29. Springer, Heidelberg (2001)
8. Arge, L.: The Buffer Tree: A Technique for Designing Batched External Data Structures. Algorithmica 37, 1–24 (2003)
9. Arge, L., Samoladas, V., Vitter, J.S.: On Two-Dimensional Indexability and Optimal Range Search Indexing. In: Proc. PODS 1999, pp. 346–357 (1999)
10. Arge, L., Vitter, J.S.: Optimal External Memory Interval Management. SIAM J. Comput. 32(6), 1488–1508 (2003)
11. Chazelle, B., Guibas, L.J.: Fractional Cascading: II. Applications. Algorithmica 1(2), 163–191 (1986)
12. Miltersen, P.B., Nisan, N., Safra, S., Wigderson, A.: On Data Structures and Asymmetric Communication Complexity. J. Comput. Syst. Sci. 57, 37–49 (1998)
13. Mortensen, C.W.: Fully Dynamic Orthogonal Range Reporting on RAM. SIAM J. Computing 35(6), 1494–1525 (2006)
14. Nekrich, Y.: A Data Structure for Multi-Dimensional Range Reporting. In: Proc. SoCG 2007, pp. 344–353 (2007)
15. Nekrich, Y.: External Memory Range Reporting on a Grid. In: Tokuyama, T. (ed.) ISAAC 2007. LNCS, vol. 4835, pp. 525–535. Springer, Heidelberg (2007)
16. Nekrich, Y.: I/O-Efficient Point Location in a Set of Rectangles. In: Laber, E.S., Bornstein, C., Nogueira, L.T., Faria, L. (eds.) LATIN 2008. LNCS, vol. 4957, pp. 687–698. Springer, Heidelberg (2008)
17. Nekrich, Y.: Dynamic Range Reporting in External Memory, arXiv: 1006.4093v1
18. Overmars, M.H.: Efficient Data Structures for Range Searching on a Grid. J. Algorithms 9(2), 254–275 (1988)
19. Pătraşcu, M.: (Data) Structures. In: Proc. FOCS 2008, pp. 434-443 (2008)
20. Pătraşcu, M., Thorup, M.: Time-space Trade-offs for Predecessor Search. In: Proc. STOC 2006, pp. 232–240 (2006)
21. Subramanian, S., Ramaswamy, S.: The P-range Tree: A New Data Structure for Range Searching in Secondary Memory. In: Proc. SODA 1995, pp. 378–387 (1995)
22. Vitter, J.S.: External Memory Algorithms and Data Structures: Dealing with Massive Data. ACM Computing Surveys 33(2), 209–271 (2001)
23. Vengroff, D.E., Vitter, J.S.: Efficient 3-D Range Searching in External Memory. In: Proc. STOC 1996, pp. 192–201 (1996)

A Cache-Oblivious Implicit Dictionary with the Working Set Property

Gerth Stølting Brodal, Casper Kejlberg-Rasmussen, and Jakob Truelsen

MADALGO*, Department of Computer Science, Aarhus University, Denmark
{gerth,jakobt,ckr}@madalgo.au.dk

Abstract. In this paper we present an implicit dictionary with the working set property i.e. a dictionary supporting insert(e), delete(x) and predecessor(x) in $\mathcal{O}(\log n)$ time and search(x) in $\mathcal{O}(\log \ell)$ time, where n is the number of elements stored in the dictionary and ℓ is the number of distinct elements searched for since the element with key x was last searched for. The dictionary stores the elements in an array of size n using *no* additional space. In the cache-oblivious model the operations insert(e), delete(x) and predecessor(x) cause $\mathcal{O}(\log_B n)$ cache-misses and search(x) causes $\mathcal{O}(\log_B \ell)$ cache-misses.

1 Introduction

In this paper we consider the problem of creating an *implicit dictionary* [9] with the *working set property*. An implicit dictionary maintains a set of n distinct keys, and encodes a data structure supporting fast insertions, deletions, predecessor queries and searches in the permutation of these keys as they are laid out in an array [9]. Between operations *no* additional space usage is allowed, while during an operation only a constant number of word registers may be used. The number of elements n is assumed externally maintained. Computation is done in a machine with a constant number of registers with a word size of $\Theta(\log n)$ bits. All operations are unit cost, similar to the RAM model. Extensive research has been done in the implicit/in-place model, from as early as binary heaps [11], to an in-place 3-d convex hull algorithm [4]. Implicit dictionaries have been the topic of several papers culminating in a dictionary supporting all operations in $\mathcal{O}(\log n)$ time [5]. For a more extensive overview see [8].

The working set property states that the time to search for an element e with key x must be $\mathcal{O}(\log \ell)$, where ℓ is the number of distinct elements searched for since e was last searched for. This property has been achieved by numerous structures. The splay tree [10], a skip list variant [2], and the working set structure [7], all achieve the property in the amortized, expected or worst-case sense. The unified access bound, which is a generalization of the working set bound, is achieved in [1]. The unified access bound states that, if $\ell(g)$ is the

---

* Center for Massive Data Algorithmics, a Center of the Danish National Research Foundation.

O. Cheong, K.-Y. Chwa, and K. Park (Eds.): ISAAC 2010, Part II, LNCS 6507, pp. 37–48, 2010.

Table 1. The operation time, and space overhead of important structures for the dictionary problem

Reference	Insert/Delete	Search	Predecessor	Additional space (words)
[5]	$\mathcal{O}(\log n)$	$\mathcal{O}(\log n)$	$\mathcal{O}(\log n)$	None
[7]	$\mathcal{O}(\log n)$	$\mathcal{O}(\log \ell)$	$\mathcal{O}(\log \ell)$	$\mathcal{O}(n)$
[3, Sec. 2]	$\mathcal{O}(\log n)$	$\mathcal{O}(\log \ell)$ exp.	$\mathcal{O}(\log n)$	$\mathcal{O}(\log \log n)$
[3, Sec. 3]	$\mathcal{O}(\log n)$	$\mathcal{O}(\log \ell)$ exp.	$\mathcal{O}(\log \ell)$ exp.	$\mathcal{O}(\sqrt{n})$
This paper	$\mathcal{O}(\log n)$	$\mathcal{O}(\log \ell)$	$\mathcal{O}(\log n)$	None

number of distinct elements accessed since g was last accessed, and $d(g,e)$ denotes the rank distance between g and e, then the search time for e must be $\mathcal{O}(\min_g \log(\ell(g) + d(g,e) + 2))$. In [3] two structures with low space overhead are presented, achieving the working set property in the expected sense, see Table 1.

The dictionary in [5] is, in addition to being implicit, also designed for the cache-oblivious model [6] where all the operations imply $\mathcal{O}(\log_B n)$ cache-misses, where B is the cache-line length that is unknown to the algorithm.

1.1   Our Results

We present an implicit dictionary with the working set property that supports insertions, deletions, and predecessor queries in $\mathcal{O}(\log n)$ time and search queries in $\mathcal{O}(\log \ell)$ time. Our result improves the construction of [3, Section 2] by requiring no additional space. Furthermore our structure is cache-oblivious and supports insert, delete and predecessor operations in $\mathcal{O}(\log_B n)$ cache-misses and search in $\mathcal{O}(\log_B \ell)$ cache misses.

Our implicit cache-oblivious dictionary makes essential use of the notion of an implicit *moveable* dictionary, i.e. a dictionary stored in a consecutive sub-array that can be moved to the left or the right, one position at a time. We construct a moveable dictionary from a constant number of the implicit and cache-oblivious dictionaries from [5], achieving a dictionary inheriting the same properties, but which is also moveable. The moveable dictionary is in itself an interesting result because it is a general transformation, that can be applied to any data structure that can be laid out in an array and grows/shrinks in one end and supporting insertions and deletions. Hence we can plug in say a binary heap, and get a moveable binary heap.

In the literature the working set property is often stated in terms of the number of operations. We note that if we perform a search for an element whenever it is inserted, we will also satisfy these kinds of bounds.

This paper is organized as follows. In Section 2 we present our implicit moveable dictionary. In Section 3 we show how our implicit working set dictionary structure is constructed by composing $\mathcal{O}(\log \log n)$ implicit moveable dictionaries.

2   A Moveable Dictionary

In this section we describe an implicit moveable dictionary which can be laid out in an array in the range $[i; j]$, where $n = j - i + 1$ is the number of elements in the dictionary. When deleting an element from the dictionary we are allowed to shrink the dictionary from the left or the right end, such that the structure now lies in the range $[i + 1; j]$ or $[i; j - 1]$, respectively. Likewise we can insert and expand the dictionary at the left or right end such that the structure now lies in the range $[i - 1; j]$ or $[i; j + 1]$, respectively. The structure also supports search and predecessor operations. All operations run in $\mathcal{O}(\log n)$ time. The moveable dictionary is implicit except for $\mathcal{O}(\log n)$ extra bits that need to be stored/encoded externally (in the D_i structures in Section 3).

The dictionary supports the following operations:

- Insert-left(e) and insert-right(e): inserts an element e into the dictionary which grows in the left and right side, respectively.
- Delete-left(x) and delete-right(x): deletes the element with key x from the dictionary which shrinks in the left and right side, respectively.
- Search(x): returns the element e with key x in the dictionary if such an element exits, otherwise **none** is returned.
- Predecessor(x): is given a key x and returns the element e in the dictionary with the largest key less than x.

An amortized solution can be obtained using two of the dictionaries by Franceschini and Grossi [5] (in the following denoted FG dictionaries). Let r be an index in the range $i \le r \le j$. One FG dictionary denoted R is located in the range $[r; j]$ and grows to the right as normal, and one FG dictionary denoted L is located in the range $[i; r - 1]$ and grows to the left, i.e. for L we have inverted all the indexes of the original FG dictionary. The insert-left and insert-right operations insert elements into L and and R, respectively. The delete-left operation searches for the element e to be deleted in L and R. If e is in L it is deleted from L and we are done. Otherwise e is deleted from R and an arbitrary element is deleted from L and inserted into R – provided L is non-empty. If L is empty we first rebuild the data structure such that L and R differ in size by at most one, by repeatedly reinserting into new L and R structures starting from the new index $r = \lceil \frac{i+j}{2} \rceil$. The delete-right operation is handled symmetrically. To search for an element with a given key we search in L and then in R; to find the predecessor element of a given key we find the predecessor in L and R and return the largest of the two. Since [5] supports all operations in $\mathcal{O}(\log n)$ time, all operations run in $\mathcal{O}(\log n)$ amortized time, which e.g. can be seen using the potential function $\Phi = |\,|L| - |R|\,|$.

In the following we describe how to deamortize the above construction using incremental rebalancing of L and R. An additional FG dictionary C is placed between L and R (see Figure 1). In the following we w.l.o.g. assume that $n \ge 24$, such that all intervals stated below are guaranteed to include an integer. If L or R get outside the range $[\frac{3}{24}n, \frac{7}{24}n]$, say L is getting too big/small, we initialize an incremental *job* to make L smaller/bigger by transferring elements to/from C.

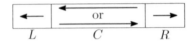

Fig. 1. We have three FG dictionaries L, C and R, where L always grows/shrinks in the left direction, and R grows/shrinks in the right, and C will change direction during the execution of the jobs to shrink or grow L or R

Each time an insert and delete operation is executed we perform a constant number of steps of the current job. While resizing L there might be a pending job waiting for resizing R, and vice versa. During the execution of a job we have a temporary FG dictionary, which can be one of either L', C' or R', depending on how far we are in the execution of the job (see Figure 2).

2.1 Methods and Jobs

The insert-left and delete-left operations, and the grow-left and shrink-left jobs described here have analogous right-versions.

Search(x). We always have the structures L, C and R, and possibly one of the structures L', C' or R'. We search each of the at most four structures. If we find an element e with key x we return e, otherwise we return **none**.

Predecessor(x). As in search we search for the predecessor in each of the structures L, C, R and possibly one of L', C' or R', and return the largest of the four candidates found.

Insert-left(e). We insert e into L. If $|L| > \frac{7}{24}n$ we initialize a shrink-left job unless a left job is already running/pending.

Delete-left(x). We delete the element with key x from L. We can do this even though the element we want to delete resides in L', C, C', R or R' by swapping the element we want to delete with one from L. We can swap elements by performing two deletions and two insertions. If $|L| < \frac{3}{24}n$ we initialize a grow-left job unless a left job is already running/pending.

Grow-left. The job consists of the following steps to be performed incrementally (see Figure 2 (left)). Notice that during the incremental work, deletions and insertions are performed on L and R by the update operations. We let n_{init} denote the size of the dictionary when the job is initialized, and assume that n_{init} is remembered when the job is initialized.

1) If C is not growing to the left then turn C around so it grows toward L. We turn C around by creating a new C' in the growing end of C which grows towards C, into which we insert all the elements of C, one element at a time.
2) Construct L' of size $\lceil \frac{2}{24} n_{\text{init}} \rceil$ at the beginning of L, growing to the right, by deleting elements from C and inserting them into L'.

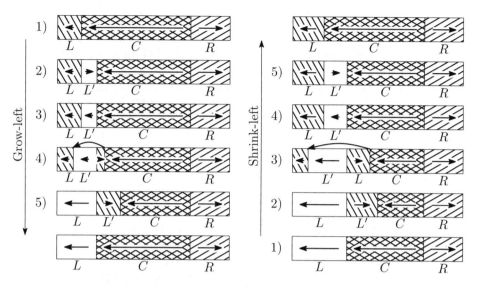

Fig. 2. The steps of the two operations grow-left and shrink-left, notice that they are almost each other's inverse. (Left) The five steps of the grow-left operation, notice that in step 4) the arrow at the top means that we have split L up into two by use of address-mapping. (Right) The five steps of the shrink-left operation, in step 3) we have again used address-mapping to split L in two.

3) Turn L' around so it faces L, like we turned C in step 1).
4) Continue deleting an element from C and inserting it into L', so L' expands into L. The element overridden in L is moved into the empty place in C where we took the element to place in L'. We do this by splitting L into two pieces by address-mapping, see steps 3) and 4) in Figure 2 (left). When we have moved L completely to the right of L', we swap the names of L and L'.
5) Merge L' back into C, by deleting an element from L' and inserting it into C until L' is empty.

Shrink-left. The job consists of the following steps (see Figure 2 (right)). Notice the similarity to grow-left.

1) If C is not growing to the left then turn C around so it grows toward L.
2) Create L' by deleting $\left\lceil \frac{5}{24} n_{\text{init}} \right\rceil$ elements from C, one element at a time and inserting them into L', which we create to the left of C.
3) Swap the names of L and L'. Delete an element from L' and insert it into C so it expands into L, then move the element overridden in L to the empty space to the left of L', do this one element at a time until L is moved completely to the left of L'.
4) Turn L' around so it faces C.
5) Merge L' back into C.

2.2   Correctness

The correctness of the search and predecessor operations follows directly from the fact that the dictionary consists of at most four FG dictionaries. Similarly the insert-left and insert-right operations insert a single new element into an FG dictionary and otherwise only moves elements between the FG dictionaries. The only operations remaining to be considered are the delete-left and delete-right operations. In the following we only consider the delete-left operation (delete-right is symmetric). The only technical detail we need to argue about is that there always is a non-empty FG dictionary L oriented to the left that has its leftmost element stored in the leftmost entry in the subarray.

In the following when considering a job, we let n_{init}, n_0, n_{finish} denote the size of the moveable dictionary: when the job was initialized, when the execution of the job started, and just after it is finished, respectively.

By performing the incremental work sufficiently fast, we will be able to perform the job during at most βn_0 moveable dictionary updates, for any constant $\beta > 0$. An upper bound on the number of primitive steps (that is movement of one element from one FG dictionary to another one, and possibly move in memory) per update is: During the execution of the job at most βn_0 insertions can take place, i.e. the dictionary always has size at most $(1 + \beta)n_0$. Therefor each of the five steps of a job require at most $(1 + \beta)n_0$ primitive steps. In total there are at most $5(1 + \beta)n_0$ primitive steps. By performing at least $5(1 + \beta)/\beta$ primitive steps per update, the job finishes within βn_0 updates.

To relate n_{init} and n_0 we make the observation that any job under execution will finish during the next βn updates, where n is the current number of elements in the dictionary. To see this, observe that a job that has run for d updates needs to be executed for at most $\beta n_0 - d \leq \beta(n_0 - d) \leq \beta n$ further updates, provided $\beta \leq 1$. From this it follows that when a job is initialized, it at most takes βn_{init} updates before the current job finishes and the new job starts being executed, i.e. $(1 - \beta)n_{\text{init}} \leq n_0 \leq (1 + \beta)n_{\text{init}}$.

Let t_{finish} denote the number of updates between the initialization of a job until it it is finished. We have $t_{\text{finish}} \leq \beta n_{\text{init}} + \beta n_0 \leq \beta n_{\text{init}} + \beta(1 + \beta)n_{\text{init}} = (\beta^2 + 2\beta)n_{\text{init}}$. We get $n_{\text{finish}} \leq n_{\text{init}} + t_{\text{finish}} \leq (1 + \beta^2 + 2\beta)n_{\text{init}}$ and $n_{\text{finish}} \geq n_{\text{init}} - t_{\text{finish}} \geq (1 - \beta^2 - 2\beta)n_{\text{init}}$.

During the lifetime of a job, i.e. between its initialization and its the time it finish, there are always at least $\frac{3}{24}n_{\text{init}} - t_{\text{finish}} \geq (\frac{3}{24} - \beta^2 - 2\beta)n_{\text{init}}$ elements that still can be deleted from the leftmost FG dictionaries which shrink the subarray from the left. By selecting β sufficiently small such that $\beta^2 + 2\beta < \frac{3}{24}$, this number is always non-zero.

What remains to be argued is that i) $\frac{3}{24}n_{\text{finish}} \leq |L| \leq \frac{7}{24}n_{\text{finish}}$ when a left job is finished, ii) $|C| \geq \lceil \frac{2}{24}n_{\text{init}} \rceil$ when a grow job starts its execution, and iii) $|L| \geq \lceil \frac{5}{24}n_{\text{init}} \rceil$ immediately before step 3) in a shrink job. We need i) to ensure that $\frac{3}{24}n \leq |L| \leq \frac{7}{24}n$ holds just before a job is initialized, and ii) and iii) to ensure that grow-left and shrink-left are well defined, respectively.

The above can be shown by the following observations: i) After a shrink or grow job $|L| \leq \frac{5}{24}n_{\text{init}} + 1 + t_{\text{finish}} \leq \frac{6}{24}n_{\text{init}} + t_{\text{finish}}$ which is less than $\frac{7}{24}n_{\text{finish}}$

for $\beta^2 + 2\beta \leq \frac{1}{31}$. Similarly after a shrink or grow job $|L| \geq \frac{5}{24}n_{init} - t_{finish}$ which is greater than $\frac{3}{24}n_{finish}$ for $\beta^2 + 2\beta \leq \frac{2}{27}$. ii) Before grow-left $|C| \geq n_0 - |L| - |R| \geq (n_{init} - \beta n_{init}) - (\frac{7}{24}n_{init} + \beta n_{init}) - \frac{7}{24}n_{init}(1+\beta)$ which is greater than $\frac{3}{24}n_{init} \geq \lceil \frac{2}{24}n_{init} \rceil$ for $\beta \leq \frac{7}{55}$. iii) In shrink-left $|L| \geq \frac{7}{24}n_{init} - t_{finish}$ which is greater than $\frac{6}{24}n_{init} \geq \lceil \frac{5}{24}n_{init} \rceil$ for $\beta^2 + 2\beta \leq \frac{1}{24}$. We note that setting $\beta = \frac{1}{63}$ will satisfy all the stated constraints.

The $\mathcal{O}(\log n)$ time bounds for the operations follow from the $\mathcal{O}(\log n)$ time bounds of the FG dictionaries. In the cache-oblivious model we notice that because the FG dictionary is cache-oblivious and we only use a constant number of FG dictionaries, where we split at most one of them into two parts by address-mapping then we only multiply the bound on the cache-misses from the FG dictionary by a constant factor. Hence all operations cause $\mathcal{O}(\log_B n)$ cache-misses.

We notice that we can make the moveable dictionary implicit such that we do not need to store $\mathcal{O}(\log n)$ bits between operations. We do this by introducing a block D of $\mathcal{O}(\log n)$ elements to the left of L which *pair-encodes* the $\mathcal{O}(\log n)$ bits. With pair-encoding we mean that each consecutive pair of elements encodes a bit. If the key of the first element is lower than the key of the second, the pair-encodes a 0 bit. If on the other hand the key of the first element is greater than the key of the second, the pair-encodes a 1 bit. As we need to read this block to get the $\mathcal{O}(\log n)$ bits, we can maintain (and possibly move) D when we perform insert-left, insert-right, delete-left and delete-right operations. From a cache-oblivious viewpoint this does also not change the asymptotic bound on the number of cache-misses.

3 Construction of the Working Set Dictionary

In the following we describe our working set dictionary archiving insertions, deletions and predecessor searches in $\mathcal{O}(\log n)$ time and searches in $\mathcal{O}(\log \ell)$. We first describe the overall structure leaving the details of the memory layout to be handled in Section 3.3. The structure is composed of $\mathcal{O}(\log \log n)$ blocks, where the i'th block B_i stores $\mathcal{O}(2^{2^i})$ elements. The main design goal is to have elements that have been searched for within the last ℓ distinct searches located in one of the first $\mathcal{O}(\log \log \ell)$ blocks.

Block B_i consists of a list D_i of size w_i where $w_i = \alpha 2^i$ for some appropriate constant α, and three implicit moveable dictionaries, L_i, C_i and R_i. We use D_i to pair-encode $\mathcal{O}(2^i)$ bits, used for memory management in the working set dictionary and storing data needed between operations in the moveable dictionaries L_i, C_i and R_i. Block B_i contains exactly $2 \cdot 2^{2^i} + w_i$ elements, except for the last block B_m that might contain less than $2 \cdot 2^{2^i} + w_i$ elements, as this is the block that grows or shrinks when we insert or delete, respectively.

When an element e is searched for it is moved from its current block B_j to the first block B_0. To make room for this in B_0, we move an element from each block B_i to B_{i+1} until we reach the block B_j where e was originally located. We move elements from R_i to L_{i+1}, for $i = 0, \ldots, j-1$ (see Figure 3). Once R_i

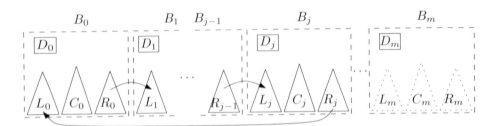

Fig. 3. Layout of the data structure. The arrows indicate the movement of elements after an element in R_j has been searched for. The dotted lines in block B_m indicate that the strucutres do not necessarily exist.

is empty we move C_i to R_i, and L_i to C_i. Doing this we can guarantee that at least 2^{2^i} distinct elements have been searched for since any element in R_i was last searched for. We can give this guarantee because an element will be located in C_i at least until searches for 2^{2^i} other elements have been performed.

3.1 Invariants

Our data structure satisfies the invariants below. Here I.1 to I.4 are about the sizes of data structures and are important for memory management. On the other hand I.5 to I.8 are about the location of elements according to when they were last searched for and are important for achieving the working set property.

I.1 $|C_i| \leq 2^{2^i}$ and $|R_i| \neq 0 \Rightarrow |C_i| = 2^{2^i}$, for all i.

I.2 $|D_i| \leq w_i$ and $|L_i| + |C_i| + |R_i| \neq 0 \Rightarrow |D_i| = w_i$, for all i.

I.3 $|L_i| + |R_i| = 2^{2^i}$, for all $i < m$, and $|L_m| + |R_m| \leq 2^{2^m}$.

I.4 $|L_i| < 2^{2^i}$, for all i.

I.5 All elements searched for since L_i was last empty are contained in L_i, D_i or B_j for some $j < i$.

I.6 For any e in some C_i either at least $|L_i|$ distinct elements have been searched for after e was last searched for or e has never been searched for.

I.7 For any e in some R_i either at least 2^{2^i} distinct elements have been searched for after e was last searched for or e has never been searched for.

I.8 For any e in D_i, L_i or C_i, for $i > 0$, either at least $2^{2^{i-1}}$ distinct elements have been searched for after e was last searched for, or e has never been searched for.

From the invariants we make the following observations:

O.1 $|D_i| = w_i$ for all $i < m$ (from I.2 and I.3).

O.2 $|R_i| > 0$ for all $i < m$ (from I.3 and I.4).

O.3 $|C_i| = 2^{2^i}$ for all $i < m$ (from I.1 and O.2).

O.4 $|B_i| = w_i + 2 \cdot 2^{2^i}$ for all $i < m$ (from O.1, O.3, and I.3).

O.5 For $i > 0$ and any e in B_i, either at least $2^{2^{i-1}}$ distinct elements have been searched for after e was last searched for or e has never been searched for (from I.7 and I.8).

3.2   Operations

Our data structure uses the operations shift and find internally, and supports the operations insert, delete, predecessor and search. Below is a detailed description of all operations.

Shift(j) handles the case when $|R_j| = 0$ and $|L_j| = 2^{2^j}$, i.e. I.4 is violated for block B_j. This is done by discarding R_j, renaming C_j to R_j, renaming L_j to C_j, and creating a new empty L_j. After shift(j) finishes I.4 also holds for B_j.

Find(x) finds the data structure S_i containing the element with key x or returns **none** if no such element exists. Here S_i will be either D_i, L_i, C_i or R_i for some i. This is done by searching for x in the blocks starting with B_0 and going in an incremental linear fashion towards B_m. Within each block, x is searched for in D_i using a linear scan, and the implicit moveable dictionaries L_i, C_i and R_i are searched for x using their built-in search operation. As soon as x is found, a reference to the data structure S_i containing the element is returned, and no further blocks are considered. In the case when x is not found in any of the blocks **none** is returned.

Predecessor(x) returns the element e in the data structure with the largest key less than x. This is done for $B_0, \ldots, B_m$ by a linear scan of D_i and invoking the built-in predecessor operation on L_i, C_i and R_i and returning the element among the results with the highest key.

Insert(e) inserts the element e into the data structure. This is done by inserting e into one of the data structures in B_m. It is inserted into D_m if $|D_m| < w_m$. Otherwise, if $|C_m| < 2^{2^m}$ it is inserted into C_m, else it is inserted into R_m. If this makes $|L_m| + |R_m| = 2^{2^m}$, then a new block B_{m+1} is initialized by incrementing m by one.

Delete(x) deletes the element with key x from the data structure. We first check if x is in the dictionary by performing a find(x) operation. If x is not found we return. Here S_j will be one of D_j, L_j, C_j or R_j. If B_m is empty, m is decremented by one. An arbitrary element e is deleted from the first of the structures R_m, C_m, L_m and D_m that is non-empty. If e has key x we return, else the element with key x is deleted from S_j and e is inserted into S_j.

Search(x) returns the element e with key x or **none** if such an element does not exist. This is done by performing a find(x) operation, finding the data structure S_j containing x. If x is not in the data structure then **none** is returned.

If x is found in a data structure S_j then the element e with key x is found by running the built-in search method on S_j. If S_j is either D_0 or L_0 we return e immediately. If $S_j = C_j$ and $|R_j| > 0$, an arbitrary element g is removed from R_j, e is removed from C_j and g is inserted in C_j. In the other case where $S_j \neq C_j$ or $|R_j| = 0$, the element e with key x is deleted from S_j.

In all cases we then proceed by deleting an arbitrary element h from R_{i-1} and inserting it into L_i, for $i = j, \ldots, 1$. In the special case where $i = j$ and $S_j = D_j$ we insert h into D_j instead of L_j.

We then insert e into L_0. Now for $i = 0, \ldots, j$ we check whether $|L_i| = 2^{2^i}$, and if this is the case we perform a shift(i) operation. Finally we return e.

3.3   Memory Management

From O.4 we know that any block except the last will contain a fixed number of elements, namely $2 \cdot 2^{2^i} + w_i$. This implies that we can lay out the blocks sequentially in the array, and then we only have to worry about memory management inside each block. The last block B_m can vary in size, and is located at the end of the array where growing and shrinking must occur.

By I.2 we know that D_i will be completely constructed before the other structures are needed, therefore we lay it out sequentially in the beginning of the block. The remaining structures will be laid out sequentially in the order: L_i, C_i, R_i.

Right before we insert an element into L_i, we move C_i and R_i one position to the right to make room. We can move L_i, C_i or R_i to the right by performing a delete-left operation on an arbitrary element e followed by an insert-right(e). This moving will take time $\mathcal{O}(2^i)$. We do the same when inserting into C_i, but here we only move R_i one position to the right. We never need to move structures to the left.

To perform queries on the substructures in a block B_i we need to store various information in D_i. We need n_{L_i}, n_{C_i} and n_{R_i}: the size of L_i, C_i and R_i, respectively. We store n_{C_i} and n_{R_i} in D_i explicitly using 2^i bits each, whereas n_{L_i} can be computed as $n_{L_i} = |B_i| - w_i - n_{C_i} - n_{R_i}$. Furthermore we store in D_i the $\Theta(2^i)$ bits we allow the moveable dictionaries L_i, C_i and R_i to maintain between operations, denoted $data_{L_i}$, $data_{C_i}$ and $data_{R_i}$.

We maintain all these bits in D_i, using pair-encoding. The fields are stored in the following order: $data_{L_i}$, n_{C_i}, $data_{C_i}$, n_{R_i}, $data_{R_i}$. Whenever we add an element to or remove an element from D_i we maintain the ordering of the pair by performing a swap if needed.

To perform an operation on block B_i we need to know the index b_i of the first element, which can be computed as $b_0 = 0$, and $b_i = b_{i-1} + 2 \cdot 2^{2^{i-1}} + w_{i-1}$. We may also need $|D_i|$ which can be computed as $|D_i| = \min(w_i, n - b_i)$, and $|B_i|$ which can be computed as $|B_i| = w_i + 2 \cdot 2^{2^i}$ if $i < m$ and $|B_m| = n - b_m$ otherwise.

Whenever we want to perform an operation on C_i, we first extract n_{L_i}, n_{C_i}, n_{R_i} and $data_{C_i}$ from the pair-encoding in D_i and put them into registers. From the sizes and the value of b_i we can compute the index of the first element in C. Using that information we can run the operation on the implicit moveable dictionary. Once that is done we $data_C$ write back to the pair-encoding in D_i. Totally this requires $\mathcal{O}(2^i)$ time. We do similarly if we perform an operation on L_i or R_i.

When performing a shift operation we override n_{R_i} and $data_{R_i}$ with n_{C_i} and $data_{C_i}$ and we override n_{C_i} and $data_{C_i}$ with n_{L_i} and $data_{L_i}$. This renames the data structures, initiating a new empty L_i before the old full one, and "deletes" the old empty R_i.

During an insert operation, when D_i increases to w_i, we initialize n_{L_i}, $data_{L_i}$, n_{C_i}, $data_{C_i}$, n_{R_i} and $data_{R_i}$. Finally we calculate m when it is needed as the minimal value where $\sum_{j=0}^{m} 2 \cdot 2^{2^j} + w_j > n$.

3.4   Analysis

To see that the invariants are maintained for the operations we need to show that each invariant is maintained for each operation. In general this is tedious but trivial. As an example below we give the proof for the shift operation to give a taste of how the proofs go. In the following S_i refers to a data structure before the shift operation and S_i' refers to the same data structure after the operation, similarly for m.

We now prove that shift is correct. We assume that B_j satisfies all invariants except I.4 before shift(j). Since the shift operation requires that $|L_j| = 2^{2^j}$ and $C_j' = L_j$ I.1 holds for j after the shift operation. Because $|L_j| = 2^{2^j}$ and I.2 holds before the operation we know that $|D_j| = w_j$. Since the shift operation did not change D_j, I.2 also holds for j after the operation. When verifying I.3 we have two cases: if $j = m$, then from I.1 we know that $|C_m| \leq 2^{2^m}$ so $|L_m'| + |R_m'| = 0 + |C_m| \leq 2^{2^m}$ so I.3 holds. Else if $j < m$ then by O.3 we know that $|C_j| = 2^{2^j}$. Now since L_j' is empty and $R_j' = C_j$, then I.3 holds for $j < m$ after the operation. Since L_j' is empty, no elements have been accessed after it was last empty thus I.5 trivially holds for j. Likewise because $|L_j'| = 0$, then I.6 is maintained for j. Because shift(j) assumed $|L_j| = 2^{2^j}$, and because $R_j' = C_j$, then I.6 immediately implies that I.7 holds for j after shift(j). Lastly, since all elements in L_j' and C_j' come from L_j, and D_j' contains the same elements as D_j then I.8 held for j since it holds before the shift(j) operation.

The core of the analysis of the running times of the predecessor and search operations stems from the find operation. Let ℓ be the number of distinct elements searched for since we last searched for some element e in some block B_j. By O.5 we know that at least $2^{2^{j-1}}$ elements have been searched for after e was last searched for so $\ell \geq 2^{2^{j-1}}$, i.e. $2^j = \mathcal{O}(\log \ell)$. For each block we use constant time to calculate b_i, $|D_i|$, and whether $i = m$. This can be done since we have already computed b_{i-1} once b_i is needed. The time used for the find operation in block B_i is $\mathcal{O}(2^i)$, plus the time for doing the linear scan in D_i, and $\mathcal{O}(\log 2^{2^i}) = \mathcal{O}(2^i)$ for doing searches in L_i, C_i and R_i from the bounds on the moveable dictionary. The total time for doing searches in all blocks $0, \ldots, j$ is then $\mathcal{O}(\sum_{i=0}^{j} 2^i) = \mathcal{O}(2^j) = \mathcal{O}(\log \ell)$ which becomes the time for the search operation. Since predecessor queries need to access all blocks, they require $\mathcal{O}(\log n)$ time. Similarly insertions and deletions require $\mathcal{O}(\log n)$ time.

From the cache-oblivious viewpoint we incur $\mathcal{O}(2^i/B)$ cache-misses when searching D_i and $\mathcal{O}(\log_B 2^{2^i}) = \mathcal{O}(2^i/\log B)$ cache-misses when searching in L_i, C_i and R_i. Since the blocks $B_0, \ldots, B_{\lfloor \log \log B \rfloor}$ in total store $\mathcal{O}(B)$ elements and are stored consecutively in the array, the accesses to these blocks imply a total of $\mathcal{O}(1)$ cache-misses. For the remaining blocks $2^{2^i} \geq B$ and in total we incur $\mathcal{O}(\sum_{i=\lfloor \log \log B \rfloor + 1}^{j} 2^i / \log B) = \mathcal{O}(\log_B \ell)$ cache-misses for the find operation. It follows that search implies $\mathcal{O}(\log_B \ell)$ cache-misses, and predecessor queries, insert and delete operations imply $\mathcal{O}(\log_B n)$ cache-misses.

Acknowledgements

We would like to thank Mark Greve and Freek van Walderveen for their help on proofreading this paper.

References

1. Bădoiu, M., Cole, R., Demaine, E.D., Iacono, J.: A unified access bound on comparison-based dynamic dictionaries. Theoretical Computer Science 382(2), 86–96 (2007)
2. Bose, P., Douïeb, K., Langerman, S.: Dynamic optimality for skip lists and B-trees. In: SODA 2008, pp. 1106–1114. SIAM, Philadelphia (2008)
3. Bose, P., Howat, J., Morin, P.: A distribution-sensitive dictionary with low space overhead. In: Dehne, F., et al. (eds.) WADS 2009. LNCS, vol. 5664, pp. 110–118. Springer, Heidelberg (2009)
4. Chan, T.M.Y., Chen, E.Y.: Optimal in-place algorithms for 3-d convex hulls and 2-d segment intersection. In: SoCG 2009, pp. 80–87. ACM, New York (2009)
5. Franceschini, G., Grossi, R.: Optimal worst-case operations for implicit cache-oblivious search trees. In: Dehne, F., Sack, J.-R., Smid, M. (eds.) WADS 2003. LNCS, vol. 2748, pp. 114–126. Springer, Heidelberg (2003)
6. Frigo, M., Leiserson, C.E., Prokop, H., Ramachandran, S.: Cache-oblivious algorithms. In: FOCS 1999, pp. 285–297. IEEE, Los Alamitos (1999)
7. Iacono, J.: Alternatives to splay trees with $\mathcal{O}(\log(n))$ worst-case access times. In: SODA 2001, pp. 516–522. SIAM, Philadelphia (2001)
8. Mortensen, C.W., Pettie, S.: The complexity of implicit and space-efficient priority queues. In: Dehne, F., López-Ortiz, A., Sack, J.-R. (eds.) WADS 2005. LNCS, vol. 3608, pp. 49–60. Springer, Heidelberg (2005)
9. Munro, J.I., Suwanda, H.: Implicit data structures for fast search and update. Journal of Computer and System Sciences 21(2), 236–250 (1980)
10. Sleator, D.D., Tarjan, R.E.: Self-adjusting binary search trees. J. ACM 32(3), 652–686 (1985)
11. Williams, J.W.J.: Algorithm 232: Heapsort. Communications of the ACM 7(6), 347–348 (1964)

The (p, q)-total Labeling Problem for Trees

Toru Hasunuma[1], Toshimasa Ishii[2], Hirotaka Ono[3], and Yushi Uno[4]

[1] Department of Mathematical and Natural Sciences, The University of Tokushima,
Tokushima 770–8502 Japan
hasunuma@ias.tokushima-u.ac.jp
[2] Department of Information and Management Science, Otaru University of Commerce,
Otaru 047-8501, Japan
ishii@res.otaru-uc.ac.jp
[3] Department of Economic Engineering, Kyushu University, Fukuoka 812-8581, Japan
hirotaka@en.kyushu-u.ac.jp
[4] Department of Mathematics and Information Sciences, Graduate School of Science,
Osaka Prefecture University, Sakai 599-8531, Japan
uno@mi.s.osakafu-u.ac.jp

Abstract. A (p, q)-total labeling of a graph G is an assignment f from the vertex set $V(G)$ and the edge set $E(G)$ to the set of nonnegative integers such that $|f(x) - f(y)| \geq p$ if x is a vertex and y is an edge incident to x, and $|f(x) - f(y)| \geq q$ if x and y are a pair of adjacent vertices or a pair of adjacent edges, for all x and y in $V(G) \cup E(G)$. A k-(p, q)-total labeling is a (p, q)-total labeling $f : V(G) \cup E(G) \rightarrow \{0, \ldots, k\}$, and the (p, q)-total labeling problem asks the minimum k, which we denote by $\lambda_{p,q}^T(G)$, among all possible assignments. In this paper, we first give new upper and lower bounds on $\lambda_{p,q}^T(G)$ for some classes of graphs G, in particular, tight bounds on $\lambda_{p,q}^T(T)$ for trees T. We then show that if $p \leq 3q/2$, the problem for trees T is linearly solvable, and give a complete characterization of trees achieving $\lambda_{p,q}^T(T)$ if in addition $\Delta \geq 4$ holds, where Δ is the maximum degree of T. It is contrasting to the fact that the $L(p, q)$-labeling problem, which is a generalization of the (p, q)-total labeling problem, is NP-hard for any two positive integers p and q such that q is not a divisor of p.

1 Introduction

In the channel/frequency assignment problems, we need to assign different frequencies to 'close' transmitters so that they can avoid interference. The $L(p, q)$-labelings of a graph have been extensively studied as one of important graph theoretical models of this problem. An $L(p, q)$-*labeling* of a graph G is an assignment f from the vertex set $V(G)$ to the set of nonnegative integers such that $|f(x) - f(y)| \geq p$ if x and y are adjacent and $|f(x) - f(y)| \geq q$ if x and y are at distance 2, for all x and y in $V(G)$. A k-$L(p, q)$-labeling is an $L(p, q)$-labeling $f : V(G) \rightarrow \{0, \ldots, k\}$, and the $L(p, q)$-*labeling problem* asks the minimum k, which we denote by $\lambda_{p,q}(G)$, among all possible assignments. Notice that we can use $k + 1$ different labels when $\lambda_{p,q}(G) = k$ since we can use 0 as a label for conventional reasons. We can find a lot of related results on $L(p, q)$-labelings in comprehensive surveys by Calamoneri [3] and by Yeh [25]. From the applicational point of view, we assume that $p \geq q \geq 1$ unless otherwise stated. Also, we assume that p and q are relatively prime, since otherwise, an $L(p, q)$-labeling is equivalent to an $L(p/\ell, q/\ell)$-labeling, where $\ell = \gcd(p, q)$.

O. Cheong, K.-Y. Chwa, and K. Park (Eds.): ISAAC 2010, Part II, LNCS 6507, pp. 49–60, 2010.

(p, q)-total labeling and a conjecture. In [24], Whittlesey et al. studied the $L(2, 1)$-labeling number of incidence graphs, where the *incidence graph* of a graph G is the graph obtained from G by replacing each edge (v_i, v_j) with two edges (v_i, v_{ij}) and (v_{ij}, v_j) after introducing one new vertex v_{ij}. Observe that an $L(p, q)$-labeling of the incidence graph of a given graph G can be regarded as an assignment f from $V(G) \cup E(G)$ to the set of nonnegative integers such that $|f(x) - f(y)| \geq p$ if x is a vertex and y is an edge incident to x, and $|f(x) - f(y)| \geq q$ if x and y are a pair of adjacent vertices or a pair of adjacent edges, for all x and y in $V(G) \cup E(G)$. Such a labeling of G is called a (p, q)-*total labeling* of G, while the case of $q = 1$ was first introduced as a $(p, 1)$-total labeling by Havet and Yu [15,16]. In particular, a k-(p, q)-total labeling is a (p, q)-total labeling $f : V(G) \cup E(G) \to \{0, \ldots, k\}$, and the (p, q)-*total labeling problem* asks the minimum k among all possible assignments. We call this invariant, the minimum k, the (p, q)-*total labeling number*, which is denoted by $\lambda_{p,q}^T(G)$.

We notice that a $(1, 1)$-total labeling of G is equivalent to a total coloring of G. Generalizing the Total Coloring Conjecture [2,22], Havet and Yu [15,16] conjectured that

$$\lambda_{p,1}^T(G) \leq \Delta + 2p - 1 \tag{1}$$

holds for any graph G, where Δ denotes the maximum degree of a vertex in G. They also investigated bounds on $\lambda_{p,1}^T(G)$ under various assumptions and some of their results are described as follows: (i) $\lambda_{p,1}^T(G) \geq \Delta + p - 1$, (ii) $\lambda_{p,1}^T(G) \geq \Delta + p$ if $p \geq \Delta$, (iii) $\lambda_{p,1}^T(G) \leq \min\{2\Delta + p - 1, \chi(G) + \chi'(G) + p - 2\}$ for any graph G where $\chi(G)$ and $\chi'(G)$ denote the chromatic number and the chromatic index of G, respectively, and (iv) $\lambda_{p,1}^T(G) \leq n + 2p - 2$ if G is the complete graph where $n = |V(G)|$. In particular, it follows by (iii) that if G is bipartite, then $\lambda_{p,1}^T(G) \leq \Delta + p$ holds (by $\chi(G) \leq 2$ and König's theorem), and if in addition, $p \geq \Delta$, then $\lambda_{p,1}^T(G) = \Delta + p$ by (ii) [1,15,16]. Also, Bazzaro et al. [1] showed that $\lambda_{p,1}^T(G) \leq \Delta + p + s$ for any s-degenerated graph (by $\chi(G) \leq s + 1$ and $\chi'(G) \leq \Delta + 1$), where an *s-degenerated graph* G is a graph which can be reduced to a trivial graph by successive removal of vertices with degree at most s, that $\lambda_{p,1}^T(G) \leq \Delta + p + 3$ for any planar graph (by the Four-Color Theorem), and that $\lambda_{p,1}^T(G) \leq \Delta + p + 1$ for any outerplanar graph other than an odd cycle (since any outerplanar graph is 2-degenerated, and any outerplanar graph other than an odd cycle satisfies $\chi'(G) = \Delta$ [9]). Also, there are many related works about bounds on $\lambda_{p,1}^T(G)$ [6,14,19,20]. From the algorithmic point of view, Havet and Thomassé [17] recently showed that for bipartite graphs, if (i) $p \geq \Delta$ or (ii) $\Delta = 3$ and $p = 2$, then the $(p, 1)$-total labeling problem is polynomially solvable and otherwise it is NP-hard.

In [7,18,21], the (r, s, t)-coloring problem which is a generalization of the (p, q)-total labeling problem was studied, while results in the cases corresponding to the (p, q)-total labeling problem (actually, the cases of $t \geq r = s$) are limited to paths, cycles, stars or the complete graph with some p and q. To our best knowledge, there are quite few studies about (p, q)-total labelings other than these. In this paper, we focus on the (p, q)-total labeling problem for some classes of graphs, especially for trees.

$L(p, q)$-labelings and (p, q)-total labelings of trees. Let T be a tree. As for the $L(2, 1)$-labeling problem, Griggs and Yeh [12] showed that $\lambda_{2,1}(T) \in \{\Delta + 1, \Delta + 2\}$. Moreover, Chang and Kuo [5] showed that $\lambda_{2,1}(T)$ can be computed in polynomial time, and recently Hasunuma et al. [13] gave a linear time algorithm for this problem. However, a characterization of trees T achieving $\lambda_{2,1}(T)$ is still open. Also, it was shown in [4] that $\lambda_{p,1}(T) \leq \min\{\Delta + 2p - 2, 2\Delta + p - 2\}$ and that $L(p, 1)$-labeling problem for trees can be solved in $O((p + \Delta)^{5.5}n) = O(\lambda_{p,1}(T)^{5.5}n)$ time by extending the algorithm in [5], where $n = |V(G)|$. On the other hand, Fiala et al. [8] showed that the $L(p, q)$-labeling problem for trees is NP-hard for any two positive integers p and q such that q is not a divisor of p. Concerning bounds on $\lambda_{p,q}(T)$, Georges and Mauro [11] gave the exact value of $\lambda_{p,q}(T)$ for the infinite regular trees T, which gives tight upper bounds on $\lambda_{p,q}(T)$; following their results, we have $\lambda_{p,q}(T) \leq p + (2\Delta - 2)q$.

As for the (p, q)-total labeling problem, the above algorithms can also be applied to the case of $q = 1$, since the incidence graph of a tree is also a tree. Moreover, due to the structure of the incidence graph of a tree, it can be observed that $\lambda_{p,1}^T(T)$ becomes much smaller than $\lambda_{p,1}(T)$. By bounds for bipartite graphs in [1,15,16], it follows that $\lambda_{p,1}^T(T) \in \{p + \Delta - 1, p + \Delta\}$, and that if $p \geq \Delta$, then $\lambda_{p,1}^T(T) = p + \Delta$. Recently, Wang and Chen [23] gave a characterization of trees T achieving $\lambda_{2,1}^T(T)$ in the case of $\Delta = 3$.

Our contributions. In this paper, we mainly focus on (p, q)-total labeling problem for trees for general p and q, and obtain the following results:

- (Upper bounds on $\lambda_{p,q}^T(T)$) If $p = q + r$ for $r \in \{0, 1, \ldots, q - 1\}$ and $\Delta > 1$ (resp., $\Delta = 1$), then $\lambda_{p,q}^T(T) \leq p + (\Delta - 1)q + r$ holds and this bound is tight (resp., $\lambda_{p,q}^T(T) = p + q$). If $p \geq 2q$, then $\lambda_{p,q}^T(T) \leq p + \Delta q$ holds and this bound is tight. In particular, if $p \geq \Delta q$, then $\lambda_{p,q}^T(T) = p + \Delta q$.
- (Lower bounds on $\lambda_{p,q}^T(T)$) If $q \leq p < (\Delta - 1)q$, then $\lambda_{p,q}^T(T) \geq p + (\Delta - 1)q$ holds and this bound is tight. If $p = (\Delta - 1)q + r$ for $r \in \{0, 1, \ldots, q - 1\}$, then $\lambda_{p,q}^T(T) \geq p + (\Delta - 1)q + r$ holds and this bound is tight. If $p \geq \Delta q$, then $\lambda_{p,q}^T(T) = p + \Delta q$.
- The (p, q)-total labeling problem with $p \leq 3q/2$ for trees can be solved in linear time. In particular, if $\Delta \geq 2$, we have $\lambda_{p,q}^T(T) \in \{p + (\Delta - 1)q, p + (\Delta - 1)q + r\}$. If $p > q$ and $\Delta \geq 4$, then $\lambda_{p,q}^T(T) = p + (\Delta - 1)q$ holds if and only if no two vertices with degree Δ are adjacent.
- In the case of $p = 2q$, the condition that no two vertices with degree Δ are adjacent is sufficient for $\lambda_{p,q}^T(T) = p + (\Delta - 1)q$, while in the case of $p > 3q/2$ and $p \neq 2q$, this condition is not sufficient.
- For any two nonnegative integers p and q, the $L(p, q)$-labeling problem for trees can be solved in polynomial time if $\Delta = O(\log^{1/3} |I|)$ where $|I| = \max\{|V(T)|, \log p\}$. Particularly, if Δ is a fixed constant, it is solved in linear time.

The first and second results provide tight upper and lower bounds on $\lambda_{p,q}^T(T)$ for all pairs (p, q) with $p \geq q$. The first statement in the third result indicates that as for the (p, q)-total labeling problem for trees, there exists a tractable case even if q is not a divisor of p, in contrast to the NP-hardness of the $L(p, q)$-labeling problem. The second and third statements in the third result completely characterize trees T achieving $\lambda_{p,q}^T(T)$ in the case of $p \leq 3q/2$ and $\Delta \geq 4$ (note that if $p = q$, we have $\lambda_{p,q}^T(T) = p + (\Delta - 1)q$ by the first and second results). This is also contrasting to the fact that no simple characterization of trees T achieving $\lambda_{2,1}(T)$ is known even for the $L(2, 1)$-labeling problem.

Organization of the paper. The rest of this paper is organized as follows. In Section 2, after giving some basic definitions, we show several properties about bounds on $\lambda_{p,q}^T(G)$. In Sections 3 and 4, we focus on the cases where a given graph is a tree. Section 3 provides tight upper and lower bounds on $\lambda_{p,q}^T(T)$ for trees T. In Section 4, we propose a linear time algorithm for solving the (p, q)-total labeling problem with $p \leq 3q/2$ for trees, and give a characterization of trees T achieving $\lambda_{p,q}^T(T)$ in the case of $p > q$ and $\Delta \geq 4$. Also, we discuss the case of $p > 3q/2$. Finally, we give concluding remarks in Section 5. Some parts of the detailed analyses are omitted due to space limitation.

2   Bounds on $\lambda_{p,q}^T(G)$

In this section, we investigate several properties on (p, q)-total labelings of a graph G. For this, we first define some terminology. A graph G is an ordered set of its vertex set $V(G)$ and edge set $E(G)$ and is denoted by $G = (V(G), E(G))$. We assume throughout this paper that a graph is undirected, simple and connected unless otherwise stated. Therefore, an edge $e \in E(G)$ is an unordered pair of vertices u and v, which are *end vertices* of e, and we often denote it by $e = (u, v)$. Let $N_G(v)$ denote the set of neighbors of a vertex v in G; $N_G(v) = \{u \in V \mid (u, v) \in E(G)\}$. The *degree* of a vertex v is $|N_G(v)|$, and is denoted by $d_G(v)$. A vertex v with $d_G(v) = k$ is called a *k-vertex*. We use $\Delta(G)$ (resp., $\delta(G)$) to denote the maximum (resp., minimum) degree of a vertex in a graph G. A $\Delta(G)$-vertex is called *major*. We often drop G in these notations if there are no confusions. For a (p, q)-total labeling $f : V(G) \cup E(G) \rightarrow \{0, 1, \ldots, k\}$ of G and an edge $e = (u, v) \in E(G)$, we may denote $f(e)$ by $f(u, v)$. Let $\bar{f}$ denote the labeling such that $\bar{f}(z) = k - f(z)$ for each $z \in V(G) \cup E(G)$. Note that $\bar{f}$ is also a (p, q)-total labeling of G.

We have the following lemmas about upper and lower bounds on $\lambda_{p,q}^T(G)$, some of which are extensions of those discussed in the case of $q = 1$ [16].

Lemma 1. *(i)* $\lambda_{p,q}^T(G) \geq p + (\Delta - 1)q$.
(ii) If G has a major vertex whose neighbors are all major, then $\lambda_{p,q}^T(G) \geq p + \Delta q$ holds for $p \geq 2q$, and $\lambda_{p,q}^T(G) \geq p + (\Delta - 1)q + r$ holds for $p = q + r$ $(r = 0, 1, \ldots, q - 1)$.
(iii) $\lambda_{p,q}^T(G) \geq p + \min\{p, \Delta q\}$. *Hence,* $\lambda_{p,q}^T(G) \geq p + (\Delta - 1)q + r$ *holds where* $r = q$ *if* $p \geq \Delta q$ *and* $r = p - (\Delta - 1)q$ *otherwise.*   □

Lemma 2. *(i)* $\lambda_{p,q}^T(G) \leq p + q(\chi(G) + \chi'(G) - 2)$.
(ii) $\lambda_{p,q}^T(G) \leq p + (2\Delta - 1)q$.
(iii) Let G be the complete graph. Then, $\lambda_{p,q}^T(G) \leq \min\{p + (2\Delta - 1)q, 2p + \Delta q - 1\}$. *In particular,* $\lambda_{p,q}^T(G) = p + (2\Delta - 1)q$ *if* $p \geq 2\Delta q + 1$ *and* $|V(G)| \geq 3$.   □

In the case where G is a path (resp., cycle), the incidence graph of G is also a path (resp., cycle). The following lemma is obtained directly from Georges and Mauro's results about $L(p, q)$-labeling of paths or cycles [10].

Lemma 3. *(i) Let G be a path. We have* $\lambda_{p,q}^T(G) = p + q$ *(resp., $p + 2q$, resp., $2p$) if* $|V| = 2$ *(resp., $|V| \geq 3$ and $p \geq 2q$, resp., $|V| \geq 3$ and $p \leq 2q$).*
(ii) Let G be a cycle. We have $\lambda_{p,q}^T(G) = p + 2q$ *if (a) $|V|$ is even and $p \geq 2q$ or (b) $2|V| \neq 0 \bmod 3$ and $p \leq 2q$,* $\lambda_{p,q}^T(G) = p + 3q$ *if $|V|$ is odd and $p \geq 3q$, and* $\lambda_{p,q}^T(G) = 2p$ *otherwise.*   □

Similarly to the arguments in Section 1, we can see that the following properties hold, where a graph is called *a series-parallel graph* or *a partial 2-tree* if it contains no subgraph isomorphic to a subdivision of the complete graph with four vertices. Notice that an outerplanar graph is series-parallel.

Corollary 1. (i) *If G is bipartite, then $\lambda_{p,q}^T(G) \leq p + \Delta q$. In particular, if $p \geq \Delta q$, then $\lambda_{p,q}^T(G) = p + \Delta q$.*
(ii) *If G is s-degenerated, then $\lambda_{p,q}^T(G) \leq p + (\Delta + s)q$.*
(iii) *If G is planar, then $\lambda_{p,q}^T(G) \leq p + (\Delta + 3)q$.*
(iv) *If G is series-parallel, then $\lambda_{p,q}^T(G) \leq p + (\Delta + 1)q$.*   □

3   Tight Bounds on $\lambda_{p,q}^T(G)$ for Trees

In this section, we show the following properties about tight upper and lower bounds on $\lambda_{p,q}^T(T)$ for trees T.

Theorem 1. *Let T be a tree. Then the following properties hold.*
(i) *If $p \geq \Delta q$, then $\lambda_{p,q}^T(T) = p + \Delta q$.*
(ii) *If $p = (\Delta - 1)q + r$ $(r = 0, 1, \ldots, q - 1)$, then $\lambda_{p,q}^T(T) \geq p + (\Delta - 1)q + r$ and this bound is tight.*
(iii) *If $p \geq 2q$, then $\lambda_{p,q}^T(T) \leq p + \Delta q$ and this bound is tight.*
(iv) *If $p = q + r$ $(r = 0, 1, \ldots, q-1)$ and $\Delta > 1$ (resp., $\Delta = 1$), then $\lambda_{p,q}^T(T) \leq p + (\Delta - 1)q + r$ and this bound is tight (resp., $\lambda_{p,q}^T(T) = p + q$).*

Since trees are bipartite, the statement (i) and the former part of the statement (iii) follow from Corollary 1 (i). The former part of the statement (ii) follows from Lemma 1 (iii), and it is not difficult to see that a star T achieves $\lambda_{p,q}^T(T) = p + (\Delta - 1)q + r$. Lemma 1 (ii) indicates that a tree T which has a major vertex whose neighbors are all major achieves $\lambda_{p,q}^T(T) = p + \Delta q$ (resp., $p + (\Delta - 1)q + r$) if $p \geq 2q$ (resp., if $p = q + r$ $(< 2q)$, $\Delta > 1$, and the former part of the statement (iv) is true). The case of $\Delta = 1$ in the statement (iv) follows from Lemma 3.

In the rest of this section, we give a proof of the former part of the case of $\Delta > 1$ in the statement (iv) to complete the proof of this theorem. For this, we assume that $\Delta \geq 2$ and give an algorithm for finding a $(p + (\Delta - 1)q + r)$-(p, q)-total labeling of T if $p = q + r$ for $r \in [0, q - 1]$. For simplicity of description, assume that $T = (V, E)$ is a tree such that all non-leaves are major.

From [10, Lemma 2.1], it follows that there exists a $\lambda_{p,q}^T(T)$-(p, q)-total labeling of T which consists of labels with form $\alpha p + \beta q$ where $\alpha, \beta \in \mathbb{Z}_+$. Here we can assume that $\lambda_{p,q}^T(T) \geq p + (\Delta - 1)q + r$ by Lemma 1(ii) and the assumption on T. Considering these two properties, we will seek a $(p + (\Delta - 1)q + r)$-(p, q)-total labeling of T with form $\alpha p + \beta q$. Then, it is not difficult to see that the candidates of labels of such a form to be assigned for each non-leaf vertex (i.e., major vertex) are 0, $p + (\Delta - 1)q + r$ $(= 2p + (\Delta - 2)q)$, or $p + iq$ for some $i \in [0, \Delta - 2]$. In particular, if a major vertex v has label $p + iq$ for $i \in [0, \Delta - 2]$, then the set of labels for edges incident to v is $\{jq \mid j \in [0, i]\} \cup \{p + jq + r \mid j \in [i+1, \Delta - 1]\}$. Based on these observations, we regard T as a rooted tree by choosing a major vertex v_r as the root, and assign labels with form $\alpha p + \beta q$ to $V \cup E$ from the root v_r in the breadth-first-search order, as shown in Algorithm (p, q)-LABEL. Actually, we use labels 0, p, $(\Delta - 1)q + r$ $(= p + (\Delta - 2)q)$, and $p + (\Delta - 1)q + r$ for vertices, and repeat applying essentially four types of labelings to each scanned vertex, its incident edges, and its children. In the description of the algorithm, $p(v)$ denotes the parent of v (if exists) and $C(v)$ denotes the set of children of v for each vertex v.

Algorithm 1. Algorithm (p, q)-LABEL

Input: A tree $T = (V, E)$ with $\Delta \geq 2$ such that all non-leaves are major, and two positive integers p and q with $p = q + r$ and $r \in [0, q - 1]$.

Output: A (p, q)-total labeling $f : V \cup E \to \{0, 1, \ldots, p + (\Delta - 1)q + r\}$ of T.

1: Assign label 0 to the root v_r; let $f(v_r) := 0$. For each $i \in [0, \Delta - 2]$, let $f(v_r, c_i(v_r)) := p + iq$ and $f(c_i(v_r)) := p + (\Delta - 1)q + r$, where $C(v_r) = \{c_i(v_r) \mid i = 0, 1, \ldots, \Delta - 1\}$ (i.e., assign labels $p + iq$ and $p + (\Delta - 1)q + r$ to the edge $(v_r, c_i(v_r))$ and the child $c_i(v_r)$ of v_r, respectively). Let $f(v_r, c_{\Delta - 1}(v_r)) := p + (\Delta - 1)q + r$ and $f(c_{\Delta - 1}(v_r)) := (\Delta - 1)q + r$.

2: **while** there exists a non-leaf $v \in V - \{v_r\}$ such that $f(v)$ has been determined but no label is assigned to any child of v where $C(v) = \{c_i(v) \mid i = 0, 1, \ldots, \Delta - 2\}$ **do**

3:     **if** (Case-1) $f(p(v), v) \in \{p + iq \mid i \in [0, \Delta - 2]\}$ and $f(v) = p + (\Delta - 1)q + r$ **then**

4:         Let $f(c_i(v)) := 0$ for each $i \in [0, \Delta - 3]$, $f(c_{\Delta - 2}(v)) := p$, and $f(v, c_{\Delta - 2}(v)) := 0$. Assign labels in $\{p + iq \mid i \in [0, \Delta - 2]\} - \{f(p(v), v)\}$ injectively to edges $\{(v, c_i(v)) \mid i \in [0, \Delta - 3]\}$.

5:     **else if** (Case-2) $f(p(v), v) = p + (\Delta - 1)q + r$ and $f(v) = (\Delta - 1)q + r$ **then**

6:         Let $f(c_i(v)) := p + (\Delta - 1)q + r$ and $f(v, c_i(v)) := iq$ for each $i \in [0, \Delta - 2]$.

7:     **else if** (Case-3) $f(p(v), v) \in \{iq \mid i \in [0, \Delta - 2]\}$ and $f(v) = p + (\Delta - 1)q + r$ **then**

8:         Let $f(c_i(v)) := (\Delta - 1)q + r$ for each $i \in [0, \Delta - 3]$, $f(c_{\Delta - 2}(v)) := 0$, and $f(v, c_{\Delta - 2}(v)) := (\Delta - 1)q + r$. Assign labels in $\{iq \mid i \in [0, \Delta - 2]\} - \{f(p(v), v)\}$ injectively to edges $\{(v, c_i(v)) \mid i \in [0, \Delta - 3]\}$.

9:     **else if** (Case-4) $f(p(v), v) \in \{iq \mid i \in [0, \Delta - 2]\}$ and $f(v) = (\Delta - 1)q + r$ **then**

10:         Let $f(c_i(v)) := p + (\Delta - 1)q + r$ for each $i \in [0, \Delta - 3]$, $f(c_{\Delta - 2}(v)) := 0$, and $f(v, c_{\Delta - 2}(v)) := p + (\Delta - 1)q + r$. Assign labels in $\{iq \mid i \in [0, \Delta - 2]\} - \{f(p(v), v)\}$ injectively to edges $\{(v, c_i(v)) \mid i \in [0, \Delta - 3]\}$.

11:     **else if** (Case-j') $\overline{f}(p(v), v))$ and $\overline{f}(v)$ satisfy the conditions of Case-j for $j \in [1, 4]$ **then**

12:         After determining labels for $f(c_i(v))$ and $f(v, c_i(v))$ according to the above (Case-j) based on $\overline{f}(p(v), v))$ and $\overline{f}(v)$, let $f(c_i(v)) := p + (\Delta - 1)q + r - f(c_i(v))$ and $f(v, c_i(v)) := p + (\Delta - 1)q + r - f(v, c_i(v))$ for each $i \in [0, \Delta - 2]$.

13:     **end if**

14: **end while**

15: Output f as a (p, q)-total labeling of T.

We prove the correctness of Algorithm (p, q)-LABEL. Note that whenever a vertex v is chosen in line 2, $f(p(v), v)$ has also been already determined. Also note that the labels assigned in each step do not violate the feasibility. Hence, it suffices to show that as a result of line 1 (resp., each iteration of the while loop in lines 2–14), each $c_i(v_r) \in C(v_r)$ (resp., $c_i(v) \in C(v)$) satisfies the conditions of Case-j or Case-j' in lines 2–14 for some $j \in \{1, 2, 3, 4\}$. As for the children of v_r, $c_i(v_r)$ for $i \in [0, \Delta - 2]$ satisfies the conditions of Case-1 and $c_{\Delta - 1}(v_r)$ satisfies those of Case-2. Also as for the children of v in each case of lines 2–14, we can prove this as follows, where Case-j', $j' \in \{1, 2, 3, 4\}$ is omitted by symmetry of labelings:

(Case-1) $c_i(v)$, $i \in [0, \Delta - 3]$ satisfies the conditions of Case-1' and $c_{\Delta-2}(v)$ satisfies those of Case-2'.

(Case-2) $c_i(v)$, $i \in [0, \Delta - 2]$ satisfies the conditions of Case-3.

(Case-3) $c_i(v)$, $i \in [0, \Delta - 3]$ satisfies the conditions of Case-4 and $c_{\Delta-2}(v)$ satisfies those of Case-1'.

(Case-4) $c_i(v)$, $i \in [0, \Delta - 3]$ satisfies the conditions of Case-3 and $c_{\Delta-2}(v)$ satisfies those of Case-3'.

Notice that in Case-1, $c_i(v)$ for $i \in [0, \Delta - 3]$ satisfies the conditions of Case-1' because $\{p + iq \mid i \in [0, \Delta - 2]\} = \{p + (\Delta - 1)q + r - (p + iq) \mid i \in [0, \Delta - 2]\}$ by $p = q + r$. Consequently, the correctness of the algorithm is proved and hence the proof of Theorem 1 is completed.

Also, we remark that Algorithm (p, q)-LABEL can be implemented to run in linear time.

4   Algorithms for (p, q)-total Labelings of Trees

In this section, we consider an algorithm for finding an optimal (p, q)-total labeling (i.e., a $\lambda^T_{p,q}(T)$-(p, q)-total labeling) of trees T. Here, we focus on the cases of $\Delta \geq 3$ and $p > q$ since the case of $\Delta \leq 2$ has been shown as Lemma 3, and in the case of $\Delta \geq 3$ and $p = q$, we have $\lambda^T_{p,q}(T) = p + (\Delta - 1)q$ by Lemma 1 (i) and Theorem 1 and such a labeling can be found in linear time by Algorithm (p, q)-LABEL. We discuss the case of $p \leq 3q/2$ in Subsection 4.1 and other cases in Subsection 4.2.

4.1   Case: $p \leq 3q/2$

Assume that $p \leq 3q/2$. We show that the problem can be solved in linear time, and we give a complete characterization of trees T with $\Delta \geq 4$ achieving $\lambda^T_{p,q}(T)$; namely, we have the following theorem.

Theorem 2. *Let T be a tree with $p \leq 3q/2$.*
(i) An optimal (p, q)-total labeling (i.e., a $\lambda^T_{p,q}(T)$-(p, q)-total labeling) of T can be found in linear time.
(ii) In the case of $\Delta \geq 2$, we have $\lambda^T_{p,q}(T) \in \{p + (\Delta - 1)q, p + (\Delta - 1)q + r\}$ where $r = p - q$.
(iii) In the case of $p > q$ and $\Delta \geq 4$, $\lambda^T_{p,q}(T) = p + (\Delta - 1)q$ if and only if

$$\text{no two major vertices are adjacent in } T. \tag{2}$$

First we consider the case of $\Delta \geq 4$. In this case, for proving Theorem 2, it suffices to show the following two lemmas. Note that the first lemma holds for an arbitrary graph.

Lemma 4. *Let G be a graph. If $p \leq 3q/2$ and $\lambda^T_{p,q}(G) < p + (\Delta - 1)q + r$ where $r = p - q$, then the condition (2) is satisfied.* $\square$

Lemma 5. *If $p \leq 3q/2$, the condition (2) is satisfied, and $\Delta \geq 4$, then $\lambda^T_{p,q}(G) = p + (\Delta - 1)q$ holds, and such a labeling can be found in linear time.*

Recall that by Lemma 1 (i) and Theorem 1, we have $p+(\Delta-1)q \leq \lambda_{p,q}^T(T) \leq p+(\Delta-1)q+$ r where $r = p - q$. Hence, Lemmas 4 and 5 indicate that either $\lambda_{p,q}^T(T) = p + (\Delta - 1)q$ or $\lambda_{p,q}^T(T) = p+(\Delta-1)q+r$ holds, and that the former case is characterized by the condition (2). Furthermore, in both cases, an optimal labeling can be found in linear time by Lemma 5 and Algorithm (p, q)-LABEL. Thus, these two lemmas show Theorem 2 in the case of $\Delta \geq 4$.

On the other hand, in the case of $\Delta = 3$, there exist instances T with $\lambda_{p,q}^T(T) >$ $p + 2q$ even if the condition (2) holds. For example, consider a tree T which contains the configuration (a) in Fig. 1 in which each major vertex is drawn by a black circle, and assume for contradiction that T admits a $(p + 2q)$-(p, q)-total labeling f. Without loss of generality, let $f(u) = 0$ and $f(u, v) = p$ (note that the set of labels for edges incident to u is $\{p, p + q, p + 2q\}$). By the feasibility of f, we have $f(v) \in [2p, p + 2q]$. Since w is major, it follows that $f(w) = 0$, however, we cannot assign any label to the edge (v, w). Similarly, we can observe that any tree which contains the configuration (b) in Fig. 1 cannot admit a $(p + 2q)$-(p, q)-total labeling. We can observe that there are many other such instances, and it seems difficult to characterize instances T achieving $\lambda_{p,q}^T(T)$ in the case of $\Delta = 3$.

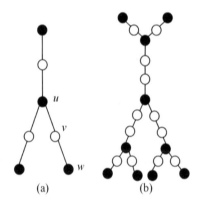

(a)   (b)

Fig. 1. Configurations that any tree T with $\lambda_{p,q}(T) = p + 2q$ does not contain in the case of $\Delta = 3$, where each major vertex is drawn by a black circle.

Nevertheless, we can prove that the case of $\Delta = 3$ is linearly solvable and $\lambda_{p,q}^T(T) \in$ $\{p + 2q, p + 2q + r\}$ in another way. The following lemma shows Theorem 2 in the case of $\Delta = 3$.

Lemma 6. *If $\Delta = 3$, then $\lambda_{p,q}^T(G) \in \{p + 2q, p + 2q + r\}$ holds, and such a labeling can be found in linear time.*   □

The latter part of Lemma 6 can be proved in a more general setting as the following theorem.

Theorem 3. *Let $|I| = \max\{|V(T)|, \log p\}$. For any nonnegative integers p, q, the $L(p, q)$-labeling problem for trees (hence, the (p, q)-total labeling problem for trees also) can be solved in polynomial time, if $\Delta = O(\log^{1/3} |I|)$ for general p, or if $\Delta = O(\log^{1/2} |I|)$ for $p = \Omega(\Delta q)$. In particular, it can be solved in linear time, if Δ is bounded by a constant.*

$\square$

Due to space limitation, we omit the proofs of Lemmas 4 and 6 and Theorem 3. We here give a proof of Lemma 5.

Proof of Lemma 5. Assume that a tree $T = (V, E)$ satisfies the condition (2) and $\Delta \geq 5$, while the case of $\Delta = 4$ is omitted due to space limitation. Also assume that the every non-major and non-leaf vertex in T is a $(\Delta - 1)$-vertex for simplicity of description. Then, we prove this lemma by showing that we can find a $(p + (\Delta - 1)q)$-(p, q)-total labeling of T according to Algorithm (p, q)-OPTLABEL$\Delta 5$. The algorithm starts with choosing a major vertex v_r as the root and assign labels to $V \cup E$ in the breadth-first-search order in a similar way to Algorithm (p, q)-TREE. Let M denote the set of major vertices.

Observe that the labelings in each step do not violate the feasibility (note that $3q - p \geq p$ and $2q \geq p$ by $p \leq 3q/2$). Hence, for proving the correctness of the algorithm, we show that as a result of line 1 (resp., the while-loop in lines 2–19), each $c_i(v_r) \in C(v_r)$ (resp., $c_i(v) \in C(v)$) satisfies the conditions of Case-j or Case-j' of lines 2–19 for some $j \in \{1, 2, \ldots, 6\}$. Notice that by the condition (2), all children of each major vertex are non-major. Hence, $c_i(v_r)$, $i \in [1, \Delta - 1]$ satisfies the conditions of Case-2 and $c_0(v_r)$ satisfies those of Case-5. Similarly, we can observe that the children of v in Case-1 of lines 2–19 satisfy the conditions of Case-2 or Case-5. As for children of v in other cases, we can prove this as follows:

(Case-2) $c_i(v)$ satisfies the conditions of Case-1, Case-3, or Case-4.
(Case-3) $c_i(v)$ satisfies the conditions of Case-1, Case-1', Case-2, or Case-2'.
(Case-4) $c_i(v)$ satisfies the conditions of Case-1, Case-1', Case-2, Case-5, or Case-6.
(Case-5) $c_i(v)$ satisfies the conditions of Case-1, Case-1', Case-2, or Case-6.
(Case-6) $c_i(v)$ satisfies the conditions of Case-1 or Case-2.

Also, it is not difficult to see that Algorithm (p, q)-OPTLABEL$\Delta 5$ can be implemented to run in linear time.

$\square$

We remark that since the procedure of Case-4 needs the assumption of $\Delta \geq 5$, Algorithm (p, q)-OPTLABEL$\Delta 5$ cannot be applied to the cases of $\Delta < 5$.

4.2   Case: $p > 3q/2$

First we consider the case of $p = 2$ and $q = 1$. In this case, the condition (2) is sufficient for $\lambda_{2,1}^T(T) = \Delta + 1$, as described in the following lemma.

Lemma 7. *Let T be a tree. If $\Delta \geq 4$ and the condition (2) is satisfied, then $\lambda_{2,1}^T(T) = \Delta + 1$ and a $(\Delta + 1)$-$(2, 1)$-total labeling of T can be found in linear time.* $\square$

This lemma can be proved in a similar way to the proof of Lemma 5.

Algorithm 2. Algorithm (p, q)-OPTLABEL$\Delta 5$

Input: A tree $T = (V, E)$ satisfying the condition (2) and $\Delta \geq 5$ such that the degree of all non-major and non-leaf vertices is $\Delta - 1$, and two integers p and q with $p \leq 3q/2$.

Output: A (p, q)-total labeling $f : V \cup E \to \{0, 1, \ldots, p + (\Delta - 1)q\}$ of T.

1: Assign label 0 to the root v_r; let $f(v_r) := 0$. For each $i \in [1, \Delta - 1]$, let $f(v_r, c_i(v_r)) := p + iq$ and $f(c_i(v_r)) := q$, where $C(v_r) = \{c_i(v_r) \mid i = 0, 1, \ldots, \Delta - 1\}$. Let $f(v_r, c_0(v_r)) := p$ and $f(c_0(v_r)) := 3q$. {Note that the root v_r is major.}

2: **while** there exists a non-leaf $v \in V - \{v_r\}$ such that $f(v)$ has been determined but no label is assigned to any child of v **do**

3:   {Let M denote the set of major vertices. Denote $C(v)$ by $\{c_i(v) \mid i = 0, 1, \ldots, |C(v)|-1\}$ such that $d(c_0(v)) \geq d(c_1(v)) \geq \cdots \geq d(c_{|C(v)|-1}(v))$ and let j be an index such that $M \cap C(v) = \{c_i(v) \mid i \in [0, j]\}$ if $M \cap C(v) \neq \emptyset$, and $j = -1$ otherwise.}

4:   **if** (Case-1) v is major, $f(p(v), v) \in \{p + iq \mid i \in [0, \Delta - 1]\}$ and $f(v) = 0$ **then**

5:     Assign labels in $\{p + iq \mid i \in [0, \Delta - 1]\} - \{f(p(v), v)\}$ injectively to edges $\{(v, c_i(v)) \mid i \in [0, \Delta - 2]\}$ so that $f(v, c_0(v)) < f(v, c_1(v)) < \cdots < f(v, c_{\Delta-2}(v))$, and let $f(c_i(v)) := q$ for each $i \in [0, \Delta - 2]$. Only if $f(v, c_0(v)) = p$, then relabel $c_0(v)$ as $f(c_0(v)) := 3q$.

6:   **else if** (Case-2) v is non-major, $f(p(v), v) \in \{p + iq \mid i \in [1, \Delta - 1]\}$ and $f(v) = q$ **then**

7:     Assign labels in $\{p + iq \mid i \in [1, \Delta - 1]\} - \{f(p(v), v)\}$ injectively to edges $\{(v, c_i(v)) \mid i \in [0, \Delta - 3]\}$ so that $f(v, c_0(v)) < f(v, c_1(v)) < \cdots < f(v, c_{\Delta-3}(v))$, let $f(c_i(v)) := 0$ for each $i \in [0, j]$ and $f(c_i(v)) := 2q$ for each $i \in [j + 1, \Delta - 3]$. Only if $f(v, c_0(v)) = p + q$ and $M \cap C(v) = \emptyset$, then relabel $c_0(v)$ as $f(c_0(v)) := 4q$.

8:   **else if** (Case-3) v is non-major, $f(p(v), v) \in \{p + iq \mid i \in [2, \Delta - 1]\}$ and $f(v) = 2q$ **then**

9:     Assign labels in $\{p + iq \mid i \in [2, \Delta - 1]\} - \{f(p(v), v)\}$ injectively to edges $\{(v, c_i(v)) \mid i \in [1, \Delta - 3]\}$, and let $f(c_i(v)) := 0$ for each $i \in [1, j]$, $f(c_i(v)) := q$ for each $i \in [\max\{1, j+1\}, \Delta-3]$, and $f(v, c_0(v)) := 0$. If $M \cap C(v) \neq \emptyset$, then let $f(c_0(v)) := p+(\Delta-1)q$ and otherwise let $f(c_0(v)) := p + (\Delta - 2)q$.

10: **else if** (Case-4) v is non-major, $f(p(v), v) = p + q$ and $f(v) = 4q$ **then**

11:    Let $f(v, c_0(v)) := 0$, $f(v, c_i(v)) := p + (i + 2)q$ for each $i \in [2, \Delta - 3]$, $f(c_i(v)) := 0$ for each $i \in [2, j]$ and $f(c_i(v)) := q$ for each $i \in [\max\{2, j + 1\}, \Delta - 3]$. If $|M \cap C(v)| \geq 2$, then let $f(c_0(v)) := f(c_1(v)) := p + (\Delta - 1)q$, and $f(v, c_1(v)) := q$. If $|M \cap C(v)| = 1$, then let $f(c_0(v)) := p + (\Delta - 1)q$, $f(c_1(v)) := 3q$, and $f(v, c_1(v)) := p$. If $M \cap C(v) = \emptyset$, then let $f(c_0(v)) := 2q$, $f(c_1(v)) := 3q$, and $f(v, c_1(v)) := p$.

12: **else if** (Case-5) v is non-major, $f(p(v), v) = p$ and $f(v) = 3q$ **then**

13:    Let $f(v, c_0(v)) := 0$, $f(v, c_i(v)) := p + (i + 2)q$ for each $i \in [1, \Delta - 3]$, $f(c_i(v)) := 0$ for each $i \in [1, j]$, and $f(c_i(v)) := q$ for each $i \in [\max\{1, j + 1\}, \Delta - 3]$. If $M \cap C(v) \neq \emptyset$, then let $f(c_0(v)) := p + (\Delta - 1)q$ and otherwise let $f(c_0(v)) := 2q$.

14: **else if** (Case-6) v is non-major, $f(p(v), v) = 0$ and $f(v) = 2q$ **then**

15:    Let $f(v, c_i(v)) := p + (i + 2)q$ for each $i \in [0, \Delta - 3]$, $f(c_i(v)) := 0$ for each $i \in [0, j]$, and $f(c_i(v)) := q$ for each $i \in [j + 1, \Delta - 3]$.

16: **else if** (Case-j') $\overline{f}(p(v), v))$ and $\overline{f}(v)$ satisfy the conditions of Case-j for $j \in [1, 6]$ **then**

17:    After determining labels for $f(c_i(v))$ and $f(v, c_i(v))$ according to the above (Case-j) based on $\overline{f}(p(v), v))$ and $\overline{f}(v)$, let $f(c_i(v)) := p + (\Delta - 1)q - f(c_i(v))$ and $f(v, c_i(v)) := p + (\Delta - 1)q - f(v, c_i(v))$ for each i.

18: **end if**

19: **end while**

20: Output f as a (p, q)-total labeling of T.

On the other hand, the condition (2) is not necessary for $\lambda_{2,1}^T(T) = \Delta + 1$, in contrast to the case of $p \leq 3q/2$. For example, a tree T which consists of two adjacent major vertices and $2(\Delta - 1)$ leaves satisfies $\lambda_{2,1}^T(T) = \Delta + 1$, and there are many other such instances.

We also remark that the case of $p = 2$ and $q = 1$ is linearly solvable by Hasunuma et al.'s algorithm for the $L(2, 1)$-labeling problem [13], while the algorithm proposed in the proof of this lemma is much simpler (though it can be applied only to some restricted cases). Here we omit the details about this algorithm.

Consider the case of $p > 3q/2$ and $p \neq 2q$. In this case, the condition (2) is not even sufficient for $\lambda_{p,q}^T(T) = p + (\Delta - 1)q$, i.e., if $p > 3q/2$ and $p \neq 2q$, then for an arbitrary Δ, there exist instances T with $\lambda_{p,q}^T(T) > p + (\Delta - 1)q$ even if the condition (2) holds. For example, a tree which contains the configuration (a′) is one of such instances, where the configuration (a′) denotes one obtained from (a) in Fig. 1 by replacing each vertex with degree 3 (resp., degree 2) with a vertex with degree $\Delta > 0$ (resp., degree $\Delta - 1$) (note that v has $\Delta - 1$ major vertices as its neighbors).

5   Concluding Remarks

In this paper, we have discussed the (p, q)-total labeling problem for general p and q. By extending known results about the case of $q = 1$, we have derived upper and lower bounds on $\lambda_{p,q}(G)$ for some classes of graphs G. In particular, we provided tight bounds on $\lambda_{p,q}(T)$ for trees T for all possible p and q. Also, in the case of $p \leq 3q/2$, we showed that the (p, q)-total labeling problem can be solved in linear time, and characterized trees T achieving $\lambda_{p,q}^T(T)$ if $\Delta \geq 4$, in contrast to the counterparts of the $L(p, q)$-labeling problem. On the other hand, in the case of $3q/2 < p \leq \Delta q - 1$ and $p \neq 2q$, it is left open whether the (p, q)-total labeling problem for trees is polynomially solvable or not.

It is also challenging to derive a tight upper bound on $\lambda_{p,q}^T(G)$ for a general graph G, where even the case of $q = 1$ is open. We here give the following conjecture, which is a generalization of Havet and Yu's conjecture (1).

Conjecture 1. $\lambda_{p,q}^T(G) \leq 2p + \Delta q - 1$.

By Lemma 2, Corollary 1, and the fact that $\lambda_{1,1}^T(G) \leq \Delta + 1$ for any series-parallel graph G [26], this conjecture is true if $p > (\Delta - 1)q$ holds or G is the complete graph, a bipartite graph, or a series-parallel graph.

Another interesting issue might be to investigate the case $p < q$. We actually obtain tight bounds on $\lambda_{p,q}^T(T)$ for trees T about the case, though we omit the details.

References

1. Bazzaro, F., Montassier, M., Raspaud, A.: $(d, 1)$-total labelling of planar graphs with large girth and high maximum degree. Discr. Math. 307, 2141–2151 (2007)
2. Behzad, M.: Graphs and their chromatic numbers. Ph.D. Thesis, Michigan State University (1965)

3. Calamoneri, T.: The $L(h,k)$-labelling problem: A survey and annotated bibliography. The Computer Journal 49, 585–608 (2006) (The $L(h,k)$-Labelling Problem: A Survey and Annotated Bibliography,
http://www.dsi.uniroma1.it/~calamo/PDF-FILES/survey.pdf, ver. October 19, 2009)

4. Chang, G.J., Ke, W.-T., Kuo, D., Liu, D.D.-F., Yeh, R.K.: On $L(d,1)$-labeling of graphs. Discr. Math. 220, 57–66 (2000)

5. Chang, G.J., Kuo, D.: The $L(2,1)$-labeling problem on graphs. SIAM J. Discr. Math. 9, 309–316 (1996)

6. Chen, D., Wang, W.: (2,1)-Total labelling of outerplanar graphs. Discr. Appl. Math. 155, 2585–2593 (2007)

7. Dekar, L., Effantin, B., Kheddouci, H.: $[r,s,t]$-coloring of trees and bipartite graphs. Discr. Math. 310, 260–269 (2010)

8. Fiala, J., Golovach, P.A., Kratochvíl, J.: Computational Complexity of the Distance Constrained Labeling Problem for Trees. In: Aceto, L., Damgård, I., Goldberg, L.A., Halldórsson, M.M., Ingólfsdóttir, A., Walukiewicz, I. (eds.) ICALP 2008, Part I. LNCS, vol. 5125, pp. 294–305. Springer, Heidelberg (2008)

9. Fiorini, S.: On the chromatic index of outerplanar graphs. J. Combin. Theory Ser. B 18, 35–38 (1975)

10. Georges, J.P., Mauro, D.W.: Generalized vertex labeling with a condition at distance two. Congr. Numer. 109, 141–159 (1995)

11. Georges, J.P., Mauro, D.W.: Labeling trees with a condition at distance two. Discr. Math. 269, 127–148 (2003)

12. Griggs, J.R., Yeh, R.K.: Labelling graphs with a condition at distance 2. SIAM J. Discr. Math. 5, 586–595 (1992)

13. Hasunuma, T., Ishii, T., Ono, H., Uno, Y.: A linear time algorithm for $L(2,1)$-labeling of trees. In: Fiat, A., Sanders, P. (eds.) ESA 2009. LNCS, vol. 5757, pp. 35–46. Springer, Heidelberg (2009)

14. Hasunuma, T., Ishii, T., Ono, H., Uno, Y.: A tight upper bound on the (2,1)-total labeling number of outerplanar graphs. In: CoRR abs/0911.4590 (2009)

15. Havet, F., Yu, M.-L.: (d,1)-Total labelling of graphs. Technical Report 4650, INRIA (2002)

16. Havet, F., Yu, M.-L.: (p,1)-Total labelling of graphs. Discr. Math. 308, 496–513 (2008)

17. Havet, F., Thomassé, S.: Complexity of (p,1)-total labelling. Discr. Appl. Math. 157, 2859–2870 (2009)

18. Kemnitz, A., Marangio, M.: $[r,s,t]$-Colorings of graphs. Discr. Math. 307, 199–207 (2007)

19. Lih, K.-W., Liu, D.D.-F., Wang, W.: On (d,1)-total numbers of graphs. Discr. Math. 309, 3767–3773 (2009)

20. Montassier, M., Raspaud, A.: (d,1)-total labeling of graphs with a given maximum average degree. J. Graph Theory 51, 93–109 (2006)

21. Villà, M.S.: [r,s,t]-colourings of paths, cycles and stars. Doctoral Thesis, TU Bergakademie, Freiberg (2005)

22. Vizing, V.G.: Some unsolved problems in graph theory. Russian Mathematical Surveys 23, 125–141 (1968)

23. Wang, W., Chen, D.: (2,1)-Total number of trees maximum degree three. Inf. Process. Lett. 109, 805–810 (2009)

24. Whittlesey, M.A., Georges, J.P., Mauro, D.W.: On the λ-number of Qn and related graphs. SIAM J. Discr. Math. 8, 499–506 (1995)

25. Yeh, R.K.: A survey on labeling graphs with a condition at distance two. Discr. Math. 306, 1217–1231 (2006)

26. Zhou, X., Matsuo, Y., Nishizeki, T.: List total colorings of series-parallel graphs. J. Discr. Algorithms 3, 47–60 (2005)

Drawing a Tree as a Minimum Spanning Tree Approximation

Emilio Di Giacomo[1], Walter Didimo[1], Giuseppe Liotta[1], and Henk Meijer[2]

[1] Dip. di Ingegneria Elettronica e dell'Informazione, Università degli Studi di Perugia
{digiacomo,didimo,liotta}@diei.unipg.it
[2] Roosevelt Academy, The Netherlands
h.meijer@roac.nl

Abstract. We introduce and study $(1 + \varepsilon)$-*EMST drawings*, i.e. planar straight-line drawings of trees such that, for any fixed $\varepsilon > 0$, the distance between any two vertices is at least $\frac{1}{1+\varepsilon}$ the length of the longest edge in the path connecting them. $(1 + \varepsilon)$-EMST drawings are good approximations of Euclidean minimum spanning trees. While it is known that only trees with bounded degree have a Euclidean minimum spanning tree realization, we show that every tree T has a $(1 + \varepsilon)$-EMST drawing for any given $\varepsilon > 0$. We also present drawing algorithms that compute $(1 + \varepsilon)$-EMST drawings of trees with bounded degree in polynomial area. As a byproduct of one of our techniques, we improve the best known area upper bound for Euclidean minimum spanning tree realizations of complete binary trees.

1 Introduction

The Euclidean minimum spanning tree of a set points in $2D$ and in $3D$ is among the most fundamental and hence most studied geometric structures (see, e.g. [14]). In their seminal paper, Monma and Suri [13] initiated the investigation of the combinatorial properties of the Euclidean minimum spanning trees in the plane. This investigation naturally leads to the following question: Which are those trees that have an *EMST drawing*, i.e. a straight-line drawing that is also a Euclidean minimum spanning tree of the set of its vertices?

Besides their relevance in geometric graph theory, EMST drawings are also interesting for graph drawing applications. Namely, an EMST drawing satisfies some aesthetic requirements that are fundamental for the readability of a tree visualization: No two edges cross each other, groups of closely related vertices visually cluster together, and less related vertices are relatively far apart from each other [1,4,7,11].

Unfortunately, not all trees have an EMST drawing. Monma and Suri [13] proved that every tree with maximum vertex degree at most five admits an EMST drawing, while no tree with a vertex of degree greater than six admits this type of representation. As for trees having maximum degree equal to six, Eades and Whitesides [5] showed that it is NP-hard to decide whether such trees admit an EMST drawing. In order to enlarge the family of representable trees, the computation of EMST drawings in three-dimensional space was initiated in [12]. The authors of [12] proved that all trees with maximum vertex degree nine admit an EMST drawing in 3D-space, while no tree with

O. Cheong, K.-Y. Chwa, and K. Park (Eds.): ISAAC 2010, Part II, LNCS 6507, pp. 61–72, 2010.

vertex degree larger than twelve has an EMST drawing. King [10] reduced the gap between upper and lower bound by showing that all trees with vertex degree up to ten admit an EMST drawing in 3D-space.

In this paper we want to compute planar straight-line drawings of trees where groups of adjacent vertices be relatively close to each other while non-adjacent vertices be relatively far apart from one another. In order to overcome the vertex degree limitations imposed by EMST drawings, we define a new type of drawing that "approximates" an EMST drawing. Given a constant $\varepsilon > 0$ and a tree T, a $(1+\varepsilon)$-EMST drawing of T is a planar straight-line drawing Γ of T that satisfies the following *proximity constraint*: For any two vertices u and v, $d(u, v) \geq \frac{1}{1+\varepsilon}|e_T(u, v)|$, where $d(u, v)$ is the Euclidean distance between u and v in Γ, and $|e_T(u, v)|$ is the maximum length of an edge of Γ in the path from u to v in T.

One of the leading motivations behind our study is to investigate the area requirements of "good approximations" of EMST-drawings. Perhaps the most longstanding open problem about EMST-drawings is due to Monma and Suri [13], who conjecture that there exists a tree T of maximum degree five and a constant $c > 1$ such that any two-dimensional EMST drawing of T requires $\Omega(c^n \times c^n)$ area. Recently, Kaufmann [8] and Frati and Kaufmann [6] have made some significant progress on this problem disproving the conjecture of Monma and Suri for vertex degree up to four. In [6,8] an area bound of $O(n^{21.252})$ is proved for trees having vertex degree at most four and of $O(n^{11.387})$ for those having vertex degree at most three. In the same papers it is also shown that these bounds can be significantly improved when the input tree has logarithmic height: For example, an area bound of $O(n^{4.3})$ is proved for EMST drawings of complete binary trees. However, the question whether all trees having vertex degree at most five admit an EMST drawing of polynomial area remains to date unanswered.

An overview the results in this paper is as follows.

- We study the relationships between $(1+\varepsilon)$-EMST drawings and Euclidean minimum spanning trees. We show that the sum of the lengths of the edges in a $(1+\varepsilon)$-EMST drawing is within a $(1+\varepsilon)$-factor of the sum of the lengths of the edges of a Euclidean minimum spanning tree of the points representing the vertices (Section 2).
- We present a drawing algorithm that, for any given $\varepsilon > 0$ and any tree T, computes a two-dimensional $(1+\varepsilon)$-EMST drawing of T (Section 3).
- We describe polynomial area approximation schemes for $(1+\varepsilon)$-EMST drawings: Any tree with n vertices and degree bounded by a constant admits a $(1+\varepsilon)$-EMST drawing whose area is $O(n^{c+f(\varepsilon)})$, where c is a positive constant and $f(\varepsilon)$ is a polylogarithmic function of ε that tends to infinity as ε tends to zero (Section 4).
- We study ad-hoc polynomial area approximation schemes for families of trees that have logarithmic height. We obtain area bounds that significantly improve the general case. These techniques are also extended to the 3D-space (Section 5.1).
- Finally, as a variant of our techniques, we are able to compute an EMST drawing of a complete binary tree of n vertices in $O(n^{3.802})$ area. This result improves the best previously known upper bound of $O(n^{4.3})$ proved by Frati and Kaufmann [6] (Section 5.2).

For reasons of space, some proofs are sketched or omitted.

2 $(1 + \varepsilon)$-EMST Drawings

Let $T = (V, E)$ be a tree and let Γ be a straight-line drawing of T. We denote by $|e|$ the length of edge e in Γ and we call $|\Gamma| = \sum_{e \in E} |e|$ the *weight* of Γ. For any pair of vertices $u, v \in V$, $d(u, v)$ is the Euclidean distance between u and v in Γ, $\pi_T(u, v)$ is the path from u to v in T, and $e_T(u, v)$ is the longest edge of Γ along the path $\pi_T(u, v)$.

Given a set of points P, a *Euclidean minimum spanning tree* of P is a geometric tree spanning all points of P and having minimum total weight. In this paper we denote by $EMST(P)$ a Euclidean minimum spanning tree of P, and by $|EMST(P)|$ its weight. Let Γ be a straight-line drawing of a tree T and let P be the set of points corresponding to the vertices of T in Γ. If Γ is a Euclidean minimum spanning tree of P, we say that Γ is an *EMST drawing*. We recall that a drawing Γ is an EMST drawing if and only if it satisfies the following condition: $\forall u, v \in V, d(u, v) \geq |e_T(u, v)|$. Also, every EMST drawing is planar, i.e., it does not contain edge crossings (see, e.g., [14]).

Let $\varepsilon > 0$ be a given constant and let T be a tree. A $(1 + \varepsilon)$-*EMST drawing* of T is a planar straight-line drawing of T such that for any two vertices u and v, $d(u, v) \geq \frac{1}{1+\varepsilon} |e_T(u, v)|$. This last condition will be referred to as the *proximity constraint* of Γ.

The next theorem establishes a relationship between $(1 + \varepsilon)$-EMST drawings and Euclidean minimum spanning trees. Its proof is omitte dfor reasons of space.

Theorem 1. *Let $\varepsilon > 0$ be a given constant. Let T be a tree, let Γ be a straight-line drawing of T, and let P be the set of points corresponding to the vertices of T in Γ. If Γ is a $(1 + \varepsilon)$-EMST drawing of T, then $|\Gamma| \leq (1 + \varepsilon)|EMST(P)|$.*

3 Computing $(1 + \varepsilon)$-EMST Drawings of General Trees

In this section we consider the problem of computing a $(1+\varepsilon)$-EMST drawing of a tree. We start by remarking that the converse of Theorem 1 does not hold, which is a major difference to take into account between the problem of computing an EMST drawing and the one of computing a $(1 + \varepsilon)$-EMST drawing. Namely, every planar straight-line drawing of a tree having minimum weight satisfies the property that $\forall u, v \in V, d(u, v) \geq |e_T(u, v)|$; for a contrast, it is not true that every planar straight-line drawing of a tree whose weight is $(1 + \varepsilon)$ times the weight of a Euclidean minimum spanning tree satisfies the proximity constraint. Hence, a simple construction like the one illustrated in Fig. 1 (suitably fix the length of and edge according to the desired approximation factor and make all other edge lengths negligible) computes a drawing whose weight can be made arbitrarily close to the one of a minimum spanning tree, but it does not guarantee that for any two vertices u and v, $d(u, v) \geq \frac{1}{1+\varepsilon} |e_T(u, v)|$. The proof of the next theorem is in fact based on a different technique.

Theorem 2. *Let $\varepsilon > 0$ be a given constant. Any tree admits a $(1 + \varepsilon)$-EMST drawing that can be computed in $O(n)$ time in the real RAM model of computation.*

Sketch of Proof: Let T be a tree. Root T at an arbitrary vertex v. We describe a recursive algorithm that computes a drawing of T contained in a disc C_v of radius r, for any given value of $r > 0$. A high-level description of the algorithm is as follows. Vertex

Fig. 1. By making edge e sufficiently long, the weight of the drawing can arbitrarily approximate the weight of a Euclidean minimum spanning tree of its vertex set. However, the resulting drawing may not be a $(1 + \varepsilon)$-EMST drawing.

v is drawn at the center of C_v; each neighbor u of v is drawn at an arbitrary point of a distinct circle centered at v and of radius smaller than r. The subtree rooted at each neighbor u of v is recursively drawn inside a sufficiently small disc C_u centered at u and such that $C_u \subset C_v$. The radii of the concentric circles hosting the neighbors of v and the radius of each C_u are chosen in such a way that the resulting drawing satisfies the statement. See also Fig. 2.

More formally, let n be the number of vertices of T, let $r > 0$ be a given value and assume that any tree with at most $n - 1$ vertices admits a planar drawing contained in a disc of radius r' for any $r' > 0$ and satisfying the proximity constraint (the base case with $n = 1$ is trivially true by representing T as the center of this disc). Denote with $\deg(v)$ the degree of the root v of T. We prove that T admits a planar drawing Γ that satisfies the proximity constraint and that is contained in a disc of radius r centered at v. Compute a real number c such that $c \geq \max\{\varepsilon, \frac{1+\varepsilon}{\varepsilon}\}$; note that, this also implies that $\frac{c-1}{c} \geq \frac{1}{1+\varepsilon}$ and $c > 1$. Choose λ to be a real number such that $\lambda c^{\deg(v)+1} = r$. Let $u_1, u_2, \ldots, u_{\deg(v)}$ be the neighbors of v. Draw v at the origin of the plane and draw u_i at polar coordinates $(\lambda c^i, (i - 1) \cdot \theta)$, where $\theta = \frac{\pi}{\deg(v)-1}$. Clearly, no two edges (v, u_i) and (v, u_j) of T overlap. The subtree rooted at each u_h ($h = 1, 2, \ldots, \deg(v)$) is recursively drawn in the disc centered at u_h and having radius $r' = \min\{\frac{\varepsilon}{1+\varepsilon}\lambda c, (d(u_2, u_1) - \lambda(c^2 - c))/2\}$. This drawing exists by inductive hypothesis. Also, it is easy to see that $(d(u_2, u_1) - \lambda(c^2 - c))/2 \leq (d(u_i, u_j) - \lambda(c^i - c^j))/2$, for $1 \leq j < i \leq \deg(v)$.

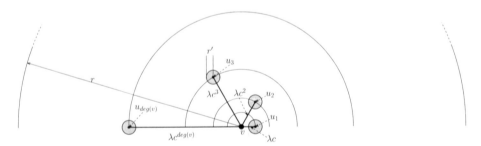

Fig. 2. The drawing construction described in the proof of Theorem 2

We prove that the computed drawing Γ of T is contained in the disc of radius r centered at v. The distance between v and the vertex that is farthest from v is at most $\lambda c^{deg(v)} + r'$. We need to prove that such a distance is at most $\lambda c^{deg(v)+1} = r$, i.e. that $\lambda c^{deg(v)+1} - \lambda c^{deg(v)} \geq r'$. We have $\lambda c^{deg(v)+1} - \lambda c^{deg(v)} = \lambda c^{deg(v)}(c-1)$ and $(c-1) \geq \frac{c}{1+\varepsilon}$. Thus, $\lambda c^{deg(v)+1} - \lambda c^{deg(v)} = \lambda c^{deg(v)}(c-1) \geq \lambda c \frac{c^{deg(v)}}{1+\varepsilon}$. By definition of c, $c^{deg(v)} > c \geq \varepsilon$; since $r' \leq \frac{\varepsilon}{1+\varepsilon}\lambda c$, we have $\lambda c^{deg(v)+1} - \lambda c^{deg(v)} \geq \lambda r \frac{c^{deg(v)}}{1+\varepsilon} \geq \lambda c \frac{\varepsilon}{1+\varepsilon} \geq r'$.

We now prove that Γ satisfies the proximity constraint. Let u and w be any two vertices of Γ and assume they are not adjacent, because otherwise the statement is trivially satisfied. If u and w are in the same subtree rooted at u_i, then they satisfy the statement by induction. If u is in a subtree rooted at u_h and w is in a subtree rooted at u_k with $h > k$ we have $d(u,w) \geq d(u_h, u_k) - 2r' \geq d(u_h, u_k) - (d(u_2, u_1) - \lambda(c^2 - c)) \geq d(u_h, u_k) - (d(u_h, u_k) - \lambda(c^h - c^k)) = \lambda(c^h - c^k) \geq \lambda(c^h - c^{h-1}) = \lambda c^h \frac{c-1}{c} \geq \frac{1}{1+\varepsilon}|e_T(u,w)|$. If u is in a subtree rooted at u_k and $w = v$ we have $d(u,w) = d(u,v) \geq d(v, u_k) - r' = \lambda c^k - r' \geq \lambda c^k - \frac{\varepsilon}{1+\varepsilon}\lambda c \geq \lambda c^k - \frac{\varepsilon}{1+\varepsilon}\lambda c^k = \frac{1}{1+\varepsilon}\lambda c^k = \frac{1}{1+\varepsilon}|e_T(u,w)|$. Since Γ is a planar drawing by construction, we conclude that Γ is a $(1+\varepsilon)$-EMST drawing.

The algorithm spends $O(deg(v))$ time for each vertex v, which implies an $O(n)$ time complexity.   □

It may be worth recalling that no tree having vertex degree higher than six can be represented as a Euclidean minimum spanning tree of a set of points and that it is NP-hard deciding whether a tree of degree six admits an EMST drawing [5,13]. Theorem 2 provides a tool to construct a drawing that is as close as possible to an EMST drawing for those trees that have degree larger than five. We observe however that the drawing algorithm of Theorem 2 may lead to drawings whose area[1] is exponential in n. For example, let T be a star-tree with n vertices (i.e., T consists of a vertex connected to $n-1$ leaves). The algorithm of Theorem 2 computes a drawing of T inside a disc whose radius is $r = \lambda c^n$, and therefore the area of the smallest axis-parallel rectangle including this disc is $O(c^{2n})$. Computing $(1 + \varepsilon)$-EMST drawings of polynomial area is the subject of the next two sections.

4   Polynomial Area Approximation Schemes for Bounded Degree Trees

In this section we show that a tree with n vertices and bounded degree admits a $(1+\varepsilon)$-EMST drawing whose area is $O(n^{c+f(\varepsilon)})$, where c is a positive constant and $f(\varepsilon)$ is a polylogarithmic function of ε that tends to infinity as ε tends to zero.

The very general idea of our approach is similar to that used in many papers that compute compact drawings of trees: Recursively compute the drawing by composing subdrawings of subtrees; if each composition increases the area of the current drawing

---

[1] The area of a drawing Γ is the area of the smallest axis-aligned rectangle enclosing Γ, for a given vertex resolution rule. A vertex resolution rule defines the minimum distance between any two vertices.

by a constant factor and if the number of recursive steps is a logarithmic function of the input size, then the total area is polynomial (see, e.g. [3,9]).

Based on this idea one could think of approaching the construction of $(1+\varepsilon)$-EMST drawings in polynomial area by using the edge-separator theorem of Valiant [15]. Namely, every tree T with n vertices and vertex degree at most Δ has an edge (called edge-separator) whose removal leaves two components, each containing at most $\frac{\Delta-1}{\Delta} \cdot n$ vertices. Therefore, one might think of drawing each of the components recursively, and add the removed edge back with a sufficient length that guarantees the proximity constraint. Because of the size of each component, the number of levels in the recursion is $O(\log_b n)$, with $b = \frac{\Delta}{\Delta-1}$ and hence the area of the resulting drawing is polynomial in n. Unfortunately, it is not clear how this simple approach could lead to drawings without edge crossings. We therefore follow a different approach.

In order to guarantee a logarithmic number of recursive steps, we decompose the tree into subtrees of smaller size by means of a *greedy path decomposition*. Let T be a rooted tree such that each vertex has at most k children, and let v_0 be the root of T. A *greedy path of T* is a path $v_0, v_1, \ldots, v_k$ connecting the root v_0 to a leaf v_k and such that v_i is the root of the largest subtree rooted at v_{i-1} ($1 \le i \le k$). A greedy path decomposition of a rooted tree T consists of recursively identifying greedy paths and on removing them so to decompose the tree into rooted subtrees of smaller size. The decomposition ends when the tree is a path (possibly consisting of a single vertex). Greedy paths decompositions of rooted trees have, for example, been used by Chan [2] to compute compact drawings of binary trees, and by Kaufmann [8] to prove polynomial area bounds for EMST drawings of ternary trees.

Let T be a tree with a given greedy path decomposition and let T' be a subtree of T. The *greedy depth* of T' (with respect to the given decomposition) is denoted as $\gamma(T')$, and defined as follows: (i) If T' is a path, $\gamma(T') = 1$; (ii) otherwise, $\gamma(T') = \max_i\{\gamma(T_i)\} + 1$, where each T_i is a tree obtained from T' by removing its greedy path for the given decomposition. Intuitively, the greedy depth of a tree for a given greedy path decomposition is the depth of the recursion in the decomposition process. If T has n vertices, the size of a subtree of T having greedy depth i is at most $\frac{n}{2^i}$. This immediately implies the following property.

Property 1. Let T be a tree with n vertices and a given greedy path decomposition. Then, $\gamma(T) \le \lceil \log_2 n \rceil$.

Theorem 3. *Let $\varepsilon > 0$ be a given constant. Let $\Delta > 2$ be a positive integer. Let T be a tree with n vertices and vertex degree at most Δ. T admits a $(1 + \varepsilon)$-EMST drawing whose area is $O(n^{4+2\log_2(\frac{c\Delta+1+c\Delta+c^2-3c}{c-1})})$, where c is a constant such that $c \ge \max\{\frac{1+\varepsilon}{\varepsilon}, \frac{1}{\sin\frac{\pi}{2(\Delta-1)}}\}$. Furthermore, such a drawing can be computed in $O(n)$ time in the real RAM model of computation.*

Sketch of Proof: We describe a drawing algorithm assuming that T is rooted and each internal vertex of T has exactly $\Delta - 1$ children. This assumption is not restrictive. Indeed, if T is a tree of degree at most Δ, we can always root T at a leaf and add to any internal vertex of degree $k < \Delta - 1$ a set of $\Delta - 1 - k$ dummy children. In this way, the number of vertices of the augmented tree is at most $(\Delta - 1)n$, and hence, still linear in n.

The drawing algorithm applies a recursive construction based on a greedy path decomposition. Let $T' = (V', E')$ be a subtree of T such that $\gamma(T') = i$ ($1 \leq i \leq \lceil \log_2 n \rceil$). Let Π be the greedy path of T'. The algorithm constructs a drawing Γ' of T' by composing the drawings of all trees obtained from T' by removing Π. Denote by $n(T')$ the number of vertices of T'. The algorithm will maintain the following invariants for drawing Γ' (see Fig. 3(a) for an illustration):

(I1) $\forall u, v \in V', d(u, v) \geq \frac{1}{1+\varepsilon} |e_{T'}(u, v)|$.
(I2) Γ' is completely contained in the north-east quadrant of a disc C' such that C' is centered at the root v' of T' and the radius of C' is $r' = n(T')(2c(b+1))^{\log_2 n(T')+1}$ where $b = (c^{\Delta-1} + 2\frac{c^{\Delta-1}-1}{c-1})$.
(I3) Γ' is planar.

We now prove that Γ' exists. The proof is by induction on the greedy depth i of T'. The base case is for $i = 1$. Since $\Delta > 2$, each internal vertex of T' has $\Delta - 1 \geq 2$ children, and therefore in the base case T' consists of a single vertex. T' is drawn as a single point centered at a disc of radius $2c(b+1)$ and thus Γ' satisfies all invariants.

By inductive hypothesis, each subtree with greedy depth $i - 1$ admits a drawing satisfying Invariants (I1), (I2), and (I3). We construct Γ' as follows (see also Fig. 3(b)).

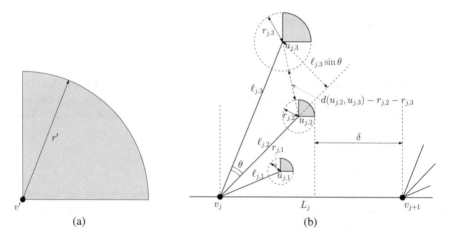

Fig. 3. The drawing construction in the proof of Theorem 3

Denote by $v_1, v_2, \ldots, v_h$ the vertices of Π, and let $u_{j,1}, u_{j,2}, \ldots, u_{j,\Delta-2}$ the children of v_j that are not in Π ($1 \leq j \leq h - 1$). The vertices $v_1, v_2, \ldots, v_h$ are drawn on a horizontal line, in this order from left to right. The distance between v_j and v_{j+1} ($1 \leq j \leq h - 1$) is denoted by L_j and its value will be specified later. Denote by $\Gamma_{u_{j,k}}$ the drawing of each subtree rooted at $u_{j,k}$ ($1 \leq j \leq h - 1, 1 \leq k \leq \Delta - 2$). By Invariant (I2) each $\Gamma_{u_{j,k}}$ is contained in a disc of suitable radius; we denote this radius as $r_{j,k}$ and we assume that the children of v_j are ordered so that $r_{j,k} < r_{j,k+1}$ ($1 \leq k \leq \Delta - 3$).

For $1 \leq k \leq \Delta - 2$, drawing $\Gamma_{u_{j,k}}$ is placed in such a way that the polar coordinates of $u_{j,k}$ with respect to the position of v_j are $(\ell_{j,k}, (k+1)\theta)$, where $\theta = \frac{\pi}{2(\Delta-1)}$ and $\ell_{j,k}$ is defined as follows:

$$\ell_{j,k} = \begin{cases} cr_{j,0} & k = 0 \\ c(\ell_{j,k-1} + r_{j,k-1} + r_{j,k}) & k > 1 \end{cases}$$

The value of L_j is set to $L_j = c(\ell_{j,\Delta-2} + \ell_{j+1,\Delta-2} + r_{j,\Delta-2} + r_{j+1,\Delta-2})$.

For reasons of space, the proof that the Invariants (I1), (I2) and (I3) are maintained is omitted. From Invariants (I1) and (I3) it follows that Γ is a $(1 + \varepsilon)$-EMST drawing of T. Also, by Invariant (I2), Γ is contained in a disc of radius $n(2c(b+1))^{\log_2 n+1} = (2c(b+1))n^{2+\log_2(c(b+1))}$. Hence, the area of the drawing is $O(n^{2+\log_2(c(b+1))})$ which, by the definition of b, is $O(n^{4+2\log_2(\frac{c^{\Delta+1}+c^{\Delta}+c^2-3c}{c-1})})$.

The algorithm spends $O(deg(v))$ time for each vertex v, i.e., $O(n)$ time in total.   □

We recall that it is not known how to draw in polynomial area a tree of degree five as a Euclidean minimum spanning tree in the plane [13]. Theorem 3 implies that, for any given constant $\varepsilon > 0$, an approximation of an EMST drawing with polynomial area exists for trees with degree five vertices.

5 Trees with Logarithmic Height

We devote this section to trees having small vertex degree and logarithmic height. We describe ad-hoc algorithms that compute $(1 + \varepsilon)$-EMST drawings of these trees by using significantly less area than the one given by Theorem 3. As a byproduct of this study, we show how to realize a complete binary tree with n vertices as a Euclidean minimum spanning tree in area $O(n^{3.802})$, which improves the best previously known upper bound of $O(n^{4.3})$ proved by Frati and Kaufmann [6].

5.1 Trees with Vertex Degree at Most Six

In the next theorem we exploit the maximum vertex degree six and the logarithmic height of the input tree to design a recursive algorithm that does not use the greedy path decomposition. In the statement, if a rooted tree T has degree Δ, each internal vertex of T has at most $\Delta - 1$ children.

Theorem 4. *Let $\varepsilon > 0$ and $h > 0$ be given constants. Let Δ be a positive integer such that $3 \leq \Delta \leq 6$. Let T be a rooted tree with n vertices, vertex degree at most Δ, and height at most $h\log_{\Delta-1} n$. T admits a $(1 + \varepsilon)$-EMST drawing whose area is $O(n^{2h\log_{\Delta-1}(c+2)})$, where $c = \frac{2+\varepsilon}{\varepsilon}$. Furthermore, such a drawing can be computed in $O(n)$ time in the real RAM model of computation.*

Sketch of Proof: We describe an algorithm that constructs a drawing of T and prove that this drawing satisfies the properties in the statement. For any vertex v of T, denote by T_v the subtree of T rooted at v. If i is the level of v ($0 \leq i \leq h\log_{\Delta-1} n$), the algorithm recursively constructs a drawing Γ_v of T_v inside a disc C_v centered at v, with a suitable radius r_i. Γ_v will be such that the following invariants hold:

(I1) $\forall u, w \in T_v$, $d(u, w) \geq \frac{1}{1+\varepsilon} |e_{T_v}(u, w)|$;

(I2) There exists a ray of C_v departing from v that does not cross any edge of Γ_v; in the following we call this ray the *free ray* of Γ_v.

(I3) Γ_v is planar.

We show how to construct Γ_v by induction on the level of v, going from the highest level of T to level 0. Level 0 is the level of the root of T.

A vertex v at the deepest level of T is a leaf; in this case T_v is drawn as a single point centered at a disc of radius 1, so that Invariants (I1)–(I3) trivially hold for Γ_v.

Suppose by induction that for any vertex v' at level $i > 0$, a drawing $\Gamma_{v'}$ of $T_{v'}$ that satisfies Invariants (I1)–(I3) exists. Let v be a vertex at level $i - 1$ and let $u_1, u_2, \ldots, u_d$ be its children. Note that, if v is not the root of T then $d \leq \Delta - 1$; if v is the root of T then $d \leq \Delta$. Drawing Γ_v is constructed by combining the drawings Γ_{u_j} of T_{u_j} ($1 \leq j \leq d$). Namely, Γ_v is drawn inside a disc C_v of radius r_{i-1}, such that v is placed at the center of C_v and the drawings Γ_{u_j} are distributed around v. More precisely, if we assume (without loss of generality) that v is placed at the origin of the plane, then each u_j is placed at a point of polar coordinates $((r_{i-1} - r_i), j\frac{2\pi}{k})$, and Γ_{u_j} is rotated so that its free rays has the direction of the segment connecting v to u_j. Finally, the radius of C_v is set to $r_{i-1} = (c + 2)r_i$, where $c = \frac{2+\varepsilon}{\varepsilon}$. See also Fig. 4(a).

We first prove that Invariant (I1) holds for Γ_v. Let u and w be two arbitrary vertices of T_v. Three cases are possible. If u and w are both in the same T_{u_j} ($1 \leq j \leq d$) then $d(u, w) \geq \frac{1}{1+\varepsilon} |e_T(u, w)|$ by the inductive hypothesis.

If $u \in T_{u_j}$ and $w \in T_{u_l}$, with $1 \leq j < l \leq d$, we have that $d(u, w) \geq d(u_j, u_l) - 2r_i$. Denote as δ be the distance between the discs containing two consecutive drawings Γ_{u_j} and $\Gamma_{u_{j+1}}$ around v (see also Fig. 4(a)). We have that $d(u, w) \geq d(u_j, u_l) - 2r_i \geq d(u_j, u_{j+1}) - 2r_i = \delta$; also $e_{T_v}(u, w) = (v, u_j)$ and therefore $|e_{T_v}(u, w)| = r_{i-1} - r_i$. Hence, it suffices to show that $\delta \geq \frac{1}{1+\varepsilon}(r_{i-1} - r_i)$. By simple trigonometry, $\frac{\delta}{2} + r_i = (c + 1)r_i \sin(\pi/\Delta)$. It follows that $\delta \geq \frac{1}{1+\varepsilon}(r_{i-1} - r_i)$ can be rewritten as $2((c + 1)\sin(\pi/\Delta) - 1)r_i \geq \frac{1}{1+\varepsilon}(c + 1)r_i$, which is verified for $c \geq \frac{1+2(1+\varepsilon)(1-\sin(\pi/\Delta))}{2(1+\varepsilon)\sin(\pi/\Delta)-1}$ and $2(1 + \varepsilon)\sin(\pi/\Delta) - 1 > 0$. Note that for $3 \geq \Delta \geq 6$ we have both $\frac{2+\varepsilon}{\varepsilon} \geq \frac{1+2(1+\varepsilon)(1-\sin(\pi/\Delta))}{2(1+\varepsilon)\sin(\pi/\Delta)-1}$ and $2(1 + \varepsilon)\sin(\pi/\Delta) - 1 > 0$.

If u coincides with v and w is in T_{u_j} ($1 \leq j \leq d$), we have that $d(u, w) \geq r_{i-1} - 2r_i$; also $e_{T_v}(u, w) = (v, u_j)$ and therefore $|e_{T_v}(u, w)| = (r_{i-1} - r_i)$. To show Invariant (I1) it suffices to prove that $r_{i-1} - 2r_i \geq \frac{1}{1+\varepsilon} r_{i-1} - r_i$. By construction, $r_{i-1} - 2r_i = cr_i$ and $r_{i-1} - r_i = (c+1)r_i$. It follows that the previous inequality can be rewritten as $cr_i \geq \frac{1}{1+\varepsilon}(c + 1)r_i$, which is verified since $c = \frac{2+\varepsilon}{\varepsilon}$.

We now prove that Invariant (I2) also holds for Γ_v. Since the distance δ between the discs containing two consecutive drawings Γ_{u_j} and $\Gamma_{u_{j+1}}$ around v is at least $\frac{1}{1+\varepsilon}(r_{i-1} - r_i)$, it follows that δ is positive. Hence, a free ray of Γ_v is any ray from v passing between the discs containing Γ_{u_j} and $\Gamma_{u_{j+1}}$.

Finally, by construction, it is easy to see that Invariant (I3) holds.

It follows that, if v is the root of T, drawing $\Gamma = \Gamma_v$ is a $(1+\varepsilon)$-EMST drawing. The bound on the area of the drawing is proved by observing that the radius r_0 of the disc containing Γ is related to the radius r_i by the following equation: $r_0 = (c+2)^i r_i$. Since the height of T is at most $h \log_{\Delta-1} n$ we have $r_0 \le (c+2)^{h \log_{\Delta-1} n} = n^{h \log_{\Delta-1}(c+2)}$, which implies an area bound of $O(n^{2h \log_{\Delta-1}(c+2)})$.

The algorithm spends $O(deg(v))$ time for each vertex v, i.e., $O(n)$ time in total.   □

It could be interesting to compare the bounds of Theorem 3 with those of Theorem 4. Suppose that T is a complete rooted tree with n vertices such that every internal vertex has five children. Assume that we wish to be within a factor $\varepsilon = 0.5$ of having an EMST drawing of T, i.e. we want to compute a 1.5-EMST drawing of T. By using the construction of Theorem 3 the area of the drawing is $O(n^{26.17})$; by using Theorem 4, we can compute a 1.5-EMST drawing of T in $O(n^{2.42})$ area.

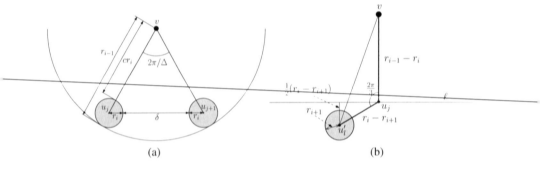

Fig. 4. (a) The drawing construction in the proof of (a) Theorem 4 (b) Theorem 6

One may wonder whether Theorem 4 can be extended to trees of degree larger than six. Notice however that the function $\frac{1+2(1+\varepsilon)(1-\sin(\frac{\pi}{\Delta}))}{2(1+\varepsilon)\sin(\frac{\pi}{\Delta})-1}$ used in Theorem 4 is finite and positive for every $\varepsilon > 0$ only when $3 \le \Delta \le 6$. If $\Delta > 6$ the argument in the proof of Theorem 4 cannot be applied. This motivates us to look at $(1 + \varepsilon)$-EMST drawings in three dimensions. The proof of the next theorem is omitted.

Theorem 5. *Let $\varepsilon > 0$ and $h > 0$ be given constants. Let Δ be a positive integer such that $3 \le \Delta \le 12$. Let T be a rooted tree with n vertices, vertex degree at most Δ, and height at most $h \log_{\Delta-1} n$. T admits a $(1 + \varepsilon)$-EMST in three dimensional space whose volume is $O(n^{3h \log_{\Delta-1}(c+2)})$, where $c = \frac{2+\varepsilon}{\varepsilon}$. Furthermore, such a drawing can be computed in $O(n)$ time in the real RAM model of computation.*

Finally, we observe that it is possible to use the drawing technique of Theorem 4 to compute $(1+\varepsilon)$-EMST drawings of trees of degree higher than six, provided that $(1+\varepsilon)$ approximation factor is not required to be arbitrarily close to 1, but it depends on the vertex degree.

5.2   *EMST* Drawings of Complete Binary Trees

Frati and Kaufmann [6] prove that a complete binary tree can be drawn as a Euclidean minimum spanning tree in area $O(n^{4.3})$. In this section we improve this bound as an application of the techniques in the proof of Theorem 4. Namely, we show that if $\Delta = 3$ the condition $cr_i \geq \frac{1}{1+\varepsilon}(c+1)r_i$ does not need to be verified for proving the correctness of the geometric construction of Theorem 4. An implication of this observation is that for $\Delta = 3$ the proof of Theorem 4 also works by setting $\varepsilon = 0$.

Theorem 6. *Let T be a rooted complete binary tree with n vertices. T admits an EMST drawing in area $O(n^{3.802})$. Furthermore, such a drawing can be computed in $O(n)$ time in the real RAM model of computation.*

Sketch of Proof: As already observed, a straight-line two-dimensional drawing of a tree T is a Euclidean minimum spanning tree of the points representing its vertices if and only if that the following property holds:

$$\forall u, w \in V, d(u, w) \geq |e_T(u, w)| \tag{1}$$

Let z be the root of T. We draw the two edges incident to z as segments of the same length and forming an angle of $\frac{2\pi}{3}$. By applying the drawing technique described in the proof of Theorem 4, we recursively construct a drawing where for each internal vertex v different from z any two consecutive edges around v form an angle of $\frac{2\pi}{3}$. Let v be an internal vertex of height $i - 1 > 0$ and let u_1 and u_2 be its children. According to our drawing technique the segment connecting v to u_1 forms an angle of $\frac{2\pi}{3}$ with the positive x-axis, and the segment connecting v to u_2 forms an angle of $-\frac{2\pi}{3}$ with the positive x-axis. Hence the positive x-axis is a free ray for T_v; this free ray is used to connect the drawing of T_v to the parent of v after suitable rotation.

It remains to prove that the computed drawing satisfies Property (1). The proof is by induction. Let u and w be any two vertices of T_v. If u and w are vertices of the same subtree T_{u_j} ($1 \leq j \leq 2$), then Property (1) holds by induction. If u is in the subtree T_{u_1} and w is in T_{u_2} then $d(u, w) \geq d(u_1, u_2) - 2r_i = \delta$ and therefore it is sufficient to guarantee $\delta \geq (r_{i-1} - r_i)$. We have that $\frac{\delta}{2} + r_i = (c+1)r_i \sin(\pi/3) = (c+1)\frac{\sqrt{3}}{2}r_i$. Thus, it must be $\sqrt{3}(c+1)r_i - 2r_i \geq (c+1)r_i$, from which we obtain $c \geq \frac{3-\sqrt{3}}{\sqrt{3}-1} > 1.732$. If u coincides with v and if w is in the subtree T_{u_j} ($1 \leq j \leq 2$), then assume that w does not coincide with u_j (otherwise Property (1) is trivially true) and let u'_1 and u'_2 be the children of u_j. Vertex w is a vertex of $T_{u'_l}$ ($1 \leq l \leq 2$). We have that $d(u, w) \geq d(v, u'_l) - r_{i+1}$ (see Fig. 4(b)). Let ℓ be the straight-line orthogonal to the segment representing edge (v, u_j) passing through u_j. The distance from u'_l to ℓ is $(r_i - r_{i+1}) \sin(\frac{\pi}{6}) = \frac{1}{2}(r_i - r_{i+1}) = \frac{1}{2}(c+1)r_{i+1}$, that is larger than r_{i+1} for any $c > 1$. Thus, for every $c > 1$, the disc containing $T_{u'_l}$ and v are on opposite sides of ℓ, which means that $d(u, w) > |e_T(u, w)| = |(v, u_j)| = r_{i-1} - r_i$.

Therefore we can choose $r_{i-1} = (c + 2)r_i$ with $c = 1.733$, which implies that the radius of the disc enclosing the whole drawing is $r_0 = (c + 2)^{\log_2 n+1} = (c + 2)n^{\log_2(c+2)} = (c + 2)n^{\log_2(3.733)} < (c + 2)n^{1.901}$ and the area of the drawing is $O(n^{3.802})$.

The algorithm spends $O(deg(v))$ time for each vertex v, i.e., $O(n)$ time in total. $\square$

6 Open Problems

The study of $(1+\varepsilon)$-EMST drawings and of their variants opens interesting perspectives both from the theoretical and from the practical point of view. For example, one can observe that drawings of trees computed by using spring embedder heuristics seem to be $(1+\varepsilon)$-EMST drawings in many cases (see, e.g., [4]). It would be nice to experimentally study this possible correlation. Also, extending the concept of "approximated drawing" to other types of proximity rules and/or to other families of graphs is a promising research subject. For example, does every triangulated planar graph admit a straight-line drawing whose weight is at most $(1+\varepsilon)$ times the weight of the Delaunay triangulation of its vertex set?

References

1. Bose, P., Di Battista, G., Lenhart, W., Liotta, G.: Proximity constraints and representable trees. In: Tamassia, R., Tollis, I.G. (eds.) GD 1994. LNCS, vol. 894, pp. 340–351. Springer, Heidelberg (1995)
2. Chan, T.M.: A near-linear area bound for drawing binary trees. Algorithmica 34(1), 1–13 (2002)
3. Di Battista, G., Eades, P., Tamassia, R., Tollis, I.G.: Graph Drawing. Prentice Hall, Upper Saddle River (1999)
4. Eades, P., Whitesides, S.: The realization problem for euclidean minimum spanning trees in NP-hard. Algorithmica 16(1), 60–82 (1996)
5. Eades, P., Whitesides, S.: The realization problem for Euclidean minimum spanning trees is NP-hard. Algorithmica 16, 60–82 (1996)
6. Frati, F., Kaufmann, M.: Polynomial area bounds for MST embeddings of trees. RT-DIA-122-2008, Dept. of Comput. Sc. Univ. Roma Tre (2008)
7. Jaromczyk, J.W., Toussaint, G.T.: Relative neighborhood graphs and their relatives. Proc. IEEE 80(9), 1502–1517 (1992)
8. Kaufmann, M.: Polynomial area bounds for MST embeddings of trees. In: Hong, S.-H., Nishizeki, T., Quan, W. (eds.) GD 2007. LNCS, vol. 4875, pp. 88–100. Springer, Heidelberg (2008)
9. Kaufmann, M., Wagner, D. (eds.): Drawing Graphs. LNCS, vol. 2025. Springer, Heidelberg (2001)
10. King, J.: Realization of degree 10 minimum spanning trees in 3-space. In: Canadian Conference on Computational Geometry (CCCG 2006) (2006)
11. Liotta, G.: Proximity drawings. In: Tamassia, R. (ed.) Handbook of Graph Drawing and Visualization. CRC Press, Boca Raton (to appear)
12. Liotta, G., Di Battista, G.: Computing proximity drawings of trees in the 3-dimemsional space. In: Sack, J.-R., Akl, S.G., Dehne, F., Santoro, N. (eds.) WADS 1995. LNCS, vol. 955, pp. 239–250. Springer, Heidelberg (1995)
13. Monma, C., Suri, S.: Transitions in geometric minimum spanning trees. Discrete Comput. Geom. 8, 265–293 (1992)
14. Preparata, F.P., Shamos, M.I.: Computational Geometry: An Introduction, 3rd edn. Springer, Heidelberg (October 1990)
15. Valiant, L.G.: Universality considerations in VLSI circuits. IEEE Trans. Computers 30(2), 135–140 (1981)

k-cyclic Orientations of Graphs

Yasuaki Kobayashi[1], Yuichiro Miyamoto[2], and Hisao Tamaki[1]

[1] Meiji University, Kawasaki, Japan 214-8571
{yasu0207,tamaki}@cs.meiji.ac.jp
[2] Sophia University, Chiyoda-ku, Tokyo, Japan 102-8554
miyamoto@sophia.ac.jp

Abstract. An *orientation* of an undirected graph G is a directed graph D on $V(G)$ with exactly one of directed edges (u, v) and (v, u) for each pair of vertices u and v adjacent in G. For integer $k \geq 3$, we say a directed graph D is *k-cyclic* if every edge of D belongs to a directed cycle in D of length at most k. We consider the problem of deciding if a given graph has a k-cyclic orientation. We show that this problem is NP-complete for every fixed $k \geq 3$ for general graphs and for every fixed $k \geq 4$ for planar graphs. We give a polynomial time algorithm for planar graphs with $k = 3$, which constructs a 3-cyclic orientation when the answer is affirmative.

1 Introduction

Let G be an undirected graph with vertex set $V(G)$ and edge set $E(G)$. An *orientation* of edge e of G between vertex u and v is a directed edge (u, v) or (v, u). An *orientation* of G is a directed graph on $V(G)$ that has exactly one of the two orientations of each edge of G.

Robbins [7] shows that G has a strongly connected orientation if and only if G is 2 edge-connected. Given this fact, it is natural to be interested in the "quality" of an orientation that we may obtain for a given graph [1,5,2,3,6]. Chvátal and Thomassen [1] show that there is a polynomial function f such that every graph with diameter d has a strongly connected orientation with directed diameter at most $f(d)$. They also show that it is NP-complete to decide, given graph G and integer d, if G has an orientation with diameter at most d, even if the diameter of G is 2. This decision problem can be solved in linear time when the given graph is planar [3]. Dankelmann *et al.* [2] study the relationship between the average distance between a pair of vertices in a graph and the average directed distance from a vertex to another in an orientation of that graph. Motivated by applications to traffic control in market places and factories, Ito *et al.* [6] study some optimization problems where, given a graph and a collection of *st*-pairs in the graph, we are to find an orientation of the graph such that the *st*-pairs are connected by short directed paths. They consider both the min-max problem, where the objective function is the maximum of the lengths of those directed paths, and the min-sum problem, where the objective function is the sum of the lengths of those directed paths.

O. Cheong, K.-Y. Chwa, and K. Park (Eds.): ISAAC 2010, Part II, LNCS 6507, pp. 73–84, 2010.

In this paper, we introduce the notion of k-cyclic orientations of graphs. For integer $k \geq 3$, we call an orientation D of graph G k-*cyclic* if the orientation of every edge of G belongs to a directed cycle of length k or smaller in D. This notion captures the local quality of an orientation as opposed to the global quality captured by directed diameters or average directed distances. Observe that D is a k-cyclic orientation of G if and only if D has a directed path form u to v of length $k-1$ or smaller for every pair of vertices u and v adjacent in G. Thus, the question of finding the minimum value of k such that G has a k-cyclic orientation is equivalent to the special case of the min-max problem of [6], where (s, t) is in the specified collection of st-pairs if and only if s and t are adjacent in G. This special case is important, especially for small values of k, since the solution for this special case is a $(k-1)$-approximate solution for an arbitrary collection of st-pairs on the same graph.

We show that the problem of deciding if a graph G has a k-cyclic orientation is NP-complete for every fixed $k \geq 3$ for general graphs. We also show that this problem remains NP-complete for planar graphs if $k \geq 4$.

On the positive side, we give a polynomial time algorithm that solves this problem for planar graphs with $k = 3$. This algorithm constructs a 3-cyclic orientation of the given graph, when the answer to the decision problem is affirmative. This algorithm is based on the following observation. Consider the special case where G is a plane embedded graph such that every cycle of G with length k or smaller bounds a face. In this case, G has a k-cyclic orientation if and only if the planar dual of G has a proper 3-coloring, using colors white, red, and blue, such that every dual vertex of degree greater than k is colored white. The correspondence between a feasible orientation and a feasible coloring can be obtained by the following rule: if a face is bounded by a cycle of length k or smaller that is oriented clockwise (counterclockwise) around the face then color the corresponding dual vertex red (blue); otherwise color the corresponding dual vertex white. This observation rather straightforwardly leads to a polynomial time algorithm for 3-cyclic orientation for this special case of planar graphs. The extension of this result to general planar graphs, however, is not simple, because of the existence of non-facial 3-cycles. We overcome the difficulty by replacing the standard planar dual of the given graph by some variant, in which the structures internal to non-facial 3-cycles are replaced by appropriate gadgets that depend on the "types" of those cycles. These types are determined by recursive applications of the main algorithm. See Subsection 3.2 for details.

The rest of this paper is organized as follows. In Section 2 we prove the NP-completenes of the problem for genral graphs. In Section 3 we study the problem for planar graphs and give the negative and positive results stated above.

2   General Graphs

In this section, we consider our orientation problem for general graphs. We first define some notation and terms that we use throughout the paper.

In this paper, all graphs are simple, unless otherwise stated. For graph G, we denote by $V(G)$ the set of vertices and $E(G)$ the set of edges of G. The set of vertices adjacent to v in G is denoted by $N_G(v)$ and the degree $|N_G(v)|$ of each vertex v of G is denoted by $d_G(v)$. For $U \subseteq V(G)$, we let $N_G(U)$ denote $\bigcup_{v \in U} \{N_G(v)\} \setminus U$. We may omit the subscript when no confusion may arise. For each vertex set $U \subseteq V(G)$, we denote by $G[U]$ the subgraph of G induced by U. We denote $G[V(G) \setminus U]$ by $G \setminus U$ and, for each $A \subseteq E(G)$, a spanning subgraph of G with edge set $E(G) \setminus A$ by $G \setminus A$.

We call a cycle C of G a k-cycle, $\leq k$-cycle, or $> k$-cycle if the number of vertices on C is k, at most k, or larger than k, respectively. Let G be a graph and D an orientation of G. For each subgraph H of G, we denote by $D|H$ the restriction of D on H, i.e., the orientation of H that is a sub-digraph of D. We say that an orientation D of G *extends* an orientation D' of subgraph H of G, if $D' = D|H$. We say that a cycle C of G is *cyclic in* D if $D|C$ is a directed cycle. We say that an edge of G is k-cyclic if e belongs to some $\leq k$-cycle of G that is cyclic in D. Thus, a k-cyclic orientation of G is an orientation in which every edge of G is k-cyclic.

Theorem 1. *The problem of deciding if a given graph has a k-cyclic orientation is NP-complete for every fixed $k \geq 3$.*

The fact that this problem is in NP is trivial. We prove the hardness by a reduction from Not All Equal 3SAT (NAE-3SAT), which is known to be NP-complete (see [4]). In NAE-3SAT, given a boolean formula ϕ in CNF with a set $X = \{x_1, x_2, \ldots, x_n\}$ of variables and a set $S = \{c_1, c_2, \cdots, c_m\}$ of clauses, each of which consists of exactly three literals, we are to decide whether ϕ has a not-all-equal assignment, that is, a truth assignment on X in which each clause of ϕ has at least one true literal and at least one false literal.

We describe the reduction for $k = 3$. A generalization for $k > 3$ is not difficult. (An alternative is to use the NP-hardness of the problem on planar graphs for $k \geq 4$ proved in the next section.) Given an instance ϕ of NAE-3SAT, we construct a graph G_ϕ as follows. For each clause c_j we have a clause gadget G_j that is isomorphic to K_4 and has vertices v_j^0, v_j^1, v_j^2, and w_j. We interpret the superscript k in v_j^k modulo 3, so that $v_j^3 = v_j^0$. Let C_j be the 3-cycle of G_j on v_j^0, v_j^1, and v_j^2. For each $0 \leq k < 3$, we say that the orientation (v_j^k, v_j^{k+1}) of edge $\{v_j^k, v_j^{k+1}\}$ is *positive* and the inverse orientation is *negative*. See Figure 1(a). A key observation in our reduction is that, for each orientation D of C_j, D can be extended into an orientation of G_j in which every edge incident with w_j is 3-cyclic if and only if D orients at least one edge of C_j positively and at least one edge of C_j negatively. This property of the clause gadget leads us to associate the kth literal of c_j with edge $\{v_j^k, v_j^{k+1}\}$, for $0 \leq k < 3$.

For each variable x_i of ϕ, we have an edge $e_i = \{y_i, z_i\}$. We call the orientation (y_i, z_i) of e_i *positive* and the other *negative*. We connect e_i with each edge e in clause gadgets that is associated with literal x_i or $\bar{x}_i$ as in Figure 1(b) or (c). Observe that, in any 3-cyclic orientation of G_ϕ, the signs of the orientaions of e_i and e are identical when the literal is positive and distinct when the literal is negative.

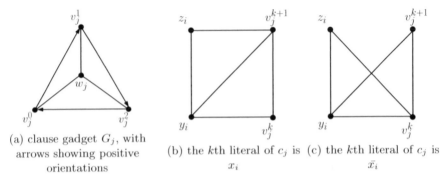

(a) clause gadget G_j, with arrows showing positive orientations

(b) the kth literal of c_j is x_i

(c) the kth literal of c_j is $\bar{x}_i$

Fig. 1. A clause gadget and its connections with a variable gadget

From this observation and the one above, it should be clear that the set of orientations of $G_\phi \setminus \{w_j \mid 1 \leq j \leq m\}$ that can be extended into a 3-cyclic orientation of G_ϕ is in one-to-one correspondence with the set of not-all-equal assignments of ϕ and therefore that the reduction is correct.

3   Planar Graphs

In this section, we consider our orientation problem for planar graphs. In the first subsection, we prove the NP-completeness for $k \geq 4$. In the next subsection, we develop a polynomial time algorithm for $k = 3$.

We need some notation and terms. Let G be a plane graph, that is, a planar graph with a fixed embedding in the plane. The *dual* of a plane graph G, for our purposes, is a graph on the set of faces of G where two faces are adjacent if and only if they share an edge in G. In this paper, we do not need the planar embedding of the dual and hence regard it as a simple graph even if two faces of G are adjacent across more than two edges. We call a cycle C of G *facial* if it bounds a face of G. We denote by $B(G)$ the unique facial cycle of G that bounds its infinite face. A face of G is a k-*face*, $\leq k$-*face*, or $> k$-*face*, if the cycle that bounds it is a k-cycle, $\leq k$-cycle, or $> k$-cycle, respectively. We call a plane graph G k-*facial*, if every $\leq k$-cycle of G is facial.

A 3-coloring χ of graph G, for our purposes, is an assignment of one of the colors red, blue, and white to each vertex of $V(G)$. We say 3-coloring χ is *proper* if it colors two vertices with different colors whenever those vertices are adjacent.

As observed in the introduction, the problem of finding a k-cyclic orientation of a given biconnected plane graph G can be formulated, provided that G is k-facial, as that of finding a proper 3-coloring of the planar dual of G that colors all $> k$-faces white.

In our NP-hardness proof for $k \geq 4$, the plane graphs we construct for the reduction are k-facial and we use the observation above in our reasoning on those graphs. In our polynomial time algorithm for $k = 3$, we extend this equivalence of the 3-cyclic orientation problem with a 3-coloring problem to a similar equivalence for general biconnected plane graphs.

3.1   NP-Completenes for $k \geq 4$

Theorem 2. *The problem of deciding if a given planar graph has a k-cyclic orientation is NP-complete for every fixed $k \geq 4$.*

Our reduction is from planar 3SAT, which is known to be NP-complete (see [4]). In planar 3SAT, we are given a formula ϕ in CNF with a set $X = \{x_1, x_2, \ldots, x_n\}$ of variables and a set $S = \{c_1, c_2, \ldots, c_m\}$ of clauses each of which contains at most three literals, such that the bipartite graph B_ϕ between X and S, in which x_i is adjacent to c_j if and only if x_i appears in c_j, is planar. The question is if ϕ has a satisfying assignment, that is, an assignment of true or false to each variable in X that makes at least one literal of each clause in S true. We assume that each variable appears positively in at least one and at most two clauses and negatively in exactly one clause. The reduction from the general planar 3SAT to this restricted form of planar 3SAT is straightforward using a standard technique.

We describe the reduction for $k = 4$. A generalization to $k > 4$ is straightforward.

Suppose we are given a planar 3SAT instance ϕ with the above restriction. We construct a plane graph G_ϕ such that G_ϕ has a 4-cyclic orientation if and only if ϕ has a satisfying assignment.

The clause gadget for clause c_j is shown in Figure 2. It has three designated edges e_j^1, e_j^2, and e_j^3, where e_j^k for each k is associated with the kth literal of c_j and is identified with a certain edge in the variable gadget representing the variable of the literal. Except for this identification of the edges associated with

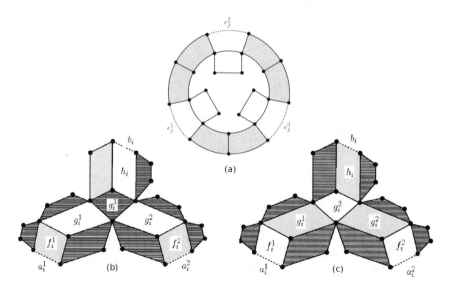

Fig. 2. (a) Clause gadget for c_j, variable gadget for x_i, (b) shown with the coloring for x_i = true and (c) for x_i = false, where shaded faces are red and hatched faces are blue

literals, each clause gadget has no connection with other parts of G_ϕ. Observe that each shaded 4-face in Figure 2 must be colored red or blue in any 3-coloring that corresponds to a 4-cyclic orientation of G_ϕ, since it contains edges incident with a $>$4-face. Therefore it is impossible to color all the 4-faces incident with e_j^1, e_j^2, and e_j^3 red or blue since then an odd dual cycle would be colored in two colors. This means that for at least one $k \in \{1, 2, 3\}$, the face across e_j^k, in a variable gadget, must be colored red or blue. The variable gadget for x_i is shown in Figure 2. It has designated edges a_i^1, a_i^2, and b_i. Edge a_i^k for each $k = 1$ or 2 is identified with an edge in a clause gadget that is associated with a positive literal x_i and edge b_i is identified with an edge associated with a negative literal $\bar{x}_i$. Since B_ϕ is planar, these identifications for all variable occurrences can be done in such a manner that the resulting graph G_ϕ is a plane graph. We also note that G_ϕ is 4-facial and every 4-cycle of G_ϕ is contained in some single clause or variable gadget.

The proof of the following lemma that states the correctness of the reduction can be found in the full paper.

~~Lemma 1.~~ *ϕ is satisfiable if and only if G_ϕ has a 4-cyclic orientation.*

3.2 Polynomial Time Algorithm for Finding 3-Cyclic Orientations of Planar Graphs

In this subsection, we develop a polynomial time algorithm for the 3-cyclic orientation problem for planar graphs. This is done by reducing our problem to a certain 3-coloring problem for graphs.

Let G be a graph. A *white-purple constraint*, or simply a *constraint* on G, is a pair (W, P) of disjoint vertex sets of G. We say a 3-coloring χ of G *respects* constraint (W, P), if χ colors each vertex in W white and each vertex in P red or blue.

We say that a constraint (W, P) on G is *strongly admissible* if every $v \in V(G)$ with $d_G(v) \geq 4$ is in W and $d_{G \setminus W}(v) \leq 2$ for every $v \in P$. As we saw earlier, the 3-cyclic orientation problem for 3-facial plane graphs can be reduced to the problem of deciding, given a graph G and a strongly admissible constraint $(W, \emptyset)$ on G, if G has a proper 3-coloring that respects this constraint. Our reduction for general planar graphs requires a slightly more general constrained 3-coloring problem.

Let (W, P) be a constraint on G and $A \subseteq V(G)$. We say (W, P) is *A-admissible* if the following conditions hold.

A1 $W \neq \emptyset$.
A2 $A \subseteq P$.
A3 $d_G(v) \leq 3$ for each $v \in V(G) \setminus W$.
A4 $d_{G \setminus W}(v) \leq 2$ for each $v \in P \setminus A$.

We say that constraint (W, P) on G is *admissible* if it is A-admissible for some $A \subseteq V(G)$ with $|A| \leq 1$. The proof of following theorem can be found in the full paper.

Theorem 3. *Given a graph G and an admissible constraint (W, P) on G, we can in polynomial time decide if G has a proper 3-coloring respecting this constraint and construct such a coloring if the answer is affirmative.*

Our reduction of the orientation problem for general plane graphs to this constrained 3-coloring problem is inductive on the nesting structure of non-facial 3-cycles.

Let G be a biconnected plane graph with an infinite 3-face and D an orientation of G. We call D *near 3-cyclic*, if every edge $e \in E(G) \setminus E(B(G))$ is 3-cyclic in D.

Let G be a biconnected plane graph. We call a cycle of G *internal* if it does not bound the infinite face of G. Let C be an internal 3-cycle of G. We denote by $G\langle C\rangle$ the subgraph of G induced by the set of vertices lying on C or drawn inside of C. Since G is simple, $G\langle C\rangle$ is obtained by removing all vertices and edges of G lying in the infinite face of C. We say that C is *relevant* if $G\langle C\rangle$ has a 3-cyclic orientation in which C is cyclic. Otherwise C is *irrelevant*. Note that each internal facial 3-cycle of G is relevant in G.

The following notion of skeletons is crucial in our reduction of the orientation problem to the constrained coloring problem. Let G be a biconnected plane graph. A *skeleton* S of G is a biconnected subgraph of G that satisfies the following conditions.

S1 $B(S) = B(G)$
S2 Every internal facial cycle of S is either a facial > 3-cycle of G or is a relevant 3-cycle of G.
S3 Every non-facial 3-cycle of S is irrelevant in G.

Every biconnected plane graph G has a unique skeleton, which can be identified as follows. Let $\mathcal{R}$ be the set of all the relevant internal 3-cycles of G and let $\mathcal{R}'$ be the set of maximal elements of $\mathcal{R}$ with respect to containment in the drawing of G. The skeleton S of G is obtained by removing vertices that lie in the inside of cycles of $\mathcal{R}'$ and thus making each cycle of $\mathcal{R}'$ facial in S. We denote by $\tilde{G}$ the skeleton of G. It will turn out that the skeleton of G can be constructed in polynomial time but this is shown in the whole inductive proof that the 3-cyclic orientation problem for a plane graph can be solved in polynomial time.

To describe the use of skeletons in our reduction, we need some more definitions.

Let G be a biconnected plane graph, C an arbitrary internal 3-cycle of G, and D an arbitrary orientation of G or of $G\langle C\rangle$. When C is not cyclic in D, D orients two edges of C in the same direction along C and the other in the opposite direction. We call the former two edges the *major edges of C* and the latter the *minor edge of C* with respect to D. Let $e \in E(C)$. We say that C is *e-majored* in D, if C is not cyclic in D and e is one of the major edges of C with respect to D. We say that C is *e-minored* in D, if C is not cyclic in D and e is the minor edge of C with respect to D. We say that C is *e-minored* in G, without reference to a particular orientation, if there is some near 3-cyclic orientation D of $G\langle C\rangle$ in which C is *e-minored*.

For each relevant internal 3-cycle C of G, let $M_G(C)$ denote the set of edges $e \in E(C)$ such that C is e-minored in G. Observe that, if C is relevant and $|M_G(C)| = 3$, then for any prescribed orientation of C, $G\langle C \rangle$ has a near 3-cyclic orientation that extends the prescribed orientation of C. Because of this, we call an internal 3-cycle C of G *universal* in G if it is relevant and $|M_G(C)| = 3$. Otherwise, C is *non-universal* in G. We call a relevant internal 3-cycle C of G *normal* in G if either C is universal or

N1 $M_G(C) \neq \emptyset$ and
N2 for each $e \in M_G(C)$, there is an e-minored near 3-cyclic orientation D of $G\langle C \rangle$ in which each $e' \in E(C) \setminus \{e\}$ is 3-cyclic.

We call a biconnected plane graph G *normal* if every internal facial 3-cycle of $\tilde{G}$ is normal in G. We later prove that every biconnected plane graph is normal. This proof is inductive and we need some lemmas that assume a given biconnected graph to be normal. These lemmas are also used in proving the main result of this subsection.

Suppose C is a universal 3-cycle of G. We call an edge e of C *internally coverable for C in G* if there is a near 3-cyclic orientation of $G\langle C \rangle$ in which C is e-majored and e is 3-cyclic. For each universal 3-cycle C of G, we denote by $I_G(C)$ the set of internally coverable edges of C.

For each biconnected plane graph G that is not a cycle, we define graph R_G as follows. R_G is similar to the planar dual of $\tilde{G}$ and our plan is to reduce the problem of finding a 3-cyclic orientation of G to a constrained coloring problem on R_G. We assume that G is normal in this construction.

To construct R_G, we first construct a graph $R_G(C)$ for each facial cycle C of $\tilde{G}$, including the cycle $B(G)$ bounding the infinite face. Here we neglect the distinction between finite and infinite faces of $\tilde{G}$ and pretend that $B(G)$ is relevant and universal if $B(G)$ is a 3-cycle, although this property has been defined only for internal 3-cycles. For each such C and an edge e of C, we also define a vertex v_C^e of $R_G(C)$, which we need in the description of the entire graph R_G. If C is a >3-cycle, a universal 3-cycle, or a non-universal cycle with $|M_G(C)| = 1$ then $R_G(C)$ consists of a single vertex v_C. We let $v_C^e = v_C$ for every $e \in E(C)$ in this case. Otherwise, i.e., if C is a non-universal cycle with $|M_G(C)| = 2$, $R_G(C)$ consists of five vertices, t_C, a_C^1, a_C^2, b_C^1, and b_C^2, and five edges $\{t_C, a_C^1\}$, $\{t_C, a_C^2\}$, $\{a_C^1, a_C^2\}$, $\{a_C^1, b_C^1\}$, and $\{a_C^2, b_C^2\}$. We let $v_C^e = t_C$ for $e \in E(C) \setminus M_G(C)$, $v_C^{e_1} = b_C^1$ for one edge $e_1 \in M_G(C)$, and $v_C^{e_2} = b_C^2$ for the other edge $e_2 \in M_G(C)$. We call the graph $R_G(C)$ in this case a *hut*, t_C the *top*, a_C^1 and a_C^2 the *eaves*, and b_C^1 and b_C^2 the *bases* of the hut. We combine these graphs into one graph R_G as follows. Let e be an arbitrary edge of $\tilde{G}$ and let C_1 and C_2 be the two facial cycles containing e. Then, R_G has an edge, denoted by e^*, between $v_{C_1}^e$ and $v_{C_2}^e$ if and only if neither of the following conditions hold.

O1 For $i = 1$ or 2, $M_G(C_i) = \{e\}$.
O2 For $i = 1$ or 2, C_i is universal in G and $e \in I_G(C_i)$.

We remark that R_G may not be simple, that is, may have parallel edges. The following simple property of a hut is essential.

Lemma 2. *Let H be a hut. Then, every proper 3-coloring of H either colors all of the top and the two bases of H red, colors all of the top and the two bases of H blue, or colors one base of H red and the other base blue. Moreover, all of these three types of proper 3-colorings of H do exist.*

The following lemma is at the heart of our reduction.

Lemma 3. *Let G be a biconnected plane graph that is not a cycle. Suppose G is normal. Then, G has a 3-cyclic orientation if and only if R_G has a proper 3-coloring that respects constraint $(W, P_1 \cup P_2)$, where W is the set of vertices v_C such that C is a facial >3-cycle of $\tilde{G}$, P_1 is the set of vertices v_C such that C is a facial 3-cycle of $\tilde{G}$ that is non-universal in G with $|M_G(C)| = 1$, and P_2 is the set of vertices b_C^1 and b_C^2 such that C is a facial 3-cycle of $\tilde{G}$ that is non-universal in G with $|M_G(C)| = 2$.*

Proof. We only prove the "only if" part. The other direction can be found in the full paper.

Suppose G has a 3-cyclic orientation D. We say that D is *skeleton-maximal* if the set of facial 3-cycles of $\tilde{G}$ that are cyclic in D is maximal subject to D being 3-cyclic. We assume that D is skeleton-maximal in the following.

We define a 3-coloring χ of R_G as follows. Let C be a facial cycle of $\tilde{G}$. If C is a >3-cycle then χ colors v_C white. Suppose C is universal in G. If C is cyclic in D, then χ colors v_C red or blue: red if the orientation is clockwise around the face C bounds and blue otherwise. Otherwise, χ colors v_C white. Suppose C is non-universal in G and $|M_G(C)| = 1$. If C is cyclic in D, then χ colors v_C in the same manner as when C is universal. Otherwise, if D orients the two major edges of C clockwise around the face C bounds, then χ colors v_C red and otherwise χ colors v_C blue.

Finally, suppose C is non-universal in G and $|M_G(C)| = 2$. Let e_0 be the edge in $E(C) \setminus M_G(C)$. By the construction of R_G, $v_C^{e_0}$ is the top of the hut $R_G(C)$ and v_C^e for each $e \in M_G(C)$ is a base of the hut. For each $e \in E(C)$, χ colors v_C^e red if D orients e clockwise around the face C bounds and blue if D orients e counterclockwise. This determines the colors of the top and the bases of the hut. Note that either the colors of the top and the bases are all identical or the colors of the two bases are distinct, since if the colors of the two bases are identical and that of the top is different, this means that C is e_0-minored in D, contrary to the assumption that $e_0 \notin M_G(C)$. The eaves of the hut are colored appropriately, one white and the other red or blue, so that χ is locally proper on this hut, which is possible due to Lemma 2. This completes the description of χ.

It is immediate from the definition of χ that χ colors every vertex in W white and every vertex in $P_1 \cup P_2$ red or blue and therefore respects constraint $(W, P_1 \cup P_2)$. We show that χ is proper. For each C such that $R_C(G)$ is a hut, that χ is proper on the hut is already insured by the coloring rule above. Let e be an arbitrary edge of $\tilde{G}$ and let C_1 and C_2 be the two facial cycles of $\tilde{G}$ containing e. We need to show that, if e^* is present in R_G, then χ colors $v_{C_1}^e$ and $v_{C_2}^e$ with different colors. Suppose χ colors both $v_{C_1}^e$ and $v_{C_2}^e$ red. We claim that at least one of C_1 and C_2 is e-minored in D. Suppose each of C_1 and C_2 is either cyclic in D

or e-majored in D. Then, since χ colors both $v^e_{C_1}$ and $v^e_{C_2}$ red, e must be oriented clockwise around both of the faces bounded by C_1 and C_2, a contradiction. So, either C_1 or C_2, say C_1, is e-minored. If $|M_G(C_1)| = 1$ then e^* is missing from R_G, so the color conflict does not arise. If $|M_G(C_1)| = 2$, that χ colors $v^e_{C_1}$ red means, from the definition of χ above, that e is oriented clockwise around the face bounded by C_1, again contradicting the orientation of e around C_2. Therefore, $v^e_{C_1}$ and $v^e_{C_2}$ cannot be colored both red as long as e^* is present in R_G. Similarly, $v^e_{C_1}$ and $v^e_{C_2}$ cannot be colored both blue as long as e^* is present in R_G.

Suppose that $v^e_{C_1}$ and $v^e_{C_2}$ are both colored white. Then, for $i = 1, 2$, either C_i is a > 3-cycle or a 3-cycle that is universal in G and not cyclic in D. Since e is 3-cyclic in D, there is some 3-cycle C of G that is cyclic in D and contains e. This C is relevant and therefore must be contained in $G\langle C_1\rangle$ or $G\langle C_2\rangle$, by the definition of the skeleton. We say C_i *covers* e, if $G\langle C_i\rangle$ contains such a C, for $i = 1, 2$. Suppose first that both C_1 and C_2 covers e. We cannot have both C_1 and C_2 e-minored in D, since if we had then we would be able to flip the orientation of e in D, replace the orientation of $G\langle C_i\rangle$ by a near 3-cyclic orientation in which C_i is cyclic, for $i = 1, 2$, and obtain a 3-cyclic orientation D' of G such that the set of facial 3-cycles of G that are cyclic in D' is a proper superset of the set of those cyclic in D, contradicting the skeleton-maximality of D. Therefore, C_i is e-majored in D for either $i = 1$ or 2, and therefore $e \in I_G(C_i)$ and e^* is missing from R_G. Next suppose that exactly one of C_1 and C_2, say C_1, covers e. Then C_1 must be e-majored in D, since otherwise D would not be skeleton-maximal similarly as above, and therefore $e \in I_G(C_1)$ and e^* is missing from R_G. Therefore, $v^e_{C_1}$ and $v^e_{C_2}$ cannot be colored both white as long as e^* is present in R_G.   □

We need the following variants of this lemma.

Lemma 4. *Let G be a biconnected plane graph that is not a cycle such that its infinite face is bounded by a 3-cycle C_0. Suppose G is normal. Then, G has a 3-cyclic orientation in which C_0 is cyclic if and only if R_G has a proper 3-coloring that respects constraint $(W, P_1 \cup P_2 \cup \{v_{C_0}\})$, where W, P_1, and P_2 are as in Lemma 3.*

Proof. Given a skeleton maximal 3-cyclic orientation of G in which C_0 is cyclic, we construct a proper 3-coloring χ of R_G that respects $(W, P_1 \cup P_2)$ as described in the proof of Lemma 3. It is immediate from the construction that χ colors v_{C_0} red or blue and hence respects $(W, P_1 \cup P_2 \cup \{v_{C_0}\})$. Given a proper 3-coloring of R_G that colors v_{C_0} red or blue, we construct a 3-cyclic orientation of G as described in the proof of Lemma 3. It is immediate from the construction that C_0 is cyclic in D.   □

We note that the constraint $(W, P_1 \cup P_2)$ in Lemma 3 is $\emptyset$-admissible and constraint $(W, P_1 \cup P_2 \cup \{v_{C_0}\})$ in Lemma 4 is $\{v_{C_0}\}$-admissible. The parameter A in the definition of admissibility, which is the source of most complications in the proof of Theorem 3, comes from the need of Lemma 4.

For each biconnected plane graph G with an infinite 3-face that is not a cycle, we consider the following variant of R_G. Let C_0 be the 3-cycle bounding the infinite face of G with $C_0 = \{e_1, e_2, e_3\}$ and let C_i be the internal cycle of $\tilde{G}$ which contains e_i for $i = 1, 2, 3$. We replace in R_G the subgraph $R_G(C_0)$, which is a single vertex v_{C_0}, by three vertices u_1, u_2, and u_3, let $v_{C_0}^{e_i} = u_i$ for $i = 1, 2, 3$, and connect $v_{C_0}^{e_i}$ with $v_{C_i}^{e_i}$ in the same manner as R_G. Let R_G' denote the resulting graph. We say that an orientation D and a 3-coloring χ of R_G' are *boundary-consistent* with each other if, for $i = 1, 2, 3$, D orients e_i clockwise (counterclockwise) around the infinite face if and only if χ colors $v_{C_0}^{e_i}$ red (blue). The following variant of Lemma 3 is the basic tool for analyzing the subgraph $G\langle C \rangle$ in our inductive proofs.

Lemma 5. *Let G be a biconnected plane graph with an infinite 3-face. Suppose G is normal. Then, for each near 3-cyclic orientation D of G, there is a proper 3-coloring χ of R_G' that respects constraint $(W, P_1 \cup P_2 \cup \{v_{C_0}^{e_1}, v_{C_0}^{e_2}, v_{C_0}^{e_3}\})$ and is boundary-consistent with D. Conversely, for each proper 3-coloring χ of R_G' that respects constraint $(W, P_1 \cup P_2 \cup \{v_{C_0}^{e_1}, v_{C_0}^{e_2}, v_{C_0}^{e_3}\})$, there is a near 3-cyclic orientation D of G that is boundary-consistent with χ.*

Proof. From a near 3-cyclic orientation D of G, we construct a coloring χ of R_G' in the same manner as in the proof of Lemma 3, except that the color of $v_{C_0}^{e_i}$, $i = 1, 2, 3$, is determined by the single edge e_i. The inverse translation is exactly the same as in the proof of Lemma 3. □

We are ready to prove the following statement announced earlier, which allows us to apply Lemmas 3, 4, and 5 without the assumption that G is normal.

Lemma 6. *Every biconnected planar graph G is normal.*

The proof of this lemma is by induction on the nesting structure of G and uses Lemma 5 for the induction step, can be found in the full paper. What we have developed so far, together with Theorem 3 allows us to decide if a given biconnected plane graph G has a 3-cyclic orientation in polynomial time, provided that we know, for each non-facial 3-cycle C of G

(1) whether C is relevant or not, so we can identify the skeleton of G,
(2) when C is relevant, for each $e \in E(C)$, whether C is e-minored or not, so we can determine $M_G(C)$ and whether C is universal or not, and
(3) when C is universal, for each $e \in E(C)$, whether e is internally coverable for C in G or not.

We call these characteristics of faces and edges of G the *skeleton characteristics* of G.

Lemma 7. *Let G be a biconnected plane graph and let $\mathcal{M}(G)$ be the set of containment maximal internal 3-cycles of G: an internal 3-cycle of C of G is in $\mathcal{M}(G)$ if and only if there is no internal 3-cycle $C' \neq C$ of G such that $G\langle C' \rangle$ contains C. Given the skeleton and the skeleton characteristic of $G\langle C \rangle$ for each $C \in \mathcal{M}(G)$, we can compute in polynomial time the skeleton and the skeleton characteristics of G.*

The proof of this lemma can be found in the full paper.

We are ready to describe our algorithm for finding a 3-cyclic orientation of a given biconnected plane graph G. It uses a recursive algorithm Analyze whose input is an internal 3-cycle C of G. The task of Analyze is to decide if C is relevant in G and, if it is, then compute the set $M_G(C)$. Moreover, if $|M_G(C)| = 3$, that is, if C is universal, then it also computes the set $I_G(C)$. If C is facial in G then this task is trivial. Otherwise, Analyze computes the set $\mathcal{M}(G\langle C\rangle)$ of containment-maximal internal 3-cycles of $G\langle C\rangle$ and recursively analyze each $C \in \mathcal{M}(G)$. This gives the skeleton and the skeleton characteristics of $G\langle C\rangle$, using which Analyze completes its task by the method described in the proof of Lemma 7. Given this procedure Analyze , our main task is simple. Given a biconnected plane graph G, we compute the skeleton and the skeleton characteristics of G applying Analyze to 3-cycles in $\mathcal{M}(G)$. Then, we decide if G has a 3-cyclic orientation applying Lemma 3 and Theorem 3. When the answer is affirmative, we can extract a 3-cyclic orientation of G using the proofs of these lemmas and theorem.

References

1. Chvátal, V., Thomassen, C.: Distances in orientations of graphs. Journal of Combinatorial Theory, Series B 24(1), 65–75 (1978)
2. Dankelmann, P., Oellermann, O.R., Wu, J.-L.: Minimum average distance of strong orientations of graphs. Discrete Appl. Math. 143(1-3), 204–212 (2004)
3. Eggemann, N., Noble, S.D.: Minimizing the oriented diameter of a planar graph. Electronic Notes in Discrete Mathematics 34(1), 267–271 (2009)
4. Garey, M.R., Johnson, D.S.: Computers and Intractability: A Guide to the Theory of NP-Completeness. W. H. Freeman and Company, New York (1979)
5. Louis Hakimi, S., Schmeichel, E.F., Young, N.E.: Orienting graphs to optimize reachability. Information Processing Letters 63, 229–235 (1997)
6. Ito, T., Miyamoto, Y., Ono, H., Tamaki, H., Uehara, R.: Route-enabling graph orientation problems. In: Dong, Y., Du, D.-Z., Ibarra, O. (eds.) ISAAC 2009. LNCS, vol. 5878, pp. 403–412. Springer, Heidelberg (2009)
7. Robbins, H.E.: A theorem on graphs, with an application to a problem of traffic control. The American Mathematical Monthly 46(5), 281–283 (1939)

Improved Bounds on the Planar Branchwidth with Respect to the Largest Grid Minor Size

Qian-Ping Gu[1] and Hisao Tamaki[2]

[1] Simon Fraser University, Burnaby BC Canada, V5A 1S6
qgu@cs.sfu.ca
[2] Meiji University, Kawasaki, Japan 214-8571
tamaki@cs.meiji.ac.jp

Abstract. For graph G, let bw(G) denote the branchwidth of G and gm(G) the largest integer g such that G contains a $g \times g$ grid as a minor. We show that bw$(G) \leq 3$gm$(G) + 1$ for every planar graph G. This is an improvement over the bound bw$(G) \leq 4$gm(G) due to Robertson, Seymour and Thomas. Our proof is constructive and implies quadratic time constant-factor approximation algorithms for planar graphs for both problems of finding a largest grid minor and of finding an optimal branch-decomposition: $(3 + \epsilon)$-approximation for the former and $(2 + \epsilon)$-approximation for the latter, where ϵ is an arbitrary positive constant. We also study the tightness of the above bound. A $k \times h$ cylinder, denoted by $C_{k,h}$, is the Cartesian product of a cycle on k vertices and a path on h vertices. We show that bw$(C_{2h,h}) = 2$gm$(C_{2h,h}) = 2h$ and therefore the coefficient in the above upper bound is within a factor of $3/2$ from the best possible.

1 Introduction

Results

Let G be a graph. An ordered pair (A, B) of edge sets of G is called an *edge separation*, (or simply *separation* in this paper) of G if $A \cup B = E(G)$ and $A \cap B = \emptyset$. The *middle set* of this separation is $V(A) \cap V(B)$, where $V(E)$ for $E \subseteq E(G)$ denotes the set of vertices incident with some edge in E. The *order* of a separation is the cardinality of its middle set. A *branch-decomposition* is a system of separations represented by a ternary tree. Formally, a branch-decomposition of G is a pair (T, ϕ), where T is a tree each internal node of which has degree 3 and ϕ is a bijection from the set of leaves of T to $E(G)$. Then, each edge of T is naturally associated with a bipartition (L_1, L_2) of the set of leaves of T and hence with a separation $(\phi(L_1), \phi(L_2))$ of G. The *width* of branch-decomposition (T, ϕ) is the maximum of the orders of the separations associated with the edges of T or 0 if $|E(G)| \leq 1$ and hence T does not have any edge. The *branchwidth* of G is the minimum width over all branch-decompositions of G. We denote by bw(G) the branchwidth of G.

Let G and H be graphs. H is a minor of G if H can be obtained from some subgraph of G through a possibly empty sequence of edge contractions. We

O. Cheong, K.-Y. Chwa, and K. Park (Eds.): ISAAC 2010, Part II, LNCS 6507, pp. 85–96, 2010.
© Springer-Verlag Berlin Heidelberg 2010

denote by gm(G) the largest integer g such that G contains a $g \times g$ grid as a minor. Since the branchwidth of a $g \times g$ grid is g and bw(G) $\geq$ bw(H) holds if H is a minor of G, gm(G) is a lower bound on bw(G). Our main result is the following:

Theorem 1. *For every planar graph G, we have* bw(G) $\leq$ 3gm(G) $+ 1$.

This improves on the previously best known bound of bw(G) $\leq$ 4gm(G) due to Robertson, Seymour and Thomas [18][1].

We also show that the coefficient in the bound cannot be improved to below 2. Let $k \geq 3$ and $h \geq 1$ be integers. A $k \times h$ cylinder, denoted by $C_{k,h}$, is the Cartesian product of a cycle on k vertices and a path on h vertices: it has a vertex (u, v) for each vertex u of the cycle and each vertex v of the path and (u, v) is adjacent with (u', v') if and only if $u = u'$ and v is adjacent with v' on the path or $v = v'$ and u is adjacent with u' on the cycle. We show that bw($C_{k,h}$) $= \min\{k, 2h\}$ and:

Theorem 2. *For every integer $h \geq 2$, we have* gm($C_{2h,h}$) $= h$.

Therefore, for $G = C_{2h,h}$, we have bw(G) $= 2$gm(G).

The proof of Theorem 1 uses an extension of some known upper bounds on the branchwidth of planar graphs and hypergraphs. These upper bounds are based on the "radius" of planar graphs [16], which roughly corresponds to the outerplanarity [1], and are first observed for the treewidth of planar graphs [16,2] and later for the branchwidth of planar graphs [21,3] and of planar hypergraphs [21]. Although our results are on planar graphs, our proof of Theorem 1 involves hypergraphs and requires a non-trivial extension of the bound of [21], which is embodied by Theorem 4 described in Section 2. This extension may be of an independent interest.

Background and consequences

The notions of branch-decompositions and branchwidth are introduced by Robertson and Seymour [17] and are related to tree-decompositions and treewidth which play central roles in their graph minor theory. For an arbitrary graph G, the treewidth tw(G) of G and the branchwidth of G are linearly related by inequalities bw(G) $\leq$ tw(G) $+ 1 \leq \lfloor \frac{3\text{bw}(G)}{2} \rfloor$ and there are simple translations between tree- and branch-decompositions that prove these inequalities.

One of the key ingredients of the graph minor theory is the fact that bw(G) is upper-bounded by some function of gm(G), which is a straightforward lower bound of bw(G). For general graphs, the known upper bound is huge (bw(G) $\leq 20^{2(\text{gm}(G))^5}$ [18]), while we have a linear bound on planar graphs as stated above. This type of relations between bw(G) and gm(G) have important algorithmic

---

[1] Their statement of the bound is that every planar graph with a tangle of order $4g - 3$ has a $g \times g$ grid minor. Equivalently, since G does not have a $g \times g$ grid minor with $g = $ gm(G)$+1$, every planar graph does not have a tangle of order $4($gm(G)$+1)-3 = 4$gm(G) $+ 1$, which implies that G admits a branch-decomposition of width 4gm(G).

applications which are extensively studied in the recent development of algorithmic graph minor theory and bidimensionality theory [5,6,7,8]. In such algorithmic applications, we typically use either a large grid minor of the given graph to find a desired structure in the graph or a small-width branch-decomposition to solve the problem by dynamic programming. The constant c in the bound $\mathrm{bw}(G) \leq c \cdot \mathrm{gm}(G)$ is important as it appears in the exponent of the running time of those algorithms.

These bounds on the branchwidth of planar graphs imply similar bounds on the treewidth through the linear relation mentioned above. In particular, the bound $\mathrm{bw}(G) \leq 4\mathrm{gm}(G)$ of [18] implies $\mathrm{tw}(G) \leq 6\mathrm{gm}(G)$. Thomas [20] and Grigoriev [10] independently improve the constant in this bound to 5. Theorem 1 gives a better constant 4.5.

Grigoriev *et al.* [11] study the tightness of these bounds for treewidth. Their best construction is a graph G for which $\frac{3}{2}\mathrm{gm}(G) - 1 \leq \mathrm{tw}(G) \leq \frac{3}{2}\mathrm{gm}(G)$ holds. They give some candidate graphs, including one similar to a cylinder, for which $\mathrm{tw}(G) \geq 2\mathrm{gm}(G) - 1$ is expected to hold. Theorem 2 verifies a variant of their expectation since $tw(C_{2k,k}) = 2k = 2\mathrm{gm}(C_{2k,k})$.

Computing a large grid minor of planar graphs have many important algorithmic applications. It is not known whether a largest $(\mathrm{gm}(G) \times \mathrm{gm}(G))$ grid minor of a planar graph can be found in polynomial time. The above mentioned bound of [18] implies a 4-approximation algorithm for this largest grid minor problem on planar graphs. Bodlaender, Grigoriev and Koster give an $O(n^2 \log n)$ time algorithm with the same approximation ratio for this problem [4], where n is the number of vertices of G. Our proof for Theorem 1 is constructive and implies an $O(n^2)$ time $(3 + \epsilon)$-approximation algorithm, where ϵ is an arbitrary positive constant.

Theorem 1 is a consequence of a slightly more general result.

Theorem 3. *Let G be a planar graph and k, h be integers with $k \geq 3$ and $h \geq 1$. Then G has either a minor isomorphic to a $k \times h$ cylinder or a branch-decomposition of width at most $k + 2h - 2$.*

Setting $h = k$ in this theorem gives Theorem 1 since a $k \times h$ grid is a subgraph of a $k \times h$ cylinder. Another interesting case is when $h = \lceil k/2 \rceil$. As we show in Section 4, the branchwidth of $C_{k,\lceil k/2 \rceil}$ is k. This motivates us to define another lower bound on the branchwidth: $\mathrm{cm}(G)$ is the largest k such that G contains $C_{k,\lceil k/2 \rceil}$ as a minor. Then, our theorem implies that $\mathrm{bw}(G) \leq 2\mathrm{cm}(G) + 1$ for planar graph G. Thus, $\mathrm{cm}(G)$ is a better lower bound on $\mathrm{bw}(G)$ than $\mathrm{gm}(G)$ in the sense that it is provably tight within a factor of approximately 2 for planar graphs.

In a related paper [13], based on the construction in Theorem 3, we develop efficient approximation algorithms for the largest grid minor and the optimal branch-decomposition of planar graphs. These algorithms give a trade-off between the running time and the approximation ratio. At one end, they run in $O(n^2)$ time and give a $(3 + \epsilon)$-approximation for the largest grid-minor and a $(2 + \epsilon)$-approximation for the optimal branch-decomposition. At the other end, they run in $O(n^{1+\epsilon})$ time and give constant-factor approximations where the constant is, roughly speaking, inversely proportional to ϵ. Although this approximation algorithm for branch-decomposition is less important than that for grid-minor construction, as an $O(n^3)$-time algorithm for exact optimal decomposition [19,12] is known, it is still significantly faster than the exact algorithm.

Organization of this paper

In the next section, we prove Theorem 3, assuming a radius-based upper bound on the branchwidth of a planar hypergraph. In Section 3, we sketch the proof of this radius-based upper bound. In Section 4, we informally discuss the branch-width and the grid minor size of the cylinder. Due to space limitations, we omit the proofs of many lemmas: they can be found in technical reports [14,15] and in the forthcoming full paper.

2   Main Result

We prove Theorem 3 by induction: to obtain a branch-decomposition of G, we find an appropriate separation, construct the branch-decompositions of the two separated parts, and then combine them into a branch-decomposition of G. As will be clear, in each of these "parts", we need to regard the other part as a hyperedge in order for this approach to work. Thus, we need to work on hypergraphs.

The definitions of separations and of branch-decompositions given earlier for graphs work for hypergraphs without any change. A plane embedding of a hypergraph is also a straightforward generalization of that of a graph, where edges are drawn as discs having vertices on its boundary.

Let G be a hypergraph and (A, B) a separation of G. We define $\partial_G(A) = \partial_G(B) = V(A) \cap V(B)$. We denote by $G|A$ the hypergraph obtained from G by replacing its subgraph induced by A by a single hyperedge $\partial_G(A)$ (and thus ignoring all the vertices in $V(A) \setminus V(B)$). The following lemma is straightforward from the definition of branch-decompositions and is implicit in the literature [19].

Lemma 1. *Let G be a hypergraph and (A, B) a separation of G. Let T_A be a branch-decomposition of $G|A$ of width w_A and T_B a branch-decomposition of $G|B$ of width w_B. Let tree T be the result of concatenating T_A and T_B by identifying the leaf $\partial_G(A)$ of T_A and the leaf $\partial_G(B)$ of T_B and then ignoring the identified node to join the two incident edges into one. Then, T is a branch-decomposition of G of width $\max\{w_A, w_B\}$.*

For the base case of the induction, we use an upper bound on the branchwidth of a hypergraph with respect to the radius of the radial graph of the hypergraph. Let G be a plane hypergraph (a planar hypergraph with a fixed plane embedding). The *radial graph* of G is a bipartite graph between the set of vertices of G and the set of faces of G, where each edge represents a vertex-face incidence of G. If G is a plane graph, then the branchwidth of G is upper-bounded roughly by the radius of the radial graph of G. This type of radius-based upper bounds are first observed for the treewidth of planar graphs [16,2], and later for the branchwidth [21,3]. Tamaki [21] generalizes the bound to plane hypergraphs and gives a linear time algorithm for constructing a branch-decomposition achieving the bound.

In this paper, we prove and use the following upper bound for hypergraphs which extends the bound of [21]. We say that a hypergraph is *totally 2-connected* if it is both 2 vertex-connected and 2 edge-connected (note that the former condition does not imply the latter for general hypergraphs). The *order* of an edge of a hypergraph is the number of vertices it is incident with.

Theorem 4. *Let $k \geq 2$ be an integer and $d \geq 2$ an even integer. Let G be a totally 2-connected plane hypergraph of maximum edge order at most k. Suppose there is an edge $e_0 \in E(G)$ such that, for every face r of G there is a face s of G incident with e_0 such that the distance between r and s in the radial graph of G is at most d. Then, we have $\mathrm{bw}(G) \leq k + d$.*

We sketch a proof of this theorem in the next section. We remark that the algorithm implied by the proof runs in linear time and is of practical value, as are the above mentioned radius-based bounds which have found many applications in approximation algorithms.

For the induction step, we are given a plane hypergraph whose radial graph has radius greater than d and look for a separation of order smaller than k. A key observation is that, when we fail, we find a cylinder minor of the original graph G that certifies $\mathrm{bw}(G) \geq k$. To formalize this observation, we need some preparations.

Let G be a plane hypergraph. We say that a curve μ on the sphere Σ is *G-normal* if μ does not intersect itself and intersects G only at its vertices. We call a closed G-normal curve that is not self-intersecting a *G-noose*. We use a G-normal curve mainly to represent a walk in the radial graph of G. In this paper (not necessarily in consistence with the literature), the *length* of a G-normal curve μ, denoted by $\mathrm{length}_G(\mu)$, is the number of incidences between the segments of $\mu \setminus V(G)$ and vertices in $\mu \cap V(G)$. Equivalently, $\mathrm{length}_G(\mu)$ is the length of the walk in the radial graph that μ represents. For vertices or faces $x, y \in V(G) \cup F(G)$, G-distance between x and y, denoted by $\mathrm{dist}_G(x, y)$, is the length of the shortest G-normal curve connecting x and y, or equivalently the length of the shortest path in the radial graph of G between x and y. We extend this notion of distance between sets of vertices/faces: for $X, Y \subseteq V(G) \cup F(G)$: $\mathrm{dist}_G(X, Y) = \min_{x \in X, y \in Y} \mathrm{dist}_G(x, y)$. We also write $\mathrm{dist}_G(x, Y)$ for $\mathrm{dist}_G(\{x\}, Y)$ and $\mathrm{dist}_G(X, y)$ for $\mathrm{dist}_G(X, \{y\})$.

For each set of faces $F \subseteq F(G)$ and an integer $i \geq 0$, we denote by $\mathrm{cont}_G(F, i)$ the subgraph of G induced by the set of edges $e \in E(G)$ such that e is incident to both a face r with $\mathrm{dist}_G(r, F) = 2i$ and a face r' with $\mathrm{dist}_G(r', F) = 2i+2$. We denote by $\overrightarrow{\mathrm{cont}}_G(F, i)$ the directed version of $\mathrm{cont}_G(F, i)$ in which each edge is oriented so that the face r' with $\mathrm{dist}_G(r', F) = 2i+2$ is on the right with respect to the orientation. We omit the proof of the following lemma in this conference version.

Lemma 2. *Let G be a plane graph, $F \subseteq F(G)$, and $i \geq 0$ an integer. Then, $\mathrm{cont}_G(F, i)$ consists of edge-disjoint cycles.*

The following lemma formalizes the above stated observation.

Lemma 3. *Let G be a plane graph and $k \geq 3$, $h \geq 2$ integers. Let r_1 and r_2 be two faces of G such that the following conditions are satisfied.*

1. *There is no separation (A, B) of G of order smaller than k such that $V_G(r_1) \subseteq V_G(A)$ and $V_G(r_2) \subseteq V_G(B)$.*
2. *$\text{dist}_G(r_1, r_2) \geq 2h$.*

Then, G contains a $k \times h$ cylinder as a minor.

Proof. For $i \geq 0$, by Lemma 2, $\text{cont}_G(\{r_1\}, i)$ consists of edge-disjoint cycles. Moreover, since $\text{dist}_G(r_1, r_2) \geq 2h$, if $i < h$ then $\text{cont}_G(\{r_1\}, i)$ is non-empty and has a unique cycle c_i that separates r_1 from r_2; one of the connected components of $\Sigma \setminus c_i$ contains r_1 and the other contains r_2. The h cycles c_i for $0 \leq i < h$ are mutually vertex-disjoint, for otherwise a vertex would be incident to two faces, say t_1 and t_2, with $\text{dist}_G(t_1, r_1) = \text{dist}_G(t_2, r_1) + 4$, which is impossible by definition.

We claim that there are k vertex-disjoint paths between $V(c_0)$ and $V(c_{h-1})$. Suppose otherwise. Then by Menger's theorem, there must be a separation (A, B) of G of order smaller than k such that $V(c_0) \subseteq V_G(A)$ and $V(c_{h-1}) \subseteq V_G(B)$ and hence $V_G(r_0) \subseteq V_G(A)$ and $V_G(r_1) \subseteq V_G(B)$, contradicting the assumption.

These k vertex-disjoint paths, together with the h vertex-disjoint cycles we have found above, constitute a minor of G isomorphic to a $k \times h$ cylinder. □

The induction would be straightforward if this lemma held not only for plane graphs but also for plane hypergraphs. This unfortunately is not the case and we need a more elaborate structure of induction in which we work on hypergraphs that are almost graphs. An appropriate generalization of Theorem 3 for an induction proof is as follows.

Theorem 5. *Let G be a plane hypergraph with at least one edge and $k \geq 3$ and $h \geq 2$ be integers. Suppose that every edge of G except for one edge e_0 is of order 2 and e_0 is of order k or smaller. Then G has either a minor isomorphic to a $k \times h$ cylinder or a branch-decomposition of width at most $k + 2h - 2$.*

To prove this theorem, we need the following technical lemma.

Lemma 4. *Let G be a plane hypergraph and d a positive integer. Let $S \subseteq F(G)$ be a set of faces of G. Let c_0 and c_1 be edge disjoint cycles of G such that, for each $i = 0, 1$, c_i contains all the faces of S and c_{1-i} in one side of it. Suppose further that, for $i = 0, 1$ and for each $v \in V(c_i)$, $\text{dist}_G(v, S) = d$.*

For $i = 0, 1$, let $(B_i, E(G) \setminus B_i)$ be a separation of G induced by some G-noose that satisfies the following conditions.

1. *$E(c_i) \subseteq B_i$ and $E_G(S) \subseteq E(G) \setminus B_i$.*
2. *The order $|\partial_G(B_i)|$ of this separation is the smallest subject to condition (1).*
3. *B_i is minimal subject to conditions (1) and (2).*

Then, either $E(c_0) \subseteq B_1$, $E(c_1) \subseteq B_0$, or $B_0 \cap B_1 = \emptyset$.

Proof (of Theorem 5). We assume that G is connected, as otherwise the proof may be applied to each connected component of G.

The proof is by induction on the number of edges of G. The base case $|E(G)| \le 2$ is trivial.

Suppose $|E(G)| > 2$. First suppose that $|V_G(e_0)| < k$. Let $e \in E(G)$ be an edge distinct from e_0 with $V_G(e) \cap V_G(e_0) \ne \emptyset$. Let $A = \{e_0, e\}$ and $X = \partial_G(A)$. Since $|V_G(e)| = 2$, $|X| \le k$. Therefore, we may apply the induction hypothesis to $G|A$ and obtain either a minor isomorphic to a $k \times h$ cylinder or a branch-decomposition T of $G|A$ of width $\le k + 2h - 2$. In the former case, we are done. In the latter case, we add two leaves e and e_0 to the leaf X of T obtaining a branch-decomposition of G of width $\le k + 2h - 2$.

Therefore, suppose $|V_G(e_0)| = k$. Let G' be the plane subhypergraph of G induced by the edge set $E(G) \setminus \{e_0\}$. By our assumption on G, G' is a graph.

Let r_0 denote the face of G' that contains e_0 of G and c_0 the cycle of G' that bounds r_0. By Lemma 2, $\text{cont}_{G'}(\{r_0\}, h - 1)$ consists of edge disjoint cycles. Let $m \ge 0$ be the number of those cycles and $c_1, \ldots, c_m$ the list of those cycles.

For each i, let (A_i, B_i) be a separation of G' of the smallest order such that $V(c_i) \subseteq V_{G'}(A_i)$ and $V(c_0) \subseteq V_{G'}(B_i)$. Suppose first that the order of separation (A_i, B_i) is $\ge k$ for some $1 \le i \le m$. Let r_i denote the connected component of $\Sigma \setminus c_i$ that does not contain r_0. Let G_i denote the graph obtained from G' by removing edges and vertices drawn in r_i. Applying Lemma 3 to G_i with faces r_0 and r_i, we obtain a minor of G isomorphic to a $k \times h$ cylinder and are done.

So, suppose $|\partial_{G'}(A_i)| < k$ for every $1 \le i \le m$. Let $\mathcal{A} = \{A_i \mid 1 \le i \le m\}$. By Lemma 4, we may choose these subsets in $\mathcal{A}$ so that, for each pair $1 \le i < j \le m$, $A_i \cap A_j = \emptyset$ unless $E(c_i) \subseteq A_j$ or $E(c_j) \subseteq A_i$. If $E(c_i) \subseteq A_j$ then we remove A_i from the collection; if $E(c_j) \subseteq A_i$ then we remove A_j from the collection. In this manner, we obtain a subcollection $\mathcal{A}'$ of subsets of $E(G')$ such that

1. for each $A \in \mathcal{A}'$, $|\partial_G(A)| < k$,
2. for each $1 \le i \le m$, there is some $A \in \mathcal{A}'$ such that $E(c_i) \subseteq A$ and $E(c_0) \subseteq E(G') \setminus A$, and
3. for each pair of distinct elements $A, B \in \mathcal{A}'$, we have $A \cap B = \emptyset$.

Let $\mathcal{A}' = \{A'_1, \ldots, A'_{m'}\}$. Let $G_0 = G$ and, for $1 \le i \le m'$, $G_i = G_{i-1}|A'_i$.

Let F_0 be the set of faces of G incident with e_0. Since these faces are contained in face r_0 of G' and hence not bounded by an edge of any A'_i, F_0 is also a set of faces of $G_{m'}$. By construction, each face r of $G_{m'}$ has $\text{dist}_{G_{m'}}(r, F_0) \le h - 1$. Therefore, $G_{m'}$ satisfies the assumptions of Theorem 4 and hence has branchwidth $\le k + 2h - 2$. By the induction hypothesis, for each i, $1 \le i \le m'$, either $G|A'_i$ has a $k \times h$ cylinder as a minor or has branchwidth $\le k + 2h - 2$. If none of them has a $k \times h$ cylinder as a minor, then by a repeated application of Lemma 1, G has branchwidth $\le k + 2h - 2$. This completes the proof of Theorem 5.

$\square$

3   An Upper Bound on the Branchwidth of a Plane Hypergraph

In this section, we sketch a proof of Theorem 4. For a technical reason, we prove a dual version of this theorem, in which the roles of vertices and faces are interchanged. In an early version of this paper [14], a slightly weaker bound of $k + d + 1$ is proved using the rat-catching characterization of the branchwidth of planar hypergraphs [19]. This previous proof is not fully constructive in the sense that, to construct a branch-decomposition achieving the bound, we need to invoke the $O(n^3)$ time optimal branch-decomposition algorithm for planar graphs [19,12]. Here we give an independent constructive proof, that leads to a linear-time algorithm for constructing a branch-decomposition achieving the bound.

The branch-decompositions we construct for the proof of Theorem 4 is of a special type, called *sphere-cut branch-decompositions* [9], which we define below.

Let G be a plane hypergraph and Σ the sphere on which G is drawn. Recall that G consists of discs representing edges and points on the boundary of edges representing vertices. Recall also the definition of G-normal curves, G-nooses, and their length.

Let ν be a G-noose. Of the two connected components of $\Sigma \setminus \nu$, we regard the one to the right of ν with respect to its orientation the *inside* of ν. Let $E_\nu(G)$ denote the set of edges that lie inside of ν. Then, ν induces a separation $(E_\nu(G), E_{\text{rev}(\nu)}(G))$, where $\text{rev}(\nu)$ denotes the reversal of ν. The middle set of this separation is the set of vertices on ν. A sphere-cut branch-decomposition of a plane hypergraph G is a branch-decomposition of G in which each separation associated with an edge of the decomposition tree is induced by some G-noose. It is known that every plane hypergraph G has an optimal branch-decomposition (of width $\text{bw}(G)$) that is a sphere-cut branch-decomposition [19,9].

Let G be a plane hypergraph. We say that a G-noose is *simple* if its intersection with each face of G is a single segment.

Let d be a positive integer, G a plane hypergraph, ν a simple G-noose, and X a set of vertices and faces of G that appear on ν consecutively. We say that ν *is d-compact for* G *with center* X, if $\text{dist}'_{G|E_{\text{rev}(\nu)}(G)}(v, X) \leq d$ for every $v \in V(E_\nu(G))$, where $\text{dist}'_H(v, X)$ is defined to be $\text{dist}_H(v, X \cap V(H))$ if $X \cap V(H) \neq \emptyset$ and $\text{dist}_H(v, X) + 1$ otherwise.

The following lemma essentially shows that the part of a plane hypergraph of maximum edge order k that is bounded by a d-compact simple G-noose, with some additional conditions, has a sphere-cut branch-decomposition of width at most $k + d$.

Lemma 5. *Let k, d be positive integers. Let G be a plane hypergraph with maximum edge order k. If there is a simple G-noose ν, that is d-compact for G with some center X with $|X| \leq k + 1$ and has length $\text{length}_G(\nu) \leq 2(d + k)$, then $G|E_{\text{rev}(\nu)}(G)$ has a sphere-cut decomposition of width at most $d + k$.*

The proof is by induction on the number of edges in $E_\nu(G)$. We omit the details. We remark that this lemma generalizes the bound in [21] and implies all previously known radius-based upper bounds, both on treewidth [16,2] and on branchwidth [21,3].

Roughly speaking, the proof of Theorem 4 consists in showing that each plane hypergraph G satisfying the condition of the theorem can be decomposed into two or three parts such that (1) each of these parts are bounded by a d-compact G-noose and hence has a sphere-cut branch-decomposition of desired width by Lemma 5 and (2) the decomposition of G into these parts can be represented by a sphere-cut branch-decomposition of desired width. Thus, applying Lemma 1 twice or thrice, we obtain a sphere-cut branch-decomposition of G of desired width.

The following lemma formalizes this idea.

Lemma 6. *Let d, k be positive integers, with $k \geq 2$. Let G be a totally 2-connected plane hypergraph, with maximum edge order k and with at least two vertices. Suppose there is an edge $e_0 \in E(G)$ such that $\mathrm{dist}_G(v, V(e_0)) \leq d$ for every vertex $v \in V(G)$. Then, there are G-nooses ν_j, $1 \leq j \leq t$, where $t = 2$ or 3, sets of vertices and faces X_j, $1 \leq j \leq t$, and an edge $e_1 \in E(G)$ such that the following conditions are satisfied.*

1. *The insides of ν_j for $1 \leq j \leq t$ are mutually disjoint.*
2. *For $1 \leq j \leq t$, the vertices and faces in X_j appear consecutively on ν_j, ν_j is d-compact for G with center X_j, $|X_j| \leq k + 1$, and $\mathrm{length}_G(\nu_j) \leq 2(k + d)$.*
3. *$E(G) = \{e_0, e_1\} \cup \bigcup_{1 \leq i \leq t} E_{\nu_i}(G)$.*
4. *Let G_j, $0 \leq j \leq t$, be defined by $G_0 = G$ and $G_j = G_{j-1}|E_{\nu_j}(G)$ for $0 < j \leq t$. Then, G_t has a sphere-cut decomposition of width at most $k + d$.*

The ideas behind the proof of this lemma are as follows. We consider the radial graph of G and construct a breadth-first search tree T_0 in this radial graph, rooted at e_0. Then we consider the *medial graph* of G [19], which is the planar dual of the radial graph of G, and construct a spanning tree T of the medial graph deleting all the edges intersected by the edges of T_0. On this tree T, we perform a fine-tuned balancing argument to find an appropriate hyperedge e_1. Given these two lemmas, the proof of Theorem 4 is fairly straightforward. We omit the details of the proofs.

The algorithm implied by this proof runs in linear time. It can be shown that the upper bound in Theorem 4 is tight.

4 The Branchwidth and Grid-Minor Size of a Cylinder

In this section, we study the branch-width and the grid-minor size of the cylinder. Recall a $k \times h$ cylinder $C_{k,h}$ is the Cartesian product of a cycle on k vertices and a path on h vertices. We call each of the h images of the cycle in $C_{k,h}$ a *row* of $C_{k,h}$ and each of the k images of the path a *column* of $C_{k,h}$.

Theorem 6. *For arbitrary integers $k \geq 3$ and $h \geq 1$, we have $\mathrm{bw}(C_{k,h}) = \min\{k, 2h\}$.*

The upper bound can be exhibited by simple "linear" branch-decompositions, whose trees is a caterpillar: leaves directly hang on internal nodes arranged in a single path ("spine"). The order of the edges of $C_{k,h}$ appearing at the leaves along the spine is either row-major or column-major, depending on whether $k \leq 2h$ or not. The lower bound can be established using a variant of the standard obstructions to branch-decompositions, namely tangles [17].

Theorem 2 states that $gm(C_{2h,h}) = h$. One direction, $gm(C_{2h,h}) \geq h$, is trivial as an $h \times h$ grid is clearly a subgraph of $C_{2h,h}$. The other direction, $gm(C_{2h,h}) \leq h$, may be the most difficult result to prove in this work. We describe some ideas here.

For contradiction, let $h \geq 2$ and suppose $C_{2h,h}$ had a minor isomorphic to an $(h + 1) \times (h + 1)$ grid. Fix the embedding of $C_{2h,h}$. Let H be the subgraph of $C_{2h,h}$ contracted into the grid minor and ψ be the mapping that maps each vertex v of H to the vertex of the grid into which v is contracted. We extend this mapping to each subgraph of H' of H: $\psi(H')$ is the subgraph of the grid into which H' is contracted. We assume an embedding of $\psi(H)$ naturally induced by the mapping ψ and the embedding of $C_{2h,h}$, where each vertex v of $\psi(H)$ is represented by a disc containing all the vertices in $\psi^{-1}(v)$ and each edge of $\psi(H)$, say between u and v, is a segment of an edge of H between some vertex in $\psi^{-1}(u)$ and some vertex in $\psi^{-1}(v)$.

Then, H must have a cycle B such that $\psi(B)$ bounds the large face of $\psi(H)$ (usually drawn as the outer face). We define the *outside* of B to be the region of the sphere bounded by B and contains the large face of $\psi(H)$; the other region bounded by B is the *inside* of B.

We call the two faces of $C_{2h,h}$ bounded by the cycles of length $2h$ at the ends of the cylinder *end faces* of $C_{2h,h}$. The easy case is when both of the two end faces of $C_{2h,h}$ lie outside of B. Then, by "unfolding" $C_{2h,h}$, we can show that a $k \times h$ grid with sufficiently large k has a subgraph isomorphic to H. This is a contradiction, since the branchwidth of such a grid is h no matter how large k is.

So suppose one of the end faces of $C_{2h,h}$ (we call it the *bottom face* and the other the *top face*) lies inside of B. We call a subpath of $\psi(B)$ a *side* of $\psi(H)$ if it is between two distinct "corner vertices" (vertices of degree 2) and does not contain any other corner vertex. Let B_0, B_1, B_2, and B_3 be the subpaths of B such that $\psi(B_0)$, $\psi(B_1)$, $\psi(B_2)$, and $\psi(B_3)$ are sides of $\psi(H)$ and concatenating them in this order comprises $\psi(B)$. Note that $V(B_i) \cap V(B_{i+1 \bmod 4})$ for each $0 \leq i < 4$ is nonempty and is mapped to a single corner vertex of $\psi(H)$. We call a pair of vertices u and v of B *antipodal* if there is some i, $0 \leq i < 4$, such that $u \in V(B_i)$ and $v \in V(B_{i+2 \bmod 4})$. A basic fact we use in order to derive a contradiction is that, for antipodal $u, v \in V(B)$, the length of each $C_{2h,h}$-normal curve between u and v that stays inside of B must be at least $2h$, for otherwise their would be a $\psi(H)$-normal curve between $\psi(u)$ and $\psi(v)$ of length strictly smaller than $2h$. This is impossible for the $(h + 1) \times (h + 1)$ grid $\psi(H)$.

We define a certain notion of visibility from the bottom face and show that, for each antipodal pair to respect the above condition, there must be some "hidden" subpath B_{hid} of B that is long in the sense that $\psi(B_{\text{hid}})$ is not contained in any two sides of the grid $\psi(H)$. Moreover, the endvertices of B_{hid} can be connected by a $C_{2h,h}$-normal curve, called a *slit*, that stays inside of B and intersects each row of $C_{2h,h}$ at most once. The slit divides the grid $\psi(H)$ into two parts: we call the one bounded by $\psi(B_{\text{hid}})$ and the slit the *hidden part* and the other one *visible part*. A crucial observation is that the hidden part is the more essential part of the grid and is almost as difficult to embed in $C_{2h,h}$ (as a minor) as the entire grid. In particular, we can show that the hidden part must contain the top face of $C_{2h,h}$ in one of its internal faces, just as we have shown above that $\psi(H)$ must contain at least one of the end faces of $C_{2h,h}$. To show this, let H_{hid} be the subgraph of H such that $\psi(H_{\text{hid}})$ is the hidden part of $\psi(H)$. Since the bottom face of $C_{2h,h}$ is contained in the visible part of $\psi(H)$ and not in the hidden part, if the top face is not contained in the hidden part either then we can unfold $C_{2h,h}$ and obtain a $k \times h$ grid for sufficiently large k that contains a subgraph isomorphic to H_{hid}. If we augment this grid by attaching the visible part of $\psi(H)$ at the slit, the resulting graph contains $\psi(H)$ as a minor. This is a contradiction since the branchwidth of the augmented grid is still h. Therefore, B contains both of the end faces in its inside.

Now we view B from the top face and find another slit. We can derive a contradiction by considering a $k \times h$ grid for sufficiently large k augmented by two subgraphs of $\psi(H)$: the part visible from the bottom face attached at the first slit and the part visible from the top face attached at the second slit. This augmented grid still has branchwidth h while containing an $(h + 1) \times (h + 1)$ grid as a minor.

References

1. Baker, B.S.: Approximation algorithms for NP-complete problems on planar graphs. J. ACM 41, 153–180 (1994)
2. Bodlaender, H.L.: A partial k-arboretum of graphs with bounded treewidth. Theoretical Computer Science 209, 1–45 (1998)
3. Bodlaender, H.L., Feremans, C., Grigoriev, A., Penninkx, E., Sitters, R., Wolle, T.: On the minimum corridor connection problem and other generalized geometric problems. Computational Geometry: Theory and Applications 42, 939–951 (2009)
4. Bodlaender, H.L., Grigoriev, A., Koster, A.M.C.A.: Treewidth lower bounds with brambles. Algorithmica 51(1), 81–98 (2008)
5. Dorn, F., Fomin, F.V., Thilikos, D.M.: Catalan structures and dynamic programming in H-minor-free graphs. In: Proc. of the 2008 Symposium on Discrete Algorithms, SODA 2008, pp. 631–640 (2008)
6. Demaine, E.D., Hajiaghayi, M.T.: Graphs excluding a fixed minor have grids as large as treewidth, with combinatorial and algorithmic applications through bidimensionality. In: Proc. of the 2005 Symposium on Discrete Algorithms, SODA 2005, pp. 682–689 (2005)
7. Demaine, E.D., Hajiaghayi, M.T.: Bidimensionality, map graphs, and grid minors, arXiv:Computer Science, DM/052070, v1 (2005)
8. Demaine, E.D., Hajiaghayi, M.T., Kawarabayashi, K.: Algorithmic graph minor theory: decomposition, approximation, and coloring. In: Proc. of the 2005 IEEE Symposium on Foundation of Computer Science, FOCS 2005, pp. 637–646 (2005)

9. Dorn, F., Penninkx, E., Bodlaender, H.L., Fomin, F.V.: Efficient exact algorithms on planar graphs: exploiting sphere cut branch decompositions. In: Brodal, G.S., Leonardi, S. (eds.) ESA 2005. LNCS, vol. 3669, pp. 95–106. Springer, Heidelberg (2005)
10. Grigoriev, A.: Tree-width and large grid minors in planar graphs (2008) (submitted for publication)
11. Grigoriev, A., Marchal, B., Usotskaya, N.: On planar graphs with large treewidth and small grid minors. Electronic Notes in Discrete Mathematics 32, 35–42 (2009)
12. Gu, Q.P., Tamaki, H.: Optimal branch decomposition of planar graphs in $O(n^3)$ time. ACM Trans. Algorithms 4(3), article No.30, 1–13 (2008)
13. Gu, Q.P., Tamaki, H.: Constant-factor approximations of branch-decomposition and largest grid minor of planar graphs in $O(n^{1+\epsilon})$ time. In: Proc. of the 2009 International Symposium on Algorithms and Computation (ISAAC 2009), pp. 984–993 (2009)
14. Gu, Q.P., Tamaki, H.: Improved bounds on the planar branchwidth with respect to the largest grid minor size, Technical Report, SFU-CMPT-TR 2009-17 (July 2009)
15. Gu, Q.P., Tamaki, H.: A radius-based linear-time-constructive upper bound on the branchwidth of planar hypergrpahs, Technical Report, SFU-CMPT-TR 2009-21 (November 2009)
16. Robertson, N., Seymour, P.D.: Graph minors III. Planar tree-width. J. of Combinatorial Theory, Series B 36, 49–64 (1984)
17. Robertson, N., Seymour, P.D.: Graph minors X. Obstructions to tree decomposition. J. of Combinatorial Theory, Series B 52, 153–190 (1991)
18. Robertson, N., Seymour, P.D., Thomas, R.: Quickly excluding a planar graph. J. of Combinatorial Theory, Series B 62, 323–348 (1994)
19. Seymour, P.D., Thomas, R.: Call routing and the ratcatcher. Combinatorica 14(2), 217–241 (1994)
20. Thomas, R.: Tree decompositions of graphs, http://www.math.gatech.edu/~thomas/SLIDE/slide2.ps, p.32
21. Tamaki, H.: A linear time heuristic for the branch-decomposition of planar graphs. In: Proc. of 11th Annual European Symposium on Algorithms, pp. 765–775 (2003)

Maximum Overlap of Convex Polytopes under Translation*

Hee-Kap Ahn[1], Siu-Wing Cheng[2], and Iris Reinbacher[1]

[1] Department of Computer Science and Engineering, POSTECH, Korea
{heekap, irisrein}@postech.ac.kr
[2] Department of Computer Science and Engineering, HKUST, Hong Kong
scheng@cse.ust.hk

Abstract. We study the problem of maximizing the overlap of two convex polytopes under translation in $\mathbb{R}^d$ for some constant $d \geq 3$. Let n be the number of bounding hyperplanes of the polytopes. We present an algorithm that, for any $\varepsilon > 0$, finds an overlap at least the optimum minus ε and reports a translation realizing it. The running time is $O(n^{\lfloor d/2 \rfloor + 1} \log^d n)$ with probability at least $1 - n^{-O(1)}$, which can be improved to $O(n \log^{3.5} n)$ in $\mathbb{R}^3$. The time complexity analysis depends on a bounded incidence condition that we enforce with probability one by randomly perturbing the input polytopes. This causes an additive error ε, which can be made arbitrarily small by decreasing the perturbation magnitude. Our algorithm in fact computes the maximum overlap of the perturbed polytopes. All bounds and their big-O constants are independent of ε.

1 Introduction

Many applications perform geometric shape matching to find a transformation of one shape in order to maximize some similarity measure with another shape. The problem of matching convex shapes has been used in tracking regions in an image sequence [13] and measuring symmetry of a convex body [11]. One robust similarity measure for two shapes is their *overlap*—the volume of their intersection. In this paper, we consider maximizing the overlap of two convex polytopes under translation in $\mathbb{R}^d$ for $d \geq 3$. The dimension d is treated as a constant and so is any value depending on d alone.

In $\mathbb{R}^2$, the maximum overlap problem has been studied for convex and simple polygons. Let n be the number of input polygon edges. De Berg et al. [3] can maximize the overlap of two convex polygons under translation in $O(n \log n)$ time. Mount et al. [14] can do the same for two simple polygons in $O(n^4)$ time. When both rotation and translation are allowed, Ahn et al. [2] can align two convex polygons with an overlap at least $1 - \varepsilon$ times the optimum for any $\varepsilon \in (0, 1)$. They get a running time of $O((1/\varepsilon) \log n + (1/\varepsilon^2) \log(1/\varepsilon))$, assuming that there are two input arrays storing the

---

* Research of Ahn was supported by Basic Science Research Program through the National Research Foundation of Korea (NRF) funded by the Ministry of Education, Science and Technology (MEST) (No. 2010-0009857). Research of Cheng was partly supported by Research Grant Council, Hong Kong, China (project no. 612109). Research of Reinbacher was partly supported by a postdoc matching fund of HKUST.

O. Cheong, K.-Y. Chwa, and K. Park (Eds.): ISAAC 2010, Part II, LNCS 6507, pp. 97–108, 2010.
© Springer-Verlag Berlin Heidelberg 2010

polygon vertices in order around the boundary. In the case of translation only, they improve the running time to $O((1/\varepsilon)\log n + (1/\varepsilon)\log(1/\varepsilon))$.

The maximum overlap problem for convex polytopes under translation in $\mathbb{R}^d$ for $d \geq 3$ has been studied by Ahn et al. [1] and Fukuda and Uno [9]. Let n be the number of hyperplanes defining the convex polytopes. Ahn et al.'s algorithm finds the maximum overlap of two convex polytopes under translation in $O(n^{(d^2+d-3)/2}\log^{d+1} n)$ expected time. Given k convex polytopes for some constant $k \geq 2$, Fukuda and Uno can translate them to give an overlap of at least opt $- \varepsilon$ for any $\varepsilon > 0$, where opt denotes the maximum overlap. They require $O(\log(\text{opt}/\varepsilon))$ calls to a subroutine returning the value and gradient of the overlap function for given translations of the polytopes. Some critical details of this subroutine are missing though. In any case, the running time does not depend on the input size n alone. They also gave an algorithm to find the maximum overlap of k possibly non-convex polytopes under translation in $O(n^{kd^2+d})$ time.

Vigneron [16] studied the optimization of algebraic functions, which can be applied to align two possibly non-convex polytopes under rigid motion. For two convex polytopes, Vigneron's method returns in $O\big(\varepsilon^{-\Theta(d^2)}n^{\Theta(d^3)}(\log\frac{n}{\varepsilon})^{\Theta(d^2)}\big)$ time an overlap under rigid motion that is at least $1 - \varepsilon$ of the optimum.

We give a new algorithm for the maximum overlap problem for two convex polytopes under translation in $\mathbb{R}^d$ for $d \geq 3$. Our model of computation is the real-RAM model in which the operations $(+, -, \times, /)$ can be performed in constant time. We also make the standard assumption that it takes $O(1)$ time to solve a system of $O(1)$ polynomials of fixed degree in $O(1)$ variables. For any $\varepsilon > 0$, we can find an overlap at least the optimum minus ε and report a translation realizing it. Our algorithm runs in $O(n^{\lfloor d/2\rfloor+1}\log^d n)$ time with probability $1 - n^{-O(1)}$, which can be improved to $O(n\log^{3.5} n)$ in $\mathbb{R}^3$. The time complexity analysis depends on a bounded incidence condition, which may fail in degenerate situations. We enforce it with probability one by randomly perturbing the input polytopes. This causes an additive error ε, which can be made arbitrarily small by decreasing the perturbation magnitude. Our algorithm in fact computes the maximum overlap of the perturbed polytopes. The running time bounds, the probability bound, and the big-O constants in these bounds are independent of ε.

2 Background

An *i-flat* is $L + v$ for some i-dimensional linear subspace L and for some point $v \in \mathbb{R}^d$, i.e., a copy of L translated by the vector v. A *hyperplane* in $\mathbb{R}^d$ is a $(d-1)$-flat. Any hyperplane and the points on one side of it form a closed *halfspace*. Given a subset $X \subset \mathbb{R}^d$, its *affine hull* $\text{aff}(X)$ is the flat of the lowest dimension containing X.

A topological space M is a *k-manifold* if any point $x \in M$ has a neighborhood homeomorphic to $\mathbb{R}^k$ or a k-dimensional closed halfspace for some k. The *interior* of M, denoted by $\text{int}(M)$, is the set of points in M with a neighborhood homeomorphic to $\mathbb{R}^k$. The *boundary* of M, denoted by $\text{bd}(M)$, is $M \setminus \text{int}(M)$. For example, a point and any open convex set have empty boundary; the boundary of a line segment consists of its two endpoints; the boundary of a polygon consists of its vertices and edges.

Let X and Y be two subsets of $\mathbb{R}^d$. The *closure* of X, denoted by $\text{cl}(X)$, is the smallest closed subset of $\mathbb{R}^d$ containing X. The *Minkowski sum* of X and Y is defined as $X \oplus Y = \{x + y : x \in X, y \in Y\}$. So $\dim(X \oplus Y) \leq \dim(X) + \dim(Y)$. For any $\alpha \in \mathbb{R}^d$, we have $X \oplus \{\alpha\} = X + \alpha$.

A convex polytope P in $\mathbb{R}^d$ is the common intersection of (closed) halfspaces. These are the *bounding halfspaces* and their boundaries the *bounding hyperplanes* of P. Assume that P has dimension d. For $0 \leq k \leq d$, a *k-face* of P is the k-dimensional common intersection of P and some bounding hyperplane(s). Taking no bounding hyperplane in the intersection gives the d-face, which is P itself. We follow the convention to call the 0-faces *vertices*, the 1-faces *edges*, and the $(d-1)$-faces *facets*. We use $\mathrm{faces}(P)$ to denote the set of k-faces of P for $0 \leq k \leq d$. The faces with dimensions less than d are called *proper faces* and they are subsets of $\mathrm{bd}(P)$. In non-degenerate situations, a k-face lies in exactly $d - k$ bounding hyperplanes. In degenerate situations, a k-face may lie in more. Each face of P is a convex polytope.

An *i-simplex* is an i-dimensional convex polytope with exactly $i + 1$ vertices.

Let $\mathcal{F}$ be a finite family of convex subsets of $\mathbb{R}^d$, each of dimension $d - 1$ or less. The *arrangement* $\mathrm{Arr}(\mathcal{F})$ of $\mathcal{F}$ is a partition of $\mathbb{R}^d$ into disjoint *cells*. A cell is either a connected component in $\mathbb{R}^d \setminus (\bigcup_{S \in \mathcal{F}} S)$, or a maximal collection of points in $\bigcup_{S \in \mathcal{F}} S$ that belong to the same elements in $\mathcal{F}$.

Lemmas 1 and 2 below state the results on the ε-net theory [10] and cuttings [6] that we use heavily. Let H be a set of hyperplanes. For any $r \in (1, |H|)$, a *$(1/r)$-cutting* for H is a collection of simplices with disjoint interiors, which together cover $\mathbb{R}^d$ such that at most $|H|/r$ hyperplanes in H intersect $\mathrm{int}(\tau)$ for any d-simplex τ in the collection.

Lemma 1. *Let H be a multiset of hyperplanes in $\mathbb{R}^d$. Let $r \in (1, |H|)$ and $\delta \in (0, 1)$ be two parameters. There is a number $j_{d,r,\delta} = \Theta(dr \log(dr/\delta))$ such that, if we draw $j_{d,r,\delta}$ hyperplanes from H uniformly at random and form an arrangement A of the hyperplanes drawn (after removing duplicates), then it holds with probability at least $1 - \delta$ that at most $|H|/r$ hyperplanes in H intersect $\mathrm{int}(\tau)$ for any d-simplex τ whose interior lies in a cell of A.*

Remark. In Lemma 1, to achieve a probability bound of $1 - |H|^{-O(1)}$, we must draw $O(r \log |H|)$ hyperplanes to guarantee that at most $|H|/r$ hyperplanes intersect $\mathrm{int}(\tau)$.

Lemma 2. *Let H be a set of hyperplanes in $\mathbb{R}^d$. For any $r \in (1, |H|)$, a $(1/r)$-cutting of H of size $O(r^d)$ can be constructed in $O(|H|r^{d-1})$ time such that each d-simplex in the cutting stores the hyperplanes in H that intersect its interior.*

3   Overview

Let P_1 and P_2 be the two convex input polytopes, specified by n distinct bounding hyperplanes. The complexity of P_j, $j \in \{1, 2\}$, is $O(n^{\lfloor d/2 \rfloor})$, the number of its faces [8]. We always translate P_1 and keep P_2 stationary. We need the following definitions.

- For any vector $\alpha \in \mathbb{R}^d$, Q_α denotes the common intersection $(P_1 + \alpha) \cap P_2$.
- For any $f \in \mathrm{faces}(P_1)$ and $g \in \mathrm{faces}(P_2)$, $\gamma_{f,g}$ denotes the set $\{\alpha \in \mathbb{R}^d : (\mathrm{int}(f) + \alpha) \cap \mathrm{int}(g) \neq \emptyset\}$, which is a single point or an open convex set. One can verify that $\gamma_{f,g} = (-\mathrm{int}(f)) \oplus \mathrm{int}(g)$.
- Γ denotes the set $\{\gamma_{f,g} : \dim(\gamma_{f,g}) < d\}$.

The dimension of $\gamma_{f,g}$ is less than d if for any $\alpha \in \mathbb{R}^d$ such that $(\mathrm{int}(f) + \alpha) \cap \mathrm{int}(g) \neq \emptyset$, we can perturb α slightly to α' such that $(\mathrm{int}(f) + \alpha') \cap \mathrm{int}(g) = \emptyset$. Thus, if we move a point α in $\mathbb{R}^d$, there is a combinatorial change in Q_α whenever the point α crosses an element of Γ. There is no combinatorial change in Q_α if the point α varies within a cell in $\mathrm{Arr}(\Gamma)$. Let $\mathrm{vol}(Q_\alpha)$ denote the volume of Q_α. The function $\mathrm{vol}(Q_\alpha)^{1/d}$ is concave over $\{\, \alpha \in \mathbb{R}^d : Q_\alpha \neq \emptyset \,\}$ [15].

We follow the high level approach in the algorithm of de Berg et al. for convex polygons [3], which we call POLYGON. One can extend POLYGON directly to higher dimensions, but this gives an $\Omega(n^{2\lfloor d/2\rfloor})$ running time in the worst case as we explain below. For $d = 2$, Γ consists of open line segments (translations that place a vertex of P_1 in the interior of an edge of P_2 and vice versa) and the endpoints of the closure of these line segments (translations that align vertices of P_1 and P_2). Let L be the set of horizontal lines through the segment endpoints in Γ. Each line in L is the set of translations that place a vertex of P_1 at the same height of some vertex of P_2. The arrangement of Γ is divided into strips by the lines in L. POLYGON locates the strip containing the solution by probing L in a binary search manner. In each probe, POLYGON solves the maximum overlap problem for P_1 and P_2 with translations restricted to a line $\ell \in L$, and decide whether the solution for the original 2D problem lies above or below ℓ. Let S be the strip obtained by the binary search. De Berg et al. showed that S is stabbed by $O(n)$ open line segments in Γ. POLYGON scans the vertices and edges of P_1 and P_2 in order from top to bottom to find the $O(n)$ vertex-edge pairs that induce the open line segments in Γ stabbing S. Then, exploiting the concavity of $\mathrm{vol}(Q_\alpha)^{1/2}$, POLYGON constructs a sequence of cuttings (Lemma 2) to prune the search space to the cell in $\mathrm{Arr}(\Gamma) \cap S$ that contains the solution for the 2D maximum overlap problem.

For $d \geq 3$, the lines in L become parallel hyperplanes and each hyperplane is the set of translations that place a vertex of P_1 at the same height as some vertex of P_2. The hyperplanes in L cut $\mathrm{Arr}(\Gamma)$ into d-dimensional slabs. One can still locate the slab S containing the solution for the maximum overlap problem by a binary search. However, for a vertex v of P_1, the translated slab $v + S$ can cross $\Theta(n^{\lfloor d/2\rfloor})$ faces of P_2, so v induces $\Theta(n^{\lfloor d/2\rfloor})$ elements of Γ that stab S. Summing over all faces of P_1, there can be $\Theta(n^{2\lfloor d/2\rfloor})$ elements of Γ that stab S. Hence, it would take $\Omega(n^{2\lfloor d/2\rfloor})$ time to construct a cutting on the elements of Γ stabbing S.

Instead of parallel slabs, we prune $\mathrm{Arr}(\Gamma)$ using the ε-net theory (Lemma 1). First, we define a set $\widehat{\Gamma}$ of hyperplanes, each containing one element of Γ. We generate a random subset $\widehat{\mathcal{E}}_0 \subset \widehat{\Gamma}$ of size $\Theta(n^{\lfloor d/2\rfloor} \log n)$. The ε-net theory ensures that $O(n^{\lfloor d/2\rfloor})$ hyperplanes in $\widehat{\Gamma}$ stab any d-simplex in a cell of $\mathrm{Arr}(\widehat{\mathcal{E}}_0)$ with high probability, in particular, the cell C that contains the solution of the maximum overlap problem. How do we locate C? As binary search no longer works, we instead construct a sequence of cuttings on $\widehat{\mathcal{E}}_0$ to prune the search space to C, or more precisely to a d-simplex $\rho_0 \subseteq C$ containing the solution. During this pruning, we recursively solve instances of the maximum overlap problem for P_1 and P_2 with translations restricted to a hyperplane in $\widehat{\mathcal{E}}_0$ in order to tell which side of this hyperplane we should step into.

The challenge is to find the elements of Γ that stab ρ_0 so that we can search in ρ_0 via cuttings. For the direct extension of POLYGON to high dimensions, we would scan the faces of P_1 and P_2 in a direction orthogonal to the slabs and extract the face pairs that induce the elements of Γ stabbing a particular slab. However, in our case scanning no longer works. We prove a characterization of the elements of Γ that stab ρ_0, which allows us to find them using linear programming on P_1 and P_2. This is the key idea to defy the $O(n^{2\lfloor d/2 \rfloor})$ bound. The speedup in $\mathbb{R}^3$ is obtained by replacing the linear programming with suitable queries using the Dobkin-Kirkpatrick structure [7].

Degeneracy in P_1, P_2 and $\mathrm{Arr}(\widehat{\Gamma})$ has a great impact on the running time. For efficiency, the linear programming step requires each face of P_1 and P_2 to be incident to $O(1)$ other faces. When pruning the search space using a cutting, we need to decide which side of a hyperplane $\ell \in \widehat{\mathcal{E}}_0$ to step into, after obtaining the translation $\alpha \in \ell$ that maximizes Q_α over ℓ. If α lies in a cell of $\mathrm{Arr}(\widehat{\Gamma})$ that is incident to many other cells, it may take a long time to decide which side of ℓ we should step into. This explains the need for the bounded incidence condition (precise definition given in the next section).

4   Algorithm

We first give some definitions and then detail the algorithm outlined in the previous section. For each element $\gamma_{f,g} \in \Gamma$, define a hyperplane $\widehat{\gamma}_{f,g}$ containing $\gamma_{f,g}$ as follows.

– Suppose that $\dim(f) + \dim(g) < d$. If $\dim(\gamma_{f,g}) = d - 1$, then $\widehat{\gamma}_{f,g} = \mathrm{aff}(\gamma_{f,g})$. Otherwise, we pick a unit vector v orthogonal to $\mathrm{aff}(\gamma_{f,g})$ uniformly at random and define $\widehat{\gamma}_{f,g}$ to be the $(d-1)$-flat through $\gamma_{f,g}$ orthogonal to v.
– Suppose that $\dim(f) + \dim(g) \geq d$. Since $\dim(\gamma_{f,g}) < d$ by the definition of Γ, there is a face h of f such that $\dim(h) + \dim(g) < d$ and $\mathrm{aff}(\gamma_{h,g}) = \mathrm{aff}(\gamma_{f,g})$. (Pick any if there are more than one such h's.) The hyperplane $\widehat{\gamma}_{h,g}$ is already defined in the previous case. We set $\widehat{\gamma}_{f,g} = \widehat{\gamma}_{h,g}$.

We define $\widehat{\Gamma}$ to be the multiset $\{\, \widehat{\gamma}_{f,g} : \gamma_{f,g} \in \Gamma \,\}$. Duplicates exist in $\widehat{\Gamma}$ if two distinct face pairs induce the same hyperplane. Both $\widehat{\Gamma}$ and Γ have $O(n^{2\lfloor d/2 \rfloor})$ elements, so we cannot afford to generate either of them completely.

Consider two quantities. First, the maximum number of faces in P_1 or P_2 that have a non-empty common intersection. Second, the maximum number of hyperplanes in $\widehat{\Gamma}$ that have a non-empty common intersection. If these quantities have a constant upper bound, the *bounded incidence condition* is satisfied, which we assume in the rest of the paper. The time complexity analysis of our algorithm depends on it, but the correctness of our algorithm does not. We show in the full version of the paper that for any $\varepsilon > 0$, we can perturb the input to enforce the bounded incidence condition with probability one and the maximum overlap for the perturbed input is at most ε less the optimum.

We call our algorithm LOCATE. Given an m-flat Π, LOCATE(Π) returns the translation $\alpha \in \Pi$ that maximizes $\mathrm{vol}(Q_\alpha)$ over Π. The original maximum overlap problem is solved by setting $m = d$. LOCATE calls a subroutine PRUNE that takes three parameters, an m-flat Π, a d-simplex τ containing the optimal translation in Π, and any subset $\widehat{\mathcal{E}} \subseteq \widehat{\Gamma}$. PRUNE$(\Pi, \tau, \widehat{\mathcal{E}})$ outputs a d-simplex $\tau' \subseteq \tau$ such that τ' contains the optimal translation in Π and $\mathrm{int}(\tau')$ lies in a cell of $\mathrm{Arr}(\widehat{\mathcal{E}})$. Fig. 1 shows the pseudocodes of LOCATE and PRUNE. Although the solution lies in the m-flat, we search the arrangement $\mathrm{Arr}(\widehat{\Gamma})$ in $\mathbb{R}^d$ for notational convenience.

LOCATE(Π) /* return the optimal translation in Π */

1. If $\dim(\Pi) = 0$, return Π; otherwise, construct a d-simplex τ_0 that contains the optimal translation in Π.
2. Sample a subset $\widehat{\mathcal{E}}_0 \subset \widehat{\Gamma}$ of $\Theta(n^{\lfloor d/2 \rfloor} \log n)$ hyperplanes.
3. $\rho_0 :=$ PRUNE($\Pi, \tau_0, \widehat{\mathcal{E}}_0$).
4. Compute a subset $\widehat{\mathcal{E}}_1 \subset \widehat{\Gamma}$ that has $O(n^{\lfloor d/2 \rfloor})$ size and contains $\{\widehat{\gamma}_{f,g} \in \widehat{\Gamma} : \gamma_{f,g} \cap \mathrm{int}(\rho_0) \neq \emptyset\}$.
5. $\rho_1 :=$ PRUNE($\Pi, \rho_0, \widehat{\mathcal{E}}_1$).
6. Return the translation $\alpha \in \rho_1 \cap \Pi$ that maximizes $\mathrm{vol}(Q_\alpha)^{1/d}$.

PRUNE($\Pi, \tau, \widehat{\mathcal{E}}$) /* return a d-simplex $\tau' \subseteq \tau$ such that τ' contains the optimal translation in Π and $\mathrm{int}(\tau')$ lies in a cell of $\mathrm{Arr}(\widehat{\mathcal{E}})$. */

1. Set $\tau' = \tau$. Let α denote the translation in Π that maximizes $\mathrm{vol}(Q_\alpha)$ over Π.
2. Compute a $\frac{1}{2}$-cutting of $\widehat{\mathcal{E}}$. Find the d-simplex τ'' in the cutting that contains α.
3. Triangulate $\tau' \cap \tau''$. Update τ' to be the d-simplex in this triangulation that contains α. Remove from $\widehat{\mathcal{E}}$ the hyperplanes that avoid $\mathrm{int}(\tau')$.
4. Return τ' if $\widehat{\mathcal{E}}$ becomes empty. Otherwise, go to step 2.

Fig. 1. Pseudocodes of LOCATE and PRUNE

4.1 How LOCATE Works

Refer to the pseudocode of LOCATE in Fig. 1. In step 1, τ_0 is constructed as follows. For $j \in \{1, 2\}$, we compute P_j and its axes-parallel bounding box B_j in $O(n^{\lfloor d/2 \rfloor} + n \log n)$ time [5]. The translations that bring B_1 and B_2 into intersection form a box B which can be computed in $O(1)$ time. We can take τ_0 to be any d-simplex containing B. By steps 2 and 3, we call PRUNE($\Pi, \tau_0, \widehat{\mathcal{E}}_0$) with a random subset $\widehat{\mathcal{E}}_0 \subset \widehat{\Gamma}$. We want the d-simplex ρ_0 returned by PRUNE to be stabbed by only few hyperplanes in $\widehat{\Gamma}$ because we will construct cuttings on them later. By the ε-net theory, a d-simplex in any cell of $\mathrm{Arr}(\widehat{\mathcal{E}}_0)$ is stabbed by $(|\widehat{\Gamma}|/|\widehat{\mathcal{E}}_0|) \log n$ hyperplanes with probability $1 - n^{-O(1)}$. We have $|\widehat{\Gamma}| = O(n^{2\lfloor d/2 \rfloor})$ and we make $|\widehat{\mathcal{E}}_0| = O(n^{\lfloor d/2 \rfloor} \log n)$ to optimize the running time of LOCATE. Lemma 3 below explains how we pick the subset $\widehat{\mathcal{E}}_0$.

Lemma 3. *We can sample in $O(n^{\lfloor d/2 \rfloor} \log^2 n)$ time a subset $\widehat{\mathcal{E}}_0 \subset \widehat{\Gamma}$ of $\Theta(n^{\lfloor d/2 \rfloor} \log n)$ size such that, with probability $1 - n^{-O(1)}$, for any d-simplex ρ whose interior lies in a cell of $\mathrm{Arr}(\widehat{\mathcal{E}}_0)$, only $O(n^{\lfloor d/2 \rfloor})$ hyperplanes in $\widehat{\Gamma}$ intersect $\mathrm{int}(\rho)$.*

Proof. Let F_j^i be the number of i-faces of P_j for $j \in \{1, 2\}$ and $i \in [0, d]$. For $k \in [0, d-1]$, let $\widehat{\Gamma}_k$ be the multiset $\{\widehat{\gamma}_{f,g} \in \widehat{\Gamma} : \dim(f) + \dim(g) = k\}$. We sample a hyperplane from $\widehat{\Gamma}_k$ uniformly at random as follows. First, pick an integer $i \in [0, k]$ with probability $F_1^i F_2^{k-i} / (\sum_{a=0}^k F_1^a F_2^{k-a})$. Second, pick an i-face of P_1 and a $(k-i)$-face of P_2 with probabilities $1/F_1^i$ and $1/F_2^{k-i}$, respectively. Repeat to pick $\Theta(n^{\lfloor d/2 \rfloor} \log n)$ face pairs that induce $\Theta(n^{\lfloor d/2 \rfloor} \log n)$ hyperplanes in $\widehat{\Gamma}_k$. The set $\widehat{\mathcal{E}}_0$ contains all hyperplanes sampled over $k \in [0, d-1]$ with duplicates removed via sorting. The time needed is $O(n^{\lfloor d/2 \rfloor} \log^2 n)$.

Take any d-simplex ρ whose interior lies in a cell of $\mathrm{Arr}(\widehat{\mathcal{E}}_0)$. It follows immediately from Lemma 1 that, with probability $1 - n^{-O(1)}$, only $O(n^{\lfloor d/2 \rfloor})$ hyperplanes in $\bigcup_{k=0}^{d-1} \widehat{\Gamma}_k$ intersects $\mathrm{int}(\rho)$. It is possible for $\mathrm{int}(\rho)$ to intersect a hyperplane $\widehat{\gamma}_{f,g}$ in $\widehat{\Gamma}$ where $\dim(f) + \dim(g) \geq d$ and so $\widehat{\gamma}_{f,g} \notin \bigcup_{k=0}^{d-1} \widehat{\Gamma}_k$. By the definition of $\widehat{\Gamma}$, we have $\widehat{\gamma}_{f,g} = \widehat{\gamma}_{h,g}$ for some face h of f where $\dim(h) + \dim(g) < d$, which implies that $\widehat{\gamma}_{h,g} \in \bigcup_{k=0}^{d-1} \widehat{\Gamma}_k$. We charge the intersection between $\mathrm{int}(\rho)$ and $\widehat{\gamma}_{f,g}$ to the intersection between $\mathrm{int}(\rho)$ and $\widehat{\gamma}_{h,g}$. By the bounded incidence condition, the intersection between $\mathrm{int}(\rho)$ and $\widehat{\gamma}_{h,g}$ is charged only $O(1)$ times. $\qquad\square$

We discuss how PRUNE works in the next section, and we defer to Section 4.3 the discussion of step 4, the generation of a subset $\widehat{\mathcal{E}}_1 \subset \widehat{\Gamma}$ that contains $\{\widehat{\gamma}_{f,g} \in \widehat{\Gamma} : \gamma_{f,g} \cap \mathrm{int}(\rho_0) \neq \emptyset\}$. After step 5, we have a d-simplex ρ_1 such that ρ_1 contains the optimal translation in Π and $\mathrm{int}(\rho_1)$ lies in a cell of $\mathrm{Arr}(\widehat{\mathcal{E}}_1)$. The property of $\widehat{\mathcal{E}}_1$ implies that $\mathrm{int}(\rho_1)$ lies in a cell of $\mathrm{Arr}(\Gamma)$. (Some $\widehat{\gamma}_{f,g}$ in $\widehat{\Gamma}$ may intersect $\mathrm{int}(\rho_1)$, but $\gamma_{f,g}$ does not.) We describe below how to find the optimal translation in step 6.

We first obtain a formula φ for $\mathrm{vol}(Q_\alpha)$ for any $\alpha \in \mathrm{int}(\rho_1)$ by defining a *canonical triangulation* T_α of Q_α as follows. The canonical triangulations of the $(d-1)$-faces of Q_α are recursively defined. Then, fix a vertex q of Q_α and connect it to every simplex in $\mathrm{bd}(Q_\alpha)$ not incident to q to get T_α. If we have the volume formulae for the d-simplices in T_α, their sum gives the formula for $\mathrm{vol}(Q_\alpha)$. The *signed volume* of a d-simplex with vertices $v_0, v_1, \ldots, v_d$ is $\frac{1}{d!} \det(v_1 - v_0, v_2 - v_0, \ldots, v_d - v_0)$, where each v_i is viewed as a column vector. Since there is no combinatorial change as α varies in $\mathrm{int}(\rho_1)$, the vertex coordinates of Q_α are fixed linear functions in α and there is no combinatorial change in T_α. So the signed volumes of the d-simplices in T_α do not change sign. We construct T_{α_0} for a fixed translation $\alpha_0 \in \mathrm{int}(\rho_1)$ to determine which d-simplices in T_α have negative volume and multiply their formulae by -1. Constructing Q_α and T_α takes $O(n^{\lfloor d/2 \rfloor} + n \log n)$ time and $|T_\alpha| = O(n^{\lfloor d/2 \rfloor})$. Note that the volume formulae for the d-simplices in T_α are polynomials of degree $O(1)$ in $O(1)$ variables, and therefore their sum is also a polynomial of degree $O(1)$ in $O(1)$ variables. We can compute a formula φ for $\mathrm{vol}(Q_\alpha)$ with $O(n^{\lfloor d/2 \rfloor})$ terms in $O(n^{\lfloor d/2 \rfloor} + n \log n)$ time.

Combinatorial changes may happen if we move α from $\mathrm{int}(\rho_1)$ to $\mathrm{bd}(\rho_1)$. Nonetheless, these possible changes are that some d-simplices in T_α may become degenerate and have zero volume. So the formula φ is valid for any $\alpha \in \rho_1$.

We convert φ to a formula ψ using the barycentric coordinates of $\alpha \in \rho_1 \cap \Pi$ as the variables. The formula ψ has $O(n^{\lfloor d/2 \rfloor})$ terms that are polynomials of degree $O(1)$ in $O(1)$ variables. The conversion takes $O(n^{\lfloor d/2 \rfloor})$ time. We maximize $\psi^{1/d}$ by standard calculus. If $\psi^{1/d}$ attains its maximum in $\mathrm{int}(\rho_1 \cap \Pi)$, we have the optimal translation. Otherwise, $\psi^{1/d}$ attains its maximum in $\mathrm{bd}(\rho_1 \cap \Pi)$ and we repeat the conversion of φ and the maximization for each face of $\rho_1 \cap \Pi$.

Lemma 4. LOCATE(Π) *runs in* $T(n, m) = T_g + T_p + O(n^{\lfloor d/2 \rfloor} \log^2 n)$ *time, where* T_g *denotes the time to generate* $\widehat{\mathcal{E}}_1$ *in step 4 and* T_p *denotes the total running time of* PRUNE *in steps 3 and 5.*

4.2   How PRUNE Works

Let α^* denote the translation in an m-flat Π that maximizes the overlap over Π. PRUNE takes parameters Π, a d-simplex τ containing α^*, and a set $\widehat{\mathcal{E}}$ of hyperplanes, and returns a d-simplex $\tau' \subseteq \tau$ such that $\alpha^* \in \tau'$ and $\text{int}(\tau')$ lies in a cell of $\text{Arr}(\widehat{\mathcal{E}})$. Assume for now an oracle that, given an $(m-1)$-flat $\ell \subset \Pi$, decides which side of ℓ contains α^*.

Refer to the pseudocode of PRUNE in Fig. 1. In step 2, we construct a $(1/2)$-cutting of $\widehat{\mathcal{E}}$ that has $O(1)$ size and can be computed in $O(|\widehat{\mathcal{E}}|)$ time by Lemma 2. Let H be the set of bounding hyperplanes of the $(d-1)$-simplices in the cutting. Running the oracle on $h \cap \Pi$ for all $h \in H$ tells us which sides of the hyperplanes in H contain α^*. This gives the d-simplex τ'' in the cutting that contains α^*. In step 3, we triangulate $\tau' \cap \tau''$ in $O(1)$ time and use the oracle as before to find the d-simplex in the triangulation that contains α^*. By Lemma 2, at least half of the hyperplanes in $\widehat{\mathcal{E}}$ are removed in step 3. Thus, steps 2–4 iterate $O(\log |\widehat{\mathcal{E}}|)$ times and PRUNE takes $O(T_o \log |\widehat{\mathcal{E}}| + |\widehat{\mathcal{E}}| + \frac{1}{2}|\widehat{\mathcal{E}}| + \frac{1}{4}|\widehat{\mathcal{E}}| + \ldots) = O(T_o \log |\widehat{\mathcal{E}}| + |\widehat{\mathcal{E}}|)$ time, where T_o is the time to run the oracle once.

We describe below how the oracle works. Let F be the restriction of $\text{vol}(Q_\alpha)^{1/d}$ to Π. For a cell C of $\text{Arr}(\Gamma)$, let F_C denote the restriction of F to $\text{cl}(C) \cap \Pi$ and let ∇F_C denote the gradient of F_C. We run LOCATE(ℓ) to find the translation $\tilde{\alpha} \in \ell$ that maximizes the overlap over ℓ. Intuitively, the gradient of F at $\tilde{\alpha}$ points to the side of ℓ containing α^*. However, this idea fails because F may not be smooth at $\tilde{\alpha}$, leaving the gradient of F undefined at $\tilde{\alpha}$. We get around this problem as follows. We call a cell C of $\text{Arr}(\Gamma)$ special if $\text{cl}(C)$ contains $\tilde{\alpha}$ and $\nabla F_C(\tilde{\alpha})$ points into C. If there is no special cell, we report that $\alpha^* = \tilde{\alpha}$. If there is a special cell C, we report the side of ℓ that $\nabla F_C(\tilde{\alpha})$ points to. We argue that our strategy is correct as follows. Take the path of steepest ascent on the graph of F from $F(\tilde{\alpha})$ to $F(\alpha^*)$ and project it to Π. If the projected path does not leave ℓ at $\tilde{\alpha}$, we have $\alpha^* = \tilde{\alpha}$, so for any cell C whose closure contains $\tilde{\alpha}$, the gradient $\nabla F_C(\tilde{\alpha})$ cannot point into C, i.e., no special cell. If the projected path leaves ℓ at $\tilde{\alpha}$, it enters a special cell C and, by the maximality of $F(\tilde{\alpha})$ over ℓ, the projected path never returns to ℓ. Thus, $\nabla F_C(\tilde{\alpha})$ points to the side of ℓ containing α^*. There cannot be two special cells; otherwise, the steepest ascent at $F(\tilde{\alpha})$ projects to a direction v in Π that points outside some special cell C. By definition, $|\nabla F_C(\tilde{\alpha})|$ is greater than the magnitude of the gradient of F_C at $\tilde{\alpha}$ in direction v, which by the concavity of the graph of F, is at least the steepest ascent at $F(\tilde{\alpha})$. But then one can ascend faster on the graph of F_C in direction $\nabla F_C(\tilde{\alpha})$, a contradiction.

The oracle requires the computation of $\nabla F_C(\tilde{\alpha})$ for each cell C of $\text{Arr}(\Gamma)$. We describe this computation in the following. Let $\mathcal{A} \subset \Gamma$ be the subset of elements whose closure contain $\tilde{\alpha}$. They are induced by the intersecting face pairs of $P_1 + \tilde{\alpha}$ and P_2, so we can compute $\mathcal{A}$ by constructing $Q_{\tilde{\alpha}}$ in $O(n^{\lfloor d/2 \rfloor} + n \log n)$ time. Let $\widehat{\mathcal{A}} = \{\widehat{\gamma}_{f,g} : \gamma_{f,g} \in \mathcal{A}\}$. We have $|\widehat{\mathcal{A}}| = O(1)$ by the bounded incidence condition as all hyperplanes in $\widehat{\mathcal{A}}$ go through $\tilde{\alpha}$. The closure of each cell of $\text{Arr}(\widehat{\mathcal{A}})$ contains $\tilde{\alpha}$. Locally at $\tilde{\alpha}$, $\text{Arr}(\widehat{\mathcal{A}})$ is a refinement of the cells of $\text{Arr}(\Gamma)$ whose closure contain $\tilde{\alpha}$. So it suffices to compute $\nabla F_C(\tilde{\alpha})$ for each cell C of $\text{Arr}(\widehat{\mathcal{A}})$, which can be done as follows. Compute the unit vector v that points into $\text{cl}(C) \cap \Pi$ in the average direction of the edges of $\text{cl}(C) \cap \Pi$. For any faces f of P_1 and g of P_2 where $(f + \tilde{\alpha}) \cap g \neq \emptyset$, we check whether $f + \tilde{\alpha} + rv$

intersects g, treating r as arbitrarily small. This gives the face lattice of $Q_{\tilde{a}+rv}$. We want to compute the formula for $\text{vol}(Q_{\tilde{a}+rv})$ as in the previous section, but there is one difference. The face lattice of $Q_{\tilde{a}+rv}$ allows us to construct the canonical triangulation $T_{\tilde{a}+rv}$ of $Q_{\tilde{a}+rv}$. This gives the signed volume formula for each d-simplex in $T_{\tilde{a}+rv}$. The unknown r is the only variable in the formula. However, since we do not know an exact value of r, we cannot evaluate the signed volumes of the d-simplices in $T_{\tilde{a}+rv}$ and flip the signs of the negative volumes in order to obtain a formula for $\text{vol}(Q_{\tilde{a}+rv})$. Instead, we decide whether a d-simplex τ in $T_{\tilde{a}+rv}$ has negative volume as follows. Let V_τ denote the signed volume formula of τ, which is a polynomial in r of fixed degree. We compute the ith derivative $\frac{d^i V_\tau}{dr^i}$ for the smallest $i \geq 0$ such that $\frac{d^i V_\tau}{dr^i}\big|_{r=0}$ is non-zero. (The 0th derivative is V_τ itself.) If $\frac{d^i V_\tau}{dr^i}\big|_{r=0}$ is positive, then τ has positive volume; otherwise, τ has negative volume. This takes $O(1)$ time per d-simplex in $T_{\tilde{a}+rv}$. Hence, for each cell C of $\text{Arr}(\widehat{\mathcal{A}})$, we can compute $\nabla F_C(\tilde{\alpha})$ in $O(n^{\lfloor d/2 \rfloor} + n \log n)$ time.

Lemma 5. PRUNE$(\Pi, \tau, \widehat{\mathcal{E}})$ runs in $O((T(n, m-1) + n^{\lfloor d/2 \rfloor} + n \log n) \log |\widehat{\mathcal{E}}| + |\widehat{\mathcal{E}}|)$ time, where $T(n, m-1)$ is the time for LOCATE to run on an $(m-1)$-flat.

4.3 The Generation of $\widehat{\mathcal{E}}_1$

The step 4 of LOCATE generates a subset $\widehat{\mathcal{E}}_1 \subset \widehat{\Gamma}$ that contains the set $\{\widehat{\gamma}_{f,g} \in \widehat{\Gamma} : \gamma_{f,g} \cap \text{int}(\rho_0) \neq \emptyset\}$. We discuss how to do this in $O(n^{\lfloor d/2 \rfloor + 1} \log n)$ time and ensure that $|\widehat{\mathcal{E}}_1| = O(n^{\lfloor d/2 \rfloor})$. Recall that the *Minkowski sum* of two subsets X and Y of $\mathbb{R}^d$ is $X \oplus Y = \{x + y : x \in X, y \in Y\}$. So $\dim(X \oplus Y) \leq \dim(X) + \dim(Y)$.

We first compute a set $\mathcal{E}_1$ of face pairs from P_1 and P_2, each inducing an element in Γ as follows. We initialize $\mathcal{E}_1$ to be empty. For each face h_1 of P_1 and for each face σ of ρ_0, we compute the vertices of $(h_1 \oplus \sigma) \cap P_2$. For each vertex computed, if it is equal to $(\text{int}(h_1) \oplus \sigma) \cap \text{int}(h_2)$ for some face h_2 of P_2, we insert into $\mathcal{E}_1$ all face pairs (f, g) where $h_1 \in \text{faces}(f)$ and $h_2 \in \text{faces}(g)$ such that $\dim(\gamma_{f,g}) < d$. (By storing with f and g the basis vectors of $\text{aff}(f)$ and $\text{aff}(g)$, we can check in $O(1)$ time whether $\dim(f \oplus g) < d$ and this suffices as $\dim(\gamma_{f,g}) = \dim(f \oplus g)$.) The vertices of $(h_1 \oplus \sigma) \cap P_2$ that are not induced by $\text{int}(h_1) \oplus \sigma$ do not trigger any insertion into $\mathcal{E}_1$. At the end, we set $\widehat{\mathcal{E}}_1 = \{\widehat{\gamma}_{f,g} : (f, g) \in \mathcal{E}_1\}$ and remove any duplicates via sorting.

Our analysis in the rest of this section is divided into three parts. First, we show that $\widehat{\mathcal{E}}_1$ contains the set $\{\widehat{\gamma}_{f,g} \in \widehat{\Gamma} : \gamma_{f,g} \cap \text{int}(\rho_0) \neq \emptyset\}$. Second, we show that $|\widehat{\mathcal{E}}_1| = O(n^{\lfloor d/2 \rfloor})$ with probability $1 - n^{-O(1)}$. Third, we show that, with probability $1 - n^{-O(1)}$, it takes $O(n^{\lfloor d/2 \rfloor + 1} \log n)$ time to compute the vertices of $(h_1 \oplus \sigma) \cap P_2$ over all faces h_1 of P_1 and all faces σ of ρ_0.

The first part. We first prove two geometric properties and then show that $\widehat{\mathcal{E}}_1$ contains the set $\{\widehat{\gamma}_{f,g} \in \widehat{\Gamma} : \gamma_{f,g} \cap \text{int}(\rho_0) \neq \emptyset\}$.

Lemma 6. *The following properties hold for each element $\gamma_{f,g} \in \Gamma$.*

(i) *Suppose that $(\text{int}(f) \oplus \sigma) \cap \text{int}(g)$ is a single point for some face σ of ρ_0. Then, $\widehat{\gamma}_{f,g} \cap \text{int}(\rho_0) \neq \emptyset$ or $\widehat{\gamma}_{f,g}$ contains a vertex of ρ_0.*
(ii) *Suppose that $\gamma_{f,g}$ intersects ρ_0. There exists a face h_1 of f, a face h_2 of g, and a face σ of ρ_0 such that $(\text{int}(h_1) \oplus \sigma) \cap \text{int}(h_2)$ is a single point.*

Proof. Since $(\mathrm{int}(f) \oplus \sigma) \cap \mathrm{int}(g) \neq \emptyset$, some translation in σ brings $\mathrm{int}(f)$ and $\mathrm{int}(g)$ into intersection. Thus, $\gamma_{f,g} \cap \sigma \neq \emptyset$ and (i) follows as σ is a face of ρ_0.

Consider (ii). Recall that $\gamma_{f,g}$ is a point or an open convex set. So $\mathrm{cl}(\gamma_{f,g})$ is a convex polytope. Among the faces of $\mathrm{cl}(\gamma_{f,g})$ that intersects ρ_0, we choose those with the lowest dimension. Among these faces, we choose a face $\mathrm{cl}(\gamma_{h_1,h_2})$ such that $\dim(h_1) + \dim(h_2)$ is minimum. Since ρ_0 does not intersect any face of $\mathrm{cl}(\gamma_{f,g})$ with dimension less than $\dim(\gamma_{h_1,h_2})$, the boundary of $\mathrm{cl}(\gamma_{h_1,h_2})$ avoids ρ_0, which implies that some face σ of ρ_0 intersects γ_{h_1,h_2} in a single point. That is, there is a unique translation $\alpha = \gamma_{h_1,h_2} \cap \sigma$ such that $(\mathrm{int}(h_1) + \alpha) \cap \mathrm{int}(h_2) \neq \emptyset$. We claim that $(\mathrm{int}(h_1) + \alpha) \cap \mathrm{int}(h_2)$ is a single point, which implies (ii). If $(\mathrm{int}(h_1) + \alpha) \cap \mathrm{int}(h_2)$ is not a single point, its closure has a vertex $(\mathrm{int}(h_1') + \alpha) \cap \mathrm{int}(h_2')$ for some $h_1' \in \mathrm{faces}(h_1)$ and $h_2' \in \mathrm{faces}(h_2)$ where h_1' is a proper face of h_1 or h_2' is a proper face of h_2. Thus, $\dim(\gamma_{h_1',h_2'}) \leq \dim(\gamma_{h_1,h_2})$ and $\gamma_{h_1',h_2'}$ intersects σ, but $\dim(h_1') + \dim(h_2') < \dim(h_1) + \dim(h_2)$. This contradicts our choice of γ_{h_1,h_2}. $\qquad\square$

Lemma 7. $\widehat{\mathcal{E}}_1$ *contains the set* $\{ \widehat{\gamma}_{f,g} \in \widehat{\Gamma} : \gamma_{f,g} \cap \mathrm{int}(\rho_0) \neq \emptyset \}$.

Proof. Choose a $\gamma_{f,g} \in \Gamma$ intersecting $\mathrm{int}(\rho_0)$. By Lemma 6(ii), $(\mathrm{int}(h_1) \oplus \sigma) \cap \mathrm{int}(h_2)$ is a single point for a face h_1 of f, a face h_2 of g, and a face σ of ρ_0. So $(\mathrm{int}(h_1) \oplus \sigma) \cap \mathrm{int}(h_2)$ is a vertex of $(h_1 \oplus \sigma) \cap P_2$. We collect this vertex and add (f, g) to $\mathcal{E}_1$. $\qquad\square$

The second part. Lemma 8 below gives an $O(n^{\lfloor d/2 \rfloor})$ bound on the number of vertices computed by our generation procedure. It follows that $|\widehat{\mathcal{E}}_1| = O(n^{\lfloor d/2 \rfloor})$.

Lemma 8. *With probability* $1 - n^{-O(1)}$, *there are* $O(n^{\lfloor d/2 \rfloor})$ *vertices in the convex polytopes* $(h_1 \oplus \sigma) \cap P_2$ *over all faces* h_1 *of* P_1 *and all faces* σ *of* ρ_0.

Proof. Each vertex is $(\mathrm{int}(f) \oplus \sigma) \cap \mathrm{int}(g)$ for a face f of P_1, a face g of P_2, and a face σ of ρ_0. If $\dim(\gamma_{f,g}) < d$, we give the vertex a blue color; if $\dim(\gamma_{f,g}) = d$, we give it a red color. It is possible for a vertex to receive both colors if it is induced by two face pairs (f, g) and (f', g') such that $\dim(\gamma_{f,g}) < d$ and $\dim(\gamma_{f',g'}) = d$. We count these two colored instances of the same vertex separately in our analysis.

Consider the blue vertices. Lemma 6(i) implies that $\widehat{\gamma}_{f,g} \cap \mathrm{int}(\rho_0) \neq \emptyset$ or $\widehat{\gamma}_{f,g}$ contains a vertex of ρ_0. By Lemma 3, with probability $1 - n^{-O(1)}$, there are $O(n^{\lfloor d/2 \rfloor})$ hyperplanes $\widehat{\gamma}_{f,g}$ in $\widehat{\Gamma}$ where $\widehat{\gamma}_{f,g} \cap \mathrm{int}(\rho_0) \neq \emptyset$. By the bounded incidence condition, any vertex of ρ_0 lies in $O(1)$ hyperplanes $\widehat{\gamma}_{f,g}$ in $\widehat{\Gamma}$. For a face σ of ρ_0, the blue vertex $(\mathrm{int}(f) \oplus \sigma) \cap \mathrm{int}(g)$ may be constructed more than once if there are other faces $f' \in \mathrm{faces}(P_1)$ and $g' \in \mathrm{faces}(P_2)$ such that $(\mathrm{int}(f') \oplus \sigma) \cap \mathrm{int}(g') = (\mathrm{int}(f) \oplus \sigma) \cap \mathrm{int}(g)$. Nevertheless, the pairs (f', g') and (f, g) are already counted separately in the above as we apply Lemma 6, Lemma 3 and the bounded incidence condition. Another factor $2^{d+1} - 1$ is needed as we go over all faces σ of ρ_0. So we compute $O(n^{\lfloor d/2 \rfloor})$ blue vertices, counting multiplicities.

Consider a red vertex $(\mathrm{int}(f) \oplus \sigma) \cap \mathrm{int}(g)$. For any translation $\alpha \in \sigma$, we have $(\mathrm{int}(f) + \alpha) \cap \mathrm{int}(g) \subseteq (\mathrm{int}(f) \oplus \sigma) \cap \mathrm{int}(g)$, which is a single point. Therefore, for any translation $\alpha \in \sigma$, if $(\mathrm{int}(f) + \alpha) \cap \mathrm{int}(g) \neq \emptyset$, then

$$(\mathrm{int}(f) \oplus \sigma) \cap \mathrm{int}(g) = (\mathrm{int}(f) + \alpha) \cap \mathrm{int}(g). \qquad (1)$$

Fix σ and a translation α_0 in σ. Divide the red vertices $(\text{int}(f) \oplus \sigma) \cap \text{int}(g)$ over all faces f of P_1 and g of P_2 into two groups, one satisfying $(\text{int}(f) + \alpha_0) \cap \text{int}(g) \neq \emptyset$ and the other satisfying $(\text{int}(f) + \alpha_0) \cap \text{int}(g) = \emptyset$. By (1), the number of red vertices in the first group is no more than the number of vertices of $(P_1 + \alpha_0) \cap P_2$, which is $O(n^{\lfloor d/2 \rfloor})$. For each red vertex $(\text{int}(f) \oplus \sigma) \cap \text{int}(g)$ in the second group, we charge it to a blue vertex as follows. Since $(\text{int}(f) \oplus \sigma) \cap \text{int}(g)$ is a single point, by continuity, $(f \oplus \sigma) \cap g$ is equal to this single point. We choose a face h_1 of f and a face h_2 of g such that $(\text{int}(h_1) \oplus \sigma) \cap \text{int}(h_2) \neq \emptyset$ and $\dim(h_1) + \dim(h_2)$ is minimized. Thus, $(\text{int}(h_1) \oplus \sigma) \cap \text{int}(h_2)$ is the single point $(f \oplus \sigma) \cap g$ and the minimization ensures that $\dim(h_1) + \dim(h_2) < d$. So $\dim(\gamma_{h_1,h_2}) \leq \dim(h_1) + \dim(h_2) < d$. It follows that $(\text{int}(h_1) \oplus \sigma) \cap \text{int}(h_2)$ is a blue vertex (it is a vertex of $(h_1 \oplus \sigma) \cap P_2$). We charge the red vertex $(\text{int}(f) \oplus \sigma) \cap \text{int}(g)$ to it. For another red vertex $(\text{int}(f') \oplus \sigma) \cap \text{int}(g')$ to charge to $(\text{int}(h_1) \oplus \sigma) \cap \text{int}(h_2)$, we must have $h_1 \in \text{faces}(f')$ and $h_2 \in \text{faces}(g')$. So the blue vertex $(\text{int}(h_1) \oplus \sigma) \cap \text{int}(h_2)$ is charged $O(1)$ times by the bounded incidence condition. It follows that $O(n^{\lfloor d/2 \rfloor})$ red vertices are induced by each face σ of ρ_0. Another factor $2^{d+1} - 1$ is needed as we go over all faces σ of ρ_0. Thus, $O(n^{\lfloor d/2 \rfloor})$ red vertices are computed, counting multiplicities. □

The third part. The next result bounds the time to generate $\widehat{\mathcal{E}}_1$.

Lemma 9. *Computing $\widehat{\mathcal{E}}_1$ takes $O(n^{\lfloor d/2 \rfloor + 1} \log n)$ time with probability $1 - n^{-O(1)}$.*

Proof. Let h_1 be a face of P_1 and let σ be a face of ρ_0. The face h_1 is the intersection of $O(n)$ halfspaces and hyperplanes. The Minkowski sum of each such halfspace or hyperplane with σ has $O(1)$ size and can be computed in $O(1)$ time. So the linear constraints defining $h_1 \oplus \sigma$ can be computed in $O(n)$ time.

We run Megiddo's linear programming algorithm to find a vertex ν of $(h_1 \oplus \sigma) \cap P_2$ in $O(n)$ time [12]. We visit the vertices adjacent to ν in two steps. First, we compute the supporting lines of edges incident to ν as follows. The point ν is dual to a $(d-1)$-flat and each bounding hyperplane through ν is dual to a point in this $(d-1)$-flat. The supporting lines of the edges incident to ν correspond to the $(d-2)$-faces of the convex hull of the dual points. By the bounded incidence condition and the constant size of σ, there are $O(1)$ such dual points, so it takes $O(1)$ time to compute their convex hull and hence the supporting lines of the edges incident to ν. Second, we shoot rays from ν along all these supporting lines and find the first hyperplane that each ray stops at by checking the linear constraints not containing ν in $O(n)$ time. These stopping points are the vertices adjacent to ν. Altogether, we can visit the vertices adjacent to ν in $O(n)$ time. Hence, it takes $O(n + k_{\sigma,h_1} n \log n)$ time to visit all vertices of $(h_1 \oplus \sigma) \cap P_2$, where k_{σ,h_1} is the number of such vertices and the $O(\log n)$ term comes from using a dictionary to record vertices visited.

A vertex of $(h_1 \oplus \sigma) \cap P_2$ is equal to $(\text{int}(h_1) \oplus \sigma) \cap \text{int}(h_2)$ for some face h_2 of P_2 if and only if that vertex lies in the translates of the bounding hyperplanes through h_1 but not in the translates of any other bounding hyperplane of P_1. Each such vertex can be found in $O(n)$ time. Hence, it takes $O(n^{\lfloor d/2 \rfloor + 1} \log n)$ time to construct $\widehat{\mathcal{E}}_1$ because P_1 has $O(n^{\lfloor d/2 \rfloor})$ faces and $\sum_{\sigma,h_1} k_{\sigma,h_1} = O(n^{\lfloor d/2 \rfloor})$ with probability $1 - n^{-O(1)}$ by Lemma 8. (We remove duplicates in $\widehat{\mathcal{E}}_1$ in $O(n^{\lfloor d/2 \rfloor} \log n)$ time via sorting.) □

Final results. By the results in Lemmas 4, 5, 8, and 9, we get the recurrence $T(n, m) = O(T(n, m-1) \log n + n^{\lfloor d/2 \rfloor+1} \log n)$ with boundary condition $T(n, 0) = O(1)$. The solution is $T(n, m) = O(mn^{\lfloor d/2 \rfloor+1} \log^m n)$. We can reduce the running time in $\mathbb{R}^3$ by replacing the linear programming with suitable queries using the Dobkin-Kirkpatrick structure [7]. The details can be found in the full version of the paper.

Theorem 1. *Let P_1 and P_2 be two convex polytopes in $\mathbb{R}^d$, $d \geq 3$, specified by n bounding hyperplanes. For any $\varepsilon > 0$, we can compute an overlap of P_1 and P_2 under translation that is at most ε less than the optimum. The running time is $O(n^{\lfloor d/2 \rfloor+1} \log^d n)$ with probability $1 - n^{-O(1)}$. In $\mathbb{R}^3$, the running time can be improved to $O(n \log^{3.5} n)$.*

References

1. Ahn, H.-K., Brass, P., Shin, C.-S.: Maximum overlap and minimum convex hull of two convex polyhedra under translation. Comput. Geom. Theory and Appl. 40, 171–177 (2008)
2. Ahn, H.-K., Cheong, O., Park, C.-D., Shin, C.-S., Vigneron, A.: Maximizing the Overlap of Two Planar Convex Sets under Rigid Motions. Comput. Geom. Theory and Appl. 37, 3–15 (2007)
3. de Berg, M., Cheong, O., Devillers, O., van Kreveld, M., Teillaud, M.: Computing the Maximum Overlap of Two Convex Polygons under Translations. Theory of Comput. Syst. 31, 613–628 (1998)
4. Chazelle, B.: An optimal algorithm for intersecting three-dimensional convex polyhedra. SIAM J. Computing 21, 671–696 (1992)
5. Chazelle, B.: An optimal convex hull algorithm in any fixed dimension. Discr. Comput. Geom. 9, 377–409 (1993)
6. Chazelle, B.: Cutting Hyperplanes for Divide-and-Conquer. Discr. Comput. Geom. 9, 145–159 (1993)
7. Dobkin, D.P., Kirkpatrick, D.G.: Determining the separation of preprocessed polyhedra – a unified approach. In: Proc. 17th Internat. Colloq. Automata Lang. Program., pp. 400–413 (1990)
8. Edelsbrunner, H.: Algorithms in Combinatorial Geometry. Springer, Heidelberg (1987)
9. Fukuda, K., Uno, T.: Polynomial time algorithms for maximizing the intersection volume of polytopes. Pacific J. Optimization 3, 37–52 (2007)
10. Haussler, D., Welzl, E.: Epsilon-nets and simplex range queries. Discr. Comput. Geom. 2, 127–151 (1987)
11. Heijmans, H.J.A.M., Tuzikov, A.V.: Similarity and symmetry measures for convex shapes using Minkowski addition. IEEE Trans. PAMI 20, 980–993 (1998)
12. Megiddo, N.: Linear programming in linear time when the dimension is fixed. J. ACM 31, 114–127 (1984)
13. Meyer, F., Bouthemy, P.: Region-based tracking in an image sequence. In: Sandini, G. (ed.) ECCV 1992. LNCS, vol. 588, pp. 476–484. Springer, Heidelberg (1992)
14. Mount, D.M., Silverman, R., Wu, A.Y.: On the area of overlap of translated polygons. Computer Vision and Image Understanding 64, 53–61 (1996)
15. Sangwine-Yager, J.R.: Mixed Volumes. In: Gruber, P.M., Wills, J.M. (eds.) Handbook on Convex Geometry, vol. A, pp. 43–71. Elsevier, Amsterdam (1993)
16. Vigneron, A.: Geometric optimization and sums of algebraic functions. In: Proc. ACM–SIAM Sympos. Alg., pp. 906–917 (2010)

Approximate Shortest Homotopic Paths in Weighted Regions*

Siu-Wing Cheng[1], Jiongxin Jin[1], Antoine Vigneron[2], and Yajun Wang[3]

[1] Department of Computer Science and Engineering, HKUST, Hong Kong
[2] INRA, UR 341 Mathématiques et Informatique Appliquées, Jouy-en-Josas, France
[3] Microsoft Research Asia, Beijing, China

Abstract. Let P be a path between two points s and t in a polygonal subdivision $\mathcal{T}$ with obstacles and weighted regions. Given a relative error tolerance $\varepsilon \in (0, 1)$, we present the first algorithm to compute a path between s and t that can be deformed to P without passing over any obstacle and the path cost is within a factor $1 + \varepsilon$ of the optimum. The running time is $O(\frac{h^3}{\varepsilon^2}kn\,\text{polylog}(k, n, \frac{1}{\varepsilon}))$, where k is the number of segments in P and h and n are the numbers of obstacles and vertices in $\mathcal{T}$, respectively. The constant in the running time of our algorithm depends on some geometric parameters and the ratio of the maximum region weight to the minimum region weight.

1 Introduction

Given a path P in the plane, the shortest homotopic path problem is to find a minimum-cost path that can be deformed to P without crossing any obstacle. The problem originates from research in VLSI (e.g. [8,11]). Forbus et al. [6] described a planning system in which a user makes a path sketch for vehicles or people and then the system generates the detailed optimized path homotopic to the sketch. It is natural to consider non-Euclidean cost models because different regions incur different costs; for example, traveling in swamps is harder than traveling on roads.

The *weighted region* model is the first non-Euclidean cost model and there has been much work on it (e.g. [1,12,13]). The environment is a polygonal subdivision, each region f has a weight w_f, and the subpath cost within a region f is w_f times the subpath length. Computing the exact shortest path seems hard and only approximation algorithms are known so far. The first algorithm of Mitchell and Papadimitriou [12] runs in $O(n^8 \log \frac{nN\rho}{\varepsilon})$ time, where n is the number of subdivision vertices, the vertices have integer coordinates in $[0, N]$, and ρ is the ratio of the maximum region weight to the minimum region weight. Subsequently, other algorithms have been proposed whose running times have a lower dependence on n. The most notable approach is to compute the shortest path in a graph obtained by discretizing the input subdivision, so as to approximate the true shortest path (e.g. [1,13]). Sun and Reif [13] gave an algorithm that runs in $O(\frac{n}{\varepsilon} \log \frac{n}{\varepsilon} \log \frac{1}{\varepsilon})$ time, where the hidden constant depends on some geometric parameters. Aleksandrov et al. [1] achieved the best dependence on n

---

* The research of Cheng and Jin was supported by the Research Grant Council, Hong Kong, China (project no. 612107).

O. Cheong, K.-Y. Chwa, and K. Park (Eds.): ISAAC 2010, Part II, LNCS 6507, pp. 109–120, 2010.

and ε with a running time of $O(\frac{n}{\sqrt{\varepsilon}} \log \frac{n}{\varepsilon} \log \frac{1}{\varepsilon})$, where the hidden constant depends on ρ and some geometric parameters. No result is known so far on the shortest homotopic path problem in weighted regions, although several results are known when the cost of a path is its length [2,4,9].

The main result in this paper is a $(1 + \varepsilon)$-approximate shortest homotopic path algorithm for any $\varepsilon \in (0, 1)$ in weighted regions. Let P be a path between two points s and t in a polygonal subdivision $\mathcal{T}$ with obstacles and weighted regions. Self-intersections in P are allowed. Given $\varepsilon \in (0, 1)$, our algorithm computes a path between s and t that can be deformed to P without passing over any obstacle and the path cost is within a factor $1 + \varepsilon$ of the optimum. The running time is $O(\frac{h^3}{\varepsilon^2} kn \operatorname{polylog}(k, n, \frac{1}{\varepsilon}))$, where k is the number of segments in P and h and n are the numbers of obstacles and vertices in $\mathcal{T}$, respectively. The constant in our running time depends on ρ and some geometric parameters. These geometric parameters and the dependence on them are of the same kind as in the work of Sun and Reif [13] as we use their result as a subroutine.

2   Preliminaries

We denote the input polygonal subdivision by $\mathcal{T}$, which consists of vertices, edges, and polygonal faces. Some polygonal faces are marked as inaccessible and each connected component of inaccessible faces forms an obstacle. The remaining polygonal faces are accessible and they are called the *regions* of $\mathcal{T}$. Each region f is associated with a positive weight $w_f > 0$. Without loss of generality, we assume that $\mathcal{T}$ is connected, every obstacle is a simple polygon, every region is a triangle, and the minimum region weight is equal to 1. We use ρ to denote the maximum region weight in $\mathcal{T}$.

Consider a line segment pq and a region f. Let $|pq|$ denote the length of pq. We use $\operatorname{int}(\cdot)$ to denote the interior of the operand. If $\operatorname{int}(pq) \subset \operatorname{int}(f)$ or pq is contained in an edge adjacent to f only, we define $\operatorname{cost}_{\mathcal{T}}(pq) = w_f |pq|$. If pq is contained in an edge shared between f and another region g, we define $\operatorname{cost}_{\mathcal{T}}(pq) = \min\{w_f, w_g\} \cdot |pq|$. A *polygonal path* Q is a polyline in $\mathcal{T}$ with finitely many segments. A *link* of Q is a maximal segment in Q that lies in a region of $\mathcal{T}$. An endpoint of a link is called a *node*. We use $|Q|$ to denote the length of Q. We use $\operatorname{cost}_{\mathcal{T}}(Q)$ to denote the sum of the costs of its links. Notice that $|Q| \leq \operatorname{cost}_{\mathcal{T}}(Q) \leq \rho|Q|$.

We use P to denote the input polygonal path. We use s and t to denote the endpoints of P and we enforce them to be vertices of $\mathcal{T}$ by splitting regions if necessary. Two paths with the same endpoints are *homotopic* if one can be deformed to the other without passing over any obstacle.

3   Overview

We present a simplified version of our strategy to highlight the main ideas. This simplified strategy cannot be turned into an effective algorithm, for instance, because no algorithm is known for computing an exact shortest path in weighted regions.

We are given a triangulated domain with obstacles, and we want to find a shortest path homotopic to a given input path P, with endpoints s and t. We first need to encode the homotopy of P. To this end, we build a spanning tree of the obstacles, with an

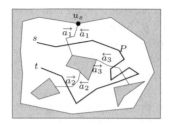 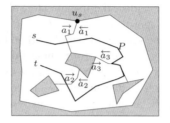

Fig. 1. The obstacles are shaded. After canceling one $\overleftarrow{a_3}$ and one $\overrightarrow{a_3}$, the path P becomes a new path P' that crosses the edge a_3 once.

extra edge connecting it to a point u_s on the outer face of our domain. The edges of this spanning tree are denoted by $a_1, a_2, \ldots, a_h$. We choose each such edge a_i to be a shortest path between two points lying on obstacles, or between u_s and a point lying on an obstacle.

We follow P from s to t to trace the edges that it crosses as well the crossing directions (determined with respect to an arbitrarily chosen orientation of the a_i's). In Fig. 1, the trace is $\overrightarrow{a_1}\overleftarrow{a_3}\overrightarrow{a_3}\overleftarrow{a_3}\overleftarrow{a_2}$, where $\overleftarrow{a_i}$ means crossing a_i from right to left and $\overrightarrow{a_i}$ means crossing a_i from left to right. If $\overleftarrow{a_i}$ and $\overrightarrow{a_i}$ appears consecutively in the trace, we can cancel the two crossings. This corresponds to making a shortcut along a_i between the two crossings as illustrated in Fig. 1. The important point is that the above cancellation does not change the homotopy of the path. When all cancellations are done, the reduced trace S_P is a unique encoding of the homotopy of P. Indeed, two paths P and Q with the same endpoints are homotopic if and only if $S_P = S_Q$.

Since the tree edges are shortest paths, a shortcut (canceling two adjacent symbols in the trace) does not increase the path cost. This is ideal because it means that for any path P, there is a shortest path P^* homotopic to P that crosses the spanning tree as dictated by S_P. The path P^* makes no redundant crossing. A natural approach to compute such a shortest path is as follows. Assume that S_P starts with $\overrightarrow{a_1}\overleftarrow{a_3}\overleftarrow{a_2}\ldots$. We know that P^* will first reach a_1 from the left. As we do not know at which point of a_1 it arrives, we can discretize a_1 by placing many vertices along it. For each of these vertices, we compute an approximate shortest path from s, treating the edges a_i of our tree as obstacles. As these paths avoid our spanning tree, they lie in a simply connected region. Thus, we do not need to consider their homotopy class and we can apply known algorithms for approximate shortest paths in weighted regions.

After crossing a_1, we know that P^* will reach a_3 from the right. So we perform a second round of approximate shortest paths computation (where the paths are not allowed to cross our spanning tree). We perform this computation with multiple sources, each source being one of the vertices placed on a_1, and each such vertex having an additive weight which is the approximate shortest distance from s to this vertex. The target points, again, are the vertices placed densely along a_3. We repeat this process for each symbol in S_P, and we obtain an approximate shortest path homotopic to P.

Our actual algorithm follows similar ideas, but there are important differences as we face several difficulties. The most obvious one is that no algorithm is known for computing an exact shortest path in weighted regions. Second, the spanning tree calls for repeated shortest path computations in order to connect the obstacles, which is rather wasteful. We replace the spanning tree above by another tree, the *anchor tree*, which is basically an approximate shortest path tree from u_s to one vertex of each obstacle. The homotopy encoding S_P is still based on the crossings between P and the anchor tree, but we change it slightly for technical convenience. Since the paths in the anchor tree are not exact shortest paths, we cannot expect a shortest path homotopic to P to cross the anchor tree as dictated by S_P. To conform to S_P, we have the reroute the optimal path along the anchor tree in the analysis. This demands a careful construction of the anchor tree so that the rerouting error is small. Another major issue is that we have to keep S_P short because the running time of our algorithm is directly related to it. Finally, to make our algorithm run faster, we will not discretize the anchor tree. We will still run one round of approximate shortest paths computation for each symbol in S_P, but in the absence of vertices on the anchor tree, multiple crossings of the anchor tree (instead of just one) may have to be taken at the end of a round. We need to do this quickly while conforming to S_P. The rest of this paper explains how to handle these difficulties.

4   The Subdivision S and the Graph H_ε

We introduce a graph H_ε which is the discretization of some subset of T based on the scheme of Sun and Reif [13]. We briefly review their construction below. Given a subdivision K with triangular regions, Sun and Reif place $O(\frac{1}{\varepsilon}\log\frac{1}{\varepsilon})$ Steiner points on each edge of K, where the hidden constant depends on some geometric parameters. The vertices of K and these Steiner points form the vertex set of a graph which we denote by $G_\varepsilon(K)$. Every two vertices p and q of $G_\varepsilon(K)$ on the boundary of a region are connected by the edge pq with weight $\text{cost}_K(pq)$. There are $O(\frac{1}{\varepsilon}|K|\log\frac{1}{\varepsilon})$ vertices and $O(\frac{1}{\varepsilon^2}|K|\log^2\frac{1}{\varepsilon})$ edges in $G_\varepsilon(K)$. So Dijkstra's algorithm returns a shortest path or a shortest path tree in $G_\varepsilon(K)$ in $O(\frac{1}{\varepsilon^2}|K|\log\frac{|K|}{\varepsilon}\log\frac{1}{\varepsilon})$ time [7]. A shortest path in $G_\varepsilon(K)$ is a $(1+\varepsilon)$-approximate shortest path in K. Sun and Reif gave a faster shortest path algorithm that avoids generating the edges of $G_\varepsilon(K)$, but we do not need this as other tasks will prove to be more time-consuming. Aleksandrov et al. [1] have a related construction with better dependence on ε, but we cannot use it due to some technical difficulties.

The graph H_ε is $G_\varepsilon(S)$ for some refinement S of a subset of T. We will run multiple rounds of Dijkstra's algorithm on a subgraph H_{alg} of H_ε to generate a $(1+\varepsilon)$-approximate shortest homotopic path. A dense enough discretization is sufficient for this purpose. We will use another graph H_{fen} whose edges are contained in H_ε to compute the anchor tree for encoding the homotopy of P. This requires extra properties as we explain below. Although H_{alg} and H_{fen} serve different purposes, a $(1+\varepsilon)$-approximate shortest homotopic path has to interact with the anchor tree, i.e., cross it. The relations among H_ε, H_{alg}, and H_{fen} facilitate the analysis.

Let L_{st} denote the length of a minimum-length path homotopic to P. Let B denote an axis-parallel box centered at s with width $4\rho L_{st}$. The cost of the shortest path homotopic to P is between L_{st} and ρL_{st}. So for any $\varepsilon \in (0,1)$, the box B contains any $(1+\varepsilon)$-approximate shortest path homotopic to P, which means that only the obstacles inside B are relevant. The restriction to B controls the costs of the paths in the anchor tree which keeps short the canonical crossing sequence of P.

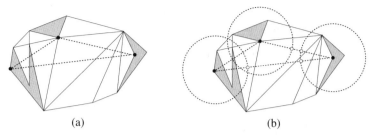

(a)   (b)

Fig. 2. The obstacles are shaded. We ignore the box B for simplicity. In (a), the black dots denote the anchors and the dashed segments form the anchor triangulation. In (b), the circles have radii δ_{fen} and the white dots are the extra vertices inserted.

For each obstacle inside B, we pick one of its vertices to be an *anchor*. We compute the *anchor triangulation*, a triangulation of the anchors as well as the four corners of B. We superimpose the anchor triangulation on $B \cap \mathcal{T}$ to obtain a subdivision $\mathcal{T}'$. Notice that an anchor triangulation edge may be split by the obstacles and the edges of $B \cap \mathcal{T}$ into several edges in $\mathcal{T}'$. Fig. 2(a) gives an illustration. The anchor triangulation edges provide shortcuts in $\mathcal{T}'$ that one can take in building the anchor tree. This controls the length of the canonical crossing sequence of P.

We need to prevent any path in the anchor tree from spiraling around the obstacles in order to keep short the canonical crossing sequence of P. For this purpose, for each edge uv in the anchor triangulation, the subset of uv within a distance $\delta_{\text{fen}} = \varepsilon L_{st}/\Theta(\rho kn)^{O(1)}$ from u or v plays a special role in building the anchor tree. Either this subset consists of two segments ux and vy or it is the edge uv. In the former case, we insert x and y as extra vertices into $\mathcal{T}'$ if they do not fall inside obstacles. Fig. 2(b) shows an example. The exact value of δ_{fen} will be specified in the proof of Theorem 1, our main result.

The subdivision $\mathcal{S}$ is the refinement of $\mathcal{T}'$ so that all regions become triangles. W.l.o.g., we assume that $\mathcal{S}$ is connected. It has $O(hn)$ vertices and $O(hn)$ edges. We construct H_ε as $G_\varepsilon(\mathcal{S})$, which has $O(\frac{h}{\varepsilon}n \log \frac{1}{\varepsilon})$ vertices and $O(\frac{h}{\varepsilon^2}n \log^2 \frac{1}{\varepsilon})$ edges.

5   Anchor Tree

We introduce an *anchor tree* $\mathcal{A}$ to connect the anchors. The crossings between $\mathcal{A}$ and P will be used to encode the homotopy of P. Let u_s be a vertex in $\mathcal{S}$ with the largest y-coordinate. The anchor tree $\mathcal{A}$ consists of two parts, a non-self-intersecting subtree in $\mathcal{S}$ that is rooted at u_s and spans all anchors, and a ray that shoots upward from u_s to infinity. So $\mathcal{A}$ is a rooted tree with the root at vertical infinity.

Let $a_1, a_2, \ldots, a_h$ be the anchors in $\mathcal{A}$. Let α_i denote the directed tree path in $\mathcal{A}$ from a_i to vertical infinity. Although the paths $\alpha_1, \alpha_2, \ldots$ may overlap, we view them as non-crossing and side by side. Fig. 3(a) shows an example. The *crossing sequence* of P is built by traversing P from s to t, appending a symbol $\overleftarrow{a_i}$ or $\overrightarrow{a_i}$ whenever P crosses α_i. We append $\overrightarrow{a_i}$ if α_i is crossed from left to right with respect to its direction. We append $\overleftarrow{a_i}$ otherwise. Fig. 3(b) shows an example. If $\overleftarrow{a_i}$ and $\overrightarrow{a_i}$ are adjacent in the crossing sequence, we can cancel them. It corresponds to a path deformation that does

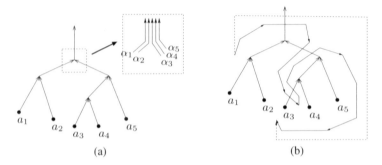

(a)   (b)

Fig. 3. In (b), the crossing sequence of the solid path is $\overrightarrow{a_1}\overrightarrow{a_2}\overrightarrow{a_3}\overrightarrow{a_4}\overrightarrow{a_5}\overleftarrow{a_5}\overleftarrow{a_4}\overleftarrow{a_3}\overrightarrow{a_3}\overleftarrow{a_3}\overrightarrow{a_3}\overrightarrow{a_4}\overrightarrow{a_5}$. It can be reduced to the crossing sequence $\overrightarrow{a_1}\overrightarrow{a_2}\overrightarrow{a_3}\overrightarrow{a_4}\overrightarrow{a_5}$ of the dashed path.

not pass over any obstacle. Repeating until no other symbol can be deleted gives the unique *canonical crossing sequence* as implied by Lemma 1 below. Cabello et al. [3] used vertical lines though obstacles to define the crossing sequence when the path cost is its length. The anchor tree generalizes this idea. The same idea of using a tree to encode homotopy was also used by Kaufmann and Mehlhorn [10].

Lemma 1. *Let H denote $\mathbb{R}^2$ minus the obstacles with anchors. Two paths in H with the same endpoints are homotopic if and only if their canonical crossing sequences are identical.*

We construct the subtree of $\mathcal{A}$ rooted at u_s as a shortest path tree in some subgraph of H_ε as follows. For edge uv of the anchor triangulation, its subset within a distance δ_{fen} from u or v consists of collinear edges in S. Due to obstacles, these collinear edges may form several connected components and we call each connected component a *fence*. Fig. 4 shows an example. To keep the canonical crossing sequence of P short, we should prevent any path in $\mathcal{A}$ from spiraling around the obstacles and hence anchors. We achieve this by making the interior of fences impenetrable. This is easily done by splitting some vertices of H_ε as follows. We split every vertex v of S in the interior of a fence into two copies, one on each side of the fence, and these two copies are not connected. Any edge incident to v is made incident to the copy of v on the same side of the fence as that edge. Notice that one can still pass through a fence at its endpoints. We use H_{fen} to denote the resulting graph. Note that each edge in H_{fen} coincides with an edge in H_ε. We compute the subtree of $\mathcal{A}$ rooted at u_s as the shortest path tree in H_{fen} from u_s to all anchors. The next result states several properties of $\mathcal{A}$.

Lemma 2. *$\mathcal{A}$ has $O(\frac{h}{\varepsilon}n\log\frac{1}{\varepsilon})$ size and can be computed in $O(\frac{h}{\varepsilon^2}n\log\frac{n}{\varepsilon}\log\frac{1}{\varepsilon})$ time. Let γ_i, $i \in [1, h]$, denote the paths in $\mathcal{A}$ between u_s and the anchors.*

(i) *$\text{cost}_{\mathcal{T}}(\gamma_i) = O(\rho^2 n L_{st})$.*
(ii) *The subpath of γ_i between any two nodes p and q has cost at most $d_{pq} + O(\rho h \delta_{\text{fen}})$, where d_{pq} is the shortest path cost in H_ε between p and q.*
(iii) *Let y be a crossing point between γ_i and an edge vw of the anchor triangulation. If $|vy| < \delta_{\text{fen}}$, then y lies on an obstacle.*

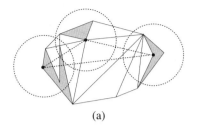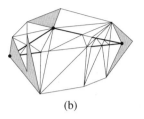

(a)   (b)

Fig. 4. The shaded regions are obstacles. We ignore the box B for simplicity. In (a), the black dots denote the anchors, the dashed segments form the anchor triangulation, and the dashed circles have radii δ_{fen}. In (b), the fences are shown as bold segments and the refined subdivision is $\mathcal{S}$. Notice that some fences consist of several edges of $\mathcal{S}$.

(iv) *Suppose that γ_i intersects an edge of the anchor triangulation at two points x and y. If xy does not intersect any obstacle, the subpath of γ_i between x and y has cost at most $\text{cost}_T(xy)$.*

A key property of the anchor tree is that it ensures that the crossing sequence of P has low dependence on n and ε.

Lemma 3. *The canonical crossing sequence S_P of P has length $O(\rho h^2 k \log \frac{\rho k n}{\varepsilon})$.*

Proof. (Sketch.) We break the k segments in P at their crossings with the vertical ray in $\mathcal{A}$. There are at most k such crossings, so P is partitioned into at most $2k$ subsegments such that each subsegment may cross the subtree of $\mathcal{A}$ rooted at u_s but not the vertical ray. Our strategy is to deform each subsegment and show an $O(\rho h \log \frac{\rho k n}{\varepsilon})$ bound on the number of crossings between the deformed subsegment and any path from u_s to an anchor in $\mathcal{A}$.

Take a segment ℓ in P and a path γ in $\mathcal{A}$ from u_s to an anchor. Let x and x' be two crossings between ℓ and γ that appear consecutively along γ. The subpath of γ between x and x' forms a simple cycle with xx'. If no obstacle lies inside this cycle, we deform ℓ by morphing xx' to a curve next to the subpath of γ between x and x' as shown in Fig. 5. This eliminates the crossing x, x', or both. The deformed ℓ is homotopic to ℓ because the deformation does not pass over any obstacle (no obstacle lies inside the cycle). The deformed ℓ has no new crossing with $\mathcal{A}$ because xx' is replaced by a curve next to a subpath in $\mathcal{A}$. Also, the deformed ℓ does not cross itself because the choices of x and x' ensure that γ does not cross ℓ between x and x'. We repeat until no more

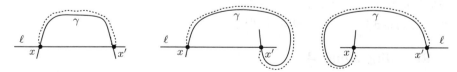

Fig. 5. Morph xx' to follow the dashed curve. This eliminates the crossings x and x' on the left, x in the middle, and x' on the right.

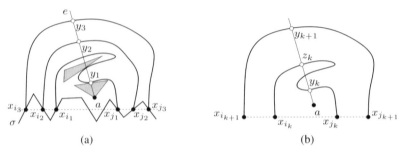

Fig. 6. (a) The shaded triangles denote the obstacles; the dashed line denotes ℓ; the polygonal curve denotes σ; the bold curves denote $\gamma(x_{i_1}, x_{j_1})$, $\gamma(x_{i_2}, x_{j_2})$ and $\gamma(x_{i_3}, x_{j_3})$. (b) z_k is the last crossing along e with $C_{i_k j_k}$ before y_{k+1}. As $z_k y_{k+1}$ avoids the obstacles, by Lemma 2(iv), $|y_k y_{k+1}| \geq |z_k y_{k+1}| \geq \mathrm{cost}_{\mathcal{T}}(\gamma(z_k, y_{k+1}))/\rho \geq |a y_k|/\rho$. So $|a y_{k+1}| \geq (1 + 1/\rho)|a y_k|$.

crossings with $\mathcal{A}$ can be eliminated. Let σ be the final deformed ℓ. By induction, we can show that σ is homotopic to ℓ, and σ does not cross itself.

We define $\gamma(p, q)$ to be the subcurve of γ between two points p and q on it. The subcurve $\sigma(p, q)$ is similarly defined. Let $x_1, x_2, \ldots$ denote the crossings between γ and σ. All these crossings lie on ℓ by our deformation. Consider the set of cycles $\{C_{ij} = \sigma(x_i, x_j) \cup \gamma(x_i, x_j) : x_i \text{ and } x_j \text{ are consecutive along } \gamma\}$. We order the subscripts of C_{ij} such that u_s is closer to x_i than x_j along γ. Each cycle is simple and it must enclose some anchors. We cluster the cycles that enclose the same anchors. The cycles in the same cluster are nested. Rotate the plane so that the subsegment ℓ is horizontal. We divide a cluster into a *left-group* and a *right-group*, depending on whether x_i lies to the left or right of x_j on ℓ. The two sets of anchors enclosed by two different cycles are either disjoint or one set is a subset of the other set. Therefore, there are at most $2h$ left- and right-groups. We show that a left-group has $O(\rho \log \frac{\rho k n}{\varepsilon})$ cycles as follows. The size of a right-group can be analyzed similarly.

There exists an edge e of the anchor triangulation that cuts through all cycles in the left-group and ends at some anchor a inside the innermost cycle. (The existence of e is ensured because we include the corners of the box B in the anchor triangulation.) Walk along e away from a. Identify the first crossing between e and each cycle in the left-group. Label these crossings as $y_1, y_2, \ldots, y_m$ at increasing distances from a. Label the cycles so that y_k lies on $C_{i_k j_k}$ for $k \in [1, m]$. It follows that $C_{i_k j_k}$ is nested in $C_{i_{k+1} j_{k+1}}$ for $k \in [1, m-1]$. See Figure 6(a) for an example. We can show that $|a y_2| \geq \delta_{\text{fen}}$ and $|a y_k| \geq (1 + 1/\rho)^{k-2}|a y_2|$ for $k \in [2, m]$ by Lemma 2 and the optimality of γ. Figure 6(b) illustrates the idea of the proof. The details are omitted. We have $(1 + 1/\rho)^{m-2}\delta_{\text{fen}} \leq (1 + 1/\rho)^{m-2}|a y_2| \leq |a y_m|$, which is at most $|\gamma(x_{i_m}, x_{j_m})| \leq \mathrm{cost}_{\mathcal{T}}(\gamma)$. Thus, $m = O\left(\frac{1}{\log(1+1/\rho)} \log \frac{\mathrm{cost}_{\mathcal{T}}(\gamma)}{\delta_{\text{fen}}}\right) = O\left(\rho \log \frac{\rho k n}{\varepsilon}\right)$ as $\mathrm{cost}_{\mathcal{T}}(\gamma) = O(\rho^2 n L_{st})$ and $\delta_{\text{fen}} = \varepsilon L_{st}/\Theta(\rho k n)^{O(1)}$. $\qquad\square$

6   Rerouting along $\mathcal{A}$

Our algorithm will run $|S_P| + 1$ rounds of shortest path computation starting from the source s in a subgraph of H_ε. In each round, $\mathcal{A}$ is treated as an obstacle. At the end of each round, we cross $\mathcal{A}$ in a way compatible with the remaining symbols in S_P. We reroute the optimal path along $\mathcal{A}$ in the analysis so that the structure of the rerouted optimum is similar to ours. Our path is as short as the rerouted optimum by construction. It is thus important to bound the rerouting error. In this section, we explain the rerouting for a path Q in H_ε with canonical crossing sequence S_Q.

Split Q into a concatenation of subpaths and edges $Q_1 \cdot u_1 v_1 \cdot Q_2 \cdot u_2 v_2 \cdots$ such that each subpath Q_i has no canonical crossing and each edge $u_i v_i$ crosses $\mathcal{A}$ at one or more canonical crossings in S_Q. We describe successive conversions from Q_i below: $Q_i \to Q_i^1 \to Q_i^2 \to Q_i^3$, such that the homotopy is preserved. All crossings between Q_i and $\mathcal{A}$ are cancellable. Canceling two adjacent symbols can be implemented by rerouting Q_i along $\mathcal{A}$. After doing all the cancellations, we get a path Q_i^1 that does not cross $\mathcal{A}$. This step is illustrated by the conversion from Fig. 7(a) to Fig. 7(b). For each path γ in $\mathcal{A}$ from u_s to some anchor, we shortcut Q_i^1 along the right side of γ between the first and last contact points of Q_i^1 on the right side of γ, and shortcut analogously along the left side of γ. The resulting path is Q_i^2. This step is illustrated by the conversion from Fig. 7(b) to Fig. 7(c).

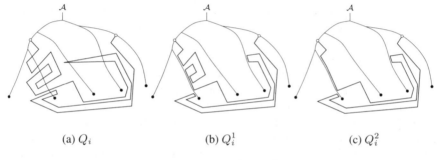

(a) Q_i (b) Q_i^1 (c) Q_i^2

Fig. 7. $Q_i \to Q_i^1 \to Q_i^2$

Finally, we convert Q_i^2 to a homotopic path Q_i^3 in H_ε as follows. Assume for now that γ_j is the only path in $\mathcal{A}$ that overlaps with Q_i^2. We snap Q_i^2 to some nodes in $Q_i^2 \cap \gamma_j$ and we use the example in Fig. 8 to illustrate this step. The white dots denote some vertices of H_ε. The path Q_i^2 starts to follow γ_j at the first contact x until Q_i^2 leaves γ_j at y. To obtain Q_i^3, we replace dx and xc by dc, and we replace uy and yv by uv.

For bounding the rerouting error, it is instructive to view the whole process as a direct conversion from Q_i to Q_i^3 by swapping subpaths between γ_j and Q_i. That is, delete ac and qu from γ_j, delete bd and pv from Q_i, and then insert ab, cd, pq, and uv. The converted Q_i is Q_i^3 and the subpath of γ_j between c and u is replaced by the subpath of Q_i between b and p. Let β_j denote the converted γ_j. Analogous to the fact that given a convex quadrilateral, the total length of its diagonals is at least the total length of any two opposite sides, we can show that $\mathrm{cost}_S(\gamma_j) + \mathrm{cost}_S(Q_i) \geq \mathrm{cost}_S(\beta_j) + \mathrm{cost}_S(Q_i^3)$. By Lemma 2(ii), we have $\mathrm{cost}_S(\gamma_j) \leq \mathrm{cost}_T(\beta_j) + O(\rho h \delta_{\mathrm{fen}})$, which implies that $\mathrm{cost}_T(Q_i^3) \leq \mathrm{cost}_S(Q_i) + O(\rho h \delta_{\mathrm{fen}})$. So far, we have only considered the rerouting of Q_i^2 along one path in $\mathcal{A}$. Rerouting along all h paths gives $\mathrm{cost}_S(Q_i^3) \leq \mathrm{cost}_S(Q_i) + O(\rho h^2 \delta_{\mathrm{fen}})$.

Fig. 8. The dashed polyline denotes $Q_i \setminus Q_i^2$

Lemma 4. *Let Q be a path in H_ε with canonical crossing sequence S_Q. We can convert Q to a homotopic path Q^3 in H_ε such that:*

(i) *Q^3 is the concatenation $Q_1^3 \cdot u_1 v_1 \cdot Q_2^3 \cdot u_2 v_2 \cdots$ such that Q_i^3 does not cross $\mathcal{A}$ and $u_i v_i$ crosses $\mathcal{A}$ at one or more canonical crossings in S_Q.*

(ii) *$\text{cost}_S(Q^3) \leq \text{cost}_S(Q) + O(\rho h^2 \delta_{\text{fen}} |S_Q|).$*

7   Main Algorithm

First, we construct $\mathcal{A}$ using Lemma 2 and superimpose it on S. Since $\mathcal{A}$ bends only at vertices of H_ε on the edges of S, no new nodes are generated, so the overlay has size $O(\frac{h}{\varepsilon} n \log \frac{1}{\varepsilon})$ by Lemma 2 and can be constructed in linear time.

Next, we obtain a subgraph H_{alg} of H_ε by deleting any edge pq that intersects $\mathcal{A}$. We intersect $\mathcal{A}$ with P by brute force to find its canonical crossing sequence S_P in $O(\frac{h}{\varepsilon} kn \log \frac{1}{\varepsilon})$ time. We run $|S_P| + 1$ rounds of shortest path computation in H_{alg}. In the initialization, for each vertex p of H_{alg}, we set a vector $p[i] = \infty$ for $i \in [0, |S_P|]$. The entry $p[i]$ will store the shortest path cost in H_{alg} from s to p subject to the constraint that the canonical crossing sequence of the path consists of the first i symbols in S_P.

In the first round, we set $s[0] = 0$ and compute shortest paths in H_{alg} from s to all other vertices. The shortest path cost of a vertex p is stored at $p[0]$ during this round. Let σ_{pq} denote the canonical crossing sequence of the segment pq. At the end of the round, for any edge pq of H_ε such that σ_{pq} is a prefix of S_P, we update $q[|\sigma_{pq}|]$ to be $\min\{q[|\sigma_{pq}|], p[0] + \text{cost}_S(pq)\}$. In general, the jth round begins with selecting vertices v of H_{alg} such that $v[j-1] \neq \infty$ and run Dijkstra's algorithm in H_{alg} from these vertices as multiple sources. This is akin to the computation of a weighted Voronoi diagram. The shortest path cost of a vertex p is stored at $p[j-1]$ during this round. Similarly, at the end of the jth round, we find all edges pq of H_ε such that σ_{pq} matches S_P from the jth to the $(j + |\sigma_{pq}| - 1)$th symbols, and update $q[j + |\sigma_{pq}| - 1]$. That is, $q[j + |\sigma_{pq}| - 1] = \min\{q[j + |\sigma_{pq}| - 1], p[j-1] + \text{cost}_S(pq)\}$. The final shortest path cost is stored at $t[|S_P|]$.

At the end of each round, we have to find all eligible edges in H_ε to update the entries $q[i]$'s. Each edge pq in H_ε lies inside a region. It may cross $O(\frac{1}{\varepsilon} \log \frac{1}{\varepsilon})$ segments in $\mathcal{A}$ and crossing one such segment corresponds to gaining up to $O(h)$ symbols. It means that $|\sigma_{pq}| = O(\frac{h}{\varepsilon} \log \frac{1}{\varepsilon})$. It is time-consuming to check every edge in H_ε. Fortunately, we can do it more efficiently by preprocessing.

Lemma 5. *We can build a data structure in $O(|S_P| \frac{h}{\varepsilon^2} n \log^2 \frac{1}{\varepsilon})$ time so as to report the eligible edges in H_ε in time proportional to their number at the end of each round.*

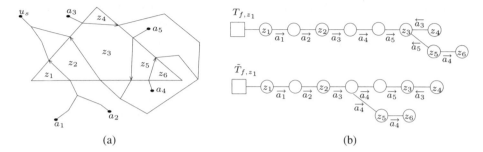

Fig. 9. (a) The division of a region into zones. (b) T_{f,z_1} and $\tilde{T}_{f,z_1}$.

Proof. Each region f of S is split by A into disjoint *zones*, each being a simple polygon. There are $O(\frac{1}{\varepsilon} \log \frac{1}{\varepsilon})$ zones in f because each zone contains some vertex of H_ε in f. We build a dual tree T_f to model the adjacency of the zones in f. Each node of T_f represents a zone and two zones are connected in T_f if they are adjacent. Building T_f takes $O(\frac{1}{\varepsilon} \log \frac{1}{\varepsilon})$ time. Fig. 9(a) shows the zones in a region f.

For each zone z in f, root T_f at z and attach z to a dummy parent. Then we expand each edge between a zone z' and its child zone z'' into $O(h)$ edges, each containing one symbol that is gained by going from zone z' to zone z''. Denote by $T_{f,z}$ the resulting rooted tree. It has $O(\frac{h}{\varepsilon} \log \frac{1}{\varepsilon})$ size. Fig.9(b) shows T_{f,z_1} for the example in Fig. 9(a). In $T_{f,z}$, we can read off the symbol sequence from any vertex p in zone z to any vertex q in another zone. But this sequence may not be canonical. We perform a BFS of $T_{f,z}$, while modifying $T_{f,z}$ on the fly. Suppose that we visit a node x from its parent x' and let ϕ be the symbol on the edge $x'x$. The path from z to x' gives a sequence of symbols $\phi_1, \phi_2, \cdots, \phi_{i-1}, \phi_i$. If ϕ does not cancel ϕ_i, we just continue with the BFS. If ϕ cancels ϕ_i, we detach x from x', make x a child of the grandparent x^* of x', and set ϕ_{i-1} to be the symbol on the edge x^*x. Then, we continue with the BFS. Basically, we are reducing the crossing sequences while generating them. Let $\tilde{T}_{f,z}$ denote the final rooted tree converted from $T_{f,z}$. $\tilde{T}_{f,z}$ is a prefix tree of canonical crossing sequences from z to all other zones in f. The bottom figure in Fig. 9(b) shows an example.

Then, we find in S_P the occurrences of all sequences in $\tilde{T}_{f,z}$ as follows. We construct a suffix tree T_S for S_P in $O(|S_P|)$ time [5]. Then, we traverse $\tilde{T}_{f,z}$ in a depth-first manner while navigating up and down T_S correspondingly. It takes $O(|\tilde{T}_{f,z}| + |S_P|)$ time to find for each sequence σ in $\tilde{T}_{f,z}$ the subtree of T_S that stores exactly the suffixes of S_P beginning with σ, which can then be traversed to output all occurrences of σ. There are $O(\frac{1}{\varepsilon} \log \frac{1}{\varepsilon})$ sequences in $\tilde{T}_{f,z}$ and each appears at most $|S_P|$ times in S_P. Therefore, the total time to find all the occurrences of the sequences in $\tilde{T}_{f,z}$ in S_P is $O(|\tilde{T}_{f,z}| + |S_P| + |S_P|\frac{1}{\varepsilon} \log \frac{1}{\varepsilon}) = O(|S_P|\frac{1}{\varepsilon} \log \frac{1}{\varepsilon})$. Repeating for all zones in all regions gives a running time of $O(|S_P|\frac{h}{\varepsilon^2}n \log^2 \frac{1}{\varepsilon})$. We use $|S_P|$ lists to store the results. The jth list contains all zone pairs (z, z') such that the canonical crossing sequence from z to z' matches S_P at the jth position. At the end of the jth round, for each zone pair (z, z') in the jth list, we report all edges pq of H_ε such that p is in z and q is in z'. ☐

Theorem 1. *Let P be a polygonal path of k segments in a weighted subdivision $\mathcal{T}$ with h obstacles and n vertices. For any $\varepsilon \in (0, 1)$, we can compute a $(1 + \varepsilon)$-approximate shortest path homotopic to P in $O(\frac{h^3}{\varepsilon^2} kn \operatorname{polylog}(k, n, \frac{1}{\varepsilon}))$ time, where the hidden constant depends on ρ and some geometric parameters.*

Proof. Let O be the shortest path in $\mathcal{T}$ homotopic to P. Using the analysis of Sun and Reif [13], the path O can be snapped to a $1 + \varepsilon$ homotopic approximation O' in H_ε. Then, O' can be converted to a path O'' that satisfies Lemma 4. Our algorithm returns a path cost at most $\operatorname{cost}_S(O'') \leq \operatorname{cost}_S(O') + O(\rho h^2 \delta_{\text{fen}} |S_P|) \leq (1 + \varepsilon) \operatorname{cost}_S(O) + O(\rho h^2 \delta_{\text{fen}} |S_P|)$. If we set $\delta_{\text{fen}} = \varepsilon L_{st}/(\rho h^2 |S_P|)$, the additive term becomes $O(\varepsilon L_{st}) = O(\varepsilon \operatorname{cost}_S(O))$. Hence, our path cost is $(1 + O(\varepsilon)) \operatorname{cost}_S(O)$. The factor $1 + O(\varepsilon)$ can be made $1 + \varepsilon$ by manipulating the constants. By Lemma 5, the preprocessing takes $O(|S_P| \frac{h}{\varepsilon^2} n \log^2 \frac{1}{\varepsilon})$ time. Consider the shortest path computation. Since H_{alg} has $O(\frac{h}{\varepsilon} n \log \frac{1}{\varepsilon})$ vertices and $O(\frac{h}{\varepsilon^2} n \log^2 \frac{1}{\varepsilon})$ edges, one round of Dijkstra takes $O(\frac{h}{\varepsilon^2} n \operatorname{polylog}(k, n, \frac{1}{\varepsilon}))$ time. We use the eligible edges pq of H_ε to update the entries $q[i]$'s at the end of each round, which takes $O(\frac{h}{\varepsilon^2} n \log^2 \frac{1}{\varepsilon})$ time. The total running time is $O(|S_P| \frac{h}{\varepsilon^2} n \operatorname{polylog}(k, n, \frac{1}{\varepsilon})) = O(\frac{h^3}{\varepsilon^2} kn \operatorname{polylog}(k, n, \frac{1}{\varepsilon}))$, where the hidden constant depends on ρ and some geometric parameters. $\square$

References

1. Aleksandrov, L., Maheshwari, A., Sack, J.-R.: Determining approximate shortest paths on weighted polyhedral surfaces. J. ACM 52, 25–53 (2005)
2. Bespamyatnikh, S.: Computing homotopic shortest paths in the plane. J. Alg. 49, 284–303 (2003)
3. Cabello, S., Liu, Y., Mantler, A., Snoeyink, J.: Testing Homotopy for Paths in the Plane. Discr. Comput. Geom. 31, 61–81 (2004)
4. Efrat, A., Kobourov, S.G., Lubiw, A.: Computing homotopic shortest paths efficiently. Comput. Geom. Theory and Appl. 35, 162–172 (2006)
5. Farach, M.: Optimal suffix tree construction with large alphabets. In: Proc. 38th Annu. Sympos. Found. Comput. Sci., pp. 137–143 (1997)
6. Forbus, K.D., Uhser, J., Chapman, V.: Qualitative spatial reasoning about sketch maps. AI Magazine 24, 61–72 (2004)
7. Fredman, M.L., Tarjan, R.E.: Fibonacci heaps and their uses in improved network optimization algorithms. J. ACM 34, 596–615 (1987)
8. Gao, S., Jerrum, M., Kaufmann, M., Kehlhorn, K., Rülling, W., Storb, C.: On continuous homotopic one layer routing. In: Proc. 4th Annu. Sympos. Comput. Geom., pp. 392–402 (1998)
9. Hershberger, J., Snoeyink, J.: Computing minimum length paths of a given homotopy class. Comput. Geom. Theory and Appl. 4, 63–98 (1994)
10. Kaufmann, M., Mehlhorn, K.: On local routing of two-terminal nets. J. Comb. Theory, Ser. B 55, 33–72 (1992)
11. Leiserson, C.E., Maley, F.M.: Algorithms for routing and testing routability of planar VLSI layouts. In: Proc. 17th Annu. Sympos. Theory of Comput., pp. 69–78 (1985)
12. Mitchell, J., Papadimitriou, C.: The weighted region problem: Finding shortest paths through a weighted planar subdivision. J. ACM 38, 18–73 (1991)
13. Sun, Z., Reif, J.: On finding approximate optimal paths in weighted regions. J. Alg. 58, 1–32 (2006)

Spanning Ratio and Maximum Detour of Rectilinear Paths in the L_1 Plane

Ansgar Grüne[1], Tien-Ching Lin[2], Teng-Kai Yu[2,3], Rolf Klein[1],
Elmar Langetepe[1], D.T. Lee[2,3], and Sheung-Hung Poon[4]

[1]Institut für Informatik I, Universität Bonn, Bonn, Germany
[2]Institute of Information Science, Academia Sinica, Nankang, Taipei, Taiwan
[3]Department of Computer Science and Information Engineering, National Taiwan University, Taipei, Taiwan
[4]Department of Computer Science, National Tsing Hua University, Hsinchu, Taiwan
ansgar.gruene@googlemail.com, kero@iis.sinica.edu.tw, tkyu@ntu.edu.tw
rolf.klein@uni-bonn.de, elmar.langetepe@informatik.uni-bonn.de
dtlee@iis.sinica.edu.tw, spoon@cs.nthu.edu.tw

Abstract. The spanning ratio and maximum detour of a graph G embedded in a metric space measure how well G approximates the minimum complete graph containing G and metric space, respectively. In this paper we show that computing the spanning ratio of a rectilinear path P in L_1 space has a lower bound of $\Omega(n \log n)$ in the algebraic computation tree model and describe a deterministic $O(n \log^2 n)$ time algorithm. On the other hand, we give a deterministic $O(n \log^2 n)$ time algorithm for computing the maximum detour of a rectilinear path P in L_1 space and obtain an $O(n)$ time algorithm when P is a monotone rectilinear path.

Keywords: rectilinear path, maximum detour, spanning ratio, dilation, L_1 metric, Manhattan plane.

1 Introduction

Given a connected graph $G = (V, E)$ embedded in a metric space M, the *detour* between any two distinct points p_i, p_j in $U = \bigcup_{e \in E} e$ is defined as

$$\delta_G(p_i, p_j) = \frac{d_G(p_i, p_j)}{\|p_i, p_j\|_M},$$

where $\|p_i, p_j\|_M$ denotes the distance between p_i and p_j in M and $d_G(p_i, p_j)$ is the shortest path between p_i and p_j on G. The *maximum detour* $\delta(G)$ of G is defined as the maximum detour over all pairs of distinct points in U, i.e.,

$$\delta(G) = \max_{p_i, p_j \in U, p_i \neq p_j} \delta_G(p_i, p_j).$$

If we restrict the points p_i, p_j to the vertex set of G, then the maximum detour is also called *spanning ratio*, *dilation* or *stretch factor* $\sigma(G)$ of G, i.e.,

$$\sigma(G) = \max_{p_i, p_j \in V, p_i \neq p_j} \delta_G(p_i, p_j).$$

O. Cheong, K.-Y. Chwa, and K. Park (Eds.): ISAAC 2010, Part II, LNCS 6507, pp. 121–131, 2010.

Given any connected graph embedded in any metric space, the spanning ratio can be computed in a straightforward manner by computing the all-pairs shortest paths of G. By using Dijkstra's algorithm [8] with Fibonacci heaps [9], we can find the spanning ratio in $O(n(m+n \log n))$ time and $O(n)$ space, where n and m are the numbers of vertices and edges, respectively.

Sometimes the geometric properties of special graph classes can be exploited to obtain a better upper bound [3,13,18]. If G is a connected graph embedded in the Euclidean space $\mathbb{R}^2$, it is easy to see that the maximum detour is infinite if G is non-planar. But if G is planar, we can compute the maximum detour by first computing shortest paths for all pairs of vertices in $O(n^2 \log n)$ time (since $|E| = O(n)$) and then using this information to find the maximum detour between each pair of edges. Wulff-Nilsen [19] recently gave an algorithm for computing the maximum detour of a planar graph in $\mathbb{R}^2$ in $O(n^{\frac{3}{2}} \log^3 n)$ expected time. The case of G being a planar polygonal chain is of particular interest. Agarwal et al. [1] gave an $O(n \log n)$ time randomized algorithm for computing the spanning ratio or maximum detour of a polygonal path in $\mathbb{R}^2$, and used it to obtain an $O(n \log^2 n)$ time randomized algorithm for computing the spanning ratio or maximum detour of cycles and trees in $\mathbb{R}^2$. They also claimed that it is possible to obtain a deterministic algorithm for computing the spanning ratio or maximum detour of a polygonal path in $O(n \log^c n)$ running time by parametric search, for some constant $c > 2$.

Ebbers-Baumann et al. [5] developed an ε-approximation algorithm that runs in $O(\frac{n}{\varepsilon} \log n)$ time for computing the maximum detour of a polygonal chain in $\mathbb{R}^2$. Narasimhan and Smid [15] studied the problem of approximating the spanning ratio of an arbitrary geometric connected graph in $\mathbb{R}^d$. They gave an $O(n \log n)$-time algorithm that computes a $(1 - \varepsilon)$-approximate value of the spanning ratio of a path, cycle, or tree in $\mathbb{R}^d$.

In this paper, we show that computing the spanning ratio of a rectilinear path P in L_1 space has a lower bound of $\Omega(n \log n)$ in the algebraic computation tree model and describe a deterministic $O(n \log^2 n)$ time algorithm. This is the first sub-quadratic deterministic algorithm for computing the spanning ratio of a polygonal path embedded in a metric space avoiding complicated parametric search methods. We also give a deterministic $O(n \log^2 n)$ time algorithm for computing the maximum detour of a rectilinear path P in L_1 space, and we obtain an optimal deterministic $O(n)$ time algorithm when P is a monotone rectilinear path.

2   Preliminaries and Problem Definition

In this section we present the preliminaries and give the formal problem definitions. In the L_1 *plane* (also called Manhattan plane), the distance of two points $p_i = (x_i, y_i)$ and $p_j = (x_j, y_j)$ is defined as $\|p_i, p_j\|_{L_1} = d_{L_1}(p_i, p_j) = |x_i - x_j| + |y_i - y_j|$. A *path* $P = (V, E)$ of $n \geq 2$ vertices is a connected undirected graph, in which every vertex has degree two, except the two end vertices of degree one. If all of the edges of a path are either horizontal or vertical, we call this path a *rectilinear path*. In this paper, we will focus on rectilinear paths in which a vertex is either an end vertex or a corner vertex. A corner vertex $v \in V$ is a common vertex of a horizontal edge and a vertical edge and has degree 2. In general, vertices may not necessarily exist only at corners or at ends.

But the existence of non-corner and non-end vertices will not affect the correctness and complexities of our algorithms. Thus the algorithms presented in this paper can solve the problem for general rectilinear paths as well. Fig.1 (a) shows an example, where we can find that apart from the two end vertices of the rectilinear path, the other vertices are placed at corners.

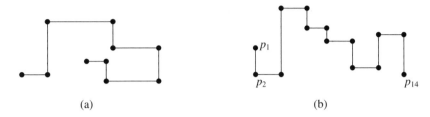

(a)   (b)

Fig. 1. (a) A rectilinear path with all vertices at corners. (b) A rectilinear path that is monotone with respect to the x-axis.

If a rectilinear path has non-decreasing x-coordinates from one of its end vertices to the other, we say that this path is *monotone* with respect to the x-axis. Monotone with respect to the y-axis can be defined similarly. Without loss of generality, we assume that monotone rectilinear paths in this paper are all monotone with respect to the x-axis. We refer to the vertices of an n-vertices monotone rectilinear path P from its left end to its right end as $p_1, p_2, ..., p_n$. Fig.1 (b) shows an example of a x-monotone rectilinear path.

Consider a connected graph $G = (V, E)$ in the L_1 plane. The distance (weight) of an edge $e \in E$ is defined as the L_1 distance of its two incident vertices, and the distance of any two points p_i and p_j on G (not necessarily in V) is defined as the length of the shortest path between them on G, denoted as $d_G(p_i, p_j)$.

In this paper we will compute the *spanning ratio* and *maximum detour* of a rectilinear path P in the L_1 plane. The rest of the paper is organized as follows. In Section 3, we show a lower bound for computing the spanning ratio of a rectilinear path P in the L_1 plane, even for the case when the path P is monotone. Section 4 gives a deterministic $O(n \log^2 n)$ time algorithm for computing the spanning ratio of P. Section 5 gives an $O(n \log^2 n)$ time algorithm for computing the maximum detour of P and an $O(n)$ time algorithm when the path is monotone. We conclude in Section 6.

3   The Lower Bound

In this section we show that computing the spanning ratio of a rectilinear path P in the L_1 plane has a lower bound $\Omega(n \log n)$ in the algebraic computation tree model. The proof follows an idea of Grüne et al's presentation [10] at EuroCG'03 that has not yet been submitted for publication.

The INTEGER ELEMENT DISTINCTNESS PROBLEM is to decide whether n integers $y_1, y_2, \ldots, y_n$ are all distinct. It is known that this problem has a lower bound of $\Omega(n \log n)$ in the algebraic computation tree model [20]. We will show that we can transform an instance $y_1, y_2, \ldots, y_n$ of INTEGER ELEMENT DISTINCTNESS PROBLEM into an instance

$P = (V, E)$ in $O(n)$ time. Let $y_{max} = \max\limits_{1 \le i \le n} y_i$ and $y_{min} = \min\limits_{1 \le i \le n} y_i$. If y_{min} is negative, we add $|y_{min}| + 1$ to every number to make all numbers positive. We set the vertex set $V = \{p_{4i-3} = (\frac{2i-2}{2n}, \hat{y} + i), p_{4i-2} = (\frac{2i-1}{2n}, \hat{y} + i), p_{4i-1} = (\frac{2i-1}{2n}, y_i), p_{4i} = (\frac{2i}{2n}, y_i) \mid i = 1, 2, \ldots, n\}$, where $\hat{y} = 3y_{max} + 2n + 1$ (the reason will be shown later), and the edge set $E = \{e_j = (p_j, p_{j+1}) \mid j = 1, 2, \ldots, 4n - 1\}$. Then $P = (p_1, p_2, \ldots, p_{4n})$ is a rectilinear path that is monotone with respect to the x-axis. We say a vertex p_i in P is a *low vertex* if its y-coordinate is smaller than $\hat{y}$, and a *high vertex* otherwise.

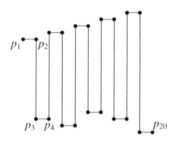

Fig. 2. Transforming an instance of the integer element distinctness problem into a rectilinear path.

Fig.2 is an example of transforming an instance $(3, 2, 4, 3, 1)$ of INTEGER ELEMENT DISTINCTNESS PROBLEM into a rectilinear path. By substituting $n = 5$, $i = 1$ and $y_1 = 3$ into the formula, we have $p_1 = (\frac{2i-2}{2n}, \hat{y} + i) = (0, \hat{y} + 1)$, and p_2, p_3 and p_4 are $(\frac{1}{10}, \hat{y} + 1)$, $(\frac{1}{10}, 3)$ and $(\frac{2}{10}, 3)$, respectively. It is easy to see that the y-coordinates of high vertices are nondecreasing from left to right, but the y-coordinates of low vertices vary according to the values of y_i's.

Lemma 1. *Let $p_i, p_{i+1}, p_j, p_{j+1}$ be four vertices in P, where p_i and p_{i+1} have the same y-coordinate, p_j and p_{j+1} have the same y-coordinate, and $i + 1 < j$. We have*

$$\delta_P(p_i, p_{j+1}) \le \delta_P(p_{i+1}, p_{j+1}) = \delta_P(p_i, p_j) \le \delta_P(p_{i+1}, p_j).$$

Proof. $\delta_P(p_i, p_j) = \frac{d_P(p_i, p_j)}{d_{L_1}(p_i, p_j)} = \frac{|\overline{p_i p_{i+1}}| + d_P(p_{i+1}, p_j)}{|\overline{p_i p_{i+1}}| + d_{L_1}(p_{i+1}, p_j)} \le \frac{d_P(p_{i+1}, p_j)}{d_{L_1}(p_{i+1}, p_j)} = \delta_P(p_{i+1}, p_j).$

$\delta_P(p_{i+1}, p_{j+1}) = \frac{|\overline{p_j p_{j+1}}| + d_P(p_{i+1}, p_j)}{|\overline{p_j p_{j+1}}| + d_{L_1}(p_{i+1}, p_j)} = \frac{|\overline{p_i p_{i+1}}| + d_P(p_{i+1}, p_j)}{|\overline{p_i p_{i+1}}| + d_{L_1}(p_{i+1}, p_j)} = \delta_P(p_i, p_j).$

$\delta_P(p_i, p_{j+1}) = \frac{d_P(p_i, p_{j+1})}{d_{L_1}(p_i, p_{j+1})} = \frac{|\overline{p_i p_{i+1}}| + d_P(p_{i+1}, p_{j+1})}{|\overline{p_i p_{i+1}}| + d_{L_1}(p_{i+1}, p_{j+1})} \le \frac{d_P(p_{i+1}, p_{j+1})}{d_{L_1}(p_{i+1}, p_{j+1})} = \delta_P(p_{i+1}, p_{j+1}).$ □

Lemma 1 shows that for any four vertices in such a situation only (p_{i+1}, p_j) can contribute to the spanning ratio. We call such a pair of vertices a *candidate pair*.

Lemma 2. *If a candidate pair (p_i, p_j) have one low vertex and one high vertex, then there exists another candidate pair of vertices, both are high vertices or low vertices, such that their detour is larger than $\delta_P(p_i, p_j)$.*

Proof. Without loss of generality, we assume that p_i is to the left of p_j, p_i is a high vertex, and p_j is a low vertex. Let vertex p_k be the next high vertex to the right of p_j. Since $\hat{y} = 3y_{max} + 2n + 1 > y_{max} + n + 1$, we have

$$\delta_P(p_i, p_j) = \frac{d_P(p_i,p_j)}{d_{L_1}(p_i,p_j)} \leq \frac{d_P(p_i,p_j)}{\hat{y}-y_{max}} < \frac{d_P(p_i,p_j)}{n+1} \leq \frac{d_P(p_i,p_j)}{d_{L_1}(p_i,p_k)} \leq \frac{d_P(p_i,p_k)}{d_{L_1}(p_i,p_k)} = \delta_P(p_i, p_k).$$

The case of p_i being a low vertex and p_j being a high vertex is similar.   □

Lemma 3. *If a candidate pair (p_i, p_j) are both high or both low vertices with different y-coordinates, then* $\delta_P(p_i, p_j) \leq \frac{4n}{3}(\frac{1}{2} + \hat{y} + n - y_{min})$.

Proof. Without loss of generality, we assume that p_i is to the left of p_j. Let the distance between p_i and p_j along the x-axis be $\frac{2m-1}{2n}$.

$$\delta_P(p_i, p_j) = \frac{d_P(p_i,p_j)}{L_1(p_i,p_j)} \leq \frac{\frac{2m-1}{2n}+2m(\hat{y}+n-y_{min})}{\frac{2m-1}{2n}+1} = \frac{2-\frac{1}{m}+2(\hat{y}+n-y_{min})}{2-\frac{1}{m}+\frac{1}{m}}$$

$$\leq \frac{\frac{1}{n}+2(\hat{y}+n-y_{min})}{\frac{3}{2n}} \leq \frac{4n}{3}(\frac{1}{2n} + \hat{y} + n - y_{min}) \leq \frac{4n}{3}(\frac{1}{2} + \hat{y} + n - y_{min})$$

Since $L_1(p_i, p_j) \geq \frac{2m-1}{2n}+1$, $d_P(p_i, p_j) \leq \frac{2m-1}{2n}+2m(\hat{y}+n-y_{min})$ and $\frac{2-\frac{1}{m}}{2n}+\frac{1}{m} \geq \frac{1}{2n}+\frac{1}{m} \geq \frac{1}{2n} + \frac{1}{n} = \frac{3}{2n}$, we have $\delta_P(p_i, p_j) \leq \frac{4n}{3}(\frac{1}{2n} + \hat{y} + n - y_{min}) \leq \frac{4n}{3}(\frac{1}{2} + \hat{y} + n - y_{min})$.   □

Lemma 4. *If a candidate pair (p_i, p_j) are both low vertices with the same y-coordinate, then* $\delta_P(p_i, p_j) \geq 2n(\hat{y} - y_{max})$.

Proof. Let the distance between p_i and p_j along x-axis be $\frac{2m-1}{2n}$. Then, $\delta_P(p_i, p_j) \geq \frac{2m(\hat{y}-y_{max})}{\frac{2m-1}{2n}} = \frac{2m(2n\hat{y}-2ny_{max})}{2m-1} \geq 2n(\hat{y} - y_{max})$.   □

Combining the above lemmas together, we now show that this problem has a lower bound of $\Omega(n \log n)$.

Theorem 1. *Computing the spanning ratio of a rectilinear path P in the L_1 plane has a lower bound of $\Omega(n \log n)$ in the algebraic computation tree model, even if the given rectilinear path is x-monotone.*

Proof. By Lemma 2, the spanning ratio must occur at a candidate pair of two high or two low vertices. Substituting $\hat{y} = 3y_{max} + 2n + 1$ into the formulas of Lemma 3 and Lemma 4, we have

$$2n(\hat{y} - y_{max}) = 2n(2y_{max} + 2n + 1) = 2n(\frac{2}{3}(\hat{y} - 2n - 1) + 2n + 1)$$
$$= 2n(\frac{2}{3}\hat{y} + \frac{2}{3}n + \frac{1}{3}) > 2n(\frac{2}{3}\hat{y} + \frac{2}{3}n + \frac{1}{3}) - \frac{4n}{3}(y_{min}) = \frac{4n}{3}(\frac{1}{2} + \hat{y} + n - y_{min}).$$

Therefore, if we choose $\hat{y} = 3y_{max} + 2n + 1$, then the spanning ratio $\delta(P) \geq 2n(2y_{max} + 2n + 1)$ if and only if there exists a candidate pair of two low vertices with the same y-coordinate. The existence of a candidate pair of two low vertices with the same y-coordinate is equivalent to the existence of two numbers y_i and y_j (with $i \neq j$) of the same value in the given instance of the INTEGER ELEMENT DISTINCTNESS PROBLEM.   □

4   Computing the Spanning Ratio of a Rectilinear Path

In this section we compute the spanning ratio of a rectilinear path P in the L_1 plane. We define that a vertex $p_i = (p_i.x, p_i.y)$ is *dominated* by another vertex $p_j = (p_j.x, p_j.y)$ if $p_i.x \leq p_j.x$ and $p_i.y \leq p_j.y$, denoted by $p_i \preceq p_j$. For a vertex p_i in V, let p_i^* be the vertex in V such that $\delta_P(p_i, p_i^*) = \max\{\delta_P(p_i, p_j) \mid p_j \in V\}$. We say that p_i^* is the *best partner* of p_i in V. Thus if we know the best partner of each vertex, then it is easy to compute the spanning ratio of P. It suffices to consider the detours from p_i to the vertices to the right of it, i.e., to find the maximum detour from p_i to the set $P_i = \{p_j \mid p_i.x \le p_j.x\}$. But the size of each P_i could be $O(n)$ and the time complexity might become $O(n^2)$ if we find the best partner of each vertex p_i in a brute force manner. In the following, we give an $O(n \log^2 n)$ time and $O(n)$ space algorithm. We divide the set P_i into two subsets: $D_i^+ = \{p_j \mid p_i \le p_j\}$ and $D_i^- = P_i \setminus D_i^+$. We denote the best partner of p_i in D_i^+ as p_i^+ and in D_i^- as p_i^-. We only focus on D_i^+ here; the case of D_i^- is similar. That is, we only need to find the p_i^+ for each p_i in D_i^+. Without loss of generality, we assume that all vertices in P are in the first quadrant.

To solve this problem, we transform the vertices of V from the L_1 plane to the L_2 plane as follows. We transform each vertex p_j in V into a point $q_j = (q_j.x, q_j.y) = (d_{L_1}(o, p_j), d_P(p_1, p_j))$ in $\mathbb{R}^2$ in a one-to-one manner, where o is the origin. For convenience, we call the original L_1 plane the *primal plane* and the transformed space the *dual plane*. In other words, in the dual plane q_j has as its x-coordinate the L_1 distance between the origin o and p_j and as its y-coordinate the path length from p_1 to p_j. The point set $Q_i^+ = \{q_j \mid p_j \in D_i^+\}$ in the dual plane corresponds to the point set D_i^+ in the primal plane. Therefore, we have $\delta_P(p_i, p_i^+) = \max_{q_j \in Q_i^+} |m(i, j)|$, where $m(i, j) = \frac{q_j.y - q_i.y}{q_j.x - q_i.x}$.

Thus the spanning ratio $\delta_P(p_i, p_i^+)$ occurs at either maximum $m(i, j)$ or minimum $m(i, j)$ among all q_j in Q_i^+. This problem now is equivalent to finding the two tangent lines from q_i to the convex hull of Q_i^+. Fig.3 shows an example. In Fig.3(a), p_i has the dominating set $D_i^+ = \{p_a, p_b, p_c, p_d, p_e\}$. In Fig.3(b), we transform p_i and p_a, p_b, p_c, p_d, p_e into the dual plane. The maximum and minimum values of $m(i, j)$ can be found by the two tangent lines from q_i to the convex hull of $Q_i^+ = \{q_a, q_b, q_c, q_d, q_e\}$.

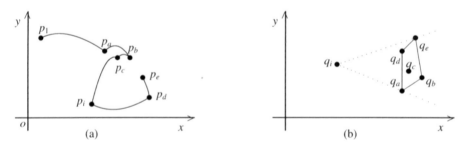

(a)   (b)

Fig. 3. (a) A vertex p_i and its $D_i^+ = \{p_a, p_b, p_c, p_d, p_e\}$. (b) Finding $\delta(p_i, p_i^+)$ in the dual plane by the two tangent lines from q_i to the convex hull of Q_i^+.

Based on this transformation, if we can find D_i^+ for each p_i, we can find p_i^+ for each p_i by making tangent queries from q_i to the convex hull of Q_i^+. We observe that the tangent query is decomposable. A query is called *decomposable* if the answer to the query over the entire set can be obtained by combining the answers to the queries to a suitable collection of subsets of the set. We will partition D_i^+ into $\log n$ subsets by the divide-and-conquer method and make the tangent queries from q_i to the convex hulls of the corresponding subsets in the dual plane and choose the one with maximum slope.

Our divide-and-conquer approach works as follows. Let p_m be the vertex in P such that $p_m.x$ is the median of the x-coordinates of all vertices in P. We divide the set P into two subsets: $P_L = \{p_i \mid p_i.x \leq p_m.x\}$ and $P_R = \{p_j \mid p_j.x > p_m.x\}$. We then sort the vertices of P_L and P_R in descending y-coordinates respectively. We iterate on each vertex in P_L in descending y-coordinate order such that we can find its best partner in P_R. Then we solve the subproblems P_L and P_R recursively. While iterating on each vertex in P_L in descending y-coordinate order, assume that after iterating on the vertex p_i in P_L we have maintained a subset $D_R^+ = \{p_k \mid p_k \in P_R, p_i.y \leq p_k.y\}$ in the primal plane and the convex hull of the corresponding subset $Q_R^+ = \{q_k \mid p_k \in D_R^+\}$ in the dual plane. For the next iterating vertex p_j in P_L, we first insert into D_R^+ those vertices in P_R whose y-coordinates are between $p_i.y$ and $p_j.y$ and their corresponding points in the dual plane into the convex hull of Q_R^+ respectively and then make a tangent query from q_j to the convex hull of Q_R^+.

Preparata [16] proposed an optimal algorithm for updating the convex hull in $O(\log n)$ time for the insertion only case. Hershberger and Suri [12] obtained an offline version of dynamic convex hull that can process a sequence of n insertion, deletion, and query instructions in total $O(n \log n)$ time and $O(n)$ space. If we implement our convex hull by either of the dynamic convex hull data structures, we can afford tangent query or insertion in $O(\log n)$ time. Therefore, the total time complexity of our algorithm is $T(n) = 2T(\frac{n}{2}) + O(n \log n) = O(n \log^2 n)$.

Theorem 2. *The spanning ratio of a rectilinear path in the L_1 plane can be found in $O(n \log^2 n)$ time and $O(n)$ space.*   □

5   Computing the Maximum Detour of a Rectilinear Path

In this section we compute the maximum detour of a rectilinear path $P = (V, E)$ in the L_1 plane. The maximum detour can occur on any two distinct points in $U = \bigcup_{e \in E} e$. In a previous work, Grüne et al. [10] presented an $O(n^2)$ algorithm for finding the maximum detour of a simple polygon P. There the detour between two points was defined as the ratio of the minimum length of all connecting paths contained in P, divided by the straight distance. They observed that linear time suffices for monotone rectilinear polygons in L_1. We also come to a linear time conclusion for monotone rectilinear paths; but for arbitrary rectilinear paths, we obtain an upper bound of only $O(n \log^2 n)$.

The following lemma is useful and will be used several times.

Lemma 5. *For any three points p, q, r in $U = \bigcup_{e \in E} e$ with $p \leq q$ and $q \leq r$, we have $\delta_P(p, r) \leq \max\{\delta_P(p, q), \delta_P(q, r)\}$.*

Proof. There are three cases to be considered, depending on which one of $\{p, q, r\}$ lies between the two others on P, see Fig.4: (a) $d_P(p, r) = d_P(p, q) + d_P(q, r)$; (b) $d_P(p, q) = d_P(p, r) + d_P(r, q)$; (c) $d_P(q, r) = d_P(q, p) + d_P(p, r)$.

For case (a), we have $\delta_P(p, r) = \frac{d_P(p,r)}{d_{L_1}(p,r)} = \frac{d_P(p,q)+d_P(q,r)}{d_{L_1}(p,q)+d_{L_1}(q,r)} \leq \max\{\frac{d_P(p,q)}{d_{L_1}(p,q)}, \frac{d_P(q,r)}{d_{L_1}(q,r)}\}$
$= \max\{\delta_P(p, q), \delta_P(q, r)\}$.

For case (b), we have $\delta_P(p,r) = \frac{d_P(p,r)}{d_{L_1}(p,r)} \leq \frac{d_P(p,r)}{d_{L_1}(p,q)} \leq \frac{d_P(p,q)}{d_{L_1}(p,q)} = \delta_P(p,q)$.

For case (c), we have $\delta_P(p,r) = \frac{d_P(p,r)}{d_{L_1}(p,r)} \leq \frac{d_P(p,r)}{d_{L_1}(q,r)} \leq \frac{d_P(q,r)}{d_{L_1}(q,r)} = \delta_P(q,r)$.   □

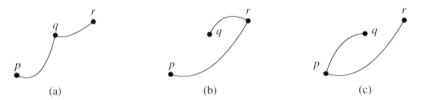

Fig. 4. (a) $d_P(p,r)=d_P(p,q)+d_P(q,r)$ (b) $d_P(p,q)=d_P(p,r)+d_P(r,q)$ (c) $d_P(q,r)=d_P(q,p)+d_P(p,r)$

First in Section 5.1, we give an $O(n)$ time and $O(n)$ space algorithm to compute the maximum detour when the rectilinear path is monotone. We then present an $O(n \log^2 n)$ time and $O(n)$ space algorithm for the general case in Section 5.2.

5.1   Monotone Rectilinear Paths

Let us assume that $d_P(p_1, p_i)$, for $i = 2, 3, ..., n$, has been computed in $O(n)$ time. For any two distinct points on P, if the *open* straight line segment connecting them has no intersection with P, we say these two points are *visible* from each other; they form a *visible pair*.

Lemma 6. *At least one of the pairs of points on P contributing to the maximum detour must be a visible pair, and these two points must have the same y-coordinate.*

Proof. By Lemma 5, if $p, q \in P$ and the open segment $\overline{pq}$ intersects P at r, then one of the two detours $\delta_P(p,r)$ and $\delta_P(r,q)$ must be no less than $\delta_P(p,q)$. Thus one of the pairs of points contributing to the maximum detour must be a visible pair.

For a visible pair of points $p, q \in P$, if $p.y \neq q.y$, then we will show that there exists a pair of points such that their detour larger than $\delta_P(p,q)$. Without loss of generality, we assume that the path on P between p and q is below the segment $\overline{pq}$ and $p \leq q$.

If there is a point r on the path between p and q such that $p \leq r$ and $r \leq q$, we have either $\delta_P(p,r) \geq \delta_P(p,q)$ or $\delta_P(r,q) \geq \delta_P(p,q)$ by Lemma 5. We then either replace point p by point r if $\delta_P(r,q) \geq \delta_P(p,q)$ or replace point q by point r if $\delta_P(p,r) \geq \delta_P(p,q)$. If we repeat the above procedure on the path between p and q until there is no point r on the path between p and q such that $p \leq r$ and $r \leq q$, we can then move point q downward to a point q' such that $q'.y = p.y$, and we have $\delta_P(p,q') \geq \delta_P(p,q)$.   □

Given the lemma above, which says that two points defining the maximum detour must be visible from each other and have the same y-coordinate, we shall call such a pair *horizontally visible*.

Lemma 7. *For any horizontally visible pair on P contributing to the maximum detour, at least one of these two points must be a vertex.*

Proof. We will show that for a horizontally visible pair $p, q \in P$, if both p and q are not a vertex, there exists a pair of points such that their detour larger than $\delta_P(p, q)$. Without loss of generality, we assume that p is to the left of q and the path on P between p and q is below $\overline{pq}$. If we move p and q upward simultaneously while keeping their L_1 distance the same, their detour $\delta_P(p, q)$ will increase as the path length from p to q on P increases. Thus we can keep moving p and q upward until one of them coincides with a vertex.   □

Thus we can restrict our search of the candidate pairs of points to horizontally visible pairs, with a vertex in each pair. Thus the number of candidate pairs is no more than the number of vertices. Fig.5 (a) shows an example of all the candidate pairs on the path P. We use a *ray-shooting* method to find all the candidate pairs. We will shoot rays from each vertex to a target point horizontally visible from the vertex. Thus we can divide the valid rays into four types, according to the four types of vertices from which we shoot the rays, i.e., top-right, bottom-right, top-left, and bottom-left corner vertices.

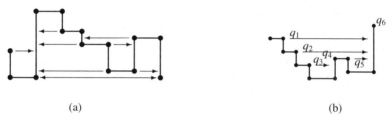

(a)   (b)

Fig. 5. (a) Find all candidate pairs by ray-shooting. (b) Shoot rays horizontally to the right from top-right vertices.

We only discuss the top-right corner case, as others are similar. Fig.5 (b) shows an example in which there are four rays shooting from four top-right corner vertices, q_1, q_2, q_3, and q_5. We use a stack S to help calculate the detours of this type of candidate pairs. We traverse path P from left to right. When we go downward and encounter a top-right vertex, we push the vertex into S. For the example in Fig.5 (b), we push q_1, q_2 and q_3 into S, respectively. When the path goes upwards and we encounter a vertex q_i, we pop the vertices lower than q_i from S and compute the detours associated with the horizontally visible pairs. For example in Fig.5 (b), when we encounter the vertex q_4, we pop q_3 and compute the detour $\delta_P(q_3, q)$ of the horizontally visible pairs (q_3, q), where q is the horizontal projection from q_3 on the vertical edge containing q_4. Since a vertex can be pushed into and popped from S only once, the total time complexity for finding the maximum detour in a monotone rectilinear path is $O(n)$, and the space complexity is obviously $O(n)$.

Theorem 3. *The maximum detour of a n-vertex monotone rectilinear path in the L_1 plane can be found in $O(n)$ time and $O(n)$ space.*   □

5.2   Non-monotone Rectilinear Path

Now we consider the case of a non-monotone rectilinear path P. The candidate pairs contributing to the maximum detour can be restricted to the following two cases. It can be proved similarly as in Lemmas 6 and 7.

Lemma 8. *Among the pairs of points contributing the maximum detour, there is one satisfies one of the following two properties: (1) it is a pair of visible vertices; (2) it is either a horizontally visible pair of points (with the same y-coordinate) or a vertically visible pair of points (with the same x-coordinate), and at least one of the two points must be a vertex.*   □

We can use the algorithm shown in Section 4 to deal with case (1), which takes $O(n \log^2 n)$ time. For case (2), we need to do both vertical and horizontal ray-shooting from V to P. The total number of rays is $O(n)$. We roughly describe the algorithm below. It can be done in $O(n \log n)$ time.

Consider shooting rays horizontally to the right from top-right and bottom-right corner vertices. We first sort the vertical edges by their x-coordinates, and then use a plane sweep method sweeping a vertical line from left to right. During the sweep, we maintain a binary search tree which consists of *active* corner vertices. An *active* corner vertex is one whose rightward ray has not yet been created. When scanning a new edge e, those vertices in the binary search tree whose y-coordinates lie between the y-coordinates of the two end vertices of e will shoot their rightward rays to e, creating horizontally visible pairs of points. We then delete those vertices from the binary search tree and insert the two end vertices of edge e, if they are top-right or bottom-right corner vertices, into the binary search tree. Obviously, this algorithm takes time $O(n \log n)$. The other types of rays, horizontally to the left, vertically upward and vertically downward, can be handled in a similar way. Thus we can find all horizontally and vertically visible pairs of points in $O(n \log n)$ time. Therefore, the theorem follows.

Theorem 4. *The maximum detour of a n-vertex rectilinear path in the L_1 plane can be found in $O(n \log^2 n)$ time and $O(n)$ space.*   □

6   Conclusion

We have shown that the problem of computing the spanning ratio of a rectilinear path P in the L_1 plane has a lower bound of $\Omega(n \log n)$ in the algebraic computation tree model and we have given a deterministic $O(n \log^2 n)$ time algorithm. We have also given a deterministic $O(n \log^2 n)$ time algorithm for computing the maximum detour of a rectilinear path P in the L_1 plane and have obtained an optimal $O(n)$ time algorithm for the monotone case.

There is still a gap between the lower bound $\Omega(n \log n)$ and upper $O(n \log^2 n)$ for the spanning ratio problem. How to bridge the gap will be of interest. As for the maximum detour problem for non-monotone rectilinear paths, we have not been able to make any use of the property that the maximum detour must be defined by a visible pair of points. Whether one can get a more efficient algorithm exploiting this or any other property is also of interest. Finally whether or not $\Omega(n \log n)$ is a lower bound for computing the maximum detour of a path remains open.

Acknowledgement

The authors would like to thank the anonymous referees for their careful reading of the paper and for their precise suggestions.

This work was supported in part by the National Science Council, Taiwan, under the Grants NSC 98-2221-E-001-007-MY3, NSC 98-2221-E-001-008-MY3, NSC 97-2221-E-007-054-MY3 and by a DFG-NSC Joint Research Project.

References

1. Agarwal, P.K., Klein, R., Knauer, C., Langerman, S., Morin, P., Sharir, M., Soss, M.: Computing the Detour and Spanning Ratio of Paths, Trees, and Cycles in 2D and 3D. Discrete Comput. Geom. 39(1-3), 17–37 (2008)
2. Agarwal, P.K., Klein, R., Knauer, C., Sharir, M.: Computing the Detour of Polygonal Curves, TRB 02-03, Freie Universität Berlin, Fachbereich Mathematik und Informatik (2002)
3. Aichholzer, O., Aurenhammer, F., Icking, C., Klein, R., Langetepe, E., Rote, G.: Generalized Self-approaching Curves. Discrete Appl. Math. 109, 3–24 (2001)
4. Alstrup, S., Holm, J.: Improved Algorithms for Finding Level Ancestors in Dynamic Trees. In: Welzl, E., Montanari, U., Rolim, J.D.P. (eds.) ICALP 2000. LNCS, vol. 1853, pp. 73–84. Springer, Heidelberg (2000)
5. Ebbers-Baumann, A., Klein, R., Langetepe, E., Lingas, A.: A Fast Algorithm for Approximating the Detour of a Polygonal Chain. Comput. Geom. Theory Appl. 27, 123–134 (2004)
6. Cormen, T.H., Leiserson, C.E., Rivest, R.L., Stein, C.: Introduction to Algorithms, 2nd edn. MIT Press and McGraw-Hill (2001)
7. de Berg, M., van Kreveld, M., Overmars, M., Schwarzkopf, O.: Computational Geometry, Second Revised Edition, pp. 105–110. Springer, Heidelberg (2000), Section 5.3: Range Trees
8. Dijkstra, E.W.: A note on two problems in connexion with graphs. Numerische Mathematik 1, 269–271 (1959)
9. Fredman, M.L., Tarjan, R.E.: Fibonacci heaps and their uses in improved network optimization algorithms. Journal of the ACM 34(3), 596–615 (1987)
10. Grüne, A.: Umwege in Polygonen. Diploma Thesis, Institute of Computer Science I, Bonn (2002)
11. Gudmundsson, J., Knauer, C.: Dilation and Detours in Geometric Networks. In: Gonzalez, T.F. (ed.) Handbook of Approximation Algorithm and Metaheuristics. Chapman & Hall/CRC (2007), Section 52
12. Hershberger, J., Suri, S.: Offline maintenance of planar configurations. J. Algorithms 21, 453–475 (1996)
13. Icking, C., Klein, R., Langetepe, E.: Self-approaching curves. Math. Proc. Camb. Philos. Soc. 125, 441–453 (1999)
14. Langerman, S., Morin, P., Soss, M.: Computing the Maximum Detour and Spanning Ratio of Planar Paths, Trees and Cycles. In: Alt, H., Ferreira, A. (eds.) STACS 2002. LNCS, vol. 2285, pp. 250–261. Springer, Heidelberg (2002)
15. Narasimhan, G., Smid, M.: Approximating the Stretch Factor of Euclidean Graphs. SIAM J. Comput. 30(3), 978–989 (2000)
16. Preparata, F.P.: An Optimal Real Time Algorithm for Planar Convex Hulls. Comm. ACM 22, 402–405 (1978)
17. Preparata, F.P., Shamos, M.I.: Computational Geometry: An Introduction. Springer, New York (1985)
18. Rote, G.: Curves with increasing chords. Math. Proc. Camb. Philos. Soc. 115, 1–12 (1994)
19. Wulff-Nilsen, C.: Computing the Maximum Detour of a Plane Graph in Subquadratic Time. In: Hong, S.-H., Nagamochi, H., Fukunaga, T. (eds.) ISAAC 2008. LNCS, vol. 5369, pp. 740–751. Springer, Heidelberg (2008)
20. Yao, A.C.-C.: Lower Bounds for Algebraic Computation Trees of Functions with Finite Domains. SIAM Journal on Computing 20(4), 655–668 (1991)

Approximation and Hardness Results for the Maximum Edge q-coloring Problem[*]

Anna Adamaszek[1] and Alexandru Popa[2]

[1] Centre for Discrete Mathematics and its Applications (DIMAP) and
Department of Computer Science, University of Warwick, UK
`A.M.Adamaszek@warwick.ac.uk`
[2] Department of Computer Science, University of Bristol, UK
`popa@cs.bris.ac.uk`

Abstract. We consider the problem of coloring edges of a graph subject to the following constraint: for every vertex v, all the edges incident to v have to be colored with at most q colors. The goal is to find a coloring satisfying the above constraint and using the maximum number of colors. This problem has been studied in the past from the combinatorial and algorithmic point of view. The optimal coloring is known for some special classes of graphs. There is also an approximation algorithm for general graphs, which in the case $q = 2$ gives a 2-approximation. However, the complexity of finding the optimal coloring was not known.

We prove that for any integer $q \geq 2$ the problem is NP-Hard and APX-Hard. We also present a 5/3-approximation algorithm for $q = 2$ for graphs with a perfect matching.

1 Introduction

We are given an integer q and a simple, undirected graph $G = (V, E)$. An assignment of colors to the edges of G is called an *edge q-coloring* if for every vertex $v \in V$ all the edges incident to v are colored with at most q different colors. Notice that the notion of coloring is different than in the classical edge coloring problem, as neighbouring edges can have the same color. An edge q-coloring that uses the maximum number of colors is called a *maximum edge q-coloring*. We consider the problem of finding a maximum edge q-coloring of a given graph.

Motivation. The edge q-coloring problem is related to some recent results in wireless mesh networks. Despite significant advances, today's wireless LAN cannot provide the same level of bandwidth as the wired ones. The bandwidth problem due to interference is crucial in single-channel wireless mesh networks.

However, wireless LAN standards allow multiple non-overlapping frequency channels to be used simultaneously to increase the bandwidth available to the users. In 2005, Raniwala and Chiueh [9] proposed a multi-channel wireless mesh

---

[*] Research supported in part by the Centre for Discrete Mathematics and its Applications (DIMAP), EPSRC award EP/D063191/1.

O. Cheong, K.-Y. Chwa, and K. Park (Eds.): ISAAC 2010, Part II, LNCS 6507, pp. 132–143, 2010.
© Springer-Verlag Berlin Heidelberg 2010

network architecture that equips each network node with multiple interface cards. Raniwala et al. show in [9,10] that even with just two interface cards on each node it is possible to improve the network throughput by a factor of 6 to 7 compared to the single-channel ad hoc network architecture.

We can analyze this setting from the theoretical point of view and ask how many channels can be used simultaneously by a given network. A network can be viewed as a graph where each computer is represented by a vertex. Therefore a channel assignment where each computer has q cards is equivalent to an edge q-coloring of a graph. The number of colors in a maximum edge q-coloring of a graph equals the number of channels that can be used simultaneously by a network.

Previous results. The number of colors used in a maximum edge q-coloring is called an *anti-Ramsey number* and has been extensively studied in the area of extremal graph theory. In general, for given graphs G and H the *anti-Ramsey number* $ar(G, H)$ is defined to be the maximum number k such that there exists an assignment of k colors to the edges of G in which every copy of H in G has at least two edges with the same color. A coloring of G is an edge q-coloring if and only if each subgraph $K_{1,q+1}$ of G (a star with $q + 1$ edges) has two edges with the same color. Therefore the number of colors in a maximum edge q-coloring of G equals $ar(G, K_{1,q+1})$.

The study of anti-Ramsey numbers started in 1975 with a paper by Erdős et al. [1] and led to many results (see [5] for a survey). The most studied case is $G = K_n$, but there are also results on computing or estimating the value of $ar(G, K_{1,q+1})$ for some classes of graphs G. In [6], Jiang shows that the number of colors in a maximum edge q-coloring of a clique K_n for $q \leq n - 2$ is either $\lfloor \frac{1}{2}n(q - 1) \rfloor + \lfloor \frac{n}{n-q+1} \rfloor$ or $\lfloor \frac{1}{2}n(q - 1) \rfloor + \lfloor \frac{n}{n-q+1} \rfloor + 1$, improving the previous estimation by Manoussakis et al. [7]. Also in [6], it is shown that the number of colors in a maximum edge q-coloring of a bipartite graph $K_{n,n}$ equals $n(q - 1) + \lfloor \frac{n}{n-q+1} \rfloor$. Montellano-Ballesteros [8] computes the values $ar(Q_n, K_{1,q+1})$ and $ar(C_m \times C_n, K_{1,q+1})$, where Q_n is a hypercube and $C_m \times C_n$ a product of cycles. The author presents also an upper bound on the value of $ar(G, K_{1,q+1})$ if the minimum degree of G is at least $q + 4$.

In this paper, our focus is on the algorithmic aspects of the problem. The problem of finding a maximum edge q-coloring of a given graph has been studied by Feng et al. [3,2,4]. They provide a 2-approximation algorithm for $q = 2$ and a $(1 + \frac{4q-2}{3q^2-5q+2})$-approximation for $q > 2$. They show that the problem is solvable in polynomial time for trees and complete graphs in the case $q = 2$, but the complexity for general graphs has been left as an open problem.

The algorithm presented by Feng et al. in [3,2,4] finds for a given graph G a maximum subgraph H with maximum degree $q - 1$. Every edge from H is colored using a unique color and each connected component of the rest of the graph is colored with a new color. To show the above approximation ratios, Feng et al. consider a *character subgraph* of G. For an edge colored graph G a character subgraph is a subgraph of G that contains exactly one edge from each color class. The number of edges in a character subgraph equals the number of colors in the optimal coloring. Comparing the number of edges in graph H with the number of edges in a character subgraph gives the approximation ratios.

Our contributions. We prove that the problem of finding a maximum edge q-coloring for a given graph is NP-Hard and APX-Hard for any integer $q \geq 2$. That solves the open problem by Feng et al. [3,2,4]. Moreover, we show that the problem of finding the number of colors in a maximum edge q-coloring is APX-Hard, and therefore computing the value of $ar(G, K_{1,q+1})$ is APX-Hard.

To prove the hardness result we first consider a more general version of the edge q-coloring problem, where each vertex has its own upper bound on the number of colors it can be incident to. We show that this problem is APX-Hard even if every vertex can be incident to either one or two colors. To prove this result we construct a reduction from the MAX-3SAT problem with an upper bound on the number of occurrences of each variable in the formula. Then we show the hardness for the maximum edge q-coloring problem.

Feng et al. [2,4] proposed Algorithm 1 to approximate a maximum edge 2-coloring. The outcome of the algorithm depends on the matching found in step 1, as in the example shown in Fig. 1. Feng et al. [2,4] show that Algorithm 1 is a 2-approximation for general graphs for the case $q = 2$.

Algorithm 1. input: graph $G = (V, E)$

1 : Find a maximum matching M in G.
2 : Assign a unique color to each edge from M.
3 : Find the connected components in the graph $(V, E \setminus M)$.
4 : Color the edges inside each connected component using a new color.

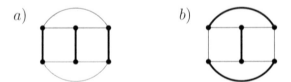

Fig. 1. Possible colorings output by Algorithm 1 for the same graph. Bold edges represent the matching edges and they are colored with unique colors. The matching found in a) yields an optimal coloring (using 5 colors). The matching found in b) yields a coloring with only 4 colors.

We analyze the performance of Algorithm 1 for graphs with a perfect matching and prove that it achieves approximation ratio $\frac{5}{3}$. This improves upon a 2-approximation due to Feng et al. [2,4].

To show the above result, we first consider the performance of the algorithm for a class of graphs which we call minimal graphs, consisting of a perfect matching and a tree. We show that for such graphs the optimal coloring uses less than

$\frac{5}{6}n + 1$ colors. As Algorithm 1 outputs a coloring with $\frac{n}{2} + 1$ colors, it gives a $\frac{5}{3}$-approximation for this class of graphs. Then we show that minimal graphs are the worst case for the algorithm and the approximation ratio for all graphs with a perfect matching is $\frac{5}{3}$.

The rest of the paper is organized as follows. In Section 2 we show that Algorithm 1 yields a $\frac{5}{3}$-approximation for graphs with a perfect matching. In Section 3 we prove the hardness results. Section 4 contains conclusions and open problems.

2 Approximation for Graphs with a Perfect Matching

In this section we focus on the case $q = 2$ and show that Algorithm 1 gives a $\frac{5}{3}$-approximation for graphs with a perfect matching.

We define $\mathrm{ALG}(G, M)$ to be the number of colors returned by Algorithm 1 for a graph G, if the matching found in the first step is M. $\mathrm{OPT}(G)$ denotes the number of colors used in an optimal solution for G.

2.1 Approximation Guarantee for Minimal Graphs

We first show the result for a smaller class of graphs with a perfect matching.

Definition 1. *A* minimal graph *is a simple graph* $G = (V, M \cup T)$, $M \cap T = \emptyset$, *consisting of a perfect matching* M *and a tree* T.

A *star* is a graph $K_{1,l}$ for some integer $l \geq 1$. A graph $G = (V, E)$ is called a 2-*star* if $|V| \geq 3$ and there are vertices $v, v', v'' \in V$ such that $E = \{vv', v'v''\} \cup \{vw : w \in V \setminus \{v, v', v''\}\}$. Therefore, a 2-star is a graph such that removing the edge $v'v''$ yields a star centered at v. Vertex v is called *center* of a 2-star, and v'' is called a *pending vertex*.

We want to show an upper bound on the number of colors in any edge 2-coloring of a minimal graph. To show it, we need the following result.

Lemma 1. *Given a minimal graph* $G = (V, M \cup T)$ *and an edge 2-coloring* c_G *of* G *we can construct a minimal graph* $G' = (V, M \cup T')$ *and an edge 2-coloring* $c_{G'}$ *of* G' *such that:*

1. *$c_{G'}$ uses the same number of colors as c_G.*
2. *For each color c used by the coloring $c_{G'}$ the set of tree edges colored with c is either a star or a 2-star. Moreover, in every 2-star there is a matching edge between the center and the pending vertex.*

Proof. We look at the graph G colored according to c_G. For a color c let $E_c = \{e \in T : c_G(e) = c\}$ (the set of tree edges of G colored with c) and denote by V_c the set of vertices incident to E_c. Assume that (V_c, E_c) does not satisfy condition 2 from the statement of the lemma. We fix an arbitrary edge $e = vv'$ in E_c and for any other edge $e' = ww'$ in E_c we perform the following operation. We can assume without loss of generality that in the tree T the vertex v is closer to w

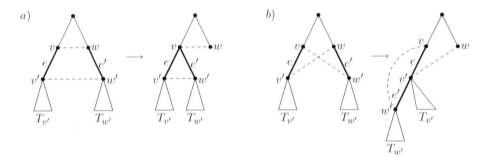

Fig. 2. Modifying the graph to make each color-component a star or a 2-star. Bold edges are sharing one color. Dashed edges represent matching edges.

than to w'. If there is no matching edge between v and w', we change the edge ww' to vw' (as in Fig. 2a). Otherwise we change the edge ww' to $v'w'$ (as in Fig. 2b).

~~Let $G_{e'}$ be the graph after~~ modifying the edge e' as described above. Observe that $G_{e'}$ is a minimal graph. The set of vertices and ~~the matching edges remain~~ the same. One of the tree edges has been modified, but the edges still form a tree. Considering the two cases above we make sure that no two neighbouring vertices in the tree are connected by a matching edge.

Coloring c_G yields a valid coloring for $G_{e'}$: all the edges are colored using the same colors as in c_G, including the modified edge. The only vertex that has an additional edge incident to it is either v or v'. The additional edge is colored with color c and the vertex is already incident to the edge e colored with c, so there are no new colors introduced for any vertex and the coloring remains feasible.

After performing these modifications for all the edges from the set $E_c \setminus \{e\}$ the modified graph is still a minimal graph and the coloring c_G yields a valid coloring for it. The set of edges colored with color c is either a star (if v was not connected by a matching edge to any of the vertices w') or a 2-star (otherwise, in this case there is a matching edge between v and the pending vertex).

Notice that this operation does not change the connected components of the tree T which are colored with colors different than c. After performing the above modifications for all the colors we get a graph G' and a coloring $c_{G'}$ that satisfy the lemma.   □

We are now ready to prove the upper bound on the number of colors used in an edge 2-coloring of a minimal graph.

Lemma 2. *Each edge 2-coloring of a minimal graph uses less than $\frac{5}{6}n+1$ colors.*

Proof. From Lemma 1 we know that it is enough to consider a minimal graph $G = (V, M \cup T)$ and a coloring c_G of G satisfying the properties: For each color c the set of edges $E_c = \{e \in T : c_G(e) = c\}$ is either a star or a 2-star. In each 2-star there is a matching edge connecting the center with the pending vertex.

Let l denote the number of leaves (vertices of degree one) in T, and m the number of monochromatic internal vertices in T (i.e. vertices which are not leaves, for which all incident tree edges are colored with the same color). The number of colors used by the edges of T is exactly $n-(l+m)+1$. The reason is that in each internal node which is not monochromatic one new color is introduced. The other color is already fixed by an edge connecting the vertex with its parent. As the root has no parent, it introduces one additional color.

Since M is a perfect matching it contains $\frac{n}{2}$ edges. We define a to be the number of edges from the matching which connect two leaves or two monochromatic vertices or a leaf and a monochromatic vertex. We also define b to be the number of edges from the matching which connect a leaf or a monochromatic vertex with another vertex. Since the matching is perfect, all the vertices are present in the matching and, therefore, $2a + b = l + m$.

We analyze how the edges from the matching can influence the coloring. If a matching edge is incident to an internal vertex which is not monochromatic, the matching edge has to be colored with one of the two colors already used by the vertex. The only edges that can introduce new colors are the a edges that connect leaves or monochromatic vertices. These edges can be colored with a new color each, therefore in an optimal coloring we have a new colors.

There are two other types of matching edges that are feasible (but do not introduce new colors):

- The b matching edges which connect a leaf or a monochromatic vertex with another vertex.
- Matching edges that connect two internal vertices sharing the same color. Let M_1 be the set of such edges. We have $|M_1| = \frac{n}{2} - a - b$.

For each color c the set (V_c, E_c) is a star or a 2-star. Edges from M_1 cannot connect neighbouring vertices in a tree, so they can only connect siblings (in a star or a 2-star) or the center of a 2-star with the pending vertex. Therefore all the matching edges from M_1 connect vertices that are at distance exactly two from each other. We want to show an upper bound for $|M_1|$.

Let $\deg(v)$ be the degree of v in T and $d(v)$ the number of matching edges from M_1 connecting two neighbours of v in T. Then $|M_1| = \sum_{v:\ \deg(v)\geq 2} d(v)$. We know that $d(v) \leq \lfloor \frac{\deg(v)}{2} \rfloor$. Moreover, if $\deg(v) = 2$ and $d(v) = 1$ then v is monochromatic. Therefore

$$\frac{n}{2} - a - b = |M_1| \leq \sum_{v:\ \deg(v)\geq 3} \lfloor \frac{\deg(v)}{2} \rfloor + m \leq \sum_{v:\ \deg(v)\geq 3} (\deg(v) - 2) + m < l + m$$

as $\sum_{v:\ \deg(v)\geq 2}(\deg(v) - 2) + 2 = l$.

We get the following bound on the number of colors used by c_G:

$$|c_G| = n - (l+m) + a + 1 < n + (a + b - \frac{n}{2}) + a + 1 = \frac{n}{2} + (l+m) + 1.$$

As $2a + b = l + m$, we get that $a \leq \frac{l+m}{2}$. That gives the second bound

$$|c_G| = n - (l+m) + a + 1 \leq n - \frac{l+m}{2} + 1.$$

The largest value of $|c_G|$ is possible when $l + m = \frac{n}{3}$ and then we get the bound $|c_G| < \frac{5}{6}n + 1$. $\qquad\qquad\square$

Theorem 1. *Let $G = (V, M \cup T)$ be a minimal graph. Then $OPT(G) < \frac{5}{3} \cdot ALG(G, M)$.*

Proof. Since the matching M has $\frac{n}{2}$ edges and after removing the matching the graph has one connected component, $ALG(G, M) = \frac{n}{2} + 1$. From Lemma 2 we know that $OPT(G) < \frac{5}{6}n + 1$. We get that $OPT(G) < \frac{5}{3} \cdot ALG(G, M)$. $\qquad\square$

The bounds in Lemma 2 and Theorem 1 are tight. For an arbitrary integer k there is a minimal graph $G = (M \cup T)$ with $6k$ vertices, which can be colored optimally using $5k$ colors, and the perfect matching M yields a coloring with only $3k + 1$ colors.

2.2   Approximation Guarantee for Graphs with a Perfect Matching

In the proof of the main theorem of this section we need the following lemma.

Lemma 3. *Let $\alpha > 1$ be a constant such that for every minimal graph $G = (V, M \cup T)$ we have $OPT(G) < \alpha \cdot ALG(G, M)$. Then for every graph $G' = (V', M' \cup F)$ consisting of a perfect matching M' and a forest F such that $M' \cap F = \emptyset$ we have $OPT(G') < \alpha \cdot ALG(G', M')$.*

Proof. Suppose that the graph G' consists of a perfect matching M' and $t > 1$ trees. If we add arbitrary $t - 1$ edges connecting the forest into a tree, we obtain a minimal graph G with the same set of vertices and the same set of matching edges M'. $OPT(G) \geq OPT(G') - (t - 1)$, since the optimal coloring of G' can be transformed into a coloring of G, coloring each added edge vw with an arbitrary color c used by v and merging c with a color used by w if necessary.

The number of colors output by the algorithm depends on the number of connected components in the graph after removing the matching edges. Therefore $ALG(G', M') = ALG(G, M') + (t - 1)$. By the assumption $OPT(G) < \alpha ALG(G, M')$. We get $OPT(G') < \alpha ALG(G, M') + (t - 1) < \alpha(ALG(G, M') + (t - 1)) = \alpha ALG(G', M')$. $\qquad\square$

Theorem 2. *Let $\alpha > 1$ be a constant such that for every minimal graph $G = (V, M \cup T)$ we have $OPT(G) < \alpha \cdot ALG(G, M)$. Then Algorithm 1 is an α-approximation for all graphs with a perfect matching.*

Proof. Proof by contradiction. Let $G = (V, E)$ be a graph with minimum number of edges such that Algorithm 1 does not give an α-approximation for it after choosing some $M \subseteq E$ as a perfect matching. Let $E' = E \setminus M$. We know that E' is not a forest (otherwise we are done by Lemma 3), so it contains some cycle $e_1, \ldots, e_m$. Let v be the vertex incident to both e_1 and e_2.

Fix some optimal coloring c of G. We can have one of the following cases:

- $c(e_1) = c(e_2)$.
- $c(e_1) \neq c(e_2)$. Let $e_v \in M$ be the matching edge incident to v. As v can be incident to edges in only two colors, we have $c(e_1) = c(e_v)$ or $c(e_2) = c(e_v)$. We can assume that $c(e_1) = c(e_v)$.

Consider $G' = (V, E \setminus \{e_1\})$. Each coloring of G induces a coloring of G'. Coloring c induces a coloring of G' with exactly the same number of colors, therefore $OPT(G') \geq OPT(G)$.

Removing the edge e_1 does not change the number of connected components in the graph $(V, E \setminus M)$, so $\text{ALG}(G', M) = \text{ALG}(G, M)$.

As $\frac{OPT(G')}{\text{ALG}(G', M)} \geq \frac{OPT(G)}{\text{ALG}(G, M)} > \alpha$, Algorithm 1 is not an α-approximation for G'. That is a contradiction with the minimality of G. We get that Algorithm 1 is an α-approximation for any graph G having a perfect matching.   □

From Theorem 1 and Theorem 2 we immediately get:

Theorem 3. *Algorithm 1 is a $\frac{5}{3}$-approximation algorithm for all graphs with a perfect matching.*

3   Hardness Results

In this section we show that the maximum edge q-coloring problem is APX-Hard for any integer $q \geq 2$. We first prove the APX-Hardness of the following problem:

Problem. Maximum edge $1, 2$-coloring
Instance. A graph $G = (V, E)$ and a function $c : V \to \{1, 2\}$
Task.   Find an edge coloring using a maximum number of colors such that for each vertex $v \in V$ the edges incident to v are colored with at most $c(v)$ distinct colors.

We consider the decision version of the problem: whether there is a feasible coloring using at least a given number of colors.

Theorem 4. *The maximum edge $1, 2$-coloring problem is APX-Hard, even for the class of graphs which admit a coloring using at least $\beta \cdot |V|$ colors for some constant $\beta > 0$.*

Proof. We present a reduction from the MAX-3SAT(13) problem, which is known to be APX-Hard (see e.g. [11]):

Problem. MAX-3SAT(13)
Instance. A CNF formula ϕ with at most three literals per clause, every variable occurs in at most 13 clauses.
Task.   Find an assignment that satisfies the maximum number of clauses.

In the decision version of the MAX-3SAT(13) problem we ask whether there exists an assignment satisfying at least a given number of clauses.

Given a formula ϕ with m clauses we construct the following graph (see Fig.3). The set of vertices and the number of colors that can be incident to them are:

- F — a special vertex with $c(F) = 1$,
- $c_1, \ldots, c_m$ — one vertex for each clause, $c(c_i) = 2$ for $i = 1, \ldots, m$,

- one vertex for each occurrence of a literal in the formula. If literal x appears a times and $\neg x$ appears b times, we have vertices $x_1, \ldots, x_a$ and $\neg x_1, \ldots, \neg x_b$. We set $c(x_i) = c(\neg x_j) = 1$,
- for each variable x for each $i \leq a$ and $j \leq b$ — a vertex x_{ij} with $c(x_{ij}) = 2$.

The set of edges is as follows:

- Fc_i for $i = 1, \ldots, m$,
- Fx_{ij} for all variables x and values i, j,
- edges connecting c_i with vertices representing the literals from the i-th clause,
- $x_i x_{ij}$ for all variables x and values i, j,
- $\neg x_j x_{ij}$ for all variables x and values i, j.

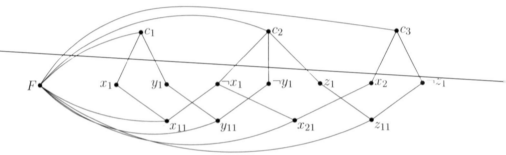

Fig. 3. A graph for the formula $\phi = (x \vee y) \wedge (\neg x \vee \neg y \vee z) \wedge (x \vee \neg z)$

We show that if there is an assignment satisfying exactly l clauses from the formula ϕ, there is a coloring of G using exactly $l+1$ colors. Given an assignment of the variables that satisfies l clauses, we use colors from the set $C = \{f, 1, \ldots, m\}$ and construct the following coloring:

- color all the edges incident to F with color f,
- for each clause color the edges connecting c_i with satisfied literals using color i and with unsatisfied literals using color f,
- color each edge $x_i x_{ij}$ with the same color as the edge connecting x_i with its clause,
- color each edge $\neg x_j x_{ij}$ with the same color as the edge connecting $\neg x_j$ with its clause.

The i-th color is used if and only if the i-th clause is satisfied by the given assignment. The color f is always used. Therefore the number of used colors equals $l + 1$. The coloring is feasible: F is incident to one color, c_i to at most two colors, x_i and $\neg x_j$ to one color, x_{ij} to at most two colors (either x_i or $\neg x_j$ is false, so two edges incident to vertex x_{ij} are colored with color f).

We prove now that if there is a coloring using $l + 1$ colors, then there is an assignment satisfying at least l clauses of the formula. Call f the only color incident to F. Edges $x_i x_{ij}$ and $\neg x_j x_{ij}$ must have the same colors as the edges connecting x_i and $\neg x_j$ to their clauses, so they do not introduce any new colors. Therefore the remaining l colors are introduced by the edges incident to vertices c_i, at most one new color for each c_i (since $c(c_i) = 2$ and c_i is already incident to f). We can assume without loss of generality that the vertices introducing new colors are $c_1, \ldots, c_l$. Let us call i the color incident to c_i for $i = 1, \ldots, l$. We can extract from the coloring an assignment satisfying clauses $c_1, \ldots, c_l$. We consider only the part of the graph representing the first l clauses. We want to satisfy each literal which is connected to its clause by an edge with color different than f (with color i if it is in the i-th clause). The assignment is feasible: we cannot set x and $\neg x$ as true (satisfied), as it would result in three colors adjacent to some x_{ij} where x_i and $\neg x_j$ represent the occurrence of these literals in clauses (we can assume that x_i and $\neg x_j$ do not appear in the same clause, as that clause would be always satisfied and could be removed). In the assignment we can set both y and $\neg y$ to false (unsatisfied) for some variable y, but we can repair it easily by setting the value of y arbitrarily. We do the same with variables that do not appear in any of the l chosen clauses. In such an assignment there are at least l satisfied clauses.

Graph G can be colored with $l + 1$ colors if and only if there is an assignment satisfying l clauses in ϕ. Therefore if we can approximate the maximum edge $1, 2$-coloring problem with an arbitrarily small constant factor $\alpha > 1$, we can do the same for the MAX-3SAT(13) problem. It shows that the maximum edge $1, 2$-coloring problem is APX-Hard.

The problem remains APX-Hard for the class of graphs which have a coloring with at least $\beta \cdot |V|$ colors for some constant $\beta > 0$. Notice that each CNF formula has an assignment that satisfies at least half of the clauses (it is either an assignment that sets all the variables to TRUE, or all to FALSE).

As each variable occurs in at most 13 places in the formula, the number of vertices in the graph satisfies $|V| < 50m$. As each formula has an assignment satisfying at least $\frac{m}{2}$ clauses, each graph created from a formula has a coloring using at least $\frac{m}{2} + 1 > \frac{|V|}{100}$ colors. Therefore the problem remains APX-Hard for a class of graphs which admit a coloring using at least $\beta \cdot |V|$ colors for $\beta = \frac{1}{100}$.
$\square$

Theorem 5. *For an arbitrary integer $q \geq 2$ the maximum edge q-coloring problem is APX-Hard.*

Proof. We prove the theorem via a reduction from the maximum edge $1, 2$-coloring problem for the class of graphs which have a coloring with at least $\beta \cdot |V|$ colors. From Theorem 4 we know that the problem is APX-Hard.

Let $G = (V, E)$ be an instance of the maximum edge $1, 2$-coloring problem with n vertices and set $k = |\{v \in V : c(v) = 1\}|$. We show that we can build an instance $G' = (V', E')$ of the maximum edge q-coloring problem such that there is a coloring of G' using $k + l + (q - 2)n$ colors if and only if there is a coloring of G using l colors. The respective colorings can be reconstructed from each other.

We create a graph G' from G by adding for each vertex $v \in V$ a set of $q - c(v)$ vertices $v_1, \ldots, v_{q-c(v)}$ and connecting each of them with v.

If there is a coloring of G using l colors, then there is a coloring of G' using $k + l + (q - 2)n$ colors. We obtain it by coloring each of the $k + (q - 2)n$ added edges with a new color. It is easy to check that such a coloring is feasible for G'.

If there is a coloring of G' using $k + l + (q - 2)n$ colors, there is a coloring of G using l colors. We can modify the coloring of G' without decreasing the number of colors in such a way that all the $k + (q - 2)n$ added edges are colored with unique colors. We do it by performing the following operation for each vertex $v \in V$:

- Let $c_1, \ldots, c_p$ ($p \leq q$) be the set of colors of the edges incident to v. We merge these colors into one color (if $c(v) = 1$ or $p < q$) or arbitrarily into two colors, in such a way that both of them are present in the subgraph induced by V (if $c(v) = 2$ and $p = q$). The number of colors decreases by at most $q - 1$ (for $c(v) = 1$) or $q - 2$ (for $c(v) = 2$).
- Recolor the added edges (vv_i) using unique colors. The number of colors increases by $q - 1$ (for $c(v) = 1$) or $q - 2$ (for $c(v) = 2$).

The number of colors used in the modified coloring is at least $k + l + (q - 2)n$. Deleting the added edges and vertices decreases the number of used colors by exactly $k + (q - 2)n$, and the coloring of remaining edges is feasible for G: we decrease the number of colors incident to a vertex v by $q - 2$ (if $c(v) = 2$) or $q - 1$ (if $c(v) = 1$), therefore there are at most two (one) colors incident to vertices with $c(v) = 2$ and $c(v) = 1$ respectively. If there are more than l colors used in the obtained coloring, we can decrease it to l by merging some colors.

The graph G has an edge $1, 2$-coloring using l colors if and only if the graph G' has an edge q-coloring using $k + l + (q - 2)n$ colors. From Theorem 4 we know it is NP-Hard to distinguish between the case when all colorings of G use at most l colors and the case when there exists a coloring using αl colors for some constant $\alpha > 1$ and for $l > \beta n$. Therefore it must be NP-Hard to distinguish between graphs G' which cannot be colored with more than $k + l + 1 + (q - 2)n$ colors and graphs admitting an edge q-coloring with at least $k + \alpha l + 1 + (q - 2)n$ colors. As $l > \beta n$ for some constant β, $k + l + 1 + (q - 2)n = O(l)$ and the maximum edge q-coloring problem is APX-Hard.   □

Corollary 1. *For any integer $q \geq 3$ the problem of finding the anti-Ramsey number $ar(G, K_{1,q})$ for a given graph G is APX-Hard.*

4   Conclusions

In this paper we show that the maximum edge q-coloring problem is APX-Hard for any integer $q \geq 2$. Moreover we show that the 2-approximation algorithm considered in [4] is a $\frac{5}{3}$-approximation for graphs with a perfect matching. A natural open problem is to find better approximation algorithms for this problem, or to prove that such algorithms do not exist.

Given the motivation of the problem via the wireless mesh networks, the following variant might be interesting to consider. Given a simple, undirected graph, color the edges of the graph in such a way that maximum number of times any color is used is minimized and the q-constraints are respected.

Acknowledgements. We would like to thank to Artur Czumaj and Nigel Smart for their useful comments. The second author is funded by an EPSRC PhD studentship.

References

1. Erdős, P., Simonovits, M., Sós, V.T.: Anti-ramsey theorems. In: Infinite and finite sets (Colloq., Keszthely, 1973; dedicated to P. Erdős on his 60th birthday), Vol. II, vol. 10, pp. 633–643. Colloq. Math. Soc. János Bolyai (1975)
2. Feng, W., Chen, P., Zhang, B.: Approximate maximum edge coloring within factor 2: a further analysis. In: ISORA, pp. 182–189 (2008)
3. Feng, W., Zhang, L., Qu, W., Wang, H.: Approximation algorithms for maximum edge coloring problem. In: Cai, J.-Y., Cooper, S.B., Zhu, H. (eds.) TAMC 2007. LNCS, vol. 4484, pp. 646–658. Springer, Heidelberg (2007)
4. Feng, W., Zhang, L., Wang, H.: Approximation algorithm for maximum edge coloring. Theor. Comput. Sci. 410(11), 1022–1029 (2009)
5. Fujita, S., Magnant, C., Ozeki, K.: Rainbow generalizations of ramsey theory: A survey. Graphs and Combinatorics (2010)
6. Jiang, T.: Edge-colorings with no large polychromatic stars. Graphs and Combinatorics 18(2), 303–308 (2002)
7. Manoussakis, Y., Spyratos, M., Tuza, Z., Voight, M.: Minimal colorings for properly colored subgraphs. Graphs and Combinatorics 12(1), 345–360 (1996)
8. Montellano-Ballesteros, J.J.: On totally multicolored stars. Journal of Graph Theory 51(3), 225–243 (2006)
9. Raniwala, A., Chiueh, T.-c.: Architecture and algorithms for an ieee 802.11-based multi-channel wireless mesh network. In: INFOCOM, pp. 2223–2234. IEEE, Los Alamitos (2005)
10. Raniwala, A., Gopalan, K., Chiueh, T.-c.: Centralized channel assignment and routing algorithms for multi-channel wireless mesh networks. Mobile Computing and Communications Review 8(2), 50–65 (2004)
11. Vazirani, V.V.: Approximation Algorithms. Springer, Heidelberg (2004)

3-Colouring AT-Free Graphs in Polynomial Time

Juraj Stacho

Wilfrid Laurier University, Department of Physics and Computer Science,
75 University Ave W, Waterloo, ON N2L 3C5, Canada
stacho@cs.toronto.edu

Abstract. Determining the complexity of the colouring problem on AT-free graphs is one of long-standing open problems in algorithmic graph theory. One of the reasons behind this is that AT-free graphs are not necessarily perfect unlike many popular subclasses of AT-free graphs such as interval graphs or co-comparability graphs. In this paper, we resolve the smallest open case of this problem, and present a polynomial time algorithm for 3-colouring of AT-free graphs.

1 Introduction

The colouring problem is one of the most studied problems on graphs. It is also one of the first problems known to be NP-hard [4]. In other words, it is unlikely that there is a polynomial time algorithm for solving this problem. This is true even in very special cases such as in planar graphs, line graphs, graphs of bounded degree or if the number of colours k is fixed and at least three. On the other hand, for $k = 2$ the problem is polynomially solvable, as is the general problem for many structured classes of graphs such as interval graphs, chordal graphs, comparability graphs, and more generally for perfect graphs [5]. In these cases, the special structure of the classes in question allows for polynomial algorithms.

We study the colouring problem in the class of *AT-free graphs*, i.e., graphs with no *asteroidal triple* (a triple of vertices such that between any two vertices of the triple there is a path disjoint from the closed neighbourhood of the third vertex). This class is a generalization of interval and co-comparability graphs as well as some non-perfect graphs such as the complements of triangle-free graphs. Unlike other standard optimization problems such as the independent set or the clique problem whose complexity on AT-free graphs is known (the former is solvable in polynomial time, while the latter is NP-hard [2]), the complexity of colouring is not known on AT-free graphs.

As a first step towards resolving this, we propose in this paper a polynomial time algorithm for the 3-colouring problem on AT-free graphs. In particular, we prove the following theorem.

Theorem 1. *There is an $O(n^4)$ time algorithm to decide, given an AT-free graph G, if G is 3-colourable and to construct a 3-colouring of G if it exists.*

O. Cheong, K.-Y. Chwa, and K. Park (Eds.): ISAAC 2010, Part II, LNCS 6507, pp. 144–155, 2010.

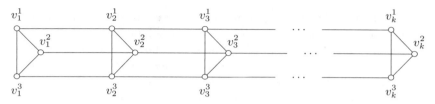

Fig. 1. The triangular strip of order k

We show this in three stages:

(1) we reduce the problem to AT-free graphs with no induced diamonds,
(2) we show how to decompose every AT-free graph with no induced diamond and no K_4 into triangular strips (see Figure 1) using stable cutsets, and
(3) we prove that we are allowed to contract minimal stable separators without changing the answer to the problem.

This reduces the problem to graphs whose blocks are triangular strips which are all clearly 3-colourable. If at any stage we encounter K_4, a clique on four vertices, we declare the graph not 3-colourable. A sketch of an algorithm resulting from this is presented below as Algorithm 1. (Note that $G/_S$ denotes the graph we obtain from G by contracting the set of vertices S into a single vertex.)

Input: An AT-free graph G.
Output: A 3-colouring of G or "G is not 3-colourable".

1   **if** G contains K_4 **then**
2   **return** "G is not 3-colourable"

    /* Now G contains no K_4 */

3   **if** G contains adjacent vertices u, v with $|N(u) \cap N(v)| \geq 2$ **then**
4   Recursively find a 3-colouring of $G/_{N(u) \cap N(v)}$.

    /* Now G contains no induced diamond and no K_4 */

5   **if** G contains a cutpoint or is disconnected **then**
6   Recursively colour all blocks of G.

    /* Now G is 2-connected and contains no induced diamond and no K_4 */

7   **if** G contains a minimal stable separator S with $|S| \geq 2$ **then**
8   Recursively find a 3-colouring of $G/_S$.

    /* Now G is a triangular strip */

9   Construct a 3-colouring of G.

Algorithm 1. Find a 3-colouring of an AT-free graph

Note that, in the above algorithm, once Line 5 is reached, the graph is guaranteed to be 3-colourable. This follows from the fact that AT-free graphs with no induced diamond and no K_4 are 3-colourable (we prove this as Theorem 3).

Hence, to obtain a decision algorithm, one can modify the procedure in Algorithm 1 to announce that "G is 3-colourable" once Line 5 is reached.

We remark that, for instance, in graphs with no induced path on five [11] or six [10] vertices the problem of 3-colouring is also known to be polynomially solvable even though to compute the chromatic number is NP-hard in both classes [8]. (In fact, in the former case, the k-colouring problem for every fixed k is polynomially solvable [7].) The main reason behind this is that in these cases we are able to reduce the problem of 3-colouring to an instance of 2-satisfiability which is solvable in polynomial time. Our approach for AT-free graphs differs from this in that it instead focuses on efficient decomposition of AT-free graphs to graphs for which 3-colouring can be decided in polynomial time.

In the following sections, we examine the main ingredients to the proof of correctness of our algorithm which are summarized in the following two theorems.

Theorem 2. *Let G be an AT-free graph with at least three vertices and no induced diamond or K_4. Then either*

(i) G is a triangular strip, or

(ii) G contains a stable cutset.

Theorem 3. *Every AT-free graph G with no induced diamond and no K_4 is 3-colourable. Moreover, if G contains a minimal stable separator S, then there is a 3-colouring of G in which all vertices of S have the same colour.*

In the final section, we explain implementation details needed to guarantee the running time $O(n^4)$.

2   Notation

In this paper, a graph is always simple, undirected, and loopless.

For a vertex v of a graph G, we denote by $N_G(v)$ the set of vertices adjacent to v in G, and write $N_G[v] = N_G(v) \cup \{v\}$. We drop the index G and write $N(v)$ and $N[v]$ whenever it is clear from context. For $X \subseteq V(G)$, we write $G[X]$ for the subgraph of G induced by X, and write $G - X$ for the subgraph of G induced by $V(G) \setminus X$. A set $X \subseteq V(G)$ is *stable*, if $G[X]$ contains no edges, and X is a *clique*, if $G[X]$ has all possible edges. As usual, K_n denotes the complete graph on n vertices, and *diamond* is the (unique) graph on four vertices with five edges.

We say that a path P of a graph G is *missed* by a vertex x if no vertex of P is adjacent to x. A triple of vertices x, y, z of a graph G is *asteroidal* if between any two vertices of the triple there exists a path missed by the third vertex.

We write $G/_S$ for the graph we obtain from G by *contracting* (i.e., identifying) all vertices in S into a single vertex. That is,

$$V(G/_S) = (V(G) \setminus S) \cup \{s\} \text{ where } s \notin V(G),$$
$$E(G/_S) = \left\{xy \in E(G) \mid x, y \notin S\right\} \cup \left\{sy \mid xy \in E(G) \wedge x \in S \wedge y \notin S\right\}.$$

A set $S \subseteq V(G)$ *disconnects* vertices a, b in G if a and b are in different connected components of $G - S$. We say that S is a *cutset* of G if it disconnects some vertices a, b. We say that S is a *minimal separator* of G if there exist vertices a and b such that S disconnects a and b, but no proper subset of S disconnects them. (Note that a minimal separator is not necessarily an inclusion-wise minimal cutset; for example, consider a 4-cycle with a pendant vertex.)

For a complete terminology, see [13].

3   Removing Diamonds

In this section, we explain how to reduce the problem to the case of AT-free graphs with no induced diamonds. We show that if we have a diamond in G, i.e., we have adjacent vertices u, v such that their common neighbourhood contains two non-adjacent vertices, then we can contract any maximal set S of pair-wise non-adjacent common neighbours of u, v and the resulting graph remains AT-free. It is also 3-colourable if and only if G is, since in any 3-colouring of G all vertices of S must have the same colour. Thus we show the following theorem.

Theorem 4. *If u, v are adjacent vertices of an AT-free graph G and S is a maximal stable set in $N(u) \cap N(v)$, then $G/_S$ is AT-free. Moverover, G is 3-colourable if and only if $G/_S$ is.*

To prove this, we use a more general tool that allows contracting specific sets in G without creating asteroidal triples. We say that a set $S \subseteq V(G)$ is *externally connected* in G, if for each $x \in V(G)$ with $N[x] \cap S = \emptyset$, the set S is contained in a (single) connected component of $G - N[x]$.

Lemma 1. *Let G be an AT-free graph and $S \subseteq V(G)$ be an externally connected set in G. Then $G/_S$ is AT-free.*

Proof. Let s denote the vertex of $G/_S$ to which we contracted the vertices of S, and suppose that $G/_S$ contains an asteroidal triple $\{x, y, z\}$. Let P be a path in $G/_S$ from y to z missed by x. If s is not on P, then P is also a path in G, and if $x = s$, then every vertex of S misses P in G. So, suppose that s belongs to P and is not one of the endpoints of P. Let u, v be the two neighbours of s on P. By the construction of $G/_S$, there exist vertices $a, b \in S$, such that $ua, vb \in E(G)$. Since $xs \notin E(G/_S)$, we have $N_G[x] \cap S = \emptyset$, and since S is externally connected in G, we conclude that a and b, and hence, u and v are in the same connected component of $G - N_G[x]$. Consequently, there is a path in G from y and z missed by x. Similarly, if s is one of the endpoints of P, say $y = s$, then we conclude that there exists a path in G missed by x between z and each vertex of S.

This proves that if s is not one of x, y, z, then x, y, z is an asteroidal triple of G, and otherwise, if say $x = s$, then a, y, z is an asteroidal triple of G for every $a \in S$, a contradiction.   □

From this lemma, we immediately obtain a proof of Theorem 4 as well as two other corollaries that we make use of later.

Lemma 2. *If G is AT-free and $G[S]$ is connected, then $G/_S$ is AT-free.*

Proof. By Lemma 1, it suffices to show that S is externally connected. This is obvious, since S induces a connected subgraph in $G - N[x]$ for $N[x] \cap S = \emptyset$. $\square$

Lemma 3. *If G is AT-free and S is a minimal separator, then $G/_S$ is AT-free.*

Proof. Again, we show that S is externally connected. Consider $x \in V(G)$ with $N[x] \cap S = \emptyset$, and let K denote the connected component of $G - S$ that contains x. Since S is a minimal separator, there exists a connected component K' of $G - S$ different from K such that each vertex of S has a neighbour in K'. Therefore, $G[K' \cup S]$ is connected, and so, S belongs to a connected component of $G - N[x]$, since clearly $N[x] \cap (K' \cup S) = \emptyset$. This proves that S is externally connected, and so the claim follows from Lemma 1. $\square$

Proof of Theorem 4. For the first part of the claim, it again suffices to prove that S is externally connected. Consider $x \in V(G)$ with $N[x] \cap S = \emptyset$. Therefore x is not adjacent to any vertex of S implying that $S \cup \{x\}$ is a stable set. By the maximality of S, x is non-adjacent to one of u, v. By symmetry, suppose that $xu \notin E(G)$. Then $S \cup \{u\}$ is in a connected component of $G - N[x]$ since $G[S \cup \{u\}]$ is connected. So, we conclude that S is indeed externally connected.

For the second part of the claim, let s be the vertex of $G/_S$ to which we contracted S. If we have a 3-colouring of $G/_S$, then we can extend this colouring of G by colouring all vertices of S with the colour of s. Conversely, if we have a 3-colouring of G, then u, v have different colours in this colouring, and hence, all vertices of S must have the same colour. So, we use this colour for s and colour all other vertices of $G/_S$ as in G. This clearly yields a 3-colouring of $G/_S$. $\square$

4   Structural Decomposition

In this section, we prove Theorem 2 asserting that every AT-free graph with no induced diamond and no K_4 decomposes into triangular strips via stable cutsets.

The *triangular strip* of order k is the graph formed by taking three disjoint paths $P^1 = v_1^1, v_2^1, \ldots, v_k^1$, $P^2 = v_1^2, v_2^2, \ldots, v_k^2$, $P^3 = v_1^3, v_2^3, \ldots, v_k^3$ and adding a triangle on v_i^1, v_i^2, v_i^3 for each $i = 1 \ldots k$. In other words, the triangular strip of order k is the cartesian product of an induced path on k vertices and a triangle. (See Figure 1 for an illustration.) We say that the triangles v_1^1, v_1^2, v_1^3 and v_k^1, v_k^2, v_k^3 of the triangular strip of order k are the *end-triangles*.

We say that G is a triangular strip if G is isomorphic to the triangular strip of order k for some k. Clearly, every triangular strip is AT-free and contains no induced diamond or K_4. Note that triangular strips have no stable cutsets; in other words, the two conditions of Theorem 2 are mutually exclusive.

Let G be an AT-free graph with $|V(G)| \geq 3$, no induced diamond, and no K_4. First, we observe that it suffices to prove Theorem 2 for 2-connected graphs G, since any cutpoint (and also the empty set) forms a stable cutset of G. Since G contains no diamond and no K_4, no two triangles of G share an edge. We show that, actually, no two triangles share a vertex provided G is 2-connected.

Lemma 4. *Let G be a 2-connected AT-free graph with no induced diamond and no K_4. Then every vertex of G is in at most one triangle.*

Proof. Let x be a vertex that belongs to two different triangles, namely, a triangle x, a, b and a triangle x, u, v. Clearly, $\{u, v\} \cap \{a, b\} = \emptyset$, since otherwise x, u, v, a, b induces a diamond or a K_4 in G. For the same reason, there is no edge between vertices u, v and a, b.

Since G is 2-connected, $G - x$ is connected, and hence, there is a path between vertices u, v and a, b in $G - x$. Let P be a shortest such path. Without loss of generality, P is a path from u to a. Let y be the second vertex on P (after u). Clearly, $yv \notin E(G)$ and $xy \notin E(G)$, since otherwise y, v, u, x induces a diamond or a K_4. Also, y is not adjacent to one of a, b, since otherwise y, a, b, x induces a diamond. In particular, $yb \notin E(G)$, since otherwise $ya \notin E(G)$ and u, y, b is a shorter path from u, v to a, b which contradicts the minimality of P.

We now show that $\{y, v, b\}$ is an asteroidal triple of G. Indeed, v, x, b is a path from v to b missed by y, and v, u, y is a path from v to y missed by b. Finally, $P' = P \setminus \{u\} \cup \{b\}$ is a path from y to b missed by v, since $vy \notin E(G)$ and v has no neighbour on $P \setminus \{u, y\}$ by the minimality of P.   □

By the above lemma, every vertex of G is in at most one triangle. If some vertex v is in no triangle, then $N(v)$ is a stable cutset of G unless $V(G) = N[v]$ in which case v is a cutpoint because G is assumed to have at least three vertices.

This implies that we may assume that every vertex of G is in exactly one triangle. In other words, G contains a triangular strip (of order 1). We show that by taking a maximal such strip, we either get the whole graph G or find a stable cutset in G, thus proving Theorem 2. To simplify the proof of this, we need the following technical lemma.

Lemma 5. *Let G be an AT-free graph with no induced diamond and no K_4, and let H be a (not necessarily induced) subgraph of G isomorphic to a triangular strip. Then (i) H is induced in G, and (ii) no vertex of H has a neighbour in $G - V(H)$ except for the vertices of the end-triangles of H.*

Proof. Let v_j^i for $i = 1, 2, 3$ and $j = 1 \ldots k$ for some k be the vertices of H. Suppose that H is not induced in G, and let $v_j^i v_{j'}^{i'}$ be an edge not in H such that $j < j'$ and $j' - j$ is smallest possible. By symmetry, we may assume that $i' = 1$, and $i \in \{1, 2\}$. Clearly, $j \neq j'$.

First, we observe that v_j^i is not adjacent to $v_{j'}^2$ and $v_{j'}^3$, since otherwise $v_j^i, v_{j'}^1, v_{j'}^2, v_{j'}^3$ induces a diamond or K_4 in G. This also implies $j' - j \geq 2$. By the choice of j, j' and the fact that $j' - j \geq 2$, we conclude that v_j^i is not adjacent to $v_{j+1}^3, v_{j+2}^3, \ldots, v_{j'}^3$, and v_{j+1}^3 is not adjacent to $v_{j'}^1$. By the same token, $v_{j'}^2$ is not adjacent to v_{j+1}^1 and v_{j+1}^3. We show that $\{v_j^i, v_{j+1}^3, v_{j'}^2\}$ is an asteroidal triple in G. Indeed, the path $v_j^i, v_{j'}^1, v_{j'}^2$ is missed by v_{j+1}^3, and the path $v_{j+1}^3, v_{j+2}^3, \ldots, v_{j'}^3, v_{j'}^2$ is missed by v_j^i. Finally, $v_{j'}^2$ is non-adjacent to at least one of v_j^1, v_j^3 otherwise $v_j^1, v_j^2, v_j^3, v_{j'}^2$ induces a diamond or K_4 in G. If $v_j^3 v_{j'}^2 \notin E(G)$, then the path v_j^i, v_j^3, v_{j+1}^3 is missed by $v_{j'}^2$. Otherwise, $v_j^1 v_{j'}^2 \notin E(G)$ in which case $v_j^i, v_j^1, v_{j+1}^1, v_{j+1}^3$ is a path (or walk) missed by $v_{j'}^2$. This proves (i).

For (ii), let $x \notin V(H)$ be a vertex adjacent to v_j^i for some $i \in \{1,2,3\}$ and $j \in \{2 \ldots k-1\}$. By symmetry, we may assume $i = 1$. Clearly, x is non-adjacent to both v_j^2 and v_j^3, otherwise x, v_j^1, v_j^2, v_j^3 induces a diamond or K_4 in G. First, suppose that x is also adjacent to v_{j+1}^1. Then x is non-adjacent to both v_{j+1}^2 and v_{j+1}^3, since otherwise $x, v_{j+1}^1, v_{j+1}^2, v_{j+1}^3$ induces a diamond or a K_4. But now $\{x, v_j^3, v_{j+1}^3\}$ is an asteroidal triple in G. Indeed, the path x, v_j^1, v_j^3 is missed by v_{j+1}^2, the path $v_j^3, v_{j+1}^3, v_{j+1}^2$ is missed by x, and the path v_{j+1}^2, v_{j+1}^1, x is missed by v_j^3. So, we may assume $xv_{j+1}^1 \notin E(G)$, and by symmetry, also $xv_{j-1}^1 \notin E(G)$.

Suppose that x is non-adjacent to both v_{j+1}^2 and v_{j-1}^3. Then $\{x, v_{j+1}^2, v_{j-1}^3\}$ is an asteroidal triple in G. Indeed, the path $x, v_j^1, v_j^2, v_{j+1}^2$ is missed by v_{j-1}^3, the path $v_{j+1}^2, v_j^2, v_j^3, v_{j-1}^3$ is missed by x, and the path $v_{j-1}^3, v_j^3, v_j^1, x$ is missed by v_{j+1}^2. So x has at least one neighbour among v_{j+1}^2, v_{j-1}^3. By the same token, x has at least one neighbour among v_{j+1}^3, v_{j-1}^2. Clearly, x cannot be adjacent to both v_{j+1}^2, v_{j+1}^3 or to both v_{j-1}^2, v_{j-1}^3, since we get an induced diamond in G on $x, v_{j+1}^1, v_{j+1}^2, v_{j+1}^3$, or on $x, v_{j-1}^1, v_{j-1}^2, v_{j-1}^3$. So, by symmetry, we may assume that x is adjacent to v_{j-1}^2 and v_{j+1}^2 and non-adjacent to v_{j-1}^3 and v_{j+1}^3. But then $\{x, v_{j-1}^3, v_{j+1}^3\}$ is an asteroidal triple in G. Indeed, the path x, v_{j+1}^2, v_{j+1}^3 is missed by v_{j-1}^3, the path $v_{j+1}^3, v_j^3, v_{j-1}^3$ is missed by x, and the path v_{j-1}^3, v_{j-1}^2, x is missed by v_{j+1}^3. That concludes the proof of (ii).   □

Now, we are finally ready to prove Theorem 2.

Proof of Theorem 2. As remarked in the discussion above, we may assume that G is 2-connected, and contains a triangle (triangular strip).

Let H be the largest triangular strip induced in G. If $V(H) = V(G)$, then G is a triangular strip, and we are done. Otherwise, there exists a vertex $v \in V(G) \setminus V(H)$ adjacent to a vertex of H. By Lemma 5, v is adjacent to a vertex c of an end-triangle of H; let a, b be the other two vertices of this triangle. Clearly, $va, vb \notin E(G)$ since otherwise v, a, b, c induces a diamond or K_4 in G.

First, we note that $N(b) \setminus \{a\}$ and $N(a) \setminus \{b\}$ are stable sets, since otherwise a or b is in two triangles which is not possible by Lemma 4. Also, the sets $N(a) \cap N(v)$ and $N(b) \cap N(v)$ are both stable sets of G, because otherwise we have an induced diamond in G. Moreover, we prove that there are no edges between the two sets. Suppose otherwise, and let $u \in N(a) \cap N(v)$ and $w \in N(b) \cap N(v)$ be adjacent. We observe that if $u \in V(H)$, then u belongs to a triangle in H and the triangle u, v, w. But these triangles are different since $v \notin V(H)$ contradicting Lemma 4. Hence, $u \notin V(H)$, and by the same token, $w \notin V(H)$. So, $G[V(H) \cup \{u, v, w\}]$ contains a spanning triangular strip which is, by Lemma 5, induced in G. This, however, contradicts the maximality of H.

Now, suppose that there are no edges between $N(b) \setminus \{a\}$ and $N(a) \cap N(v)$. In other words, $S = (N(b) \setminus \{a\}) \cup (N(a) \cap N(v))$ is a stable set. We show that S is a stable cutset of G separating a from v. Suppose otherwise, and let P be a shortest path in $G - S$ from a to v. Let z be the second vertex on P (after a). Since $N(a) \cap N(v) \subseteq S$, we conclude $zv \notin E(G)$. Also, $zc \notin E(G)$, since otherwise a, b, c, z induces a diamond or K_4 in G. By the same token, $zb \notin E(G)$. We show that $\{b, v, z\}$ is an asteroidal triple in G which will yield a contradiction. Indeed, the path v, c, b is missed by z, the path z, a, b is missed by v, and $P \setminus \{a\}$ is a path from z to v missed by b, since all neighbours of b except for a are in S.

Similarly, if there are no edges between $N(a) \setminus \{b\}$ and $N(b) \cap N(v)$, we conclude that G contains a stable cutset. So, we may assume that there exists $x \in N(a) \cap N(v)$ adjacent to some $y \in N(b) \setminus \{a\}$, and $x' \in N(b) \cap N(v)$ adjacent to some $y' \in N(a) \setminus \{b\}$. We show that this is impossible. Clearly, $y, y' \notin N(v)$, since there are no edges between $N(a) \cap N(v)$ and $N(b) \cap N(v)$. Also, y is not adjacent to any of a, c, x', since otherwise b is in two triangles which is impossible by Lemma 4. By the same token, y' is not adjacent to any of b, c, x. We show that G contains an asteroidal triple. Suppose that $yy' \notin E(G)$. Then $\{y, y', v\}$ is an asteroidal triple of G. Indeed, the path y, x, v is missed by y', the path y', x', v is missed by y, and the path y, b, a, y' is missed by v. So, $yy' \in E(G)$ in which case $\{x, c, x'\}$ is an asteroidal triple of G. Indeed, the path x, a, c is missed by x', the path c, b, x' is missed by x, and the path x, y, y', x' is missed by c.

That concludes the proof.   □

5   Proof of Theorem 3

The proof is by induction on $|V(G)|$. Let G be an AT-free graph with no induced diamond and no K_4. If G has at most 2 vertices, the claim is trivially satisfied.

Therefore, we may assume $|V(G)| \geq 3$. If G has a stable cutset, then by (possibly) removing some of its vertices, we can find a minimal stable separator in G. So, if G has no minimal stable separator, then it must be, by Theorem 2, a triangular strip with vertices v_j^i for $i = 1, 2, 3$ and $j = 1 \ldots k$ for some k. We obtain a 3-colouring of G by assigning each v_j^i the colour $((i + j) \bmod 3) + 1$.

So, we may assume that G contains a minimal stable separator S. If S is empty, then G is disconnected and we obtain a 3-colouring of G by independently 3-colouring its connected components by induction. If S has one element, then G has a cutpoint and we obtain a 3-colouring of G by 3-colouring its blocks by induction, and permuting the colours in blocks so that they match on cutpoints. In both cases, all vertices in S have the same colour. So, we may assume $|S| \geq 2$.

To prove the claim, it now suffices to show that for every connected component K of $G - S$, there exists a 3-colouring of $G[K \cup S]$ in which all vertices of S have the same colour.

Let K be a (fixed) connected component of $G - S$. Let S' denote the set of vertices of S with at least one neighbour in K. If $S' \neq S$, then S' is a minimal stable separator in $G' = G - (S \setminus S')$. By induction, there exists a 3-colouring of G' in which all vertices of S' have the same colour. By restricting this colouring to $K \cup S$ and colouring the vertices of $S \setminus S'$ with the common colour of the vertices of S', we obtain the required 3-colouring. (Note that the vertices of $S \setminus S'$ are isolated in $G[K \cup S]$.)

Hence, we may assume that every vertex of S has a neighbour in K. Further, since S is a minimal separator, there exists a connected component K' of $G - S$ different from K such that each vertex of S also has a neighbour in K'. Let G' denote the graph $G[K \cup K' \cup S]/_{K'}$, and let x be the vertex of G' to which we contracted K'. By Lemma 2, G' is AT-free. Moreover, G' contains no induced diamond or K_4, since any such subgraph is either in G, or contains x, but x belongs to no triangle of G'. Also, S is a minimal separator in G'. Hence, if G' has fewer vertices than G, then, by induction, there exists a 3-colouring of G' in which all vertices of S have the same colour. This colouring when restricted to the vertices $K \cup S$ yields the required 3-colouring.

It follows that we may assume that $G - S$ has exactly two connected components, one of which is K, the other consists of a single vertex x, and every vertex of S is adjacent to x and has a neighbour in K.

Now, let S^* be a smallest subset of S such that $\bigcup_{u \in S^*} N(u) = \bigcup_{u \in S} N(u)$. Suppose that S^* contains three distinct vertices u, v, w. By the minimality of S^*, there exist vertices u', v', w' such that $u' \in N(u) \setminus (N(v) \cup N(w))$, $v' \in N(v) \setminus (N(u) \cup N(w))$ and $w' \in N(w) \setminus (N(u) \cup N(v))$. Clearly, $u', v', w' \in K$ since S is a stable set and u, v, w are adjacent to x. Suppose that $u'v' \notin E(G)$. Then $\{u', v', x\}$ is an asteroidal triple of G. Indeed, the path u', u, x is missed by v', the path v', v, x is missed by u', and x misses any path in K between u' and v'. Hence, we must conclude $u'v' \in E(G)$, and by the same token, $u'w', v'w' \in E(G)$. However, then $\{u, v, w\}$ is an asteroidal triple in G. Indeed, the path u, u', v', v is missed by w, the path u, u', w', w is missed by v, and the path v, v', w', w is missed by u. We therefore conclude $|S^*| \leq 2$.

If $S^* \neq S$, then we consider the graph $G' = G - (S \setminus S^*)$. Clearly, S^* is a minimal separator in G', and therefore, there exists, by induction, a 3-colouring of G' in which all vertices of S^* have the same colour. We extend this colouring to G by colouring all vertices of $S \setminus S^*$ with the common colour of the vertices of S^*. By the definition of S^*, this yields the required 3-colouring.

Hence, we may assume that S consists of exactly two vertices u and v. We let $A = N(u) \setminus N(v)$, $B = (N(u) \cap N(v)) \setminus \{x\}$, and $C = N(v) \setminus N(u)$. By the minimality of S^*, we have $A \neq \emptyset$ and $C \neq \emptyset$. Moreover, each vertex $a \in A$ is adjacent to every vertex $c \in C$, since otherwise $\{a, c, x\}$ is an asteroidal triple of G. Indeed, the path a, u, x is missed by c, the path c, v, x is missed by a, and any path between a and c in K is missed by x. Furthermore, A is a stable set in G, since any adjacent $a, a' \in A$ yield an induced diamond u, a, a', c for any vertex $c \in C$. By the same token, C is a stable set. Finally, B is a stable set, since any adjacent $b, b' \in B$ yield an induced diamond b, b', u, v in G.

Suppose that there is $b \in B$ adjacent to some $a \in A$, and let $c \in C$. We show that $N(b) \setminus \{u\} \subseteq N(c)$. Suppose otherwise and let $w \in N(b) \setminus \{u\}$ be such that $wc \notin E(G)$. Clearly, $bc \notin E(G)$ since otherwise u, a, b, c induces a diamond in G. Also, $wu, wa \notin E(G)$ since otherwise w, a, b, u induces a diamond or K_4 in G. Finally, $wx \notin E(G)$, since w is not one of u, v and x is only adjacent to u, v. It follows that $\{w, x, c\}$ is an asteroidal triple in G. Indeed, the path w, b, a, c is missed by x, the path c, a, u, x is missed by w, and the path x, u, b, w is missed by c. This proves that $N(b) \setminus \{u\} \subseteq N(c)$. Now, by induction, there exists a 3-colouring of $G - b$ in which u and v have the same colour. We extend this colouring to G by assigning b the colour of c. Clearly, b and u have different colours in this colouring, since otherwise c, u, v have the same colour, impossible since cv is an edge in $G - b$. Also, b has colour different from its other neighbours, since $N(b) \setminus \{u\} \subseteq N(c)$. So, this gives the required 3-colouring.

It follows that we may assume that there are no edges between A and B. In other words, $A \cup B$ is a stable set. It is also a minimal separator of $G[K \cup S]$ separating u from v, since u, a, c, v and u, b, v are paths from u to v for each $a \in A$, $b \in B$, and $c \in C$. So, by induction, there is a 3-colouring of $G[K \cup S]$ in which all vertices of $A \cup B$ have the same colour. If $B \neq \emptyset$, then by recolouring u with the colour of v, we obtain the required 3-colouring. So, we conclude $B = \emptyset$.

Now, suppose that there is $a \in A$ with $N(a) \subseteq C \cup \{u\}$. If $|A| \geq 2$, then u, v is a minimal separator in $G - a$, and hence, there exists, by induction, a 3-colouring of $G - a$ in which u, v have the same colour. Recall that $N(a') \supseteq C \cup \{u\}$ for all $a' \in A$. So, by assigning a the colour of any vertex in $A \setminus \{a\}$, we obtain the required 3-colouring. Hence, we may assume $A = \{a\}$, and we observe that C is a minimal separator of $G - u$ separating a from v. By induction, there is a 3-colouring of $G - u$ in which all vertices of C have the same colour. To obtain the required 3-colouring, we colour u with the colour of v and recolour a with the colour different from the colour of u and the common colour of the vertices of C.

Hence, we may assume that there exists $w \in N(a) \setminus (C \cup \{u\})$ for some $a \in A$, and by symmetry, we also have $z \in N(c) \setminus (A \cup \{v\})$ for some $c \in C$. We show that G contains an asteroidal triple which will lead to a contradiction. Clearly, w and z are both different from and non-adjacent to all of u, v, x. If $wc \in E(G)$ or $wz \in E(G)$, then $\{w, u, v\}$ is an asteroidal triple. Indeed, the path w, a, u is missed by v, the path w, c, v or w, z, c, v is missed by u, and the path u, x, v is missed by w. So, $wc, wz \notin E(G)$, and by symmetry, $za \notin E(G)$. But now $\{z, w, x\}$ is an asteroidal triple. Indeed, the path w, a, c, z is missed by x, the path w, a, u, x is missed by z, and the path z, c, v, x is missed by w.

That concludes the proof. □

6   The Algorithm

In this section, we finally prove Theorem 1 by showing that Algorithm 1 is correct and can be implemented to run in time $O(n^4)$.

The correctness follows easily from Theorems 2, 3, 4 and Lemma 3. We therefore focus on the details of $O(n^4)$ implementation.

First, we note that the complexity is easily seen to be polynomial, since all the tests in Algorithm 1 are polynomial including the test in Line 7 which follows from [1]. Also, the algorithm makes at most n recursive calls, since each call reduces the graph by at least one vertex. So, to get the running time $O(n^4)$, it suffices to explain how to implement each test in time $O(n^3)$.

The test in Line 3 has a straightforward implementation of complexity $O(n^3)$. Similarly, the test in Line 5 can be clearly implemented in time $O(n^2)$ by the standard algorithm of [12]. Also, we can recognize and colour triangular strips in Line 9 in time $O(n^2)$ by iteratively removing triangles on vertices of degree 3.

For the test in Line 1, we do the following. If we execute Line 1 for the first time, we test if G contains a K_4 by trying all possible 4-sets of vertices in time $O(n^4)$. If we reach Line 1 by recursion to $G/_S$ and s is the vertex of $G/_S$ to which we contracted S, then we only test if the neighbourhood of s in $G/_S$ contains a triangle. This requires only $O(n^3)$ time, and it is enough to verify that $G/_S$ contains no K_4, since before contracting S, the graph G was assumed to contain no K_4 (because we have reached at least Line 3 before the recursive call).

Therefore, it remains to show that we can implement the test in Line 7 in time $O(n^3)$. This requires a little more work. As remarked earlier, if the neighbourhood of some vertex x is a stable set, then either $N(x)$ is a stable cutset of G, or x is a cutpoint of G, or $|V(G)| \leq 2$. It turns out that a partial converse of this is also true as shown in the following lemma.

Lemma 6. *If S is a minimal stable separator of an AT-free graph G, then there exists a vertex $x \in V(G)$ with $N(x) \supseteq S$.*

Proof. Let S be a counterexample to the claim and let S^* be a smallest subset of S for which there is no vertex x with $N(x) \supseteq S^*$. Clearly, $|S^*| \geq 2$.

First, suppose that $|S^*| = 2$. Hence, $S^* = \{x, y\}$ for some vertices x, y. Since S is a minimal separator, there are connected components K, K' of $G - S$ such that each vertex of S has a neighbour in both K and K'. In particular, we have $u \in N(x) \cap K$ and $v \in N(x) \cap K'$. Clearly, $uy, vy \notin E(G)$ by the minimality of S^*. This implies that $\{u, v, y\}$ is an asteroidal triple of G. Indeed, the path u, x, v is missed by y. Also, since S is a minimal separator, we have a path P in $G[K \cup \{y\}]$ from y to u, and a path P' in $G[K' \cup \{y\}]$ from y to v. Clearly, P is missed by v and P' is missed by u.

We therefore conclude $|S^*| \geq 3$, and let x, y, z be any three vertices of S^*. By the minimality of S^*, there exist vertices a, b, c such that $N(a) \supseteq S^* \setminus \{x\}$, $N(b) \supseteq S^* \setminus \{y\}$, and $N(c) \supseteq S^* \setminus \{z\}$, and also $ax, by, cz \notin E(G)$. Therefore, $\{x, y, z\}$ is an asteroidal triple of G. Indeed, the path x, c, y is missed by z, the path y, a, z is missed by x, and the path z, b, x is missed by y.

That concludes the proof. ☐

We further need the following observation which is easy to check.

Observation 7. *If S is a stable cutset of a connected graph G, and $S' \supseteq S$ is a stable set, then S' is also a stable cutset of G.* ☐

Now, if G contains a minimal stable separator S, then all we have to do, by Lemma 6, is to find a vertex x with $N(x) \supseteq S$. Since G is assumed to have no induced diamond and no K_4 in Line 7, $N(x)$ contains, by Lemma 4, at most two maximal stable sets one of which contains S. But then this set is also a stable cutset of G by Observation 7. So, to find a minimal stable separator in G, we test for each vertex x if $N(x)$ or $N(x) \setminus \{u\}$ or $N(x) \setminus \{v\}$ is a stable cutset where uv (if exists) is the unique edge in $G[N(x)]$. This can be accomplished in time

$O(n^2)$ by a standard graph search, so altogether $O(n^3)$ for all x. If a stable cutset S is found, we reduce it to a minimal stable separator by iteratively removing vertices of S and testing if the resulting set is a still a cutset. Again, $O(n^3)$ time, since it suffices to test each vertex only once.

7   Conclusion

In this paper, we have shown how to find in polynomial time a 3-colouring of a given AT-free graph if one exists. To this end, we used a nice structural decomposition of AT-free graphs without diamonds. Note that similar structural results are also known for other restrictions of AT-free graphs [3,6].

Finally, after submitting the paper for review, we learned that Haiko Müller et al. announced that for every fixed k, the k-colouring problem on AT-free graphs is solvable in polynomial time [9]. Their result is yet to be published.

Acknowledgements

The author would like to thank Derek Corneil and anonymous referees for useful suggestions, and also Kathie Cameron and Chính Hoàng for financial support.

References

1. Brandstädt, A., Dragan, F.F., Le, V.B., Szymczak, T.: On stable cutsets in graphs. Discrete Applied Mathematics 105, 39–50 (2000)
2. Broersma, H., Kloks, T., Kratsch, D., Müller, H.: Independent sets in asteroidal triple-free graphs. SIAM Journal on Discrete Mathematics 12, 276–287 (1999)
3. Corneil, D., Stacho, J.: The structure and recognition of C_4-free AT-free graphs (2010) (manuscript)
4. Garey, M.R., Johnson, D.S.: Computers and Intractability: A Guide to the Theory of NP-Completeness. W. H. Freeman, New York (1979)
5. Grötschel, M., Lovász, L., Schrijver, A.: The ellipsoid method and its consequences in combinatorial optimization. Combinatorica 1, 169–197 (1981)
6. Hemper, H., Kratsch, D.: On claw-free asteroidal triple-free graphs. Discrete Applied Mathematics 121, 155–180 (2002)
7. Hoàng, C.T., Kaminski, M., Lozin, V.V., Sawada, J., Shu, X.: Deciding k-colorability of P_5-free graphs in polynomial time. Algorithmica 57, 74–81 (2010)
8. Král, D., Kratochvíl, J., Tuza, Z., Woeginger, G.J.: Complexity of coloring graphs without forbidden induced subgraphs. In: Brandstädt, A., Le Van, B. (eds.) WG 2001. LNCS, vol. 2204, pp. 254–262. Springer, Heidelberg (2001)
9. Müller, H.: Personal communication
10. Randerath, B., Schiermeyer, I.: 3-colorability $\in \mathcal{P}$ for P_6-free graphs. Discrete Applied Mathematics 136, 299–313 (2004)
11. Sgall, J., Woeginger, G.J.: The complexity of coloring graphs without long induced paths. Acta Cybernetica 15, 107–117 (2001)
12. Tarjan, R.E.: Depth first search and linear graph algorithms. SIAM Journal on Computing 1, 146–160 (1972)
13. West, D.B.: Introduction to Graph Theory, 2nd edn. Prentice Hall, Englewood Cliffs (2000)

On Coloring Graphs without Induced Forests[*]

Hajo Broersma, Petr A. Golovach, Daniël Paulusma, and Jian Song

School of Engineering and Computing Sciences, Durham University,
Science Laboratories, South Road, Durham DH1 3LE, England
{hajo.broersma,petr.golovach,daniel.paulusma,jian.song}@durham.ac.uk

Abstract. The ℓ-COLORING problem is the problem to decide whether a graph can be colored with at most ℓ colors. Let P_k denote the path on k vertices and $G + H$ and $2H$ the disjoint union of two graphs G and H or two copies of H, respectively. We solve a known open problem by showing that 3-COLORING is polynomial-time solvable for the class of graphs with no induced $2P_3$. This implies that the complexity of 3-COLORING for graphs with no induced graph H is now classified for any fixed graph H on at most 6 vertices. The VERTEX COLORING problem is the problem to determine the chromatic number of a graph. We show that VERTEX COLORING is polynomial-time solvable for the class of triangle-free graphs with no induced $2P_3$ and for the class of triangle-free graphs with no induced $P_2 + P_4$. This solves two open problems of Dabrowski, Lozin, Raman and Ries and implies that the complexity of VERTEX COLORING for triangle-free graphs with no induced graph H is now classified for any fixed graph H on at most 6 vertices. Our proof technique for the case $H = 2P_3$ is based on a novel structural result on the existence of small dominating sets in $2P_3$-free graphs that admit a k-coloring for some fixed k.

1 Introduction

Graph coloring involves the labeling of the vertices of some given graph by integers called colors such that no two adjacent vertices receive the same color. The corresponding ℓ-COLORING problem is the problem to decide whether a graph can be colored with at most ℓ colors. The related VERTEX COLORING problem is the problem of determining the smallest number of colors a graph can be colored with. Due to the fact that ℓ-COLORING is NP-complete for any fixed $\ell \geq 3$, there has been considerable interest in studying its complexity when restricted to certain graph classes. Without doubt one of the most well-known results in this respect is that ℓ-COLORING is polynomially solvable for perfect graphs. More information on this classic result and on the general motivation, background and related work on coloring problems restricted to special graph classes can be found in several surveys [18, 19] on this topic.

We continue the study of the computational complexity of the ℓ-COLORING and VERTEX COLORING problem restricted to graphs in which a small forest F is forbidden as an induced subgraph (such graphs are called F-free). This

---

[*] This work has been supported by EPSRC (EP/G043434/1).

O. Cheong, K.-Y. Chwa, and K. Park (Eds.): ISAAC 2010, Part II, LNCS 6507, pp. 156–167, 2010.

problem has been studied in many papers by different groups of researchers [2–5, 9–11, 13–15, 17, 19, 20].

The focus on forests as forbidden induced subgraph can be justified by the following result. Kamiński and Lozin [10] showed that 3-COLORING is NP-complete for the class of graphs of girth (the length of a shortest induced cycle) at least p for any fixed $p \geq 3$. This immediately implies that 3-COLORING is NP-complete for the class of H-free graphs, if H contains a cycle.

The 3-COLORING problem is also NP-complete for the class of claw-free graphs (graphs with no induced 4-vertex star $K_{1,3}$), even for the subclass of claw-free graphs that are also diamond-free and 4-regular, as shown by Kochol, Lozin and Randerath [12]. This immediately implies that 3-COLORING is NP-complete for the class of F-free graphs, if F is a forest that contains a vertex with degree at least 3. In our previous paper [3] we showed that 3-COLORING is polynomial-time solvable for F-free graphs if F is a linear forest on at most 6 vertices, except for the case that $F = 2P_3$, i.e., F consists of two disjoint paths on 3 vertices. In Section 2, we settle this remaining case by proving that the 3-COLORING problem is also polynomial-time solvable for the class of $2P_3$-free graphs. Our approach to proving this case involves novel structural results on the existence of small dominating sets in graphs that admit a k-coloring. We explain these new results in Section 2 as well. Our new result together with all other above results lead to the following theorem.

Theorem 1. *Let H be a fixed graph on at most 6 vertices. Then 3-COLORING for H-free graphs is polynomial-time solvable if H is a linear forest; otherwise it is NP-complete.*

The complexity status of the 3-COLORING problem restricted to F-free graphs is open for many forests F on seven or more vertices, in particular for paths. It is even unknown whether there exists a fixed integer $k \geq 7$ such that 3-COLORING is NP-complete for P_k-free graphs. This indicates how difficult these complexity questions are, and puts the results of this paper in the right perspective. For larger values of ℓ, more is known on the complexity status of the ℓ-COLORING problem restricted to P_k-free graphs. The currently sharpest known results are that 4-COLORING is NP-complete for P_8-free graphs [3] and that 6-COLORING is NP-complete for P_7-free graphs [2]. It is unknown whether there exists an integer ℓ such that ℓ-COLORING is NP-complete for P_6-free graphs.

Hoàng et al. [9] showed that ℓ-COLORING for any fixed integer ℓ is polynomial-time solvable for P_5-free graphs. In contrast, Král' et al. [13] proved that VERTEX COLORING is NP-hard on P_5-free graphs. In fact, they give a complete complexity classification of the VERTEX COLORING problem restricted to graphs in which one fixed graph is forbidden as an induced subgraph. In particular, this problem is NP-hard for triangle-free graphs (also called K_3-free graphs). This motivated a study by Kamiński and Lozin [10] on the computational complexity of the VERTEX COLORING problem on triangle-free graphs with one extra forbidden subgraph H. They show that VERTEX COLORING is NP-hard for triangle-free H-free graphs for any fixed graph H that is not a forest. A very recent paper of Dabrowski et al. [5] deals with the computational complexity of this problem in triangle-free F-free graphs where F is a forest on at most 6 vertices. They conclude that the problem is NP-hard when $F = K_{1,5}$ and polynomial-time solvable in all other cases, except when $F = P_2 + P_4$ and $F = 2P_3$. They leave these two cases as open problems. They call the case $F = 2P_3$ a "challenging" open problem, because the class of triangle-free $2P_3$-free graphs has unbounded clique-width, whereas there is still some hope that the clique-width of $(P_2 + P_4)$-free triangle-free graphs is bounded. We solve the two mentioned open problems in Section 4 by presenting two polynomial-time algorithms that solve VERTEX COLORING for triangle-free $2P_3$-free graphs and for triangle-free $(P_2 + P_4)$-free graphs, respectively. All above results yield the following theorem.

Theorem 2. *Let H be a fixed graph on at most 6 vertices. Then* VERTEX COLORING *for triangle-free H-free graphs is polynomial-time solvable if H is a forest and $H \neq K_{1,5}$; otherwise it is* NP*-hard.*

We would like to emphasize that our algorithm for the case $F = 2P_3$ relies on our polynomial-time result on the 3-COLORING problem for $2P_3$-free graphs in Section 2 and another new polynomial-time result, namely on the 4-COLORING problem for triangle-free $2P_3$-free graphs in Section 3. The latter result is another application of our key results on dominating sets of Section 2. We think these results are interesting in their own right and might lead to future applications.

In order to present our results, we start by introducing some additional terminology and notations.

Notations and terminology. We only consider finite undirected graphs without loops and without multiple edges. Let $G = (V, E)$ be a graph. For $u \in V$ let $N_G(u) = \{v \mid uv \in E\}$ denote the *neighborhood* of u and let $d_G(u) = |N_G(u)|$ denote the *degree* of u. Let U be a subset of V. Then we define $N_G(U) = \{v \in V \setminus U \mid uv \in E \text{ for some } u \in U\}$. We write $G[U]$ to denote the subgraph of G *induced by* the vertices in U, i.e., the subgraph of G with vertex set U and an edge between two vertices $u, v \in U$ whenever $uv \in E$. Furthermore, U is called a *dominating set* of G if every vertex of G is in U or adjacent to a vertex of U, and U is called an *independent set* if there is no edge between any two vertices in U. If $G[U]$ is a *complete* graph, i.e., if there is an edge between any two vertices of U, then U is called a *clique*.

We use K_n and P_n to denote the complete graph and path on n vertices, respectively. The disjoint union of two graphs G and H is denoted $G + H$, and the disjoint union of k copies of G is denoted kG. A *linear forest* is the disjoint union of a collection of paths. Let $\{H_1, \ldots, H_p\}$ be a set of graphs. Then we say that a graph G is $(H_1, \ldots, H_p)$-*free* if G has no induced subgraph isomorphic to a graph in $\{H_1, \ldots, H_p\}$; if $p = 1$, we sometimes use H_1-free instead of (H_1)-free.

A *(vertex) coloring* of a graph $G = (V, E)$ is a mapping $c : V \to \{1, 2, \ldots\}$ such that $c(u) \neq c(v)$ whenever $uv \in E$. Here $c(u)$ is referred to as the *color* of u. An ℓ-*coloring* of G is a coloring c of G with $c(V) \subseteq \{1, \ldots, \ell\}$. Here we use the notation $c(U) = \{c(u) \mid u \in U\}$ for $U \subseteq V$. We let $\chi(G)$ denote the *chromatic number* of G, i.e., the smallest ℓ such that G has an ℓ-coloring. We say that a graph G is ℓ-*chromatic* if $\chi(G) = \ell$ and ℓ-*colorable* if $\chi(G) \leq \ell$. The problem ℓ-COLORING is the problem to decide whether a given graph admits an ℓ-coloring. The VERTEX COLORING problem is the problem of determining the chromatic number of a given graph.

A *list-assignment* of a graph $G = (V, E)$ is a function L that assigns a list $L(u)$ of so-called *admissible* colors to each $u \in V$. We say that a coloring $c : V \to \{1, 2, \ldots\}$ *respects* L if $c(u) \in L(u)$ for all $u \in V$. In this case we also call c a *list-coloring*.

In *pre-coloring extension* we assume that a (possibly empty) subset $W \subseteq V$ of G is pre-colored with $c_W : W \to \{1, 2, \ldots\}$ and the question is whether we can extend c_W to a coloring of G. If c_W is restricted to $\{1, 2, \ldots, \ell\}$ and we want to extend it to an ℓ-coloring of G, we say we deal with the *pre-coloring extension version of ℓ-COLORING*. This problem is relevant for us in the following sense. In our coloring algorithms we sometimes color the vertices of a subset $W \subseteq V$ in every possible way (in order to start some branching). In that case we say that we *pre-color W*, and then we must solve the pre-coloring extension version.

2   Key Ingredients for Coloring $2P_3$-Free Graphs

Our polynomial-time algorithms heavily rely on a number of structural properties of k-colorable $2P_3$-free graphs. We present these properties in this section, together with some other useful observations.

We start off with the following well-known observation, the proof of which follows from the fact that the decision problem in this case can be modeled and solved as an instance of the 2-SATISFIABILITY problem. This approach has been introduced by Edwards [6] and is folklore now.

Observation 1 ([6]). *Let G be a graph in which every vertex has a list of admissible colors of size at most 2. Then checking whether G has a coloring respecting these lists is solvable in polynomial time.*

Let $G = (V, E)$ be a $2P_3$-free graph. Let I be an independent set in G, and let X be a subset of $V \backslash I$. We write $I(X) := N_G(X) \cap I$ and $I(\overline{X}) := I \backslash N_G(X)$, so $I = I(X) \cup I(\overline{X})$ and $I(X) \cap I(\overline{X}) = \emptyset$. If every vertex in $N_G(I) \backslash X$ has exactly one neighbor in $I(\overline{X})$ then we say that X *pseudo-dominates I*. An example of a set X that pseudo-dominates a set I is illustrated in Figure 1.

The proofs of the following two lemmas are omitted due to page restrictions.

Lemma 1. *Let I be an independent set in a $2P_3$-free graph $G = (V, E)$. Then $G[V \backslash I]$ contains a clique X that pseudo-dominates I.*

Lemma 2. *Let G be a $2P_3$-free graph that contains a set X and an independent set I, such that X pseudo-dominates I. Let $k \geq 1$. If $I(\overline{X})$ contains more than k vertices with degree at least k in G, then G is not k-colorable.*

The following lemma states a useful relationship between k-colorability of $2P_3$-free graphs with minimum degree at least k and the existence of a dominating set, the size of which is bounded by a quadratic function in k. Its proof uses Lemmas 1 and 2.

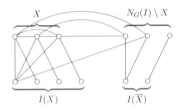

Fig. 1. A set X that pseudo-dominates a set I

Lemma 3. *Let G be a connected $2P_3$-free graph with minimum degree at least k for some integer k. If G is k-colorable, then G contains a dominating set of size at most $2k^2 + 3$.*

Proof. Let G be a connected $2P_3$-free graph with minimum degree at least k for some integer k. Assume that G is k-colorable. Then G is K_{k+1}-free. If G is a complete graph, then G contains a dominating set of size 1, and the statement of the lemma holds. Suppose G is not a complete graph. Then G contains an induced path uvw. If $\{u, v, w\}$ is a dominating set of G, then the statement of the lemma holds. Suppose $\{u, v, w\}$ is not a dominating set of G.

Let G' be the graph obtained from G after removing u, v, w, and all vertices in $N_G(\{u, v, w\})$. Because G is $(K_{k+1}, 2P_3)$-free and u, v, w form an induced P_3, we find that G' is (K_{k+1}, P_3)-free. Hence, every component of G' is isomorphic to a graph from $\{K_1, \ldots, K_k\}$.

We partition the vertices of G' into at most k independent sets $I_1, \ldots, I_k$ as follows. First we form I_1 by taking exactly one vertex from each component of G'. We remove I_1 from G' and repeat the above step to obtain I_2 if there were any vertices of G' left. We proceed in this way until all vertices of G' have been used. This will happen after at most k steps, because every component of G' has at most k vertices at the start of this procedure.

We apply Lemma 1 to each I_h in order to find a clique X_h in G that pseudo-dominates I_h. Because G is K_{k+1}-free, $|X_h| \leq k$ for $h = 1, \ldots, k$.

We apply Lemma 2 to G and each I_h in order to find that $|I_h(\overline{X_h})|$ has size at most k for $h = 1, \ldots, k$. Then $D = \{u, v, w\} \cup X_1 \cup \cdots \cup X_k \cup I_1(\overline{X_1}) \cup \cdots \cup I_k(\overline{X_k})$ has at most $3 + k^2 + k^2 = 2k^2 + 3$ vertices. Since D is a dominating set in G, this completes the proof of Lemma 3. $\qquad\square$

Remark 1. If in Lemma 3 the graph G is a $(K_3, 2P_3)$-free graph, then every component in the graph G' has size at most two. Hence, there is no I_h, and consequently, no X_h for $h \geq 3$. Moreover, in that case, X_1 and X_2 have size at most two. This means that the dominating set D in the statement of Lemma 3 has size at most $3 + 2 + 2 + k + k = 2k + 7$. We also observe that in this case the size of a maximum independent set in $G[D]$ is at most $4 + 2k$.

We observe that the graph in Lemma 3 is required to have minimum degree at least k. This is not a problem due to the following well-known procedure. Let G be a graph. Remove all vertices with degree at most $k - 1$ from G. Propagate this until we obtain a graph with minimum degree at least k, denoted as $G_{\geq k}$. We observe the following.

Observation 2. *Let k be a fixed integer. Then $G_{\geq k}$ can be obtained in polynomial time. Furthermore, $G_{\geq k}$ is k-colorable if and only if G is k-colorable.*

As an immediate consequence of our results we will now show that testing whether a given $2P_3$-free graph is 3-colorable can be done in polynomial-time.

Theorem 3. *The 3-COLORING problem can be solved in polynomial time for the class of $2P_3$-free graphs.*

Proof. Let $G = (V, E)$ be a $2P_3$-free graph on n vertices. By Observation 2 we may assume that G has minimum degree at least 3. We first check in $O(n^4)$ time whether G is K_4-free. If not, then G is not 3-colorable. Suppose G is K_4-free.

Suppose G is disconnected. Then G contains at most one component that is not isomorphic to a complete graph on at most three vertices, because G is $2P_3$-free. Hence, we may assume that G is connected. We give each vertex u of G a list $L(u) = \{1, 2, 3\}$ of admissible colors. We now search for a dominating set of at most 21 vertices. We can do this in $O(n^{21})$ time by brute force. If we do not find such a set, then G is not 3-colorable, due to Lemma 3. Otherwise, let D be the dominating set with $|D| \leq 21$ that we found.

We pre-color the vertices of D and adjust the list of each vertex in $V \setminus D$ by removing the colors of its neighbors in D. Because D is a dominating set of G, we find that every vertex in G has a list of admissible colors of size at most 2. Then checking whether G has a coloring respecting these lists is solvable in polynomial time by Observation 1. Because there are at most $3^{|D|} \leq 3^{21}$ different pre-colorings of D, our algorithm runs in polynomial time.   □

3   Coloring $(K_3, 2P_3)$-Free Graphs with at Most Four Colors

We next present an algorithm that solves the 4-COLORING problem for the class of $(K_3, 2P_3)$-free graphs.

3.1   Outline of the Algorithm

Our algorithm first assigns a list with colors $1, 2, 3, 4$ to every vertex of the input graph G. Our goal is to reduce the list of every vertex to a list with at most two admissible colors such that Observation 1 can be used. For this purpose our algorithm first preprocesses G, thereby reducing the lists of admissible colors of every vertex by at least one. This preprocessing heavily relies on Lemma 3 and is explained in detail in Section 3.2. After the preprocessing stage, we either find that G has no 4-coloring, or else we find a constant-bounded number of so-called suitable list-assignments of G. Due to the preprocessing, every list in every suitable list-assignment is a proper subset of $\{1, 2, 3, 4\}$, thus of size at most three. We show that it suffices to find a coloring of G respecting one of the *suitable* list-assignments of G. However, for a suitable list-assignment L', we might not be able to apply Observation 1 immediately, because there might still be vertices u with $|L'(u)| = 3$. In Section 3.3 we apply a polynomial-time branching algorithm that reduces the size of such lists, thereby enabling the use of Observation 1.

We note that during the execution of the algorithm some vertices may get an empty list of admissible colors at some moment. In that case our algorithm can immediately output No. We do not write this explicitly in the description of our algorithm, because such a case will be spotted anyway, namely at the moment we apply Observation 1.

3.2   The Preprocessing

Let G be a $(K_3, 2P_3)$-free graph, every vertex u of which has a list $L(u) = \{1, 2, 3, 4\}$ of admissible colors. If G is disconnected, then G contains at most one component that is not isomorphic to a complete graph on at most two vertices, because G is $2P_3$-free. Henceforth, we assume that G is connected. By Observation 2, we may assume that G has minimum degree at least 4. We preprocess G in three phases.

Phase 1. Reduce the list sizes by at least 1
The algorithm checks if G has a dominating set D of size at most $2 \cdot 4 + 7 = 15$. If not, it outputs No. Suppose G has such a dominating set D. Then the algorithm pre-colors every vertex of D with a color from $\{1, 2, 3, 4\}$ and adjusts the lists of every other vertex by removing the colors of its neighbors in D.

After Phase 1, we can partition the set of vertices of G into five sets A, B_1, B_2, B_3, B_4, some of which may be empty. They are defined as follows. We let A consist of all vertices with a list of at most two admissible colors. Observe that we have not removed the vertices of D. Because these have been pre-colored, they have a list of exactly one admissible color. Hence, by definition, $D \subseteq A$. For $i = 1, \ldots, 4$, we let B_i consist of all vertices with list $\{1, 2, 3, 4\} \setminus \{i\}$. We note that each $G[B_i]$ contains at most one component on more than two vertices, due to our assumption that G is $(K_3, 2P_3)$-free. We denote this component by H_i if it exists. We write F_i for the subgraph of $G[B_i]$ induced by the vertices of the edge-components (components isomorphic to K_2). So, each F_i is the disjoint union of a number of edges.

Phase 2. Pre-color each F_i
We pre-color every vertex of F_i for $i = 1, \ldots, 4$ respecting its list of admissible colors. Afterwards, the algorithm adjusts the lists of every other vertex in $B_1 \cup B_2 \cup B_3 \cup B_4$ by removing the colors of its neighbors in $F_1 \cup F_2 \cup F_3 \cup F_4$. We redefine A, B_1, B_2, B_3, B_4 by moving every vertex in $B_1 \cup B_2 \cup B_3 \cup B_4$ that has a new list of at most two admissible colors to A.

Phase 3. Reduce the list sizes in each H_i by at least one
For $i = 1, \ldots, 4$ the algorithm acts as follows. It checks if H_i has a dominating set D_i of size at most $2 \cdot 3 + 7 = 13$. If not it outputs No. Suppose H_i has

such a dominating set D_i. Then the algorithm pre-colors every vertex of D_i respecting its list of admissible colors and adjusts the lists of every other vertex in $B_1 \cup B_2 \cup B_3 \cup B_4$ by removing the colors of its neighbors in D_i. We denote the resulting list-assignment by L' and call L' a *suitable* list-assignment of G.

After Phase 3, we redefine A, B_1, B_2, B_3, B_4 by moving every vertex in $B_1 \cup B_2 \cup B_3 \cup B_4$ that has a new list of at most two admissible colors to A. Because of Phases 2 and 3, each B_i now induces a set of isolated vertices in G. However, a vertex from B_i may be adjacent to a vertex from B_j for some i, j with $i \neq j$.

Before we continue with the description of our algorithm, we need to show the following two lemmas, the proofs of which have been omitted. The first lemma shows that we can restrict ourselves to suitable list-assignments of G. Note that G has no suitable list-assignment if our algorithm has outputted No in Phase 1, 2 or 3. Otherwise, the number of suitable list-assignments depends on the number of different pre-colorings in Phase 1, 2 and 3. Hence, G may have many suitable list-assignments. However, the second lemma shows that the number of suitable list-assignments is bounded by a constant and that we can find all of them in polynomial time.

Lemma 4. *Let G be a connected $(K_3, 2P_3)$-free graph with minimum degree at least four. Then G has a 4-coloring if and only if there exists a suitable list-assignment L' such that G has a coloring respecting L'.*

Lemma 5. *Let G be a connected $(K_3, 2P_3)$-free graph with minimum degree at least four. Then the number of suitable list-assignments of G is constant-bounded and can be obtained in polynomial time.*

Due to Lemma 4 our algorithm is left with the following task:

Check for each suitable list-assignment L' whether G has a coloring that respects L'.

Due to Lemma 5 our algorithm runs in polynomial time if it performs the above task in polynomial time for every suitable list-assignment. In Section 3.3 we consider a single suitable list-assignment L' of G and show that this is indeed the case.

3.3   Reducing the Lists of Size Three in a Suitable List-Assignment

Let L' be a suitable list-assignment created from a connected $(K_3, 2P_3)$-free graph G. Recall that, due to the preprocessing, $V(G) = A \cup B_1 \cup B_2 \cup B_3 \cup B_4$ with the following four properties; also recall that D is the dominating set that got pre-colored in Phase 1.

P1. $|L'(u)| \leq 2$ for every $u \in A$;

P2. $L'(v) = \{1, 2, 3, 4\} \backslash \{i\}$ for every $v \in B_i$ and for every $1 \leq i \leq 4$;

P3. B_i is an independent set for every $1 \leq i \leq 4$;

P4. Every vertex of B_i is adjacent to at least one vertex of D that has color i.

Our algorithm now starts a branching procedure in order to reduce the number of admissible colors in the list of every vertex in each B_i by at least one, thereby enabling the use of Observation 1. This is described below.

Phase 4. The branching
Our algorithm first considers B_1, then B_2, then B_3, and finally B_4 by applying the following code for each B_i. Recall that $B_i(X) = N_G(X) \cap B_i$ and that $B_i(\overline{X}) = B_i \backslash N_G(X)$.

(i) Determine a clique X in $G[V \backslash D]$ with $X = \{x\}$ or $X = \{x_1, x_2\}$ that pseudo-dominates B_i.

(ii) If $X = \{x\}$, then do as follows for every pair $p, q \in \{1, 2, 3, 4\} \backslash \{i\}$ with $p \neq q$:

    1. Set $L'(u) := \{p, q\}$ for every $u \in B_i(X)$.
    2. Remove all vertices in $B_i(\overline{X})$ with at most two neighbors in $V \backslash D$.
    3. Pre-color all remaining vertices in $B_i(\overline{X})$ respecting L'.
    4. If $i \leq 3$, then start Phase 4 with B_{i+1}; otherwise apply Observation 1.

If the above branching does not lead to a coloring of G respecting L', then G might still have such a coloring. However, in that case all three colors from $\{1, 2, 3, 4\} \backslash \{i\}$ must occur on $B_i(X)$. This means that our algorithm must do as follows: pre-color x by i, remove x from G and repeat Phase 4 with set B_i.

(iii) If $X = \{x_1, x_2\}$, then do as follows for all 4-tuples (p, q, r, s) with $p, q, r, s \in \{1, 2, 3, 4\} \backslash \{i\}$, $p \neq q$ and $r \neq s$:

    1. Set $L'(u) := \{p, q\}$ for every $u \in B_i(\{x_1\})$.
    2. Set $L'(v) := \{r, s\}$ for every $v \in B_i(\{x_2\})$.
    3. Remove all vertices in $B_i(\overline{X})$ with at most two neighbors in $V \backslash D$.
    4. Pre-color all remaining vertices in $B_i(\overline{X})$ respecting L'.
    5. If $i \leq 3$, then start Phase 4 with B_{i+1}; otherwise apply Observation 1.

If the above branching does not lead to a coloring of G respecting L', then G might still have such a coloring. However, in that case, all three colors from $\{1, 2, 3, 4\} \backslash \{i\}$ must occur on $B_i(\{x_1\})$, implying that x_1 has at least three neighbors in B_i and must get color i, or they must all three occur on $B_i(\{x_2\})$, implying that x_2 has at least three neighbors in B_i and must get color i. Note that this is an "either or" situation, because x_1 and x_2 are adjacent, and as such they cannot both be colored with color i. We branch in either direction.

    First we check if x_2 has at least three neighbors in B_i. If so, then do as follows for every $p \in L'(x_1) \backslash \{i\}$:

    6. Set $L'(x_1) = p$.
    7. Set $L'(u) := L'(u) \backslash \{p\}$ for every $u \in B_i(\{x_1\})$.
    8. Set $L'(x_2) := \{i\}$.
    9. Remove x_1, x_2 from G, set $B_i := B_i \backslash B_i(\{x_1\})$, and repeat Phase 4 with B_i.

If the above branching does not lead to a coloring of G respecting L', or if x_2 has at most two neighbors in B_i, then check if x_1 has at least three neighbors in B_i. If not, then pre-color every vertex in $B_i(X)$. Then start Phase 4 with set B_i if $i \leq 3$ or apply Observation 1 if $i = 4$. Otherwise do as follows for every $r \in L'(x_2)\backslash\{i\}$:

10. Set $L'(x_2) = r$.
11. Set $L'(v) := L'(v)\backslash\{r\}$ for every $v \in B_i(\{x_2\})$.
12. Set $L'(x_1) := \{i\}$.
13. Remove x_1, x_2 from G, set $B_i := B_i\backslash B_i(\{x_2\})$, and repeat Phase 4 with B_i.

(iv) If all calls to Observation 1 yield no coloring, then G has no coloring respecting L', and the algorithm outputs No. Otherwise, if there is a call to Observation 1 that yields a coloring c, then the algorithm extends c to a coloring of G that respects L' by coloring the vertices it has removed from G in the reverse order of their removal, in such a way that L' is respected.

The proof of the following lemma has been omitted.

Lemma 6. *Phase 4 is correct and runs in polynomial time.*

By Lemmas 4–6 we immediately obtain the following result.

Theorem 4. *The 4-COLORING problem can be solved in polynomial time for the class of $(K_3, 2P_3)$-free graphs.*

4   Determining the Chromatic Number

We present two polynomial-time algorithms that solve the VERTEX COLORING problem for $(K_3, 2P_3)$-free graphs and for $(K_3, P_2 + P_4)$-free graphs, respectively. We need the following theorem. We omit the proof due to page restrictions. However, the second statement has been observed by Dabrowski et al. [5] without proof (because the result is easy to prove).

Theorem 5. *Every $(K_3, 2P_3)$-free graph can be colored with at most 5 colors, and every $(K_3, P_2 + P_4)$-free graph can be colored with at most 4 colors.*

Theorem 6. *The VERTEX COLORING problem can be solved in polynomial time for the class of $(K_3, 2P_3)$-free graphs and for the class of (K_3, P_2+P_4)-free graphs.*

Proof. We present two algorithms. Each algorithm just checks for the input graph G whether it is k-colorable for increasing k up to a certain value k^*, which is equal to 4 in case G is $(K_3, 2P_3)$-free, and which is equal to 3 in case G is $(K_3, P_2 + P_4)$-free. If no coloring of G has been found then $\chi_G = k^* + 1$ is outputted. The correctness of these algorithms follows immediately from Theorem 5. Below we show that they run in polynomial time.

We first note that a graph can be colored with at most one color if and only if it consists of isolated vertices only. Secondly, a graph can be colored with at most two colors if and only if it is bipartite.

Suppose our input graph G is $(K_3, 2P_3)$-free. Then, by Theorem 3 and 4, we can test in polynomial time whether G is 3-colorable or 4-colorable, respectively.

Suppose G is $(K_3, P_2 + P_4)$-free. In our previous paper [3] we showed that 3-COLORING can be solved in polynomial time even for $(P_2 + P_4)$-free graphs.

From the above we conclude that both our algorithm for coloring a $(K_3, 2P_3)$-free graph and our algorithm for coloring a $(K_3, P_2 + P_4)$-free graph run in polynomial time. □

5   Future Research

One can explore various directions to extend the polynomial-time results in this paper, and determining the complexity of the following problems is still open.

1. 3-COLORING for (K_3, sP_3)-free graphs for any fixed $s \geq 3$;
2. k-COLORING for $2P_3$-free graphs for any fixed $k \geq 4$;
3. 3-COLORING for (K_3, P_7)-free graphs.

We expect that our key results in Section 2 will be useful for solving Problems 1 and 2. As an aside, we mention that Dabrowski et al. [5] showed that the VERTEX COLORING problem is polynomial-time solvable for the class of (K_3, sK_2)-free graphs for any fixed $s \geq 2$.

Another interesting research direction is to characterize the class of $(K_3, 2P_3)$-free graphs with chromatic number 4 or 5. Such a characterization could lead to a certifying algorithm, just as the ISAAC 2009 paper of Bruce, Hoàng, and Sawada [4] successfully shows for 3-COLORABILITY of P_5-free graphs. We have no examples of $(K_3, 2P_3)$-free graphs with chromatic number 5 and expect such graphs to have a large number of vertices. We note that, even in the possible case that such graphs do no exist at all, our polynomial-time algorithm in Section 3 for solving 4-COLORING is still useful, because it produces a 4-coloring of a 4-chromatic $(K_3, 2P_3)$-free graph. An infinite class of 4-chromatic $(K_3, 2P_3)$-free graphs can for example be obtained from the Grötzsch graph (see, e.g., [1]) by replacing the unique vertex v of degree 5 by a set of vertices, all adjacent to the five neighbors of v. These graphs are also $(K_3, P_2 + P_4)$-free.

References

1. Bondy, J.A., Murty, U.S.R.: Graph Theory. Springer Graduate Texts in Mathematics 244 (2008)
2. Broersma, H.J., Fomin, F.V., Golovach, P.A., Paulusma, D.: Three complexity results on coloring P_k-free graphs. In: Fiala, J., Kratochvíl, J., Miller, M. (eds.) IWOCA 2009. LNCS, vol. 5874, pp. 95–104. Springer, Heidelberg (2009)

3. Broersma, H.J., Golovach, P.A., Paulusma, D., Song, J.: Narrowing down the gap on the complexity of coloring P_k-free graphs. In: Proceedings of the 36th International Workshop on Graph-Theoretic Concepts in Computer Science (WG 2010). LNCS (to appear, 2010)

4. Bruce, D., Hoàng, C.T., Sawada, J.: A certifying algorithm for 3-colorability of P_5-free graphs. In: Dong, Y., Du, D.-Z., Ibarra, O. (eds.) ISAAC 2009. LNCS, vol. 5878, pp. 594–604. Springer, Heidelberg (2009)

5. Dabrowski, K., Lozin, V., Raman, R., Ries, B.: Colouring vertices of triangle-free graphs. In: Proceedings of the 36th International Workshop on Graph-Theoretic Concepts in Computer Science (WG 2010). LNCS (to appear, 2010)

6. Edwards, K.: The complexity of coloring problems on dense graphs. Theoret. Comput. Science 43, 337–343 (1986)

7. Garey, M.R., Johnson, D.S.: Computers and Intractability: A Guide to the Theory of NP-Completeness. Freeman, San Francisco (1979)

8. Grötschel, M., Lovász, L., Schrijver, A.: The ellipsoid method and its consequences in combinatorial optimization. Combinatorica 1, 169–197 (1981)

9. Hoàng, C.T., Kamiński, M., Lozin, V., Sawada, J., Shu, X.: Deciding k-colorability of P_5-free graphs in polynomial time. Algorithmica 57, 74–81 (2010)

10. Kamiński, M., Lozin, V.V.: Coloring edges and vertices of graphs without short or long cycles. Contributions to Discrete Math. 2, 61–66 (2007)

11. Kamiński, M., Lozin, V.V.: Vertex 3-colorability of Claw-free Graphs. Algorithmic Operations Research 21 (2007)

12. Kochol, M., Lozin, V.V., Randerath, B.: The 3-Colorability Problem on Graphs with Maximum Degree Four. SIAM J. Comput. 32, 1128–1139 (2003)

13. Král', D., Kratochvíl, J., Tuza, Z., Woeginger, G.J.: Complexity of coloring graphs without forbidden induced subgraphs. In: Brandstädt, A., Le Van, B. (eds.) WG 2001. LNCS, vol. 2204, p. 254. Springer, Heidelberg (2001)

14. Kratochvíl, J.: Precoloring extension with fixed color bound. Acta Math. Univ. Comen. 62, 139–153 (1993)

15. Le, V.B., Randerath, B., Schiermeyer, I.: On the complexity of 4-coloring graphs without long induced paths. Theor. Comp. Science 389, 330–335 (2007)

16. Lozin, V., Mosca, R.: Independent sets in extensions of $2K_2$-free graphs. Discrete Appl. Math. 146, 74–80 (2005)

17. Randerath, B., Schiermeyer, I.: 3-Colorability $\in$ P for P_6-free graphs. Discrete Appl. Math. 136, 299–313 (2004)

18. Randerath, B., Schiermeyer, I.: Vertex colouring and forbidden subgraphs - a survey. Graphs and Combin. 20, 1–40 (2004)

19. Tuza, Z.: Graph colorings with local constraints - a survey. Discuss. Math. Graph Theory 17, 161–228 (1997)

20. Woeginger, G.J., Sgall, J.: The complexity of coloring graphs without long induced paths. Acta Cybern. 15, 107–117 (2001)

On the Approximability of the Maximum Interval Constrained Coloring Problem[*]

Stefan Canzar[1], Khaled Elbassioni[2], Amr Elmasry[2,3], and Rajiv Raman[4]

[1] Centrum Wiskunde & Informatica, Amsterdam, The Netherlands
[2] Max-Planck-Institut für Informatik, Saarbrücken, Germany
[3] Datalogisk Institut, University of Copenhagen, Denmark
[4] DIMAP and Department of Computer Science, University of Warwick, UK

Abstract. In the MAXIMUM INTERVAL CONSTRAINED COLORING problem, we are given a set of intervals on a line and a k-dimensional requirement vector for each interval, specifying how many vertices of each of k colors should appear in the interval. The objective is to color the vertices of the line with k colors so as to maximize the total weight of intervals for which the requirement is satisfied. This $\mathcal{NP}$-hard combinatorial problem arises in the interpretation of data on protein structure emanating from experiments based on hydrogen/deuterium exchange and mass spectrometry. For constant k, we give a factor $O(\sqrt{|\text{OPT}|})$-approximation algorithm, where OPT is the *smallest-cardinality* maximum-weight solution. We show further that, even for $k = 2$, the problem remains APX-hard.

1 Introduction

The INTERVAL CONSTRAINED COLORING (ICC) problem was introduced recently by Althaus et al. [1,2] as the mathematical abstraction of a problem appearing in the interpretation of experimental data in biochemistry. Monitoring exchange rates via mass spectrometry is a method used to obtain information about the 3-dimensional structure of proteins. ICC captures the problem of increasing the resolution of the exchange data from peptic fragments to single residues. We refer the interested reader to Althaus et al. [2] for more on the biochemical background. INTERVAL CONSTRAINED COLORING is a decision problem that asks for a coloring of an ordered sequence of n vertices $V = [n]$ using k colors such that a given set of requirements is satisfied. Each requirement is made up of a closed interval $I \subseteq [n]$, and a complete specification of how many elements in I should be colored with each color.

More formally, let $\mathcal{I}$ be a set of m intervals defined on $V = [n]$, let $[k]$ be a set of color classes, and $r : \mathcal{I} \times [k] \to \mathbb{Z}_+$ be a requirement function such that

$$\sum_{c \in [k]} r(I, c) = |I| \quad \text{for all } I \in \mathcal{I}. \tag{1}$$

[*] Part of the work was done while the first, third and forth authors were members of Max-Planck Institute. A. Elmasry is on leave from Alexandria University of Egypt. R. Raman's research is supported by the Centre for Discrete Mathematics and its Applications (DIMAP) at University of Warwick, EPSRC award EP/D063191/1.

O. Cheong, K.-Y. Chwa, and K. Park (Eds.): ISAAC 2010, Part II, LNCS 6507, pp. 168–179, 2010.
© Springer-Verlag Berlin Heidelberg 2010

The given set of intervals $\mathcal{I}$ is said to be *feasible* (or *colorable*) if there exists a coloring $\chi : V \to [k]$ such that for every $I \in \mathcal{I}$ we have $N_\chi(I, c) = r(I, c)$, for all $c \in [k]$, where $N_\chi(I, c) = |\{i \in I : \chi(i) = c\}|$ denotes the number of vertices in I assigned color c by χ. Clearly, not all sets of intervals are colorable. In the biochemical application from which our problem arises, the requirement function models exchange data collected in real experiments, which usually contain some noise. Hence, we might not obtain a colorable instance even if the real underlying instance is colorable. This motivates the study of the problem of extracting the largest subset $\mathcal{I}' \subseteq \mathcal{I}$ that is colorable. More precisely, we consider a more general version of the problem where each interval also has a weight, in addition to its color requirements, given by $w : \mathcal{I} \to \mathbb{R}_+$, and we wish to find a maximum-weight colorable subset of the intervals, as well as produce a feasible coloring. We define the problem formally below.

Definition 1 (MAX-FEASIBLE-COLORING (MFC)). *Given a set of intervals $\mathcal{I}$ with non-negative weights $w : \mathcal{I} \to \mathbb{R}_+$, a requirement function $r : \mathcal{I} \times [k] \to \mathbb{Z}^+$ satisfying (1), the* MAX-FEASIBLE-COLORING (MFC) *problem asks for finding a maximum weight colorable subset $\mathcal{I}' \subseteq \mathcal{I}$.*

The MFC problem can also be cast as a problem on linear systems, as observed by Byrka et al. [3]. A 0/1 matrix A has the consecutive 1's property if in each row of A the 1's appear consecutively. Consider the following system, where $A \in \{0, 1\}^{m \times n}$ is the consecutive 1's matrix derived in the natural manner from the MFC problem, and $r^{(i)} \in \mathbb{Z}_+^m$ is a column vector of requirements for the i-th color from the m intervals. Let I be the $n \times n$ identity matrix.

$$\begin{bmatrix} A & 0 & \dots & 0 \\ 0 & A & \dots & 0 \\ \vdots & & \ddots & \vdots \\ 0 & \dots & & A \\ I & \dots & & I \end{bmatrix} \begin{bmatrix} x^{(1)} \\ x^{(2)} \\ \vdots \\ x^{(k)} \end{bmatrix} = \begin{bmatrix} r^{(1)} \\ r^{(2)} \\ \vdots \\ r^{(k)} \\ 1 \end{bmatrix}$$

The ICC problem asks for a feasible solution to the above system with $x^{(i)} \in \{0, 1\}^n$, $i = 1, \dots, k$. The MFC problem can then be stated as the problem of computing the maximum subset of rows of the matrix A for which the above system has a feasible solution. In this light, the problem is similar to the MAXIMUM FEASIBLE SUBSYSTEM (MFS) problem, where an infeasible linear system $Ax = b$ is given and we wish to find the largest subsystem that has a feasible solution. However, in the MFS problem we may allow a rational vector x.

1.1 Our Results

We study the approximability of MFC and present the first non-trivial approximation algorithm for the problem. In particular, we give a factor $O(\sqrt{|\text{OPT}|})$-approximation algorithm, where we denote throughout by $\text{OPT} \subseteq \mathcal{I}$ an optimal set of intervals, that is a maximum-weight colorable subset, with *smallest cardinality*. The main technique that we use is to decompose the problem into simpler instances using Dilworth's Theorem [4], and then show how we can solve these simpler instances in polynomial time. A similar technique was used earlier [5] to obtain an $O(\sqrt{|OPT|}\log n)$-approximation algorithm for the following MAXIMUM FEASIBLE SUBSYSTEM problem: Given an infeasible linear system $l \leq Ax \leq u, x \geq 0$, where A is a consecutive 1's matrix, the problem is to find the largest subsystem for which there is a feasible solution satisfying the non-negativity constraints $x \geq 0$. However, the same decomposition does not work for the MFC problem. In particular, one needs to use a different poset definition when applying Dilworth's Theorem. As it turns out, the new decomposition can also be used to save the $\log n$-factor in the above approximation factor for MFS.

Note that the INTERVAL CONSTRAINED COLORING problem for $k = 2$ colors can be solved in polynomial time via linear programming. In contrast, we show that the maximization version of the problem, MFC, is APX-hard for $k = 2$. This is akin to the situation for 2SAT and MAX2SAT. We thus improve on the result in [3], which shows that MFC is APX-hard for $k = 3$.

1.2   Related Work

The ICC problem has been introduced in [2] by Althaus et al. who formulated the problem of improving the resolution of exchange data as a linear minimization problem subject to integer linear constraints (ILP), and proposed a branch-and-bound approach to enumerate all optimal colorings (see also [6,7]). In [1], Althaus et al. studied the problem from a theoretical point of view. They showed it to be NP-hard in general and developed algorithms that, given a *fractionally* feasible instance, find a coloring that satisfies all the requirements within ± 1 of the prescribed value. Furthermore, they considered the maximization variant of the problem, MFC, and showed that if one is allowed to relax the coloring requirements by a small factor of $(1 + \epsilon)$, then there is an algorithm that finds a coloring satisfying (with violations) the optimal number of intervals, and running in quasi-polynomial time if the number of colors is constant.

Komusiewicz et al. [8] showed that the problem is fixed-parameter tractable with respect to parameters such as the maximum interval length and the maximum number of intervals containing a given vertex. Very recently, the decision variant of the problem (ICC) was shown to be NP-hard even when $k = 3$ [3]; this settles the complexity of the problem since, for $k = 2$, the problem is polynomially solvable [2]. However, prior to the current paper, both inapproximability of MFC for $k = 2$, and non-trivial approximations, for constant k, were not known.

A related or specialized version of the INTERVAL CONSTRAINED COLORING problem also arises in discrete tomography. The goal in these problems is to reconstruct an image from the partial information that is available. In particular, we have an $m \times n$ matrix A whose entries have to be colored with k colors. The entries of the matrix are integers in the range $\{1, \ldots, k\}$. Further, we are given k row vectors of dimension m, and k column vectors of dimension n, which tell us the number of entries of each color in each row and column. The problem is then to reconstruct the coloring of the matrix A with the information in the row and column vectors. This is a special case of a 2-dimensional version of our problem. See [9] for a recent survey, and [10] where the authors study the problem when the path in our case is replaced by a general graph. We believe that our techniques would be helpful in discrete tomography applications as well.

2   Preliminaries

In this section we recall some basic facts that we will use in our algorithm, and introduce some definitions. Let $(\mathcal{P}, \preceq)$ be a *partially ordered set (poset)*. A *chain* (respectively, anti-chain) is a set of pairwise comparable (respectively, incomparable) elements. The well-known Dilworth Theorem [4] states that, in any poset $\mathcal{P}$, the maximum size of anti-chain is equal to the minimum number of chain-covers (that is, a partition of the poset into chains). An immediate corollary is that $\mathcal{P}$ either contains a chain or an anti-chain of size at least $\sqrt{|\mathcal{P}|}$. Applying this recursively we obtain the following decomposition.

Lemma 1. *Let $(\mathcal{P}, \preceq)$ be any poset. Then, $\mathcal{P}$ can be decomposed into $k \leq 2\sqrt{|\mathcal{P}|}$ sets $\mathcal{P}_1, \ldots, \mathcal{P}_k$ such that, for each i, the induced subposet $(\mathcal{P}_i, \preceq)$ forms either a chain or an anti-chain.*

Proof. Let $\mathcal{P}_1$ be a chain or anti-chain of size $\sqrt{|\mathcal{P}|}$. Recurse on $(\mathcal{P} - \mathcal{P}_1, \preceq)$. The recurrence we get for the number of iterations is $f(p) \leq 1 + f(p - \sqrt{p})$, where $p = |\mathcal{P}|$. This is satisfied with $f(p) = 2\sqrt{p}$. $\square$

Corollary 1. *Let $(\mathcal{P}, \preceq)$ be a poset and $w : \mathcal{P} \to \mathbb{R}_+$ be a weight function. Then there is either a chain or an anti-chain in $\mathcal{P}$ of size at least $\frac{w(\mathcal{P})}{2\sqrt{|\mathcal{P}|}}$.*

Our approximation algorithm is based on decomposing the problem into simpler instances, which can be solved in polynomial time. The instances are defined on special classes of intervals, described more precisely in the following definitions: A set of intervals $\mathcal{I}$ such that for every pair of intervals $I, I' \in \mathcal{I}$ either $I \subseteq I'$ or $I' \subseteq I$ will be called a *tower*. A set of intervals $\mathcal{I}$ such that for every pair of intervals $I, I' \in \mathcal{I}$ neither $I \subseteq I'$ nor $I' \subseteq I$ will be called an *anti-tower*. An anti-tower in which all the intervals intersect will be called a *staircase*.

We will use the following notation: For a set of intervals $\mathcal{I} \subseteq 2^V$, let $\Phi(\mathcal{I})$, $\Psi(\mathcal{I}, u, v)$ and $\Upsilon(\mathcal{I})$ denote respectively the set of towers, the set of anti-towers starting at $u \in V$ and ending at $v \in V$, and the set of independent sets (that is, pairwise disjoint intervals) from $\mathcal{I}$.

An optimal tower $\mathcal{I}' \subseteq \mathcal{I}$ (respectively, anti-tower, or staircase) is one for which there exists a feasible coloring such that $w(\mathcal{I}') \overset{\text{def}}{=} \sum_{I \in \mathcal{I}'} w(I)$ is maximized among all such towers (respectively, anti-towers, or staircases). We will derive our main theorem in Section 3 form the following two lemmas, both of which assume that k is a constant.

Lemma 2. *Given any instance of* MFC*, we can find an optimal tower in polynomial time.*

Lemma 3. *Given any instance of* MFC *and two vertices u and v, we can find an optimal staircase starting at u and ending at v in polynomial time.*

Two staircases are independent if the end point of one is less than the starting point of the other, i.e., all the intervals of one are independent (i.e., disjoint) of the intervals of the other. We will also need the following decomposition.

Proposition 1 ([11]). *Any anti-tower can be partitioned into two subsets of intervals, each of which is a set of independent staircases.*

3   The Approximation Algorithm

The algorithm proceeds as follows. We first find the largest-weight colorable tower (step 1). This can be done easily by dynamic programming, sketched in Section 3.1. We then find, for every pair of vertices $u, v \in V$, $u < v$, an optimal staircase starting at u and ending at v, and define a new weight function w' on every possible interval of V (step 3). This can be done using the dynamic program presented in Section 3.2. Using this weight function, we compute a maximum wight independent set of intervals (step 5); we emphasize that we find this independent set *among all possible intervals* in V, not only form $\mathcal{I}$. This can also be done by dynamic programming (see e.g. [12]). Finally, the algorithm returns the larger among the two weights computed at steps 1 and 5. Note that it is straightforward to modify the algorithm to also return a coloring with the same weight that satisfies a subset of $\mathcal{I}$.

Algorithm 1. Find-MFC$(V, \mathcal{I}, r, w)$

1: $W_1 = \max_{\mathcal{I}' \in \Phi(\mathcal{I})} w(\mathcal{I}')$ (cf. Section 3.1)
2: **for** every $u, v \in V$ s.t. $u < v$ **do**
3:      $w'_{u,v} = \max_{\mathcal{I}' \in \Psi(\mathcal{I}, u, v)} w(\mathcal{I}')$ (cf. Section 3.2)
4: **end for**
5: Let $W_2 = \max_{\mathcal{I}' \in \Upsilon(\{[u,v]:u,v \in V,\ u<v\})} w'(\mathcal{I}')$
6: **return** $\max\{W_1, W_2\}$

Theorem 1. *Algorithm Find-MFC outputs a feasible coloring for a subset of weight at least $\frac{w(\text{OPT})}{4\sqrt{\text{OPT}}}$.*

Proof. Consider the intervals in a minimum-cardinality optimal solution OPT. Define a poset $\mathcal{P} = (\text{OPT}, \subseteq)$ by the containment relation on these intervals. A chain in such a poset is a tower, and an anti-chain is an anti-tower. By corollary 1, there is a tower or an anti-tower of weight at least $\frac{w(\text{OPT})}{2\sqrt{|\text{OPT}|}}$. If there is such an anti-tower, then, by Proposition 1, there is an independent set of staircases of weight at least $\frac{w(\text{OPT})}{4\sqrt{|\text{OPT}|}}$. Note that step 5 of the algorithm outputs the maximum weight of an independent set of staircases. Thus, in all cases, the total weight of intervals colored by the algorithm is as claimed.   □

3.1   Finding an Optimal Tower; Proof of Lemma 2

Let $\mathcal{I} = \{I_1, \ldots, I_m\}$ be a given set of intervals, where we assume that $I_i = [a_i, b_i]$ for $i \in [m]$. Given the requirement function $r : \mathcal{I} \times [k] \to \mathbb{Z}_+$ satisfying (1), we construct a partially ordered set $(\mathcal{P}, \preceq)$, where $\mathcal{P} \subseteq \mathbb{Z}^{k+2}$ is defined as follows: there is one-to-one correspondence between $\mathcal{P}$ and $\mathcal{I}$; interval $I_i \in \mathcal{I}$ is mapped to the point $(-a_i, b_i, r(I_i, 1), \ldots, r(I_i, k))$, and for two points $P, P' \in \mathcal{P}$, $P \preceq P'$ if and only if at each coordinate P is at most the value of P'. The algorithm for finding an optimal tower is based on the following observation.

Observation 1. $\mathcal{I}' \subseteq \mathcal{I}$ *is a colorable tower if and only if it is a chain in* $\mathcal{P}$.

Thus, finding an optimal tower is equivalent to finding a maximum-weight chain, which can be done in polynomial time (see, e.g., [12]).

3.2   Finding an Optimal Staircase; Proof of Lemma 3

Let $I_1, I_2, \ldots, I_m$ be the sequence of intervals in $\mathcal{I}$, sorted in the order of their left endpoints, i.e., if $i < j$, then $a_i \leq a_j$. Assuming this ordering, let I_s be an interval with $a_s = u$, where u is the starting vertex of the optimal staircase we are looking for. Clearly, non of the intervals I_ℓ with $\ell < s$ can be contained in the optimal staircase starting at u. For simplicity, and w.l.o.g., let us therefore assume for the remainder of this section, that we want to find the optimal staircase starting at a_1. Similarly, if I_t is an interval with $b_t = v$, non of the intervals I_ℓ with $\ell > t$ can be contained in the optimal staircase ending at v. W.l.o.g. we therefore assume in the following that we want to find the optimal staircase ending at b_m. Furthermore, we will assume that we have removed, in a preprocessing step, all intervals from the instance which do not form a staircase with I_1, i.e., intervals I_ℓ with $b_\ell \leq b_1$. Note that if such intervals I_s or I_t do not exist, the weight of an optimal staircase for this choice of u and v is defined to be $-\infty$ in the algorithm in Section 3. Similarly, if $I_s \cap I_t = \emptyset$ no staircase starting at a_s and ending at b_t exists and again its weight is defined to be $-\infty$. However, if I_s or I_t is not uniquely defined, we have to maximize over all possible choices of I_s and I_t with $I_s \cap I_t \neq \emptyset$.

We denote by $\mathcal{I}_t$ the set containing the subsequence $I_1, I_2, \ldots, I_t$, for $t \leq m$, and by OPT_t an optimal staircase to the instance induced by interval set $\mathcal{I}_t$ that contains intervals I_1 and I_t. If such a solution does not exist, we set $\text{OPT}_t \stackrel{\text{def}}{=} \emptyset$. Recall that for a given coloring χ and an arbitrary interval I defined on V, $N_\chi(I, c)$ counts the number of vertices in I colored c by χ. Accordingly, we define vector $\mathbf{N}_\chi(I) = (N_\chi(I, c))_{c \in [k]}$. Note that we denote vectors by boldface characters. For a vector $\bar{\mathbf{r}} \in \mathbb{R}^k$, $\text{OPT}_t(\bar{\mathbf{r}}, I)$ further constrains an optimal solution OPT_t to be satisfiable by a coloring χ with $\mathbf{N}_\chi(I) = \bar{\mathbf{r}}$. For such a coloring to exist, the coloring requirement $\bar{\mathbf{r}}$ imposed on interval I must be *valid* in the sense that $\|\bar{\mathbf{r}}\|_1 = |I|$. Since we will constrain optimal solutions OPT_t only by valid requirements on a tail subinterval of I_t, we will denote $\text{OPT}_t(\bar{\mathbf{r}}, I)$ simply by $\text{OPT}_t(\bar{\mathbf{r}})$. Note that such a solution $\text{OPT}_t(\bar{\mathbf{r}})$ for a valid $\bar{\mathbf{r}}$ might not exist, in which case again $\text{OPT}_t(\bar{\mathbf{r}}) \stackrel{\text{def}}{=} \emptyset$.

Intuitively, the dynamic program exploits the following optimality property Consider a set $\mathcal{I}' \subseteq \mathcal{I}_{t'}$ with $I_1, I_{t'} \in \mathcal{I}'$ that can be satisfied by a coloring χ'. Then the colors assigned by χ' to vertices within $[a_{t'}, b_1]$ can be shuffled arbitrarily without causing any interval in $\mathcal{I}'$ to be not satisfied. Therefore, for $\mathcal{I}' \cup \{I_t\}$ to be satisfiable it suffices to require that there exists a coloring χ satisfying both $I_{t'}$ and I_t such that $\mathbf{N}_\chi([b_1, b_{t'}]) = \mathbf{N}_{\chi'}([b_1, b_{t'}])$. In particular, rearranging the colors of χ' within $[a_{t'}, b_1]$ such that it satisfies I_t (under an appropriate coloring of the remaining vertices in $[b_{t'} + 1, b_t]$) does not affect the intervals in $\mathcal{I}'$, since none of them starts or ends in this interval. Furthermore, the remainder of interval $I_{t'}$, namely $[b_{t'} + 1, b_t]$, does not intersect any interval in $\mathcal{I}'$. In other words, if $I_{t'}$ is the predecessor of I_t in OPT_t, realized by a coloring χ, then the intervals in $\text{OPT}_t \setminus \{I_t\}$ must form a solution to $\mathcal{I}_{t'}$ of maximum weight among those solutions that can be realized by a coloring χ'' that conforms with χ' in $[b_1, b_{t'}]$, i.e., for which $\mathbf{N}_{\chi''}([b_1, b_{t'}]) = \mathbf{N}_{\chi'}([b_1, b_{t'}])$ holds. An equivalent composition of optimal solutions can be established if we condition on the colorig within interval $[p, b_{t'}]$, $a_t \leq p \leq b_1$, i.e., when we extend the left boundary of $[b_1, b_{t'}]$ in the above observation to any $p \geq a_t$. More formally, an optimal solution exhibits the following optimal substructure.

Theorem 2. (Optimal Substructure)
Assume that the optimal staircase OPT *starts at interval* I_1 *and ends at interval* I_t, *i.e.,* $I_1, I_t \in \text{OPT}$ *and* $I_\ell \notin \text{OPT}$ *for all* $\ell > t$. *Further assume that* $I_{t'}$ *is the predecessor of* I_t *in* OPT, *i.e.,* $I_{t'} \in \text{OPT}$, *for some* $1 \leq t' < t$, *and for all* I_ℓ *with* $t' < \ell < t$, $I_\ell \notin \text{OPT}$. *Let* χ *be a coloring satisfying intervals in* OPT. *Then for any* $a_t \leq p \leq b_1$, $w(\text{OPT} \setminus \{I_t\}) = w(\text{OPT}_{t'}(\mathbf{N}_\chi([p, b_{t'}])))$.

Proof. Let χ be a coloring satisfying intervals in OPT and let $a_t \leq p \leq b_1$ be fixed. We show that for any subset $\mathcal{I}' \subseteq \mathcal{I}_{t'}$ with $I_1, I_{t'} \in \mathcal{I}'$, that can be satisfied by a coloring χ' s.t. $\mathbf{N}_{\chi'}([p, b_{t'}]) = \mathbf{N}_\chi([p, b_{t'}])$, set $\mathcal{I}' \cup \{I_t\}$ is colorable. For that, we construct a coloring $\tilde{\chi}$ as follows.

$$\tilde{\chi}(u) = \begin{cases} \chi(u) & \text{if } u \in [b_{t'} + 1, b_t] \text{ or } u \in [a_{t'}, p - 1] \\ \chi'(u) & \text{if } u \in [p, b_{t'}] \text{ or } u \in [a_1, a_{t'} - 1] \end{cases}$$

We show that coloring $\tilde{\chi}$ satisfies $\mathcal{I}' \cup \{I_t\}$. First, consider interval I_t. The only vertices spanned by I_t whose coloring differs from χ lie in $[p, b_{t'}]$. These are colored according to χ', which satisfies $\mathbf{N}_{\chi'}([p, b_{t'}]) = \mathbf{N}_\chi([p, b_{t'}])$. Therefore $\mathbf{N}_{\tilde{\chi}}(I_t) = \mathbf{r}(I_t)$ and the interval is satisfied by $\tilde{\chi}$. Analogously, it can be shown that interval $I_{t'}$ is satisfied by $\tilde{\chi}$. Now consider an arbitrary interval $I_\ell \in \mathcal{I}'$ with $\ell < t'$. The only vertices spanned by I_ℓ whose coloring differs from χ' lie in $[a_{t'}, p - 1]$. Since both χ and χ' satisfy $I_{t'}$, and $\mathbf{N}_\chi([p, b_{t'}]) = \mathbf{N}_{\chi'}([p, b_{t'}])$, we have $\mathbf{N}_\chi([a_{t'}, p - 1]) = \mathbf{N}_{\chi'}([a_{t'}, p - 1])$ and thus $\mathbf{N}_{\tilde{\chi}}(I_\ell) = \mathbf{r}(I_\ell)$. $\square$

Let $[p, b_t]$ be a tail subinterval of I_t with $a_t \leq p \leq b_1$, and let vector $\bar{\mathbf{r}}$ be a valid coloring requirement with $\bar{\mathbf{r}} \leq \mathbf{r}(I_t)$ on that subinterval. To obtain an optimal solution to $\mathcal{I}_t$ containing I_1 and I_t that can be satisfied by a coloring χ with $\mathbf{N}_\chi([p, b_t]) = \bar{\mathbf{r}}$, the following recurrence "guesses" the predecessor $I_{t'}$ and the best allocation of $\bar{\mathbf{r}}$ to $I_\alpha = [p, b_{t'}]$ and the remaining interval $I_\beta = [b_{t'} + 1, b_t]$, and recursively solves the resulting subproblem optimally.

$$w(\mathrm{OPT}_t(\bar{\mathbf{r}})) = \max_{\substack{1 \le t' < t : \; a_{t'} < a_t \text{ and } b_{t'} < b_t \\ \bar{\bar{\mathbf{r}}} \in \mathbb{R}^k : \; \|\bar{\bar{\mathbf{r}}}\|_1 = b_{t'} - p + 1 \text{ and } \bar{\bar{\mathbf{r}}} \le \bar{\mathbf{r}}, \\ \mathbf{r}(I_{t'}) - \bar{\bar{\mathbf{r}}} \ge \mathbf{r}(I_t) - \bar{\mathbf{r}}}} w(\mathrm{OPT}_{t'}(\bar{\bar{\mathbf{r}}})). \qquad (2)$$

If there is no coloring satisfying both I_1 and I_t under the restriction $\mathbf{N}_\chi([p, b_t]) = \bar{\mathbf{r}}$, we set $\mathrm{OPT}_t(\bar{\mathbf{r}}) \stackrel{\text{def}}{=} \emptyset$ and define $w(\emptyset) \stackrel{\text{def}}{=} -\infty$.

The recurrence considers only possible predecessors $I_{t'}$ that together with I_t form a staircase. The tail subinterval $[p, b_t]$ is divided into interval I_α overlapping with $I_{t'}$ and the remaining interval I_β. A valid requirement $\bar{\bar{\mathbf{r}}}$ on I_α is chosen in such a way that there exists a coloring χ satisfying both I_α and I_β under the constraints $\mathbf{N}_\chi([p, b_t]) = \bar{\mathbf{r}}$ and $\mathbf{N}_\chi([p, b_{t'}]) = \bar{\bar{\mathbf{r}}}$. Such a coloring χ exists if and only if requirement $\bar{\bar{\mathbf{r}}}$ on $[p, b_{t'}]$ is consistent with requirement $\bar{\mathbf{r}}$ on $[p, b_t]$ (i.e., $\bar{\bar{\mathbf{r}}} \le \bar{\mathbf{r}}$) and at the same time the resulting requirement on $[a_t, p-1]$ of a coloring satisfying I_t, namely $\mathbf{r}(I_t) - \bar{\mathbf{r}}$, is consistent with the resulting requirement on $[a_{t'}, p-1]$ of a coloring satisfying $I_{t'}$, namely $\mathbf{r}(I_{t'}) - \bar{\bar{\mathbf{r}}}$ (i.e., $\mathbf{r}(I_{t'}) - \bar{\bar{\mathbf{r}}} \ge \mathbf{r}(I_t) - \bar{\mathbf{r}}$).

We compute the recurrence relation by a dynamic programming approach as follows. An entry in the dynamic program matrix $\mathbf{D}$ is indexed by an interval index t, the left boundary p of a tail subinterval of I_t, and a valid coloring requirement vector $\bar{\mathbf{r}}$ on this subinterval $[p, b_t]$. Then table entry $\mathbf{D}[t, p, \bar{\mathbf{r}}]$ stores the weight of an optimal solution to $\mathcal{I}_t$ that contains I_1 and I_t and that can be satisfied by a coloring χ with $\mathbf{N}_\chi([p, b_t]) = \bar{\mathbf{r}}$, more precisely

$$\mathbf{D}[t, p, \bar{\mathbf{r}}] = w(\mathrm{OPT}_t(\bar{\mathbf{r}})).$$

Concerning the base case $t = 1$, we simply have to decide, for each $p = a_1, a_2, \ldots, a_m$ and each valid coloring requirement $\bar{\mathbf{r}}$ on tail subinterval $[p, b_1]$ of I_1, whether interval I_1 is satisfiable under $\bar{\mathbf{r}}$:

$$\mathbf{D}[1, p, \bar{\mathbf{r}}] = \begin{cases} w(\{I_1\}) & \text{if } \bar{\mathbf{r}} \le \mathbf{r}(I_1), \\ -\infty & \text{otherwise.} \end{cases}$$

For all $2 \le t \le m$, $p = a_t, a_3, \ldots, a_m$, and valid coloring requirements $\bar{\mathbf{r}}$ on $[p, b_t]$, we simply "guess" the predecessor of I_t in $\mathrm{OPT}_t(\bar{\mathbf{r}})$ according to (2):

$$\mathbf{D}[t, p, \bar{\mathbf{r}}] = \max_{\substack{1 \le t' < t : \; a_{t'} < a_t \text{ and } b_{t'} < b_t \\ \bar{\bar{\mathbf{r}}} \in \mathbb{R}^k : \; \|\bar{\bar{\mathbf{r}}}\|_1 = b_{t'} - p + 1 \text{ and } \bar{\bar{\mathbf{r}}} \le \bar{\mathbf{r}}, \\ \mathbf{r}(I_{t'}) - \bar{\bar{\mathbf{r}}} \ge \mathbf{r}(I_t) - \bar{\mathbf{r}}}} \mathbf{D}[t', p, \bar{\mathbf{r}}],$$

where $I_\alpha = [p, b_{t'}]$ and $I_\beta = [b_{t'} + 1, b_t]$, as above. The desired value of the optimal staircase starting at a_1 and ending at b_m is $\mathbf{D}[m, a_m, \mathbf{r}(I_m)]$.

4   APX-Hardness

We show that MFC is APX-hard even with $k = 2$ colors. The reduction is from
MAX2ESAT which was shown to be APX-hard by Håstad [13]. The input to the
MAX2ESAT problem is a boolean formula ϕ in conjunctive normal form with
variables $x_1, \ldots, x_n$, and clauses $C_1, \ldots, C_m$, where each clause C_i is a disjunc-
tion of exactly 2 literals. The problem asks for an assignment of boolean values
to the variables that satisfies the maximum number of clauses(where a clause
is satisfied if it evaluates to 1). For a variable i, we let m_i denote the number
of clauses that the variable appears in. The reduction is inspired by the APX-
hardness proof for the maximum feasible subsystem on interval matrices [5]. We
construct gadgets for each variable and each clause. Let P be a path on $8n+2$ vertices, where the vertices are labeled $\{v_{-4n-1}, v_{-4n}, v_{-4n+1}, \ldots, v_{-1}, v_1, v_2, \ldots,$
$v_{4n}, v_{4n+1}\}$. In the figures, we represent the path having vertices on integer
points on the number line. However, the origin is not a vertex of P. Let the
two colors we use be BLACK and WHITE. We start with a description of the
variable gadgets.

Variable Gadget: The gadget for variable x_i consists of $2m_i$ copies of the following
basic gadget that we describe, i.e., each interval of the basic gadget for x_i is
replicated $2m_i$ times to obtain the variable gadget for x_i.

A basic gadget for variable x_i, BG_i consists of 4 sets of intervals L_i, R_i, C_i,
and B_i, nemonic for *left, right, center* and *boundary* respectively. The sets L_i, R_i
and C_i consist of 2 intervals each, labeled L_i^0 and L_i^1, and so on, and the set B_i
consists of two intervals labeled B_i' and B_i''. Figure 1 gives a graphic representation of the gadget along with the requirements and coordinates of the intervals.
The intervals in set L_i, for $i = 1, \ldots, n$ have their right end-point at -1, and
the intervals of set R_i have their left end-point at 1. The set C_i is symmetric
around 0. A variable gadget consists of $2m_i$ copies of the basic gadget BG_i for
variable x_i. Assume we fix the colors of the vertices v_{-1} and v_1 so that they
have distinct colors. Then, in any solution that satisfies the maximal number
of intervals of BG_i, the two sets of vertices $S_A = \{v_{-4i-1}, v_{-4i}, v_{4i}, v_{4i+1}\}$, and
$S_B = \{v_{-4i+1}, v_{-4i+2}, v_{4i-2}, v_{4i-1}\}$ receive two consecutive BLACK and two
consecutive WHITE colors in the same order, i.e., there are two possible ways
in which this happens encoding TRUE and FALSE.

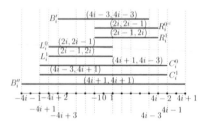

Fig. 1. The basic gadget for variable x_i

Lemma 4. *The maximum number of intervals of a basic gadget BG_i that can be satisfied simultaneously is 5. If we assume that the vertices v_{-1} and v_1 recieve distinct colors, then there are exactly two colorings of the sets S_A and S_B (modulo the colors on v_{-1} and v_1).*

Proof. Consider a basic gadget BG_i. Each set L_i, R_i and C_i consist of a pair of mutually exclusive intervals, and we can satisfy at most one interval from each set. This leaves us with the two intervals in the set B_i. Hence, we can satisfy at most 5 intervals simultaneously, viz. one each from L_i, R_i, C_i, and both intervals from B_i. Note that such a set can always be satisfied as follows. Color the vertices in the range of B'_i arbitrarily with equal numbers of each color. Now, suppose we color the vertices in the sets S_A the same color (either BLACK or WHITE), and the set S_B the same color. This gives a feasible coloring satifying 5 intervals. Note that any optimal solution must select both intervals in the set B_i and one interval from each of the remaining sets for an optimal solution.

Now consider a coloring that assigns distinct colors to the vertices -1 and 1. We claim that there is no optimal solution that simultaneously satisfies both R_i^0 and L_i^1, and similarly there is no optimal solution simultaneously satisfying both the intervals R_i^1 and L_i^0. Since we have assumed that v_{-1} and v_1 recieve distinct colors, the set of vertices $S = \{v_{-4i+3}, v_{-4i+2}, \ldots, v_{-3}, v_{-2}\} \cup \{v_2, v_3, \ldots, v_{4i-3}\}$ must satisfy the color requirement $(4i - 4, 4i - 4)$ in order to satisfy B'_i. Further, assume that we can satisfy R_i^0. Then, the range $\{v_2, \ldots, v_{4i-2}\}$ satisfy the requirement $(2i - 1, 2i - 2)$. Now, if we satisfy L_i^1, the range $\{v_{-4i+2}, \ldots, v_{-2}\}$ have color requirements $(2i - 2, 2i - 1)$. Summing them up, the range $S' = S \cup \{v_{-4i+2}, v_{4i-2}\}$ has colors $(4i - 3, 4i - 3)$. But, $S' \cup \{v_{-1}, v_1\}$ has colors $(4i - 2, 4i - 2)$, and we can not satisfy any interval in the set C_i. Hence, we can not have both R_i^0 and L_i^1 in an optimal solution. A symmetric argument holds for the case where we have R_i^1 and L_i^0 together in an optimal solution.   □

For a variable gadget then, we can satisfy $10m_i$ intervals simultaneously. We abuse notation and use L_i, L_i^0, etc. to refer either to the intervals of the basic gadget, or to the $2m_i$ copies of the variable gadget. An optimal solution for a variable gadget consisting of $\{B_i, L_i^0, R_i^0, C_i^0\}$ is said to be in a FALSE configuration, and an optimal solution for a variable gadget consisting of $\{B_i, L_i^1, R_i^1, C_i^1\}$ is said to be in a TRUE configuration.

We can show easily by induction on n that the variable gadgets can each be in TRUE or FALSE configuration independently, i.e., for any choice of an optimal configuration for each variable gadget corresponding to variables $x_1, \ldots, x_n$, there exists a coloring that satisfies all of them simultaneously. This is encoded in Lemma 5.

Lemma 5. *The variable gadgets can be independently set to a TRUE or FALSE configuration, with v_{-1} and v_1 receiving distinct colors.*

Clause Gadgets: We have 4 different clause gadgets corresponding to the 4 different types of clauses. If a clause C_p is of the form $(x_i \vee x_j)$ or $(\overline{x_i} \vee \overline{x_j})$, then the gadget for this clause consists of two intervals I_α^p, I_β^p. If C_p is of the form

$(\overline{x_i} \vee x_j)$ or $(x_i \vee \overline{x_j})$ the gadget consists of 3 intervals I_α^p, I_β^p and I_γ^p. Table 1 shows the gadgets for the 4 different types of clauses. The intervals corresponding to a clause gadget are mutually exclusive, and satisfy the property that exactly one interval will be satisfied if and only if the corresponding interval is satisfied. We show this in the next lemma.

Table 1. The Clause gadgets corresponding to the 4 different types of clauses. We assume that $i < j$.

Clause	Intervals	Requirement
$C_p = (x_i \vee x_j)$	$I_\alpha^p = [v_{-4j+2}, v_{4i-2}]$	$(2i + 2j - 3, 2i + 2j - 1)$
	$I_\beta^p = [v_{-4j+2}, v_{4i-2}]$	$(2i + 2j - 2, 2i + 2j - 2)$
$C_p = (\overline{x_i} \vee x_j)$	$I_\alpha^p = [v_{-4j+1}, v_{4i-2}]$	$(2i + 2j, 2i + 2j - 3)$
	$I_\beta^p = [v_{-4j+1}, v_{4i-2}]$	$(2i + 2j - 2, 2i + 2j - 1)$
	$I_\gamma^p = [v_{-4j+1}, v_{4i-2}]$	$(2i + 2j - 3, 2i + 2j)$
$C_p = (x_i \vee \overline{x_j})$	$I_\alpha^p = [v_{-4j+1}, v_{4i-2}]$	$(2i + 2j, 2i + 2j - 3)$
	$I_\beta^p = [v_{-4j+1}, v_{4i-2}]$	$(2i + 2j - 1, 2i + 2j - 2)$
	$I_\gamma^p = [v_{-4j+1}, v_{4i-2}]$	$(2i + 2j - 3, 2i + 2j)$
$C_p = (\overline{x_i} \vee \overline{x_j})$	$I_\alpha^p = [v_{-4j+2}, v_{4i-2}]$	$(2i + 2j - 1, 2i + 2j - 3)$
	$I_\beta^p = [v_{-4j+2}, v_{4i-2}]$	$(2i + 2j - 2, 2i + 2j - 2)$

Lemma 6. *Let C be a clause, and assume that the gadgets for all the variables are all optimally satisfied with v_{-1} and v_1 receiving distinct colors. Then, exactly one interval of C is satisfied if and only if the corresponding clause is satisfied with the truth assignment implied by the satisfied variable gadgets.*

Assume that all the variable gadgets are optimally satisfied. Then, the lemma follows directly by counting the colors of the vertices contained in the intervals corresponding to the clause gadget; the details are omitted.

The final part of the reduction, we show how we can ensure the assumption in Lemmas 4, 5 and 6 that v_{-1} and v_1 recieve distinct colors. We do this by the addition of a set of $24m$ Z intervals spanning $[v_{-1}, v_1]$, with requrirement $(1, 1)$. This completes the reduction. We are now ready to prove the main theorem.

Theorem 3. MFC *is APX-hard with $k \geq 2$ colors.*

Proof. (Sketch) Given an MAX2ESAT instance with n variables and m clauses that has an assignment satisfying k clauses, we construct a solution to the corresponding MFC as described in the previous paragraph, encoding TRUE and FALSE that satisfies at least $44m + k$ intervals. We can show that in the reverse direction that in any optimal solution to the MFC instance, all Z intervals are satisfied, and all intervals corresponding to each variable gadget are satisfied optimally. Hence, correspoding to each satisfied clause, exactly one interval corresponding to the clause gadget is satsified. Since the instance has at most $O(m)$ intervals, the reduction is gap preserving. □

The reduction above heavily uses mutually exclusive intervals with the same end-points in order to achieve the gap preserving reduction. However, we believe that we can modify the gadgets, albeit making the reduction more complicated so that there are no mutually exlusive intervals with the same end-points, and yet the reduction is gap preserving.

References

1. Althaus, E., Canzar, S., Elbassioni, K.M., Karrenbauer, A., Mestre, J.: Approximating the interval constrained coloring problem. In: Gudmundsson, J. (ed.) SWAT 2008. LNCS, vol. 5124, pp. 210–221. Springer, Heidelberg (2008)
2. Althaus, E., Canzar, S., Emmett, M.R., Karrenbauer, A., Marshall, A.G., Meyer-Bäse, A., Zhang, H.: Computing h/d-exchange speeds of single residues from data of peptic fragments. In: SAC, pp. 1273–1277 (2008)
3. Byrka, J., Karrebauer, A., Sanità, L.: Hardness of interval constrained coloring. In: Proceedings of the 9th Latin American Theoretical Informatics Symposium, pp. 583–592 (2009)
4. Dilworth, R.: A decomposition theorem for partially ordered sets. Annals of Mathematics 51, 161–166 (1950)
5. Elbassioni, K.M., Raman, R., Ray, S., Sitters, R.: On the approximability of the maximum feasible subsystem problem with 0/1-coefficients. In: SODA, pp. 1210–1219 (2009)
6. Althaus, E., Canzar, S., Ehrler, C., Emmett, M.R., Karrenbauer, A., Marshall, A.G., Meyer-Bäse, A., Tipton, J., Zhang, H.: Discrete fitting of hydrogen-deuterium-exchange-data of overlapping fragments. In: Proceedings of the 4th International Conference on Bioinformatics & Computational Biology, pp. 23–30 (2009)
7. Canzar, S., Elbassioni, K., Mestre, J.: A polynomial delay algorithm for enumerating approximate solutions to the interval constraint coloring problem. In: ALENEX, pp. 23–33. SIAM, Philadelphia (2010)
8. Komusiewicz, C., Niedermeier, R., Uhlmann, J.: Deconstructing intractability: A case study for interval constrained coloring. In: Kucherov, G., Ukkonen, E. (eds.) CPM 2009. LNCS, vol. 5577, pp. 207–220. Springer, Heidelberg (2009)
9. de Werra, D., Costa, M.C., Picouleau, C., Ries, B.: On the use of graphs in discrete tomography. 4OR **6** (2008) 101–123
10. Bentz, C., Costa, M.C., de Werra, D., Picouleau, C., Ries, B.: On a graph coloring problem arising from discrete tomography. Networks 51, 256–267 (2008)
11. Albertson, M., Jamison, R., Hedetniemi, S., Locke, S.: The subchromatic number of a graph. Discrete Mathematics 74, 33–49 (1989)
12. Schrijver, A.: Combinatorial Optimization: Polyhedra and Efficiency, Algorithms and Combinatorics, vol. 24. Springer, New York (2003)
13. Håstad, J.: Some optimal inapproximability results. J. ACM 48, 798–859 (2001)

Approximability of Constrained LCS[*]

Minghui Jiang

Department of Computer Science, Utah State University, Logan, UT 84322, USA
mjiang@cc.usu.edu

Abstract. The problem CONSTRAINED LONGEST COMMON SUBSEQUENCE
is a natural extension to the classical problem LONGEST COMMON SUBSE-
QUENCE, and has important applications to bioinformatics. Given k input sequences $A_1, \ldots, A_k$ and l constraint sequences $B_1, \ldots, B_l$, C-LCS(k,l) is the
problem of finding a longest common subsequence of $A_1, \ldots, A_k$ that is also a
common supersequence of $B_1, \ldots, B_l$. Gotthilf et al. gave a polynomial-time
algorithm that approximates C-LCS$(k,1)$ within a factor $\sqrt{\hat{m}|\Sigma|}$, where $\hat{m}$
is the length of the shortest input sequence and $|\Sigma|$ is the alphabet size. They
asked whether there are better approximation algorithms and whether there exists a lower bound. In this paper, we answer their questions by showing that their
approximation factor $\sqrt{\hat{m}|\Sigma|}$ is in fact already very close to optimal although a
small improvement is still possible:

1. For any computable function f and any $\epsilon > 0$, there is no polynomial-time
   algorithm that approximates C-LCS$(k,1)$ within a factor $f(|\Sigma|) \cdot \hat{m}^{1/2-\epsilon}$
   unless NP = P. Moreover, this holds even if the constraint sequence is unary.
2. There is a polynomial-time randomized algorithm that approximates C-LCS
   $(k,1)$ within a factor $|\Sigma| \cdot O(\sqrt{\text{OPT} \cdot \log\log\text{OPT}/\log\text{OPT}})$ with high
   probability, where OPT is the length of the optimal solution, OPT $\leq \hat{m}$.

For the problem over an alphabet of arbitrary size, we show that

3. For any $\epsilon > 0$, there is no polynomial-time algorithm that approximates
   C-LCS$(k,1)$ within a factor $\hat{m}^{1-\epsilon}$ unless NP = P.
4. There is a polynomial-time algorithm that approximates C-LCS$(k,1)$ within
   a factor $O(\hat{m}/\log\hat{m})$.

We also present some complementary results on exact and parameterized algorithms for C-LCS(k,l).

1 Introduction

LONGEST COMMON SUBSEQUENCE (LCS) is a fundamental problem in computer science. Given two input sequences A_1 and A_2 and one constraint sequence B, CON-
STRAINED LONGEST COMMON SUBSEQUENCE (C-LCS) is the problem of finding a
longest common subsequence of A_1 and A_2 that is also a supersequence of B. The
problem C-LCS is a natural extension to the classical problem LCS, and has application to computing the homology of two biological sequences with a specific or putative
structure in common [14].

We review some standard notations. For a sequence S, let $S[i]$ denote the letter of
S at position i, let $S[i,j]$ denote the subsequence of S starting at position i and ending

---

[*] Supported in part by NSF grant DBI-0743670.

O. Cheong, K.-Y. Chwa, and K. Park (Eds.): ISAAC 2010, Part II, LNCS 6507, pp. 180–191, 2010.

at position j (the subsequence is empty when $i > j$), and let $|S|$ denote the length of S. For two sequences S and T, let ST denote the concatenation of S and T, and write $S \preceq T$ if S is a subsequence of T. For a letter σ and a positive integer w, let σ^w denote a unary sequence consisting of w repetitions of σ.

The problem C-LCS can be easily generalized to a problem C-LCS(k, l) for an arbitrary number k of input sequences and an arbitrary number l of constraint sequences [6]:

Problem C-LCS(k, l)

Instance: k input sequences $A_1, \ldots, A_k$ and l constraint sequences $B_1, \ldots, B_l$ over an alphabet Σ, where $k \geq 2$, $l \geq 1$, and $|\Sigma| \geq 2$.
Problem: Find a longest sequence C such that $C \preceq A_i$ for each i, $1 \leq i \leq k$, and $B_j \preceq C$ for each j, $1 \leq j \leq l$.

Here the input size n is the total length of the k input sequences and the l constraint sequences.

For C-LCS$(2, 1)$, the most basic version of the problem C-LCS on two input sequences A_1 and A_2 and one constraint sequence B, there are dynamic programming algorithms running in $O(|A_1| \cdot |A_2| \cdot |B|)$ time [2,5]; see also [9,1,4,3] for some related results. The problem C-LCS(k, l) becomes intractable, however, when either the number k of input sequences or the number l of constraint sequences is unbounded.

An early result of Middendorf [12] on consistent sequences of type (Super, Sub) implies that even if the input and constraint sequences are over a binary alphabet, it is already NP-hard to decide whether a given instance of C-LCS$(2, l)$ has a valid solution; see also [13]. Recently, Gotthilf et al. [6] showed that if the sequences are over an arbitrary alphabet, then even if all constraint sequences have length 1, it is again NP-hard to decide whether a given instance of C-LCS$(2, l)$ has a valid solution. On the other hand, Gotthilf et al. [6] observed that C-LCS$(k, 1)$ is NP-hard because it generalizes the classical problem LCS on an arbitrary number k of input sequences, which is known to be NP-hard even if the input sequences are over a binary alphabet [11].

C-LCS$(k, 1)$ is perhaps the most interesting variant of C-LCS because of its biological applications, hence it will be the focus of this paper. Let $A_1, \ldots, A_k$ be the k input sequences, and B be the single constraint sequence. Without loss of generality, we assume that the constraint sequence B has length at least one and is a common subsequence of the k input sequences $A_1, \ldots, A_k$. Put $\hat{m} = \min_{1 \leq i \leq k} |A_i|$ and $b = |B|$. Then $\hat{m} \geq b \geq 1$.

Gotthilf et al. [6] gave a polynomial-time algorithm that approximates C-LCS$(k, 1)$ within a factor $\sqrt{\hat{m}|\Sigma|}$, and asked whether there are better approximation algorithms and whether there exists a lower bound. In the following two theorems, we show that their approximation factor $\sqrt{\hat{m}|\Sigma|}$ is in fact already very close to optimal but nevertheless a small improvement is still possible:

Theorem 1. *For any computable function f and any $\epsilon > 0$, there is no polynomial-time algorithm that approximates C-LCS$(k, 1)$ within a factor $f(|\Sigma|) \cdot \hat{m}^{1/2-\epsilon}$ unless NP $=$ P. Moreover, this holds even if the constraint sequence is unary.*

Theorem 2. *There is a polynomial-time randomized algorithm that approximates C-LCS$(k, 1)$ within a factor $|\Sigma| \cdot O(\sqrt{\text{OPT} \cdot \log \log \text{OPT}/ \log \text{OPT}})$ with high probability, where OPT is the length of the optimal solution, OPT $\leq \hat{m}$.*

For an alphabet of arbitrary size, we can have $|\Sigma| = \Theta(\hat{m})$. Then the approximation factor of Gotthilf et al.'s algorithm [6] becomes $\sqrt{\hat{m}|\Sigma|} = \Theta(\hat{m})$. In the following two theorems, we show that again this approximation factor $\Theta(\hat{m})$ is very close to optimal but nevertheless a small improvement is possible:

Theorem 3. *For any $\epsilon > 0$, there is no polynomial-time algorithm that approximates* C-LCS$(k, 1)$ *within a factor $\hat{m}^{1-\epsilon}$ unless* NP $=$ P.

Theorem 4. *There is a polynomial-time algorithm that approximates* C-LCS$(k, 1)$ *within a factor $O(\hat{m}/\log \hat{m})$.*

Although the focus of this paper is on approximability, we also obtain some complementary results on exact and parameterized algorithms. The following theorem shows that C-LCS(k, l) is fixed-parameter tractable with both the alphabet size $|\Sigma|$ and the optimal solution length OPT as parameters, and is polynomially solvable if both k and l are constants:

Theorem 5. C-LCS(k, l) *admits an exact algorithm running in time $O(|\Sigma|^{\text{OPT}+1} \cdot n)$, where* OPT *is the length of the optimal solution, and admits an exact algorithm running in time $O(\prod_{i=1}^{k}(|A_i| + 1) \cdot \prod_{j=1}^{l}(|B_j| + 1) \cdot (k + l))$.*

2 Approximation Lower Bounds for C-LCS$(k, 1)$

2.1 Proof of Theorem 1

We prove the inapproximability of C-LCS$(k, 1)$ by a reduction from MAX-CLIQUE. Our construction is inspired by Middendorf [12, Theorem 2(b)].

Let G be a graph with n vertices and m edges. We construct a C-LCS$(k, 1)$ instance consisting of

$$k = \binom{n}{2} - m + 1$$

input sequences $A_1, \ldots, A_k$ over a binary alphabet and a single constraint sequence B.

The constraint sequence B is a unary sequence of $n - 1$ zeros:

$$B = 0^{n-1}.$$

The last input sequence A_k consists of n^2 ones and $n - 1$ zeros:

$$A_k = 1^n (01^n)^{n-1}.$$

Let $V = \{1, \ldots, n\}$ be the set of vertices of the graph G. Let $\bar{E}$ be the $\binom{n}{2} - m$ pairs of vertices of the graph G that are not edges. For each pair of vertices $\bar{e}_j = \{u, v\} \in \bar{E}$, $1 \le j \le \binom{n}{2} - m$ and $1 \le u < v \le n$, we construct a corresponding input sequence A_j of $n^2 - n$ ones and n zeros:

$$A_j = (1^n 0)^{u-1} \; 0(1^n 0)^{v-u} \; (01^n)^{n-v}.$$

For comparison, observe that

$$A_k = (1^n 0)^{u-1}\ 1^n (01^n)^{v-u}\ (01^n)^{n-v}.$$

We use the term *one-block* to refer to a substring of n consecutive ones in the input sequences. Note that each input sequence A_j for $1 \le j \le \binom{n}{2} - m$ consists of $n-1$ one-blocks and n zeros, while the last input sequence A_k consists of n one-blocks and $n-1$ zeros. Thus $\hat{m} = (n-1)n + n = n^2$. This completes the construction. We refer to Figure 1 for an example.

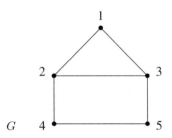

$A_1 =$		0	11111	0	11111	0	11111	0	0	11111
$A_2 =$		0	11111	0	11111	0	11111	0	11111	0
$A_3 =$	11111	0	0	11111	0	11111	0	11111	0	
$A_4 =$	11111	0	11111	0	0	11111	0	0	11111	
$A_5 =$	11111	0	11111	0	11111	0	11111	0	11111	
$B\ \ =$				0	0	0	0			
$C\ \ =$		11111	0	11111	0	11111	0	0		

Fig. 1. A graph G with $n = 5$ vertices and $m = 6$ edges. The C-LCS$(k,1)$ instance of $k = \binom{5}{2} - 6 + 1 = 5$ input sequences $A_1, \ldots, A_5$ and a single constraint sequence B. The four input sequences A_1, A_2, A_3, A_4 correspond to the 4 non-edges $\{1,4\}, \{1,5\}, \{2,5\}, \{3,4\}$. The sequence C corresponds to the clique $\{1, 2, 3\}$.

Lemma 1. *There is a clique K of q vertices in the graph G if and only if there is a sequence C of length $\ell = (q + 1)n - 1$ that is a subsequence of each input sequence A_i and is a supersequence of the constraint sequence B.*

Proof. We first prove the direct implication. Suppose there is a clique K of q vertices in the graph G. Let $i_1, \ldots, i_q$ be the q vertices in K, where $i_1 < \ldots < i_q$. We construct a sequence C of qn ones and $n - 1$ zeros as follows:

$$C = 0^{i_1 - 1}\ 1^n\ 0^{i_2 - i_1}\ 1^n\ \cdots\ 0^{i_q - i_{q-1}}\ 1^n\ 0^{n - i_q}.$$

Note that C can be obtained from A_k by deleting all one-blocks except those with block indices $i_1, \ldots, i_q$. Clearly, C is a subsequence of A_k and is a supersequence of B. For each input sequence A_j, $1 \le j \le \binom{n}{2} - m$, the two vertices u and v of the corresponding non-edge $\bar{e}_j$ cannot be both in the clique K. Consider two cases:

1. If $u \notin K$, then C is a subsequence of the following common subsequence of A_j and A_k

$$(1^n 0)^{u-1} \ (01^n)^{v-u} \ (01^n)^{n-v},$$

   which can be obtained either from A_j by deleting a zero, or from A_k by deleting the one-block with block index u.
2. If $v \notin K$, then C is a subsequence of the following common subsequence of A_j and A_k

$$(1^n 0)^{u-1} \ (1^n 0)^{v-u} \ (01^n)^{n-v},$$

   which can be obtained either from A_j by deleting a zero, or from A_k by deleting the one-block with block index v.

In either case, C is a subsequence of A_j.

We next prove the reverse implication. Suppose there is sequence C of length $\ell = (q+1)n - 1$ that is a subsequence of each input sequence A_i and is a supersequence of the constraint sequence B. Then C must contain exactly qn ones and exactly $n-1$ zeros, because A_k and B have the same number $n-1$ of zeros. Note that the ones in the input sequences are grouped into one-blocks. When selecting the common subsequence C from each input sequence, we can select the ones from left to right in each one-block and add the remaining ones of a one-block if it is only partially selected. In this way, we obtain a sequence C' that is a supersequence of C and is still a subsequence of each input sequence. Moreover, C' consists of at least q one-blocks and exactly $n-1$ zeros. Let K be the set of vertices corresponding to the block indices of these one-blocks in A_k. We claim that K is a clique in the graph G, that is, for each non-edge $\bar{e}_j = \{u, v\}$, either u or v is not in K.

We prove this claim by contradiction. Suppose that both vertices u and v of some non-edge $\bar{e}_j$ are in K. Then the corresponding one-blocks with block indices u and v in A_k are selected in C'. Since all $n-1$ zeros in A_k are selected in C', C' contains exactly $u-1$ zeros before the one-block u, exactly $n-v$ zeros after the one-block v, and exactly $v-u$ zeros between them. Observe that for any two one-blocks in A_j, if there are at least $u-1$ zeros before the left one-block and there are at least $n-v$ zeros after the right one-block, then these two one-blocks must both come from the middle part $0(1^n 0)^{v-u}$ of A_j, and hence have at most $v-u-1$ zeros between them. Thus C' cannot be a subsequence of A_j. This is a contradiction. □

We now prove the approximation lower bounds for C-LCS$(k, 1)$. Suppose there is a polynomial-time algorithm that approximates C-LCS$(k, 1)$ within a factor $f(|\Sigma|) \cdot \hat{m}^{1/2-\epsilon}$ for some computable function f and some $\epsilon > 0$. Then we can obtain a polynomial-time algorithm that approximates MAX-CLIQUE (on a graph G of n vertices) within a factor $n^{1-\epsilon}$ as follows:

1. If $n < (2f(2))^{1/\epsilon}$, use a brute-force algorithm to find a maximum clique in G, then return the clique.
2. Construct a C-LCS$(k, 1)$ instance as in our reduction, use the $f(|\Sigma|) \cdot \hat{m}^{1/2-\epsilon}$-approximation algorithm to find a subsequence of length ℓ, then obtain a clique of size q in G following the reverse implication of Lemma 1. If $q \geq 1$, return the clique of size q. Otherwise, return any single vertex in G as a clique of size 1.

The algorithm clearly finds an optimal solution in constant time if it returns in Step 1. Now assume that $n \geq (2f(2))^{1/\epsilon}$, and proceed to Step 2. Let q^* be the maximum size of a clique in G. Let ℓ^* be the maximum length of a constrained common subsequence for the reduced C-LCS$(k, 1)$ instance. By Lemma 1, we have $\ell^* = (q^* + 1)n - 1$. The algorithm finds a subsequence of length

$$\ell \geq \frac{\ell^*}{f(2) \cdot \hat{m}^{1/2-\epsilon}} = \frac{(q^* + 1)n - 1}{f(2) \cdot \hat{m}^{1/2-\epsilon}} \geq \frac{q^* + 1}{f(2) \cdot \hat{m}^{1/2-\epsilon}} n - 1,$$

then obtains a clique of size

$$q \geq \frac{q^* + 1}{f(2) \cdot \hat{m}^{1/2-\epsilon}} - 1.$$

Recall that $\hat{m} = n^2$ and $n \geq (2f(2))^{1/\epsilon}$. It follows that

$$\max\{q, 1\} \geq \frac{q+1}{2} \geq \frac{q^* + 1}{2f(2) \cdot \hat{m}^{1/2-\epsilon}} > \frac{q^*}{2f(2) \cdot n^{1-2\epsilon}} = \frac{n^\epsilon}{2f(2)} \cdot \frac{q^*}{n^{1-\epsilon}} \geq \frac{q^*}{n^{1-\epsilon}}.$$

Let us recall the following result of Zuckerman which improves an earlier result of Håstad [8]:

Theorem 6 (Zuckerman 2007 [15]). *For any $\epsilon > 0$, there is no polynomial-time algorithm that approximates* MAX-CLIQUE *within a factor $n^{1-\epsilon}$ unless* NP = P.

By our reduction, it follows that for any computable function f and any $\epsilon > 0$, there is no polynomial-time algorithm that approximates C-LCS$(k, 1)$ within a factor $f(|\Sigma|) \cdot \hat{m}^{1/2-\epsilon}$ unless NP = P. The proof of Theorem 1 is now complete.

2.2 Proof of Theorem 3

It is easy to check that the reduction that Jiang and Li [10] used to prove the inapproximability of LCS over an arbitrary alphabet is an L-reduction from MAX-CLIQUE. Thus, in conjunction with the result of Zuckerman [15], this L-reduction actually implies the following theorem although it is not explicitly stated in their paper:

Theorem 7 (Jiang and Li 1995 [10]). *For any $\epsilon > 0$, there is no polynomial-time algorithm that approximates* LCS *over an arbitrary alphabet within a factor $\hat{m}^{1-\epsilon}$ unless* NP = P, *where $\hat{m}$ is the length of the shortest input sequence.*

Since C-LCS$(k, 1)$ over an arbitrary alphabet includes LCS over an arbitrary alphabet as a special case, Theorem 3 immediately follows.

3 Improved Approximation Algorithms for C-LCS$(k, 1)$

3.1 Proof of Theorem 2

Let C^* be a constrained longest common subsequence. Since $B \preceq C^*$, we can embed B inside C^* in some fixed way such that $C^* = C_0 B[1] C_1 \ldots B[b] C_b$, then assign each index a between 0 and b (which we call a *slot*) a *value* that is the length of the subsequence C_a.

The previous approximation algorithm of Gotthilf et al. [6] is essentially a greedy algorithm that composes a constrained common subsequence from two parts: the constraint sequence B itself and a $|\Sigma|$-approximation of the subsequence C_a for a slot a of the highest value. Our improved algorithm for C-LCS$(k, 1)$ uses Gotthilf et al.'s greedy algorithm [6] as the first step, then supplements it with a brute-force algorithm and a random procedure. Instead of betting on a single large slot, the random procedure guesses a large number s of slots of high value. Intuitively, the random procedure and the greedy algorithm complement each other in their respective worst cases. Then an improved approximation ratio can be obtained by balancing them with suitably chosen parameters.

Algorithm A1

1. Run Gotthilf et al.'s algorithm [6] to find a constrained common subsequence:

   (a) For each slot a, $0 \le a \le b$, do the following:

       i. Partition the constraint sequence $B \rightarrow B[1, a]\, B[a + 1, b]$.
       ii. Partition each input sequence $A_i \rightarrow L_{i,a}\, M_{i,a}\, R_{i,a}$ such that $B[1, a] \preceq L_{i,a}$, $B[a + 1, b] \preceq R_{i,a}$, and $M_{i,a}$ is maximal.
       iii. Find a longest unary sequence M_a that is a common subsequence of $M_{i,a}$, $1 \le i \le k$.

   (b) Compose a sequence $B[1, a]\, M_a\, B[a + 1, b]$ for a slot a such that $|M_a|$ is maximum.

2. Let z be the smallest positive integer[1] such that for all integers $\ell \ge z$,

$$\ell \ge 16\lceil \log \ell / \log \log \ell \rceil^3. \tag{1}$$

   For each integer ℓ, $b + 1 \le \ell < z$, use brute force to find a constrained common subsequence of length ℓ if it exists: enumerate all $|\Sigma|^\ell$ candidate sequences of length ℓ, and for each candidate sequence check whether it is a supersequence of the constraint sequence B and is a common subsequence of the input sequences A_i, $1 \le i \le k$.

3. For each integer ℓ, $\max\{b + 1, z\} \le \ell \le \hat{m}$, set the parameters

$$s = \lceil \log \ell / \log \log \ell \rceil, \quad l = \lfloor (\ell \cdot \log \log \ell / \log \ell)^{1/2} \rfloor - 1, \quad \text{and} \quad w = \left\lceil \frac{l}{|\Sigma|} \right\rceil,$$

   then for some tunable constant r (which controls the probability), repeat the following random procedure for $r(4s)^s$ rounds to find a constrained common subsequence of length $b + sw$:

[1] A calculation shows that $z = 324$. Our choice of this value is somewhat arbitrary. We set the parameter z to a concrete value here mostly for convenience, so that later in the analysis we can prove a concrete approximation ratio $\lambda \le |\Sigma|\sqrt{\text{OPT} \cdot \log \log \text{OPT}/ \log \text{OPT}}$ without using the big-O notation. In actual implementation, we can set z to a smaller value, which results in a reduced running time of Step 2 at the cost of an increased approximation ratio λ that is still $O(|\Sigma|\sqrt{\text{OPT} \cdot \log \log \text{OPT}/ \log \text{OPT}})$.

(a) Randomly select (sample with replacement) s slots between 0 and b. Sort the s slots in ascending order: $0 \leq b_1 \leq \ldots \leq b_s \leq b$. Randomly select a letter $\sigma_i \in \Sigma$ for each slot b_i, $1 \leq i \leq s$.

(b) If the s slots are all distinct, that is, $0 \leq b_1 < \ldots < b_s \leq b$, compose a candidate constrained sequence

$$B[1, b_1] \, \sigma_1^w \, \ldots \, B[b_{s-1}+1, b_s] \, \sigma_s^w \, B[b_s+1, b],$$

and check whether it is a common subsequence of the input sequences A_i, $1 \leq i \leq k$.

4. Return the longest constrained sequence found.

Approximation Ratio. Let OPT be the length of the constrained longest common subsequence C^*. Let APX_1, APX_2, and APX_3, respectively, be the maximum length of a constrained common subsequence found in Step 1, Step 2, and Step 3 of the algorithm. Put $\lambda_1 = \text{OPT}/APX_1$, $\lambda_2 = \text{OPT}/APX_2$, $\lambda_3 = \text{OPT}/APX_3$, and $\lambda = \min\{\lambda_1, \lambda_2, \lambda_3\}$.

We clearly have $b \leq \text{OPT} \leq \hat{m}$. If $\text{OPT} = b$, then $APX_1 = \text{OPT}$ and $\lambda_1 = 1$. Also, if $b+1 \leq \text{OPT} < z$, then $APX_2 = \text{OPT}$ and $\lambda_2 = 1$. So suppose that

$$\max\{b+1, z\} \leq \text{OPT} \leq \hat{m}.$$

Then OPT is equal to ℓ for some iteration in Step 3.

Put

$$f = \sqrt{(\log \text{OPT}/\log\log \text{OPT})/|\Sigma|}, \tag{2}$$

We will show that

$$\lambda \leq \sqrt{\text{OPT} \cdot |\Sigma|}/f,$$

hence

$$\lambda \leq |\Sigma|\sqrt{\text{OPT} \cdot \log\log \text{OPT}/\log \text{OPT}} \leq |\Sigma|\sqrt{\hat{m} \cdot \log\log \hat{m}/\log \hat{m}}.$$

We first look at Step 1. Let h be the highest value of a slot. Clearly,

$$h \geq \left\lceil \frac{\text{OPT} - b}{b+1} \right\rceil \geq 1.$$

Gotthilf et al.'s algorithm [6] finds a constrained common subsequence of length

$$APX_1 \geq b + \left\lceil \frac{h}{|\Sigma|} \right\rceil \geq \max\left\{ b+1, \frac{\text{OPT}}{(b+1)|\Sigma|} \right\},$$

thus

$$\lambda_1 \leq \min\left\{ \frac{\text{OPT}}{b+1}, (b+1)|\Sigma| \right\}.$$

If $h \geq \sqrt{\text{OPT} \cdot |\Sigma|} \cdot f$, then $APX_1 \geq \sqrt{\text{OPT}/|\Sigma|} \cdot f$ and $\lambda_1 \leq \sqrt{\text{OPT} \cdot |\Sigma|}/f$. So suppose that

$$h \leq \sqrt{\text{OPT} \cdot |\Sigma|} \cdot f. \tag{3}$$

If $b+1 \geq \sqrt{OPT/|\Sigma|} \cdot f$ or $b+1 \leq \sqrt{OPT/|\Sigma|}/f$, then again $\lambda_1 \leq \sqrt{OPT \cdot |\Sigma|}/f$. So suppose that

$$\sqrt{OPT/|\Sigma|}/f \leq b+1 \leq \sqrt{OPT/|\Sigma|} \cdot f. \tag{4}$$

Now proceed to Step 3. Consider the iteration where $\ell = OPT$. From (2) we have

$$s = \lceil \log OPT/\log\log OPT \rceil = \lceil f^2|\Sigma| \rceil, \tag{5}$$

and

$$l = \lfloor (OPT \cdot \log\log OPT/\log OPT)^{1/2} \rfloor - 1 = \lfloor \sqrt{OPT/|\Sigma|}/f \rfloor - 1. \tag{6}$$

Also, from (1) we have

$$OPT \geq 16s^3. \tag{7}$$

Let t be the number of slots of value at least l. Then the number of slots of value less than l is $b+1-t$. Since

$$OPT \leq b + t \cdot h + (b+1-t) \cdot l,$$

we have

$$t \geq \frac{OPT - b - (b+1)l}{h-l} \geq \frac{OPT - (b+1)(l+1)}{h}$$
$$\geq \frac{OPT - \sqrt{OPT/|\Sigma|} \cdot f \cdot \sqrt{OPT/|\Sigma|}/f}{\sqrt{OPT \cdot |\Sigma|} \cdot f}$$
$$= (1 - 1/|\Sigma|)\sqrt{OPT/|\Sigma|}/f \geq \frac{1}{2}\sqrt{OPT/|\Sigma|}/f, \tag{8}$$

where the third inequality follows from (3), (4), and (6). Then, from (8), (7), and (5) we have

$$t \geq \frac{1}{2}\sqrt{\frac{OPT}{f^2|\Sigma|}} \geq \frac{1}{2}\sqrt{\frac{16s^3}{s}} = 2s, \tag{9}$$

and from (9) and (5) we have

$$b \geq \frac{b+1}{2} \geq \frac{t}{2} \geq s \geq 2f^2. \tag{10}$$

If a constrained common subsequence of length $b + sw$ is found in this iteration of Step 3, then from (6) and (10) we have

$$APX_3 \geq b + f^2|\Sigma| \cdot \frac{l}{|\Sigma|} \geq b + \sqrt{OPT/|\Sigma|} \cdot f - 2f^2 \geq \sqrt{OPT/|\Sigma|} \cdot f,$$

thus

$$\lambda_3 \leq \sqrt{OPT \cdot |\Sigma|}/f.$$

Probability. We now estimate the probability that a constrained common subsequence is found in Step 3 in the iteration where $\ell = OPT$. First consider the probability p that a constrained common subsequence is found in one round of the random procedure. Since the random procedure always finds a constrained common subsequence if it guesses correctly s distinct slots of value at least l, and guesses correctly the dominating letter for each of the s slots, we have

$$p \geq \frac{t!/(t-s)!}{(b+1)^s} \cdot \frac{1}{|\Sigma|^s}. \tag{11}$$

From (9), we have $t!/(t-s)! \geq (t/2)^s$. From (8) and (4), we have $t/(b+1) \geq 1/(2f^2)$. Also recall (5) that $s = \lceil f^2|\Sigma| \rceil$. Thus

$$p \geq \left(\frac{t/2}{(b+1)|\Sigma|} \right)^s \geq \left(\frac{1}{4f^2|\Sigma|} \right)^s \geq \left(\frac{1}{4s} \right)^s.$$

Put $x = (4s)^s$. Then each round of the random procedure finds a constrained common subsequence with probability at least $1/x$. Since the random procedure is repeated for rx rounds, the probability that a constrained common subsequence is not found in rx consecutive rounds is at most

$$(1 - 1/x)^{rx} \leq 1/e^r,$$

which can be made arbitrarily small by choosing the constant r sufficiently large.

Time Complexity.   Step 1 and Step 2 are clearly polynomial. For Step 3 to be polynomial, it is sufficient that $s = O(\log n / \log \log n)$ so that $(4s)^s = \text{poly}(n)$, where n is the input size. This is clearly satisfied since $s = \lceil \log \ell / \log \log \ell \rceil$ and $\ell \leq \hat{m} \leq n$.

3.2   Proof of Theorem 4

We obtain an $O(\hat{m}/\log \hat{m})$ approximation for C-LCS$(k, 1)$ over an arbitrary alphabet using Halldórsson's partitioning technique [7]. Assume without loss of generality that $\hat{m} \geq 2$.

Algorithm A2

1. Find a shortest input sequence $\hat{A}$, which has length $\hat{m}$, and partition it into $q \leq \lceil \hat{m}/\log \hat{m} \rceil$ substrings $\hat{A} \rightarrow S_1 \dots S_q$ such that each substring S_p, $1 \leq p \leq q$, has length at most $\lceil \log \hat{m} \rceil$.
2. For each pair of indices u and v, $0 \leq u < v \leq b+1$, and for each subsequence T of each substring S_p, $1 \leq p \leq q$, compose a candidate constrained sequence

$$B[1, u] \, T \, B[v, b],$$

   and check whether it is a supersequence of the constraint sequence B and is a common subsequence of the input sequences A_i, $1 \leq i \leq k$.
3. Return the longest constrained sequence found.

Approximation Ratio.   Let C^* be a constrained longest common subsequence. Since $\hat{A} = S_1 \ldots S_q$ and $C^* \preceq \hat{A}$, we can partition C^* into q substrings $C^* \to C_1 \ldots C_q$ such that $C_p \preceq S_p$ for $1 \leq p \leq q$. Similarly, since $C^* = C_1 \ldots C_q$ and $B \preceq C^*$, we can partition B into q substrings $B \to T_1 \ldots T_q$ such that $T_p \preceq C_p$ for $1 \leq p \leq q$.

By the Pigeonhole principle, at least one of the q substrings of C^*, say C_p, has length at least $1/q$ times the length of C^*. This substring is enumerated by the algorithm as some subsequence T of S_p. Let $B[1, u] = T_1 \ldots T_{p-1}$ and $B[v, b] = T_{p+1} \ldots T_q$. Then

$$B[1, u] \, T \, B[v, b] = T_1 \ldots T_{p-1} \, C_p \, T_{p+1} \ldots T_q \preceq C_1 \ldots C_{p-1} \, C_p \, C_{p+1} \ldots C_q = C^*.$$

The length of $B[1, u] \, T \, B[v, b]$ is at least the length of T, which is at least $1/q \geq 1/\lceil \hat{m}/\log \hat{m} \rceil$ times the length of C^*.

Time Complexity.   The dominating step of the algorithm is Step 2. There are $O(b^2)$ pairs of indices u and v, $\lceil \hat{m}/\log \hat{m} \rceil$ substrings S_p, and at most $2^{\lceil \log \hat{m} \rceil} = O(\hat{m})$ subsequences T of each substring S_p. Thus the total number of candidate constrained sequences is $O(b^2 \hat{m}^2 / \log \hat{m})$. For each candidate constrained sequence, it takes $O(n)$ time to check whether it is valid. The overall running time of the algorithm is polynomial.

4   Exact Algorithms for C-LCS(k, l)

In this section we prove Theorem 5 by presenting two exact algorithms for C-LCS(k, l).

Our first exact algorithm, which runs in $O(|\Sigma|^{\text{OPT}+1} \cdot n)$ time, is a trivial brute-force algorithm: for $\ell = 1, \ldots, \hat{m}$, enumerate all $|\Sigma|^\ell$ sequences of length ℓ, then for each candidate sequence check in $O(n)$ time whether it is a constrained common subsequence; stop the iteration if for some ℓ no candidate sequence of length ℓ is a constrained common subsequence.

Our second exact algorithm is based on dynamic programming, and achieves a running time of $O(\prod_{i=1}^{k}(|A_i| + 1) \cdot \prod_{j=1}^{l}(|B_j| + 1) \cdot (k + l))$. For simplicity, we only compute the maximum length of a constrained common subsequence (or report that the problem has no solution). By standard techniques, an actual constrained common subsequence of the maximum length can be found (if it exists) within the same running time.

Denote by $L(a_1, \ldots, a_k; b_1, \ldots, b_l)$ the maximum length of a constrained common subsequence for the subproblem with partial input sequences $A_1[1, a_1], \ldots, A_k[1, a_k]$ and partial constraint sequences $B_1[1, b_1], \ldots, B_l[1, b_l]$, where $0 \leq a_i \leq |A_i|$ and $0 \leq b_j \leq |B_j|$ for $1 \leq i \leq k$ and $1 \leq j \leq l$. We use the value $-\infty$ to indicate that a subproblem has no solution. The desired entry is $L(|A_1|, \ldots, |A_k|; |B_1|, \ldots, |B_l|)$.

The base cases are

$$L(0, \ldots, 0; 0, \ldots, 0) = 0,$$

and

$$L(a_1, \ldots, a_k; b_1, \ldots, b_l) = -\infty, \quad \text{if } \min_{1 \leq i \leq k} a_i < \max_{1 \leq j \leq l} b_j.$$

The recurrence is

$$L(a_1, \ldots, a_k; b_1, \ldots, b_l)$$

$$= \max \begin{cases} \max\limits_{1 \le i \le k} L(a_1, \ldots, a_{i-1}, a_i - 1, a_{i+1}, \ldots, a_k; b_1, \ldots, b_l) \\ L(a_1 - 1, \ldots, a_k - 1; b'_1, \ldots, b'_l) + 1, \end{cases}$$

where the second case applies only if $A_1[a_1] = \ldots = A_k[a_k] = \sigma$ for some $\sigma \in \Sigma$; then we let $b'_j = b_j - 1$ if $B_j[b_j] = \sigma$ and let $b'_j = b_j$ if $B_j[b_j] \ne \sigma$.

References

1. Ann, H.-Y., Yang, C.-B., Tseng, C.-T., Hor, C.-Y.: Fast algorithms for computing the constrained LCS of run-length encoded strings. In: Proceedings of the 2009 International Conference on Bioinformatics & Computational Biology (BIOCOMP 2009), pp. 646–649 (2009)
2. Arslan, A., Eğecioğlu, Ö.: Algorithms for the constrained longest common subsequence problems. International Journal of Foundations of Computer Science 16, 1099–1109 (2005)
3. Bonizzoni, P., Della Vedova, G., Dondi, R., Pirola, Y.: Variants of constrained longest common subsequence. Information Processing Letters 110, 877–881 (2010)
4. Chen, Y.-C., Chao, K.-M.: On the generalized constrained longest common subsequence problems. Journal of Combinatorial Optimization, doi:10.1007/s10878-009-9262-5.
5. Chin, F.Y.L., De Santis, A., Ferrara, A.L., Ho, N.L., Kim, S.K.: L Ho, and S. K. Kim. A simple algorithm for the constrained sequence problems. Information Processing Letters 90, 175–179 (2004)
6. Gotthilf, Z., Hermelin, D., Lewenstein, M.: Constrained LCS: hardness and approximation. In: Ferragina, P., Landau, G.M. (eds.) CPM 2008. LNCS, vol. 5029, pp. 255–262. Springer, Heidelberg (2008)
7. Halldórsson, M.M.: Approximation via partitioning. Technical Report IS-RR-95-0003F, School of Information Science, Japan Advanced Institute of Science and Technology, Hokuriku (1995)
8. Håstad, J.: Clique is hard to approximate within $n^{1-\epsilon}$. Acta Mathematica 182, 105–142 (1999)
9. Iliopoulos, C.S., Rahman, M.S.: New efficient algorithms for the LCS and constrained LCS problems. Information Processing Letters 106, 13–18 (2008)
10. Jiang, T., Li, M.: On the approximation of shortest common supersequences and longest common subsequences. SIAM Journal on Computing 24, 1122–1139 (1995)
11. Maier, D.: The complexity of some problems on subsequences and supersequences. Journal of the ACM 25, 322–336 (1978)
12. Middendorf, M.: On finding minimal, maximal, and consistent sequences over a binary alphabet. Theoretical Computer Science 145, 317–327 (1995)
13. Middendorf, M., Manlove, D.F.: Combined super-/substring and super-/subsequence problems. Theoretical Computer Science 320, 247–267 (2004)
14. Tsai, Y.-T.: The constrained longest common subsequence problem. Information Processing Letters 88, 173–176 (2003)
15. Zuckerman, D.: Linear degree extractors and the inapproximability of Max Clique and Chromatic Number. Theory of Computing 3, 103–128 (2007)

Approximation Algorithms for the Multi-Vehicle Scheduling Problem

Binay Bhattacharya* and Yuzhuang Hu

School of Computing Science, Simon Fraser University, Burnaby, Canada, V5A 1S6
{binay,yhu1}@cs.sfu.ca

Abstract. In this paper we investigate approximation algorithms for the multi-vehicle scheduling problem (MVSP). In MVSP we are given a graph $G = (V, E)$, where each vertex u of V is associated with a job $j(u)$, and each edge e has a non-negative weight $w(e)$. There are m identical vehicles available to service the jobs. Each job $j(u)$ has its own release time $r(u)$ and handling time $h(u)$. A job $j(u)$ can only be serviced by one vehicle after its release time $r(u)$, and the handling time $h(u)$ represents the time needed to finish processing $j(u)$. The objective is to find a schedule in which the maximum completion time of the jobs, i.e. the makespan, is minimized. In this paper we present a 3-approximation algorithm for MVSP on trees, and a $(5 - \frac{2}{m})$-approximation algorithm for MVSP on general graphs.

1   Introduction

In the vehicle routing problem (VRP) [5], we are given m vehicles and a complete graph G where each edge is associated with non-negative edge cost satisfying the triangle inequality. The objective is to find a separate tour for each vehicle such that the total cost of the tour is minimized. In the typical setting of VRP, a depot node o is also given, and in each tour the vehicle should start from and end at the depot.

VRP is NP-hard as it is a generalization of the famous traveling salesman problem (TSP) [13]. VRP in fact represents one of the most challenging tasks in the area of optimization, and has been extensively investigated by many researchers. Various tools, such as integer programming, local search, and clustering, have been used to solve this class of problems [17]. As a well-recognized approach of dealing with NP-hard problems, good approximation algorithms for VRP are of great theoretical and practical importance. In this paper, we study approximation algorithms for a variant of VRP. More specifically, we investigate approximation algorithms for the multi-vehicle scheduling problem (MVSP) [9].

1.1   Multi-Vehicle Scheduling Problem (MVSP)

In MVSP we are given a graph $G = (V, E)$, where each vertex u of V is associated with a job $j(u)$, and each edge e has a non-negative weight $w(e)$. There are m

---

* Research was partially supported by MITACS and NSERC.

O. Cheong, K.-Y. Chwa, and K. Park (Eds.): ISAAC 2010, Part II, LNCS 6507, pp. 192–205, 2010.

identical vehicles available to service the jobs. Each job $j(u)$ has its own release time $r(u)$ and handling time $h(u)$. A job $j(u)$ can only be serviced by one vehicle after its release time $r(u)$, and the handling time $h(u)$ represents the time needed to finish processing $j(u)$. This implies that when the vehicle arrives earlier, the vehicle may have to wait until the time $r(u)$ to service $j(u)$. The vehicle may also choose to move to other vertices and come back to u later to service $j(u)$. The objective of MVSP is to find a schedule in which the maximum completion time of the jobs, i.e., the makespan, is minimized.

MVSP is a variation of the classical VRP with time windows (VRPTW). In VRPTW each vertex u in the graph is associated with a full time window $[r(u), d(u)]$, where $r(u)$ and $d(u)$ denote the release time and the deadline for the job $j(u)$ respectively. Time windows are called *soft*, if they are considered non-biding, i.e., the vehicle is allowed to arrive earlier or later at a customer node. However, when the time window constraint is violated, a separate penalty cost of earliness or lateness will incur. Time windows are called *hard*, if the customer nodes must be serviced within the specified time intervals. A survey for VRPTW can be found in [17]. For approximation algorithms, Yehuda et al. gave an $O(\log n)$-approximation algorithm for VRPTW on a line in a geometric space [3]; Blum et al. presented an $O(\log^2 n)$-approximation algorithm for VRPTW on general graphs [2].

MVSP can be viewed as a relaxation of VRPTW, as in MVSP only release times are associated with the customer nodes. The current research status on MVSP is as follows. A substantial amount of research work [4] [7] [14] has been devoted to a special case of MVSP, called SVSP, where only one vehicle is available to service the customers. Nagamochi et al. [9] introduced MVSP in the general sense and gave a 2-approximation algorithm for MVSP on paths. For MVSP on trees, Nagamochi et al. [15] [16] gave a $(5 - 2/(m+1))$-approximation algorithm. There are also two polynomial time approximation schemes (PTAS) [1] [8] for MVSP on trees under certain restrictions. For MVSP on general graphs, a $(9 + \epsilon)$-approximation can be derived from Even et al. [6].

1.2   Our Results and Solution Techniques

We summarize our results as follows.

1. For MVSP on trees, we design a 3-approximation algorithm.
2. For MVSP on general graphs, we present a $(5 - \frac{2}{m})$-approximation algorithm. This algorithm is based on the 3-approximation algorithm we propose for MVSP on trees.

Our results are obtained by introducing an appropriate relaxation problem for MVSP. We apply dynamic programming to give a 3-approximation for VRP on trees. The main idea of this algorithm, is to use dynamic programming to indirectly decompose the original problem P into a set of disjoint subproblems. Approximating these subproblems separately gives us the solution for P with the desired approximation ratio.

Dynamic programming is one of the most fundamental and powerful tools for designing efficient algorithms. However, its application in approximation algorithms for VRP, is relatively new. The existing way of using dynamic programming for VRP, e.g., in [2] [9] and [10], works as follows. First a set of disjoint NP-hard subproblems is defined for the original problem P, and an approximation algorithm A with ratio α is designed for these subproblems. Typically these subproblems have good properties, and the algorithm A can approximate them well. In this method, dynamic programming is used as a master algorithm to locate a set of subproblems with cost bounded by $\beta \cdot OPT(P)$, where $OPT(P)$ denotes the optimal solution cost of P. During this process all the configurations of the subproblems are tried, and the algorithm A is applied to each of the configurations. The one with the smallest cost is chosen to be the final solution. It is easy to see that this solution is an $(\alpha\beta)$-approximation for P.

This approach relies on the fact that all the configurations of the subproblems of P can be examined in polynomial time by dynamic programming. Therefore it is only applicable when the underlying graph is a path [9], or when some ordering can be found for the underlying graph (as in the $O(\log^2 n)$-approximation algorithm for VRPTW [2], where the dynamic programming proceeds based on an ordering of the vertices). However, for VRP on trees, we do not know how to deploy this method, since for a vertex u, the number of possible configurations of the subproblems containing u is exponential in the number of children of u.

We apply dynamic programming to obtain a 3-approximation algorithm for MVSP on trees in the following way. Our algorithm consists of two steps. In the first step, we use dynamic programming to decompose the original problem P into a set of disjoint subproblems. However, as it is not possible to try all the configurations of the subproblems by directly obtaining solutions for P, we instead find a relaxation problem P' and locate a set S of disjoint subproblems. In the second step we work on approximation algorithms for the subproblems in S. We show that a good approximation for P can be obtained by approximating these subproblems well.

The two steps in our algorithm are highly connected. For MVSP on trees, the problem P' solved in the first step in fact comes from the approximation algorithm used in the second step. We locate P' in the following way. Define an edge e to be a gap if and only if in the optimal solution no vehicle passes through e to service customers. Define a gapless subproblem to be a subproblem whose underlying subgraph contains no gaps. For a gapless subproblem P_1, we design an approximation algorithm which produces a solution with cost function $f(P_1)$. Let $S(P)$ denote the set of gapless subproblems of P. We treat $\max_{P_1 \in S(P)} f(P_1)$ as a bound, and define P' to be the problem of locating a set of subproblems with the smallest possible bound $\max_{P_1 \in S(P)} f(P_1)$. More details of this algorithm can be found in Section 2 of this paper.

The rest of the paper is organized as follows. In Section 2 we present a 3-approximation algorithm for MVSP on trees. In Section 3 we show that a $(5 - \frac{2}{m})$-approximation is possible for MVSP on general graphs. In Section 4 we conclude the paper and give figure research directions.

2   3-Approximation for MVSP on Trees

In this section we show the 3-approximation algorithm for MVSP on trees. The following notation is used in describing our algorithms. Assume we are given a rooted tree T. For a node u in T, we denote the parent of u by $p(u)$, and the subtree rooted at u (including u) by T_u. Here $p(u)$ is null if u is the root. We use $w(e)$ to denote the weight or cost of edge e.

2.1   Defining the Problem P' for MVSP on Trees

To define the relaxation problem P' for MVSP on trees, we first introduce several lower bounds for a gapless subproblem of an MVSP instance on trees. These lower bounds are also used to design approximation algorithms for MVSP on paths [9]. For a gapless subproblem P_1, we define two lower bounds for the makespan:

$$LB_1(P_1) = \max_{u \in V(P_1)}\{r(u) + h(u)\}, \qquad LB_2(P_1, m') = \frac{W(P_1) + H(P_1)}{m'}$$

where $V(P_1)$ and $E(P_1)$ are the respective vertex set and edge set involved in P_1, m' represents the number of vehicles used to service the jobs in P_1, and $W(P_1)$ and $H(P_1)$ are the total edge weights of $E(P_1)$ and the total handling times of the jobs associated with the vertices in $V(P_1)$ respectively.

A collection of disjoint sets $S_1, S_2, \cdots, S_t$ is called a *partition* of the graph $G = (V, E)$, if $S_i \subseteq V$ ($1 \leq i \leq t$) and the vertices of S_i ($1 \leq i \leq t$) induce a connected subgraph in G. We define a *vehicle configuration* to be a partition of the tree where each connected component C_i in the partition is associated with a positive integer m_i and $\sum_i m_i = m$. Given a vehicle configuration VC, we call an edge a *cut edge* under VC, if its two ends are in different subsets of vertices in the partition corresponding to VC.

For a vertex u, we denote $C_{VC}(u)$ to be the set of connected components in T_u under VC. Let $S_{VC}(u)$ be the set of subproblems defined on the corresponding connected components in $C_{VC}(u)$. We define a new bound $B(G, m)$ to be $\min_{VC} \max_{P_2 \in S_{VC}(o)}(B(P_2, m') = LB_1(P_2) + (2W(P_2) + H(P_2))/m')$, where o is the root of the tree and m' is the number of vehicles allocated for P_2 under VC. We similarly define $B(T_u, m')$ where $m' \in (0, m]$ is the number of vehicles used to service the vertices in the subtree T_u. Our relaxation problem P' for MVSP is just to find the set of subproblems corresponding to the bound $B(G, m)$. We first solve P', namely compute $B(G, m)$, in strongly polynomial time by dynamic programming. By doing so we find a set of cut edges and the original problem is correspondingly decomposed to a set S of disjoint subproblems. In the second step we figure out a feasible schedule for the subproblems in S with a makespan of at most $B(G, m)$. Note that the cut edges we located might be different from the gaps that are defined on the optimal MVSP solution.

We first justify the second step of our algorithm in Lemma 1. The proof is constructive.

Lemma 1. *For MVSP on trees the subproblems corresponding to $B(G, m)$ induce a feasible MVSP solution with cost at most $B(G, m)$.*

Proof. Let VC be the vehicle configuration corresponding to $B(G,m)$. Consider a subproblem P_2 defined on a connected component under VC. Assume m' vehicles are involved in P_2 under VC. In the optimal solution for P_2, an edge might be traversed by more than one vehicles. We double the tree edges and obtain a hamiltonian path containing all the jobs (by a depth first traversal of the tree). We traverse the path from the root, and associate the handling times with the jobs when they are visited for the first time. Therefore we transformed P_2 for MVSP on trees to another subproblem P'_2 of MVSP on paths with $W(P'_2) = 2W(P_2)$, $H(P'_2) = H(P_2)$ and $LB_1(P'_2) = LB_1(P_2)$. Using the algorithm in [9], we obtain a schedule for P'_2 with cost bounded by $LB_1(P'_2) + LB_2(P'_2, m')$. This schedule is also feasible for P_2 and has a cost of at most $LB_1(P_2) + (2W(P_2) + H(P_2))/m' = B(P_2, m')$. The lemma then holds after applying this analysis to every other connected component under VC. □

Lemma 2. *For MVSP on trees the bound $B(G,m)$ corresponds to a feasible MVSP solution with cost bounded by $3 \cdot OPT(P)$.*

Proof. The gapless subproblems will be encountered when computing $B(G,m)$. The same operation in the proof of Lemma 1 can be applied on the gapless subproblems. Since $B(G,m)$ is the smallest possible, we have that $B(G,m) \leq 3 \cdot OPT(P)$. Due to Lemma 1 there exists a feasible solution with cost at most $B(G,m)$. This completes the proof. □

The above proof also shows how to solve the subproblems once they are located, therefore in the following we only focus on how to locate the appropriate subproblems.

2.2 Locating the Subproblems for MVSP on Trees

The core of our solution is an algorithm to solve the following simple feasibility decision problem D: Given a real number λ, is $B(G,m) \leq \lambda$?

Solving the Decision Problem. In the following we assume that only the leaves of the tree are associated with handling times. The general case can be easily transformed to this setting by creating a pseudo vertex u' for each vertex u in the graph and connecting u and u' by a zero cost edge. We set $h(u')$ to $h(u)$. It is easy to see that the optimal MVSP solution remains the same after this transformation.

We solve the feasibility problem also by dynamic programming. Given a vehicle configuration VC, we denote $P_{VC}(u)$ to be the subproblem defined on the connected component containing u in $C_{VC}(u)$, and $P'_{VC}(u)$ to be the set of subproblems defined on the connected components not containing u in $C_{VC}(u)$. Recall that $C_{VC}(u)$ consists of all the connected components of T_u under VC.

To solve the decision problem, we maintain a variable $obj(u)$ and a table $table(u)$ for each vertex u. The variable $obj(u)$ records the minimum number m' of vehicles needed to guarantee that $B(T_u, m') \leq \lambda$. The variable $table(u)$ has two dimensions. Assume the minimum value of an entry $table(u)[i_1][i_2]$ of $table(u)$

is obtained under the vehicle configuration VC. Then this value represents $2 \cdot W(P_{VC}(u)) + H(P_{VC}(u))$, given that i_1 equals $LB_1(P_{VC}(u))$, and i_2 vehicles have been used to service the jobs in the subproblems of $P'_{VC}(u)$. Here $LB_1(T_u)$ represents the first lower bound LB_1 for all the jobs in T_u. In other words, given the constraints of i_1 and i_2, the value of the entry $table(u)[i_1][i_2]$ equals the minimum of $2 \cdot W(P_{VC}(u)) + H(P_{VC}(u))$, for any vehicle configuration VC in T_u. For each vertex u, we initialize all the table entries of $table(u)$ to $+\infty$. Given an input of λ, the algorithm maintains that an entry $table(u)[i_1][i_2]$ is not $+\infty$ if and only if there exists a vehicle configuration VC under which the B bound of each subproblem in $P'_{VC}(u)$ is less than or equal to λ.

We show the pseudo code of generating the tables in Figure 1.

For a leaf u, since there must be a vehicle to service u, and there are no other jobs in T_u, the entry $table(u)[r(u) + h(u)][0]$ is set to $h(u)$. This is for the case where the edge $(p(u), u)$ is not a cut edge. The edge cost of $(p(u), u)$ will be considered later when merging $table(u)$ to $table(p(u))$. When $(p(u), u)$ is a cut edge, again as u needs to be serviced, we set $obj(u)$ to 1.

When u is a non-leaf node, the dynamic programming proceeds as follows. We assume that all the tables and variables of its children have already been computed. Let the children of u be $c_1, c_2, \cdots, c_t$ (in an arbitrary order). The algorithm scans this list of children from left to right, and incorporates the tables of the children into $table(u)$ one at a time. The variable $obj(u)$ is updated after all the children of u are considered and $table(u)$ becomes available. The computation of $obj(u)$ can be expressed as

$$obj(u) = \min_{i_1, i_2}(i_2 + \lceil table(u)[i_1][i_2]/(\lambda - i_1)\rceil).$$

where $1 \leq i_1 \leq LB_1(G)$, $0 \leq i_2 \leq m$.

For an entry $table(u)[i_1][i_2]$, assume its value is obtained under the vehicle configuration VC. Then $table(u)[i_1][i_2]$ represents $2 \cdot W(P_{VC}(u)) + H(P_{VC}(u))$. Recall that i_2 records the number of vehicles used in the subproblems of $P'_{VC}(u)$. Therefore the total number of vehicles used in T_u corresponding to the entry $table(u)[i_1][i_2]$ is $i_2 + \lceil table(u)[i_1][i_2]/(\lambda - i_1)\rceil$. As we assume $e = (p(u), u)$ is a cut edge, and i_1 represents $LB_1(P_{VC}(u))$, we can then get the minimum for $obj(u)$ after trying all possible values of i_1 and i_2.

The updating of $table(u)$ for a vertex u is crucial for solving the decision problem D. Let v be a child of u. Assume $table(v)$ is already available before we start to incorporate $table(v)$ into $table(u)$. Let $table_v(u)$ be the resulting new table for u. Every entry of $table_v(u)$ is initialized to $+\infty$, and $table(u)$ will be set to $table_v(u)$ after processing v. Note that we need to work on this new table, because additional vehicles may be needed to service all the vertices in T_v. Consider the following cases:

Case 1: (u, v) is a cut edge. In this case $table_v(u)$ is updated as

$$table_v(u)[i_1][i_2 + obj(v)] = table(u)[i_1][i_2], \text{ given that } i_2 + obj(v) \leq m.$$

Here $1 \leq i_1 \leq LB_1(G)$, $0 \leq i_2 \leq m$. The entry $table_v(u)[i_1][i_2 + obj(v)]$ will be updated if it is greater than $table(u)[i_1][i_2]$.

Algorithm MVSP-decision(T_u, m, λ)
Input: an MVSP instance P defined on a tree T_u, an integer m and a real number λ.
Output: a table showing a bound $B(T_u, m)$ (at least λ).

```
1  if u is a leaf then
2      obj(u) = 1
3      table(u)[r(u) + h(u)][0] = h(u)
4      return table(u)
5  endif
6
7  for each child v of u do
8      table(v) = MVSP-decision(Tᵥ, m, λ);
9      tmp ← a new table with each entry set to +∞
10     for i₁ = 0 to LB₁(G) do
11         for i₂ = 0 to m do
12             Comment: when e = (u, v) is a cut edge
13             if table(u)[i₁][i₂] < tmp[i₁][i₂ + obj(v)] and obj(v) ≤ m − i₂ then
14                 tmp[i₁][i₂ + obj(v)] = table(u)[i₁][i₂]
15             endif
16             Comment: when e = (u, v) is not a cut edge
17             for j₁ = 0 to LB₁(G) do
18                 for j₂ = 0 to m − i₂ do
19                     t₁ = max{i₁, j₁}
20                     t₂ = table(u)[i₁][i₂] + table(v)[j₁][j₂] + 2w((u, v))
21                     if t₂ < tmp[t₁][i₂ + j₂]) then
22                         tmp[t₁][i₂ + j₂] = t₂
23                     endif
24     table(u) ← tmp
25
26 Comment: consider the case when e = (p(u), u) is a cut edge
27 for i₁ = 0 to LB₁(G) do
28     for i₂ = 0 to m do
29         t₁ = ⌈table(u)[i₁][i₂]/(λ − i₁)⌉
30         if t₁ + i₂ < obj(u) then
31             obj(u) = t₁ + i₂
32         endif
33
34 return table(u)
```

Fig. 1. Solving the decision problem for MVSP on trees

Note that an entry $table(u)[i_1][i_2]$ has a meaningful value (not $+\infty$), if and only if under the corresponding vehicle configuration VC, $B(P_2, m') \leq \lambda$ holds for every subproblem P_2 in $P'_{VC}(u)$. Here m' denotes the number of vehicles allocated for P_2 under VC. This is implemented by line 13 in the algorithm MVSP-decision in Figure 1. Recall that $obj(u)$ records the minimum number m' of vehicles to guarantee that $B(T_u, m') \leq \lambda$. Therefore if $i_2 + obj(u)$ exceeds m, then we know that VC is not a feasible vehicle configuration and there is no need to complete the rest of the computation for VC.

Case 2: (u, v) is not a cut edge. In this case $table_v(u)$ is updated as

$$table_v(u)[t][i_2 + j_2] = table(u)[i_1][i_2] + table(v)[j_1][j_2] + 2 \cdot w((u, v)).$$

where $1 \leq i_1, j_1 \leq LB_1(G)$, $0 \leq i_2 + j_2 \leq m$, and $t = \max\{i_2, j_2\}$. The entry $table_v(u)[t][i_2 + j_2]$ is updated when $table(u)[i_1][i_2] + table(v)[j_1][j_2] + 2w((u, v))$ is smaller.

Under this case we need to merge the two components C_u and C_v containing u and v respectively under the current settings for the two entries of $table(u)$ and $table(v)$. The new component has LB_1 equal to the maximum of that of C_u and C_v, so we update the first dimension of the new entry to be $t = \max\{i_2, j_2\}$. After the merging, the connected components not containing u and v remain unchanged, therefore we set the second dimension to $i_2 + j_2$. Finally we need to add $2 \cdot w((u, v))$ to this entry, as (u, v) now becomes part of the new connected component containing both u and v.

The above algorithm runs in strongly polynomial time. Recall that $LB_1(P_1) = \max_{u \in V(P_1)} \{r(u) + h(u)\}$, therefore for a vertex u the second dimension of $table(u)$ takes at most $|V|$ distinct values. The time complexity of this algorithm is dominated by Case 2 when updating $table(u)$ for each vertex u. The algorithm runs in time $O(m^2 |V|^3)$.

It is not difficult to see that this algorithm can be used to solve a disguised problem D' of D: Given a real number λ, compute the smallest possible bound $B(G, m)$ (at least λ) subject to the constraint that, under the corresponding vehicle configuration VC the bound $B(C, m')$ of each connected component C, not including the root o in $C_{VC}(o)$, is at most λ. Here m' is the number of vehicles used for the customers in C. We will show in the next subsection that a strongly polynomial time algorithm for the optimization problem P' can be obtained by solving D'.

Solving the Optimization Problem. In this subsection we present a polynomial time algorithm called MVSP-optimization (Figure 2) to solve our optimization problem. This algorithm is in the same spirit as that of the parametric searching technique. Developed by Megiddo in [11] and [12], the parametric searching technique works as follows. Let λ^* be the optimal solution of an optimization problem P_1. Assume that for P_1 we have a decision problem $D_1(\lambda)$ which is monotone in λ, in the sense that we can decide whether $\lambda < \lambda^*$, $\lambda = \lambda^*$ or $\lambda > \lambda^*$. Assume that we have an algorithm A for the decision problem $D_1(\lambda)$. To solve the optimization problem P_1, Megiddo's idea is to run A generically. It seems somewhat strange to run an algorithm when the input is still unknown. The core in Megiddo's method is to maintain an open interval I where λ^* lies throughout the execution of the algorithm A. More specifically, I is initialized to $(-\infty, +\infty)$, and at each step of running A with unknown input, a critical value t_1 is computed and the concrete version of A is executed with parameter t_1. Therefore after the first step the interval I is shrunk to either $(-\infty, t_1]$ or $[t_1, +\infty)$. When the generic version of A is terminated, we either find λ^* or an interval with its lower end equals λ^*.

It is crucial to generate the critical values in the parametric search framework. These critical values in fact discretize the problem P_1 and make it possible to compute P_1 optimally in polynomial time. Intuitively the critical values, or the steps of A represent all the tests which the optimal solution must pass. On the other hand, if a solution passes all such tests, then it is a candidate of the optimal solution. Note that as long as all the critical values are tested, the master optimization algorithm need not necessarily to be the same with A, e.g., in many cases sorting all the critical values suffices for solving the optimization problem.

The algorithm MVSP-optimization in Figure 2 shares the same spirt with the parametric searching technique, however, the details of MVSP-optimization are different from the methods in [11] and [12]. Instead of running the algorithm for the decision problem generically, we take advantage of the structure of the tree and perform the tests bottom up along the path from the vertex being visited to the root of the tree. Note that the idea of exploring parallelism for serial algorithms as described in [12] does not seem to apply in this context.

Recall that P' is to compute the bound $B(T, m)$ for a tree T and a given integer m. We first define the discrete events, or the critical values, needed by our method. Given a subtree T_u of T and a number $0 < m' \le m$, we define the critical value at u to be the bound $B(T_u, m')$ under the constraint that m' vehicles are assigned to service the jobs in T_u. Corresponding to this definition, our algorithm runs in the bottom up fashion: the computation of the bound $B(T, m)$ starts from the leaves of the tree, and the optimal value of LB for a subtree T_u will be available after all the vertices in T_u have been processed.

The reason we choose $B(T_u, m')$ $(0 < m' \le m)$ to be the critical value at a vertex u is as follows. Let $F^*(T)$ be the optimal forest corresponding to the bound $B(T, m)$. Then $B(T, m)$ is determined by a particular connected component C of $F^*(T)$. Assume u is the root of C and T_u is allocated m' vehicles in total in the optimal solution. Then we can get the optimal solution by applying the algorithm MVSP-decision in Figure 1 with the parameter $B(T_u, m')$. Given the values $B(T_u, m')$ for every subtree T_u of T and every integer $0 < m' \le m$, it is easy to see that we can compute $B(T, m)$ by $m \cdot |V|$ applications of the algorithm MVSP-decision.

For a leaf u of T, it is trivial to compute $B(T_u, m')$ $(0 < m' \le m)$. In our algorithm, whenever a critical value $B(T_u, m')$ $(u \in T, 0 < m' \le m)$ is known, we propagate the test on this value along the path from u to the root o of T. In other words, for every vertex u' on the path from u to o, we apply algorithm MVSP-decision on $T_{u'}$ with $m'' \in [1, m]$ vehicles and the feasibility parameter $B(T_u, m')$.

For an arbitrary vertex u, a critical value $B(T_u, m')$ $(u \in T, 0 < m' \leq m)$ is computed if and only if for every vertex u' (other than u) in T_u, the test of $B(T_{u'}, m')$ $(0 < m' \leq m)$ has already been taken on T_u. It is easy to see that the optimal solution can be computed after $m^2|V|^2$ calls to the algorithm MVSP-decision. Therefore the time complexity of the algorithm is $O(m^4|V|^5)$. The pseudo code of the above algorithm is listed in Figure 2.

Algorithm MVSP-optimization(T, m, o)
Input: an MVSP instance P defined on a tree T, an integer m and a root o.
Output: the bound $B(T, m)$.

```
 1 if u is a leaf then
 2    B(T_u, 1) = r(u) + 2h(u)
 3 endif
 4
 5 for each child v of u do
 6    call MVSP-optimization(T_v, m, o)
 7
 8 for every vertex u' on the path from u to the root o do
 9    for every integer m' = 0 to m do
10       for every integer m'' = 0 to m do
11          call MVSP-decision(T_u', m'', B(T_u, m'))
```

Fig. 2. Locating the subproblems for MVSP on trees

Lemma 3. *The algorithm MVSP-optimization computes the bound $B(G, m)$ exactly. The running time of the algorithm is $O(m^4|V|^5)$.* □

Proof. In the algorithm MVSP-optimization, by computing $B(T_u, m')$ $(u \in V, 1 \leq m' \leq m)$, we are in fact assuming that the edge $(p(u), u)$ is a cut edge. The lemma is trivially true when $|V| = 1$. Assume the lemma holds when $|V| < n$ $(n > 1)$. Assume the graph $G = (V, E)$ has n vertices and VC is the vehicle configuration corresponding to $B(G, m)$. Assume $e_1 = (p(u_1), u_1), e_2 = (p(u_2), u_2), \cdots, e_t = (p(u_t), u_t)$ are all the cut edges where $p(u_1), p(u_2), \cdots, p(u_t)$ are in the connected component on which $P_{VC}(o)$ is defined. Assume $m_1, m_2, \cdots, m_t$ vehicles are allocated for the vertices in $T_{u_1}, T_{u_2}, \cdots, T_{u_t}$ respectively under VC. According to our assumption, $B(T_{u_1}, m_1), B(T_{u_2}, m_2), \cdots, B(T_{u_t}, m_t)$ can be computed exactly in polynomial time. In MVSP-optimization the critical value tests with feasibility parameters $B(T_{u_1}, m_1), B(T_{u_2}, m_2), \cdots, B(T_{u_t}, m_t)$ will be performed against the root o. The lemma then holds for G due to the definition of the disguised decision problem D' and the fact that $B(G, m)$ is determined by either $P_{VC}(o)$ or a subproblem in $P'_{VC}(o)$.

By Lemmas 1, 2 and 3 we establish

Theorem 1. *There is a 3-approximation for MVSP on trees.* □

3 $(5 - \frac{2}{m})$-Approximation for MVSP on General Graphs

The algorithm MVSP-optimization can be further utilized to get a $(5 - \frac{2}{m})$-approximation for MVSP on general graphs. We simply apply MVSP-optimization on a minimum spanning tree of the underlying graph G. We then obtain a feasible MVSP solution by the strategy mentioned in the proof of Lemma 1. In the following we prove that the cost of this solution is at most $(5 - \frac{2}{m}) \cdot OPT(P)$.

It is well known that minimum spanning trees have the property stated in Lemma 4.

Lemma 4. *Let C be the cycle introduced by adding an edge e to a minimum spanning tree T. Then the cost of e is no less than that of any edge on C.*

Lemma 4 together with Lemma 5 are crucial for our analysis. In the following we define $F^*(P)$ to be the subgraph on which the gapless subproblems are defined for an MVSP problem P.

Lemma 5. *For an MVSP problem P the cost of any edge in $F^*(P)$ is a lower bound on $OPT(P)$.*

Assume we already know $F^*(P)$ and the optimal cost $OPT(P)$. We describe in the following an algorithm MVSP1 based on $F^*(P)$, which is designed for analysis only. In MVSP1, we first remove any edge with cost more than $OPT(P)$ from the original graph G. Let the resulting graph be G'. We then compute minimum spanning trees for each of the connected components of G'. Finally we apply the strategy mentioned in the proof of Lemma 1 to the computed minimum spanning trees. As shown in Lemma 6, the cost of such a solution is at most $(5 - \frac{2}{m}) \cdot OPT(P)$.

Lemma 6. *Given the subgraph $F^*(P)$ for an MVSP problem P, the cost of the solution produced by MVSP1, is at most $(5 - \frac{2}{m}) \cdot OPT(P)$.*

Proof. Let G' be the resulting graph after removing edges with costs more than $OPT(P)$ from the original graph G. Due to Lemma 5, all the edges of $F^*(P)$ still remain in the connected components of G'. Consider an arbitrary connected component CC of G'. Let $CC_1, CC_2, \cdots, CC_t$ be the connected components of $F^*(P)$ in CC. We may assume that each CC_i $(1 \le i \le t)$ is a tree, as otherwise some edges can be removed from a connected component of $F^*(P)$ to make it a tree. We construct a new graph $G'' = (V'', E'')$ where each vertex $v_i \in V''$ $(1 \le i \le t)$ corresponds to CC_i $(1 \le i \le t)$, and the cost of an edge $(v_i, v_j) \in E''$ $(1 \le i, j \le t)$ equals the minimum edge cost of any edge between a vertex in CC_i and a vertex in CC_j in G'.

Let $e_1, e_2, \cdots, e_{t-1}$ be the minimum spanning tree edges in G''. Adding these edges to $F^*(P)$ in G' will produce a connected tree T' spanning all the vertices in CC. Let P_0 and P_i $(1 \le i \le t)$ be the associated subproblems defined on T' and CC_i respectively. Let m' and m_i $(1 \le i \le t)$ be the vehicles used in T' and CC_i respectively. We have the following inequalities.

(1) $t \leq m'$ (2) $W(P_i) + H(P_i) \leq m_i \cdot OPT(P)$ (3) $w(e_i) \leq OPT(P)$

where $1 \leq i \leq t$. The first inequality holds, because at least one vehicle should be allocated for each C_i $(1 \leq i \leq t)$. The second inequality is due to $(W(P_i) + H(P_i))/m_i \leq OPT(P)$ $(1 \leq i \leq t)$. The third inequality is implied by the algorithm MVSP1.

Therefore we have

$$B(T', m') = LB_1(P_0) + (2 \cdot W(P_0) + H(P_0))/m'$$
$$\leq OPT(P) + 2 \cdot \sum_{i=1}^{t}(W(P_i) + H(P_i))/m' + 2 \cdot \sum_{i=1}^{t-1} w(e_i)/m'$$
$$\leq OPT(P) + 2 \cdot \sum_{i=1}^{t} m_i \cdot OPT(P)/m' + 2 \cdot (t-1) \cdot OPT(P)/m'$$
$$\leq (5 - \frac{2}{m}) \cdot OPT(P),$$

The lemma then holds after applying the same analysis to other connected components of G'.   □

Consider another algorithm MVSP2, where we first compute the minimum spanning tree F of G and then remove the edges with costs more than $OPT(P)$ from F. In the following we show that MVSP1 and MVSP2 are essentially equivalent.

Lemma 7. *Given $F^*(P)$ for an MVSP problem P, MVSP1 and MVSP2 produce the same solution or two solutions with the same cost.*

Proof. Assume CC is a connected component of G'. We claim that the vertices in CC appear *contiguously* in any minimum spanning tree, say F, of G. This *contiguity* is in the sense that any vertex not in CC is adjacent to at most one vertex in CC by the edges of F.

Assume the vertices in CC are in two disjoint subtrees T_1 and T_2 of F. As CC is a connected component, there is one edge e in CC that connects T_1 and T_2 and has cost at most $OPT(P)$. Adding e to F will create a cycle. As T_1 and T_2 are disjoint in F, on this cycle there must exist two edges e_1 and e_2 of F, each of which connects a vertex of CC to a vertex not in CC. According to MVSP1, both e_1 and e_2 have costs larger than $OPT(P)$. A contradiction to Lemma 4. □

Given a minimum spanning tree F of G, our algorithm MVSP-optimization gives the best partition of the tree to minimize $B(F, m)$. We proved in Lemma 6 and Lemma 7 that there exists a partition of F such that the corresponding LB bound is at most $(5 - \frac{2}{m}) \cdot OPT(P)$. Further due to Lemma 1, we establish

Theorem 2. *There is a $(5 - \frac{2}{m})$-approximation for MVSP on general graphs.*
□

4   Conclusions and Future Work

In this paper we obtained a 3-approximation for MVSP on trees, and a $(5 - \frac{2}{m})$-approximation for MVSP on general graphs. We showed that by finding appropriate relaxation problems, dynamic programming can be applied in the design of approximation algorithms for VRP on trees. Our $(5 - \frac{2}{m})$-approximation for MVSP on general graphs is just based on the 3-approximation algorithm for MVSP on trees. In the future we would like to further improve the approximation ratio for MVSP on general graphs, possibly by investigating new ways of extending the use of dynamic programming for VRP to general graphs.

References

1. Augustine, J.E., Seiden, S.: Linear time approximation schemes for vehicle scheduling problems. Theoretical Computer Science 324(2-3), 147–160 (2004)
2. Bansal, N., Blum, A., Chawla, S., Meyerson, A.: Approximation algorithms for deadline-TSP and vehicle routing with time-windows. In: Proceedings of the 36th Annual ACM Symposium on Theory of Computing, pp. 166–174 (2004)
3. Bar-Yehuda, R., Even, G., Shahar, S.: On approximating a geometric prize-collecting traveling salesman problem with time windows. In: Proceedings of the 11th Annual European Symposium on Algorithms, pp. 55–66 (2003)
4. Bhattacharya, B.K., Carmi, P., Hu, Y., Shi, Q.: Single Vehicle Scheduling Problems on Path/Tree/Cycle Networks with Release and Handling Times. In: Proceedings of the 19th International Symposium on Algorithms and Computation, pp. 800–811 (2008)
5. Dantzig, G.B., Ramser, R.H.: The truck dispatching problem. Management Science 6, 80–91 (1959)
6. Even, G., Garg, N., Konemann, J., Ravi, R., Sinha, A.: Min-max tree covers of graphs. Operations Research Letters 32(4), 309–315 (2004)
7. Karuno, Y., Nagamochi, H.: Better approximation ratios for the single-vehicle scheduling problems on line-shaped networks. Networks 39(4), 203–209 (2002)
8. Karuno, Y., Nagamochi, H.: A polynomial time approximation scheme for the multi-vehicle scheduling problem on a path with release and handling times. In: Proceedings of the 12th International Symposium on Algorithms and Computation, pp. 36–47 (2001)
9. Karuno, Y., Nagamochi, H.: A 2-Approximation Algorithm for the Multi-vehicle Scheduling Problem on a Path with Release and Handling Times. In: Proceedings of the 9th Annual European Symposium on Algorithms, pp. 218–229 (2001)
10. Korula, N., Chekuri, C.: Approximation algorithms for orienteering with timewindows (September 2007) (manuscript)
11. Megiddo, N.: Combinatorial optimization with rational objective functions. Math. Oper. Res. 4, 414–424 (1979)
12. Megiddo, N.: Applying parallel computation algorithms in the design of serial algorithms. Journal of ACM 30(4), 852–865 (1983)
13. Menger, K.: Reminiscences of the vienna circle and the mathematical colloquium. Dortmund (1994)
14. Nagamochi, H., Mochizuki, K., Ibaraki, T.: Complexity of the Single Vehicle Scheduling Problem on Graphs. Inform. Systems Oper. Res. 35(4), 256–276 (1997)

15. Nagamochi, H., Okada, K.: A faster 2-approximationalgorithm for the minmax p-traveling salesmenproblem on a tree. Discrete Applied Mathematics 140(1-3), 103–114 (2004)
16. Nagamochi, H., Okada, K.: Polynomial time 2-approximation algorithms for the minmax subtree cover problem. In: Ibaraki, T., Katoh, N., Ono, H. (eds.) ISAAC 2003. LNCS, vol. 2906, pp. 138–147. Springer, Heidelberg (2003)
17. Toth, P., Vigo, D. (eds.): The vehicle routing problem. SIAM Monographs on Discrete Mathematics and Applications, vol. 9. SIAM, Phaladelphia (2002)

On Greedy Algorithms for Decision Trees[*]

Ferdinando Cicalese[1], Tobias Jacobs[2], Eduardo Laber[3], and Marco Molinaro[4]

[1] University of Salerno, Italy
[2] National Institute of Informatics, Japan
[3] PUC – Rio de Janeiro, Brazil
[4] Carnegie Mellon University, USA

Abstract. In the general search problem we want to identify a specific element using a set of allowed tests. The general goal is to minimize the number of tests performed, although different measures are used to capture this goal. In this work we introduce a novel greedy approach that achieves the best known approximation ratios simultaneously for many different variations of this identification problem. In addition to this flexibility, our algorithm admits much shorter and simpler analyses than previous greedy strategies. As a second contribution, we investigate the potential of greedy algorithms for the more restricted problem of identifying elements of partially ordered sets by comparison with other elements. We prove that the latter problem is as hard to approximate as the general identification problem. As a positive result, we show that a natural greedy strategy achieves an approximation ratio of 2 for tree-like posets, improving upon the previously best known 14-approximation for this problem.

1 Introduction

The problem of efficiently searching in a discrete set is a fundamental one in computer science [20] and as such it appears in many diverse variants and in a surprisingly wide range of areas, e.g., database [5, 9], learning [7], parallel assembly of multipart products [10], image processing [3], data compression [14], and more generally, theory of algorithms [1, 2, 4, 8, 19–26]. In this paper we try to contribute to the large literature on searching by considering very general search problems and analyzing the performance of a simple novel greedy approach. We show that this novel approach matches most of the best known bounds to date and, remarkably, allows for a very direct and less involved analysis as compared to the state of the art.

Problem Definition. Let U be a finite set of objects and $n = |U|$. An initially marked but unknown object $u^* \in U$ has to be identified. A weight function $w : U \to \mathbb{N}$ is given, which indicates for each object $u \in U$ the likelihood that u is the marked object to be identified. The identification is done by adaptively performing tests from a given finite set T. We denote by m the cardinality of T.

---

[*] This work was supported by a fellowship within the Postdoc-Programme of the German Academic Exchange Service (DAAD).

O. Cheong, K.-Y. Chwa, and K. Park (Eds.): ISAAC 2010, Part II, LNCS 6507, pp. 206–217, 2010.

For each $t \in T$ there is an associated partition, $\mathcal{G}_t = \{G_t^1, \ldots, G_t^k\}$ of the universe U. The output of the test is the index j of the set in the partition $\mathcal{G}_t$ containing the object u^* which has to be identified, i.e., the j such that $u^* \in G_t^j$. In such case, we say that the objects in G_t^j satisfy (agree with) the test performed. For each test t there is a cost c_t that has to be paid in order to perform the test.

An identification strategy is typically represented by a decision tree D where each internal node ν of D maps to a test t_ν and has $|\mathcal{G}_{t_\nu}|$ children, indexed according to the elements of the partition $\mathcal{G}_{t_\nu}$. The tree has exactly n leaves, which are in one-to-one correspondence with elements in U. Each leaf ℓ is associated with an element in U which is the unique object satisfying the tests on the path from the root of D to the leaf ℓ. Given such a decision tree, the corresponding strategy is to start performing the test associated to the root of D; if j is the outcome of the test, then the test associated to the jth child is to be performed, and so on, until a leaf is reached, indicating the marked object for the given instance. Implicit in the definition of decision trees given here is that we only allow instances of the problem where each object $u \in U$ has a set of tests which uniquely identifies u.

Given a decision tree D, the cost $cost^D(u)$ of identifying an object u following the strategy defined by D is defined as the sum of the costs of the tests associated to the nodes on the unique path from the root of D to the leaf associated to u.

We consider two different measures of performance for a decision tree: The *average identification cost* of D is defined as the expected costs of identifying an objects chosen in accordance to the likelihood $w(.)$ when we use the strategy associated to D, i.e., in formulae:

$$avgcost(D) = \sum_{u \in U} w(u) cost^D(u).$$

We remark that w is not a probability distribution.

We also consider the *worst identification cost* of D, defined as the maximum over all $u \in U$ of the cost of identifying u using the decision tree D, i.e., in formulae:

$$worstcost(D) = \max_{u \in U} cost^D(u).$$

Existing Results. The problem of finding the decision tree with minimum average cost is a direct generalization of the Binary Identification Problem [12], which coincides with the particular case when each test defines a bipartition. Since there has been a substantial amount of work in variations of the above identification problem, we will specify the type of instance we refer to by means of a three field notation, inspired by scheduling problems. More precisely, we will use the notation $aCbW|cr|Obj$, where the first field is used for indicating numerical restrictions on the input, the second field for indicating combinatorial restrictions, and the third field is used to specify the objective function. In the first field, $aCbW$ with $a, b \in \{u, n\}$ indicates whether the costs and weights are uniform or not (e.g., $uCnW$ indicates a problem with uniform costs and non-uniform weights). As for the second field, the combinatorial restrictions we consider are $cr \in \{B, M, P, T\}$, respectively for Binary tests, Multiway or k-ary tests, Poset, Tree-like poset. Finally, we consider the objective functions described above, i.e., $Obj \in \{A, W\}$, for average-case, and worst-case, respectively.

The problem of finding a decision tree with minimum worst-case cost does not admit an $o(\log n)$-approximation algorithm unless P=NP, even when costs are uniform [24]. The same lower bound holds for the $uCnW|B|A$ version [5].

On the algorithmic side, it is known that a simple greedy algorithm achieves an $O(\log n)$-approximation for minimizing the worst-case cost with binary tests and uniform costs [3]. The problem of finding a decision tree with minimum average cost has received much more attention. For the $uCnW|B|A$ version, Kosaraju et al. [21] and DasGupta [7] independently proposed greedy algorithms with an $O(\log W/w_{min})$ approximation factor, where W is the sum of the weights of all objects and w_{min} is the weight of the object with minimum weight. In addition, Kosaraju et. al. showed how to modify their greedy approach to attain an $O(\log n)$-approximation. Still in the case of binary tests, Adler and Heeringa [2] present a $(\ln n + 1)$-approximation for the $nCuW|B|A$ version.[1]

In [5], Chakaravarthy et al. started the study of multiway tests. They present an $O(\log^2 n)$ approximation for the $uCnW|M|A$ problem. In a later paper, Chakaravarthy et al. [6] managed to cut a $\log n$ factor for uniform weight instances, showing that the $uCuW|M|A$ problem admits a $O(\log n)$-approximation.

Recently Guillory and Blimes [15] simultaneously extended many of the above results by showing that a natural extension of the algorithm proposed in [7] achieves approximation ratio of $12 \log W/w_{min}$ for the $nCnW|M|A$ problem. Finally, Gupta et al. [16] proved that such a problem admits an $O(\log n)$-approximation, matching the $\Omega(\log n)$ lower bound up to constant factors.

Our Results. One of the aim with which we embarked in our investigation was to better understand the potentiality of the greedy approach pursuing a more direct analysis and trying to disclose the problem's crucial structure. In fact in all the works mentioned in the above long historical excursus [2, 6, 7, 15, 16, 21], the analyses are quite lengthy and involved, requiring several pages to prove the approximations. This might be a result of the two specific greedy criteria considered in those paper, namely the *shrinkage-cost ratio* and the *minimization of the heaviest group*. In the paper considering the former criterium, at each step the algorithm selects the test i which maximizes the *shrinkage-cost ratio* defined by $\frac{\Delta_i(S,w)}{c_i} = \frac{1}{c_i}\left(w(S) - \sum_{j=1}^{k} \frac{w(G_i^j \cap S)^2}{w(S)}\right)$, where S is the set of objects consistent with the tests performed so far [2, 7, 15]. In the case when only uniform weights are considered the weights are substituted by cardinalities of the corresponding sets. In the papers considering the latter criterium, greedy means selecting the test that generates a partition of S whose heaviest group is as light as possible [6, 21]. We remark that the approach taken in [16] is different, but somehow less transparent to the actual identification problem structure, as it relies on much heavier machinery such as submodular optimization and variants of the TSP problem.

---

[1] Note that in the paper, the authors use the term weight to refer to the cost of a test.

Our first contribution is an alternative greedy strategy which, in order to select the next test to perform, takes into account simultaneously the cardinality and the weight of the sets in the partitions induced by the possible available tests. This novel greedy strategy allows to achieve (or improve) the best known results for different types of instances considered in literature and is prone to a short, and in our opinion, neater analysis. More specifically, we give a short proof that our algorithm attains a $4 \ln(W/w_{\min})$-approximation for $nCnW|M|A$. This improves by a constant factor the $12 \ln(W/w_{\min})$ analysis of [15]. For $uCnW|M|A$ our algorithm achieves a $O(\log n)$-approximation. Finally, for instances with non-uniform costs and uniform weights it simultaneously achieves an $O(\log n)$ approximation for both the average-case and worst-case cost objective functions.

In the second part of the paper we investigate the potential of greedy approaches for restricted classes of the average-case multiway identification problem that have been studied in the literature. We consider the class where the set U is a poset and for each element $u \in U$ there exists a binary test which reports whether the marked object is smaller than or equal to u [25]. It turns out that there is little hope for better approximations for this class since we show an $\Omega(\log n)$ hardness of approximation. Thus, we consider the class of instances where the poset is a rooted tree, the tests have uniform costs and objects may have non-uniform weights. This corresponds to the NP-complete problem of searching in trees [18, 23]. For this class, we show that the greedy strategy which always selects the test which induces the most balanced partition attains a 2-approximation, improving over the 14-approximation given in [23].

2   A Novel Greedy Procedure

Recall the problem definition and the notation given in the introduction. We say that test t splits the objects into groups $G_t^1, \ldots, G_t^k$. Let $n_t^j = |G_t^j|$ and let W_t^j be the sum of the weights of the objects in G_t^j. Let $j^* = \text{argmax}_j\{n_t^j W_t^j\}$. In order to simplify the notation, we use the shorthands $n_t = n_t^{j^*}$ and $W_t = W_t^{j^*}$.

Now we describe the greedy approach we propose, dubbed GREEDY. Let $I = (U, T, w)$ be an instance of an identification problem, where $|U| = n$ and $W = \sum_{u \in U} w(u)$. Let $t \in T$ be the test which minimizes $c_t/(nW - n_t W_t)$. Then the root of the decision tree of GREEDY is associated to the test t and its children are decision trees obtained recursively by applying GREEDY to the instances $I_1, \ldots, I_k$, where $I_i = (G_t^i, T, w)$.

The intuition behind GREEDY's criterium is that it penalizes a test whether it has a high cost or it induces a partition of U that contains a set that is either large or heavy.

We will use the following subadditivity property on the cost of an optimal decision tree for the identification problem. This property was observed in [6] for the average-case model. Similar arguments show that it also holds for the worst-case model.

Proposition 1. *[6] Let U be the set of objects in an instance of the identification problem and let D^* and E^* be optimal decision trees with respect to the average-case* *model and worst-case model, respectively. Let $U_1, \ldots, U_k$ be disjoint subsets of U, and, for $i = 1, \ldots, k$, let D_i^* (E_I^*) be an optimal decision tree for the subinstance of I defined on the set of objects U_i for the average-case (worst-case) model. Then*

$$\sum_{i=1}^{k} avgcost(D_i^*) \leq avgcost(D^*)$$

and

$$\max_{i=1}^{k}\{worstcost(E_i^*)\} \leq worstcost(E^*).$$

2.1 Multiway Tests, Non-uniform Weights and Non-uniform Costs

We start our analysis of the approximation provided by GREEDY considering the case of $nCnW|M|A$ instances. We are going to show that in this case GREEDY provides a solution with cost which is at most $2 \log nW/w_{\min}$ times the cost of an optimal decision tree, where $w_{\min}$ is the minimum weight assigned to an object. Without loss of generality, in the following analysis we assume $w_{\min} = 1$.

Let $cost(I)$ denote the cost of the decision tree produced by GREEDY on instance I and let $OPT(I)$ be the cost of the optimal decision tree for the same instance.

Let τ denote the first test performed by GREEDY. Also, for each $\ell = 1, \ldots, k$, let I_ℓ be the instance associated with the set of objects G_τ^ℓ. Then, the cost incurred by GREEDY is given by $cost(I) = Wc_\tau + \sum_\ell cost(I_\ell)$. Moreover, using Proposition 1, we have $OPT(I) \geq \sum_{\ell=1}^{k} OPT(I_\ell)$. Given any lower bound LB on $OPT(I)$, we can bound the approximation ratio attained by GREEDY as follows:

$$\frac{cost(I)}{OPT(I)} \leq \frac{Wc_\tau + \sum_\ell cost(I_\ell)}{\max\{LB, \sum_\ell OPT(I_\ell)\}} \leq \frac{Wc_\tau}{LB} + \max_\ell \frac{cost(I_\ell)}{OPT(I_\ell)} \tag{1}$$

We now focus on devising a suitable lower bound LB on the cost of an optimal decision tree D^* for the instance I. For each test $t \in T$ we define α_t as the sum of the weights of the objects associated with leaves that are descendants of some node associated with test t in D^*.[2] Clearly we have $OPT(I) = \sum_t c_t \alpha_t$. The greedy rule implies that $c_\tau/(nW - n_\tau W_\tau) \leq c_t/(nW - n_t W_t)$ for each $t \in T$. It follows that

$$OPT(I) \geq \frac{c_\tau}{nW - n_\tau W_\tau} \sum_t (nW - n_t W_t)\alpha_t \tag{2}$$

We now note that we can interpret $\sum_t (nW - n_t W_t)\alpha_t$ as the cost of a decision tree for a modified version of instance I where the cost of test t is changed from c_t to $nW - n_t W_t$. We can then use the following result.

Claim. *Let $\tilde{I}$ be an instance obtained from I by changing the costs of the tests so that for each t, it holds that $c_t \geq nW - n_t W_t$. Then $OPT(\tilde{I}) \geq nW^2/2$.*

---

[2] Note that the leaves contributing to α_t do not necessarily induce a subtree of D^*, as t can appear in more than one node of D^*.

Proof of Claim. We use induction on n. If $n = 2$ then any test that splits the two objects has cost at least $2W - W'$, where W' is the weight of the heaviest object. Thus, the cost of the optimal tree is at least $W(2W - W') \geq W^2$, which establishes the base case.

Suppose now that the instance $\tilde{I}$ comprises $n > 2$ objects. Let $\tilde{t}$ be the test at the root of an optimal tree for $\tilde{I}$ and notice that $n_{\tilde{t}} < n$. If $n_{\tilde{t}} = 1$ then $OPT(\tilde{I}) \geq W(nW - W_{\tilde{t}}) \geq nW^2/2$. If $n_{\tilde{t}} > 1$ then $OPT(\tilde{I}) \geq W(nW - n_{\tilde{t}}W_{\tilde{t}}) + n_{\tilde{t}}W_{\tilde{t}}^2/2$ because: (i) $W(nW - n_{\tilde{t}}W_{\tilde{t}})$ is a lower bound on the contribution of the test $\tilde{t}$ to $OPT(\tilde{I})$ and (ii) by induction, $n_{\tilde{t}}W_{\tilde{t}}^2/2$ is a lower bound for the instance associated with the objects of the group induced by test $\tilde{t}$ that has $n_{\tilde{t}}$ objects with total weight $W_{\tilde{t}}$. Since $W^2/2 \geq W_{\tilde{t}}(W - W_{\tilde{t}}/2)$, collecting the terms with $n_{\tilde{t}}$ gives $W(nW - n_{\tilde{t}}W_{\tilde{t}}) + n_{\tilde{t}}W_{\tilde{t}}^2/2 \geq nW^2/2$, which establishes the result.   □

From the above claim and equation (2) we get that

$$OPT(I) \geq \frac{c_\tau nW^2}{2(nW - n_\tau W_\tau)}.$$

Then replacing LB by this lower bound in equation (1) gives

$$\frac{cost(I)}{OPT(I)} \leq \frac{2(nW - n_\tau W_\tau)}{nW} + \max_\ell \frac{cost(I_\ell)}{OPT(I_\ell)}.$$

Thus, assuming by induction on the number of objects that for each instance I_ℓ the approximation ratio attained by GREEDY is at most $2\ln(W_\tau^\ell n_\tau^\ell)$, which by the definition of W_τ and n_τ is at most $2\ln(W_\tau n_\tau)$, we have

$$\frac{cost(I)}{OPT(I)} \leq 2\left(1 - \frac{n_\tau W_\tau}{nW} + \ln(n_\tau W_\tau)\right) \leq 2\ln nW \leq 4\ln W, \qquad (3)$$

where the second inequality uses the fact that $\ln x \leq x - 1$ for all $x > 0$ and the last inequality uses the fact $w_{min} = 1$.

In general, dropping the assumption $w_{min} = 1$, the above analysis yields the approximation

$$\frac{cost(I)}{OPT(I)} \leq 2\ln n \frac{W}{w_{min}}, \qquad (4)$$

2.2   Uniform Costs and Non-uniform Weights

In the case of $uCnW|M|A$ instance, we can strengthen the above result and show that GREEDY attains a $1 + 4\ln n$ approximation. Note that, for the particular case when the costs are uniform GREEDY selects a test t with smallest $n_t W_t$.

Let D be the tree constructed by GREEDY and for each node $v \in D$ let D_v denote its subtree rooted at v and $w(D_v)$ denote the sum of the weights of the leaves in D_v. Define the set $V = \{v : w(D_v) \leq W/n\}$, that is, the set of nodes in GREEDY's tree such that the weight of the subtree below it is at most W/n. It follows that these nodes contribute with at most W for the cost of D. Then the approximation ratio can be bounded by

$$\frac{cost(D)}{OPT} = \frac{\sum_{v \in D-V} w(D_v)}{OPT} + \frac{\sum_{v \in V} w(D_v)}{OPT} \leq \frac{\sum_{v \in T-V} w(D_v)}{OPT} + 1, \quad (5)$$

where the inequality follows from the fact that $OPT \geq W$.

We can now estimate the ratio $\frac{\sum_{v \in T-V} w(D_v)}{OPT}$ in terms of an instance where each object has weight at least W/n. We are coalescing objects that appear in the decision tree of GREEDY in subtrees whose leaves have total cost $\leq W/n$.

Thus, by using the analysis of the previous section, from (4) we have

$$\frac{\sum_{v \in T-V} w(D_v)}{OPT} \leq 2 \ln \left(n \frac{W}{W/n} \right) = 4 \ln n,$$

which, together with (5), gives the desired result.

2.3   GREEDY is Bi-Criteria for Non-uniform Costs and Uniform Weights

We now show that for instances with uniform weights GREEDY attains an $O(\log n)$-approximation simultaneously for both average-case and worst-case objectives. The first part of the claim follows directly from the general result of section 2.1, hence we now prove the approximation for the worst-case objective.

Consider an instance I with arbitrary costs and uniform weights and let D be the decision tree produced by GREEDY on I. Let t be the first test performed by GREEDY and D_ℓ be the decision tree produced by GREEDY on the instance I_ℓ induced by G_t^ℓ. From the recursive nature of the algorithm we have

$$worstcost(D) = c_t + \max_\ell \{worstcost(D_\ell)\}. \quad (6)$$

Now, let D^* denote an optimal decision tree for the instance I and D_ℓ^* be the optimal decision tree for the instance I_ℓ. Using Proposition 1 we can again bound $worstcost(D^*)$ as

$$worstcost(D^*) \geq \max\{LB, \max_\ell \{worstcost(D_\ell^*)\}\}, \quad (7)$$

where LB is any lower bound on $worstcost(D^*)$.

We now turn to the issue of defining a suitable lower bound LB on the cost of D^*. In the following we identify nodes of D^* with the tests they map to. We choose a root-to-leaf path $v_1, v_2, \ldots, v_p$ in D^* as follows. First, v_1 is the root of D^*. Then, assuming we have already chosen $v_1, \ldots, v_i$, we choose v_{i+1} as the child of v_i that is used to split the objects that lie in the largest set of the partition of G_{v_i}. If the largest set has only one object, it corresponds to a leaf ℓ, then we set $v_{i+1} = \ell$. The process stops when we reach the leaf v_p. We set $LB = \sum_{i=1}^{p-1} c_{v_i}$. Our goal is now to bound this quantity.

Let u_i be the number of objects that are in the subtree of D^* rooted at v_i but not in the subtree rooted at v_{i+1}. By definition of the v_i's, we have that

$u_i + n_{v_i} \leq n$ holds for $i = 1, \ldots, p-1$. Then, the greedy criterium allows us to write

$$\frac{c_t}{n^2 - n_t^2} \leq \frac{c_{v_i}}{n^2 - (n-u_i)^2} = \frac{c_{v_i}}{2nu_i - u_i^2} \quad \text{for } i = 1, \ldots, p-1.$$

Finally, adding up the costs of the nodes $v_1, \ldots, v_{p-1}$ and using the fact that $\sum_{j=1}^{p-1} u_j = n-1$, we get

$$LB \geq \frac{c_t}{n^2 - n_t^2} \sum_{i=1}^{p-1} (2nu_i - u_i^2) = \frac{c_t}{n^2 - n_t^2} \left(2n(n-1) - \sum_{i=1}^{p-1} u_i^2 \right) \geq \frac{c_t(n^2 - 1)}{n^2 - n_t^2}.$$

Now we can bound the approximation ratio of GREEDY as the ratio of (6) and (7), which gives

$$\frac{n^2 - n_t^2}{n^2 - 1} + \max_\ell \frac{cost(D_\ell)}{cost(D_\ell^*)}.$$

Assuming, by induction on the size of the instances, that on I_ℓ GREEDY provides a solution at most $\ln((n_t^\ell)^2 - 1)$ away from the optimal one, we have that the approximation ratio becomes

$$\frac{n^2 - n_t^2}{n^2 - 1} + \ln(n_t^2 - 1) = 1 - \frac{n_t^2 - 1}{n^2 - 1} + \ln(n_t^2 - 1) \leq \ln(n^2 - 1).$$

3   Partially Ordered Sets

In this section we consider the binary identification problem where the elements of U form a partially ordered set and there is one test $t(u)$ for each $u \in U$ which indicates whether $u^* \leq u$ or not. That is, $\mathcal{G}_{t(u)} = \{G_{t(u)}^1, G_{t(u)}^2\}$ with $G_{t(u)}^1 = \{u' \in U : u' \leq u\}$ and $G_{t(u)}^2 = U \setminus G_{t(u)}^1$. We represent the poset by its Hasse diagram, that is, the digraph F with node-set U that contains the arc (u_i, u_j) whenever u_j covers u_i (i.e., $u_i < u_j$ and there is no u_ℓ in U such that $u_i < u_\ell < u_j$). Given a decision tree for this problem, it is useful to order its nodes such that, for a node associated to the test t, its right child corresponds to the part G_t^1 and its left child corresponds to the part G_t^2.

We show next that, for every $\epsilon > 0$, the problem $uCnW|P|A$ cannot be approximated in polynomial time within a ratio of $(0.25 - \epsilon) \log n$ unless $NP \subseteq TIME(n^{O(\log \log n)})$. This is accomplished via a reduction to Set Cover, which is hard to approximate within a factor of less than $\log n$ under this complexity assumption [11]. We remark that this reduction is similar to the one used in [5].

Consider a Set Cover instance $(X, \mathcal{S})$, where $X = \{x_1, \ldots, x_n\}$ is the non-empty ground set and $\mathcal{S} \subseteq 2^X$ with $\bigcup_{S \in \mathcal{S}} S = X$ is the family of covering sets. Moreover, we assume that $|\mathcal{S}| = O(n^2)$; this is without loss of generality since these instances are still hard to approximate [11].

Given a Set Cover instance $(X, \mathcal{S})$, we construct the following instance (F, w) to our identification problem. The digraph F has a node v_i for each $x_i \in X$, a node s_i for each set $S_i \in \mathcal{S}$ and one extra node r. The arcs in this graph are (v_i, r) for each v_i and arcs (v_i, s_j) for $x_i \in S_j$. The weight function w assigns weight 1 to r and weight 0 to every other node.

Now we relate the solution for these instances. Consider a cover $C = \{S_{i_1}, S_{i_2}, \ldots, S_{i_k}\}$ for the Set Cover instance $(X, \mathcal{S})$. We construct a decision tree D for (F, w) with cost $|C| + 1$ as follows. First, make the leftmost path of D contain the tests $t(s_{i_1}), t(s_{i_2}), \ldots, t(s_{i_k}), t(r)$ in this order; then complete D to form a valid decision tree. To analyze the cost of D notice that the fact that $\{S_{i_1}, S_{i_2}, \ldots, S_{i_k}\}$ is a cover implies that the node in D associated to $t(r)$ has no descendant leaf associated to a v_i. Then we have that the right child of the node associated to $t(r)$ is a leaf associated to r and hence $cost(D) = |C| + 1$.

Now let D be a solution for the instance (F, w). We claim that D gives a cover C for the Set Cover instance with $|C| = cost(D)$. For this, let P be the path of D from its root to the its leaf associated to r. The crucial property is that for each element $x_i \in X$, there is a node in P which either: (i) corresponds to the test $t(v_i)$ or (ii) corresponds to a test $t(s_j)$ such that $(v_i, s_j) \in F$. To construct the cover C, we consider each $x_i \in X$ and if it falls in case (i) we add to C any ~~set in $\mathcal{S}$ which contains~~ x_i and if x_i falls in case (ii) we add to C the set S_j. By construction, C is a cover which ~~satisfies $|C| = cost(D)$.~~

To conclude our reduction, let OPT_{SC} and OPT_{ID} be respectively the optimal value for the Set Cover and the identification problem instances. Let $n(F) = n + O(n^2) + 1$ denote the number of nodes in F. Using the previous properties, we get that an $\alpha \log(n(F))$-approximate solution for the identification problem gives a solution for Set Cover with size at most

$$\alpha \log(n(F)) OPT_{ID} \leq \alpha \log(n(F)) (OPT_{SC} + 1) \leq 2\alpha(2 \log n + O(1)) OPT_{SC},$$

where in the last inequality we used the fact that $OPT_{SC} \geq 1$. If $\alpha \leq 0.25 - \epsilon$ for some $\epsilon > 0$, then for large enough n the right hand side is at most $(1 - \epsilon)(\log n) OPT_{SC}$, hence we obtain an approximation for Set Cover with factor better than $\log n$. Given the aforementioned hardness of Set Cover, there cannot exist a $(0.25 - \epsilon) \log(n(F))$-approximation with $\epsilon > 0$ for the identification problem $uCnW|P|A$ unless $NP \subseteq TIME(n^{O(\log \log n)})$.

4   Tree-Like Posets

Having shown that finding a good search strategy for partial orders is essentially as hard as the general identification problem, we now turn our attention towards a special class of partial orders. Namely, we consider posets whose Hasse diagram F is a tree with arcs directed towards the root r, i.e. r is the unique maximum element. The test asking whether or not $u^* \leq r$ will never be applied by a reasonable search strategy and is therefore assumed not to exist in the following. Any other test "is $u^* \leq u$?" will split the tree into the the two connected components F_1, F_2 induced by the removal of the unique outgoing edge of u. We can therefore associate tests from T with edges and objects $u \in U$ with nodes of F. For simplicity we do not distinguish between tree nodes (edges) and the elements of U (of T) associated with them.

After test t has revealed that the searched element u^* is in F_1, no reasonable search strategy will perform a test corresponding to an edge in F_2, and vice versa. Therefore, in any reasonable search tree, each edge $t \in T$ appears at most once. It also always holds that $|T| = |U| - 1$. As any search tree D for F has exactly $|U| - 1$ internal nodes, it follows that each test $t \in T$ appears *exactly* once in D.

We consider here the case $uCnW|T|A$ of non-uniform weights and uniform costs. That problem version is known to be NP-hard [18] and admits a linear time algorithm achieving an approximation ratio of 14 [23]. In this section we give a simple analysis showing that a natural greedy algorithm attains a 2-approximation. We assume here that there are no elements of weight zero. In the full paper we will show that this assumption can be removed by using a more careful tie breaking strategy.

Our greedy algorithm always selects a test edge t such that the two subtrees F_1, F_2 obtained by the removal of t are as even as possible in terms of weight, i.e. the algorithm always maximizes $\min\{w(F_1), w(F_2)\}$, breaking ties arbitrarily.

In order to prove that this algorithm results in a 2-approximation, we describe a procedure for turning any search tree D^*, including the optimal one, into the greedy search tree D computed by our algorithm, and we show that during the transformation the cost increases by no more than $cost(D)/2$.

Let t be the test associated with the root of the greedy search tree D, and let F_1, F_2 be the two subtrees obtained after removing t from F. Furthermore, let D_1^* be the search tree for F_1 which is obtained from D^* by simply skipping all the tests associated with edges not in F_1, and let D_2^* be defined analogously. The transformation from D^* to D proceeds as follows: (a) Construct a search tree D' for F with test t at the root, and with the left and right subtree under the root being equal to D_1^* and D_2^*, respectively. (b) Recursively turn D_1^* and D_2^* into greedy search trees D_1 and D_2 for F_1 and F_2, respectively.

Lemma 1. *Step (a) increases the cost by at most $W/2$.*

Using this lemma, we can show by simple induction on the number of nodes in F that the transformation increases the cost by at most $cost(D)/2$. The basic case, $|T| = 1$, is trivial. From the induction hypothesis we know that step (b) of the transformation increases the cost by at most $cost(D_1)/2 + cost(D_2)/2$, so the claim follows from the fact that $cost(D) = W + cost(D_1) + cost(D_2)$.

Proof (of Lemma 1). Assume w.l.o.g. that the root t_1 of D^* is associated with an edge in the subtree F_1. Consider in D^* the path $t_1, t_2, \ldots$ that leads from t_1 to test t. Let t_k be the first test on that path which is not located in F_2, so either $t_k = t$, or t_k is associated with an edge in F_2. Furthermore, for $i = 1, \ldots, k - 1$, among the two subtrees obtained by removing t_i from F, let G_i be the one not containing t.

As $t_1, \ldots, t_{k-1}$ are in F_1 and thus are skipped in D_2^*, we have that the search path to any node from F_2 is by at least $k - 1$ shorter in D_2^* than it is in D^*. Search tree D_1^* is obtained from D^* by skipping certain tests, so the search path in D_1^* to any node in F_1 is at most as long as in D^*. For the set of nodes in the subtree $G' := F_1 - (G_1 \cup \ldots \cup G_{k-1})$ we make a stronger statement: As in D^* the search path to those nodes contains test $t_k \notin F_1$, the search path to them in D_1^* is by at least one shorter. Summarizing the findings from this paragraph, the difference $d := cost(D^*) - (cost(D_1^*) + cost(D_2^*))$ is

$$d \geq (k-1)w(F_2) + w\left(F_1 - \left(\bigcup_{i=1}^{k-1} G_i\right)\right) \geq (k-1)w(F_2) + w(F_1) - \sum_{i=1}^{k-1} w(G_i) .$$

The greedy criterion ensures that for $i = 2, \ldots, k-1$ the total weight $w(G_i)$ of all nodes in G_i is at most $\min\{w(F_2), W/2\}$. $w(G_i) \leq W/2$ holds because otherwise, as t is in $F - G_i$, greedy would rather choose t_i than t (note that here the assumptions of non-zero weights is essential). But this means that $w(G_i) \leq w(F - w(G_i))$, and therefore $w(G_i) > w(F_2)$ would again mean that t_1 is a better greedy choice than t. We charge $G_2, \ldots, G_{k-1}$ against $(k-2)$ times $w(F_2)$:

$$d \geq w(F_2) + w(F_1) - w(G_1) = w(F - G_1) \geq W/2 .$$

Now the Lemma is established by

$$cost(D') = W + cost(D_1^*) + cost(D_1^*) = W + cost(D^*) - d < cost(D^*) + W/2 .$$

5   Conclusion

We presented a new greedy approach which can be employed for different variants of the multiway identification problem (aka active learning problem). We demonstrated that this novel algorithm can be easily analyzed in several problem cases considered in the literature. We believe that our greedy approach deserves some more investigation in order to fully understand its potential and, possibly, its applicability in other contexts of search where costs of the tests and weights are simultaneously considered. With respect to our analysis, the following questions are worth considering: (1) can the analysis in Section 2.1. be improved or can its tightness be shown? (2) An interesting generalization of the result in Section 2.3 would be to prove that the cost of the costliest path is at most $\log nW$ times the cost of the costliest path in the tree that minimizes the worst-case.

We also investigated the restriction of the decision tree minimization problem to poset instances, showing that for the average-case objective general poset instances are as hard to approximate as unrestricted instances. For the case of tree-like posets, however, we proved that a greedy strategy can attain 2-approximation. In this direction, a further open problem concerns the limits of approximability of tree-like instances, e.g., existence of a PTAS vs. APX-hardness.

References

1. Abrahams, J.: Code and parse trees for lossless source encoding. In: Compression and Complexity of Sequences 1997, pp. 145–171 (1997)
2. Adler, M., Heeringa, B.: Approximating optimal binary decision trees. In: Goel, A., Jansen, K., Rolim, J.D.P., Rubinfeld, R. (eds.) APPROX and RANDOM 2008. LNCS, vol. 5171, pp. 1–9. Springer, Heidelberg (2008)

3. Arkin, E., Meijer, H., Mitchell, J., Rappaport, D., Skiena, S.: Decision trees for geometric models. International Journal of Computational Geometry and Applications 8(3), 343–364 (1998)
4. Carmo, R., Donadelli, J., Kohayakawa, Y., Laber, E.: Searching in random partially ordered sets. Theoretical Computer Science 321(1), 41–57 (2004)
5. Chakaravarthy, V., Pandit, V., Roy, S., Awasthi, P., Mohania, M.: Decision trees for entity identification: Approximation algorithms and hardness results. In: PODS, pp. 53–62 (2007)
6. Chakaravarthy, V., Pandit, V., Roy, S., Sabharwal, P.: Approximating decision trees with multiway branches. In: Albers, S., Marchetti-Spaccamela, A., Matias, Y., Nikoletseas, S., Thomas, W. (eds.) ICALP 2009. LNCS, vol. 5555, pp. 210–221. Springer, Heidelberg (2009)
7. Dasgupta, S.: Analysis of a Greedy Active Learning Strategy. In: NIPS (2007)
8. Daskalakis, C., Karp, R., Mossel, E., Riesenfeld, S., Verbin, E.: Sorting and selection in posets. In: SODA, pp. 392–401 (2009)
9. Dereniowski, D., Kubale, M.: Efficient parallel query processing by graph ranking. Fundamenta Informaticae 69 (2008)
10. Dereniowski, D.: Edge ranking and searching in partial orders. Discrete Applied Mathematics 156(13), 2493–2500 (2008)
11. Feige, U.: A threshold of ln n for approximating set cover. Journal of the ACM 45(4), 634–652 (1998)
12. Garey, M.: Optimal binary identification procedures. SIAM Journal on Applied Mathematics 23(2), 173–186 (1972)
13. Garey, M., Graham, R.: Performance bounds on the splitting algorithm for binary testing. Acta Informatica 3, 347–355 (1974)
14. Golin, M., Kenyon, C., Young, N.: Huffman coding with unequal letter costs. In: STOC, pp. 785–791 (2002)
15. Guillory, A., Bilmes, J.: Average-Case Active Learning with Costs. In: The 20th Intl. Conference on Algorithmic Learning Theory (2009)
16. Gupta, A., Krishnaswamy, R., Nagarajan, V., Ravi, R.: Approximation algorithms for optimal decision trees and adaptive TSP problems. In: ICALP (2010)
17. Hyafil, L., Rivest, R.: Constructing obtimal binary decision trees is NP-complete. Information Processing Letters 5, 15–17 (1976)
18. Jacobs, T., Cicalese, F., Laber, E., Molinaro, M.: On the Complexity of Searching in Trees: Average-Case Minimization. In: ICALP (2010)
19. Knight, W.: Search in an ordered array having variable probe cost. SIAM Journal on Computing 17(6), 1203–1214 (1988)
20. Knuth, D.: Optimum binary search trees. Acta. Informat. 1, 14–25 (1971)
21. Kosaraju, R., Przytycka, T., Borgstrom, R.: On an optimal split tree problem. In: Dehne, F., Gupta, A., Sack, J.-R., Tamassia, R. (eds.) WADS 1999. LNCS, vol. 1663, pp. 157–168. Springer, Heidelberg (1999)
22. Laber, E., Milidiú, R., Pessoa, A.: On binary searching with non-uniform costs. In: SODA, pp. 855–864 (2001)
23. Laber, E., Molinaro, M.: An Approximation Algorithm for Binary Searching in Trees. Algorithmica, doi: 10.1007/s00453-009-9325-0
24. Laber, E., Nogueira, L.: On the hardness of the minimum height decision tree problem. Discrete Applied Mathematics 144(1-2), 209–212 (2004)
25. Lipman, M., Abrahams, J.: Minimum average cost testing for partially ordered components. IEEE Transactions on Information Theory 41, 287–291 (1995)
26. Mozes, S., Onak, K., Weimann, O.: Finding an optimal tree searching strategy in linear time. In: SODA, pp. 1096–1105 (2008)

Single and Multiple Device DSA Problem, Complexities and Online Algorithms[*]

Weiwei Wu[1,2], Wanyong Tian[1,2], Minming Li[2],
Chun Jason Xue[2], and Enhong Chen[1]

[1] School of Computer Science, University of Science and Technology of China
[2] Department of Computer Science, City University of Hong Kong
{wweiwei2,twanyong2}@student.cityu.edu.hk, minmli@cs.cityu.edu.hk,
jasonxue@cityu.edu.hk, cheneh@ustc.edu.cn

Abstract. We study the single-device Dynamic Storage Allocation (DSA) problem and multi-device Balancing DSA problem in this paper. The goal is to dynamically allocate the job into memory to minimize the usage of space without concurrency. The SRF problem is just a variant of DSA problem. Our results are as follows,

- The NP-completeness for 2-SRF problem, 3-DSA problem, and DSA problem for jobs with agreeable deadlines.
- An improved 3-competitive algorithm for jobs with agreeable deadlines on single-device DSA problem. A 4-competitive algorithm for jobs with agreeable deadlines on multi-device Balancing DSA problem.
- Lower bounds for jobs with agreeable deadlines: any non-clairvoyant algorithm cannot be $(2 - \epsilon)$-competitive and any clairvoyant algorithm cannot be $(1.54 - \epsilon)$-competitive.
- The first $O(\log L)$-competitive algorithm for general jobs on multi-device Balancing DSA problem without any assumption.

1 Introduction

This paper studies Dynamic Storage Allocation (DSA) problem, which is a classic problem in computer science. The problem description is as follows. Given a set of n jobs $\mathcal{J}$, each job J_i is characterized by a *size* s_i, an *arrival time* r_i, and a *departure time/deadline* d_i. Each job should be allocated in a contiguous location at its arrival time. Once placed into address $a(J_i)$, the job occupies the allocated space from address $a(J_i)$ to $a(J_i) + s_i - 1$ (with size s_i) in the whole time interval $[r_i, d_i]$. The occupied space is available for other jobs after time d_i. Two jobs i, j are assigned with *conflict* if $[r_i, d_i] \cap [r_j, d_j] \neq \emptyset$ and $[a(J_i), a(J_i) + s_i - 1] \cap [a(J_j), a(J_j) + s_j - 1] \neq \emptyset$. We need to place all the jobs into the memory without conflict at any time. The objective is to minimize the

[*] This work was supported in part by grants from the Research Grants Council of the Hong Kong Special Administrative Region, China [Project No. CityU 117408 and 123609], National Natural Science Foundation of China (grant no. 60775037), and Research Fund for the Doctoral Program of Higher Education of China (20093402110017).

O. Cheong, K.-Y. Chwa, and K. Park (Eds.): ISAAC 2010, Part II, LNCS 6507, pp. 218–229, 2010.
© Springer-Verlag Berlin Heidelberg 2010

memory space used by all jobs. Another interpretation of DSA is to consider the following variant of Strip Packing Problem. Packing into a strip can be interpreted as packing into a 2-D geometric plane $X \times Y$ with fixed width X. In the Strip Packing Problem, each rectangle is a $(d_i - r_i + 1) \times s_i$ area. The objective is to pack all the rectangles into a strip that has a fixed width, while minimizing the maximum height of this strip. Obviously, DSA is a variant of Strip Packing Problem (we would not survey the work in this field here) where the position of rectangles in X-axis is immovable. DSA has received much attention since 1950s. Knuth proposed some basic methods in [12]. Stockmeyer proved its NP-completeness by a reduction from the 3-partition problem in 1976. He also proved DSA to be strongly NP-complete even when sizes of jobs are restricted to $\{1,2\}$ (The proof was given in Larry Stockmeyer's private communication with David Johnson and included in [3]). For *offline* version of DSA, the first constant-factor approximation algorithm—an 80-approximation first-fit mechanism was proposed by Kierstead in [10] where jobs are sorted according to their sizes. Whereafter Kierstead reduced the approximation ratio to 6 in [11]. Subsequently, Gergov presented a 5-approximation algorithm in [7] and then improved to a 3-approximation algorithm in [8]. Narayanaswamy gave another 3-approximation algorithm in [16]. The algorithm with approximation ratio $2 + \epsilon$ is proposed by Buchsbaum in [2]. For the special case of DSA with only two job sizes, Gergov [7] presented a 2-approximation algorithm. When job sizes are restricted to small values, Li et al. [13] proposed a $\frac{4}{3}$-approximation algorithm when the sizes are $\{1, 2\}$ and a 1.7-approximation algorithm when the sizes are $\{1, 2, 3\}$. For *online* version of DSA, Robson [17] showed that first-fit algorithm had a competitive ratio of $O(\log(S_{max}))$ where S_{max} is the maximum size among all jobs. Then Luby et al. [14] proved that first-fit algorithm could achieve the competitive ratio of $O(\min\{\log(S_{max}), \log(\chi)\})$, where χ denotes the maximum number of concurrent jobs for all time instants. This is an improvement over Robson [17] since generally S_{max} and χ are incomparable. They also studied the multi-device Balancing DSA problem (minimizing the maximum occupied space on m devices) and gave an optimal competitive algorithm under the assumption that $m \leq \frac{\omega^*}{S_{max}}$ where ω^* represents the maximum total size of alive jobs at any time, which is also denoted as $LOAD(\mathcal{J})$ in this paper. Gergov [7] improved the competitive ratio of first-fit strategy of online DSA to $\Theta(\max\{1, \log(S_{max} \cdot \chi / \omega^*)\})$. For cases that the live-ranges (which equals the difference of deadline and arrival time) of jobs are known when they arrive (this scenario is also referred to as *clairvoyant scheduling*), Naor et al. [15] provided an online 8-competitive algorithm for jobs with agreeable deadlines (where later released jobs have no earlier deadlines) and an $O(\min\{(\log(L), \log(\tau))\})$-competitive algorithm for general jobs , where L is the ratio between the longest and the shortest live-ranges of the jobs, and τ denotes the maximum number of concurrent jobs that have different live-ranges. [9] provides a lower bound $\Omega(\frac{\log x}{\log \log x})$-competitive for online algorithm where x can be n, S_{max}, χ, τ or $\log L$.

DSA is investigated in many applications such as memory or register allocation in operating systems and bandwidth allocation in communication networks etc.

In operating systems, compiler algorithms should decide where to place items (*arrays* or *matrices etc.*) in memory. To deal with register allocation problems or DSA problems, another conventional mechanism is to use graph coloring. Graph coloring has been extensively investigated and specifically studied for register allocation since Chaitin [4]. From [4] on, many register allocation algorithms based on graph coloring have been proposed such as [5][1][6]. Our study of DSA is motivated by stream register file (SRF) allocation problem. SRF is a software-managed on-chip memory. Since it is a critical resource, optimizing SRF utilization becomes crucial. Recently in [18], by dividing streams (can be referred to as *jobs* here) into two types (short streams with live-ranges $l \leq 2$ and long streams with live-ranges $l \geq 3$), they modeled the problem into comparability graph coloring. They considered a special case where there are at most two long items at any time instant and gave a near-optimal solution, without showing the complexity of this problem. We will refer to their model as *2-SRF problem*.

We study both the complexities and algorithms for DSA problem and its variants. Note that the SRF allocation problem studied in [18] can be transformed into DSA notation. When considering the duration of the jobs, one interesting question is that how large the value L should be in order to make the problem hard. We will refer to the problem where every job has live-range at most L as *L-DSA problem*. For the general DSA problem, we improve the performance for the single-device DSA problem studied in [15] and gives the first competitive algorithm for multi-device Balancing DSA problem without any assumption. We start by investigating the jobs with *agreeable deadlines*, which was also studied in [15]. This kind of jobs has received much concern in other domains of scheduling problems. We derive the complexity for this type of jobs for completeness. Moreover, although our method leads to the same performance as [15] for the offline setting, it is proved more extendable to the online setting both for single-device and multi-device. Our results are as follows,

- The NP-completeness for 2-SRF problem, 3-DSA problem, and DSA problem for jobs with agreeable deadlines.
- An improved 3-competitive algorithm over [15] for jobs with agreeable deadlines on single-device DSA problem. A 4-competitive algorithm for jobs with agreeable deadlines on multi-device Balancing DSA problem.
- Lower bounds for jobs with agreeable deadlines: any non-clairvoyant algorithm cannot be $(2-\epsilon)$-competitive and any clairvoyant algorithm cannot be $(1.54-\epsilon)$-competitive.
- The first $O(\log L)$-competitive algorithm for general jobs on multi-device Balancing DSA problem without any assumption.

The rest of the paper is organized as follows. In Section 2, we review the model for DSA problem and multi-device Balancing DSA problem. In Section 3, we settle the complexities for L-DSA problem and 2-SRF problem. In Section 4, we investigate the single-device DSA problem. We propose the laminar decomposition for jobs with agreeable deadlines. The complexity for jobs with agreeable deadlines is settled. We have a simple 2-approximation algorithm through laminar decomposition. This decomposition can be used to design an improved 3-competitive algorithm (for jobs with agreeable deadlines). In Section 5, we design competitive algorithms for multi-device Balancing DSA problem. A 4-competitive algorithm for jobs with agreeable deadlines is proposed and also an $O(\log L)$-competitive algorithm for general jobs. Due to the space limit, some of the proofs are omitted in this version.

2   Preliminaries

In operating systems, memory allocation algorithms need to consider where to place a set of items (often arrays, matrices etc.) so that the overall usage of memory is minimized. In DSA (Dynamic Storage Allocation) notation, we are given jobs $\mathcal{J}$ where each job i (or J_i) has a *size* s_i and a *live-range/length* starting from the *arrival time* r_i and departing at *deadline* d_i, i.e. $J_i = (r_i, d_i, s_i)$. A job i is said *alive at time* t if $t \in [r_i, d_i]$. The time is partitioned into units. For each unit of time t, if we say i has *arrival time* t, we mean job i is released at the beginning of time t. Correspondingly, if we say i has *deadline* t, we mean job i expires at the end of time t. Upon arrival of each job, we need to allocate a contiguous memory location for it. We assume that the memory is starting from address 1 instead of address 0. By $a(J_i)$ we denote that J_i is assigned or allocated to address $a(J_i)$ throughout its live-range. Once placed into address $a(J_i)$, J_i occupies the space from address $a(J_i)$ to $a(J_i) + s_i - 1$ (with size s_i) in the whole time interval $[r_i, d_i]$. The occupied space is available for other jobs after time d_i. Two jobs i, j are assigned with *conflict* if $[r_i, d_i] \cap [r_j, d_j] \neq \emptyset$ and $[a(J_i), a(J_i) + s_i - 1] \cap [a(J_j), a(J_j) + s_j - 1] \neq \emptyset$. We need to place all the jobs into the memory without conflict at any time. Once allocated, the job cannot be moved or deleted until its deadline. Only when the job leaves, its occupied memory location can be available for other jobs. The objective is to minimize the overall used memory size. For the multi-device *Balancing DSA problem*, we are given m devices and the goal is to allocate the job in a balanced manner such that the maximum occupied space on the m devices is minimum. By S_{max} we denote the maximum size over all jobs. Let $LOAD(t)$ be the total size of alive jobs at time t. We use $LOAD(\mathcal{J})$ to represent the maximum total size over all time t. Obviously $\max\{LOAD(\mathcal{J}), S_{max}\}$ is a lower bound for the optimal solution for single-device DSA problem, while multi-device Balancing DSA problem has a lower bound $\max\{\frac{LOAD(\mathcal{J})}{m}, S_{max}\}$.

Observing that in the applications such as SRF problem each item usually has a short live-range, we use *L-DSA problem* to denote the restricted DSA problem where all jobs have live-ranges at most L. The jobs with *agreeable deadlines* which receive much concern in scheduling literature are defined to be jobs where later released jobs always have no earlier deadlines.

We use $OPT(\mathcal{J})$ to denote the optimal memory space occupied by the optimal solution for jobs $\mathcal{J}$. An offline algorithm is said to be c-approximation if it outputs a solution which occupies memory space at most $c \cdot OPT(\mathcal{J})$. In the online setting, the algorithm should make decision upon the arrival of jobs. There are two versions. In the *clairvoyant* scheduling, the algorithm knows the deadline when a job is released, while in the *non-clairvoyant* scheduling the algorithm does not know this information. We say an online algorithm is *c*-competitive if it always outputs a solution within *c* times the optimal offline solution.

3 Settling the Complexity for Special Cases of DSA/SRF Problem

The decision version of *Partition problem* is as follows. Given finite set $U = \{u_1, u_2, \ldots, u_n\}$ with $\sum_{i=1}^{n} u_i = 2B$. The question is to find a subset $U' \subset U$ such that the sum of the elements in U' is exactly B. In the following we will show that the 2-SRF problem and L-DSA problem are NP-complete. The construction is reduced from Partition problem.

The problem defined in [18] is referred to as 2-SRF problem in this paper. After transforming their paradigm to the DSA notation, the problem can be considered equivalently as follows. There are two kinds of jobs, *short* jobs and *long* jobs. All the short jobs have live-ranges at most 2. The long jobs have live-ranges larger than 2, but each time there are at most two such long alive jobs. The justification of such a problem is simple because we can split the long live-range stream (job) into streams with short live-ranges by live-range splitting technique. This allows us to break long jobs into jobs limited to some constant length L. They discussed this problem by five sub-cases, among which four of these cases can be solved optimally by their proposed algorithm, while the fifth case is shown to be within the optimal solution plus 2 times the maximum size of the long jobs. Thus the complexity of this problem still remains open. The following theorem answers this question.

Theorem 1. *The decision version of 2-SRF problem is NP-complete even when all jobs have live-ranges at most $L = 3$.*

Note that the reduction in Theorem 1 uses jobs that have live-ranges at most 3, thus 3-DSA problem is also NP-complete. Simple extension of this construction can show the NP-completeness for every $L \geq 3$.

Corollary 1. *The decision version of L-DSA problem is NP-complete for $L \geq 3$.*

4 Better Online Algorithms for DSA Problem

In this section, we start by studying jobs that have *agreeable deadlines*. In this setting, all the later released jobs will have no earlier deadlines. This kind of jobs has also received much attention in other domains of scheduling problems. We will first show that the problem (with agreeable deadlines) is still NP-complete in the offline setting. Then we design a 2-approximation algorithm for jobs with agreeable deadlines in the offline version. This matches the result in [15]. However, we will show that the concept we use is more extendable to the online problem both in single-device and multi-device setting.

4.1 Approximation Algorithm through Laminar Decomposition

The following reduction shows that the problem is still NP-complete for jobs with agreeable deadlines. The reduction uses a symmetric structure of the jobs.

Theorem 2. *DSA problem for jobs with agreeable deadlines is NP-complete.*

To design an approximation algorithm, we introduce the *laminar jobs*. Laminar jobs at time t is composed of jobs $laminar(t) = \{j : t \in [r_j, d_j]\}$. The concept for our algorithm is to first decompose the original jobs into several small job sets, with jobs in the same set forming laminar jobs. These sets are then divided into two groups. Moreover, our decomposition and grouping ensure that jobs in different sets but the same group are never concurrent with each other. Thus this allows us to concentrate on the separated single set of laminar jobs. The decomposition step loops as follows:

(1) Set $t = d_1$. Let $S_1 = laminar(t)$.

(2) Assume j to be the first job in the remaining jobs. Let $t = d_j$ and $S_2 = laminar(t)$.

(3) Update j to be the first job in the remaining jobs and repeat to find all sets $S_3, S_4, \ldots, S_b$ until no jobs are left.

For the grouping step, we group the sets with odd index into group $G_1 = \{S_1, S_3, S_5, \ldots, \}$ while the others form group G_2. Note that both the decomposition and grouping can be operated in online manner since we know both the release times and deadlines of the jobs when they arrive. An offline 2-approximation algorithm can be easily obtained.

Lemma 1. *DSA problem for agreeable jobs has a 2-approximation algorithm.*

Proof. An important observation of such a grouping strategy is that the sets in the same group are pairwise non-concurrent at any time. Take the first three sets S_1, S_2, S_3 for example. Let j be the first job in S_2. Let t_a be the largest deadline of jobs in S_1 and t_b be the smallest arrival time of jobs in S_3. We note that $d_j \geq t_a$ according to the decomposition step (2). Moreover, the first job in S_3 has release time larger than d_j because otherwise this job would be grouped into S_2, hence $t_b > d_j$. Therefore we have $t_a < t_b$ and S_1, S_3 are not concurrent. Similarly this can be extended to every two sets in the same group.

Getting this, for all sets in G_1 (or G_2), we can solve each set optimally by FirstFit policy since laminar jobs is crossing at least one common time. The jobs in group G_1 (or G_2) can also be solved optimally due to the pairwise independence property of the sets. Finally, by applying $OrderedFit(\cdot)$ to $G_1 \cup G_2$ and Lemma 2, the new algorithm framework uses memory at most $2OPT(\mathcal{J})$.

Lemma 2. *(Technical Lemma) If a job set $\mathcal{J}$ is decomposed into k disjoint subsets $G_1, \ldots, G_k$, and every set G_i can be solved within $r \cdot OPT(G_i)$ by some strategy STR (i.e for each i we have $STR(G_i) \leq r \cdot OPT(G_i)$), then by applying OrderedFit to $\mathcal{J}$ we have $OrderedFit(STR, \mathcal{J}) \leq rk \cdot OPT(\mathcal{J})$.*

Algorithm 1. OrderedFit(STR,$\mathcal{J}$)

---

for each round $r = 1, 2, \ldots, k$ **do**

1. Simulate $STR(G_r)$ in virtual device r. Denote the maximum occupied address on device r to be M_r.

2. Assign all jobs in the practical memory as if they are combined by the allocations in k virtual devices. Instead of starting from address 1 on the virtual device, device r $(r \geq 2)$ starts its allocation from address $\sum_{i=2}^{r} M_{i-1} + 1$ in the practical single memory.

end for

Note that our new algorithm matches the current best algorithm in [15]. However, our decomposition step not only simplifies the structure of the jobs, but will also be proved more extendable to online single-device and multi-device setting.

4.2   Better Online Algorithm through Laminar Decomposition

We note that our decomposition step can be performed in an online manner. We will show how to achieve 3-competitive clairvoyant online algorithm by using laminar decomposition. This improves the ratio 8 by algorithms in [15].

Our strategy is to assign the second decomposed group pessimistically to ensure that we waste at most one gap in the memory, which is shown in Algorithm 2. Let S_i be generated by laminar jobs at time t_i, i.e. $S_i = laminar(t_i)$. We have $LOAD(S_i) \leq LOAD(t_i)$ by the decomposition step.

Algorithm 2. PessiOnline

---

1. Jobs are decomposed into two groups $G_1 = \{S_1, S_3, S_5, \ldots\}, G_2 = \{S_2, S_4, S_6, \ldots\}$ online by laminar property. For all jobs in the same set, the job that is earlier released will be allocated to lower address (ties are broken by assigning jobs with earlier deadline to lower address). We apply different strategies for the two groups as follows.

for each round $i = 1, 2, 3, \ldots$ **do**

2. For set S_{2i-1} in G_1, upon arrival of the jobs, it will be placed on the lowest feasible address (FirstFit). Let the maximum address used by this set be M_{2i-1}.

3. For set S_{2i} in G_2, we pessimistically assign the first one to address $M_{2i-1} + 1$. All the later jobs in the same set will be assigned one by one to the address immediately (consecutively) after the prior one, without violating the rule in Step 1.

end for

Theorem 3. *There is a 3-competitive algorithm for single-device DSA problem where jobs have agreeable deadlines.*

Proof. Obviously the occupied space is maximized at some time t_i. The analysis is based on discussing the possible cases of placing job set S_{2k-1} where $k = 1, 2, \ldots$. Three observations are the key advantages of such an allocation strategy. First, when we consider jobs in S_{2k-1} (determine the allocation upon the arrival of a job), the determination should never worry about the possible conflict with S_{2k-3}, since these two sets are separated by time t_{2k-2} and thus never concurrent. Second, because of the pessimistic allocation in Step 3 and the rule in Step 1, the jobs in S_{2k-2} is consecutively allocated and only leave at most one gap (it can only be the empty space starting from address 1) at every time in interval $[t_{2k-2}, t_{2k-1}]$ (before jobs S_{2k-1} are released). Third, because we apply the FirstFit policy to set S_{2k-1} and also the two former observations, the jobs in S_{2k-1} can be placed in at most two contiguous memory spaces (Namely, divided by a gap of empty space). For example, S_3, S_7 are assigned to two contiguous memory spaces, while S_5 is assigned to one contiguous memory space. Furthermore, the gap is generated only because there is a job with larger size being released. For the case that S_{2k-1} is separately allocated, the gap is created because the space of the gap (let the size be s_g) is less than the job (let its size be s_j) attempting to fit into this space. Namely, we have $s_g < s_j < LOAD(t_{2k-1})$. The occupied space at time t_{2k-1} is at most $LOAD(t_{2k-1}) + s_g \leq 2LOAD(t_{2k-1}) \leq 2LOAD(\mathcal{J}) \leq 2OPT(\mathcal{J})$. Since we assign the later set S_{2k} pessimistically, the space used at time t_{2k} is at most $LOAD(t_{2k-1}) + s_g + LOAD(t_{2k}) \leq 3OPT(\mathcal{J})$. For the case that S_{2k-1} is not separately allocated, the occupied space by assigning S_{2k-1} at time t_{2k-1} is $LOAD(S_{2k-1}) \leq LOAD(t_{2k-1})$. S_{2k} will not conflict with S_{2k-2} and is then consecutively allocated after S_{2k-1}, thus at time t_{2k} the occupied space is at most $LOAD(t_{2k-1}) + LOAD(t_{2k}) \leq 2LOAD(\mathcal{J}) \leq 2OPT(\mathcal{J})$. Combining these two cases, the algorithm is 3-competitive.

4.3   Lower Bounds on the Online Algorithm

For jobs with agreeable deadlines, we show the following lower bounds of online algorithm.

Theorem 4. *Any non-clairvoyant online algorithm for DSA problem with agreeable jobs cannot be $(2 - \epsilon)$-competitive. Any online clairvoyant algorithm for DSA problem with agreeable deadlines cannot be $(1.54 - \epsilon)$-competitive.*

5   Multi-device Balancing DSA Problem

In this section, we settle the multi-device Balancing DSA problem for both agreeable jobs and general jobs. In this setting, there are m devices and the objective is to balance the usage of the memories. Accurately, we aim at minimizing the maximum memory size used in the m memories. We observed that [14] studied this problem under the assumption that "$m \leq \frac{\omega^*}{S_{max}}$". With their assumption, [14] showed that their algorithm is online optimal competitive. We restudy this problem by dropping their assumption. For the multi-device DSA problem, we give a 4-competitive algorithm for jobs with agreeable deadlines and an $O(\log L)$-competitive algorithm for general jobs. The method of the online algorithm is still based on the laminar decomposition.

We also start by designing a $O(1)$-competitive algorithm for jobs with agreeable deadlines.

Balancing DSA Problem for Jobs with Agreeable Deadlines. Assume we decompose the jobs $\mathcal{J}$ to $G_1 = \{S_1, S_3, S_5, \ldots\}, G_2 = \{S_2, S_4, S_6, \ldots\}$ where $S_i = laminar(t_i)$ through laminar decomposition in an online manner as in Section 4.2. The new algorithm for multi-device is shown in Algorithm 3.

Algorithm 3. MultiOnline

---

1. Jobs are decomposed into two groups $G_1 = \{S_1, S_3, S_5, \ldots\}, G_2 = \{S_2, S_4, S_6, \ldots\}$ online by laminar property. We apply different strategies for the two groups. For all jobs in the same set, the job that is earlier released will be allocated to lower address (ties are broken by assigning jobs with earlier deadline to lower address). In the following, by *top address* at time t we denote the maximum occupied address at time t that is allocated in the current round.

for each round $i = 1, 2, 3, \ldots$ **do**

  2. For set S_{2i-1} in G_1, upon arrival of the jobs j, we simulate on every device by assigning j to the lowest feasible address (FirstFit policy). If j can be allocated to the empty space that is below the top address of the device l, then we assign j to l. If not, then we assign j to the device with minimum top address. Assign j to the selected device by FirstFit policy.

  3. For set S_{2i} in G_2, we assign the jobs pessimistically. Upon the arrival of job $j \in S_{2i}$, we suppose that l is the current device that has the minimum top address (let the address be M_l). Assign j to address $M_l + 1$. All the later jobs in the same set that is allocated to l will be assigned consecutively without violating the rule in Step 1.

end for

Theorem 5. *Algorithm MultiOnline is 4-competitive for multi-device Balancing DSA problem on jobs with agreeable deadlines.*

Proof. Through laminar decomposition, one advantage is that the load of set of laminar jobs S_i at time t_i is exactly the summation of sizes over all its jobs. We first note that the lower bounds for multi-device Balancing DSA problem are $OPT \geq \frac{LOAD(\mathcal{J})}{m}$ and $OPT \geq S_{max}$. Observing this, our allocation strategy in Algorithm 3 tries to balance the allocation for jobs in the same set S_i. Assume that the minimum (maximum) occupied address in the m devices is M_{min} (M_{max}), the allocation strategy in Step 2 and Step 3 ensures that $M_{max} - M_{min} \leq S_{max}$. First, when allocating S_1, the load at time t_1 is allocated in a balanced manner to the m devices. The maximum address used in the m devices is at most $\frac{LOAD(t_1)}{m} + S_{max}$ since the minimum address used in the m devices is at most $\frac{LOAD(t_1)}{m}$ and the difference of every two devices is at most S_{max}. When we allocate S_2, the jobs are pessimistically assigned to the current top address on the selected device (which is the device that has the minimum occupied address). The total space occupied in t_2 is exactly $LOAD(S_1)+LOAD(S_2)$ which is at most $LOAD(t_1)+LOAD(t_2)$ since assigning S_1 did not generate any empty space at time t_1. Because we assign S_2 in a balanced way, the device with maximum occupied address at time t_2 is at most $\frac{LOAD(t_1)+LOAD(t_2)}{m} + S_{max}$. Note that the optimal solution OPT uses at least $\max\{\frac{LOAD(t_1)}{m}, \frac{LOAD(t_2)}{m}, S_{max}\}$. Thus the occupied space at time t_2 is at most $3 \cdot OPT(\mathcal{J})$.

We will prove that the occupied address is at most $3 \cdot OPT(\mathcal{J})$ at time t_{2i-1} and at most $4 \cdot OPT(\mathcal{J})$ at time t_{2i}. As in the proof of Theorem 3, we discuss the possible cases of allocation for set S_{2i-1} where $i \geq 2$. Note that we allocate set S_{2i-1} by FirstFit for selected device. Consider the case that the current job has size larger than the remaining empty space that is below the selected device's top address. The allocation creates at most one gap (empty space) that has size less than S_{max}. This property holds because the strategy on a single device is the same as what we used in Algorithm 2. This job will be allocated to the top address of the selected device.

We start by the simple case that every device is at least assigned one job of S_{2i-1} to its top address (denote such a device as *saturated-device*). That is, each gap generated on the m devices is at most S_{max}. After assigning S_{2i-1}, the total occupied space in these devices at time t_{2i-1} is at most $LOAD(t_{2i-1})+m\cdot S_{max}$. In this case, we have $M_{min} \leq \frac{LOAD(t_{2i-1})}{m}+S_{max}$. Thus $M_{max} \leq M_{min}+S_{max} \leq 3 \cdot OPT(\mathcal{J})$. When we assign S_{2i} later, we need not worry about the possible conflict with S_{2i-2}. Due to the pessimistic strategy in Step 3, the jobs in S_{2i} are assigned to the top address consecutively and no space is wasted. Thus after assigning S_{2i}, the total occupied space in these devices at time t_{2i} is at most $LOAD(t_{2i-1}) + LOAD(t_{2i}) + m \cdot S_{max}$. Hence, in this case we have $M_{max} \leq M_{min} + S_{max} \leq \frac{LOAD(t_{2i-1})+LOAD(t_{2i})}{m} + S_{max} + S_{max} \leq 4 \cdot OPT(\mathcal{J})$.

We say a device is *saturated* (or *unsaturated*) if the current job is (or not) infeasible to be assigned to the empty space that is below the top address of the selected device at Step 2. Now we consider the case that some devices are *saturated* while others are not when assigning S_{2i-1}. The existence of saturated device implies that the current job is infeasible to be allocated to the empty space on the unsaturated devices according to Step 2. Thus the empty space both on the unsaturated devices and the saturated devices has size less than S_{max}. The total occupied space at time t_{2i-1} is still $LOAD(t_{2i-1}) + m \cdot S_{max}$. Thus $M_{max} \leq \frac{LOAD(t_{2i-1})}{m} + 2S_{max} \leq 3OPT(\mathcal{J})$. Similarly at time t_{2i}, we have $M_{max} \leq \frac{LOAD(t_{2i-1})+LOAD(t_{2i})}{m} + 2S_{max} \leq 4OPT(\mathcal{J})$.

It remains to consider the case that all devices are unsaturated after assigning S_{2i-1}. All the jobs of S_{2i-1} are assigned below their top addresses of the selected devices. Note that the occupied top address for these unsaturated-devices is generated at time t_{2i-2}. This value can be bounded by $4OPT(\mathcal{J})$ by the analysis above. Thus the maximum occupied space at time t_{2i-1} is at most $4 \cdot OPT(\mathcal{J})$ in this situation. Considering assigning S_{2i} later, all jobs in S_{2i-2} have departed at this time by the independence property of laminar decomposition. Thus we need not to worry about the possible conflict between S_{2i} and S_{2i-2}. Since all the load is assigned from address 0 consecutively at time t_{2i-1}. The total occupied space at time t_{2i} will be $LOAD(S_{2i-1}) + LOAD(S_{2i})$ which is at most $LOAD(t_{2i-1}) + LOAD(t_{2i})$. Thus we have $M_{max} \leq \frac{LOAD(t_{2i-1}) + LOAD(t_{2i})}{m} + S_{max} \leq 3 \cdot OPT(\mathcal{J})$ at time t_{2i} in this situation. Therefore, the competitive ratio of Algorithm 3 is at most 4.

Extending to General Jobs. Now we are ready for the competitive algorithm on multi-device for general jobs. The extension will follow some extending procedure for single-device in [15] with further modifications. We will derive a $O(logL)$-competitive algorithm without aiming at reducing its constant factor. Every job with length $(2^{i-1}, 2^i]$ is rounded to length 2^i. Then the rounded jobs with length 2^i can be grouped into a set of type i. We will make a loss of factor 2 in this step. All the jobs in the same group has the nice property that they have agreeable deadlines. Let $\mathcal{J}'$ be all the jobs that are already released at current time t and correspondingly S'_{max} be the largest job size until time t. Let lower bound $LB(t) = \max\{\frac{LOAD(\mathcal{J}')}{m}, S'_{max}\}$. According to the proof in Theorem 5, we only need to assign a slot with size $4 \cdot LB(t)$ for the set of type i which ensures the feasibility to allocate jobs $\mathcal{J}'$. There are $O(\log L)$ such sets of different types. We only need to open a new slot (starting from the maximum occupied address thus far) if the arrival job cannot be matched to a slot for its type. Note that the value $LB(t)$ will change over time, thus we need to update the slot size as follows. For the current value $LB(t)$, we allocate a slot with size $8LB(t)$ for each type i. When this value (the lower bound over time $LB(t)$) is doubled due to later released jobs, we then double the slot size that is newly opened for jobs of type i. The doubling procedure will ensure that the total size of slot that is allocated is $O(\log L \cdot LB(t))$. Thus finally the competitive ratio is $O(\log L)$.

Theorem 6. *There is an $O(\log L)$-competitive algorithm for multi-device Balancing DSA problem for general jobs.*

6   Conclusion

In this paper, we improve the algorithms for DSA problem(s) by introducing the online laminar decomposition. It not only simplifies the analysis, but is also proved more extendable both for single-device and multi-device setting.

References

1. Briggs, P., Cooper, K.D., Torczon, L.: Improvements to graph coloring register allocation. ACM Transactions on Programming Languages and Systems 16(3), 428–455 (1994)
2. Buchsbaum, A.L., Karloff, H., Kenyon, C., Reingold, N., Thorup, M.: OPT versus LOAD in dynamic storage allocation. In: Proceedioings of the 35th Annual ACM Symposium on Therory of Computing (STOC), pp. 556–564 (2003)
3. Buchsbaum, A.L., Efrat, A., Jain, S., Venkatasubramanian, S.: Restricted strip covering and the sensor cover problem. The Conference version appears in Proceedings of the 18th ACM-SIAM Symposium on Discrete Algorithms (SODA), pp. 1056–1063 (2007), The full version is at
http://arxiv.org/PS_cache/cs/pdf/0605/0605102v1.pdf

4. Chaitin, G.J.: Register allocation & spilling via graph coloring. In: Proceedings of the SIGPLAN Symposium on Compiler Construction, pp. 98–105. ACM Press, New York (1982)
5. Chow, F.C., Hennessy, J.L.: The priority-based coloring approach to register allocation. ACM Transactions on Programming Languages and Systems 12(4), 501–536 (1990)
6. George, L., Appel, A.W.: Iterated register coalescing. ACM Transactions on Programming Languages and Systems 18(3), 300–324 (1996)
7. Gergov, J.: Approximation algorithms for dynamic storage allocation. In: Díaz, J. (ed.) ESA 1996. LNCS, vol. 1136, pp. 52–61. Springer, Heidelberg (1996)
8. Gergov, J.: Algorithms for compile-time memory optimization. In: Proceedings of the 10th ACM-SIAM Symposium on Discrete Algorithms (SODA), pp. S907–S908 (1999)
9. Kalyanasundaram, B., Pruhs, K.R.: Dynamic Spectrum Allocation: The Impotency of Duration Notification. In: Kapoor, S., Prasad, S. (eds.) FST TCS 2000. LNCS, vol. 1974, pp. 421–428. Springer, Heidelberg (2000)
10. Kierstead, H.A.: The linearity of first-fit colorings of interval graphs. SIAM Journal on Discrete Mathematics 1(4), 526–530 (1988)
11. Kierstead, H.A.: A polynomial time approximation algorithm for dynamic storage allocation. Discrete Mathematics 88, 231–237 (1991)
12. Knuth, D.E.: Foundamental algorithms, 2nd edn., vol. 1. Addison-Wesley, Reading (1973)
13. Li, S.C., Leong, H.W., Quek, S.K.: New approximation algorithms for some dynamic storage allocation problems. In: Proceedings of the 10th Annual International Computing and Combinatorics Conference, pp. 339–348 (2004)
14. Luby, M.G., Naor, J., Orda, A.: Tight bounds for dynamic storage allocation. In: Proceedings of the 5th Annual ACM-SIAM Symposium on Discrete Algorithms, pp. 724–732 (1994)
15. Naor, J., Orda, A., Petruschka, Y.: Dynamic storage allocation with known durations. In: Proceedings of the 5th Annual European Symposium on Algorithms (ESA), pp. 378–387 (1997); The journal version appreas in Discrete Applied Mathematics 100(3), 203–213 (2000)
16. Narayanaswamy, N.S.: Dynamic storage allocation and on-line colouring interval graphs. In: Proceedings of the 10th Annual International Computing and Combinatorics Conference, pp. 329–338 (2004)
17. Robson, J.M.: Worst case fragmentation of first-fit and best fit storage allocation strategies. Computer Journal 20, 242–244 (1977)
18. Yang, X., Wang, L., Xue, J., Deng, Y., Zhang, Y.: Comparability graph coloring for optimizing utilization of stream register files in stream processors. In: Proceedings of the 14th ACM SIGPLAN Symposium on Principles and Practice of Parallel Programming (PPoPP), pp. 111–120 (2009)

The Onion Diagram: A Voronoi-Like Tessellation of a Planar Line Space and Its Applications[*],[**]

(Extended Abstract)

Sang Won Bae[1] and Chan-Su Shin[2]

[1] Department of Computer Science, Kyonggi University, Suwon, Korea
swbae@kgu.ac.kr
[2] Department of Digital Information and Engineering,
Hankuk University of Foreign Studies, Yongin, Korea
cssin@hufs.ac.kr

Abstract. Given a set S of weighted points in the plane, we consider two problems dealing with planar lines in $\mathbb{R}^2$ under the weighted Euclidean distance: (1) Preprocess S into a data structure that efficiently finds a nearest point among S of a query "line". (2) Find an optimal "line" that maximizes the minimum of the weighted distance to any point of S. We introduce a unified approach to both problems based on a new geometric transformation that maps lines in the plane into points in a line space. It turns out that our transformation, together with its target space, well describes the proximity relations between given weighted points S and every planar line in $\mathbb{R}^2$. We define a Voronoi-like tessellation on the line space and investigate its geometric, combinatorial, and computational properties. As its applications, we obtain several new results on the two problems.

1 Introduction

Let $S \subset \mathbb{R}^2$ be a set of n points, called *sites*, with weights assigned. For any point $x \in \mathbb{R}^2$, the distance $d_p(x)$ to a site $p \in S$ is defined to be their Euclidean distance times the weight $w_p > 0$ of p; that is, $d_p(x) := w_p \cdot d(p, x)$, where $d(\cdot, \cdot)$ denotes the Euclidean distance function. Consider two popular geometric problems:

The nearest-neighbor query problem. Preprocess S into a data structure that efficiently reports $\arg\min_{p \in S} d_p(q)$ for a query point $q \in \mathbb{R}^2$.

The max-min location problem. Find an optimal point $x^* \in \operatorname{conv}(S)$ that maximizes the minimum distance to sites; that is, $x^* = \arg\max_{x \in \operatorname{conv}(S)} \min_{p \in S} d_p(x)$, where $\operatorname{conv}(S)$ denotes the convex hull of S.

When the weights w_p of all $p \in S$ are equal (or without weights), both problems can be solved efficiently by the ordinary Voronoi diagram; the nearest-neighbor query

---

[*] This work is dedicated to our advisor, Professor Kyung-Yong Chwa, on the occasion of his honorable retirement.

[**] Work by S.W.Bae was supported by National Research Foundation of Korea(NRF) grant funded by the Korea government(MEST) (No. 2010-0005974). Work by C.-S.Shin was supported by National Research Foundation of Korea(NRF) grant funded by the Korea government(MEST) (No. 2010-0016416), and the HUFS Research Fund.

O. Cheong, K.-Y. Chwa, and K. Park (Eds.): ISAAC 2010, Part II, LNCS 6507, pp. 230–241, 2010.

can be answered in $O(\log n)$ time by the Voronoi diagram of S using $O(n)$ space and $O(n \log n)$ preprocessing time, and the max-min location problem, also known as the *largest empty circle problem*, is solved by traversing all vertices and edges of the Voronoi diagram [14]. In the general weighted case, one can exploit the *multiplicatively weighted Voronoi diagram* of given weighted points S with $O(n^2)$ construction time [2].

This paper aims to extend the above idea, exploiting Voronoi diagrams of a certain type, in solving a variant of the problems where "lines" in $\mathbb{R}^2$ are dealt with instead of "points" in $\mathbb{R}^2$:

The nearest-neighbor line query problem. Preprocess S to a data structure that efficiently reports $\arg\min_{p \in S} d_p(l)$ for a query *line* $l \subset \mathbb{R}^2$.

The max-min line location problem. Find an optimal *line* $l^* \subset \mathbb{R}^2$ intersecting $\mathrm{conv}(S)$ that maximizes the minimum distance to sites.

Note that for any line $l \subset \mathbb{R}^2$, the distance $d_p(l)$ to a site $p \in S$ is naturally extended as $d_p(l) := \min_{x \in l} d_p(x)$. In the nearest-neighbor line query problem, shortly the *line query problem*, the query object becomes a line in $\mathbb{R}^2$; in the max-min line location problem, we would like to find an optimal line. The max-min line location problem is a geometric facility location problem, where a facility of line shape to be placed is "obnoxious" so that each site in S want to stay as far away from it as possible. Both problems have been studied in computational geometry and operations research community, motivated by their natural applications.

It is known that the unweighted version of both problems can be efficiently solved commonly by the geometric dual transformation that maps a line $y = ax + b$ in the primal plane into a point $(a, -b)$ in the dual plane, and vice versa. Lee and Chiang [12] showed the line query can be processed in $O(\log n)$ time using $O(n^2)$ space and preprocessing time based on the geometric duality, while the same performance had been achieved earlier by Cole and Yap [5]. Afterwards, Nandy et al. [13] improved the required storage into $O(n^2/\log n)$ and extended it to k-nearest neighbor queries, again exploiting the duality. The max-min line location problem for unweighted points is equivalent to the problem of finding a *widest empty corridor*. The currently best algorithm solves the widest empty corridor problem in $O(n^2)$ time and $O(n)$ space by topological sweeping in the dual plane [11].

On the other hand, the geometric duality is seemingly hard to extend to the general (weighted) version of the problems. To our best knowledge, the line query problem for weighted points has not been considered in the literature. The max-min line location problem is also known as the *obnoxious line location problem*, shortly the OLL problem, and was first studied by Drezner and Wesolowsky [8] with an $O(n^3)$-time algorithm. The first near-quadratic $O(n^2 \log^3 n)$-time algorithm was recently presented by Díaz-Báñez et al. [7]; later, Chen and Wang [3] improved it to $O(n^2 \log n)$ time. Both algorithms are achieved via the parametric search technique.

The parametric search technique is known as a powerful tool to solve geometric optimization problems, resulting in theoretically efficient algorithms, while their implementation is an entirely challenging task because the simulation by a parallel decision algorithm is fairly complicated. Though several attempts to make it more practical have been suggested [4,16], its practical implementation still remains a challenge. Therefore,

an efficient algorithm avoiding the parametric search technique would be more attempting, even though its theoretical running time is slightly sub-optimal.

   This paper introduces a new geometric transformation that maps lines in the primal
plane into points in a parameterized space $\mathbb{L}$. It turns out that our mapping, together
with its target space $\mathbb{L}$, provides nice geometric insights that make easy to look into the
proximity relations between planar lines and given weighted points S. Further, we define a Voronoi-like subdivision $\mathcal{O}(S)$ on $\mathbb{L}$ generated by weighted points S, named the
onion diagram, and investigate its geometric, combinatorial and computational properties. As the nearest-neighbor query problem and the max-min location problem can be
solved by exploiting the multiplicatively weighted Voronoi diagram, the onion diagram
is applied to their line version, resulting in several new algorithmic results:

(1) We present a data structure of size $O(n^2)$ that answers a line query among n
    weighted points in $O(\log n)$ time after $O(n^2 \log n)$ preprocessing time. This is the
    first nontrivial result on the line query problem among weighted points.
(2) We present the first near-quadratic $O(n^2 \log n)$-time/$O(n^2)$-space algorithm for the
    OLL problem for weighted points that avoids the parametric search. The bounds
    match the best known ones achieved via the parametric search [3]. We also present
    an $O(n^2 \log^2 n)$-time/$O(n)$-space algorithm, again avoiding the parametric search.
(3) Our approach extends to weighted "polygon" sites, where we are given m polygons
    with n total corners. We show that the OLL problem for weighted polygons can
    be solved in $O(nm + m^2 \log m \log n)$ time using $O(nm)$ space, and that the line
    query problem for weighted polygons can be solved using $O(nm)$ space, $O(nm + m^2 \log m \log n)$ preprocessing time, and $O(\log n)$ query time.

   The previously best algorithm for the OLL problem runs in $O(nm + n \log^2 n + m^2 \log m)$ time and $O(nm)$ space [3] via the parametric search technique, and our
result on the line query problem among polygons shows the first nontrivial bound.

2   The Onion Diagram: Definition and Properties

In this section, we introduce our transformation, give the definition of the onion diagram, and then reveal its several useful properties. Due to lack of space, some proofs
are omitted but will be found in a longer version.

2.1   The Line Space $\mathbb{L}$

We first describe our coordinate system for planar lines, which introduces a new param-
eterized space $\mathbb{L}$ of planar lines in $\mathbb{R}^2$. Let $o \in \mathbb{R}^2$ be the origin (or the reference point)
in the plane. Any line $l \subset \mathbb{R}^2$ can be described by a pair (θ, λ) of two parameters with
$\theta \in [0, \pi)$ and $\lambda \in \mathbb{R}$ as follows: θ denotes the orientation of l and λ the *signed* distance
of l from o. More precisely, letting $\ell_\theta \subset \mathbb{R}^2$ be the directed line through o with direction
$\theta - \frac{\pi}{2}$, the intersection point $\ell_\theta \cap l$ is expressed as $o + \lambda \cdot (\cos(\theta - \frac{\pi}{2}), \sin(\theta - \frac{\pi}{2}))$; that
is, λ can be seen as the coordinate of point $\ell_\theta \cap l$ on ℓ_θ.

   Define $\mathbb{L} := [0, \pi) \times \mathbb{R}$ to be the set of all possible pairs of parameters. Observe
that any point in $\mathbb{L}$ corresponds to a unique line l in $\mathbb{R}^2$. In order to clearly distinguish

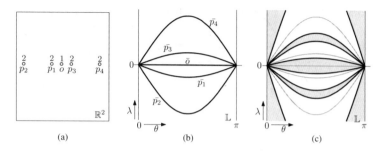

Fig. 1. An example of the onion diagram induced by $S = \{o, p_1, p_2, p_3, p_4\}$, as shown in (a)with $w_o = 1$ and $w_{p_i} = 2$ for $i = 1, \ldots 4$. (a) the five weighted points in the primal plane $\mathbb{R}^2$, (b) $\bar{p}$ for every $p \in S$, (c) the onion diagram $\mathcal{O}(S)$, where the thick curves are bisectors between o and p_i and the region $R(o)$ of $o \in S$ is depicted by the shaded region.

things, we shall call the plane $\mathbb{R}^2$ the *primal plane* and $\mathbb{L}$ the *line space*; for any line $l \subset \mathbb{R}^2$ in the primal plane, let $\bar{l} = (\theta, \lambda) \in \mathbb{L}$ be its corresponding point in the line space. In $\mathbb{L}$, we regard the θ-axis as horizontal and the λ-axis as vertical.

Note that our mapping $l \mapsto \bar{l} = (\theta, \lambda)$ is hardly said to be a geometric dual transformation, since a point $p \in \mathbb{R}^2$ in the primal plane is not mapped into a line in $\mathbb{L}$. Instead, we consider any point $p \in \mathbb{R}^2$ in the primal plane to be a function $\bar{p}(\theta)$ over $\theta \in [0, \pi)$ based on the orthogonal projection onto the line $\ell_\theta \subset \mathbb{R}^2$ with orientation $\theta - \pi/2$: let $\bar{p}(\theta)$ be the λ-coordinate of the line through p with orientation θ. Then, $\bar{p}(\theta)$ is described as a sinusoidal function with angular frequency 1:

$$\bar{p}(\theta) = \rho_p \cdot \sin(\phi_p + \theta),$$

where $\rho_p := d(p, o)$ and $\phi_p \in [0, 2\pi)$ denotes the direction from o towards p. By an abuse of notation, we denote by $\bar{p}$ the graph of $\bar{p}(\theta)$ over $\theta \in [0, \pi)$ embedded in the line space $\mathbb{L}$. Note that $\bar{o}(\theta) = 0$ for any $\theta \in [0, \pi)$ and thus $\bar{o}$ is the horizontal line through $(0, 0) \in \mathbb{L}$. See Fig. 1(a) and (b).

2.2 Definition of the Onion Diagram

We are now given a set $S \subset \mathbb{R}^2$ of n weighted point sites in the primal plane; for any $p \in S$, let $w_p > 0$ be its assigned weight. Recall that the weighted distance from a line $l \in \mathbb{R}^2$ to $p \in S$ is represented as $w_p \cdot d(p, l)$. For any $\bar{l} = (\theta, \lambda) \in \mathbb{L}$, we define $d_p(\bar{l}) := w_p \cdot |\bar{p}(\theta) - \lambda|$. Then, it is easy to check that $d_p(\bar{l}) = w_p \cdot d(p, l)$.

Now, we can define a Voronoi-like diagram on the line space $\mathbb{L}$ that is generated by the set S of sites in the primal plane $\mathbb{R}^2$ under the distance function d_p for any $p \in S$. For any $p, q \in S$ with $p \neq q$, the line space $\mathbb{L}$ is split into three regions $R(p, q)$, $B(p, q)$ and $R(q, p)$ as follows: let $B(p, q) := \{\bar{l} \in \mathbb{L} \mid d_p(\bar{l}) = d_q(\bar{l})\}$ and $R(p, q) := \{\bar{l} \in \mathbb{L} \mid d_p(\bar{l}) < d_q(\bar{l})\}$. We call $B(p, q)$ the *bisector between p and q*. The *region $R(p)$ of p with respect to S* is defined to be $R(p) := \bigcap_{q \in S \setminus \{p\}} R(p, q)$. By its definition, it is obvious that the union of the (closure of) $R(p)$ covers the whole line space $\mathbb{L}$. Consequently, the regions $R(p)$ form a planar subdivision of $\mathbb{L}$. We shall call

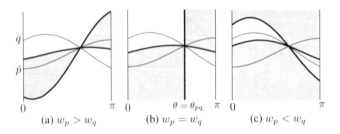

(a) $w_p > w_q$ (b) $w_p = w_q$ (c) $w_p < w_q$

Fig. 2. The shape of the bisector $B(p, q)$ with various weights w_p and w_q. In each of three figures, thin black curves are $\bar{p}$ and $\bar{q}$, thick curves are $B^{\text{in}}(p, q)$ and $B^{\text{out}}(p, q)$, and the shaded region depicts $R(p, q)$. (a) $w_p = 2$, $w_q = 1$; (b) $w_p = w_q = 1$; (c) $w_p = 1$, $w_q = 3$.

this subdivision of the line space $\mathbb{L}$ the *onion diagram* induced by the set S of sites in the primal plane, denoted by $\mathcal{O}(S)$. The term "onion" is inspired by its geometric shape; for some cases, the diagram $\mathcal{O}(S)$ resembles a cross section of an onion as shown in Fig. 1(c).

As a subdivision of $\mathbb{L}$, the onion diagram $\mathcal{O}(S)$ consists of a number of vertices, edges, and cells. Each vertex v of $\mathcal{O}(S)$ is determined by three (or more) sites $p, q, r \in S$ such that $d_p(\bar{v}) = d_q(\bar{v}) = d_r(\bar{v}) = \min_{p' \in S} d_{p'}(\bar{v})$. Each edge e of $\mathcal{O}(S)$ is a maximally connected portion of $B(p, q)$ for some $p, q \in S$ such that $d_p(\bar{l}) = d_q(\bar{l}) < d_r(\bar{l})$ for any $\bar{l} \in e$ and any $r \in S$. Each cell of $\mathcal{O}(S)$ is a maximally connected set of points $l \in \mathbb{L}$ that are closest to exactly one $p \in S$; thus, each cell of $\mathcal{O}(S)$ is a connected component of $R(p)$ for some $p \in S$.

2.3   Geometric and Combinatorial Properties

We start with an interesting relation between the onion diagram and the multiplicatively weighted Voronoi diagram.

Lemma 1. *The cross section of the onion diagram $\mathcal{O}(S)$ by a vertical line $\ell \subset \mathbb{L}$ at $\theta = \theta'$ coincides with the 1-dimensional multiplicatively weighted Voronoi diagram on the line $\ell_{\theta'}$ for $\bar{S}(\theta')$, where $\bar{S}(\theta')$ denotes the set of the orthogonal projections of each $p \in S$ onto $\ell_{\theta'}$. Therefore, ℓ intersects at most $O(n)$ cells and edges of $\mathcal{O}(S)$.* ▫

Hence, the onion diagram $\mathcal{O}(S)$ can be seen as the trace of the 1-dimensional weighted Voronoi diagram on ℓ_θ as θ increases continuously from 0 to π. This gives an idea of a plane sweep algorithm for computing $\mathcal{O}(S)$, which will be presented in Section 3. About the 1-dimensional weighted Voronoi diagram, we refer to Aurenhammer [1].

Now, we look into the shape of the bisector $B(p, q)$.

Lemma 2. *The bisector $B(p, q) \subset \mathbb{L}$ between any two sites $p, q \in S$ consists of two simple curves $B^{\text{in}}(p, q)$ and $B^{\text{out}}(p, q)$. Moreover, if $w_p \neq w_q$, then*

$$B^{\text{in}}(p, q) : \lambda = \frac{w_p \bar{p}(\theta) + w_q \bar{q}(\theta)}{w_p + w_q}, \quad B^{\text{out}}(p, q) : \lambda = \frac{w_p \bar{p}(\theta) - w_q \bar{q}(\theta)}{w_p - w_q};$$

if $w_p = w_q$, $B^{\text{in}}(p, q)$ is described as above and $B^{\text{out}}(p, q)$ is a vertical line $\{\theta = \theta_{pq}\}$, where $\theta_{pq} \in [0, \pi)$ is the unique orientation such that $\bar{p}(\theta_{pq}) = \bar{q}(\theta_{pq})$ holds.

Thus, each of $B^{\text{in}}(p,q)$ and $B^{\text{out}}(p,q)$ is either a vertical line in $\mathbb{L}$ or the graph of the sum of two sinusoidal functions, $c_1 \sin(\phi_p + \theta) + c_2 \sin(\phi_q + \theta)$ for constants c_1 and c_2 depending on w_p, w_q, p, and q. See Fig. 2.

From now, we bound the combinatorial complexity of the onion diagram $\mathcal{O}(S)$, which counts the total number of vertices, edges, and cells of $\mathcal{O}(S)$.

Lemma 3. *Any two distinct $B^{\square}(p,q)$ and $B^{\triangle}(r,s)$ intersect at most once for $\square, \triangle \in \{\text{in}, \text{out}\}$ and $p, q, r, s \in P$.*

In spite of the seemingly favorable behaviors of the bisectors $B(p,q)$ as shown in Lemma 3, we show that the region $R(p)$ of a site $p \in S$ can have a quadratic complexity $\Omega(n^2)$.

Lemma 4. *Let $S \subset \mathbb{R}^2$ be any set of n points in the primal plane in which no three are collinear. Then, there exists an assignment of weights such that the region $R(p)$ for some $p \in S$ has complexity $\Omega(n^2)$.*

Lemma 4 directly implies that the onion diagram $\mathcal{O}(S)$ has complexity $\Omega(n^2)$ in the worst case. An easy upper bound for the complexity of $\mathcal{O}(S)$ is $O(n^3)$ because each region $R(p)$ for $p \in S$ is the union of some cells in the arrangement of $B(p,q)$ whose complexity is $O(n^2)$ by Lemma 3. In the following, we show that the right bound is $\Theta(n^2)$.

For $p \in S$, we split $S \setminus \{p\}$ into two subsets S_p^{in} and S_p^{out}, where S_p^{in} is the set of points in $S \setminus \{p\}$ whose weights are at most w_p and $S_p^{\text{out}} := (S \setminus \{p\}) \setminus S_p^{\text{in}}$. Then, the region $R(p)$ is also represented by $R(p) = R^{\text{in}}(p) \cap R^{\text{out}}(p)$, where $R^{\text{in}}(p) := \bigcap_{q \in S_p^{\text{in}}} R(p,q)$ and $R^{\text{out}}(p) := \bigcap_{q \in S_p^{\text{out}}} R(p,q)$. We are more interested in the complexity of $R^{\text{in}}(p)$; it will be shown that the complexity of $\mathcal{O}(S)$ is not more than the sum of those of the $R^{\text{in}}(p)$ for all $p \in S$.

Our task is thus to tightly bound the complexity of $R^{\text{in}}(p)$. For the purpose, we observe the following:

Lemma 5. *Let $\Gamma_p := \{B^{\text{in}}(p,q), B^{\text{out}}(p,q) \mid q \in S \setminus \{p\}\}$ and $\mathcal{A}(\Gamma_p)$ be the arrangement of Γ_p. Then, $R^{\text{in}}(p)$ is the union of cells of $\mathcal{A}(\Gamma_p)$ intersected by $\bar{p}$.*

Now, we are ready to prove the linear bound on the complexity of $R^{\text{in}}(p)$.

Lemma 6. *The combinatorial complexity of $R^{\text{in}}(p)$ is at most $O(|S_p^{\text{in}}|)$.*

Proof. Let Γ_p and $\mathcal{A}(\Gamma_p)$ be defined as in Lemma 5. By Lemma 5, we are done by bounding the complexity of the union of all cells of $\mathcal{A}(\Gamma_p)$ intersected by $\bar{p}$. We cut each curve $\gamma \in \Gamma_p$ by $\bar{p}$ at the unique intersection point (its uniqueness is easily checked by equations) into two curves γ^+ and γ^-, where γ^+ lies above $\bar{p}$ and γ^- lies below $\bar{p}$. Let $\Gamma_p^+ := \{\gamma^+ \mid \gamma \in \Gamma_p\}$ and $\Gamma_p^- := \{\gamma^- \mid \gamma \in \Gamma_p\}$. Then, the boundary of $R^{\text{in}}(p)$ is exactly the union of the lower envelope $\mathcal{L}(\Gamma_p^+)$ of Γ_p^+ and the upper envelope $\mathcal{U}(\Gamma_p^-)$ of Γ_p^-. See Fig. 2(a) for illustration to $B(p,q)$ and $R(p,q)$ when $w_p > w_q$.

Now, we show that $\mathcal{L}(\Gamma_p^+)$ is corresponding to the Davenport-Schinzel sequence of order 2. Recall that any pair of two curves in Γ_p^+ intersects at most once by Lemma 3. Informally speaking, the curve segments in Γ_p^+ act almost like the half-lines; one end of each $\gamma^+ \in \Gamma_p^+$ reaches the vertical line $\{\theta = 0\}$ or $\{\theta = \pi\}$ and the other lies on $\bar{p}$. Thus, attaching a steep ray at $\gamma^+ \cap \bar{p}$ is enough to reform it as the graph of a function defined totally on $[0, \pi)$. Further, the number of pairwise intersections among such extended curves are at most two; the steep extension intersects at most once with any other curve. This implies that the lower envelope $\mathcal{L}(\Gamma_p^+)$ of Γ_p^+ corresponds to the Davenport-Schinzel sequence of order 2 and $O(|S_p^{in}|)$ symbols, whose length is at most $O(|S_p^{in}|)$ [15].

The identical argument applies to $\mathcal{U}(\Gamma_p^-)$, completing the proof. We note that a similar technique — attaching steep rays — can be found in showing the lower envelope of straight line segments corresponds to the Davenport-Schinzel sequence of order 3. ⊟

Finally, we show the tight bound on the complexity of the onion diagram.

Theorem 1. *The combinatorial complexity of the onion diagram $\mathcal{O}(S)$ induced by n weighted points S is $O(n^2)$. This bound is asymptotically tight in the worst case.*

Proof. Since the onion diagram $\mathcal{O}(S)$ is a planar subdivision and its vertices are incident to at least three edges, we are done by showing that the number of vertices of $\mathcal{O}(S)$ is $O(n^2)$. For each vertex $\bar{v} \in \mathbb{L}$ of $\mathcal{O}(S)$, let $p, q, r \in S$ with $w_p \geq w_q \geq w_r$ be the three sites defining $\bar{v}$. Then, the vertex $\bar{v}$ of $\mathcal{O}(S)$ also appears as a vertex of $R^{in}(p)$; that is, a vertex of $\mathcal{O}(S)$ is also a vertex of $R^{in}(p)$ for some $p \in S$. This implies that the number of vertices of $\mathcal{O}(S)$ does not exceed the sum of the number of vertices of $R^{in}(p)$ for all $p \in S$. By Lemma 6, we thus have

$$|\mathcal{O}(S)| \leq \sum_{p \in S} |R^{in}(p)| = \sum_{p \in S} O(|S_p^{in}|) = \sum_{p \in S} O(n) = O(n^2),$$

as claimed. The tightness of the bound is shown by Lemma 4. ⊟

We remark that the quadratic bound remains still tight even if the assigned weights w_p for $p \in S$ are all equal. In this (unweighted) case, we have $R(p) = R^{in}(p)$ and therefore the complexity of each region $R(p)$ is $O(n)$ by Lemma 6. On the other hand, one can easily see that, for any $p, q \in S$ with $p \neq q$, there is an edge of $\mathcal{O}(S)$ that is a connected portion of $B(p, q)$, provided that no three of S are collinear. Consequently, the worst-case complexity of $\mathcal{O}(S)$ is $\Theta(n^2)$.

3   Computing the Onion Diagram

In this section, we present a near-optimal algorithm for computing the onion diagram $\mathcal{O}(S)$. Our algorithm follows a standard plane sweep technique: let $\ell \subset \mathbb{L}$ be the vertical line in the line space at θ, namely the *sweep line*, and we maintain the intersection between ℓ and the cells of $\mathcal{O}(S)$ as θ increases continuously from 0 to π. Each cell C of $\mathcal{O}(S)$ is regarded to be *labeled* by $p \in S$ with $C \subseteq R(p)$. The combinatorial structure of the intersections is described by the order of labels of cells along the sweep line ℓ, and is stored in a balanced binary search tree $\mathcal{T}$.

We need the following observation for correct computation.

Lemma 7. *Each cell C of $\mathcal{O}(S)$ is θ-monotone. Moreover, the leftmost and the rightmost point of C are vertices of $\mathcal{O}(S)$.*

By Lemma 7, we know that each cell of $\mathcal{O}(S)$ intersects ℓ in an interval and that the combinatorial structure changes whenever ℓ touches a vertex of $\mathcal{O}(S)$. We call an orientation $\theta \in [0, \pi)$ an *event orientation* if ℓ touches a vertex of $\mathcal{O}(S)$ at θ. It is also said that an *event* occurs at θ and a vertex $\bar{v} \in \mathbb{L}$ of $\mathcal{O}(S)$ is associated with an event. There are two sorts of events: *merge events* and *split events*. When a merge event occurs at θ, two edges of $\mathcal{O}(S)$ are merged at the associated vertex $\bar{v}$ into one; for a split event, an edge of $\mathcal{O}(S)$ is split into two at the corresponding vertex $\bar{v}$.

Our algorithm runs in two phases: First, we predict and gather all possible events with their associated vertices and then sweep the line space $\mathbb{L}$ by the sweep line ℓ, constructing the diagram $\mathcal{O}(S)$.

Prediction phase. In the prediction phase, we compute a set V of $O(n^2)$ points in $\mathbb{L}$ which includes all vertices of $\mathcal{O}(S)$. Since each event corresponds to a vertex of $\mathcal{O}(S)$, computing V is sufficient to collect all candidates of events.

More precisely, V is determined to be the union of all vertices of $R^{\text{in}}(p)$ for $p \in S$. Then, as shown in the proof of Theorem 1, V is a superset of the set of vertices of $\mathcal{O}(S)$. Furthermore, $|V| = O(n^2)$ by Lemma 6 and it can be computed in $O(n^2 \log n)$ time by the following lemma.

Lemma 8. *For any $p \in S$, all vertices of $R^{\text{in}}(p)$ can be identified in $O(n \log n)$ time.*

Sweeping phase. In the sweeping phase, we first initialize $\mathcal{T}$ for $\theta = 0$ by computing the intersection of the sweep line ℓ and $\mathcal{O}(S)$, which coincides with a 1-dimensional weighted Voronoi diagram by Lemma 1. This initialization can be done in $O(n \log n)$ time by Aurenhammer [1]. We then put the points in V associated with the event candidates into a priority queue $\mathcal{Q}$ indexed by their θ-coordinates, and run the main loop. In the main loop, we extract the next upcoming event from $\mathcal{Q}$ and process it accordingly, updating the combinatorial structure $\mathcal{T}$.

Let $\bar{v} = (\theta, \lambda) \in V$ be a point in the line space associated with the next event at θ to be processed. There are the three sites $p, q, r \in S$ defining $\bar{v}$; that is, we have $d_p(\bar{v}) = d_q(\bar{v}) = d_r(\bar{v})$. Recall that V contains a number of "fake" events. In order to filter them out, we test the validity of the current event using the following observation: If the current event is a *true* split event, then a pair of labels $\{p, q, r\}$ must be consecutive in $\mathcal{T}$; otherwise, if it is a *true* merge event, then a permutation of labels $\{p, q, r\}$ must be consecutive in $\mathcal{T}$. The validity test of each event can be done in $O(\log n)$ time by searching the binary tree $\mathcal{T}$.

Once the validity test is passed, we update $\mathcal{T}$ accordingly, making up the tree $\mathcal{T}$ locally by insertion or deletion. This also can be done in $O(\log n)$ time. Since we process $O(n^2)$ events and each of them is handled in $O(\log n)$ time, the sweeping phase is done in $O(n^2 \log n)$ time. Once the sweeping phase is completed, we know the combinatorial structure of the onion diagram $\mathcal{O}(S)$ and thus we can build it by tracing its edges and vertices.

Theorem 2. *The onion diagram $\mathcal{O}(S)$ induced by a given set $S \subset \mathbb{R}^2$ of n weighted points can be computed in $O(n^2 \log n)$ time using $O(n^2)$ space.*   ⬜

4 Applications

In this section, we present several new algorithmic results on the line query problem and the max-min (or obnoxious) line location problem, all of which are based on the observations on the onion diagram made in previous sections.

4.1 Querying a Line for Nearest-Neighbor among Weighted Points

This is a direct application of the onion diagram. With an aid of a standard point location structure [6] on the onion diagram $\mathcal{O}(S)$, a query can be processed in $O(\log n)$ time. We hence conclude the following.

Theorem 3. *Given a set S of weighted points in $\mathbb{R}^2$, one can preprocess S into a data structure of size $O(n^2)$ in time $O(n^2 \log n)$ that can answer a nearest point among S of a query line in $O(\log n)$ time.* $\quad\square$

It is worth noting that Theorem 3 presents the first nontrivial bound on the nearest-neighbor line query problem among weighted points.

4.2 The Max-Min Line Location Problem among Weighted Points

This problem asks to find an optima line that maximizes the minimum of the weighted distances to given points S. A key observation, which was known by previous results [8, 7], is rephrased as follows in terms of the onion diagram $\mathcal{O}(S)$:

Lemma 9. *Let $\Phi(\bar{l})$ be the minimum of $d_p(\bar{l})$ over all $p \in S$; that is, $\Phi(\bar{l}) := \min_{p \in S} d_p(\bar{l})$. Suppose that $\bar{l}^* \in \mathbb{L}$ is a local maximum of function Φ. Then, one of the following cases holds: (1) $\bar{l}^*$ is a vertex of the onion diagram $\mathcal{O}(S)$; or (2) $\bar{l}^*$ lies on an edge e of $\mathcal{O}(S)$, and if $e \subset B(p,q)$ for some $p, q \in S$, the θ-coordinate of $\bar{l}^*$ is $\theta_{pq} + \pi/2$ (modulo π), where $\theta_{pq} \in [0, \pi)$ denotes the orientation from p towards q.* $\quad\square$

Lemma 9 tells us a way of finding an optimal obnoxious line l^* in the onion diagram $\mathcal{O}(S)$. Since the combinatorial complexity of $\mathcal{O}(S)$ is $O(n^2)$ by Theorem 1, it is done by traversing every vertex and edge of $\mathcal{O}(S)$ in the same time bound, once we obtain the diagram $\mathcal{O}(S)$. The total time complexity is dominated by $O(n^2 \log n)$ by Theorem 2.

Theorem 4. *Given a set S of n weighted points, the obnoxious line location (OLL) problem can be solved in $O(n^2 \log n)$ time and $O(n^2)$ space, without using the parametric search technique.* $\quad\square$

Remark that our asymptotic bounds exactly match those by Chen and Wang [3]. Nonetheless, our main contribution here is found in removing the technical difficulty depending on the parametric search and thus making its practical implementation easier. Computing the onion diagram can be done by Theorem 2 and the algorithm is relatively easy to implement; it exploits standard data structures and the plane sweep technique. Moreover, our approach based on the onion diagram provides a more geometric insight to the problem, resulting in an alternative algorithm using linear space:

Theorem 5. *Given a set S of n weighted points, the obnoxious line location (OLL) problem can be solved in $O(n^2 \log^2 n)$ time and $O(n)$ space, without using the parametric search technique.*

5   Extension to Weighted Polygonal Sites

In this section, we extend our discussion to the polygonal site case; thus, here let S be a set of m weighted simple polygons, possibly being overlapping, with n total number of corners. For each polygon $P \in S$, a weight $w_P > 0$ is assigned as before, and thus the distance from any line $l \subset \mathbb{R}^2$ to $P \in S$ is defined to be $d_P(\bar{l}) := w_P \cdot \min_{p \in P} d(p, l)$. Since we consider only the distance from lines, we can assume that each $P \in S$ is *convex*; even though $P \in S$ is not convex, its convex hull $\mathrm{conv}(P)$ can be computed in linear time [9].

As in Section 2, consider the orthogonal projection $\bar{P}(\theta)$ of $P \in S$ onto the line $\ell_\theta \subset \mathbb{R}^2$ through the origin $o \in \mathbb{R}^2$ with orientation $\theta - \pi/2$. Obviously, $\bar{P}(\theta)$ is an interval on ℓ_θ, described by two λ-coordinates $(\bar{P}^+(\theta), \bar{P}^-(\theta))$. Each of $\bar{P}^+(\theta)$ and $\bar{P}^-(\theta)$ is a continuous and piecewise sinusoidal function, consisting of $|P|$ breakpoints corresponding to orientations $\theta_{pp'}$ in which two consecutive corners p, p' of P are aligned. By an abuse of notation, we denote by $\bar{P}^+$ and $\bar{P}^-$ the graphs of functions $\bar{P}^+(\theta)$ and $\bar{P}^-(\theta)$ over $\theta \in [0, \pi)$, and by $\bar{P}$ the region bounded by $\bar{P}^+$ and $\bar{P}^-$. See Fig. 3.

Redefinition of the onion diagram. Observe that the set $\{\bar{l} \in \mathbb{L} \mid d_P(\bar{l}) = d_Q(\bar{l})\}$ contains a two-dimensional region $\bar{P} \cap \bar{Q}$, corresponding to the set of lines l in the primal plane $\mathbb{R}^2$ that intersect both P and Q; thus, $d_P(\bar{l}) = d_Q(\bar{l}) = 0$. Since we would like to have one-dimensional bisectors to define the onion diagram, we take an arbitrary linear order $\prec$ on S and redefine the bisector $B(P, Q)$ as follows: For any $P, Q \in S$ with $P \prec Q$, we let $R(P, Q) := \{\bar{l} \in \mathbb{L} \mid d_P(\bar{l}) \leq d_Q(\bar{l})\}$; $R(Q, P) := \{\bar{l} \in \mathbb{L} \mid d_P(\bar{l}) > d_Q(\bar{l})\}$; $B(P, Q) := \partial R(P, Q)$, the boundary of $R(P, Q)$. We call the set $B(P, Q) \subset \mathbb{L}$ the *bisector* between two weighted polygons P and Q. The *region* $R(P)$ of P is defined to be the intersection of $R(P, Q)$ for any $Q \in S$ with $Q \neq P$. Then, the line space $\mathbb{L}$ is decomposed by the regions $R(P)$ and the onion diagram $\mathcal{O}(S)$ induced by S is the corresponding planar subdivision of $\mathbb{L}$. See Fig. 3(d).

Complexity bound. The vertices, edges, and cells of $\mathcal{O}(S)$ are induced accordingly. Since a vertex $\bar{v}$ of $\mathcal{O}(S)$ is determined by some $P, Q, Q' \in S$ with $d_P(\bar{v}) = d_Q(\bar{v}) = d_{Q'}(\bar{v})$, three edges of $\mathcal{O}(S)$ are incident to $\bar{v}$ which are portions of $B(P, Q)$, $B(P, Q')$, and $B(Q, Q')$, respectively. On the other hand, the complexity of an edge of $\mathcal{O}(S)$ is

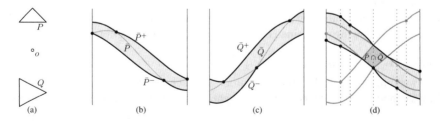

Fig. 3. (a) Two given polygons P and Q. (b)(c) $\bar{P}$ and $\bar{Q}$; black dots are breakpoints on their boundaries $\bar{P}^+$, $\bar{P}^-$, $\bar{Q}^+$, and $\bar{Q}^-$. (d) The bisector $B(P, Q)$ (black thick curves) and the breakpoints (black dots) on it, when $w_P = 8$, $w_Q = 1$ and $P \prec Q$. The gray region, including $\bar{P} \cap \bar{Q}$, depicts $R(P, Q)$. All figures are precisely drawn under scaling.

generally more than constant: each edge of $\mathcal{O}(\mathcal{S})$ consists of a number of "breakpoints". Thus, in order to correctly count the combinatorial complexity of $\mathcal{O}(\mathcal{S})$, we also need to consider the breakpoints on its edges.

Observe that a breakpoint on $B(P, Q)$ corresponds to an orientation $\theta_{pp'}$ in which two consecutive corners p, p' of P or of Q are aligned; recall that the description of function $\bar{P}(\theta)$ or $\bar{Q}(\theta)$ changes at such orientations $\theta_{pp'}$ only. Since we have only n such orientations, the number of distinct θ-coordinates of all breakpoints are at most n. Also, Lemma 1 extends to the polygon case and thus any vertical line $\ell \subset \mathbb{L}$ intersects at most $O(m)$ cells and edges of $\mathcal{O}(\mathcal{S})$. This implies that there are at most $O(nm)$ breakpoints on the edges of $\mathcal{O}(\mathcal{S})$. Fig. 3 also illustrates the breakpoints on the bisector $B(P, Q)$; one can check that every breakpoint is corresponding to such an orientation $\theta_{pp'}$.

Bounding the number of vertices of $\mathcal{O}(\mathcal{S})$ is done in a similar fashion as Lemma 6 and Theorem 1. Detailed proofs will be presented in a longer version of the paper.

Theorem 6. *The combinatorial complexity of the onion diagram $\mathcal{O}(\mathcal{S})$ induced by a given set $\mathcal{S}$ of m polygons with n total number of corners is $O(nm)$. Moreover, there are at most $O(m^2)$ vertices and $O(nm)$ breakpoints on the edges in $\mathcal{O}(\mathcal{S})$. All these bounds are asymptotically tight in the worst case.*

Algorithm. The overall framework of our algorithm computing $\mathcal{O}(\mathcal{S})$ is almost identical to that for weighted points described in Section 3. One special care is necessary to handle the breakpoints on the edges. Here, we handle three types of events: split, merge, and breakpoint events. Split and merge events are defined as in the weighted point case and breakpoint events occur whenever the sweep line ℓ touches a breakpoint on an edge of $\mathcal{O}(\mathcal{S})$.

As discussed above, breakpoint events occur at at most n different orientations and a breakpoint event can cause $O(m)$ changes on edges intersected by the current sweep line ℓ. Thus, when handling a breakpoint event determined by two consecutive corners of $P \in \mathcal{S}$, it suffices to do a linear search on $\mathcal{T}$ to find all intervals labeled by P and update them accordingly, taking $O(m)$ time per breakpoint event.

For the other two types of events, we first observe the following:

Lemma 10. *Computing the intersection $B(P, Q) \cap B(P, Q')$ for $P, Q, Q' \in \mathcal{S}$ can be done in $O(\log(|P| + |Q| + |Q'|))$ time.*

This is possible because the corners of each polygon in $\mathcal{S}$ are given sorted. The above lemma provides a primitive operation of computing a superset V of the set of all vertices of $\mathcal{O}(\mathcal{S})$, thus collecting all split and merge events, in the prediction phase. Consequently, the following is obtained by an analogous method to Lemma 8 for each $P \in \mathcal{S}$.

Lemma 11. *All split and merge events can be collected in $O(m^2 \log m \log n)$ time.*

The sweeping phase of the algorithm processes each event in $O(\log m)$ time since $\mathcal{T}$ always contains $O(m)$ intervals as aforementioned. Finally, we conclude the following.

Theorem 7. *The onion diagram $\mathcal{O}(\mathcal{S})$ induced by a set $\mathcal{S}$ of m polygons with n total corners can be computed in $O(nm + m^2 \log m \log n)$ time using $O(nm)$ space.*

As direct applications of the onion diagram $\mathcal{O}(\mathcal{S})$ induced by a given set $\mathcal{S}$ of weighted polygons, we state following new algorithmic results.

Theorem 8. *Given a set S of m weighted polygons with n total corners, one can pre-process S into a data structure of size $O(nm)$ in time $O(nm + m^2 \log m \log n)$ that can answer a nearest polygons among S of a query line in $O(\log n)$ time.* ⊟

To the best of our knowledge, the above theorem presents the first nontrivial result on the nearest-neighbor query among weighted polygons.

Theorem 9. *Given a set S of m weighted polygons with n total corners, the obnoxious line location (OLL) problem among S can be solved in $O(nm + m^2 \log m \log n)$ time and $O(nm)$ space, without using the parametric search technique.* ⊟

Remark that the previously best algorithm for the obnoxious line location problem for weighted polygons runs in $O(nm + n \log^2 n + m^2 \log m)$ time using $O(nm)$ space via the parametric search [3]. Our algorithm again avoids the parametric search, while its performance is as good as the previous one: ours even outperforms it for $m = o(\log^2 n)$.

References

1. Aurenhammer, F.: The one-dimensional weighted Voronoi diagram. Inf. Process. Lett. 22(3), 119–123 (1986)
2. Aurenhammer, F., Edelsbrunner, H.: An optimal algorithm for constructing the weighted Voronoi diagram in the plane. Pattern Recognition 17(2), 251–257 (1984)
3. Chen, D.Z., Wang, H.: Locating an obnoxious line among planar objects. In: Proc. 20th Int. Sympos. Algo. Comput. (ISAAC), pp. 740–749 (2009)
4. Cole, R.: Parallel merge sort. SIAM J. Comput. 17(4), 770–785 (1988)
5. Cole, R., Yap, C.: Geometric retrieval problems. In: Proc. 24th IEEE Sympos. Foundation of Computer Science (FOCS), pp. 112–121 (1983)
6. de Berg, M., van Kreveld, M., Overmars, M., Schwarzkopf, O.: Computational Geometry: Algorithms and Applications, 2nd edn. Springer, Heidelberg (2000)
7. Díaz-Báñez, J.M., Ramos, P.A., Sabariego, P.: The maximin line problem with regional demand. European Journal of Operational Research 181(1), 20–29 (2007)
8. Drezner, Z., Wesolowsky, G.: Location of an obnoxious route. Journal of Operational Research Society 40(11), 1011–1018 (1989)
9. Graham, R.L., Yao, F.F.: Finding the convex hull of a simple polygon. J. Algorithms 4(4), 324–331 (1983)
10. Hershberger, J.: Finding the upper envelope of n line segments in $o(n \log n)$ time. Inf. Process. Lett. 33(4), 169–174 (1989)
11. Janardan, R., Preparata, F.P.: Widest-corridor problems. Nordic J. of Computing 1(2), 231–245 (1994)
12. Lee, D., Chiang, Y.: The power of geometric duality revisited. Inform. Process Lett. 21, 117–122 (1985)
13. Nandy, S.C., Das, S., Goswami, P.P.: An efficient k nearest neighbors searching algorithm for a query line. Theoretical Computer Science 299, 273–288 (2003)
14. Preparata, F., Shamos, M.: Computational Geometry: An Introduction. Springer, Heidelberg (1985)
15. Sharir, M., Agarwal, P.K.: Davenport-Schinzel Sequences and Their Geometric Applications. Cambridge University Press, New York (1995)
16. van Oostrum, R., Veltkamp, R.C.: Parametric search made practical. Comput. Geom: Theory and Appl. 28, 75–88 (2004)

Improved Online Algorithms for 1-Space Bounded 2-Dimensional Bin Packing

Yong Zhang[1,2,*], Jingchi Chen[2], Francis Y.L. Chin[2,**], Xin Han[3,***],
Hing-Fung Ting[2,†], and Yung H. Tsin[4,‡]

[1] College of Mathematics and Computer Science, Hebei University, China
[2] Department of Computer Science, The University of Hong Kong, Hong Kong
{yzhang,jchen,chin,hfting}@cs.hku.hk
[3] School of Software, Dalian University of Technology, China
hanxin.mail@gmail.com
[4] School of Computer Science, University of Windsor, Canada
peter@uwindsor.ca

Abstract. In this paper, we study 1-space bounded 2-dimensional bin packing and square packing. A sequence of rectangular items (square items, respectively) arrive over time, which must be packed into square bins of size 1×1. $90°$-rotation of an item is allowed. When an item arrives, we must pack it into an active bin immediately without any knowledge of the future items. The objective is to minimize the total number of bins used for packing all the items in the sequence. In the 1-space bounded variant, there is only one active bin for packing the current item. If the active bin does not have enough space to pack the item, it must be closed and a new active bin is opened.

Our contributions are as follows: For 1-space bounded 2-dimensional bin packing, we propose an online packing strategy with competitive ratio 5.155, surpassing the previous 8.84-competitive bound. The lower bound of competitive ratio is also improved from 2.5 to 3. Furthermore, we study 1-space bounded square packing, which is a special case of the bin packing problem. We give a 4.5-competitive packing algorithm, and prove that the lower bound of competitive ratio is at least 8/3.

1 Introduction

The bin packing problem [1–9, 11–24] has been well studied for more than thirty years. In the general bin packing problem, a sequence of items are packed into bins without overlapping. The objective is to minimize the number of bins used for packing all the items in the sequence.

* Supported by Shanghai Key Laboratory of Intelligent Information Processing, China. Grant No. IIPL-2010-010.
** Research supported by HK RGC grant HKU-7113/07E and the William M.W. Mong Engineering Research Fund.
*** Partially supported by the Fundamental Research Funds for the Central Universities.
† Research supported by HK RGC grant HKU-7171/08E.
‡ Research supported by NSERC under grant NSERC 7811-2009.

O. Cheong, K.-Y. Chwa, and K. Park (Eds.): ISAAC 2010, Part II, LNCS 6507, pp. 242–253, 2010.
© Springer-Verlag Berlin Heidelberg 2010

Most previous studies do not impose a limit on the number of bins available for packing the items (called *active bins*). We call this model *unbounded space*. There is another model called *bounded space*, which is more realistic in many applications. In the *bounded space* model, the number of active bins is bounded by a *constant*, and each item can only be packed into one of the active bins. If none of the active bins has enough space to pack an item, one of the current active bins is closed and a new active bin is opened to pack that item.

In this paper, we consider *1-space bounded* 2-dimensional bin packing and square packing, which are interesting variants of bin packing. In the 1-space bounded variant, the number of active bins is only *one*. If an item cannot be packed into the active bin, we have to close this bin and open a new one to pack the item. In the 1-space bounded 2-dimensional bin packing problem, each item is rectangular in shape and its width and height are no more than 1. The items must be packed into square bins of size 1×1. 90°-rotation of any item is allowed, otherwise, the competitive ratio is unbounded [12]. 1-space bounded square packing is a special case of 1-space bounded 2-dimensional bin packing, where each item is a square with edge length no more than 1. Again, the objective is to minimize the number of square bins used.

For example, as shown in Fig. 1(a), there are four items to be packed into unit square bins, and the arrival order is A, B, C and D. After the packing position of A is fixed, we have two choices to pack B: rotation and without rotation. If we pack B without rotation in the same bin with A as shown in Fig. 1(b), when item C arrives, we have to open a new bin since the current active bin does not have enough space for packing C. In the optimal solution, these four items can be packed into one bin (Fig. 1(c)), since item B, C and D can be rotated and the free space in the bin can accommodate all of them in their order of arrival.

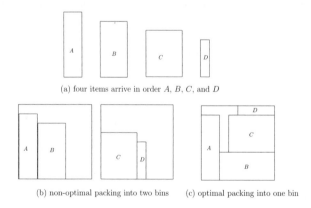

(a) four items arrive in order A, B, C, and D

(b) non-optimal packing into two bins   (c) optimal packing into one bin

Fig. 1. Example of optimal packing and non-optimal packing

We focus on the online version of 1-space bounded 2-dimensional bin packing and square packing, where the items arrive over time, and when packing the current item, we have no information of the future items and the position of the packed items in the bin cannot be changed. To measure the performance of online bin packing, the general method is to use the *asymptotic competitive ratio*. Consider an online algorithm A and an optimal offline algorithm OPT. For any sequence S of items, let $A(S)$ be the cost (number of square bins used) incurred by algorithm A and $OPT(S)$ be the corresponding optimal cost incurred by algorithm OPT. The *asymptotic competitive ratio* for algorithm A is:

$$R_A^\infty = \lim_{k \to \infty} \sup_S \{\frac{A(S)}{OPT(S)} | OPT(S) = k\}.$$

Related works

Both the offline and online version of the bin packing problem have been well studied. For the offline version of two-dimensional bin packing. Chung et. al. [5] presented an approximation algorithm with an asymptotic performance ratio of 2.125. Caprara [4] improved the upper bound to 1.69103. Bansal et al. [2] devised a randomized algorithm with an asymptotic performance ratio of at most 1.525. As for the offline lower bound of the approximation ratio, Bansal et al. [1] showed that the two-dimensional bin packing problem does not admit any asymptotic polynomial time approximation scheme.

For online one-dimensional bin packing, Johnson et al. [16] showed that the First Fit algorithm (FF) has an asymptotic competitive ratio of 1.7. Yao [24] improved the algorithm to obtain a better upper bound of 5/3. Lee et al. [17] introduced the class of Harmonic algorithms, and showed that an asymptotic competitive ratio of 1.63597 is achievable. Ramanan et al. [21] further improved the upper bound to 1.61217. The best known upper bound is that of the Super Harmonic algorithm of Seiden [22] which is 1.58889. As for the lower bound of the competitive ratio, Yao [24] showed that no online algorithm can have an asymptotic competitive ratio less than 1.5. The best known lower bound to date is 1.54014 [23].

For two-dimensional online bin packing, the best known lower bound is 1.907 [3] while the best known upper bound is 2.5545 [13].

For bounded space online bin packing, Csirik and Johnson [6] presented an 1.7-competitive algorithm (K-Bounded Best Fit algorithms (BBF_K)) for one dimensional bin packing using K active bins, where $K \geq 2$. Epstein et al. [8] gave a 1.69103^d-competitive algorithm using $(2M - 1)^d$ active bins, where $M \geq 10$ is an integer such that $M \geq 1/(1 - (1 - \varepsilon)^{1/(d+2)}) - 1$, $\varepsilon > 0$ and d is the dimension of the bin packing problem. For the *1-space bounded* variant, Fujita [12] gave an $O((\log \log m)^2)$-competitive algorithm, where m is the width of the square bin and the size of each item is $a \times b$, where a, b are integers no larger than m. He also proved that the competitive ratio for the 1-bounded space variant is at least 23/11. Recently, Chin et al. [7] proposed an 8.84-competitive packing strategy. Moreover, they proved that the lower bound of the competitive ratio is at least 2.5.

For the special case where the items are squares, there are also many results [9–11, 14, 15, 18–20]. For bounded space online square packing, Epstein and van Stee [10] gave a 2.3692-competitive algorithm, they also proved that the lower bound of the competitive ratio is at least 2.36343. Januszewski and Lassak [15] proved that any sequence of square items with a total area of at most 5/16 can be packed into a unit bin. Han et al. [14] studied a variant in which any packed item can be removed so as to guarantee a good competitive ratio and presented a packing algorithm that is 3-competitive. Note that in the above two studies, there is *only* one bin to pack the square items.

The remaining part of this paper is organized as follows. In Section 2, we propose a 5.155-competitive algorithm for the 1-space bounded bin packing problem; we also prove that the lower bound of the competitive ratio is at least 3. In Section 3, we consider 1-space bounded square packing and give a 4.5-competitive algorithm, which is the first result in this variant. Moreover, we prove that the lower bound of this variant is at least 8/3. In Section 4, we summarize our results and give some future research directions.

2   1-Space Bounded 2-Dimensional Bin Packing

In this section, we propose a packing strategy for 1-space bounded 2-dimensional bin packing with competitive ratio 5.155. In the previous 8.84-competitive algorithm, the items are classified into three types according to their sizes, and the unit bin is partitioned into two parts: the upper and lower part. The upper part only accommodates items from the two types of the larger sizes, while the lower part only accommodates items from the type of the smallest size. In our new approach, there is no partition in the unit bin which means that an item can be placed at any available position within the bin. This new approach is more flexible than the previous one and can thus achieve better performance.

Since 90°-rotation is allowed, we shall assume that for each rectangular item, the width is no less than the height. We classify the rectangular items into three classes A, B and C according to the width x as follows:

$$A = \{(x,y)|x \geq 1/2\},$$
$$B = \{(x,y)|1/4 \leq x < 1/2\}, and$$
$$C = \{(x,y)|x < 1/4\}.$$

For simplicity, let A-item denote an item belonging to class A. B-item and C-item are defined similarly.

In our packing strategy, A-items are packed into the active bin using a top-down approach while B-items and C-items are packed into the bin using a bottom-up approach. If an item cannot be packed into the bin using the strategy, we close the active bin and open a new one to pack the item.

We further divide the C-items into subclasses C_0, C_1, C_2, ... An item (x,y) belongs to subclass $C_i, i \geq 0$, if $2^{-i-1}/4 \leq x < 2^{-i}/4$. Let w_i denote the maximal possible width of items from subclass C_i. Then, $w_0 = 1/4$, $w_1 = 1/8$, ... Each item belonging to subclass C_{2j-1} or C_{2j} $(j > 0)$ is packed into a row with height w_{2j-1} and width $1/2$. The items from subclasses C_{2j-1} are packed from left to right while the items from subclass C_{2j} are packed from right to left in two subrows (upper and lower) keeping the lengths of the two subrows balanced at all times (that means a new item is always packed into the shorter subrow). Note that the C_j-items $(j > 0)$ are packed with a 90°-rotation. Figure 2 depicts a row with packed items from subclass C_{2j-1} and C_{2j}. When handling an item from subclass C_{2j-1} (or C_{2j}), a new row of height w_{2j-1} must be created if the existing rows of height w_{2j-1} cannot accommodate this item.

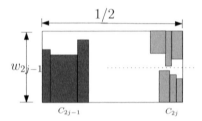

Fig. 2. Packing C_{2j-1} (or C_{2j})-items ($j > 0$) into a row

Fact 1. *For any non-last row of height w_{2j-1}, the occupation ratio is at least 5/16.*

Proof. Consider the packing configuration shown in Figure 2. Assuming that the length of the left occupied area (the total height of the packed items from subclass C_{2j-1} in this row) is y, the lengths of the upper and lower subrows in the right occupied area are y_1 and y_2 respectively, w.l.o.g., $y_1 \geq y_2$.

If this configuration cannot accommodate the next item from subclass C_{2j-1} or C_{2j}, we have $y + y_1 + (height\ of\ the\ item) > 1/2 \Rightarrow y + y_1 + 1/8 > 1/2$. Since the lengths of the two subrows in the right occupied area are balanced, we have $y_1 - y_2 \leq w_{2j} \leq 1/16$. Therefore, the total occupied area in this row is at least $(y + y_2) \times w_{2j-1}/2$. Since the total area of this row is $w_{2j-1}/2$, the occupation ratio is thus at least $y + y_2 \geq 5/16$. $\qquad\square$

The following is a detailed description of the above packing strategy.

Algorithm PS1: packing strategy for 1-space bounded 2-dimensional bin packing
  1: The A-items are packed in a top-down order with the vertical symmetry axis of each item aligning with the vertical symmetry axis of the square bin.
  2: The B-items and C-items are packed in a bottom-up order along both the left and right side of the square bin keeping the heights of these two sides balanced at all times, i.e., a new B-item (C_0-item) or newly created row of C_j-item with $j > 0$, is always packed on the side of smaller height.
  3: If there is insufficient space to pack a new item (A-item, B-item, or C_0-item) or a new row (C_j-item with $j > 0$), the bin is closed and a new bin is opened to pack the new item or row.

For example, given the current configuration in Figure 3, the height of the packed A-items is y, the left and right sides of the packed B-items and C-items are of height y_1 and y_2 respectively.

Consider packing the C_j-items with $j > 0$. If there are more than one rows for the subclass C_j, the last row could be almost empty while the occupation ratio of the other rows is at least 5/16 by Fact 1. Since the last row of each subclass could be almost empty, the total area of such rows is at most $(w_1 + w_3 + w_5 + ...) \times 1/2 = (1/8 + 1/32 + ...) \times 1/2 \approx 1/12$.

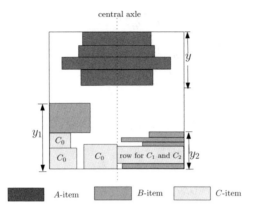

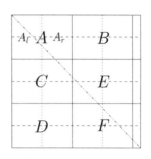

Fig. 3. Packing items into a square bin   **Fig. 4.** Partition of unit bin

Theorem 1. *The competitive ratio of the packing strategy PS1 is at most 5.155.*

Proof. For a given sequence of items, suppose the number of bins used by the packing strategy PS1 is n. Let o_A^i, o_B^i and o_C^i be the occupied space of A-, B- and C- items in the i-th bin respectively. The average occupation for all the bins is $\sum_{i=1}^n (o_A^i + o_B^i + o_C^i)/n$.

Consider the packing configuration of the i-th bin as shown in Figure 3. We have $o_A^i \geq y/2$, $o_B^i \geq (y_1 + y_2 - \sum_{j \geq 1} w_{2j-1} - m)/4 \geq (y_1 + y_2 - 1/6 - m)/4$, $o_C^i \geq m \cdot \min\{1/8, \min_{j>0}\{1/2 - w_{2j-1}\}\} \geq m/8$, where m is the total height of C_0-items and non-last rows of C_j-items with $j > 0$. Let $p_C^i = m/8$, $q_C^i = o_C^i - p_C^i$.

When n is very large, we have

$$\frac{\sum_{i=1}^n (o_A^i + o_B^i + o_C^i)}{n} \geq \min_{1 \leq i < n} \{o_A^i/2 + o_B^i/2 + p_C^i + o_A^{i+1}/2 + o_B^{i+1}/2 + q_C^{i+1}\} \quad (1)$$

W.l.o.g., assume $y_1 \geq y_2$.

- If the next A-item with height u cannot be packed into this bin, we have $y + y_1 + u > 1$ and $o_A^{i+1} \geq u/2$. Thus,

  - If $y_1 - y_2 \leq 1/4$

    $$o_A^i/2 + o_B^i/2 + p_C^i + o_A^{i+1}/2 + o_B^{i+1}/2 + q_C^{i+1}$$
    $$\geq y/4 + (y_1 + y_2 - 1/6 - m)/8 + u/4 + m/8$$
    $$\geq y/4 + (y_1 + y_2)/8 + u/4 - 1/48$$
    $$> (1 - y_1)/4 + (y_1 + y_2)/8 - 1/48$$
    $$= 11/48 - (y_1 - y_2)/8$$
    $$\geq 19/96$$

  - If $y_1 - y_2 > 1/4$, that means the top item in the bottom-left occupied area is a B-item. Thus, $o_B^i \geq (y_2 + y_2 - 1/6 - m)/4 + (y_1 - y_2)^2$. It follows that

$$o_A^i/2 + o_B^i/2 + p_C^i + o_A^{i+1}/2 + o_B^{i+1}/2 + q_C^{i+1}$$
$$\geq y/4 + (y_2 + y_2 - 1/6 - m)/8 + (y_1 - y_2)^2/2 + u/4 + m/8$$
$$\geq (y + u)/4 + y_2/4 - 1/48 + (y_1 - y_2)^2/2$$
$$> (1 - y_1)/4 + y_2/4 - 1/48 + (y_1 - y_2)^2/2$$
$$= 11/48 + (y_1 - y_2)^2/2 - (y_1 - y_2)/4$$
$$\geq 19/96$$

- If the next B-item with height u cannot be packed into this bin, we have $y + y_2 + u > 1$ and $o_B^{i+1} \geq u^2$. Thus

$$o_A^i/2 + o_B^i/2 + p_C^i + o_A^{i+1}/2 + o_B^{i+1}/2 + q_C^{i+1}$$
$$\geq y/4 + (y_1 + y_2 - 1/6 - m)/8 + u^2/2 + m/8$$
$$= y/4 + (y_1 + y_2)/8 + u^2/2 - 1/48$$
$$\geq y/4 + y_2/4 + u^2/2 - 1/48$$
$$> (1 - u)/4 + u^2/2 - 1/48$$
$$\geq 19/96$$

- If the next C-item with height u cannot be packed into this bin,
  - if this item belongs to subclass C_i $(i > 0)$, then as the width of w_i $(i > 0)$ is at most $1/8$, we must have $y + y_2 + 1/8 > 1$. Thus,

$$o_A^i/2 + o_B^i/2 + p_C^i + o_A^{i+1}/2 + o_B^{i+1}/2 + q_C^{i+1}$$
$$\geq y/4 + (y_1 + y_2 - 1/6 - m)/8 + m/8$$
$$= y/4 + (y_1 + y_2)/8 - 1/48$$
$$\geq y/4 + y_2/4 - 1/48$$
$$> (1 - 1/8)/4 - 1/48$$
$$= 19/96$$

  - if the next item with height u belongs to subclass C_0, we have $y + y_2 + u > 1$. Note that as this C_0-item will be packed into the next bin, $q_C^{i+1} \geq u^2 - u/8$. We thus have,

$$o_A^i/2 + o_B^i/2 + p_C^i + o_A^{i+1}/2 + o_B^{i+1}/2 + q_C^{i+1}$$
$$\geq y/4 + (y_1 + y_2 - 1/6 - m)/8 + m/8 + (u^2 - u/8)$$
$$= y/4 + (y_1 + y_2)/8 - 1/48 + (u^2 - u/8)$$
$$\geq y/4 + y_2/4 - 1/48 + (u^2 - u/8)$$
$$> (1 - u)/4 - 1/48 + (u^2 - u/8)$$
$$\geq 149/768$$

Combining all of the above cases, we conclude that the competitive ratio of this packing strategy is at most $768/149 < 5.155$. □

Theorem 2. *The lower bound of competitive ratio for 1-space bounded 2-dimensional bin packing is at least 3.*

Proof. Consider a sequence of items: $S = \{X_1, X_2, ..., X_{2n}, Y_1, Y_2, Z_1, Y_3, Y_4, Z_2, ..., Y_{n-1}, Y_n, Z_{n/2}, Y_1, Y_2, Z_1, Y_3, Y_4, Z_2, ..., Y_n\}$, in which $\epsilon = o(1/n^2)$, and

$$\begin{aligned}
X_{2i-1} &= (1/2 + i \cdot \epsilon, 1/2 + i \cdot \epsilon) \\
X_{2i} &= (1/2 - (i-1) \cdot \epsilon, 1/2 - (i-1) \cdot \epsilon) \\
Y_i &= (1/2 + i \cdot \epsilon, 1/2 - i \cdot \epsilon) \\
Z_i &= (1, (2i+2) \cdot \epsilon)
\end{aligned}$$

In the first part of the item sequence containing all the X_i items, no online algorithm can pack any two consecutive items into one unit square bin because the sum of the edge lengths of any two consecutive X-items is larger than 1. Thus, at least $2n$ bins are used for packing all the X_i items.

For the remaining part of the item sequence containing all the Y_i and Z_i items, no three consecutive Y_i items with an intervening Z_i item can be packed into the same bin. As a result, at least n bins are needed to pack this part of the item sequence.

For optimal offline packing, since X_{2i-1}, X_{2i+2}, Y_i, Y_i ($1 \le i \le n-1$) can be packed into one bin; X_2, Y_n, Y_n can be packed into one bin; X_{2n-1} can be packed into one bin, and all the Z_i items can be packed into one bin, the minimum number of bins needed for packing all the items is at most $n+2$.

From the above analysis, we conclude that no online algorithm can achieve a competitive ratio less than 3 for 1-space bounded 2-dimensional bin packing. □

3   1-Space Bounded Square Packing

In 1-space bounded square packing, a sequence of square items is to be packed into bins, where there is only one active bin at any time. If a newly arrived square item cannot be packed into the active bin, we close the active bin and open a new one for packing that item and subsequent items. The packing strategy in [14] can be used for this variant directly, leading to a 6-competitive algorithm for 1-space bounded square packing.

Most of the previous studies on square packing use the method of packing square in brick where a *brick* is a rectangle with aspect ratio $\sqrt{2}$. A brick can be partitioned into two smaller congruent bricks of the same size. Thus, packing a square into a brick can be done recursively. Given a square Q, we use $S(Q)$ to denote the smallest brick which can contain Q. Let $|R|$ denote the area of rectangle R.

We briefly describe the algorithm in [15] for packing a square Q in a brick B.

- If there is no empty brick in B of size greater than or equal to $S(Q)$, then give up packing Q in B.
- Else pack Q into B as follows:
  - if there is an empty brick congruent to $S(Q)$, then pack Q into it,
  - else partition the smallest empty brick P that is larger than $S(Q)$ into a sequence of bricks of area $|P|/2$, $|P|/4$,..., $2|S(Q)|$, $|S(Q)|$, $|S(Q)|$, respectively. Then pack Q into one of the two bricks of area $|S(Q)|$.

The following lemma is proved in [15].

Lemma 2. *If the above algorithm cannot pack an item Q in a brick B, then all empty bricks in B are smaller than $S(Q)$. Furthermore, there is at most one empty brick with area $|S(Q)|/2^i$ for each $i = 1, 2, ...,$ and the total area of the empty bricks is less than $|S(Q)|$.*

Fact 3. *If Q is packed in a brick congruent to $S(Q)$, then at least $1/(2\sqrt{2})$ of this brick is occupied.*

3.1   A 4.5-Competitive Algorithm

We partition each unit bin as shown in Figure 4. Note that bricks A to F are of the same size $(1/3, \sqrt{2}/3)$. Each brick can be further partitioned into two congruent bricks. For instance, brick A can be partitioned into A_l and A_r. Our packing strategy is as follows.

Algorithm PS2: For 1-space bounded square packing

1: For a square item s with edge length no greater than $\sqrt{2}/6$, we search A_l, A_r, B_l, B_r, C_l, D_l, C_r, D_r, E, F, in the listed order, for an $S(s)$ to pack the square s at its top-left corner.
2: Else, for a square item s with edge length no greater than $1/3$, we search A, B, C, D, E, F, in the listed order, for an $S(s)$ to pack s at its top-left corner.
3: Else, for a square item s with edge length no greater than $\sqrt{2}/3$, we search CD, EF, in the listed order, for an $S(s)$ to pack s at its bottom-left corner. Note that CD (or EF) is one brick for packing the item in this case.
4: Else, for a square item with edge length greater than $\sqrt{2}/3$, we pack it at the bottom-right corner of the unit bin.

Fact 4. *In executing Algorithm PS2, if an item s with edge length no greater than $\sqrt{2}/6$ is packed in C_r or D_r, the total area of the packed items is larger than $1/6$; if an item s' with edge length no greater than $\sqrt{2}/3$ is packed in E or F, the total area of the packed items is larger than $2/9$.*

Proof. According to Algorithm PS2, items with edge length no greater than $\sqrt{2}/6$ are packed into the bricks A_l, A_r, B_l, B_r, C_l, D_l, C_r, D_r in the listed order. If s is packed into C_r or D_r, then using Lemma 2, it is easily verified that there is at most one empty brick of area $|S(s)|/2^i (i > 0)$ in the preceding bricks. Therefore, by Fact 3, the total occupied area in this bin is at least $(|A| + |B| + |C_l| + |D_l| - \sum_{i>0} |S(s)|/2^i)/2\sqrt{2} + |s| > 1/6$. Similarly, if an item s' with edge length no greater than $\sqrt{2}/3$ is packed into E or F, the total occupied area is no less than $2/9$. $\qquad\square$

Theorem 3. *The competitive ratio of the packing strategy PS2 is at most 4.5.*

Proof. To prove this theorem, we will show, based on PS2, that either the total occupied area in a bin is at least $2/9$, or the total occupied area in two consecutive bins is at least $4/9$. Therefore, the average occupied area in each bin is at least $2/9$. Consider the item sequence $s_1, s_2, ..., s_{k-1}, s_k, s_{k+1}, ...,$ where s_1 is the first item packed into bin $\mathcal{I}$. Suppose s_1 to s_{k-1} are all packed into $\mathcal{I}$ by PS2, and s_{k+1} is not packed into $\mathcal{I}$. We analyze the situation in packing s_k. Let e_k denote the edge length of s_k.

- $e_k \geq 2/3$

  The area of s_k is at least $4/9$, no matter where s_k is packed, the average occupied area in $\mathcal{I}$ and the next bin is thus at least $2/9$.
- $1 - \sqrt{2}/3 \leq e_k < 2/3$

  In this case, if s_k can be packed into $\mathcal{I}$, the occupied area is larger than $2/9$. Otherwise, there must be an item in $\mathcal{I}$ preventing s_k from being packed into the bottom-right corner.

  - If the item is of edge length larger than $\sqrt{2}/3$ (it must occupy the bottom-right corner and cover part of brick EF), then the occupied area is at least $(\sqrt{2}/3)^2 = 2/9$.
  - If the item is of edge length no larger than $\sqrt{2}/3$ and resides in E or F, then according to Fact 4, the occupied area in $\mathcal{I}$ is at least $2/9$.
  - If the item is of edge length larger than $\sqrt{2}/6$ and resides in C or D, then from the packing strategy, this item is packed at the top-left corner in C or D whereas s_k is packed at the bottom-right corner of $\mathcal{I}$. Therefore, the total edge length of these two items is more than 1, and the total area of these two items is no less than $(\sqrt{2}/6 + x)^2 + (1 - \sqrt{2}/6 - x)^2 \geq 1/2$, where $\sqrt{2}/6 + x$ is the edge length of the item. Consequently, the average occupied area is at least $1/4$, which is larger than $2/9$.
  - Otherwise, the item is of edge length no greater than $\sqrt{2}/6$ and resides in C_r or D_r. From Fact 4, the total area of the packed items is at least $1/6$. The total area of items s_1 to s_k is thus at least $1/6 + (1 - \sqrt{2}/3)^2 > 0.446$, and the average occupied area is larger than $2/9$.
- $\sqrt{2}/3 < e_k < 1 - \sqrt{2}/3$

  If s_k can be packed into $\mathcal{I}$, the occupied area in this bin is at least $(\sqrt{2}/3)^2 = 2/9$. Otherwise, packing s_k at the bottom-right corner of the bin results in s_k overlapping some packed item in brick E or F. If one of the packed items is of edge length no less than $\sqrt{2}/3$, then the total occupied area in the bin is at least $2/9$. Otherwise, from Fact 4, the total occupied area contributed by the items $s_1, s_2, \ldots, s_{k-1}$ is at least $2/9$.
- $e_k \leq \sqrt{2}/3$

  If s_k cannot be packed into this bin, then similar to the above case, the occupied area in this bin is at least $2/9$.

  Otherwise, s_k can be packed into $\mathcal{I}$. But s_{k+1} cannot be packed into this bin. Again, similar to the above analysis, the average occupied area in $\mathcal{I}$ is at least $2/9$.

Since the average occupied area in each bin is at least $2/9$, the competitive ratio of our packing strategy PS2 is thus at most 4.5.   □

3.2   Lower Bound of the Competitive Ratio

Now we shall derive a lower bound of the competitive ratio for 1-space bounded square packing. Roughly speaking, we use three types of items to derive the lower bound.

- Type-X has area slightly larger than $1/4$.
- Type-Y has area slightly smaller than $1/4$.
- Type-Z has sufficiently small area.

The high level idea underlying the lower bound proof is as follows: We construct a sequence with n items of type-X, $3n$ items of type-Y and $2n/3$ items of type-Z such that we need n bins to pack the first n items of type-Y, n bins to pack the n items of type-X, and $2n/3$ bins to pack the remaining $2n$ items of type-Y and $2n/3$ items of type-Z. But for the optimal packing strategy, $n + 2$ bins is sufficient. Specifically, one bin for packing the $2n/3$ items of type-Z, $n + 1$ bins for packing all the type-X and type-Y items; in most of these bins three type-Y items and one type-X item are packed together in one bin.

Theorem 4. *There is no online algorithm with a competitive ratio less than $8/3$ for 1-space bounded square packing.*

Proof. Consider a sequence of items Y_1, X_1, Y_2, X_2, ..., Y_n, X_n, Y_n, Y', Y', Z, $3Y'$, Z, $3Y'$, Z ..., containing n type-X items, $3n$ type-Y items and $2n/3$ type-Z items. Let $\epsilon = o(1/n^2)$.

$$X_i = (1/2 + (n + 2 - i) \cdot \epsilon, 1/2 + (n + 2 - i) \cdot \epsilon)$$
$$Y_i = (1/2 - (n + 1 - i) \cdot \epsilon, 1/2 - (n + 1 - i) \cdot \epsilon)$$
$$Y' = (1/2 - n \cdot \epsilon, 1/2 - n \cdot \epsilon)$$
$$Z = (3n \cdot \epsilon, 3n \cdot \epsilon)$$

For the first part of the item sequence containing $2n$ items, any two consecutive items cannot be packed together into one bin. Thus, any online algorithm will use at least $2n$ bins for the first part.

For the remaining part containing $2n$ type-Y items and $2n/3$ type-Z items, any four consecutive type-Y items cannot be packed together into one bin because of the intervening type-Z item. Thus, any online algorithm will use at least $2n/3$ bins to pack the remaining items.

The total number of bins used by any online algorithm is at least $8n/3$.

For offline optimal packing, Y_i, X_{i+1}, Y', Y' $(i < n)$ can be packed together into one bin. X_1 can be packed into the n-th bin, while Y_n, Y_n, Y' can be packed into the $(n + 1)$-th bin. The type-Z items can all be packed into the $(n + 2)$-th bin. Thus, the minimum number of bins used for packing all the items is at most $n + 2$.

Hence, the lower bound of competitive ratio is at least $8/3$.   □

4   Concluding Remarks

We have studied 1-space bounded 2-dimensional bin packing and presented online algorithms for rectangle packing and square packing. For rectangle packing, we derived an upper bound of 5.155 and a lower bound of 3. For square packing, the corresponding bounds we derived are 4.5 and 2.667. These bounds surpass the previously best known bounds. We feel that the gap between the upper and lower bound is quite big in both cases. We thus propose closing these gaps as open problems for future research.

References

1. Bansal, N., Correa, J.R., Kenyon, C., Sviridenko, M.: Bin Packing in Multiple Dimensions: In-approximability Results and Approximation Schemes. Mathematics of Operations Research 31(1), 31–49 (2006)
2. Bansal, N., Caprara, A., Sviridenko, M.: Improved approximation algorithm for multidimensional bin packing problems. In: FOCS 2006, pp. 697–708 (2006)
3. Blitz, D., van Vliet, A., Woeginger, G.J.: Lower bounds on the asymptotic worst-case ratio of on-line bin packing algorithms (1996) (unpublished manuscript)
4. Caprara, A.: Packing 2-dimensional bins in harmony. In: FOCS 2002, pp. 490–499 (2002)
5. Chung, F.R.K., Garey, M.R., Johnson, D.S.: On packing two-dimensional bins. SIAM J. Algebraic Discrete Methods 3(1), 66–76 (1982)
6. Csirik, J., Johnson, D.S.: Bounded Space On-Line Bin Packing: Best is Better than First. Algorithmica 31, 115–138 (2001)
7. Chin, F.Y.L., Ting, H.-F., Zhang, Y.: 1-Space Bounded Algorithms for 2-Dimensional Bin Packing. In: Dong, Y., Du, D.-Z., Ibarra, O. (eds.) ISAAC 2009. LNCS, vol. 5878, pp. 321–330. Springer, Heidelberg (2009)
8. Epstein, L., van Stee, R.: Optimal Online Algorithms for Multidimensional Packing Problems. SIAM Jouranl on Computing 35(2), 431–448 (2005)
9. Epstein, L., van Stee, R.: Online square and cube packing. Acta Inf. 41(9), 595–606 (2005)
10. Epstein, L., van Stee, R.: Bounds for online bounded space hypercube packing. Discrete Optimization 4, 185–197 (2007)
11. Ferreira, C.E., Miyazawa, E.K., Wakabayashi, Y.: Packing squares into squares. Pesquisa Operacional 19, 223–237 (1999)
12. Fujita, S.: On-Line Grid-Packing with a Single Active Grid. Information Processing Letters 85, 199–204 (2003)
13. Han, X., Chin, F., Ting, H.-F., Zhang, G., Zhang, Y.: A New Upper Bound on 2D Online Bin Packing (manuscript)
14. Han, X., Iwama, K., Zhang, G.: Online removable square packing. Theory of Computing Systems 43(1), 38–55 (2008)
15. Januszewski, J., Lassak, M.: On-line packing sequences of cubes in the unit cube. Geometriae Dedicata 67, 285–293 (1997)
16. Johnson, D.S., Demers, A.J., Ullman, J.D., Garey, M.R., Graham, R.L.: Worst-Case performance bounds for simple one-dimensional packing algorithms. SIAM Journal on Computing 3(4), 299–325 (1974)
17. Lee, C.C., Lee, D.T.: A simple on-line bin packing algorithm. J. Assoc. Comput. Mach. 32, 562–572 (1985)
18. Leung, J.Y.-T., Tam, T.W., Wong, C.S., Young, G.H., Chin, F.Y.L.: Packing squares into a square. J. Parallel Distrib. Comput. 10, 271–275 (1990)
19. Kohayakawa, Y., Miyazawa, F.K., Raghavan, P., Wakabayashi, Y.: Multidimensionalcube packing. Algorithmica 40(3), 173–187 (2004)
20. Meir, A., Moser, L.: On packing of squares and cubes. Journal of Combinatorial Theory 5, 126–134 (1968)
21. Ramanan, P.V., Brown, D.J., Lee, C.C., Lee, D.T.: On-line bin packing in linear time. Journal of Algorithms 10, 305–326 (1989)
22. Seiden, S.S.: On the online bin packing problem. J. ACM 49, 640–671 (2002)
23. van Vliet, A.: An improved lower bound for on-line bin packing algorithms. Information Processing Letters 43, 277–284 (1992)
24. Yao, A.C.-C.: New Algorithms for Bin Packing. Journal of the ACM 27, 207–227 (1980)

On the Continuous CNN Problem

John Augustine* and Nick Gravin

School of Physical and Mathematical Sciences
Nanyang Technological University
Singapore 637371
jea@ics.uci.edu, ngravin@gmail.com

Abstract. In the (discrete) CNN problem, online requests appear as points in $\mathbb{R}^2$. Each request must be served before the next one is revealed. We have a server that can serve a request simply by aligning either its x or y coordinate with the request. The goal of the online algorithm is to minimize the total L_1 distance traveled by the server to serve all the requests. The best known competitive ratio for the discrete version is 879 (due to Sitters and Stougie).

We study the continuous version, in which, the request can move continuously in $\mathbb{R}^2$ and the server must continuously serve the request. A simple adversarial argument shows that the lower bound on the competitive ratio of any online algorithm for the continuous CNN problem is 3. Our main contribution is an online algorithm with competitive ratio $3+2\sqrt{3} \approx 6.464$. Our analysis is tight. The continuous version generalizes the discrete orthogonal CNN problem, in which every request must be x or y aligned with the previous request. Therefore, our result improves upon the previous best known competitive ratio of 9 for the discrete orthogonal CNN problem (due to Iwama and Yonezawa).

1 Introduction

The k-server problem has been influential in the development of online algorithms [3]. We have k servers that can move around a metric space. Requests arrive in an online manner on various locations in the metric space. After each request arrives, one of the k servers must move to the request location. The online algorithm must make this decision without any knowledge of the future requests. The objective is to minimize the sum of the distances traveled by the k servers.

A natural variant of the k-server problem, the (discrete) CNN problem, was introduced by Koutsoupias and Taylor [4]. The name derives from the following illustrative example: consider a sequence of newsworthy events that occur in street intersections in Manhattan. A CNN news crew must cover these events with minimal movement. Since they have powerful zoom lenses, they only need to be at some point on either one of the two cross streets. More formally, we are

---

* Work done in part while at Tata Research Development and Design Centre, Pune, India

O. Cheong, K.-Y. Chwa, and K. Park (Eds.): ISAAC 2010, Part II, LNCS 6507, pp. 254–265, 2010.

given a sequence of requests as points from $\mathbb{R}^2$ that appear in an online manner. We have one server that can move around in $\mathbb{R}^2$. To serve a request, the server must merely align itself to the x or y coordinate of the request. The objective is to minimize the total distance traveled by the server in L_1 norm.

There is an equivalent alternative definition that is also used in literature in which, instead of a single server that can move in 2D, we have two independent servers with one restricted to move along the x-axis, while the other is restricted to move along the y-axis. Given an online request at (a, b), either the x-axis server must move to $x = a$ or the y-axis server must move to $y = b$. The objective is to minimize the sum of distances moved by either servers. Notice that the two independent servers in different dimensions are equivalent to a single server that can move around in both dimensions. For this reason, the CNN problem is also called sum of two 1-server problems [4].

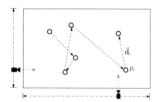

Fig. 1. Illustration for the two server definition of the continuous CNN problem

We introduce the continuous version of the CNN problem. We use the alternative two server definition to illustrate the continuous version. Consider the problem of covering the activities of a soccer match (see Figure 1). For the sake of simplicity in our illustration, let us have two cameras on rails, one along the length (i.e., the x-axis server) and the other along the breadth (the y-axis server) of the field. Their orientations are fixed perpendicular to the direction of movement (of course, pointing into the field). As the ball is kicked around, at least one of the two cameras must track the ball continuously. Informally, the input is a request point moving along a continuous trajectory that is revealed in an online manner and a server must *continuously* align itself to the x or y coordinate of the request.

We say that two points are x-aligned (respectively, y-aligned) if they share the same x (respectively, y) coordinate. Also, we say that two points are aligned if they are either x-aligned or y-aligned. We are now ready to formally define the continuous CNN problem. For this formal definition (and for the rest of the paper) we have a single server that can move around in 2D space. Our input is an online sequence of pairs $r_i = (p_i, d_i)$, where p_i is a point on $p_{i-1} + td_{i-1}$, $t \geq 0$, and d_i is a unit vector in some arbitrary direction. (In the soccer illustration, p_i is the point on the previous trajectory where the ball is intercepted and d_i is the new direction in which it is kicked.) Without loss of generality, the first point is assumed to be the origin. The server also starts at the origin. When an input pair (p_i, d_i) is revealed, the server and p_i are already aligned. The online algorithm must then commit to a continuous trajectory $T_i(t)$ of the server parameterized by

t such that for all $t \geq 0$, $T_i(t)$ is aligned with $p_i + t\boldsymbol{d}_i$. After the online algorithm commits, the next request $(p_{i+1}, \boldsymbol{d}_{i+1})$ arrives, the online server moves to the point on T_i that aligns with p_{i+1} along the trajectory T_i. The objective is to minimize the total distance traveled by the server in L_1 norm.

History of CNN problems: The discrete version of the CNN problem was discussed in several conferences and seminars in the late 1990s without any breakthroughs[1,2]. It was formally introduced by Koutsoupias and Taylor [4][1]. They conjectured that this problem has a competitive algorithm along with a lower bound of $6 + \sqrt{17}$ on the competitive ratio of any deterministic online algorithm. Their conjecture was proved affirmatively in [5] by Sitters, Stougie, and de Paepe, albeit, with an algorithm that was 10^5-competitive. For a fascinating discussion of the prevailing understanding of this problem in 2003, see [1]. Eventually, Sitters and Stougie [6] made further improvements and provided a 879-competitive algorithm. In fact, their work focussed on the *generalized k-server problem* which can be characterized as the sum of several 1-server problems on arbitrary metric spaces. The orthogonal CNN problem was introduced by Iwama and Yonezawa [2]. Each request (except the first one) must either share the x coordinate or the y coordinate with the previous request. With this restriction, they were able to improve the competitive ratio dramatically to 9.

Our Contribution: We focus on the continuous CNN problem, which is a generalization of the orthogonal CNN problem. We formalize this in the following Claim (with proof deferred to the full version).

Claim 1. *Any c-competitive algorithm $\mathcal{A}$ for the continuous CNN problem can be applied to the orthogonal CNN problem in a manner that preserves the competitive ratio.*

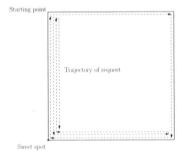
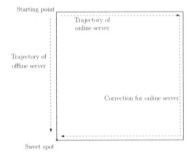

Fig. 2. Illustration for Claim 2. The figure on the left shows the request trajectory. The figure on the right shows the trajectory of online and offline servers.

Claim 2. *If there is a c-competitive algorithm for the continuous CNN problem, then $c \geq 3$ even when the request trajectory is restricted to lie on the boundary of a unit square.*

---

[1] Conference version appeared in STACS 2000.

Proof (Sketch). We now provide an example that informally illustrates how we get a lower bound of 3 on the competitive ratio of the continuous CNN problem; see Figure 2. Consider the unit square with both the optimal offline server and the online server at the top-left corner. In this adversarial example, the request moves to the bottom-right corner so that the online server is forced to choose between either a clockwise or counter-clockwise direction. Assume, without loss of generality, that it chooses the clockwise direction and moves to the top-right. The offline server, however, makes a single move down to the bottom-left. Suppose now the request moves around repeatedly in the left and bottom edges of the unit square, i.e., it makes a repeated "L" shaped move. Clearly, the offline server is already at a "sweet spot" and therefore stays unmoved. The online server, however, must correct its position and move to the sweet spot to offset its disadvantage. Notice that the online server moved three units of distance while the optimal offline server just needed one.   □

The significant contribution of our paper is an online algorithm for the continuous CNN problem with a competitive ratio of $3 + 2\sqrt{3} = 6.464$. In light of Claim 1, our result improves upon the 9-competitive algorithm for the orthogonal CNN problem [2]. Our algorithm alternates between two phases, namely, the *bishop phase* and the *rook phase*. Hence, we call it the Bishop-Rook algorithm or just the BR algorithm. Our analysis using a non-decreasing potential function is non-trivial. Finally, we show that our analysis is tight by constructing input instances for which the competitive ratio is realized.

In Section 2 we present the BR Algorithm for the continuous CNN problem. We analyze the BR algorithm in Section 3 and show that it has a competitive ratio of $(3 + 2\sqrt{3}) \approx 6.464$.

2   The BR Algorithm for the Continuous CNN Problem

We now turn our attention to the main problem that we address in this paper — the continuous CNN problem. Recall that we formally defined the input as an online sequence of pairs (p_i, d_i). Informally, we treat the request as a point starting at the origin and moving to each subsequent p_i in straight line segments whose direction is given by the vector d_i. So we use the term *request trajectory* to refer to the path traversed by the request. The server's trajectory must stay aligned with request trajectory at all times. In this section, we describe the Bishop-Rook algorithm or just the BR algorithm that alternates between two phases, namely, the Bishop phase and the Rook phase. As the name implies, the server moves diagonally during the Bishop phase. In the Rook phase, we treat the horizontal and vertical components of the server separately, leading to movements that mimic Rooks in Chess. The algorithm switches between the phases when appropriate conditions (described subsequently for each phase) are met.

The key intuition behind the algorithm is the following. Suppose the offline server manages to get to a "sweet spot" from which it can align with the request trajectory with little or no movement. Then, the online server also must home into that spot. Iwama and Yonezawa [2] also exploit this idea. They get closer to a potential sweet spot using "L" shaped moves — hence, one can call it the Knight algorithm. To achieve this homing effect in the BR algorithm, we define an **offset** vector at the end of the bishop phase that, when added to the online server's position, will point to our candidate sweet spot. In the rook phase, we use the **offset** vector to guide the online server to the sweet spot.

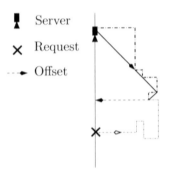

Fig. 3. Bishop phase

Bishop Phase: During the bishop phase, as the name implies, the server moves diagonally making a 45° angle with the axes. Without loss of generality, let the point p_i be at $(0,0)$ and the online server be on the non-negative part of y-axis at $(0,h)$, so $h \geq 0$; see Figure 3. Throughout the bishop phase, the server moves in a manner that maintains x-alignment with the request trajectory. Notice that this defines the x component of the server movement. To ensure the diagonal movement of the bishop phase, the server also moves in the $-y$ direction. For every maximal δx that the server moves in either the $+x$ or $-x$ direction, the server simultaneously moves a distance $|\delta x|$ in the $-y$ direction. If (and when) the position of server and request trajectory coincide, we terminate the bishop phase and switch to the rook phase. Let (s_x, s_y) be the coordinates of the point at which they coincide. Then, the **offset** vector $o = -s_x \boldsymbol{x}$, where $\boldsymbol{x}$ is the unit vector in the positive x direction.

Rook Phase: At the beginning of the rook phase, positions of server and request trajectory coincide. Without loss of generality we assume that **offset** is in the $-x$ direction. We maintain two invariants throughout the rook phase. However, in so doing, we are judicious with the L_1 distance traveled by the server.

y-**alignment:** The server and request trajectory are always y-aligned. This fully defines the movement of the server along the y direction because the server maintains the same y coordinate as that of the request.

x **coordinate inequality:** The x coordinate of the server is always less than or equal to the x coordinate of the request trajectory. This invariant is more subtle. When the x coordinate of the request trajectory is strictly greater than that of the server, the server's x coordinate stays unchanged — this is to ensure that we are judicious with the L_1 distance traveled. When the x coordinates coincide and the request trajectory is moving in the $-x$ direction, then the server moves along with the request trajectory.

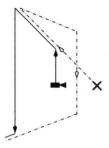

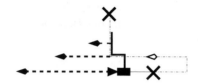

Fig. 4. Rook phase without **offset** update

Fig. 5. Rook phase showing **offset** update

During the rook phase, the **offset** vector o decreases whenever the server moves. The rate of decrease depends on the horizontal and vertical components of the movement. The rate at which $|o|$ decreases is given by:

$$|o| \leftarrow \begin{cases} |o| - (1 + \sqrt{3})|t| & \text{if server and request move a distance } t \text{ vertically} \\ |o| & \text{if request moves but server does not} \\ |o| - t & \text{if server and request move a distance } t \text{ horizontally} \end{cases}$$

When $|o|$ reaches 0, we switch to the bishop phase. Fig. 4 depicts the working of the rook phase, but does not show the change in **offset** . Fig. 5 shows how the **offset** shrinks as the phase progresses.

3   Analysis of the BR Algorithm

To simplify the analysis, we assume that we are working on an instance of the continuous *orthogonal* CNN problem, i.e, all the direction vectors d_i are orthogonal with respect to the axes. This does not affect our analysis because any straight line of arbitrary angle can be approximated by a series of infinitesimally small x and y components.

Before we proceed with the analysis, we make a simple observation that allows us to insert artificial points into the input sequence. Suppose we are given a sequence of input requests $I = ((p_1, d_1), \ldots, (p_i, d_i), (p_{i+1}, d_{i+1}), \ldots)$. Consider the sequence $I' = (p_1, d_1), \ldots, (p_i, d_i), (p_i', d_i), (p_{i+1}, d_{i+1}), \ldots$, where p_i' lies on the line segment between p_i and p_{i+1}. Then any server trajectory for serving the request sequence I will also serve I' and vice versa.

Our analysis uses a potential function that is non-decreasing throughout the execution of the algorithm. We define a cycle to be the combination of a bishop phase and the subsequent rook phase. Recall that at the start of a cycle, the **offset** is 0. We re-orient our view such that the next outstanding request is at the origin and the online server is at $(0, h)$, where $h \geq 0$. When re-orienting our view, we ensure that the potential remains unchanged. This is shown formally in Remark 1. The potential function Φ is a function of the **offset** and the parameters defined as follows:

ℓ^{opt} **and** ℓ^{on} are the distances traveled by the optimal offline server and the online server, respectively,

p^{opt} **and** p^{on} are the positions of the optimal offline server and the online server, respectively.

The potential function is given by

$$\Phi = (3 + 2\sqrt{3})\ell^{opt} - 3d(p^{on} + o, p^{opt}) - \ell^{on} - |o| + f(|o|, p^{opt}, p^{on}), \quad (1)$$

where $d(p, q)$ is the L_1 distance between points p and q. To define f, we first define $h = p_y^{opt} - p_y^{on}$, where p_y^{opt} and p_y^{on} are the y coordinates of p^{opt} and p^{on}. Now,

$$f(o, p^{opt}, p^{on}) = \begin{cases} 0 & \text{if } h \le 0 \\ (6 - 2\sqrt{3})h & \text{if } 0 \le h \le |o| \\ (6 - 2\sqrt{3})|o| & \text{if } |o| \le h \end{cases}$$

Theorem 3. Φ *is non-decreasing throughout the execution of the* BR *algorithm and this implies a competitive ratio of* $(3 + 2\sqrt{3})$.

We first provide a series of lemmas that lead to the proof of Theorem 3.

Lemma 1. *If the online server stays still,* Φ *does not decrease.*

Proof. Note that o remains unchanged when the online server stays still. Also, the optimal server either (i) does not move, (ii) moves horizontally (arbitrary distance) or (iii) moves vertically the same distance that the request moves. In all three cases, Φ does not decrease. □

Corollary 1. *From Lemma 1, it follows that, in the bishop phase,* Φ *does not decrease when request moves vertically.*

We define the offset halfplane to be the halfplane $x \le p_x^{on}$. Naturally, its complement is $x > p_x^{on}$. Since p_x^{on} can change as the online server moves, the offset halfplane also changes accordingly.

Corollary 2. *From Lemma 1, it follows that, in the rook phase,* Φ *does not decrease when request moves horizontally in the complement of the offset halfplane.*

Remark 1. At the start of each cycle, the axes of the euclidean plane can be redrawn (orthogonally) without changing Φ.

Proof. At the start of each cycle, **offset** is 0. Therefore, only the first three terms of Equation 1 are non-zero. Those three terms do not change if the axes are redrawn orthogonal to the previous axes. □

In the rest of the lemmas, since we can insert new points into the request sequence, we show that Φ does not decrease for small ϵ moves of the request in the direction specified.

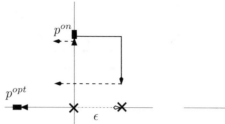

Fig. 6. Case: p^{opt} is un-changed

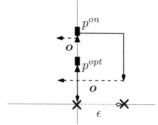

Fig. 7. Case: p^{opt} moves vertically

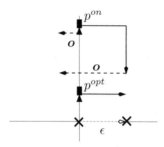

Fig. 8. Case: $p_y^{opt} \leq p_y^{on}$

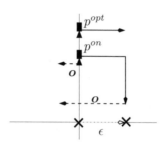

Fig. 9. Case: $p_y^{opt} \geq p_y^{on}$

Lemma 2. *In the bishop phase, Φ does not decrease when the request moves a distance ϵ in the horizontal direction.*

Proof. We treat this proof in cases based on the behavior of the optimal offline algorithm.

Case: p^{opt} is unchanged. This is only possible if p^{opt} and request are y-aligned. ℓ^{opt} is unchanged. ℓ^{on} increased by 2ϵ. $|o|$ has changed by at most ϵ. $f = 0$ because $h \leq 0$. If the request moves in the same direction as o, then $|o|$ decreases by ϵ. $d(p^{on} + o, p^{opt})$ decreased by ϵ. Overall, Φ does not decrease (see Fig. 6).

Case: p^{opt} moves vertically and aligns with request. This is a composition of Lemma 1 and the previous case (see Fig. 7).

Case: $p_y^{opt} \leq p_y^{on}$ and p^{opt} and request are x-aligned for the duration of the move. ℓ^{opt} increases by ϵ. If request moves in the same direction as o, then $|o|$ decreases by ϵ and $d(p^{on} + o, p^{opt})$ decreases by 2ϵ, otherwise, $|o|$ increases by ϵ and $d(p^{on} + o, p^{opt})$ is unchanged. ℓ^{on} increases by 2ϵ. f remains at 0. Therefore, Φ does not decrease (see Fig. 8).

Case: $p_y^{opt} \geq p_y^{on}$ and p^{opt} and request are x-aligned for the duration of the move. ℓ^{opt} increases by ϵ. If request moves in the same direction as o, then $|o|$ decreases by ϵ and $d(p^{on} + o, p^{opt})$ remains unchanged. Otherwise,

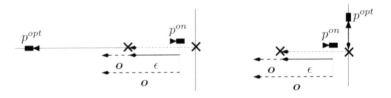

Fig. 10. Case: p^{opt} stays still   **Fig. 11.** Case: p^{opt} moves vertically

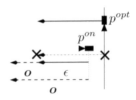

Fig. 12. Case: p^{opt} makes an x-aligned move

$|o|$ increases by ϵ and $d(p^{on} + o, p^{opt})$ increases by 2ϵ. ℓ^{on} increases by 2ϵ. h in f increased by ϵ (see Fig. 9).

The easy case is when $|o|$ decreases. We assume that either $|o| \geq h$ or $|o| \leq h$. Otherwise, we can insert a request when the change happens. With either option, the change in f term is positive and since $|o|$ decreases, one can work out that Φ increases.

When $|o|$ increases, the analysis tightens. The f term increases by $(6 - 2\sqrt{3})\epsilon$ because $|o|$ and h also increase by ϵ. Therefore, $\Delta\Phi = (3 + 2\sqrt{3})\epsilon - 6\epsilon - 2\epsilon - \epsilon + (6 - 2\sqrt{3})\epsilon = 0$.

Case: p^{opt} and request are x-aligned for the duration of the move. In this case, we are not restricting the relative locations of p^{opt} and p^{on}. In particular, $p^{on} \geq p^{opt}$ first, then after some point, the inequality is interchanged. If we insert a request at that point, then, this case breaks into the previous two cases. $\square$

Lemma 3. *In the rook phase, Φ does not decrease when request moves horizontally into the offset halfplane.*

Proof. As the request moves a distance ϵ, the online server goes with it. (So, the request does not enter the offset halfplane, but rather pushes it by a distance ϵ.) Therefore, $|o|$ decreases by ϵ.

Case: p^{opt} stays still. Clearly, p^{opt} is y-aligned with the request. ℓ^{opt} and $d(p^{on} + o, p^{opt})$ are unchanged, but ℓ^{on} increases by ϵ. Since p^{opt} and p^{on} are y-aligned, $f = 0$. Recall that $|o|$ decreases by ϵ. Therefore, Φ is unchanged (see Fig. 10).

Case: p^{opt} makes a vertical move after which, p^{opt} and request are y-aligned. We can assume that p^{opt} made the jump first before p^{on} moved

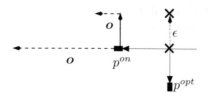

Fig. 13. Case: $p_y^{opt} \leq p_y^{on}$ and p^{opt} is x-aligned

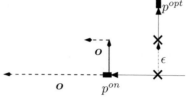

Fig. 14. Case: $p_y^{opt} \geq p_y^{on}$ and p^{opt} is x-aligned. p^{on} moves up

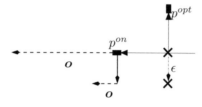

Fig. 15. Case: $p_y^{opt} \geq p_y^{on}$ and p^{opt} is x-aligned. p^{on} moves down.

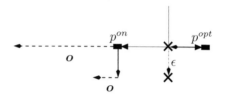

Fig. 16. Case: p^{opt} x-aligns with a horizontal move

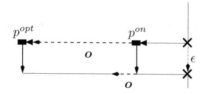

Fig. 17. Case: p^{opt} moves vertically

along with the request. From Lemma 1, Φ does not decrease when p^{opt} jumped. p^{on} moving along with the request is handled by the previous case (see Fig. 11).

Case: p^{opt} **makes an** x-**aligned move.** ℓ^{opt} and ℓ^{on} increase by ϵ. $d(p^{on} + o, p^{opt})$ decreased by ϵ. Since $|o|$ decreased by ϵ, f decreases at most by $(6 - 2\sqrt{3})\epsilon$ (see Fig. 12). Therefore,

$$\Delta(\Phi) \geq (3 + 2\sqrt{3})\epsilon + 3\epsilon + \epsilon - \epsilon - (6 - 2\sqrt{3})\epsilon \geq 0. \qquad \square$$

Lemma 4. *In the rook phase, Φ does not decrease when request moves vertically.*

Proof. Note that all vertical moves of ϵ distance in the rook phase decrease $|o|$ by $(1 + \sqrt{3})\epsilon$.

Case: $p_y^{opt} \leq p_y^{on}$ **and** p^{opt} **is** x-**aligned and therefore does not move.** ℓ^{opt} is obviously unchanged, but ℓ^{on} increases by ϵ. $d(p^{on} + o, p^{opt})$ decreased by at least $(1 + \sqrt{3})\epsilon - \epsilon = \sqrt{3}\epsilon$. Since $h \leq 0$, $\Delta(f) = 0$. Therefore, $\Delta(\Phi) \geq 3\sqrt{3}\epsilon + \sqrt{3}\epsilon > 0$ (see Fig. 13).

Case: $p_y^{opt} \geq p_y^{on}$ and p^{opt} is x-aligned, so it does not move. p^{on} and request move up by ϵ. As in the previous case, ℓ^{opt} remains unchanged, but ℓ^{on} increases by ϵ. Also, $|o|$ decreased by $(1 + \sqrt{3})\epsilon$. $d(p^{on} + o, p^{opt})$ decreased by $(1 + \sqrt{3})\epsilon + \epsilon = 2\epsilon + \sqrt{3}\epsilon$. Both h and $|o|$ decreased, so f decreased as well by at most $(1 + \sqrt{3})(6 - 2\sqrt{3})\epsilon$ (see Fig. 14). Therefore,

$$\Delta(\Phi) \geq 3(2 + \sqrt{3})\epsilon - \epsilon + (1 + \sqrt{3})\epsilon - (6 - 2\sqrt{3})(1 + \sqrt{3})\epsilon \geq 0$$

Case: p^{opt} is still, but p^{on} and request start below p^{opt}, move up and cross over to above p^{opt}. This is simply a composition of the above two cases, so Φ does not decrease.

Case: $p_y^{opt} \geq p_y^{on}$ and p^{opt} is x-aligned, so it does not move. p^{on} and request move down by ϵ. ℓ^{opt} is unchanged, but ℓ^{on} increases by ϵ. $|o|$ decreased by $(1 + \sqrt{3})\epsilon$. $d(p^{on} + o, p^{opt})$ decreased by $(1 + \sqrt{3})\epsilon - \epsilon = \sqrt{3}\epsilon$. While h increases, $|o|$ decreased. Therefore, f might decrease, but at most by $(1+\sqrt{3})(6-2\sqrt{3})\epsilon$. Therefore, $\Delta(\Phi) \geq 3\sqrt{3}\epsilon + \sqrt{3}\epsilon - (1+\sqrt{3})(6-2\sqrt{3})\epsilon = 0$ (see Fig. 15).

Case: p^{opt} starts out y-aligned, but it x-aligns itself to the request with a horizontal move. This case can be viewed as the composition of two parts. p^{opt} moves first and Φ does not decrease (by Lemma 1). Then, p^{opt} stays still, but request and server move up. This is the previous case. Hence, Φ does not decrease (see Fig. 16).

Case: p^{opt} moves vertically (up or down) and stays y-aligned. ℓ^{opt} and ℓ^{on} increase by ϵ. $|o|$ decreases by $(1 + \sqrt{3})\epsilon$, but $d(p^{on} + o, p^{opt})$ increases by at most $(1 + \sqrt{3})\epsilon$. Finally, f remains at 0. Therefore, $\Delta(\Phi) \geq (3 + 2\sqrt{3})\epsilon - 3(1 + \sqrt{3})\epsilon - \epsilon + \epsilon + \sqrt{3}\epsilon = 0$ (see Fig. 17). □

Proof (of Theorem 3). Φ started at 0 and, from Lemmas 1, 2, 3, 4, and Corollary 2, we know that it only increased. Without loss of generality, we can assume that we terminate at the end of the rook phase, at which point, the f function will evaluate to 0. If we terminate at some other point in the cycle, f might be non-zero. For the purpose of analysis, we can perform a simple trick to bring f to zero without increasing ℓ^{opt}. In particular, we artificially move the request repeatedly in an "L" shaped manner with p^{opt} at the corner. p^{on} will home in on this corner point as well and once it coincides with the corner, f will become zero without incurring any increase in ℓ^{opt}. Since $\Phi = (3 + 2\sqrt{3})\ell^{opt} - 3d(p^{on} + o, p^{opt}) - \ell^{on} - |o| \geq 0$, and $d(p^{on} + o, p^{opt})$ and $|o|$ are non negative, $(3 + 2\sqrt{3})\ell^{opt} \geq \ell^{on}$. □

Remark 2. The analysis of our algorithm is tight, i.e. there are infinite sequences of requests for which $\ell^{on} = (3 + 2\sqrt{3})\ell^{opt}$.

We defer the proof of Remark 2 to the full version. In fact, we provide two tight examples to indicate how Φ balances between multiple scenarios. We feel that minor adjustments to Φ will not reduce the competitive ratio.

Acknowledgment

We are grateful to Ning Chen, Edith Elkind, Sachin Lodha, Srinivasan Iyengar, Sasanka Roy and Dilys Thomas for useful discussions and ideas.

References

1. Chrobak, M.: Sigact news online algorithms column 1. SIGACT News 34(4), 68–77 (2003)
2. Iwama, K., Yonezawa, K.: The orthogonal CNN problem. Inf. Process. Lett. 90(3), 115–120 (2004)
3. Koutsoupias, E.: The k-server problem. Computer Science Review 3(2), 105–118 (2009)
4. Koutsoupias, E., Taylor, D.S.: The CNN problem and other k-server variants. Theor. Comput. Sci. 324(2-3), 347–359 (2004)
5. Sitters, R., Stougie, L., de Paepe, W.: A competitive algorithm for the general 2-server problem. In: ICALP 2003: 29th International Colloquium on Automata, Languages and Programming, Malaga, Spain, pp. 624–636 (2003)
6. Sitters, R.A., Stougie, L.: The generalized two-server problem. J. ACM 53(3), 437–458 (2006)

Policies for Periodic Packet Routing

Britta Peis, Sebastian Stiller, and Andreas Wiese[*]

Technische Universität Berlin,
Straße des 17. Juni 136, 10623 Berlin, Germany
{peis,stiller,wiese}@math.tu-berlin.de

Abstract. In the periodic packet routing problem each of a set of tasks repeatedly emits packets over an infinite time horizon. These have to be routed along their fixed path through a common network. A schedule must resolve the resource conflicts on the arcs, such that the maximal delay any packet experiences can be bounded. The scheduling policies themselves must be simple enough to be executed in real-time, i.e., with low computational overhead. We compare the potential of two natural classes of policies, namely, template schedules and priority schedules by giving algorithms and lower bounds.

1 Introduction

Packet routing has become a standard model for data transfer. The basic model features a graph of servers as vertices and links as arcs. Each packet travels through the network along a simple path from its start vertex to its destination vertex. For the time a packet crosses an arc, it blocks this link (or part of its bandwidth) for other packets. Each server can store several packets at the same time. In our model we assume infinite vertex capacity. Further, we assume the paths of the packets to be given. Therefore, the task is to find synchronized scheduling decisions to settle the resource conflicts on the arcs. Algorithms for these decisions are the central subject of this paper.

Usually, the data transfer between each pair of start and destination vertex consists of a huge quantity of packets. Typical examples are the real-time transfer of a video content or of a conversation. To execute complicated, individual scheduling decisions for each of these packets would create an unacceptable high computational overhead.

Therefore, a suitable model combines standard packet routing with the paradigm of classical real-time scheduling: In the *Periodic Packet Routing Problem* (PPRP) we are given a graph and a set of tasks. Each task, e.g., a connection for a voice communication, is defined by a pair of start and destination vertex. Once the connection is established along a certain routing path the task repeatedly emits packets over a long time interval. Given several such tasks in the network, one needs a simple policy to decide at each arc which packet is send first via this arc. Such a policy shall ensure a certain quality of service (QoS), i.e., that each packet is delayed along its path by at most a certain, tolerable small time span.

As in classical real-time scheduling, we distinguish two task models, the strict periodic and the sporadic task model. A strict periodic task emits one new packet exactly

---

[*] This work was partially supported by the DFG-research center MATHEON and the DFG-focus program 1307.

O. Cheong, K.-Y. Chwa, and K. Park (Eds.): ISAAC 2010, Part II, LNCS 6507, pp. 266–278, 2010.

every p time units. In the sporadic case the period p is only a lower bound on the separation time: a task can emit packets at any time, but the release dates of two packets of the same task are at least p time units apart. Strict periodic tasks model a steady demand of data transfer, e.g., for a video content. But for the broadcast of conversations a sporadic task can be a better model.

As the total duration of a connection is usually much larger than the transfer time for a single packet, we assume the time horizon of each task to be infinite. Note that an instance of the sporadic PPRP consists of an infinite set of scenarios, for each of which the policy must guarantee the desired QoS. One of these scenarios is equivalent to the strict periodic instance with the same input data.

In this work, firstly, we require as QoS that the maximum delay for every packet is finitely bounded. We prove a necessary and sufficient condition that a schedule with this QoS exists. This result holds for arbitrary graphs. For the rest of the paper we restrict to trees. For trees, secondly, we construct and analyze policies that achieve higher levels in QoS.

As stated before, the scheduling policies must be simple enough to work in real-time. A widely used class of policies in classical real-time machine scheduling are fixed priority policies. Here each task has a (distinct) priority assigned. Whenever, jobs (in our case: packets) are in conflict, the first job of the task with highest priority is served first (i.e., traverses an edge). For periodic packet routing we study *global-priority* and *edge-priority schedules*. For the former there is a global priority assignment for the tasks which is identical for every part (i.e., edge) of the system. Such a priority assignment is simple and therefore easy to implement. Edge-priority schedules are more flexible. There, each edge has again a priority for the tasks which use it. However, this list is not necessarily identical for each edge. This additional freedom allows better schedules as we will see in the sequel and does not loose too much flexibility.

A class of schedules specifically designed [1] for PPR problems are template schedules. Here at each point in time an arc is open to transfer packets of exactly one task. This exclusive openness permutes cyclically in time over all tasks that send their packets along the arc. To construct a template schedule one has to find for each arc the order in which the tasks are permuted and the offsets between the cyclic permutations of all arcs.

We show how schedules of the two types can be constructed. Further, we show how much can be gained in QoS by using template schedules instead of priority schedules. Before we give a detailed overview of our results, we summarize some definitions.

1.1 Definitions

Let $G = (V, A)$ be a directed tree. Let T denote a set of tasks $\tau_i = (s_i, t_i)$ with $s_i, t_i \in V$ such that in G there is a directed path from s_i to t_i. Let $p \in \mathbb{N}$ denote a period length. We call $I = (G, T, p)$ an instance of the *Periodic Packet Routing Problem* (PPRP). We assume a discrete time model. Each task τ_i repeatedly creates new packets which have to be transported from s_i to t_i by a routing schedule. We assume unit transit times (i.e., each packet needs one timestep to traverse an arc), unlimited storage in each vertex, and unit bandwidths (each arc can be used by at most one packet at a time).

We distinguish between the *strict* and the *sporadic PPRP*. For instances of the strict PPRP each task generates a new packet every p timesteps, starting at time $t = 0$. In instances of the sporadic PPRP it is not known a priori when the tasks emit their packets. We require only that in each time interval of length p each task emits at most one packet. We call a specification of the release times for the packets a *realization*.

For each task τ_i we denote by P_i the arcs on the unique path from s_i to t_i and define $D_i := |P_i|$. For an arc e let T_e denote the set of tasks which use e. For general graphs we assume that we are given the paths of the tasks P_i explicitly.

We also consider undirected trees, where the paths of different tasks may include the same edge in opposite direction. Still, for *undirected* trees, the edge can only be used by one packet at time. In case an edge can be used at a time by one packet in each direction, we speak of a *bidirected* tree.

Given an instance, respectively a realization, a *schedule* S must feasibly and for an infinite time horizon assign to each arc for each point in time which packet is routed on it. We say a *limit* for a task τ_i in a schedule S is a value k such that each packet which is ever created by τ_i needs at most k timesteps to reach t_i after it has been created. We denote by c the congestion, that is the maximum number of tasks which use an arc (or the maximum number of tasks which use an undirected edge in one direction). We say an instance I is feasible if there is a schedule for I such that for each task there is a finite limit. We call a schedule a *direct schedule* if no packet waits in a vertex different from its start vertex (and its destination vertex). A schedule is called *indirect* if it is not necessarily direct. We now define the two main classes of schedules under consideration.

Definition 1 (Template schedules). *Let $I = (G, T, p)$ be an instance of the sporadic PPRP or the strict PPRP. A schedule for I is a* template *schedule if there exists an integer $\bar{p} \leq p$ and a map $task : E \times \{0, ..., \bar{p} - 1\} \rightarrow T \cup \{none\}$ such that in the schedule an arc $e = (u, v)$ is used at time t by a packet M created by a task τ if and only if $task(e, t \bmod \bar{p}) = \tau$, M is located on u at time t, and no packet created by τ before M is located on u.*

Note that each map $task : E \times \{0, ..., \bar{p} - 1\} \rightarrow T \cup \{none\}$ yields a template schedule, if its restriction to every arc e is surjective on T_e. Also note that a template schedule might delay a packet even though there is no conflicting packet waiting. Apart from template schedules, this peculiarity is inevitable in a direct periodic schedule.

We consider two types of priority schedules: Global-priority schedules have a global priorization of the tasks (which is valid for each edge), edge-priority schedules might have different priorizations for each edge.

Definition 2 (Edge-priority schedule). *Let $I = (G, T, p)$ be an instance of the sporadic or strict PPRP. A schedule for I is an* edge-priority *schedule if there exists a total order $\prec_e \subseteq T \times T$ for each edge e in G such that the following holds: If in the schedule a packet M of task τ traverses arc $e = (a, b)$ between time t and $t+1$, then each packet located at time t in node a either belongs to tasks τ' with $\tau \prec_e \tau'$ or belongs to task τ and has a later release date than M.*

Definition 3 (Global-priority schedule). *Let* $I = (G, T, p)$ *be an instance of the sporadic or strict PPRP. A schedule for* I *is a* global-priority *schedule if it is an edge-priority schedule and all relations* $\prec_e$ *for the edges are identical.*

Both types of priority schedules are simpler to apply than template schedules: While priority schedules can be executed fully locally, template schedules require a 'global clock'. Edge-priority schedules are a strengthened version of global-priority schedules. We also briefly discuss two other ways to strengthen priority schedules, which we define on the fly.

1.2   Related Work

The non-periodic packet routing problem is widely studied. In a celebrated paper Leighton et al. [5] show that there is always a schedule of length $O(C + D)$ (where C denotes the maximum number of packets using an edge and D is the length of the longest path of a packet). In [6] Leighton et al. also present an algorithm which finds such schedules. Often the packet routing problem is studied on special (simple) topologies like trees [8,10] or grid graphs [7,11]. Busch et al. [2] present algorithms computing direct schedules, i.e., schedules which delay packets only in their start vertex. Also, simple greedy algorithms [9] and the complexity of the problem [4,10] have been studied.

The result by Leighton et al. [5] is extended to the periodic setting by Andrews et al. [1] guaranteeing a bound of $O(D_i + 1/r_i)$ for each session i with a packet injection rate of r_i (corresponding to a task with a period length in our notation). In particular, they introduce the template schedules which are also studied in this paper.

Our task models are borrowed from classical real-time scheduling. Here one studies real-time executable algorithms for distributing and scheduling (computational) jobs on a processor platform. However, as there is no graph involved the techniques are quite different. For an overview cf. [3].

Note that with arbitrary period lengths for the tasks (rather than all being identical to p) the PPRP with only one edge is identical to real-time scheduling on one machine. Since the latter is already very complicated, in particular in the case of priority schedules, for the PPRP we restrict to tasks with identical period lengths.

1.3   Our Contributions

We give a comprehensive characterization of the periodic packet routing problem in the various settings (direct/indirect schedules, bidirected/directed trees, template/priority schedules), see Tables 1 and 2 for a complete overview. We present algorithms and prove limitations and relations of the different types of schedules. Most importantly, we show the following results:

▶ For bidirected trees we give an algorithm to construct template schedules that guarantee a maximal delay for each packet of $2c - c(1/2)^{\lceil diam(G)/2 \rceil - 1} - 1$. This is achieved by carefully distributing the necessary delays among the tasks.

▶ For directed trees we give an edge-priority schedule that guarantees a maximal delay of $1.5c - 1$ for each packet (under mild conditions on the congestion). We give a non trivial construction that yields a lower bound tightly matching the quality achieved by the algorithm.

▶ Finally, in Section 4.3 we follow a more direct approach to compare the power of template schedules and priority schedules. We show that whenever a priority schedule achieves a certain quality of service, one can construct a template schedule imitating the priority schedule well enough to achieve the same or almost the same quality. Key to these results is to prove that priority schedules after some time show a periodic behavior.

In the following Section 2 we prove a necessary and sufficient condition on the existence of a schedule. The affirmative part of the theorem rests on a structural insight for template schedules that allows to bound the backlog of the packets.

All our algorithms are designed for trees in the setting of equal period lengths p. This is so because we show that already on chain graphs (which have a quite simple structure) no edge-priority schedule can always guarantee non-trivial limits. Also, we prove that with arbitrary period lengths already on a path no edge-priority schedule can guarantee a delay for each packet which is bounded by a constant times the period length of its respective task.

To round up the picture on PPRP we also state minor results for which we defer the proofs to our technical report[12]. Among these are complexity results and results on two other ways to strengthen the concept of priority schedules: First, we allow the priority relation to be not necessarily transitive. Under certain conditions this gives a well-defined and fruitful class of so called quasi-priority schedules. Secondly, we briefly mention the effect of an initial swing-in period in which the schedule can deviate from its priority rules.

Table 1. Overview of our results for template schedules

	Template Schedules				
	Indirect Schedules			Direct Schedules	
	Limit	Bound on c		Limit	Bound on c
Directed Trees	$c + D_i - 1$		p	$c + D_i - 1$	p
Bidirected Trees	$2c - c(1/2)^{\lceil diam(G)/2 \rceil - 1} + D_i - 1$		p	$2c + D_i - 2$	$p/2$
Undirected Trees	$4c - 2c(1/2)^{\lceil diam(G)/2 \rceil - 1} + D_i - 1$		$p/2$	$4c + D_i - 3$	$p/4$

Table 2. Overview of our results for priority schedules. All bounds hold for both the strict periodic and the sporadic setting.

	Priority Schedules			
	Type	Limit	Bound on c	Lower Bound
Bidirected Trees	Global	$2c + D_i - 1$	$p/3$	–
Directed Trees	Global	–	–	$2c + D_i - 1$
Directed Trees	Edge	$1.5c + D_i - 1$	$2p/5$	$1.5c + D_i - 1$

2 Necessary and Sufficient Bound for Feasibility

Theorem 1. *An instance $I = (G, T, p)$ of the strict or sporadic PPRP on an arbitrary graph G is feasible if and only if $c \leq p$. Also, there is always a template schedule for it which guarantees a limit of $p \cdot D_i$ for each task τ_i.*

The theorem can be shown with the following technique. For instances in which an arc e is used by more than p tasks in the long run there are arbitrarily many packets waiting to use e. With careful technical analysis we can show that some of them will have an arbitrarily large delay. For instances with $c \leq p$ we show that any template schedule such that $\{task(e, k)|0 \leq k < \bar{p}\} = T_e$ guarantees a limit of $p \cdot D_i$ for each task τ_i. In order to prove the limit we use the following insight for template schedules.

Lemma 1. *Let I be an instance of the sporadic PPRP and let S be a template schedule for I with periodicity $\bar{p}$. Assume that for each arc e we have $\{task(e, k)|0 \leq k < \bar{p}\} = T_e$. If a packet M created by a task τ_i arrives on a vertex v at time t then no packet created by τ_i arrives on v during the time interval $[t + 1, t + \bar{p})$.*

Proof. Let τ_i be a task. We prove the claim by induction over P_i. Since $\bar{p} \leq p$ the claim holds for s_i (due to our definition of the sporadic setting). Now let $\{s_i = v_0, v_1, ...v_{k-1}, v_k = t_i\}$ be the vertices on P_i. We assume by induction that the claim holds for the vertices $v_1, ..., v_\ell$. Assume that at time t a packet M created by τ_i arrives on v_ℓ. The induction hypothesis implies that no other packet created by τ_i is located on v_ℓ at time t. Since S is a template schedule with periodicity $\bar{p}$ there must be a timestep $t' \geq t$ with $t' < t + \bar{p}$ at which M traverses the arc $e_\ell = (v_\ell, v_{\ell+1})$. In particular, this implies that during the time interval $[t + \bar{p}, t' + \bar{p})$ no packet created by τ_i traverses e_ℓ (due to the periodicity of S). We conclude that M arrives on $v_{\ell+1}$ at time $t' + 1$ and no packet created by τ_i arrives at $v_{\ell+1}$ during the time interval $[t' + 2, t' + \bar{p} + 1)$.   □

For further technical we refer to our technical report [12].

3   Template Schedules

In the sequel we will study template schedules on directed trees, bidirected trees and undirected trees. Note that the case of directed trees is a special case of bidirected trees.

3.1   Directed Trees

Let $I = (G, T, p)$ be an instance of the sporadic periodic packet routing problem on a directed tree G. We present an algorithm which constructs a direct template schedule guaranteeing a limit of $c + D_i - 1$ for each task τ_i. It transfers ideas from [10] to the periodic setting. The given bound is best possible: There are instances in which there can be no better limit for every task (e.g., consider an instance on a directed path in which all tasks have identical paths). Also, in the strict periodic setting it is NP-hard to determine whether there is a template schedule which guarantees a limit of $c + D_i - 2$ for an instance on a directed tree (see technical report [12]).

Now we present the algorithm. First, we compute a feasible path-coloring for the paths of the tasks (i.e. a coloring such that any two paths sharing an edge have different colors). A proof given in [10, Section 2.1] shows that on directed trees there is always a valid coloring for the paths of the tasks $f : T \rightarrow \{0, ..., c - 1\}$ with at most c colors and this coloring can be computed in polynomial time. Then we define

a time-dependent edge-coloring $g : E \times \{0, ..., c - 1\} \rightarrow \{0, ..., c - 1\}$ which has
the property that for two consecutive arcs $e = (u, v)$ and $e' = (v, w)$ we have that
$g(e, i) = g(e', (i + 1) \bmod c)$ for $0 \leq i < c$. This yields a periodic template schedule with $\bar{p} := c$ by defining $task(e, k) := \tau_i$ if $f(\tau_i) = g(e, k)$ for all arcs e and
all $k \in \{0, ..., c - 1\}$. If $g(e, k) \notin f(T_e)$ we define $task(e, k) := none$. Denote by
$DTREE(I)$ the resulting schedule.

Theorem 2. *Let I be an instance of the sporadic periodic packet routing problem on a
directed tree with $c \leq p$. The schedule $DTREE(I)$ is a direct template schedule which
guarantees a limit of $c + D_i - 1$ for each tasks τ_i.*

Proof. The property of g is passed on to $task$: For two consecutive arcs $e = (u, v)$ and
$e' = (v, w)$ we have that $task(e, k) = task(e', (k + 1) \bmod c)$ for all $k \in \{0, ..., c -
1\}$. Therefore, once a packet has left its start vertex it is never delayed until it reaches
its destination vertex. Each packet has to wait for at most $c - 1$ timesteps in its start
vertex. We conclude that for each task τ_i it holds that $c + D_i - 1$ is a valid limit.   □

3.2   Bidirected Trees

Before we present our algorithms for the bidirected tree we introduce some structure
for the tree which we will use for all algorithms in the sequel.

Definition 4 (Tree-structure). *Let G be a directed or bidirected tree. We define a vertex v_r to be the root vertex such that for each vertex v we have that its height $h(v)$
(i.e., the distance between v_r and v) is bounded by $\lceil diam(G)/2 \rceil$ (where $diam(G)$ denotes the diameter of G). We call an arc e an up-arc if it is oriented towards v_r and a
down-arc if it is oriented away from v_r. For each task τ_i we define the vertex v_i which
is closest to v_r to be the peak vertex of τ_i. For a task τ_i we define its height $h(\tau_i)$ by
$h(\tau_i) := h(v_i)$. We say a packet moves up if it is using an up-arc. A packet moves down
if it uses a down-arc.*

Now we describe our indirect schedule $BTREE(I)$ which guarantees a limit of $2c -
c \left(\frac{1}{2}\right)^{\lceil diam(G)/2 \rceil - 1} + D_i - 1$ for each task τ_i (see [12] for formal definitions). It has
the property that on the way up we delay packets only in their start vertex (restricted
to the way up the problem is essentially PPR on a directed tree). For the way down we
use the following key observation: Assume there are two sets of tasks T and T' using
arcs $e = (u, v)$ and $e' = (u', v)$, respectively, and a common arc $\bar{e} = (v, w)$. Assume
that the values of the map $task$ for the arcs e and e' have already been defined. Then
we can assign the values of $task$ for the arc $\bar{e}$ such that no task in T is delayed on v and
each task in T' is delayed at most $|T|$ times on v by assigning the free slots in a suitable
manner to T'.

We define $\bar{p} := c$. Our algorithm considers the vertices sorted by their height and
defines the values of $task$ for their respective adjacent arcs. We start with the root vertex
v_r and define $task$ such that no packet is delayed on v_r. This can be achieved since the
graph induced by v_r and its neighbors behaves like a directed tree (to compute the
values for $task$ see $DTREE(I)$). Now let v be a vertex with $h(v) > 0$ and assume that
the algorithm has already computed the values of $task$ for arcs (u, w) with $h(u) < h(v)$

or $h(w) < h(v)$. Our schedule ensures that tasks moving up through v and tasks with peak vertex v do not delay each other (such an assignment can be found by interpreting the respective subgraph as a directed tree and finding a schedule which fits to the up-arc adjacent to v). Note that these are the only tasks which use up-arcs adjacent to v. It remains to define the prioritization of the tasks on the down-arcs adjacent to v. Denote by $T_e^{(down)}$ the tasks whose path uses a down-arc $e = (v, w)$ and which have a peak vertex different than v (i.e., above v). Denote by $T_e^{(peak)}$ the tasks which use e and which have v as peak vertex. There are two cases: If $|T_e^{(down)}| \geq c(1 - \left(\frac{1}{2}\right)^{h(v)})$ then – according to the key observation stated above – we give priority to $T_e^{(peak)}$ causing a delay of at most $c\left(\frac{1}{2}\right)^{h(v)}$ for each task in $|T_e^{(down)}|$. On the other hand, if $|T_e^{(down)}| < c(1 - \left(\frac{1}{2}\right)^{h(v)})$ we give priority to $|T_e^{(down)}|$ causing a delay of at most $c(1 - \left(\frac{1}{2}\right)^{h(v)})$ for each task in $T_e^{(peak)}$.

Theorem 3. *Let I be an instance of the sporadic PPRP on a bidirected tree with $c \leq p$. The schedule $BTREE(I)$ is a template schedule which guarantees a limit of $2c - c\left(\frac{1}{2}\right)^{\lceil diam(G)/2 \rceil - 1} + D_i - 1$ for each task τ_i.*

Proof. Let τ_i be a task with peak vertex v_i. Let M_i be a packet created by τ_i. From the definition of $task$ it follows that on its way up M_i is delayed only in its start vertex (at most $\bar{p} - 1 = c - 1$ times). On v_i the packet M_i is delayed at most $c(1 - \left(\frac{1}{2}\right)^{h(v_i)})$ times. Now let $e = (v, w)$ be a down-arc on P_i which is not adjacent to v_i (i.e., $v \neq v_i$). By construction, M_i is delayed at most $c\left(\frac{1}{2}\right)^{h(v)}$ times on v. Denote by $P_i^{\downarrow}$ all vertices on the way down of τ_i, excluding v_i and t_i. We calculate that in total M_i is delayed at most $c(1 - \left(\frac{1}{2}\right)^{h(v_i)}) + \sum_{v \in P_i^{\downarrow}} c\left(\frac{1}{2}\right)^{h(v)}$ times. With the geometric sequence and using that $h(v) \leq \lceil diam(G)/2 \rceil - 1$ for all $v \in P_i^{\downarrow}$ we obtain a limit of $2c - c\left(\frac{1}{2}\right)^{\lceil diam(G)/2 \rceil - 1} + D_i - 1$ for each packet created by τ_i. $\qquad\square$

Our analysis of $BTREE(I)$ always guarantees a bound of $2c + D_i - 2$. We provide an instance with $p = c = 2$ where no template schedule can guarantee a better bound for every task (see technical report [12]).

Direct Schedule. In contrast to the schedule $DTREE(I)$ for directed trees, $BTREE(I)$ is not a direct schedule. It is NP-hard to decide whether a direct template schedule exists or not. Furthermore, there are in fact instances, even with $c \leq 3p/4$, for which no direct schedule exists, see [12] for both results. We now present a direct schedule $BRTEE_{dir}(I)$ for any instance I of the sporadic PPRP on a bidirected tree with $c \leq p/2$, guaranteeing a limit of $2c + D_i - 2$ for each task τ_i. It transfers techniques from the non-periodic case presented in [2, Theorem 3.4].

We sort the tasks descendingly by their height $h(\tau_i)$ (the distance between v_r and P_i). W.l.o.g. let $\tau_1, \tau_2, ..., \tau_{|T|}$ be this order. Define $\bar{p} := 2c$. We iterate over the tasks with $i = 1$ to $|T|$. Consider the i-th iteration. Let $P_i = \{v_0, v_1, ..., v_{|P_i|-1}\}$ be the path of τ_i and let $e_j = (v_j, v_{j+1})$ for all $j \in \{0, ..., |P_i| - 2\}$. Let w_i be the smallest positive integer such that $task(e_j, w_i + j \mod \bar{p}) = none$ for all relevant values of j. We assign τ_i the initial start offset w_i and define $task(e_j, w_i + j \mod \bar{p}) = \tau_i$ for all respective values of j. We sorted the tasks by height and hence only the tasks which use the two edges on P_i adjacent to v_i can interfere with a task τ_i. Since we required that $c \leq p/2$ there is always a feasible value for w_i with $0 \leq w_i \leq 2c - 2 < \bar{p}$. We denote by $BTREE_{dir}(I)$ the resulting schedule.

Theorem 4. *Let I be an instance of the sporadic PPRP on a bidirected tree with $c \leq p/2$. The schedule $BTREE_{dir}(I)$ is a direct template schedule which guarantees a limit of $2c + D_i - 2$ for each task τ_i.*

3.3   Undirected Trees and Randomized Algorithms

Randomized Algorithm. We further establish a randomized algorithm which interprets a bidirected tree as two directed trees and combines the two schedules (computed like in $DTREE(I)$) with a random offset. For each task τ_i it guarantees a limit of $c + D_i - 1$ in expectation in the strict periodic case, a limit of $1.5c + D_i - 1$ in expectation in the sporadic case, and a limit of $2c + D_i - 2$ independently from the outcome of the random experiment, for details see [12].

Undirected Trees. The techniques presented for bidirected trees can be adapted to the case of undirected trees. It turns out that in this setting we need to require that $c \leq p/2$ since otherwise for any $\alpha, \beta > 0$ we can construct an instance for which there can be no limit of the form $\alpha c + D_i + \beta$ for every task τ_i and there is not necessarily a direct schedule. We derived algorithms which compute schedules with limits $4c - 2c \left(\frac{1}{2}\right)^{\lceil diam(G)/2 \rceil - 1} + D_i - 1$ (indirect schedule) and $4c + D_i - 3$ (direct schedule if we also require that $c \leq p/4$), see [12].

4   Priority Schedules

In this section we investigate edge- and global-priority schedules. It turns out that a bound on c stricter than just $c \leq p$ is necessary for edge-priority schedules (and hence for global-priority schedules) in order to guarantee a limit of the form $\alpha \cdot D_i$ for every instance as we will see in the sequel. We first describe the edge-priority schedule $EPRIO(I)$ guaranteeing a limit of $3c/2 + D_i - 2$ if $c \leq 2p/5$.

4.1   Edge-Priority schedule

Let I be an instance of the sporadic PPRP on a directed tree. We need to define a prioritization for each arc separately. First, we define that $\tau \prec_e \tau'$ if $h(\tau) < h(\tau')$ for each arc e. So now we focus on the tie-breaking for tasks τ, τ' with $h(\tau) = h(\tau')$. Consider a vertex v. Let $e_1, ..., e_r$ be the ingoing up-arcs of v and let $e'_1, ..., e'_s$ be the outgoing down-arcs of v. Denote by E_v the set of all these arcs. In case that v_r is not the root vertex let $\bar{e}$ be the remaining arc which is adjacent to v. We define the prioritization for all tasks with peak vertex v.

We define T_v to be the set of tasks which use v. Let T'_v be the tasks in T_v which do not use $\bar{e}$. We compute a minimum path coloring for the paths of the tasks T_v. This can be done by a reduction to edge-coloring of a bipartite multigraph. Exactly c colors are needed, for details see [10]. Assume that the paths are colored with colors $\{1, 2, ..., c\}$. If there is an arc $\tilde{e} \in E_v$ such that more than $c/2$ tasks use both $\tilde{e}$ and $\bar{e}$ then we assume w.l.o.g. that the paths which use $\bar{e}$ and $\tilde{e}$ use the colors $1, ..., m$ if $\bar{e}$ is an up-arc and the colors $c - m + 1, ..., c$ if $\bar{e}$ is a down-arc (assuming that there are m such paths). Note that there can be at most one arc $\tilde{e}$ with this property. Denote by $f(\tau)$ the color of the path of a task τ. Let $\tau, \tau' \in T'_v$ be a pair of tasks with peak vertex v. For each up-arc e we define $\tau \prec_e \tau'$ if $f(\tau) < f(\tau')$. For each down-arc e we define $\tau \prec_e \tau'$ if $f(\tau) > f(\tau')$. Denote by $EPRIO(I)$ the resulting schedule.

Theorem 5. *Let I be an instance of the sporadic PPRP on a directed tree with $c \leq 2p/5$. The schedule $EPRIO(I)$ guarantees a limit of $3c/2 + D_i - 1$.*

Proof. Let τ_i be a task colored with color $f(\tau_i)$. We say a task τ_j *interferes* with τ_i if P_i and P_j have a common arc on which P_j has a higher priority than P_i. We observe that at most $f(\tau_i) - 1$ tasks with peak vertex v_i interfere with τ_i on its way up and there are at most $c - f(\tau_i)$ such tasks which interfere with τ_i on its way down. By the choice of the colors for tasks in $T_v \setminus T'_v$ we have that in total at most $c/2 + (f(\tau_i) - 1) + (c - f(\tau_i)) = 3c/2 - 1$ tasks interfere with τ_i.

We require that $c \leq 2p/5$ in order to ensure that two packets created by a task with higher priority than τ_i have a certain minimum time distance when arriving at v_i. This allows to show that each packet created by τ_i is delayed at most once by each task with higher priority. For details see [12]. □

Now we show that the bound guaranteed by $EPRIO(I)$ is best possible.

Theorem 6. *For each period length p and any even value for c there exist instances of the sporadic PPRP on directed trees for which no edge-priority schedule can guarantee a better limit than $3c/2 + D_i - 1$ for every task τ_i.*

Proof. Consider a very long path along which $c/2$ tasks send their packets. Denote by $\hat{I}$ these tasks. Assume on the contrary that we have a schedule S which guarantees a better limit than $3c/2 + D_i - 1$ for each task τ_i.

At every other vertex on the path we introduce a gadget. We have a star with the vertices v_1, v_2, v_3, v_4, v_r and the arcs $(v_1, v_r), (v_2, v_r), (v_r, v_3), (v_r, v_4)$. The tasks in $\hat{I}$ use the arcs (v_2, v_r) and (v_r, v_4). We introduce $c/2$ tasks with the path (v_2, v_r, v_3), $c/2$ tasks with the path (v_1, v_r, v_3), and $c/2$ tasks with the path (v_1, v_r, v_4). We denote these tasks by I_1, I_2, and I_3, respectively. We introduce $3c^2/2$ of these gadgets. For every task $\hat{\tau} \in \hat{I}$ there can be at most $3c$ gadgets in which a task $\tau \notin \hat{I}$ has a higher priority than $\hat{\tau}$. Since $\left|\hat{I}\right| = c/2$ by the pigeon hole principle there must be a gadget F in which every task in $\hat{I}$ has a higher priority than the tasks traversing F which are not in $\hat{I}$. Hence, in F the delays of the packets have to be distributed among the tasks in $I_1 \cup I_2 \cup I_3$. One can show that then one of these tasks can be delayed up to $3c/2 - 1$ times, see [12] for details. □

We would also like to comment that without a bound on the congestion c the schedule $EPRIO(I)$ cannot guarantee the limits stated in the theorems. Even more, we can show the following proposition (which of course also holds for global-priority schedules since they are special cases of edge-priority schedules), proven in [12].

Proposition 1. *Let $\alpha > 0$. There is an instance $I_\alpha = (G, T, p)$ of the sporadic PPRP on a directed tree G with $c = p$ such that for any edge-priority schedule ES there is a task $\tau \in T$ for which ES cannot guarantee any limit k with $k \leq \alpha \cdot c + D_i$.*

Investigating edge-priority schedules only on trees might look restrictive at first glance. However, we show in the following theorem that already on chain graphs – which have a quite simple structure – no edge-static schedule can guarantee non-trivial limits for all tasks. We assume that the paths are given as part of the input.

Theorem 7. *For any $D > 0$ and any $c, p \in \mathbb{N}$ there is an instance $I = (G, T, p)$ of the sporadic PPRP with given paths on a chain graph G such that $D_i = D$ for each task $\tau_i \in T$, all tasks have the same start and the same destination vertex, each arc is used by c tasks, and for every edge-priority schedule there is at least one task $\tau \in T$ which has a limit in $\Omega(c \cdot D_i)$.*

The proof of the theorem is given in our technical report [12]. There, we also prove that if the tasks have arbitrary period lengths p_i, in general we cannot obtain schedules guaranteeing good limits for all tasks like in $EPRIO(I)$.

~~**Theorem 8.** *For every $\alpha \geq 0$ there is an instance $I = (G, T)$ of the sporadic PPRP with arbitrary period lengths on a path such that for any edge-periodic schedule there is a task $\tau_i \in T$ whose limit is at least $\alpha \cdot p_i + D_i$.*~~

4.2   Global-Priority Schedule and Strict Periodic Setting

Global-priority schedule. We define a global-priority schedule $GPRIO(I)$. Let I be an instance of the sporadic PPRP on a bidirected tree. We define a task τ_i to have a higher priority than a task τ_j if $h(\tau_i) < h(\tau_j)$. The priorization of tasks with the same peak vertex is defined arbitrarily.

Theorem 9. *Let I be a instance of the sporadic PPRP on a bidirected tree with $c \leq p/3$. The schedule $GPRIO(I)$ guarantees a limit of $2c + D_i - 2$ for each task τ_i.*

Proof. Consider a task τ_i. On its way up it is delayed only up to $c - 1$ times. This holds since there it can only be delayed by tasks which use the last up-arc on P_i and excluding τ_i there are at most $c - 1$ such tasks. Similarly, on the way down τ_i can only be delayed on v_i by tasks which use the first down-arc of P_i. We require that $c \leq p/3$ in order to ensure that each packet created by τ_i is delayed at most once by each task with higher priority. □

It is easy to see that there are instances where $GPRIO(I)$ delays a packet $2c - 2$ times. In fact, using an instance on a directed star graph we can show this applies for any global static schedule (full proof see [12]).

Theorem 10. *For each period length p and any value for c there exist instances of the sporadic PPRP on directed trees for which no global-priority schedule can guarantee a better limit than $2c + D_i - 2$ for every task τ_i.*

There are instances where even in the strict periodic setting no global-priority schedule can a guarantee better limit than $\Omega(c \cdot D_i)$ for every task τ_i, see [12].

Priority Schedules for Strict Periodic Packet Routing. For the strict periodic setting one can show that $GPRIO(I)$ and $EPRIO(I)$ can guarantee the stated bounds already if we require only that $c \leq p/2$. Also, in the strict periodic setting we can weaken the notion of global-priority schedules to *quasi-global-priority schedules*: we no longer require a relation $\prec$ which is globally a total order. Instead, we consider the restriction of $\prec$ to the tasks whose packets wait for using an arc at any timestep t. If we require only that this restriction is a total order then we can obtain a schedule which guarantees a limit of $p/a + D_i - 1$ for each task τ_i if $c \leq p/a$ for an integer $a \geq 2$.

Also, if we allow a carefully controlled initial swing-in phase after which we start executing an edge-priority schedule we can find schedules on directed and bidirected trees which guarantee limits of $p + D_i - 1$ and $2p + D_i - 1$, respectively, while allowing any congestion $c \leq p$. For details we refer to our technical report [12].

4.3 Imitation Theorems

We show that global- and edge-priority schedules can be imitated by template schedules with almost the same quality.

In the strict periodic setting we say that a template schedule S *imitates* a priority schedule PS if after a certain time t_0 the schedule S transfers a packet created by a task τ along an edge e if and only if PS does so. Since the strict period setting is a special realization of the sporadic setting, if a priority schedule PS guarantees a limit of k_i for a task τ_i then any template schedule S imitating PS guarantees a limit of $k_i + p$ for τ_i.

Theorem 11. *Every global-priority schedule can be imitated by a template schedule. Every edge-priority schedule on a bidirected tree can be imitated by a template schedule.*

This theorem can be shown using a structural insight of independent interest: After a certain time global-priority schedules on arbitrary graphs and edge-priority schedules on bidirected trees behave periodically. See [12] for details.

Acknowledgments. We would like to thank Martin Niemeier for fruitful discussions on the topic.

References

1. Andrews, M., Fernández, A., Harchol-Balter, M., Leighton, F., Zhang, L.: General dynamic routing with per-packet delay guarantees of O (distance + 1/session rate). SIAM Journal of Computing 30, 1594–1623 (2000)
2. Busch, C., Magdon-Ismail, M., Mavronicolas, M., Spirakis, P.: Direct routing: Algorithms and complexity. Algorithmica 45, 45–68 (2006)
3. Buttazzo, G.C.: Hard Real-time Computing Systems: Predictable Scheduling Algorithms And Applications. Real-Time Systems Series. Springer, Santa Clara (2004)
4. di Ianni, M.: Efficient delay routing. Theoretical Computer Science 196, 131–151 (1998)
5. Leighton, F.T., Maggs, B.M., Rao, S.B.: Packet routing and job-scheduling in $O(congestion + dilation)$ steps. Combinatorica 14, 167–186 (1994)
6. Leighton, F.T., Maggs, B.M., Richa, A.W.: Fast algorithms for finding $O(congestion + dilation)$ packet routing schedules. Combinatorica 19, 375–401 (1999)

7. Leighton, F.T., Makedon, F., Tollis, I.G.: A $2n - 2$ step algorithm for routing in an $n \times n$ array with constant size queues. In: Proceedings of the 1st Annual Symposium on Parallel Algorithms and Architectures, pp. 328–335 (1989)
8. Leung, J.Y.-T.: Handbook of Scheduling: Algorithms, Models and Performance Analysis (2004)
9. Mansour, Y., Patt-Shamir, B.: Greedy packet scheduling on shortest paths. Journal of Algorithms 14 (1993)
10. Peis, B., Skutella, M., Wiese, A.: Packet routing: Complexity and algorithms. In: Proceedings of the 7th Workshop on Approximation and Online Algorithms (2009)
11. Peis, B., Skutella, M., Wiese, A.: Packet routing on the grid. In: Proceedings of the 9th Latin American Theoretical Informatics Symposium (2010)
12. Peis, B., Stiller, S., Wiese, A.: Periodic packet routing on trees. Technical Report 008-2010, Technische Universität Berlin (April 2010)

Increasing Speed Scheduling and Flow Scheduling

Sebastian Stiller and Andreas Wiese*

Technische Universität Berlin,
Straße des 17. Juni 136, 10623 Berlin, Germany
{stiller,wiese}@math.tu-berlin.de

Abstract. Network flows and scheduling have been studied intensely, but mostly
separately. In many applications a joint optimization model for routing and scheduling is desireable. Therefore, we study flows over time with a demand split into
jobs. Our objective is to minimize the weighted sum of completion times of these
jobs. This is closely related to preemptive scheduling on a single machine with a
processing speed increasing over time. For both, flow scheduling and increasing
speed scheduling, we provide an EPTAS. Without release dates we can prove a
tight approximation factor of $(\sqrt{3}+1)/2$ for Smith's rule, by fully characterizing
the worst case instances. We give exact algorithms for some special cases and a
dynamic program for speed functions with a constant number of speeds. We can
prove a competitive ratio of 2 for the online version. We also study the class of blind
algorithms, i.e., those which schedule without knowledge of the speed function.

1 Introduction

Scheduling and network flows are two corner stones of combinatorial optimization.
These topics are intensely treated, and rich both in theory and applicability to real-
world optimization problems. Still, many real-world applications in logistic, traffic, and
telecommunication require a coupled optimization of scheduling and routing decisions.
As an example consider the container terminal of a modern harbor where containers are
carried from the storage area to the loading cranes by automatically guided vehicles.
One has to make a scheduling decision on the order in which the containers are brought
to the ships, and a routing decision for the vehicles through the small area between
storage and cranes. These applications usually surpass the algorithmic means developed
separately for scheduling and flows. Thus, one either has to reside to general purpose
methods, in particular IP models, or to decouple the optimization of scheduling and
routing. In this work we consider a basic step towards joint, combinatorial optimization
of flows and schedules.

Given an s-t-network, static capacities for the arcs' in-flow rates, and static transit
times for the arcs. Further, given a demand comprised of k jobs, each characterized by
its own flow demand and weight. The goal is to find a feasible flow over time minimizing the sum of weighted completion times, i.e., the points in time when the complete
demand of a job has reached the sink.

Using *multiple deadline flows*—a slight extensions of known, though elaborate techniques for earliest arrival flows—flow scheduling boils down to an intriguing scheduling

* This work was partially supported by the DFG-research center MATHEON and the DFG-focus
  program 1307.

O. Cheong, K.-Y. Chwa, and K. Park (Eds.): ISAAC 2010, Part II, LNCS 6507, pp. 279–290, 2010.

problem. This subproblem is simple enough to be stated in one sentence: Given jobs with processing demand and weights and a single machine with increasing speed: which schedule minimizes the weighted sum of completion times?

Increasing speed scheduling forms an interesting specialization of scheduling with varying speed and a wide generalization of scheduling with rejection. Though flow scheduling only leads to scheduling with stepwise increasing speed, we also treat increasing speed scheduling for arbitrary (integrable) speed functions.

Related work. To the best of our knowledge *flow scheduling* has not been considered so far. It has some far resemblance to flow shop problems. There is a close relation to dynamic multi-commodity, earliest arrival, and universally maximal flows.

For the single source, single sink case an earliest arrival flow (EAF) always exist [5], and can be found by a pseudopolynomial successive shortest path algorithms [12,17]. There are instances where these algorithm take exponentially many steps in a binary encoded input [18]. For multiple sinks and sources EAFs need not exist [4]. In [9] a fully polynomial-time approximation scheme for the earliest arrival *s-t*-problem is given. These results have been extended in [2] to solve EAFs with multiple sources. The solution for an instance of the flow scheduling problem is a multicommodity flow over time where all commodities have common source and common sink, optimizing a for flows unusual objective function, namely, weighted sum of completion times. Multicommodity flow over time is NP-hard even in the fractional case, and even for strongly restricted graph classes [6,7]. See also reference therein for a survey on flows over time in general.

Increasing speed scheduling with release dates clearly contains $1|r_i, pmtn| \sum w_j C_j$ as an NP-hard special case. For this an EPTAS is known [1]. For scheduling with arbitrary varying speed, in particular, when machines stop, we can argue that the $1|pmtn| \sum w_j C_j$ problem is weakly NP-hard by a reduction from the PARTITION-problem similar to that in [10]. Already [11] focus on machines with nondecreasing speed. They restrict to unit weights but treat multiple, identical machines. In contrast [13] consider parallel machines with arbitrary weight but unit length jobs. In [3] Epstein et al. consider the problem of minimizing the weighted sum of completion times on a machine with increasing and decreasing speed without knowledge of the machine, like the blind algorithms discussed in this paper.

Another closely related problem is scheduling with rejection (cf. [8] and references there within), i.e., jobs can be excluded from the schedule at a fixed penalty cost. This is weakly NP-hard for a single machine (reduction from knapsack). Moreover, the case of unit weights and the case of unit lengths are shown to be polynomial. The case where rejection costs are proportional to job weights is open. This is equivalent to a special case of increasing speed scheduling, notably, when the machine has constant speed until time t, and infinite (or sufficiently large) speed after t.

Definitions. This work treats the following two problems:

Definition 1 (Flow scheduling problem). *Consider a directed graph G with two distinct nodes s and t. For each arc we are given a static inflow capacity and a static transit time. Also, we are given a set of jobs $\mathscr{J}$ where each job J_i has a weight w_i and* *a demand ℓ_i. The goal is to find a multi-commodity flow over time from s to t with $|\mathcal{J}|$ commodities such that $\sum_{J_i \in \mathcal{J}} w_i C_i$ is minimized where C_i denotes the time when flow value corresponding to commodity i has reached ℓ_i.*

For the definition of dynamic flows over time we refer to [2,9].

Definition 2 (Increasing speed scheduling (ISS) problem). *Given a machine whose speed is given as a integrable, weakly monotonically increasing function $s : \mathbb{R}^+ \to \mathbb{R}^+$ and k jobs $J_i = (w_i, \ell_i)$. Compute a schedule which minimizes the weighted sum of completion times, i.e., we look for k integrable indicator functions $\chi_i : \mathbb{R}^+ \to \{0,1\}$ with $\chi_i(x) \cdot \chi_j(x) = 0$ for all $x \in \mathbb{R}^+$ and $i \neq j$ such that $C_i := \inf_{T \in \mathbb{R}^+} \{ \int_0^T \chi_i(x)s(x)dx \geq \ell_i \}$ exists and $\sum_{J_i \in \mathcal{J}} w_i C_i$ is minimized.*

For a job (w_i, ℓ_i) we call w_i / ℓ_i its *Smith's ratio*. Note that in the literature often the inverse is referred to as Smith's ratio. A schedule processing the jobs successively with non-increasing Smith's ratio is called a *Smith's rule algorithm*. Recall that an EPTAS is a family of $(1 + \varepsilon)$-algorithms for all $\varepsilon > 0$ with running time in $\mathcal{O}(f(\varepsilon) \cdot poly(n))$ for a function f depending only on ε.

Our contribution. In Section 2 we establish the connection between flow scheduling and increasing speed scheduling. We show that flows that are maximal for a given set of deadlines can be found in polynomial time (Theorem 1). Next, we extend the EPTAS of [1] for preemptive single machine scheduling with release dates to ISS with release dates. Together with Theorem 1 this yields an EPTAS for the flow scheduling problem. In Section 4 we give exact, polynomial time algorithms for the ISS problem in similar special cases as considered in [8] for scheduling with rejection. Moreover, we device a dynamic program in case the speed function is a step function with constantly many steps.

The most important result of this work is found in Section 5. We show that Smith's rule is a $(\sqrt{3} + 1)/2$ approximation for ISS, and that this is a tight analysis. We achieve the tightness, because we constructively characterize worst instances for Smith's rule. While at first glance one might even expect Smith's rule to be optimal, to our mind the tight analysis of its approximation factor yields the deepest insight to the behavior of the ISS problem.

In the final sections we study online algorithms and algorithms that have no knowledge of the speed function (blind algorithms). For both cases there is a lower bound. For the online case we achieve a competitive ratio of 2. A blind algorithm with approximation factor α yields an α-approximation for the flow scheduling problem. So we get a $(\sqrt{3} + 1)/2$ approximation for the flow scheduling problem.

An intriguing question that we have to leave open is, whether the ISS problem without release dates is NP-hard. Recall that it is NP-hard with release dates. Also, it would be interesting to find out whether the flow scheduling problem is NP-hard.

2 From Flows to Scheduling

We consider flows over time with single source and sink[1]. For these it is known that an earliest arrival flow (EAF) exists [5], i.e., a flow over time that has a maximal flow

---

[1] As we want to be brief on this please cf. [2,9] for basic definitions and properties. As the ISS-part is self-contained, readers with no interest in EAFs may also skip this section.

value at every point in time. In particular, one can compute in pseudo-polynomial time its inflow rate into the sink s_{EAF} which is a non-decreasing, stepwise constant function of time. For some instances this function has (in the input size of the network) pseudo-polynomially many break-points [18].

Therefore, any (single source, single sink) flow scheduling problem has an optimal solution with the following structure. Let I be the smallest interval in time such that in the complement of I no flow arrives at the sink. During I the inflow rate is always positive and the interval can be partitioned into k consecutive intervals such that during each of these intervals $[T_i, T_i + 1)$ all inflow to the sink belongs to the same job. Interpret the inflow rate to the sink as the speed of a machine. Then we can rephrase: To minimize the weighted sum of completion times (without release dates) it is best to process the jobs without preempting and in a certain order on that machine.

An EAF is by definition maximal at all points in time, thus to solve the flow scheduling problem one may calculate an EAF and solve the ISS problem with its speed function s_{EAF}. As for EAFs no strongly polynomial encoding is known (and deemed unlikely to exist) this leads to over-complicated flow schedules. The pseudo-polynomial blow up is unnecessary for an optimal solution: The flow value in any optimal solution for a flow scheduling instance needs to equal that of an EAF only at the completion time of each job. So, assuming that the optimal order of the jobs is known one can in strongly polynomial time find an optimal flow schedule by the following concept.

Definition 3 (Multiple Deadline Flows (MDF)). *Given an s-t-digraph G with non-negative, constant transit times τ and capacities u on the arcs, and a finite set of deadlines $\mathcal{T} = \{T_1, \ldots, T_k\}$. A flow over time x for transit times τ respecting at any point in time u as inflow rates of the arc is called a Multiple Deadline Flow (MDF), if for $1 \leq i \leq k$ its value up to time T_i is maximal among all feasible flows over time on (G, τ, u).*

Recall that the *value* of a flow over time is the inflow (minus the outflow, which can be assumed to be zero) at the sink node: $\sum_{(v,t) \in A(G)} \int_0^{T_i - \tau_{(v,t)}} x_{(v,t)}(\theta) d\theta - \sum_{(t,v) \in A(G)} \int_0^{T_i} x_{(t,v)}(\theta) d\theta$.

Theorem 1. *An MDF for $G, \tau, u,$ and $\mathcal{T}$ can be found in time polynomial in its input length.*

Proof. First we calculate in polynomial time the value of a maximal flow over time for each deadline T_i. Then computing the MDF is equivalent to computing a quickest dynamic transshipment in the following graph: Replace the sink t by k sink nodes t_i each connected to t by its own arc of transit time $T_k - T_i$ and infinite capacity. The demand at each sink equals the maximal flow value for the corresponding deadline T_i minus the maximal flow value for T_{i-1}. A dynamic transshipment can be found in polynomial time [9].   □

A direct algorithm given in [16] computes an MDF even in strongly polynomial time.

To solve flow scheduling, in case the optimal (or some fixed) order of the jobs for a flow scheduling problem is given by the index of the jobs, calculate the minimal time horizons T_i by quickest flows for the successive sums of demands $\sum_{1 \leq j \leq i} \ell_j$ of the first

i jobs. Then solve the MDF problem for the set of deadlines $\{T_i\}$. In the resulting flow assign the first ℓ_1 flow units that reach the sink to the first job, the following ℓ_2 flow units to the second job, and so on. This yields an optimal flow schedule (for the fixed order).

To summarize: We are in a chicken and egg problem. Given an optimal order of the jobs, one can find an optimal flow schedule in strongly polynomial time. Without the optimal order, one can use the pseudo-polynomially sized inflow rate of an EAF as speed function of an ISS problem that returns the optimal order of the jobs for the flow scheduling problem. To solve this dilemma one can *either* approximate the earliest arrival flow by a flow with a bounded number of speed changes and work with the machine given by this flow *or* calculate an order which fulfills some approximation factor independent of the actual machine speed (blind algorithms). We will pursue both of these approaches, the first leading to an EPTAS and the second to an exact approximation factor of $\frac{\sqrt{3}+1}{2}$. For both approaches one could restrict to the ISS problem with step functions as speed functions. We even treat the ISS problem with general, integrable speed functions.

3   EPTAS

In this section we present an EPTAS for the increasing speed scheduling problem with arbitrary release dates. We show later that this yields also an EPTAS for the flow scheduling problem.

Let $0 < \varepsilon < 1$. We describe an algorithm which guarantees an approximation ratio of $1 + \mathcal{O}(\varepsilon)$ and has a running time in $\mathcal{O}\left(2^{poly(1/\varepsilon)}n + n\log n\right)$. (By abuse of notation whenever we use the term $\mathcal{O}(\varepsilon)$ we refer to a function bounded by $k \cdot \varepsilon$ for some positive k.) W.l.o.g. we assume that at any time the speed of the machine is at least 1.

First, we derive a couple of properties depending on ε which we can assume for the instance without losing more than a factor of $1 + \mathcal{O}(\varepsilon)$ in the objective function in comparison to the optimum (denoted by "with $1 + \varepsilon$ loss" or "with $1 + \mathcal{O}(\varepsilon)$ loss"). Here we can extend techniques from [1] to increasing speed. Then, we show how to compute the optimal schedule with these properties by a dynamic program.

We define $R(w)$ to be the timestep when the total work that the machine has done so far equals w. As short notation we use $R_x := R((1+\varepsilon)^x)$. Note that since we assume that the machine always runs with at least unit speed we have that $R_{x+1} \leq (1+\varepsilon)R_x$. We split the time scale into intervals of the form $I_x := [R_x, R_{x+1})$. In order to simplify notation we will use the notion I_x for the interval as well as for the work that the machine does within I_x.

Lemma 1. *With $1 + \mathcal{O}(\varepsilon)$ loss, we can assume for each job J_i that $r_i \geq R(\varepsilon\ell_i)$ and that ℓ_i is a power of $(1+\varepsilon)$.*

Proof. Scaling an optimal schedule in time by a factor of $1 + \varepsilon$ the work the work of any interval $[x, y]$ is moved to the interval $[(1 + \mathcal{O}(\varepsilon))x, (1 + \mathcal{O}(\varepsilon))y]$. Since in the new interval more work can be done, we gain slack which can be used for the desired properties. Scaling increases the objective function by at most a factor of $1 + \mathcal{O}(\varepsilon)$. For details cf. [16]. □

The following lemmata are proven using a technique which we call *interval-hopping*: for every interval I_x we move the work which is done in I_x to the interval I_{x+1}. The gained free space is then used to ensure certain properties of the instance or the schedule.

Lemma 2. *Assume the adjustments of Lemma 1. With $1 + \varepsilon$ loss, we can additionally assume that each job is released at a time R_x.*

Proof. After interval-hopping set the release time of each job J_i to the largest value R_x before J_i is processed. □

In order to simplify the complexity of the problem we calculate the objective function as if each job finishes at the end of an interval. So, for the remainder of this section we do not consider the objective function $\sum_{J_j \in \mathscr{J}} w_j C_j$ but the objective function $\sum_{J_j \in \mathscr{J}} w_j \min \{R_x : C_j \leq R_x\}$. As an effect, we can assume that the machine has constant speed within each interval.

Lemma 3. *Assume the adjustments of the previous lemmata. For completion times C_i resulting from any schedule we have that*
$$\sum_{J_i \in \mathscr{J}} w_i C_i \leq \sum_{J_i \in \mathscr{J}} w_i \min \{R_x : C_i \leq R_x\} \leq (1 + \varepsilon) \sum_{J_i \in \mathscr{J}} w_i C_i.$$

Proof. The first inequality is obvious. The second inequality holds since $\min \{R_x : C_i \leq R_x\} \leq (1 + \varepsilon) C_i$ for all values C_i. □

A job J_i is called *small* if it is released at a time R_x such that $\ell_i \leq \varepsilon I_x$. Otherwise it is *large*. We denote by H_x and T_x the large and small jobs, respectively, which were released at time R_x. For a set of jobs $\mathscr{J}' \subseteq \mathscr{J}$ we denote by $p(\mathscr{J}')$ their total demand.

Lemma 4. *Assume the adjustments of the previous lemmata. With $1 + \mathcal{O}(\varepsilon)$ loss, we can additionally assume that*

- *no small job is ever preempted,*
- *no small job is processed in more than one interval,*
- *the order in which the small jobs are executed obeys Smith's rule,*
- *each large job $J_i \in H_x$ is preempted only if there is an integer $k \leq \frac{1}{\varepsilon^3}$ such that a fraction of exactly $k \cdot \frac{\varepsilon I_x}{\ell_i}$ of the job has already been processed,*
- *at any point in time in each set H_x there is at most one job which has already been processed but which has not been finished yet,*
- *the number of distinct job sizes in H_x is bounded by $|H_x| \leq 3 \log_{1+\varepsilon} \frac{1}{\varepsilon} + 1$,*
- *the number of jobs in each distinct size is bounded by $1/\varepsilon$, and*
- *$p(T_x) \leq I_x$.*

Proof. We use interval-hopping once. Then, in each interval I_{x+1} there is a free space of εI_x. We can assume that in I_{x+1} there is at most one small job J_i which is preempted. Since J_i was originally scheduled in I_x we have that $\ell_i \leq \varepsilon I_x$. Thus, in the gained free space in I_{x+1} we can finish J_i. The other claims can be shown similarly with interval-hopping or by using the properties derived so far. For details cf. [16]. □

The following lemma gives an upper bound on the time a job has to wait before it completes.

Lemma 5. *Assume the adjustments of the previous lemmata. With $1 + \varepsilon$ loss, we can additionally assume that each job which is released at time R_x finishes in the interval $I_{x+s(\varepsilon)}$ the latest, where $s(\varepsilon)$ is a constant depending only on ε.*

Proof. We use interval-hopping and shift the work of each interval I_x to the interval I_{x+1}. Lemma 4 implies a bound on the total demand on the jobs in $T_x \cup H_x$. With $s(\varepsilon) = \left\lceil \log_{1+\varepsilon} \left(\frac{1}{\varepsilon} + \frac{1}{\varepsilon^4} \left(3\log_{1+\varepsilon} \frac{1}{\varepsilon} + 1 \right) \right) \right\rceil + 1$ the gained free space in $I_{x+s(\varepsilon)}$ suffices to process all jobs of $T_x \cup H_x$. □

Now we partition the ordered list of the jobs in each set T_x into at most $2/\varepsilon^2$ packs, each with size at most $\varepsilon^2 \cdot I_x$. Denote by $P_{x,i}$ the ith pack of small jobs which are released at time R_x.

Lemma 6. *Assume the adjustments of the previous lemmata. With $1 + \varepsilon$ loss, we can additionally assume that in each interval I_x either all or none of the jobs in a pack $P_{x',i}$ are scheduled, each job which is released at time R_x finishes in the interval $I_{x+s(\varepsilon)+2}$ the latest, and the ordering of the small jobs does not necessarily obey Smith's rule anymore.*

Proof. We use interval-hopping and shift each interval I_x to I_{x+2}. This gives us a free space of εI_{x+1} in each interval I_{x+2}. Due to the original ordering by Smith's rule at most one pack $P_{x',i}$ per value x' is scheduled partially but not completely in an interval I_x. The total demand of these packs is upper bounded by $\sum_{i=x-s+1}^{x} \varepsilon^2 \cdot I_i \leq \varepsilon \cdot I_{x+1}$. Thus, in the gained free space we can finish all these packs. □

Next, we describe a *dynamic program* which finds the best solution with the above properties. Each table entry is identified by a combination of an interval I_x and for each interval I_y with $x - s(\varepsilon) \leq y < x$, we have the subset of jobs in H_y which have already been fully processed, a job $J_i \in H_y$ and an integer $k \leq \frac{1}{\varepsilon^3}$ such that a fraction of exactly $k \cdot \frac{\varepsilon l_x}{\ell_i}$ of J_i has been processed, and the subset of the packs $P_{y,i}$ which have already been fully processed.

Since we need to consider at most $s(\varepsilon) \cdot n$ intervals in total, the number of table entries is bounded by $\mathcal{O}\left(2^{poly(1/\varepsilon)} n \right)$. We obtain the following theorem:

Theorem 2. *There is a polynomial time approximation scheme for the increasing speed scheduling problem with release dates with running time $\mathcal{O}\left(2^{poly(1/\varepsilon)} n + n \log n \right)$.*

In order to do the computation in the dynamic program it is not necessary to know the exact speed function. It is sufficient to know the points in time when a total demand of $(1 + \varepsilon)^x$ has already been processed for the relevant values of x. Recall that at most $s(\varepsilon) \cdot n$ intervals are relevant for us. Thus, we obtain the following theorem:

Corollary 1. *There is an efficient polynomial time approximation scheme for the flow scheduling problem.*

Proof. Follows from Theorems 1 and 2. □

4   Tractable Cases of ISS

In this section we analyze the structure of the increasing speed scheduling problem. We identify some properties which allow efficient algorithms for certain special cases. Moreover, we provide the necessary insights for our analysis of the Smith's rule algorithm in Section 5. Throughout this section we assume that all jobs are released at time $t = 0$. Accordingly, we can restrict ourselves to non-preemptive schedules.

Proposition 1. *The ISS problem has the following properties:*

- *In any optimal schedule subsequent jobs are ordered by demand if the job with less demand has the better Smith's ratio.*
- *If in the instance there is an ordering $J_1, J_2, ..., J_{|\mathcal{J}|}$ for the jobs such that $\frac{w_i}{\ell_i} \geq \frac{w_{i+1}}{\ell_{i+1}}$ and $\ell_i \leq \ell_{i+1}$ for each i then it is optimal to order the jobs ascendingly by demand.*
- *If in an instance I all jobs have the same Smith's ratio then*
  - *there is an optimal schedule which orders the jobs ascendingly by demand and*
  - *there is a worst possible schedule which orders the jobs descendingly by demand.*
- *If all jobs have the same demand then it is optimal to order the jobs descendingly by their weight. This is still true if the speed of the machine can decrease and increase.*

Proof. All claims follow from exchange arguments, i.e., one swaps two adjacent jobs and calculates the change in the objective function. □

For machines without speed change there is a well known exchange argument showing that in an optimal schedule the jobs are ordered non-decreasingly by their Smith's factors [15]. In our setting, this argument can easily be applied to jobs starting and finishing within an interval A in which the machines has constant speed. We show that the statement even holds for the set of all jobs which end in such an interval A (and not necessarily start in A).

Lemma 7. *For an interval A with constant machine speed the jobs finishing in A are ordered according to Smith's rule in any optimal schedule.*

Proof. Straight forward calculations show the claim for all jobs which start and end in A, and for the first two jobs if the smaller job has the better Smith's ratio. If both is not the case extend A to the left (i.e., change the speed function) such that the first job starts in A. One can show the schedule remains optimal for the changed machine. For details cf. [16]. □

Now we consider step functions for the speed. We denote by s_i the different speeds of the machine and by A_i the corresponding intervals, i.e., $A_i := s(s_i)^{-1}$. The above properties allow for a *dynamic program* in case the number of speeds is bounded by a constant: It gets a list of the jobs ordered by Smith's rule. It successively removes a job J_j from the list and chooses the interval $A_i = [a_i, a_{i+1})$ in which J_j finishes. Inside A_i, the job J_j is scheduled right after the last job finishing within A_i. If J_j is the first job assigned to A_i we try all start offsets less or equal a_i for which J_j finishes within A_i. Thus, in the dynamic programming table we need to encode how many jobs have already been removed from the list and how much space (at the beginning and at the end) of each interval is already occupied by jobs.

Theorem 3. *If the number of different values for the speed function is bounded by a constant there is an exact, pseudopolynomial dynamic program for the ISS problem.*

Proof. Computing an entry in the dynamic programming table can be done in pseudopolynomial time. The number of entries in the dynamic programming table is bounded by $n \cdot L^{2k}$. Lemma 7 implies that our procedure finds the optimal solution. □

Note that this pseudopolynomial algorithm cannot be combined with the pseudopolynomial algorithm for earliest arrival flows [12,17] to achieve an exact, pseudopolynomial algorithm for the entire problem. An EAF corresponds to a machine with pseudopolynomially many speeds. Our result requires a *constant* number of speeds.

5 Tight Analysis of Smith's Rule

In this section we present a tight analysis which shows that the Smith's rule algorithm is exactly a $\frac{\sqrt{3}+1}{2}$-approximation. We achieve tightness in this result because we constructively characterize worst case instances. Let $I = (\mathcal{J}, M)$ be an instance of the increasing speed scheduling problem. We denote by $SR(I)$ the worst possible schedule which obeys Smith's rule (i.e., tie-breaking decisions are taken such that the cost of the schedule is maximized). We show how to transform I into an instance with a special structure without decreasing $SR(I)/OPT(I)$. Then we show that on instances with this structure the Smith's rule algorithm is exactly a $\frac{\sqrt{3}+1}{2}$-approximation.

Lemma 8. *For any instance I there is an instance $I' = (\mathcal{J}', M)$ such that $w_i/\ell_i = 1$ for all jobs $J_i' \in \mathcal{J}'$ and $SR(I')/OPT(I') \geq SR(I)/OPT(I)$. Moreover, if in I the demands of all jobs are integral then in I' the demands of all jobs are integral as well.*

Proof. First, we define a partition of the jobs in I into classes $\mathcal{C}_i$ such that two jobs belong to the same class if and only if they have the same Smith's ratio. Let $\tilde{I}$ be an instance with as few classes $\mathcal{C}_i$ as possible such that $SR(\tilde{I})/OPT(\tilde{I}) \geq SR(I)/OPT(I) =: \alpha$. If in $\tilde{I}$ there is only one class then scaling job weights completes the proof. So now assume that there are at least two classes in $\tilde{I}$. Denote by $SR(\mathcal{C}_i)$ and $OPT(\mathcal{C}_i)$ the amount that a class $\mathcal{C}_i$ contributes $SR(\tilde{I})$ and $OPT(\tilde{I})$, respectively. Since $SR(\tilde{I})/OPT(\tilde{I}) \geq \alpha$ there must be a class $\mathcal{C}_k$ such that $SR(\mathcal{C}_k)/OPT(\mathcal{C}_k) \geq \alpha$. If $\mathcal{C}_k$ is the class with highest Smith's ratio then we can remove all other classes and $SR(I)/OPT(I)$ will not decrease. So now assume that $\mathcal{C}_k$ is not the class with the highest Smith's ratio. Then we increase the weights of the jobs in $\mathcal{C}_k$ until they all have the Smith's ratio of $\mathcal{C}_{k-1}$. Denote by I' the resulting instance. Let $\beta > 1$ denote the factor by which we increased the weights of the jobs in $\mathcal{C}_k$. Then we calculate that $OPT(I') \leq OPT(\tilde{I}) + (\beta - 1)OPT(\mathcal{C}_k)$ and $SR(I') = SR(\tilde{I}) + (\beta - 1)SR(\mathcal{C}_k) \geq \alpha(OPT(\tilde{I}) + (\beta - 1)OPT(\mathcal{C}_k))$. This yields $SR(I')/OPT(I') \geq SR(I)/OPT(I)$. Applying this reasoning until there is only one class completes the proof. □

Recall from Proposition 1 that if all Smith's ratios are identical then we can assume that $OPT(I)$ orders the jobs ascendingly by demand and the worst Smith's rule schedule $SR(I)$ orders the jobs descendingly by demand. Now we want to study the demands of the jobs. Assume that in $\mathscr{J} = \{J_1, J_2, ..., J_{|\mathscr{J}|}\}$ the jobs are ordered ascendingly by their demand. Let k denote the largest integer such that $\sum_{i=1}^{k} \ell_i \leq \frac{1}{2} \sum_{i=1}^{|\mathscr{J}|} \ell_i$. We define $\mathscr{J}_{small} := \{J_1, J_2, ..., J_k\} \subset \mathscr{J}$. For an instance I' we denote the respective set by $\mathscr{J}'_{small}$.

Lemma 9. *Let $\varepsilon > 0$. For any instance I such that all jobs in $\mathscr{J}$ have a Smith's ratio of 1 there is an instance I' such that*

- *all jobs in I' have a Smith's ratio of 1,*
- *$\mathscr{J}'_{small}$ consist only of jobs of demand 1,*
- *$|\mathscr{J}' \setminus \mathscr{J}'_{small}| = 1$ (i.e., there is exactly one job with demand larger than 1),*
- *M' has at most one speed change which occurs when the large job is finished in $SR(I')$, and*
- *$\frac{SR(I')}{OPT(I')} \geq (1 - \varepsilon) \frac{SR(I)}{OPT(I)}$.*

The transformation from I to I' stated in the lemma is achieved by an elaborate but nowhere surprising procedure of merging and splitting jobs and carefully adjusting the speed of the machine (see [16] for details). Now we prove the approximation ratio of $\frac{\sqrt{3}+1}{2}$ for the Smith's rule algorithm.

Lemma 10. *For any instance I it holds that $SR(I)/OPT(I) \leq \frac{\sqrt{3}+1}{2}$.*

Proof. Let $\varepsilon > 0$. Using the lemmata above we derive an instance $I' = (\mathscr{J}', M')$ with the respective properties. We define $L := \sum_{J'_i \in \mathscr{J}'} \ell'_i$. Let s_1 and s_2 denote the two speed values of M' with $s_1 \leq s_2$. We set $k := L - \ell'_m$ (i.e., we have k small jobs of demand 1) and calculate that $SR(I')/OPT(I') \leq \frac{\ell'_m \cdot k + (\ell'_m)^2}{(\ell'_m)^2 + \frac{k^2}{2}}$ using that $k \leq \ell'_m$. For fixed ℓ'_m we define

$f(k) := \frac{\ell'_m \cdot k + (\ell'_m)^2}{(\ell'_m)^2 + \frac{k^2}{2}}$. The function f attains its maximum for $[0, \ell'_m]$ in $k = (-1 + \sqrt{3}) \cdot \ell'_m$. We conclude that $\frac{SR(I')}{OPT(I')} \leq \frac{\sqrt{3}+1}{2}$. Hence, for any $\varepsilon > 0$ we can show that $\frac{SR(I)}{OPT(I)} \leq \frac{1}{1-\varepsilon} \cdot \frac{SR(I')}{OPT(I')} \leq \frac{1}{1-\varepsilon} \cdot \frac{\sqrt{3}+1}{2}$. This implies that $\frac{SR(I)}{OPT(I)} \leq \frac{\sqrt{3}+1}{2}$. $\square$

Lemma 11. *For any $\varepsilon > 0$ there are instances I_ε such that $SR(I_\varepsilon)/OPT(I_\varepsilon) \geq (1 - \varepsilon) \frac{\sqrt{3}+1}{2}$.*

The instances I_ε are constructed such that the function f in the proof of Lemma 10 attains its maximum. Moreover, the weights of their jobs can be perturbed such that *any* algorithm which obeys Smith's rule outputs the same solution. Thus, we obtain the following theorem:

Theorem 4. *Any algorithm respecting Smith's rule is a $\frac{\sqrt{3}+1}{2}$-approximation, and none of these algorithms can achieve a better approximation factor for all instances.*

6   Blind and Online Algorithms

We note that the Smith's rule algorithm orders the jobs without knowledge of the speed function. We call algorithms with this property *blind* algorithms. A blind algorithm for the ISS problem allows to solve the flow scheduling problem without computing the full earliest arrival pattern. It suffices to calculate (in a second step) an MDF for the job ordering computed by the blind ISS algorithm (of the first step). With Theorem 1 we have the following corollary.

Corollary 2. *There is an approximation algorithm for the flow scheduling problem with approximation factor $\frac{\sqrt{3}+1}{2}$ which runs in polynomial time.*

Since a blind algorithm does not have knowledge of the machine it cannot compute the optimal solution for every instance. In particular, a lower bound – communicated to us by an anonymous referee – shows that a blind algorithm can achieve at best an approximation ratio of $(19+3\sqrt{65})/(22+2\sqrt{65}) \approx 1.1328$. The basic strategy for lower bounds is to confront the algorithm with two different machines. One machine runs with unit speed and another machine runs with unit speed up to a certain time and then becomes arbitrarily fast. For either ordering of the jobs, on one of the machines the achieved approximation ratio is bad.

Now we consider the increasing speed scheduling problem in an online setting. We assume the following online model: each job J_i has a release time r_i, the existence and all data of a job become known at its release time, and at time t the speed of the machine up to time t is known, the speed of the machine *after* time t is not known.

Note that neither the defined class of online algorithms nor the class blind algorithms contains the other:

We consider the following algorithm which we call the online Smith's rule algorithm, $SR_{online}(I)$: always processes the job which has the largest Smith's factor among all available jobs.

Theorem 5. The online Smith's rule algorithm has a competitive ratio of 2.

Proof. Like in Section 5 w.l.o.g. all jobs have equal Smith's ratios. We establish a lower bound instance I' by replacing each job J_i by ℓ_i/ε jobs, each with demand ε and weight $\frac{w_i}{\ell_i}\varepsilon$ such that $SR_{online}(I')$ does not preempt any job. Similarly to Proposition 1 one can show that $SR_{online}(I') = OPT(I')$ and $OPT(I') \leq OPT(I)$. The fact that the speed of the machine never decreases implies that $SR_{online}(I) \leq 2 \cdot SR_{online}(I')$ and thus $SR_{online}(I) \leq 2 \cdot OPT(I)$. For the full proof we refer to [16]. $\square$

Note that our analysis is tight since there are examples (even for the case that the speed of the machine does not change) where the online Smith's rule algorithm does not perform better than 2 [14].

The lower bound of ≈ 1.1328 for the approximation factor of a blind algorithm carries over to a lower bound for online algorithms. It actually also holds if all jobs are released at time $t = 0$ and only the information about the machine becomes available online.

For the problem $1|r_i, pmtn|\sum C_j$ the shortest remaining processing time algorithm (SRPT) is optimal: at each point in time, we process the available job which has the shortest remaining demand. Ties are broken arbitrarily. This can be generalized to our problem and proven by an exchange argument.

Theorem 6. *If all jobs in an instance I have the same weight then $SRPT(I)$ is optimal.*

References

1. Afrati, F., Bampis, E., Chekuri, C., Karger, D., Kenyon, C., Khanna, S., Milis, I., Queyranne, M., Skutella, M., Stein, C., Sviridenko, M.: Approximation schemes for minimizing average weighted completion time with release dates. In: Proceedings of the 40th Annual Symposium on Foundations of Computer Science (FOCS 1999), pp. 32–44. IEEE, Los Alamitos (1999)
2. Baumann, N., Skutella, M.: Earliest arrival flows with multiple sources. Mathematics of Operations Research 34, 499–512 (2009)
3. Epstein, L., Levin, A., Marchetti-Spaccamela, A., Megow, N., Mestre, J., Skutella, M., Stougie, L.: Universal sequencing on a single machine. In: Eisenbrand, F., Shepherd, F.B. (eds.) Integer Programming and Combinatorial Optimization. LNCS, vol. 6080, pp. 230–243. Springer, Heidelberg (2010)
4. Fleischer, L.: Faster algorithms for the quickest transshipment problem with zero transit times. In: Proceedings of the 9th Annual Symposium on Discrete Algorithms (SODA 1998), pp. 147–156 (1998)
5. Gale, D.: Transient flows in networks. Michigan Mathematical Journal 6, 59–63 (1959)
6. Hall, A., Hippler, S., Skutella, M.: Multicommodity flows over time: Efficient algorithms and complexity. In: Baeten, J.C.M., Lenstra, J.K., Parrow, J., Woeginger, G.J. (eds.) ICALP 2003. LNCS, vol. 2719, pp. 397–409. Springer, Heidelberg (2003)
7. Hall, A., Hippler, S., Skutella, M.: Multicommodity flows over time: Efficient algorithms and complexity. Theoretical Computer Science 379, 387–404 (2007)
8. Hoogeveen, H., Skutella, M., Woeginger, G.J.: Preemptive scheduling with rejection. In: Paterson, M. (ed.) ESA 2000. LNCS, vol. 1879, pp. 268–277. Springer, Heidelberg (2000)
9. Hoppe, B., Tardos, É.: Polynomial time algorithms for some evacuation problems. In: Proceedings of the 5th Annual Symposium on Discrete Algorithms (SODA 1994), pp. 433–441 (1994)
10. Labetoulle, J., Lawler, E.L., Lenstra, J.K., Rinnooy Kan, A.H.G.: Preemptive scheduling of uniform machines subject to release dates. In: Progress in Combinatorial Optimization, pp. 245–261. Academic Press, London (1984)
11. Meilijson, I., Tamir, A.: Minimizing flow time on parallel identical processors with variable unit processing time. Operations Research 32(2), 440–448 (1984)
12. Minieka, E.: Maximal, lexicographic, and dynamic network flows. Operations Research 21, 517–527 (1973)
13. Queyranne, M., Schulz, A.S.: Scheduling unit jobs with compatible release dates on parallel machines with nonstationary speed. In: Balas, E., Clausen, J. (eds.) IPCO 1995. LNCS, vol. 920, pp. 307–320. Springer, Heidelberg (1995)
14. Schulz, A.S., Skutella, M.: The power of α-points in preemptive single machine scheduling. Journal of Scheduling, 121–133 (2002)
15. Smith, W.E.: Various optimizers for single-stage production. Naval Research and Logistics Quarterly, 59–66 (1956)
16. Stiller, S., Wiese, A.: Increasing speed scheduling and flow scheduling. Technical Report 007-2010, Technische Universität Berlin (February 2010)
17. Wilkinson, W.L.: An algorithm for universal maximal dynamic flows in a network. Operations Research 19, 1602–1612 (1971)
18. Zadeh, N.: A bad network problem for the simplex method and other minimum cost flow algorithms (1973)

A Tighter Analysis of Work Stealing

Marc Tchiboukdjian[1], Nicolas Gast[1], Denis Trystram[1],
Jean-Louis Roch[1], and Julien Bernard[2]

[1] Grenoble University
firstname.lastname@imag.fr
[2] Université de Franche-Comté
julien.bernard@lifc.univ-fcomte.fr

Abstract. Classical list scheduling is a very popular and efficient technique for scheduling jobs in parallel platforms. However, with the increasing number of processors, the cost for managing a single centralized list becomes prohibitive. The objective of this work is to study the extra cost that must be paid when the list is distributed among the processors. We present a general methodology for computing the expected makespan based on the analysis of an adequate potential function which represents the load unbalance between the local lists. A bound on the deviation from the mean is also derived. Then, we apply this technique to show that the expected makespan for scheduling W unit independent tasks on m processors is equal to W/m with an additional term in $3.65 \log_2 W$. Moreover, simulations show that our bound is very close to the exact value, approximately 50% off. This new analysis also enables to study the influence of the initial repartition of tasks and the reduction of the number of steals when several thieves can simultaneously steal work in the same processor's list.

1 Introduction

List scheduling is one of the most popular technique for scheduling the tasks of a parallel program. This algorithm has been introduced by Graham [1]. Its principle is to build a list of ready tasks and schedule them as soon as there exist available resources. List schedules are low-cost (greedy) algorithms that are not too far from optimal solutions. Most proposed list algorithms always consider a centralized management of the list. However, today parallel platforms involve more and more processors. Thus, the time needed for managing such a centralized data structure can not be ignored anymore [2]. Practically, implementing such schedulers induces synchronization overheads when several processors access the list concurrently. Such overheads involve low level synchronization mechanisms.

A suitable approach to reduce the contention is to distribute the list among the processors: each processor manages its own list of tasks. When a processor becomes idle, it randomly chooses another processor and steals some work, *i.e.* it transfers some tasks from the victim's list to its own list. Such a strategy is called *work-stealing* (WS). WS has been implemented in many languages and parallel libraries including Cilk [3], Intel TBB [4] and KAAPI [5].

O. Cheong, K.-Y. Chwa, and K. Park (Eds.): ISAAC 2010, Part II, LNCS 6507, pp. 291–302, 2010.
© Springer-Verlag Berlin Heidelberg 2010

Related works. WS has been analyzed in a seminal paper of Blumofe and Leiserson [6] where they show that the expected makespan of series-parallel precedence graphs with unit tasks is bounded by $\mathbb{E}\left[C_{\max}\right] \leq W/m + O(T_\infty)$ where W is the number of tasks, T_∞ is the critical path of the graph and m is the number of processors. This analysis has been improved in [7] using a proof based on a potential function. However the precedence graph is constrained to have only one source and out-degree at most 2 which does not model the basic case of independent tasks. Simulating independent tasks with a binary tree of dependencies gives a bound of $W/m + O(\log W)$ as a complete binary tree of W nodes has a depth of $T_\infty \leq \log_2 W$. Our new analysis allows to directly devise a result for the independent tasks case.

Notice that there exist other ways to analyze work stealing where the work generation is probabilistic and that target steady state results [8,9,10,11].

Our analysis shows some similarities with the work of Berenbrink *et al.* [12]. It is based on computing the expected decrease of a potential function. However, to simplify the analysis, we introduce an adversary that controls one parameter of the model, the number of steal requests at each time step.

Contributions. We present a new methodology for studying distributed list scheduling algorithms. The bound obtained in [7] is asymptotically optimal up to a constant factor but their analysis shows big constant factors. Based on the analysis of the load balancing between two processors during a steal request, we compute the expected number of steal requests and a bound on the makespan is derived. Thanks to our analysis, the constant factors are greatly reduced and are less than 50% away from values obtained by simulation. Moreover, our analysis enables to evaluate the impact of two modifications of the WS algorithm : the influence of the initial repartition of tasks and the reduction of the number of steal requests when several thieves can simultaneously steal work in the same processors list.

Roadmap. After presenting the model and notations in Section 2, we give the principle of the analysis in Section 3. We apply this analysis to the case of unit independent tasks and study the influence of the initial repartition of tasks in Section 4. Section 5 quantifies the reduction of steals when several thieves are allowed to steal the same victim simultaneously. We analyze simulation results in Section 6.

2   Model of the Distributed List

In this section, we give properties a distributed list implementation should follow.

We consider a parallel platform composed of m identical processors. At time t, let $w_i(t)$ represent the amount of work on processor i. When $w_i(t) > 0$, processor i is active and executes some work: $w_i(t+1) \leq w_i(t)$. When $w_i(t) = 0$, processor i is idle and intends to steal a random processor j. If processor j has no work, *i.e.* $w_j(t) = 0$, the steal fails and processor i will steal again at the next time slot. Otherwise, a certain amount of work is transfered from processor j to processor

i: $w_i(t+1) + w_j(t+1) \leq w_j(t)$. Processor i will resume execution at time $t+1$. The execution terminates when all the processors are idle, *i.e.* $\forall i, w_i(t) = 0$. We also denote the total amount of work on all processors by $w(t) = \sum_{i=1}^{m} w_i(t)$ and the number of active processors by $\alpha(t) \in [0, m]$. Thus, between time t and $t+1$, there are $m - \alpha(t)$ steal requests. Notice that when $\alpha(t) = 0$, all queues are empty and thus the execution is complete.

To model the contention on the queues, no more than one steal request per processor can succeed in the same time slot. If several requests target the same processor, a random one succeeds and all the others fail. This assumption will be relaxed in Section 5.

This is a high level model of a distributed list. We will show in Section 4 how these properties accurately model the case of independent tasks. We justify here some choices of this model. There is no explicit communication cost as WS algorithms most often target shared memory platforms. A steal request is done in constant time independently of the amount of work transfered. This assumption is not restrictive: for the case of independent tasks, the description of a large number of tasks can be very short. For instance a whole subpart of an array of tasks can be represented in a compact way by the range of the corresponding indices, each cell containing the effective description of a task (a STL transform in [13]). For more general cases with dependencies, it is usually enough to transfer a task which represents a part of the graph [7].

3   Principle of the Analysis and Main Theorem

This section presents the principle of the analysis. The main result is Theorem 1 that gives bounds on the expectation of the steal requests done by the schedule as well as the probability that the number of requests exceeds this bound.

The main idea of our analysis is that we study the decrease of a potential function $\Phi(t)$, instead of studying directly the number of processors that will run out of work and become idle. The definition of $\Phi(t)$ varies depending on the scenario (see Sections 4 to 5). The diminution of the potential depends on the number of steal requests, $m - \alpha(t)$. Since $\alpha(t)$ is a complicated random process, we tackle this problem by assuming that an adversary is choosing the number of active processors $\alpha(t)$ at each step of the schedule. At the beginning, the adversary starts with $\Phi(0)$ potential. Each time, when she chooses its $\alpha(t)$, $m - \alpha(t)$ steal requests are generated and diminishes the potential. The more work requests she creates, the more the potential decreases. She tries to maximize the number of steal requests before running out of potential.

In the actual system, $\alpha(t)$ is determined by the evolution of the system and cannot be chosen at time t. The introduction of an adversary provides an upper-bound on the number of steal requests and has two main advantages. First, its simplicity makes it applicable in several scenarios, such as the ones presented in Sections 4 to 5. Moreover, we show in Section 6 that the gap between the obtained bound and the values obtained by simulation is small.

The analysis of the scenarios of sections 4 to 5 will be done in three steps.

1. First, we define a potential function and we compute the potential decrease $\delta_i^k(t)$ when the processor i receives k work requests.
2. Then we compute the expected decrease of the potential between step t and $t + 1$, $\Delta\Phi(t) \stackrel{\text{def}}{=} \Phi(t) - \Phi(t + 1)$. By linearity of expectation,

$$\mathbb{E}\left[\Delta\Phi(t)\right] = \sum_{i=1}^{m} \sum_{k=0}^{m-1} \mathbb{E}\left[\delta_i^k | i \text{ receives } k \text{ requests}\right] \mathbb{P}\left\{i \text{ receives } k \text{ requests}\right\},$$

where $\mathbb{E}\left[X|Y\right]$ denotes the expectation of X knowing Y. Using the properties of $\delta_i^k(t)$, we show that there exists a function $h(\alpha) \in (0; 1]$ such that

$$\mathbb{E}\left[\Delta\Phi(t)|\Phi(t) = \Phi, \alpha(t) = \alpha\right] \geq h(\alpha)\Phi$$

3. Finally, we obtain a bound on the expected number of steal requests $\mathbb{E}\left[R\right]$ using Theorem 1 presented in this section. An upper bound on the expected makespan $\mathbb{E}\left[C_{\max}\right]$ can be obtained using the bound on the number of steal.

The following theorem gives an upper bound on the number of steal requests using a lower bound on the expected decrease of the potential in one step.

Theorem 1. *Assume that the potential function $\Phi(t)$ satisfies:*

- *There exists a constant $d > 0$ such that $d\Phi(t) \in \mathbb{N}$.*
- *$\Phi(t)$ is non-increasing.*
- *There exists a function $h(\alpha) \in (0, 1]$ such that if $\alpha \in [1, m-1]$ processors are active at time t and $\Phi(t) = \Phi$, then the potential decreases on average by*
$$\mathbb{E}\left[\Phi(t) - \Phi(t + 1)|\Phi(t) = \Phi, \alpha(t) = \alpha\right] \geq h(\alpha) \cdot \Phi.$$

Let $\lambda = \max_{1 \leq \alpha \leq m-1} \frac{m-\alpha}{-m \log_2(1-h(\alpha))}$ and $\Phi(0)$ be the potential at $t = 0$. Then

(i) the expected number of steal requests R until $\Phi(t) \leq 1$ is bounded by
$$\mathbb{E}\left[R\right] \leq \lambda \cdot m \cdot \log_2 \Phi(0);$$

(ii) The deviation from the mean can be bounded by:
$$\mathbb{P}\left\{R \geq \lambda \cdot m \cdot \log_2 \Phi(0) + u\right\} \leq 2^{-u/(\lambda m)}$$

Proof. Without loss of generality and to simplify the notation, we assume $d = 1$.

Let T be the random variable indicating the end of the schedule: $T = \min\{t|\Phi(t) \leq 1\}$. The number of steal requests is equal to the number of idle processors at each time step. The number of steal requests *after* time t is $R(t)$:

$$R(t) = \sum_{s=t}^{T-1} m - \alpha(s)$$

The total number of steals is $R \stackrel{\text{def}}{=} R(0)$.

The sequence $\alpha(t)$ is difficult to study since it depends on the number of processors at time $t - 1$ with 0 or 1 tasks, but also the successful or unsuccessful steals. Therefore, we perform the analysis assuming a worst-case scenario: at each time t, an adversary can choose $\alpha(t)$ knowing the history of the system but not the future random choices. This can be seen as a Markov decision process with total reward criteria, see [14] for more details about Markov decision processes.

We prove by induction on Φ that for all t,

$$\mathbb{E}\left[R(t)|\Phi(t) = \Phi\right] \leq \lambda m \log_2(\Phi)$$

For $\Phi = 1$, this is clearly true since in that case $T \leq t$ and $R(t) = 0$. Assume that (3) holds for all t and all $\phi < \Phi$. $\mathbb{E}\left[R(t)|\Phi(t) = \Phi\right]$ is equal to:

$$
\begin{aligned}
\mathbb{E}\left[R(t)|\Phi(t) = \Phi\right] &= \mathbb{E}\left[m - \alpha(t) + R(t+1)|\Phi(t) = \Phi\right] \\
&= m - \alpha(t) + \mathbb{E}\left[R(t+1)|\Phi(t) = \Phi\right]
\end{aligned}
\tag{1}
$$

By definition of $\Delta\Phi(t)$, if $\Phi(t) = \Phi$, then $\Phi(t+1) = \Phi - \phi$ with probability $\mathbb{P}\left\{\Delta\Phi(t) = \phi|\Phi(t) = \Phi\right\}$. Since $\Phi(t)$ is non-increasing, $\Delta\Phi(t) \geq 0$. Therefore:

$$\mathbb{E}\left[R(t+1)|\Phi(t)=\Phi\right] = \sum_{\phi=0}^{\Phi} \mathbb{E}\left[R(t+1)|\Phi(t+1)=\Phi - \phi\right] \mathbb{P}\left\{\Delta\Phi(t) = \phi|\Phi(t)=\Phi\right\}.$$

Let us denote $p_0 \stackrel{\text{def}}{=} \mathbb{P}\left\{\Delta\Phi(t) = 0|\Phi(t) = \Phi\right\}$. Using the induction hypothesis, and the fact that $\mathbb{E}\left[R(t+1)|\Phi(t+1) = \Phi\right] = \mathbb{E}\left[R(t)|\Phi(t) = \Phi\right]$, we get from (1)

$$(1-p_0)\mathbb{E}\left[R(t)|\Phi(t)=\Phi\right] \leq m - \alpha(t) + \sum_{\phi=1}^{\Phi} \lambda m \log_2(\Phi - \phi)\mathbb{P}\left\{\Delta\Phi(t)=\phi|\Phi(t)=\Phi\right\}$$

$$= m - \alpha(t) + \lambda m \mathbb{E}\left[\log_2(\Phi - \Delta\Phi(t))|\Phi(t)=\Phi\right] - \lambda m \log_2(\Phi)p_0 \tag{2}$$

where we used the fact that $\sum_{\phi=1}^{\Phi}(\ldots) = \sum_{\phi=0}^{\Phi}(\ldots) - \lambda m \log_2(\Phi)p_0$.

Moreover, by Jensen's inequality (log is concave), we have:

$$
\begin{aligned}
\mathbb{E}\left[\log_2(\Phi - \Delta\Phi(t))|\Phi(t) = \Phi\right] &\leq \log_2(\Phi - \mathbb{E}\left[\Delta\Phi(t)|\Phi(t) = \Phi\right]) \\
&\leq \log_2(\Phi - h(\alpha(t))\Phi)
\end{aligned}
\tag{3}
$$

Combining equations (2) and (3), we get:

$$(1 - p_0)\mathbb{E}\left[R(t)|\Phi(t) = \Phi\right] \leq (1 - p_0)\lambda m \log_2(\Phi) + m - \alpha(t) + \lambda m \log_2(1 - h(\alpha(t))).$$

If $\alpha(t) = m$, the sum of the two last terms of the equation is negative since $1 - h(\alpha) \leq 1$. If $\alpha(t) = 0$, the schedule is finished. If $0 < \alpha(t) < m$, the sum of the two last terms is negative by definition of λ (λ corresponds to the worst choice of $\alpha(t)$). Dividing on both sides by $1 - p_0$ concludes the proof of (i).

The proof of (ii) is quite similar to the proof of (i). We prove by induction on Φ that $\mathbb{E}\left[2^{R(t)/(\lambda m) - \log_2(\Phi)}|\Phi(t) = \Phi\right] \leq 1$. It clearly holds for $\Phi = 1$ since in that case it is equal to 0. $\mathbb{E}\left[2^{R(t)/(\lambda m) - \log_2(\Phi)}|\Phi(t) = \Phi\right]$ is equal to

$$\sum_{\phi=0}^{\Phi} \mathbb{E}\left[2^{R(t)/(\lambda m) - \log_2(\Phi)}|\Phi(t+1)=\Phi - \phi\right] \mathbb{P}\left\{\Delta\Phi(t)=\phi|\Phi(t)=\Phi\right\}$$

Using $R(t+1) = m - \alpha + R(t)$ and introducing $\log_2(\Phi - \phi) - \log_2(\Phi - \phi)$, this equals

$$\sum_{\phi=1}^{\Phi} 2^{\frac{m-\alpha}{\lambda m} + \log_2(1 - \frac{\phi}{\Phi})} \mathbb{E}\left[2^{\frac{R(t+1)}{\lambda m} - \log_2(\Phi - \phi)}|\Phi(t+1)=\Phi-\phi\right] \mathbb{P}\left\{\Delta\Phi(t)=\phi|\Phi(t)=\Phi\right\}$$

$$+ 2^{\frac{m-\alpha}{\lambda m}} \mathbb{E}\left[2^{\frac{R(t+1)}{\lambda m} - \log_2(\Phi)}|\Phi(t+1)=\Phi\right] p_0$$

$$\leq \sum_{\phi=1}^{\Phi} 2^{\frac{m-\alpha}{\lambda m}+\log_2(1-\frac{\phi}{\Phi})} \mathbb{P}\left\{\Delta\Phi(t)=\phi|\Phi(t)=\Phi\right\} + 2^{\frac{m-\alpha}{\lambda m}} \mathbb{E}\left[2^{\frac{R(t)}{\lambda m}-\log_2(\Phi)}|\Phi(t)=\Phi\right] p_0$$

where we used the induction hypothesis for the inequality.

Then, adding and subtracting the first term of the sum $\sum_{\phi=1}^{\Phi}$, this leads to

$$(1 - 2^{\frac{m-\alpha}{\lambda m}} p_0)\mathbb{E}\left[2^{\frac{R(t)}{\lambda m}-\log_2(\Phi)}|\Phi(t)=\Phi\right] \leq 2^{\frac{m-\alpha}{\lambda m}} \mathbb{E}\left[1 - \Delta\Phi/\Phi|\Phi(t)=\Phi\right] - 2^{\frac{m-\alpha}{\lambda m}} p_0$$

$$\leq 2^{\frac{m-\alpha}{\lambda m}+\log_2(1-h(\alpha))} - 2^{\frac{m-\alpha}{\lambda m}} p_0$$

$$\leq 1 - 2^{\frac{m-\alpha}{\lambda m}} p_0.$$

where we used the definition of λ to show that the first term is less than one. This shows that $\mathbb{E}\left[2^{\frac{R(t)}{\lambda m}-\log_2(\Phi)}|\Phi(t)=\Phi\right] \leq 1$. Therefore by Markov's inequality:

$$\mathbb{P}\left\{R(t) \geq \lambda m \log_2 \Phi + u|\Phi(t)=\Phi\right\} = \mathbb{P}\left\{2^{\frac{R(t)}{\lambda m}-\log_2 \Phi} \geq 2^{\frac{u}{\lambda m}}|\Phi(t)=\Phi\right\} \leq 2^{-\frac{u}{\lambda m}}.$$

4 Unit Independent Tasks

We apply the analysis presented in the previous section for the case of independent unit tasks. In this case, each processor i maintains a local queue Q_i of tasks to execute. At every time slot, if the local queue Q_i is not empty, processor i picks a task and executes it. When Q_i is empty, processor i sends a steal request to a random processor j. If Q_j is empty or contains only one task (currently executed by processor j), then the request fails and processor i will have to send a new request at the next slot. If Q_j contains more than one task, then i is given half of the tasks (after that the task executed at time t by processor j has been removed from Q_j). The amount of work on processor i at time t, $w_i(t)$, is the number of tasks in $Q_i(t)$. At the beginning of the execution, $w(0) = W$ and tasks can be arbitrarily spread among the queues.

Applying the method presented in Section 3, the first step of the analysis is to define the potential function and compute the potential decrease when a steal occurs. For this example, $\Phi(t)$ is defined by:

$$\Phi(t) = \sum_{i=1}^{m}\left(w_i(t) - \frac{w(t)}{m}\right)^2 = \sum_{i=1}^{m} w_i(t)^2 - \frac{w^2(t)}{m}.$$

This potential represents the load unbalance in the system. If all queues have the same $w_i(t) = w(t)/m$, then $\Phi(t) = 0$. $\Phi(t) \leq 1$ implies that there is at most one processor with at most one more task than the others. In that case, there will be no steal until there is just one processor with 1 task and all others idle. Moreover, the potential function is maximal when all the work is concentrated on a single queue. That is $\Phi(t) \leq w(t)^2 - w(t)^2/m \leq (1 - 1/m)w^2(t)$.

Assume that at time t, the queue i has $w_i(t) \geq 1$ tasks. If it receives one or more steal requests, it chooses a processor j among the thieves. At time $t + 1$, i has executed one task and the rest of the work is split between i and j.

Therefore, $w_i(t+1) = \lceil (w_i(t) - 1)/2 \rceil$ and $w_j(t+1) = \lfloor (w_i(t) - 1)/2 \rfloor$. Thus $w_i(t+1)^2 + w_j(t+1)^2 = \lceil (w_i(t)-1)/2 \rceil^2 + \lfloor (w_i(t)-1)/2 \rfloor^2 \le w_i(t)^2/2 - w_i(t) + 1$. Therefore, this generates a difference of potential of

$$\delta_i^k(t) = \delta_i^1(t) \ge w_i(t)^2/2 + w_i(t) - 1. \tag{4}$$

If i receives zero steal requests, it potential goes from $w_i(t)^2$ to $(w_i(t) - 1)^2$, generating a potential decrease of $2w_i(t) - 1$. The last event contributing to the change of the potential is that $(\sum_{i=1}^m w_i(t))^2/m$ goes from $w(t)^2/m$ to $w(t+1)^2 = (w(t) - \alpha(t))^2/m$, generating a potential increase of $2\alpha(t)w(t)/m - \alpha(t)^2/m$.

Recall that at time t, there are $\alpha(t)$ active processors and therefore $m - \alpha(t)$ processors that send steal requests. A processor i receives zero steal requests if the $m - \alpha(t)$ thieves choose another processor. Each of these events is independent and happens with probability $(m - 2)/(m - 1)$. Therefore, the probability for the processor to receive one or more steal requests is $p_r(\alpha(t))$:

$$p_r(\alpha(t)) = 1 - \left(1 - \frac{1}{m-1} \right)^{m-\alpha(t)}.$$

If $\Phi(t)=\Phi$ and $\alpha(t)=\alpha$, by summing the expected decrease on each active processor δ_i^1, the expected potential decrease is greater than:

$$\sum_{i/w_i(t)>0} \left[p_r(\alpha) \left(\frac{w_i(t)^2}{2} + w_i(t) - 1 \right) + (1 - p_r(\alpha))(2w_i(t) - 1) \right] - 2w(t)\frac{\alpha}{m} + \frac{\alpha^2}{m}$$

$$= \frac{p_r(\alpha)}{2} \Phi + \frac{p_r(\alpha)}{2} \left(\frac{w(t)^2}{m} - 2w(t) + 2\frac{m-\alpha}{mp_r(\alpha)} (2w(t) - \alpha) \right) \tag{5}$$

$$\ge \frac{p_r(\alpha)}{2} \left(\Phi + \frac{w(t)^2}{m} - 2\frac{w(t)}{m} + 2\left(1 - \frac{1}{m}\right)(w(t) - \alpha) \right) \ge \frac{p_r(\alpha)}{2} \Phi$$

The details of the computation of (5) can be found in Appendix A.

Using Theorem 1 of the previous section, we conclude the analysis by the following theorem.

Theorem 2. *Let C_{max} be the makespan of W unit independent tasks scheduled by work stealing. Then:*

(i) $\mathbb{E}[C_{max}] \le \dfrac{W}{m} + \dfrac{2}{1 - \log_2(1 + \frac{1}{e})} \cdot \log_2 W + 1$

(ii) $\mathbb{P}\left\{ C_{max} \ge \dfrac{W}{m} + \dfrac{2}{1 - \log_2(1 + \frac{1}{e})} \cdot \left(\log_2 W + \log_2 \dfrac{1}{\epsilon} \right) + 1 \right\} \le \epsilon$

These bounds are optimal up to a constant factor in $\log_2 W$.

Proof. Using Equation (5), $\mathbb{E}[\Delta\Phi(t)|\Phi(t) = \phi, \alpha(t) = \alpha] \le h(\alpha)\Phi$, with $h(\alpha) = p_r(\alpha)/2$. Using Theorem 1 (i) and the fact that $\Phi(0) \le W^2$, the expected number of steal requests before $\Phi(t) \le 1$ is bounded by:

$$\mathbb{E}[R] \le \lambda m \log_2(W^2) = 2\lambda m \log_2(W),$$

with $\lambda = \max_{1 \le \alpha \le m-1}(m-\alpha)/(-m \log_2(1-h(\alpha)))$. A direct computation shows that $(m - \alpha)/(-m \log_2(1 - h(\alpha)))$ is decreasing in α. Therefore, its minimum is attained for $\alpha = 1$. This shows that $\lambda \le 1/(1 - \log_2(1 + \frac{1}{e}))$.

As said before, when $\Phi(t) \leq 1$, there is at most one processor with at least one more task than the others. Therefore, there will be a steal request only when this processor will have one task and the others zero. This happens only once and generates at most $m - 1$ steal requests.

At each time step of the schedule, a processor is either computing one task or stealing work. This shows that $m \cdot C_{\max} = W + R$. Thus:

$$\mathbb{E}\left[C_{\max}\right] \leq \frac{W}{m} + \frac{2}{1 - \log_2(1 + \frac{1}{e})} \log_2 W + 1$$

The proof of the *(i)* applies *mutatis mutandis* to prove the bound in probability *(ii)* using Theorem 1 *(ii)*.

We now give a lower bound for this problem. Consider $W = 2^{k+1}$ tasks and $m = 2^k$ processors, all the tasks being on the same processor at the beginning. In the best case, all steal requests target processors with highest loads. In this case the makespan is $C_{\max} = k + 2$: the first $k = \log_2 m$ steps for each processor to get some work; one step where all processors are active; and one last step where only one processor is active. In that case, $C_{\max} \geq \frac{W}{m} + \log_2 W - 1$.

This theorem shows that the factor before $\log_2 W$ is bounded by 1 and $2/(1 - \log_2(1 + 1/e)) < 3.65$. Simulations reported in Section 6 seem to indicate that the factor of $\log_2(W)$ is slightly less. This shows that the constants obtained by our analysis are sharp.

Initial repartition of tasks. In the worst case, all tasks are in the same queue at the beginning of the execution. Using bounds in terms of Φ_0, one can show that a more balanced initial repartition leads to fewer steal requests on average. Suppose that we take a balls-and-bins assignment as the initial repartition: for each task, we choose a processor at random and put the task in its queue. The expected value of Φ_0 is:

$$\mathbb{E}\left[\Phi_0\right] = \sum_i \mathbb{E}\left[w_i^2\right] - \frac{W^2}{m} = \sum_i \left(\mathrm{Var}\left[w_i\right] + \mathbb{E}\left[w_i\right]^2\right) - \frac{W^2}{m} = \left(1 - \frac{1}{m}\right) \cdot W$$

as w_i follows a binomial distribution. Since the number of work requests is proportional to $\log_2 \Phi_0$, this initial distribution of tasks reduces the number of steal requests by a factor of 2 on average.

5   Cooperation among Thieves

In this section, we modify the protocol for managing the distributed list. Previously, when $k > 1$ steal requests were sent on the same processor, only one of them could be served due to contention on the list. We now allow the k requests to be served in unit time. This model has been implemented in the middleware Kaapi [5]. When k steal requests target the same processor, the work is divided into $k + 1$ pieces. In practice, allowing concurrent thieves increase the cost of a steal request but we neglect this additional cost here. We assume that the k concurrent steal requests can be served in unit time. We study the influence of this new protocol on the number of steal requests in the case of unit independent tasks.

We use the potential function[1] $\Phi(t) = \sum_{i=1}^{m} w_i(t)^2$. Let us first compute the decrease of the potential when processor i receives $k \geq 1$ steal requests. If $w_i(t) > 0$, it can be written $w_i(t) = (k+1)q + r + 1$ with $0 \leq r < k+1$. After one time step and k steal requests, the work will be divided in r parts with $q+1$ tasks and $k+1-r$ parts with q tasks. By a direct computation, the potential generated by these steal requests at time $t+1$ can be bounded by:

$$r(q+1)^2 + (k+1-r)q^2 = (k+1)q^2 + r(2q+1) \leq \frac{1}{k+1}\left((k+1)q + r\right)^2 \leq \frac{w_i(t)^2}{k+1}.$$

If $m - \alpha$ processors send steal requests, the probability for an active processor to receive k steal requests is

$$p_k(\alpha) = \binom{m-\alpha}{k} \frac{1}{(m-1)^k} \left(\frac{m-2}{m-1}\right)^{m-\alpha-k}$$

The expected diminution of the potential caused by the steals on processor i is equal to $\sum_{k=0}^{m-\alpha} \delta_i^k p_k(\alpha)$. By a direct computation, this is bounded by

$$\sum_{k=0}^{m-\alpha} \delta_i^k p_k(\alpha) \geq \sum_{k=0}^{m-\alpha} \left(1 - \frac{1}{k+1}\right) w_i(t)^2 p_k(\alpha)$$

$$= w_i(t)^2 \left(1 - \frac{m-1}{m-\alpha+1}\left(1 - \left(\frac{m-2}{m-1}\right)^{m-\alpha+1}\right)\right)$$

This shows that $\mathbb{E}\left[\Delta\Phi(t)|\Phi(t) = \Phi|\alpha(t) = \alpha\right] \leq h(\alpha)\Phi$ where

$$h(\alpha) = 1 - \frac{m-1}{m-\alpha+1}\left(1 - \left(\frac{m-2}{m-1}\right)^{m-\alpha+1}\right)$$

Deriving with respect to α shows that $(m-\alpha)/-\log_2(1-h(\alpha))$ is decreasing. Thus $\lambda = \max_{1 \leq \alpha \leq m}(m-\alpha)/-m\log_2(1-h(\alpha)) = (m-1)/-m\log_2(1-h(1))$. A direct computation shows that $\lambda \leq 1/-\log_2(1-1/e)$. Therefore we can copy *mutatis mutandis* the proof of Theorem 2 to show that:

Theorem 3. *The makespan $C_{\max}^{\text{coop}}$ of W unit independent tasks scheduled with cooperative work stealing satisfies:*

(i) $\mathbb{E}\left[C_{\max}^{\text{coop}}\right] \leq \dfrac{W}{m} + \dfrac{2}{-\log_2(1-\frac{1}{e})} \cdot \log_2 W + 1$

(ii) $\mathbb{P}\left\{C_{\max}^{\text{coop}} \geq \dfrac{W}{m} + \dfrac{2}{-\log_2(1-\frac{1}{e})} \cdot \log_2 W + 1 \geq \dfrac{2}{-\log_2(1-\frac{1}{e})}\log_2(\epsilon)\right\} \leq \epsilon$

Compared to the situation with no cooperation among thieves, the number of steal requests is reduced by a factor $\frac{1 - \log_2(1+1/e)}{-\log_2(1-1/e)} \approx 1.20$. We will see in Section 6 that this is close to the value obtained by simulation.

---

[1] The same potential function as in Section 4 could be used but leads to more complex computations.

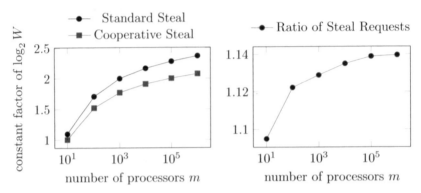

Fig. 1. (Left) Constant factor of $\log_2 W$ against the number of processors for the standard steal and the cooperative steal. (Right) Ratio of steal requests (standard/cooperative).

6   Experimental Study

Theorem 1 provides upper bound on the expected value of the makespan for the models considered in Sections 4 and 5. In this section, we experimentally study the constant factor of the $\log_2 W$ term and show that it is close to the theoretical result.

We developed a simulator that strictly follows the model of Sections 4 and 5. At the beginning, all the tasks are given to processor 0 in order to be in the worst case, *i.e.* when the initial potential Φ_0 is maximum. Each pair (m, W) is simulated 10000 to get accurate results, with a coefficient of variation about 2%.

We computed the constant factor 2λ of the $\log_2 W$ term for various number of processors and tasks. The value goes to a limit between 2 and 3 (cf. Fig. 1). This gives a constant $2\lambda \approx 2.37$ for unit independent tasks with standard steal and $2\lambda_{coop} \approx 2.08$ for unit independent tasks with cooperative steal. The theoretical values of 3.65 (standard steal) and 3.02 (cooperative steal) are close, only 50% greater than the simulation values. The difference can be explained by the use of an adversary in Theorem 1. Moreover, the analysis is fine enough to predict the advantage of the cooperative steal with a gain of 20% over the standard steal, close to the experimental gain of 14%.

7   Concluding Remarks

We have presented in this paper a new analysis of work stealing. The main result is to prove that the expected makespan to schedule a workload of W on m processors is no more than the best possible absolute lower bound W/m plus an additive term in $3.65 \log_2 W$ very close to the value obtained by simulation.

References

1. Graham, R.L.: Bounds on multiprocessing timing anomalies. SIAM Journal on Applied Mathematics 17, 416–429 (1969)
2. Hoffmann, R., Korch, M., Rauber, T.: Performance evaluation of task pools based on hardware synchronization. In: Proc. of Supercomputing (2004)
3. Frigo, M., Leiserson, C.E., Randall, K.H.: The implementation of the Cilk-5 multithreaded language. In: Proceedings of PLDI (1998)
4. Robison, A., Voss, M., Kukanov, A.: Optimization via reflection on work stealing in TBB. In: Proceedings of IPDPS, pp. 1–8 (2008)
5. Gautier, T., Besseron, X., Pigeon, L.: KAAPI: A thread scheduling runtime system for data flow computations on cluster of multi-processors. In: Proceedings of PASCO, pp. 15–23 (2007)
6. Blumofe, R.D., Leiserson, C.E.: Scheduling multithreaded computations by work stealing. Journal of the ACM 46(5), 720–748 (1999)
7. Arora, N.S., Blumofe, R.D., Plaxton, C.G.: Thread scheduling for multiprogrammed multiprocessors. Theory of Computing Systems 34(2), 115–144 (2001)
8. Berenbrink, P., Friedetzky, T., Goldberg, L.A.: The natural work-stealing algorithm is stable. SIAM Journal of Computing 32(5), 1260–1279 (2003)
9. Mitzenmacher, M.: Analyses of load stealing models based on differential equations. In: Proceedings of SPAA, pp. 212–221 (1998)
10. Hendler, D., Shavit, N.: Non-blocking steal-half work queues. In: Proceedings of PODC (2002)
11. Gast, N., Gaujal, B.: A Mean Field Model of Work Stealing in Large-Scale Systems. In: Proceedings of SIGMETRICS (2010)
12. Berenbrink, P., Friedetzky, T., Goldberg, L.A., Goldberg, P.W., Hu, Z., Martin, R.: Distributed selfish load balancing. SIAM Journal on Computing 37(4) (2007)
13. Traoré, D., Roch, J.L., Maillard, N., Gautier, T., Bernard, J.: Deque-free workoptimal parallel STL algorithms. In: Luque, E., Margalef, T., Benítez, D. (eds.) Euro-Par 2008. LNCS, vol. 5168, pp. 887–897. Springer, Heidelberg (2008)
14. Puterman, M.L.: Markov Decision Processes : Discrete Stochastic Dynamic Programming. Wiley, Chichester (2005)

A   Proof of Inequality (5) of Theorem 2

Recall that the expectation of the potential decrease is greater than:

$$\sum_{i/w_i(t)>0} \left[p_r(\alpha) \left(\frac{w_i(t)^2}{2} + w_i(t) - 1 \right) + (1 - p_r(\alpha))(2w_i(t) - 1) \right] - 2w(t)\frac{\alpha}{m} + \frac{\alpha^2}{m}$$

$$= \frac{p_r(\alpha)}{2} \left(\sum w_i(t)^2 - \frac{w(t)^2}{m} \right) + p_r(\alpha) \left(\frac{w(t)^2}{2m} + w(t) - \alpha \right)$$

$$+ (1 - p_r(\alpha))(2w(t) - \alpha) - 2w(t)\frac{\alpha}{m} + \frac{\alpha^2}{m}$$

where we used that $\sum_{i/w_i(t)>0} w_i(t) = w(t)$ and that $\sum_{i/w_i(t)>0} 1 = \alpha$ since α is the number of active processors. A direct computation shows that this is equal to

$$\frac{p_r(\alpha)}{2}\Phi + p_r(\alpha)\left(\frac{w(t)^2}{2m} + w(t) - \alpha - 2w(t) + \alpha\right) + 2w(t) - \alpha - 2w(t)\frac{\alpha}{m} + \frac{\alpha^2}{m}$$

$$= \frac{p_r(\alpha)}{2}\Phi + \frac{p_r(\alpha)}{2}\left(\frac{w(t)^2}{m} - 2w(t) + \frac{2}{p_r(\alpha)}\left(1 - \frac{\alpha}{m}\right)(2w(t) - \alpha)\right)$$

$$= \frac{p_r(\alpha)}{2}\Phi + \frac{p_r(\alpha)}{2}\left(\frac{w(t)^2}{m} - 2w(t) + \frac{2}{m}\frac{m-\alpha}{p_r(\alpha)}(2w(t) - \alpha)\right). \qquad (6)$$

Let define f by $f(\alpha) \overset{\text{def}}{=} (m - \alpha)/p_r(\alpha)$. A direct computation shows that the derivative f' is negative. Thus, f is non increasing for $1 \le \alpha \le m - 1$ and $\min_{1\le\alpha\le m-1} f(\alpha) = f(m-1) = m - 1$. Therefore, (6) is greater than:

$$\frac{p_r(\alpha)}{2}\Phi + \frac{p_r(\alpha)}{2}\left(\frac{w(t)^2}{m} - 2w(t) + 2\frac{m-1}{m}(2w(t) - \alpha)\right)$$

$$= \frac{p_r(\alpha)}{2}\Phi + \frac{p_r(\alpha)}{2}\left(\frac{w(t)^2}{m} - 2w(t) + 2w(t)\left(1 - \frac{1}{m}\right) + 2\left(1 - \frac{1}{m}\right)(w(t) - \alpha)\right)$$

$$= \frac{p_r(\alpha)}{2}\Phi + \frac{p_r(\alpha)}{2}\left(\frac{w(t)^2}{m} - \frac{2w(t)}{m} + 2\left(1 - \frac{1}{m}\right)(w(t) - \alpha)\right).$$

As $w(t) - \alpha(t) \ge 0$ (an active processor has at least one task) the last term is positive. Moreover, for all $w(t) > 1$, the second term is positive. Thus, this is greater than $\frac{p_r(\alpha)}{2}\Phi$ which concludes the proof of the inequality (5).

B   Computation of λ for the Unit Tasks

In this section, we compute the constant λ for the unit tasks. We first show that the quantity $(m - \alpha)/(- \log_2(1 - p_r(\alpha)/2))$ is decreasing in α. Then we bound the value $(m - 1)/(- \log_2(1 - p_r(1)/2))$ by $(m - 1)/(1 - \log_2(1 + 1/e))$.

Let $g(\alpha) \overset{\text{def}}{=} - \log_2(1 - p_r(\alpha)/2)$ and $f(\alpha) \overset{\text{def}}{=} (m - \alpha)/g(\alpha)$. By definition of $p_r(\alpha)$, $g(\alpha)$ can be written:

$$g(\alpha) = - \log_2\left(\frac{1}{2} + \frac{1}{2}\left(1 - \frac{1}{m-1}\right)^{m-\alpha}\right) = 1 - \log_2\left(1 + \left(1 - \frac{1}{m-1}\right)^{m-\alpha}\right).$$

Denoting $p \overset{\text{def}}{=} 1 - 1/(m - 1)$ and $x \overset{\text{def}}{=} p^{m-\alpha}$, the derivative of f w.r.t. α is:

$$f'(\alpha) = \frac{(1 + x)\ln(1 + x) - x\ln x - (1 + x)\ln 2}{(1 + x)g(\alpha)^2 \ln 2}$$

The derivative of $(1 + x)\ln(1 + x) - x\ln x - (1 + x)\ln 2$ w.r.t. x is $\ln(1 + x) - \ln(x) - \ln 2 = \ln(1 + 1/x) - \ln 2 > 0$. As $x < 1$, this shows that $(1 + x)\ln(1 + x) - x\ln x - (1 + x)\ln 2 < (1 + 1)\ln(1 + 1) - 1\ln 1 - (1 + 1)\ln 2 = 0$. This shows that $f(\alpha)$ is decreasing. Therefore, $\lambda = \max_{1\le\alpha\le m-1} \frac{1}{m}f(\alpha) = \frac{1}{m}f(1)$. Using the fact that for $m \ge 2$, $(1 - \frac{1}{m-1})^{m-1} \le 1/e$, we get that $1 - \log_2\left(1 + (1 - 1/(m - 1))^{m-1}\right) \ge 1 - \log_2(1 + 1/e)$ which shows that $\lambda \le \frac{1}{1-\log_2(1+1/e)}$.

Approximating the Traveling Tournament Problem with Maximum Tour Length 2

Clemens Thielen and Stephan Westphal

Department of Mathematics, University of Kaiserslautern, Paul-Ehrlich-Str. 14,
D-67663 Kaiserslautern, Germany
{thielen,westphal}@mathematik.uni-kl.de

Abstract. We consider the traveling tournament problem, which is a well-known benchmark problem in tournament timetabling. The most important variant of the problem imposes restrictions on the number of consecutive home games or away games a team may have. We consider the case where at most two consecutive home games or away games are allowed. We show that the well-known independent lower bound for this case cannot be reached and present an approximation algorithm that has an approximation ratio of $3/2 + \frac{6}{n-4}$, where n is the number of teams in the tournament. In the case that n is divisible by 4, this approximation ratio improves to $3/2 + \frac{5}{n-1}$.

Keywords: traveling tournament problem, timetabling, approximation algorithm.

1 Introduction

Professional sports leagues exist all over the world. Popular leagues are often of huge economic importance due to the enormous revenues generated by selling tickets and broadcasting rights for the games. Hence, the planning of these leagues is of major importance. An important aspect is the generation of a timetable for the tournaments that specifies the order in which the teams play each other during the season and the venue of each game. A well-studied variant of this problem is the traveling tournament problem (TTP), which was formally introduced by Easton et al.[1] in 2001. Given the number of teams and the pairwise distances between their home venues, TTP asks for a timetable of a double round robin tournament that minimizes the sum of the distances traveled by the teams during the season. This problem is quite important in practice, for example in the US, where the distances between two teams' home venues are often quite large, so minimizing travel distance becomes a major issue.

The variant of TTP most relevant in practice imposes restrictions on the number of consecutive home games or away games a team may have. The schedules of many major sports leagues, e.g., the Major League Baseball (MLB) in the US, contain such restrictions. The case that was studied most so far is that the number of consecutive home or away games is upper bounded by three. The case

O. Cheong, K.-Y. Chwa, and K. Park (Eds.): ISAAC 2010, Part II, LNCS 6507, pp. 303–314, 2010.
© Springer-Verlag Berlin Heidelberg 2010

TTP(2), where only two consecutive home or away games are allowed, was studied in a classical paper by Campbell and Chen [2], who scheduled a basketball conference of ten teams with a solution method based on matching techniques. Their method, however, only yields a *relaxed* tournament, which needs two time slots more than necessary. Moreover, no upper bound on the number of consecutive home games was considered and the schedules constructed by their method violate the upper bound of at most two consecutive away games for some teams.

We now formally define the traveling tournament problem (TTP) and introduce our notation. We are given a set of n teams, where $n \geq 4$ is even. An $(n \times n)$-*distance matrix* $D = (d_{ij})$ specifies the distances between the home venues of the teams, i.e., $d_{ij} \geq 0$ is the distance between the home venues of teams i and j. The distances are assumed to be symmetric (i.e., $d_{ij} = d_{ji}$ for all i, j) and satisfy $d_{ii} = 0$ for all i as well as the triangle inequality (i.e., $d_{ij} + d_{jk} \geq d_{ik}$ for all i, j, k). A *game* is an ordered pair of teams, where the first team is the home team and the second the away team. A sequence of consecutive away games of a team is called a *road trip*, and sequence of consecutive home games is called a *home stand*. A *double round robin tournament* is a collection of games in which every team plays every other team once at home and once away (i.e., at the other team's home venue). Hence, exactly $2n - 2$ time slots are necessary for a double round robin tournament. Before the tournament, each team is assumed to stay at its home venue and it has to return there after the tournament in case that its last game is an away game. Between two consecutive away games, a team travels directly from the venue of the first opponent to the venue of the second opponent. With this terminology, the traveling tournament problem for a positive integer $k \geq 2$ is defined as follows:

Definition 1 (The Traveling Tournament Problem (TTP(k)) [1])
INSTANCE: The set of n teams and the distance matrix $D = (d_{ij})$.
TASK: Compute a feasible double round robin tournament of the teams satisfying the following conditions:

> *(a) The length of any home stand is at most k.*
> *(b) The length of any road trip is at most k.*
> *(c) Game j at i is not followed immediately by game i at j.*
> *(d) The sum of the distances traveled by the teams is minimized.*

1.1   Previous Work

Since the proposal of TTP by Easton et al. [1], many approximation algorithms and heuristics have been designed for the problem (cf., for example, [3,4,5,6,7]). The first algorithm with a constant approximation ratio was the $(2 + (9/4)/(n-1))$-approximation algorithm for TTP(3) proposed by Miyashiro et al. [5]. Recently, Westphal and Noparlik [7] presented the first constant factor approximation for $k > 3$, which achieves an approximation ratio of at most $5 + 3/n + 3/(2k)$. The only approximation results on TTP(2) we are aware of are the ones due to Campbell and Chen [2] already mentioned above. The complexity of TTP was recently settled in [8] by showing that the problem is strongly

NP-hard to solve. A modified version of TTP without restrictions on the number of consecutive home games or away games was shown to be strongly NP-hard in [9]. Surveys on round robin scheduling and TTP can be found in [10,11].

1.2   Our Contribution

We show that the independent lower bound for TTP(2) obtained by matching techniques in [2] can in fact not be reached in general without violating some constraints of the TTP. Instead, we use this bound to construct an approximation algorithm that always outputs a tournament with overall distance traveled at most $3/2 + \frac{5}{n-1}$ times optimal in the case that $n/2$ is even, and at most $3/2 + \frac{6}{n-4}$ times optimal in the case that $n/2$ is odd.

2   The Lower Bound

In this section, we present the lower bound for TTP(2) obtained by Campbell and Chen [2] and show that this lower bound cannot be reached in general.

The basic idea of the lower bound is that the optimal trips for a given team i can be obtained by computing a minimum weight perfect matching in the complete undirected graph G on the set of teams with the weight of the edge from team j to team k given as the distance d_{jk} between the home venues of j and k. Figure 1 illustrates the construction. Since team i has to visit each of the other teams and may visit at most two teams in one trip, it has to use each of the dotted edges $[i, j]$, $j \neq i$, at least once. Thus, we may ignore these edges when looking for an optimal set of trips. Moreover, the number of trips of length 2 must be maximized in order to minimize the travel distance. Hence, an optimal set of trips corresponds to a partition of the teams into pairs such that the sum of the distances between the paired teams is minimized. Each pair in the pairing is then visited by team i in a single trip, and the team that is paired with i itself is visited in a trip of length 1. Such a pairing is exactly a minimum weight perfect matching in G. In particular, the optimal pairing is independent of the team i for which the travel distance is minimized.

Using this lower bound on the distance traveled by a single team for each of the teams independently yields the following lower bound on the overall distance

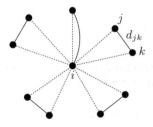

Fig. 1. Optimal trips for team i obtained from a minimum weight matching

traveled in any feasible tournament: Writing $\Delta := \sum_{i \neq j} d_{ij}$, the overall distance traveled in any feasible tournament is at least

$$\sum_{i=1}^{n}\left(d(M) + \sum_{j \neq i} d_{ij}\right) = \Delta + n \cdot d(M), \tag{1}$$

where $d(M)$ is the weight of a minimum weight perfect matching in G.

The reason that this lower bound cannot be reached in general is that the optimal trips for the teams given by the perfect matching cannot be synchronized to yield a feasible tournament. To see why, suppose that we were given a schedule in which all teams travel the optimal trips given by the minimum weight perfect matching as above. Consider two teams t_1, t_2 that are paired in the matching and look at the time slot l in the tournament in which team t_1 visits team t_2. Since t_1 uses the trips given by the matching, the trip in which it visits t_2 has length 1, so t_1 has home games in the slots $l-1, l+1$ adjacent to l. Using this, Figure 2 shows why we already obtain a contradiction to Constraint (c) of the TTP: If some team $t_3 \neq t_2$ visits t_1 in slot $l-1$, team t_3 must have visited t_2 in the previous slot $l-2$ as it travels according to the perfect matching and $[t_1, t_2]$ is a matching edge. But this implies that t_2 has a trip of length 1 in slot $l-1$, which can only be the case if it visits t_1 in this slot. t_1 was, however, already assumed to play against t_3 in slot $l-1$. Hence, the only possibility would be that t_2 visits t_1 in slot $l-1$, which contradicts Constraint (c) of the TTP as t_1 visits t_2 in the adjacent slot l.

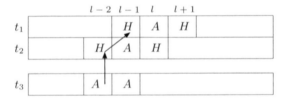

Fig. 2. No team $t_3 \neq t_2$ can visit t_1 in slot $l-1$

Even without Constraint (c) we would obtain a contradiction since, by symmetry, the same argumentation yields that only t_2 can visit t_1 in slot $l+1$, so t_2 would have to visit t_1 twice.

3   Approximation Algorithms

A very simple way to construct a 2-approximation for TTP(2) is to take an arbitrary schedule with the minimum possible number of breaks. In such a schedule, $n-2$ of the n teams have one trip containing two games, while all other trips contain exactly one game. Because of the triangle inequality, the distance traveled on each trip consisting of two games is not longer than the distance traveled when visiting the opponents of these games separately. Thus, the total length of the trips corresponding to this schedule is not greater than 2Δ. Since the cost of any feasible tournament is at least Δ as shown in the previous section, this schedule yields a 2-approximation. In the rest of the paper, we present an algorithm that constructs a tournament T of total cost at most $3/2 + \frac{6}{n-1}$ times optimal, which yields an approximation ratio less than 2 for $n \geq 14$.

3.1   Construction of the Tournament T

We assume that the n teams are numbered such that the edges $(1, 2), (3, 4), \ldots,$ $(n-1, n)$ form a minimum weight perfect matching M in the graph G. For each $i = 1, \ldots, n$, we denote the sum of the distances of team i to all other teams by $s(i)$, i.e.,

$$s(i) := \sum_{j \neq i} d_{ij}.$$

As $\sum_{i=1}^{n} s(i) = \Delta$, we can choose the numbering in such a way that $s(n-1)$ $+ s(n) \leq 2 \cdot \Delta/n$. Furthermore, we may order the two teams on each matching edge such that $s(i) \geq s(i+1)$ for every odd $i \leq n/2$, and $s(i) \leq s(i+1)$ for every odd $i > n/2$. Hence, we obtain

$$\sum_{i=1, \, i \text{ odd}}^{n/2} s(i+1) + \sum_{i=n/2+1, \, i \text{ odd}}^{n-2} s(i) \leq \frac{\Delta}{2} - (s(n-1) + s(n))/2. \qquad (2)$$

In order to schedule the matches between the teams, we apply the canonical tournament introduced by de Werra [12]. This way, we make sure that each team plays against every other team exactly once. This initial canonical schedule can be obtained by assigning the teams to the vertices of a special graph as displayed in Figure 3 for $n = 20$.

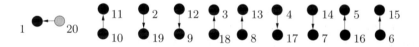

Fig. 3. Games at the first time slot for $n = 20$

The matches of the first time slot correspond to the pairs of vertices being adjacent to each other and a game always takes place at the venue of the team assigned to the head of the corresponding arc. The second day's matches can be obtained by changing the assignment of the teams to the black vertices in counterclockwise direction as shown in Figure 4. The schedules for the other time slots are derived analogously. The only difference is that the orientation of the arc incident to team 20 changes every second time, thus making sure that the road trips and home stands of team 20 do not get longer than 2.

Clearly, the first half of the tournament obtained this way has no road trip or home stand longer than two. The second half is derived from the first half by

Fig. 4. Games at the second time slot for $n = 20$

repeating the matches of the first half with changed home field advantage. As the first half does not contain any road trip or home stand longer than two matches, the second half will not contain any road trips or home stands being too long either. We only need to make sure that the connection of the two halves does not yield any such event. If we started the second half with the same match as the first one, we might obtain road trips of length four. Team 13 would encounter such a situation in the tournament stated in Figure 3 as it would end the first half of the season with a road trip containing teams 6 and 7 and start the second half with two away games against 8 and 9, thus obtaining a road trip of length four. Therefore, we start the second half with the last two games of the first half. This way, we make sure that no two games against the same opponent are played consecutively (Definition 1, Condition (c)) and additionally take care that no road trips or home stands at the connection of the two halves become too long. We can see that the latter holds by considering the four possible cases for the games at the connection point. In case the first half of some team t ends with two away games against teams t_1 and t_2, the second half starts with one home game against t_1 which is followed by another home game against t_2. We denote such a sequence as $AAHH$. The other possible sequences read $HHAA$, $AHHA$, and $HAAH$, and none of these sequences yield a road trip or home stand longer than 2.

3.2 Costs of the Tournament T for $n/2$ Even

In this section, we prove an upper bound for the total length of the tours defined by the tournament T constructed above in the case that the number n of teams is divisible by 4, i.e., $n/2$ is even. In the next section, we will derive a slightly larger bound for the case that $n/2$ is odd.

We assume that every team t having an away game against team n drives home before driving to team n's venue and drives home again after having played that match. By construction, t has a home game before or after that game anyway, so we just added one more visit home. By the triangle inequality, this can only increase the total cost of the tournament. Furthermore, we apply the triangle inequality a second time by assuming that every team drives home after the last game of the first half of the season if it is not already at home. We now estimate the distances related to the constructed tournament T separately.

C_h - *The costs related to home games of team n:* Every other team plays one away game against team n. As we can assume by application of the triangle inequality that all teams come from their home venues to play against team n and return to their home venues after this game, we know that the cost incurred thereby is at most

$$C_h \le \sum_{i=1}^{n-1} (d(i,n) + d(n,i)) = 2s(n).$$

C_a - *The costs related to away games of team n:* Analogously to the estimation of the home games of team n, we can upper bound the costs incurred by the away games of team n by first assuming that team n always returns home after each away game. This way, we derive the same upper bound of $2s(n)$ for the costs C_a incurred by the away games.

C_s - *The costs related to the first days of the season halves and the costs of returning home after the last days of the season halves:* At the first day of the season, $n/2$ teams have to travel to their opponents. We do not consider the games that the teams $n-1$ and n are involved in, as we have already taken care of these costs above. Hence, there are $n/2 - 2$ distances traveled left, which correspond to all but one of the vertical arcs in Figure 3. After the games of day $n-1$, the first half of the season is over and we assume that all teams drive home. The second half of the season starts with the matches that have already taken place at day $n-2$ and it ends with the second leg of the game of day $n-3$.

Observe that the orientation of the arcs does not have an effect on the total distance traveled. It only affects the question who is driving, which is not of interest here. In the example mentioned above, for team 13, the matches to consider are those against the teams $8, 7, 6$, and 5. If team 13 did not start the season this way but with a match against team 9 say, we would need to consider the distances to the teams $9, 8, 7$, and 6. Overall, there are $n-1$ different choices for the first and last trips of the two halves of the season and each edge of $(\{1, \ldots, n-1\} \times \{1, \ldots, n-1\})$ is part of at most four of these choices. Hence, summing up the distances of the $n-1$ different possible choices, we obtain a total of at most

$$\sum_{i=1}^{n-1} \sum_{j=i+1}^{n-1} 4d_{ij} = 2\Delta - 4s(n) \le 2\Delta.$$

Consequently, there has to be a choice for which $C_s \le 2\Delta/(n-1)$.

C_o - *The other costs:* The opponent schedule for the first half of the tournament in the example is shown in Table 1. Here, the ith row displays the opponents of team i in the order they are faced in the course of the first half of the season. We now consider the cost of each row separately. By adding an additional drive from the last opponent to the first opponent of team i in the first half of the season, we obtain a cyclic order of the opponents and the cost only increases. Observe that we already considered the distances incurred by playing against team n. The costs of the remaining trips in the cyclic order are exactly the same as the costs for playing against the opponents in order $i+1, i+2, \ldots, n-1, 1, 2, \ldots, i-1$, like for team 10 in the example from Table 1. The same can be done for the second half of the season, where we obtain the same order of games as for the first one.

As the road trips of the second half are exactly the home stands of the first half and vice verse, it suffices to consider, for each team i, a sequence of away games against the teams $i+1, i+2, \ldots, n-1, 1, 2, \ldots, i-1$ with only the first and the last trip having length one and all other trips having length two.

Table 1. Opponent schedule for the teams in the example. Games contained in trips following edges of M in the cyclic order are colored gray.

1	20	2	3	4	5	6	7	8	9	10	11	12	13	14	15	16	17	18	19
2	19	1	20	3	4	5	6	7	8	9	10	11	12	13	14	15	16	17	18
3	18	19	1	2	20	4	5	6	7	8	9	10	11	12	13	14	15	16	17
4	17	18	19	1	2	3	20	5	6	7	8	9	10	11	12	13	14	15	16
5	16	17	18	19	1	2	3	4	20	6	7	8	9	10	11	12	13	14	15
6	15	16	17	18	19	1	2	3	4	5	20	7	8	9	10	11	12	13	14
7	14	15	16	17	18	19	1	2	3	4	5	6	20	8	9	10	11	12	13
8	13	14	15	16	17	18	19	1	2	3	4	5	6	7	20	9	10	11	12
9	12	13	14	15	16	17	18	19	1	2	3	4	5	6	7	8	20	10	11
10	11	12	13	14	15	16	17	18	19	1	2	3	4	5	6	7	8	9	20
11	10	20	12	13	14	15	16	17	18	19	1	2	3	4	5	6	7	8	9
12	9	10	11	20	13	14	15	16	17	18	19	1	2	3	4	5	6	7	8
13	8	9	10	11	12	20	14	15	16	17	18	19	1	2	3	4	5	6	7
14	7	8	9	10	11	12	13	20	15	16	17	18	19	1	2	3	4	5	6
15	6	7	8	9	10	11	12	13	14	20	16	17	18	19	1	2	3	4	5
16	5	6	7	8	9	10	11	12	13	14	15	20	17	18	19	1	2	3	4
17	4	5	6	7	8	9	10	11	12	13	14	15	16	20	18	19	1	2	3
18	3	4	5	6	7	8	9	10	11	12	13	14	15	16	17	20	19	1	2
19	2	3	4	5	6	7	8	9	10	11	12	13	14	15	16	17	18	20	1
20	1	11	2	12	3	13	4	14	5	15	6	16	7	17	8	18	9	19	10

Thus, for every team i with i being even, we see that the trips visiting teams $1, 2, \ldots, i-2$ include edges that are part of M. Hence, by the triangle inequality, the distance $c(i)$ traveled by i is not greater than

$$\sum_{j=1}^{i-2} d_{ij} + \sum_{j=1,\, j\text{ odd}}^{i-3} d_{j,j+1} + 2d_{i,i-1} + 2d_{i,i+1} + 2d_{i,n-1} + \sum_{j=i+2}^{n-2} 2d_{ij}$$

$$= \sum_{j=1}^{i-2} d_{ij} + \sum_{j=1,\, j\text{ odd}}^{i-3} d_{j,j+1} + \sum_{j=i-1}^{n-1} 2d_{ij} = \sum_{j=1,j\text{ odd}}^{i-3} d_{j,j+1} + s(i) - d_{in} + \sum_{j=i-1}^{n-1} d_{ij}.$$

Analogously, for every team i with i being odd, we see that the trips visiting teams $i+2, \ldots, n-2$ include edges that are part of M. Hence, the distance $c(i)$ traveled by i is at most

$$\sum_{j=i+2}^{n-2} d_{ij} + \sum_{j=i+2,\, j\text{ odd}}^{n-3} d_{j,j+1} + 2d_{i,i-1} + 2d_{i,i+1} + 2d_{i,n-1} + \sum_{j=1}^{i-2} 2d_{ij}$$

$$= \sum_{j=i+2,\, j\text{ odd}}^{n-3} d_{j,j+1} + d_{i,n-1} + s(i) - d_{in} + \sum_{j=1}^{i+1} d_{ij}.$$

Hence, the sum $c(i) + c(i+1)$ for any odd $i < n - 2$ amounts to

$$\sum_{j=i+2,\, j\text{ odd}}^{n-3} d_{j,j+1} + d_{i,n-1} + s(i) - d_{in} + \sum_{j=1}^{i+1} d_{ij}$$

$$+ \sum_{j=1,\, j\text{ odd}}^{i-2} d_{j,j+1} + s(i+1) - d_{i+1,n} + \sum_{j=i}^{n-1} d_{i+1,j}$$

$$\leq d(M) + d_{i,n-1} + s(i) - d_{in} + s(i+1) - d_{i+1,n} + \underbrace{\sum_{j=1}^{i+1} d_{ij} + \sum_{j=i}^{n-1} d_{i+1,j} - d_{i,i+1}}_{=:f(i)}.$$

As $s(i) \geq s(i+1)$ for all odd $i \leq n/2$, we estimate $f(i)$ in this case as

$$f(i) = \sum_{j=1}^{i+1} d_{ij} + \sum_{j=i}^{n-1} d_{i+1,j} - d_{i,i+1} \qquad = \sum_{j=1}^{i-1} d_{ij} + \sum_{j=i}^{n-1} d_{i+1,j}$$

$$\leq \sum_{j=1}^{i-1}(d_{i,i+1} + d_{i+1,j}) + \sum_{j=i}^{n-1} d_{i+1,j} \qquad = \sum_{j=1}^{i-1} d_{i,i+1} + \sum_{j=1}^{i-1} d_{i+1,j} + \sum_{j=i}^{n-1} d_{i+1,j}$$

$$= (i-1){\cdot}d_{i,i+1} + s(i+1) - d_{i+1,n} \overset{i\leq n/2}{\leq} (n/2 - 1){\cdot}d_{i,i+1} + s(i+1) - d_{i+1,n}.$$

Analogously, for all odd $i > n/2$, we have that $s(i) \leq s(i+1)$, so we estimate

$$f(i) = \sum_{j=1}^{i+1} d_{ij} + \sum_{j=i}^{n-1} d_{i+1,j} - d_{i,i+1} \qquad = \sum_{j=1}^{i+1} d_{ij} + \sum_{j=i+2}^{n-1} d_{i+1,j}$$

$$\leq \sum_{j=1}^{i+1} d_{ij} + \sum_{j=i+2}^{n-1}(d_{i+1,i} + d_{ij}) \qquad = \sum_{j=1}^{n-1} d_{ij} + \sum_{j=i+2}^{n-1} d_{i+1,i}$$

$$= \sum_{j=1}^{n-1} d_{ij} + (n-i-2) \cdot d_{i+1,i} \overset{i>n/2}{\leq} s(i) - d_{in} + (n/2 - 1) \cdot d_{i+1,i}.$$

Overall, we can conclude that $\sum_{i=1}^{n-2} c(i)$ amounts to

$$\sum_{i=1,\, i\text{ odd}}^{n-2} (c(i) + c(i+1))$$

$$\leq \sum_{i=1,\, i\text{ odd}}^{n-2} \left(d(M) + d_{i,n-1} + s(i) - d_{in} + s(i+1) - d_{i+1,n} + f(i) \right)$$

$$\leq \sum_{i=1,\, i\text{ odd}}^{n-2} \left(d(M) + d_{i,n-1} + s(i) - d_{in} + s(i+1) - d_{i+1,n} \right)$$

$$+ \sum_{i=1,\, i\text{ odd}}^{n/2} \left((n/2 - 1) \cdot d_{i,i+1} + s(i+1) - d_{i+1,n} \right)$$

$$+ \sum_{i=n/2+1,\ i\ \text{odd}}^{n-2} \left(s(i) - d_{in} + (n/2 - 1) \cdot d_{i+1,i} \right)$$

$$\overset{Eq.(2)}{\leq} \sum_{i=1,\ i\ \text{odd}}^{n-2} \left(d(M) + d_{i,n-1} \right) \quad + \Delta - s(n-1) - s(n) - s(n)$$

$$+ (n/2 - 1) \cdot d(M) + \Delta/2 - (s(n-1) + s(n))/2$$

$$\leq (n/2 - 1) \cdot d(M) + 3/2 \cdot \Delta - 2s(n) + (n/2 - 1) \cdot d(M) - (s(n-1) + s(n))/2$$

$$= (n - 2) \cdot d(M) + 3/2 \cdot \Delta - 2 \cdot s(n) - (s(n-1) + s(n))/2.$$

Moreover, team $n - 1$ has to travel to all of its opponents separately incurring a cost of at most $2 \cdot s(n-1)$. Finally, the total cost can be estimated as

$$C_h + C_a + C_s + C_o$$
$$\leq 2s(n) + 2s(n) + 2\Delta/(n-1)$$
$$\quad + (n - 2) \cdot d(M) + 3/2 \cdot \Delta - 2s(n) + 2s(n-1) - (s(n-1) + s(n))/2$$
$$= 3/2 \cdot (s(n-1) + s(n)) + 2\Delta/(n-1) + (n-2) \cdot d(M) + 3/2 \cdot \Delta$$
$$\leq 3/2 \cdot (2\Delta/n) + 2\Delta/(n-1) + (n-2) \cdot d(M) + 3/2 \cdot \Delta$$
$$\leq 5\Delta/(n-1) + (n-2) \cdot d(M) + 3/2 \cdot \Delta.$$

As we have a lower bound of $\Delta + n \cdot d(M)$ for the total cost of any feasible tournament, this yields an overall approximation ratio of $\frac{3}{2} + \frac{5}{n-1}$.

3.3 Costs of the Tournament T for $n/2$ Odd

Recall that, after having played against team n, team i has to play against the teams $i + 1, i + 2, \ldots, n - 1, 1, 2, \ldots, i - 1$. In the case that n is not a multiple of 4, the estimation of the cost $c(i)$ changes slightly as there might occur an additional drive home between the $(n/2 - 1)$th game g_1 and the $n/2$th game g_2 of the sequence $i+1, i+2, \ldots, n-1, 1, 2, \ldots, i-1$, which might have been counted as a road trip before. We distinguish three cases:

Case 1: $i + n/2 \leq n - 1$. The game g_1 is against team $i + n/2 - 1$, and g_2 is the game against team $i + n/2$. For even i, we have applied the triangle inequality in this case anyway, so that $c(i)$ does not change. Otherwise, it changes by $d_{i,i+n/2-1} + d_{i,i+n/2} - d_{i+n/2-1,i+n/2}$.

Case 2: $i + n/2 = n$. The game g_1 is against team $n - 1$, and g_2 is the game against team 1. In this case, we have applied the triangle inequality anyway, so that $c(i)$ does not change.

Case 3: $i + n/2 > n$. The game g_1 is against team $i - n/2$, and g_2 is the game against team $i - n/2 + 1$. For odd i, we have applied the triangle inequality in this case anyway, so that $c(i)$ does not change. Otherwise, it changes by $d_{i,i-n/2} + d_{i,i-n/2+1} - d_{i-n/2,i-n/2+1}$.

The additional costs for all three cases sum up to

$$\tilde{C} := \sum_{\substack{i=1,\ i\ \text{odd}}}^{n/2-1} \left(d_{i,i+n/2-1} + d_{i,i+n/2} - d_{i+n/2-1,i+n/2}\right)$$

$$+ \sum_{\substack{i=n/2+1,\ i\ \text{even}}}^{n-2} \left(d_{i,i-n/2} + d_{i,i-n/2+1} - d_{i-n/2,i-n/2+1}\right)$$

$$\leq \sum_{\substack{i=1,\ i\ \text{odd}}}^{n/2-1} \left(d_{i,i+n/2-1} + d_{i,i+n/2}\right) + \sum_{\substack{i=n/2+1,\ i\ \text{even}}}^{n-2} \left(d_{i,i-n/2} + d_{i,i-n/2+1}\right)$$

$$\leq \sum_{\substack{i=1,\ i\ \text{odd}}}^{n/2-1} \left(d_{i,i+n/2-1} + d_{i,i+n/2}\right) + \sum_{\substack{i=1,\ i\ \text{odd}}}^{n/2-2} \left(d_{i+n/2,i} + d_{i+n/2,i+1}\right)$$

$$\leq \sum_{\substack{i=1,\ i\ \text{odd}}}^{n/2-2} \left(d_{i,i+n/2-1} + 2 \cdot d_{i,i+n/2} + d_{i+1,i+n/2}\right).$$

We already assumed that the teams are numbered such that the edges $(i, i+1)$ for i odd form a minimum weight perfect matching M in G and that $s(n-1) + s(n) \leq 2 \cdot \Delta/n$. We are, however, still free to choose the mapping of the edges in $\tilde{M} := M \setminus \{(s(n-1), s(n))\}$ to the pairs $(i, i+1)$. In the following, we will choose this mapping in a way that minimizes the additional costs $\tilde{C}$.

We consider the complete undirected graph G' on $\tilde{M}$. Each vertex (u, v) of G' corresponds to a pair of teams u and v that form an edge $(u, v) \in \tilde{M}$ and which we ordered such that $s(u) \leq s(v)$. Using this ordering for every vertex of G', we define a weight function w on the edges of G' by

$$w((u, v), (w, z)) := d_{uw} + d_{uz} + d_{wu} + d_{wv} \quad \text{for all } (u, v), (w, z) \in \tilde{M}.$$

The sum of the weights of the edges of G' is then

$$\sum_{((u,v),(w,z)) \in E[G']} w((u, v), (w, z)) = 1/2 \cdot \sum_{(u,v) \neq (w,z) \in \tilde{M}} (d_{uw} + d_{uz} + d_{wu} + d_{wv})$$

$$\leq 1/2 \cdot \left(\sum_{(u,v) \in \tilde{M}} s(u) + \sum_{(w,z) \in \tilde{M}} s(w) \right) = \sum_{(u,v) \in \tilde{M}} s(u) \leq \Delta/2,$$

where the last inequality holds since $s(u) \leq s(v)$ for all $(u, v) \in \tilde{M}$.

Since every complete graph is 1-factorable, there exists a decomposition of the edges of G' into $n/2 - 2$ perfect matchings. Thus, the weight of the minimum weight perfect matching M' with respect to w in G' can be estimated as

$$w(M') \leq \frac{\Delta/2}{n/2 - 2} = \frac{\Delta}{n - 4}.$$

Given $M' =: \{((u_i, v_i), (w_i, z_i)) : i = 1, 3, 5, \ldots, n/2 - 2\}$, we assign the teams to the numbers $1, 2, \ldots, n - 2$ by giving team u_i the number i, team v_i the number $i + 1$, team w_i the number $i + n/2$, and team z_i the number $i + n/2 - 1$.

Observe that this assignment violates none of the assumptions on the numbering of the teams made so far. Moreover, we can conclude after straightforward calculations that $\tilde{C} \leq \frac{\Delta}{n-4}$, which, for the case that $n/2$ is odd, leads to an overall approximation ratio of $3/2 + \frac{5}{n-1} + \frac{1}{n-4} \leq 3/2 + \frac{6}{n-4}$.

References

1. Easton, K., Nemhauser, G., Trick, M.: The traveling tournament problem description and benchmarks. In: Walsh, T. (ed.) CP 2001. LNCS, vol. 2239, pp. 580–584. Springer, Heidelberg (2001)
2. Campbell, R.T., Chen, D.S.: A minimum distance basketball scheduling problem. In: [13], pp. 15–25
3. Easton, K., Nemhauser, G., Trick, M.: Solving the travelling tournament problem: A combined integer programming and constraint programming approach. In: Burke, E.K., De Causmaecker, P. (eds.) PATAT 2002. LNCS, vol. 2740, pp. 100–109. Springer, Heidelberg (2003)
4. Anagnostopoulos, A., Michel, L., Van Hentenryck, P., Vergados, Y.: A simulated annealing approach to the travelling tournament problem. Journal of Scheduling 9(2), 177–193 (2006)
5. Miyashiro, R., Matsui, T., Imahori, S.: An approximation algorithm for the traveling tournament problem. In: Proceedings of the 7th International Conference on the Practice and Theory of Automated Timetabling (PATAT) (2008)
6. Yamaguchi, D., Imahori, S., Miyashiro, R., Matsui, T.: An improved approximation algorithm for the traveling tournament problem. In: Dong, Y., Du, D.-Z., Ibarra, O. (eds.) ISAAC 2009. LNCS, vol. 5878, pp. 679–688. Springer, Heidelberg (2009)
7. Westphal, S., Noparlik, K.: A 5.875-approximation for the traveling tournament problem. In: Proceedings of the 8th International Conference on the Practice and Theory of Automated Timetabling (PATAT) (2010)
8. Thielen, C., Westphal, S.: Complexity of the Traveling Tournament Problem. Theoretical Computer Science (2010) (online first), doi:10.1016/j.tcs.2010.10.001
9. Bhattacharyya, R.: A note on complexity of traveling tournament problem. Optimization Online (2009)
10. Kendall, G., Knust, S., Ribeiro, C., Urrutia, S.: Scheduling in sports: An annotated bibliography. Computers and Operations Research 37(1), 1–19 (2010)
11. Rasmussen, R., Trick, M.: Round robin scheduling - a survey. European Journal of Operations Research 188, 617–636 (2008)
12. de Werra, D.: Scheduling in sports. In: [14], pp. 381–395
13. Machol, R.E., Ladany, S.P., Morrison, D. (eds.): Management Science in Sports. Studies in the Management Sciences, vol. 4. North-Holland Publishing Company, Amsterdam (1976)
14. Hansen, P.: Studies on Graphs and Discrete Programming. Annuals of Discrete Mathematics, vol. 11. North-Holland Publishing Company, Amsterdam (1981)

Alphabet Partitioning for Compressed Rank/Select and Applications

Jérémy Barbay[1], Travis Gagie[1,*], Gonzalo Navarro[1,*], and Yakov Nekrich[2]

[1] Department of Computer Science, University of Chile
{jbarbay,tgagie,gnavarro}@dcc.uchile.cl
[2] Department of Computer Science, University of Bonn
yasha@cs.uni-bonn.de

Abstract. We present a data structure that stores a string $s[1..n]$ over the alphabet $[1..\sigma]$ in $nH_0(s) + o(n)(H_0(s)+1)$ bits, where $H_0(s)$ is the zero-order entropy of s. This data structure supports the queries **access** and **rank** in time $\mathcal{O}(\lg\lg\sigma)$, and the **select** query in constant time. This result improves on previously known data structures using $nH_0(s) + o(n\lg\sigma)$ bits, where on highly compressible instances the redundancy $o(n\lg\sigma)$ cease to be negligible compared to the $nH_0(s)$ bits that encode the data. The technique is based on combining previous results through an ingenious partitioning of the alphabet, and practical enough to be implementable. It applies not only to strings, but also to several other compact data structures. For example, we achieve (i) faster search times and lower redundancy for the smallest existing full-text self-index; (ii) compressed permutations π with times for $\pi()$ and $\pi^{-1}()$ improved to log-logarithmic; and (iii) the first compressed representation of dynamic collections of disjoint sets.

1 Introduction

Search queries on strings have many important applications, to the point that one is willing to sacrifice some additional space to index the string in order to support the queries in less time. The most important queries serve as primitives to implement many other operations, in particular pattern matching in full-text databases (see, e.g., [18,7,14,19] for recent discussions): given a string s, $s.\mathsf{access}(i)$ returns the ith character of s, which we denote $s[i]$; $s.\mathsf{rank}_a(i)$ returns the number of occurrences of the character a up to position i; and $s.\mathsf{select}_a(i)$ returns the position of the ith occurrence of a in s.

Wavelet trees [11] represent a string $s[1..n]$ over alphabet $[1..\sigma]$ within $n\lg\sigma + o(n\lg\sigma)$ bits, where lg denotes the logarithm in base two. The indexing space in $o(n\lg\sigma)$ is considered asymptotically "negligible" compared to the $n\lg\sigma$ bits required to hold the main data, while providing support for the queries in time $\mathcal{O}(\lg\sigma)$. Later results [10] improved the times to $\mathcal{O}(\lg\lg\sigma)$.

Regularities in the string permit further reductions in the space, from $n\lg\sigma$ bits down to $nH_k(s)$ bits, where $H_k(s)$ denotes the kth-order empirical entropy

---

* Funded in part by the Millennium Institute for Cell Dynamics and Biotechnology (ICDB), Grant ICM P05-001-F, Mideplan, Chile.

O. Cheong, K.-Y. Chwa, and K. Park (Eds.): ISAAC 2010, Part II, LNCS 6507, pp. 315–326, 2010.

Table 1. Recent bounds and our new ones for data structures supporting access, rank and select. The first row holds for $\sigma = \mathcal{O}\left(\mathrm{polylog}(n)\right)$ and the second for $\sigma = o(n)$. The space bound in the sixth row holds for $k = o(\log_\sigma n)$. The times of our Thm. 1 can be refined into a more complicated formula (see also Cor. 1).

	space (bits)	access	rank	select
[8, Thm. 3.2]	$nH_0(s) + o(n)$	$\mathcal{O}(1)$	$\mathcal{O}(1)$	$\mathcal{O}(1)$
[8, Cor. 3.3]	$nH_0(s) + o(n \lg \sigma)$	$\mathcal{O}\left(1 + \frac{\lg \sigma}{\lg \lg n}\right)$	$\mathcal{O}\left(1 + \frac{\lg \sigma}{\lg \lg n}\right)$	$\mathcal{O}\left(1 + \frac{\lg \sigma}{\lg \lg n}\right)$
[10, Thm. 2.2]	$n \lg \sigma + o(n \lg \sigma)$	$\mathcal{O}(\lg \lg \sigma)$	$\mathcal{O}(\lg \lg \sigma)$	$\mathcal{O}(1)$
[10, Thm. 2.2]	$n \lg \sigma + o(n \lg \sigma)$	$\mathcal{O}(1)$	$\mathcal{O}(\lg \lg \sigma \lg \lg \lg \sigma)$	$\mathcal{O}(\lg \lg \sigma)$
[3, Lem. 4.1]	$nH_0(s) + o(n \lg \sigma)$	$\mathcal{O}(\lg \lg \sigma)$	$\mathcal{O}(\lg \lg \sigma)$	$\mathcal{O}(1)$
[3, Thm. 4.2]	$nH_k(s) + o(n \lg \sigma)$	$\mathcal{O}(1)$	$\mathcal{O}\left((\lg \lg \sigma)^2 \lg \lg \lg \sigma\right)$	$\mathcal{O}(\lg \lg \sigma \lg \lg \lg \sigma)$
Thm 1	$nH_0(s) + o(n)(H_0(s) + 1)$	$\mathcal{O}(\lg \lg \sigma)$	$\mathcal{O}(\lg \lg \sigma)$	$\mathcal{O}(1)$
Thm 1	$nH_0(s) + o(n)(H_0(s) + 1)$	$\mathcal{O}(1)$	$\mathcal{O}(\lg \lg \sigma \lg \lg \lg \sigma)$	$\mathcal{O}(\lg \lg \sigma)$

of s (i.e., the minimum self-information of s with respect to a kth-order Markov source; see Manzini [15] for a definition and discussion). The challenge of compressing the string *while still supporting the queries efficiently* was also achieved, using as little as $nH_0(s) + o(n \lg \sigma)$ [11,8,3] and even $nH_k(s) + o(n \lg \sigma)$ bits [3] (for any $k = o(\log_\sigma n)$) while retaining the time complexities.

One problem with such space is that, on highly compressible data, the $o(n \lg \sigma)$ bits of the index are not always negligible compared to the space used to encode the compressed data. Hence the challenge is to retain the efficient support for the queries *while compressing the index redundancy* as well. In this paper we solve this challenge in the case of zero-order entropy compression, that is, the redundancy of our data structure is asymptotically negligible compared to the zero-order entropy of the text, plus $o(n)$ bits.

For comparison, the representation by Golynski et al. [10] does not compress[1] s and uses additional $\mathcal{O}\left(\frac{n \lg \sigma}{\lg \lg \sigma}\right) = o(n \lg \sigma)$ bits, but offers log-logarithmic times for the queries. Ferragina et al.'s wavelet tree [8] achieves zero-order compression plus $\mathcal{O}\left(\frac{n \lg \sigma \lg \lg n}{\lg n}\right) = o(n \lg \sigma)$ bits, supporting the queries in $\mathcal{O}\left(1 + \frac{\lg \sigma}{\lg \lg n}\right)$ time. Barbay et al. [3] obtain zero-order space and log-logarithmic times, but their redundancy is still $o(n \lg \sigma)$. See Table 1 for a summary of our bounds and previous ones.[2]

In Section 2 we show how to combine the strengths of these data structures, obtaining not only zero-order compressed space and log-logarithmic times, but

---

[1] In terms of the usual entropy measures. It compresses to the k-th order entropy of a different sequence (A. Golynski, personal communication).

[2] When we write $o(n \lg \sigma)$ bits we mean $o(n) \lg \sigma$. Although in some cases [10,3] the results are actually $n \, o(\lg \sigma)$, we point out that this can be taken as $o(n) \lg \sigma$ because, if $\sigma = \mathcal{O}\left(\mathrm{polylog}(n)\right)$, one can use a structure by Ferragina et al. [8, Thm. 3.2] that solves access, rank, and select in constant time using $nH_0(s) + o(n)$ bits. Thus one can assume $\sigma = \omega(1)$ at the very least. See also Footnote 6 of Barbay et al. [3].

also compressed redundancy. The technique can be summarized as partitioning the alphabet into sub-alphabets according to the characters' frequencies in s, storing in a multiary wavelet tree [8] the string that results from replacing the characters in s by identifiers of their sub-alphabets, and storing separate strings, each the projection of s to the characters of s belonging to each sub-alphabet, this time using Golynski et al.'s [10] structure for large alphabets. We achieve a data structure that stores a string $s[1..n]$ in $nH_0(s) + o(n)(H_0(s)+1)$ bits, thus guaranteeing that the redundancy stays negligible even when the text is very compressible. It supports queries in the times shown in Table 1 (rows 7 and 8 give two alternatives).

Then we consider various extensions and applications of our main result. In Section 3 we show how our result can be used to improve an existing text index that achieves k-th order entropy [8,3], so as to improve its redundancy and query times. This way we achieve the first self-index with space bounded by $nH_k(s) + o(n)(H_k(s)+1)$ bits, able of counting and locating pattern occurrences and extracting any substring of s, within the time complexities achieved by either of its predecessors. In Sections 4 and 5, respectively, we show how to apply our data structure to store a compressed permutation, a compressed function and a compressed dynamic collection of disjoint sets, while supporting a rich set of operations on those. This improves or gives alternatives to the best previous results [4,17,12]. We have approached these applications in such a way that an improvement to our main result, however achieved, translates into improved bounds for them as well.

2   Alphabet Partitioning

Let $s[1..n]$ be a sequence over effective alphabet $[1..\sigma]$, i.e., every character appears in s, so $\sigma \le n$. (At the end of the section we handle the case of large alphabets.) The zero-order entropy of s is $H_0(s) = \sum_{a \in [1..\sigma]} \frac{|s|_a}{n} \lg \frac{n}{|s|_a}$, where $|s|_a$ is the number of occurrences of the character a in s. Note that by convexity we have $nH_0(s) \ge (\sigma - 1) \lg n + (n - \sigma + 1) \lg(n/(n - \sigma + 1))$, a property we will use later.

Our results are based on the following alphabet partitioning scheme. Let $m[1..\sigma]$ be the sequence assigning to each character $a \in [1..\sigma]$ the value

$$m[a] = \lceil \lg(n/|s|_a) \lg n \rceil \le \lceil \lg^2 n \rceil .$$

Let $t[1..n]$ be the string over $\left[1.. \lceil \lg^2 n \rceil \right]$ obtained from s by replacing each occurrence of a by $m[a]$, for $1 \le a \le \sigma$. For $0 \le \ell \le \lceil \lg^2 n \rceil$, let σ_ℓ be the number of occurrences of ℓ in m or, equivalently, the number of distinct characters of s replaced by ℓ in t. Finally, let $s_\ell[1..|t|_\ell]$ be the string over $[1..\sigma_\ell]$ defined by $s_\ell[t.\mathrm{rank}_\ell(i)] = m.\mathrm{rank}_\ell(s[i])$.

Notice that, if both a and b are replaced by the same number in t, then $\lg(n/|s|_b) - \lg(n/|s|_a) < 1/\lg n$ and so $|s|_a/|s|_b < 2^{1/\lg n}$. It follows that, if a is replaced by ℓ in t, then $\sigma_\ell < 2^{1/\lg n}|s|_\ell/|s|_a$ (by fixing a and summing over all those b replaced by ℓ). Since

$$\sum_{\lceil \lg(n/|s|_a) \lg n \rceil = \ell} |s|_a = |s_\ell| \qquad \text{and} \qquad \sum_a |s|_a = \sum_\ell |s_\ell| = n,$$

we have

$$nH_0(t) + \sum_\ell |s_\ell| \lg \sigma_\ell$$

$$< \sum_\ell |s_\ell| \lg(n/|s_\ell|) + \sum_\ell \sum_{\lceil \lg(n/|s|_a) \lg n \rceil = \ell} |s|_a \lg \left(2^{\frac{1}{\lg n}} |s_\ell|/|s|_a\right)$$

$$= \sum_a |s|_a \lg(n/|s|_a) + n/\lg n$$

$$= nH_0(s) + o(n).$$

In other words, if we represent t with $H_0(t)$ bits per symbol and each s_ℓ with $\lg \sigma_\ell$ bits per symbol, we achieve a good overall compression. Thus we can obtain a very compact representation of a string s by storing a compact representation of t and storing each s_ℓ as an "uncompressed" string over an alphabet of size σ_ℓ.

Now we show how our approach can be used to obtain a fast and compact rank/select data structure. Suppose we have a data structure T that supports access, rank and select queries on t; another structure M that supports the same queries on m; and data structures $S_1, \ldots, S_{\lceil \lg^2 n \rceil}$ that support the same queries on $s_1, \ldots, s_{\lceil \lg^2 n \rceil}$. With these data structures we can implement

$$s.\text{access}(i) = m.\text{select}_\ell(s_\ell.\text{access}(t.\text{rank}_\ell(i))), \text{ where } \ell = t.\text{access}(i);$$

$$s.\text{rank}_a(i) = s_\ell.\text{rank}_c(t.\text{rank}_\ell(i)), \text{ where } \ell = m.\text{access}(a) \text{ and } c = m.\text{rank}_\ell(a);$$

$$s.\text{select}_a(i) = t.\text{select}_\ell(s_\ell.\text{select}_c(i)) \text{ where } \ell = m.\text{access}(a) \text{ and } c = m.\text{rank}_\ell(a).$$

We implement T and M as multiary wavelet trees [8]; we implement each S_ℓ as either a multiary wavelet tree or an instance of Golynski et al.'s [10, Thm. 2.2] access/rank/select data structure, depending on whether $\sigma_\ell \leq \lg n$ or not. The wavelet tree for T uses at most $nH_0(t) + \mathcal{O}\left(\frac{n(\lg \lg n)^2}{\lg n}\right)$ bits and operates in constant time, because its alphabet size is polylogarithmic. If S_ℓ is implemented as a wavelet tree, it uses at most $|s_\ell|H_0(s_\ell) + \mathcal{O}\left(\frac{|s_\ell| \lg |s_\ell| \lg \lg n}{\lg n}\right)$ bits[3] and operates in constant time for the same reason; otherwise it uses at most $|s_\ell| \lg \sigma_\ell + \mathcal{O}\left(\frac{|s_\ell| \lg \sigma_\ell}{\lg \lg \sigma_\ell}\right) \leq |s_\ell| \lg \sigma_\ell + \mathcal{O}\left(\frac{|s_\ell| \lg \sigma_\ell}{\lg \lg \lg n}\right)$ bits (the latter because $\sigma_\ell > \lg n$). Thus in either case the space for s_ℓ is bounded by $|s_\ell| \lg \sigma_\ell + \mathcal{O}\left(\frac{|s_\ell| \lg |s_\ell|}{\lg \lg \lg n}\right)$ bits. Finally, since M is a sequence of length σ over an alphabet of size $\lceil \lg^2 n \rceil$, the wavelet tree for M takes $\mathcal{O}(\sigma \lg \lg n)$ bits. Because of the property we referred to in the beginning of this section, $nH_0(s) \geq (\sigma - 1) \lg n$, this space is

---

[3] This is achieved by using block sizes of length $\frac{\lg n}{2}$ and not $\frac{\lg |s_\ell|}{2}$, at the price of storing universal tables of size $\mathcal{O}(\sqrt{n} \operatorname{polylog}(n)) = o(n)$ bits. Therefore all of our $o(\cdot)$ expressions involving n and other variables will be asymptotic in n.

$H_0(s) \mathcal{O}\left(\frac{n \lg \lg n}{\lg n}\right)$. By these calculations, the space for T, M and the S_ℓ's adds up to $nH_0(s) + o(n)H_0(s) + o(n)$, where the $o(n)$ term is $\mathcal{O}\left(\frac{n}{\lg \lg \lg n}\right)$.

Depending on which time tradeoff we use for Golynski et al.'s data structure, we obtain the results of Table 1. We can refine the time complexity by noticing that the only non-constant times are due to operating on some string s_ℓ, where the alphabet is of size $\sigma_\ell < 2^{1/\lg n}|s_\ell|/|s|_a$, where a is the character in question, thus $\lg \lg \sigma_\ell = \mathcal{O}(\lg \lg \min(\sigma, n/|s|_a))$.

Theorem 1. *We can store $s[1..n]$ over effective alphabet $[1..\sigma]$ in $nH_0(s) +$ $o(n)(H_0(s) + 1)$ bits and support* access, rank *and* select *queries in $\mathcal{O}(\lg \lg \sigma)$, $\mathcal{O}(\lg \lg \sigma)$, and $\mathcal{O}(1)$ time, respectively (variant (i)). Alternatively, we can support* access, rank *and* select *queries in $\mathcal{O}(1)$, $\mathcal{O}(\lg \lg \sigma \lg \lg \lg \sigma)$ and $\mathcal{O}(\lg \lg \sigma)$ time, respectively (variant (ii)). Any of the σ terms in these time complexities is actually $\min(\sigma, n/|s|_a)$, where a stands for $s[i]$ in the time of the* access *query, and for the character argument in the time of the* rank *and* select *query.*

Moreover, by implementing S_ℓ as a wavelet tree whenever $\sigma_\ell \le (\lg n)^{\lg \lg \lg n}$, we ensure to achieve the complexities of wavelet trees if those are better than the ones given above. That is, for example, $\mathcal{O}\left(\min\left(1 + \frac{\lg \sigma_\ell}{\lg \lg n}, \lg \lg \sigma_\ell\right)\right)$ instead of just $\mathcal{O}(\lg \lg \sigma_\ell)$. We can similarly match the complexity $\mathcal{O}(\lg \lg \sigma_\ell \lg \lg \lg \sigma_\ell)$. Note that, if we do this, the complexities that were $\mathcal{O}(1)$ become $\mathcal{O}\left(1 + \frac{\lg \sigma_\ell}{\lg \lg n}\right)$.

Corollary 1. *All the time complexities up to $\mathcal{O}(\lg \lg \sigma)$ in variants (i) or (ii) of Theorem 1 can be made $\mathcal{O}\left(\min\left(1 + \frac{\lg \sigma}{\lg \lg n}, \lg \lg \sigma\right)\right)$. Alternatively, all time complexities in variant (ii) can be made $\mathcal{O}\left(\min\left(1 + \frac{\lg \sigma}{\lg \lg n}, \lg \lg \sigma \lg \lg \lg \sigma\right)\right)$. As in Theorem 1, the σ term is actually $\min(\sigma, n/|s|_a)$.*

In the most general case, s is a sequence over an alphabet Σ which is not an effective alphabet, and σ symbols from Σ occur in s. Let Σ' be the set of elements that occur in s; we can map characters from Σ' to elements of $[1..\sigma]$ by replacing each $a \in \Sigma'$ with its rank in Σ'. All elements of Σ' are stored in the indexed dictionary data structure described by Raman et al. [20], so that the following queries are supported in constant time: for any $a \in \Sigma'$ its rank in Σ' can be found (for any $a \notin \Sigma'$ the answer is -1); for any $i \in [1..\sigma]$ the i-th smallest element in Σ' can be found. The indexed dictionary of Raman et al. [20] uses $\sigma \lg(e\mu/\sigma) + o(\sigma) + \mathcal{O}(\lg \lg \mu)$ bits of space, where e is the base of the natural logarithm and μ is the maximal element in Σ'; the value of μ can be specified with additional $\mathcal{O}(\lg \mu)$ bits. We replace every element in s by its rank in Σ', and the resulting string is stored using Theorem 1. Hence, in the general case the space usage is increased by $\sigma \lg(e\mu/\sigma) + o(\sigma) + \mathcal{O}(\lg \mu)$ bits and the asymptotic time complexity of queries remains unchanged.

3   Reduced Redundancy on Self-indexes

Our result can be readily carried over self-indexes. These also represent a sequence, but they support other operations related to text searching. A well known self-index [8] achieves k-th order entropy space by partitioning the Burrows-Wheeler transform [6] of the sequence and encoding each partition to its zero-order entropy. Those partitions must support queries access and rank. By using Theorem 1(i) to represent such partitions, we achieve the following result, improving previous ones [8,10,3].

Theorem 2. *Let $s[1..n]$ be a string over alphabet[4] $[1..\sigma]$. Then we can represent s using $nH_k(s) + o(n)(H_k(s) + 1)$ bits of space, for any $k \le (\alpha \log_\sigma n) - 1$ and constant $0 < \alpha < 1$, while supporting the following queries: (i) count the number of occurrences of a pattern $p[1..m]$ in s, in time $\mathcal{O}(m \lg \lg \sigma)$; (ii) locate any such occurrence in time $\mathcal{O}(\lg n \lg \lg \lg \lg n \lg \lg \lg \sigma)$; (iii) extract $s[l,r]$ in time $\mathcal{O}((r - l) \lg \lg \sigma + \lg n \lg \lg \lg \lg n \lg \lg \sigma)$. The $\lg \lg \sigma$ times can be reduced to $\mathcal{O}\left(1 + \frac{\lg \sigma}{\lg \lg n}\right)$ if convenient.*

For these particular locating and extracting times we are sampling one out of every $\lg n \lg \lg \lg n$ text positions, which maintains our lower-order space term $o(n)$ at $\mathcal{O}(n/ \lg \lg \lg n)$. Compared to Theorem 4.2 of Barbay et al. [3], we reduce the redundancy from $o(n) \lg \sigma$ to $o(n)(H_k(s) + 1)$. Our improved locating times, however, just owe to the denser sampling, which they could also use.

4   Compressing Permutations

We now show how to use access/rank/select data structures to store a compressed permutation. We follow Barbay and Navarro's notation [4] and improve their space and, especially, their time performance. They measure the compressibility of a permutation π in terms of the entropy of the distribution of the lengths of *runs* of different kinds. Let π be covered by ρ runs (using any of the previous definitions of runs [13,4,16]) of lengths $\mathsf{runs}(\pi) = \langle n_1, \ldots, n_\rho \rangle$. Then $H(\mathsf{runs}(\pi)) = \sum \frac{n_i}{n} \lg \frac{n}{n_i} \le \lg \rho$ is called the *entropy* of the runs (and, because $n_i \ge 1$, it also holds $nH(\mathsf{runs}(\pi)) \ge (\rho - 1) \lg n$). We first consider permutations which are interleaved sequences of increasing or decreasing values as first defined by Levcopoulos et al. [13] for adaptive sorting, and later on for compression [4], and then give improved results for more specific classes of runs. In both cases we consider first the application of the permutation $\pi()$ and its inverse, $\pi^{-1}()$, to show later how to extend the support to the iterated applications of the permutation, $\pi^k()$, extending and improving previous results [17].

Theorem 3. *Let π be a permutation on n elements that consists of ρ interleaved increasing or decreasing runs, of lengths $\mathsf{runs}(\pi)$. We can store π in $2nH(\mathsf{runs}(\pi)) + o(n)(H(\mathsf{runs}(\pi)) + 1)$ bits and perform $\pi()$ and $\pi^{-1}()$ queries in $\mathcal{O}\left(\min\left(1 + \frac{\lg \rho}{\lg \lg n}, \lg \lg \rho\right)\right)$ time.*

---
[4] Again, $[1..\sigma]$ does not need to be the effective alphabet (see paragraph after Thm. 1).

Proof. We first replace all the elements of the rth run by r, for $1 \leq r \leq \rho$. Let s be the resulting string and let s' be s permuted according to π, that is, $s'[\pi(i)] = s[i]$. We store s and s' using Theorem 1(i) and store ρ bits indicating whether each run is increasing or decreasing. Notice that, if $\pi(i)$ is part of an increasing run, then $s'.\mathrm{rank}_{s[i]}(\pi(i)) = s.\mathrm{rank}_{s[i]}(i)$, so

$$\pi(i) = s'.\mathrm{select}_{s[i]}\left(s.\mathrm{rank}_{s[i]}(i)\right) \ ;$$

if $\pi(i)$ is part of a decreasing run, then $s'.\mathrm{rank}_{s[i]}(\pi(i)) = s.\mathrm{rank}_{s[i]}(n) + 1 - s.\mathrm{rank}_{s[i]}(i)$, so

$$\pi(i) = s'.\mathrm{select}_{s[i]}\left(s.\mathrm{rank}_{s[i]}(n) + 1 - s.\mathrm{rank}_{s[i]}(i)\right) \ .$$

A $\pi^{-1}()$ query is symmetric. The space of the bitmap is $\rho \in o(n)H(\mathrm{runs}(\pi))$ because $nH(\mathrm{runs}(\pi)) \geq (\rho - 1)\lg n$.   □

We now consider the case of runs restricted to be strictly incrementing $(+1)$ or decrementing (-1), while still letting them be interleaved: such runs were not directly considered before.

Theorem 4. *Let π be a permutation on n elements that consists of ρ interleaved strictly incrementing or decrementing runs. For any constant $\epsilon > 0$, we can store π in $nH(\mathrm{runs}(\pi)) + o(n)(H(\mathrm{runs}(\pi)) + 1) + \mathcal{O}(\rho n^\epsilon)$ bits and perform $\pi()$ queries in $\mathcal{O}\left(\min\left(1 + \frac{\lg \rho}{\lg \lg n}, \lg \lg \rho\right)\right)$ time and $\pi^{-1}()$ queries in $\mathcal{O}(1/\epsilon)$ time.*

Proof. We first replace all the elements of the rth run by r, for $1 \leq r \leq \rho$, considering the runs in order by minimum element. Let $s \in \{1, \ldots, \rho\}^n$ be the resulting string. We store s using Theorem 1(i); we also store an array containing the runs' lengths, directions (incrementing or decrementing), and minima, in order by minimum element; and store a predecessor data structure containing the runs' minima as keys with their positions in the array as auxiliary information. The predecessor data structure is based on Lemma 4 of Andersson's paper [1]. It is an n^ϵ-ary trie where the keys are sought considering $\epsilon \lg n$ bits per trie node, and hence found in $\mathcal{O}(1/\epsilon)$ time. Each of the ρ elements may require $\mathcal{O}((1/\epsilon)n^\epsilon \lg n)$ bit space for the n^ϵ-size children arrays along its $\mathcal{O}(1/\epsilon)$-length path. By slightly adjusting ϵ the space is $\mathcal{O}(\rho n^\epsilon)$ bits. With these data structures, we can retrieve a run's data given either its array index or any of its elements.

If $\pi(i)$ is the jth element in an incrementing run whose minimum element is m, then $\pi(i) = m + j - 1$; on the other hand, if $\pi(i)$ is the jth element of a decrementing run of length l whose minimum element is m, then $\pi(i) = m + l - j$. It follows that, given i, we can compute $\pi(i)$ by using the query $j = s.\mathrm{rank}_{s[i]}(i)$ and then an array lookup at position $s[i]$ to find m, l and the direction, finally computing $\pi(i)$ from them. Also, given $\pi(i)$, we can compute i by first using a predecessor query to find the run's array position r, then an array lookup to find m, l and the direction, then computing $j = \pi(i) - m + 1$ (increasing) or $j = m + l - \pi(i)$ (decreasing), and finally using the query $i = s.\mathrm{select}_r(j)$.   □

Notice that, if π consists of ρ contiguous increasing or decreasing runs, then π^{-1} consists of ρ interleaved incrementing or decrementing runs. Therefore, Theorem 4 applies to such permutations as well, with the time bounds for $\pi()$ and $\pi^{-1}()$ queries reversed, which yields the following corollary:

Corollary 2. *Let π be a permutation on n elements that consists of ρ contiguous increasing or decreasing runs. For any constant $\epsilon > 0$, we can store π in $nH(\mathrm{runs}(\pi)) + o(n)(H(\mathrm{runs}(\pi)) + 1) + \mathcal{O}(\rho n^\epsilon)$ bits and perform $\pi()$ queries in $\mathcal{O}(1/\epsilon)$ time and $\pi^{-1}()$ queries in $\mathcal{O}\left(\min\left(1 + \frac{\lg \rho}{\lg \lg n}, \lg \lg \rho\right)\right)$ time.*

If π's runs are both contiguous and incrementing or decrementing, then so are the runs of π^{-1}. In this case we can store π in $\mathcal{O}(\rho n^\epsilon)$ bits and answer $\pi()$ and $\pi^{-1}()$ queries in $\mathcal{O}(1)$ time. To do this, we use two predecessor data structures: for each run, in one of the data structures we store the position j in π of the first element of the run, with $\pi(j)$ as auxiliary information; in the other, we store $\pi(j)$, with j as auxiliary information. To perform a query $\pi(i)$, we use the first predecessor data structure to find the starting position j of the run containing i, and return $\pi(j)+i-j$. A $\pi^{-1}()$ query is symmetric. Decreasing runs are handled as before.

Corollary 3. *Let π be a permutation on n elements that consists of ρ contiguous incrementing or decrementing runs. For any constant $\epsilon > 0$, we can store π in $\mathcal{O}(\rho n^\epsilon)$ bits and perform $\pi()$ and $\pi^{-1}()$ queries in $\mathcal{O}(1/\epsilon)$ time.*

We now show how to achieve exponentiation ($\pi^k(i)$, $\pi^{-k}(i)$) within compressed space. Munro et al. [17] reduced the problem of supporting exponentiation on a permutation π to the support of the direct and inverse application of another permutation, related but with quite distinct runs than π. Expressing their result as a succinct index and combining it with any of our results does yield a compression, but one where the space depends of the lengths of both the runs and cycles of π. The following construction, extending the technique from Munro et al. [17], retains the compressibility properties of π by building a companion data structure that uses small space to support the exponentiation, thus allowing the compression of the main data structure with any of our results.

Theorem 5. *Suppose we have a data structure D that stores a permutation π on n elements and supports queries $\pi()$ in time $g(\pi)$. Then for any $t \leq n$, we can build a succinct index that takes $\mathcal{O}((n/t)\lg n)$ bits and, when used in conjunction with D, supports $\pi^k()$ and $\pi^{-k}()$ queries in $\mathcal{O}(t\,g(\pi))$ time.*

Proof. We decompose π into its cycles and, for every cycle of length at least t, store the cycle's length and an array containing pointers to every tth element in the cycle, which we call 'marked'. We also store a compressed binary string, aligned to π, indicating the marked elements. For each marked element, we record to which cycle it belongs and its position in the array of that cycle.

To compute $\pi^k(i)$, we repeatedly apply $\pi()$ at most t times until we either loop (in which case we need apply $\pi()$ at most t more times to find $\pi^k(i)$ in the loop) or we find a marked element. Once we have reached a marked element, we use its array position and cycle length to find the pointer to the last marked element in the cycle before $\pi^k(i)$, and the number of applications of $\pi()$ needed to map that to $\pi^k(i)$ (at most t). A π^{-k} query is similar (note that it does not need to use $\pi^{-1}()$). □

As an example, given a constant $\epsilon > 0$ and a value $t \le n$, we can combine Corollary 2 and Theorem 5 to obtain a data structure that stores Sadakane's Ψ function [21] for s in $nH_0(s) + o(n)(H_0(s) + 1) + \mathcal{O}\left(\sigma n^\epsilon + (n/t)\lg n\right)$ bits and supports $\Psi^k()$ and $\Psi^{-k}()$ queries in $\mathcal{O}\left(1/\epsilon + t\right)$ time; these queries are useful when working on compressed suffix arrays and trees.

5 Compressing Functions and Dynamic Collections of Disjoint Sets

Hreinsson, Krøyer and Pagh [12] recently showed how, given $X = \{x_1, \ldots, x_n\} \subseteq [U]$ and $f : [U] \to [1..\sigma]$, where $[U]$ is the set of numbers whose binary representations fit in a machine word, they can store f restricted to X in compressed form with constant-time evaluation. Their representation uses at most $(1 + \delta)nH_0(f) + n\min(p_{\max} + 0.086, 1.82(1 - p_{\max})) + o(\sigma)$ bits, where $\delta > 0$ is a given constant and $p_{\max}$ is the relative frequency of the most common function value. We note that this bound holds even when $\sigma \gg n$.

Notice that, in the special case where $X = [1..n]$ and $\sigma \le n$, we can achieve constant-time evaluation and a better space bound using Theorem 1. We can also find all the elements in $[1..n]$ that f maps to a given element in $[1..\sigma]$ (using select), find an element's rank among the elements with the same image, or the size of the preimage (using rank), etc.

Theorem 6. *Let $f : [1..n] \to [1..\sigma]$ be a surjective function.[5] We can represent f using $nH_0(f) + o(n)(H_0(f) + 1)$ bits so that any $f(i)$ can be computed in $\mathcal{O}(1)$ time. Moreover, each element of $f^{-1}(a)$ can be computed in $\mathcal{O}(\lg \lg \sigma)$ time, and $|f^{-1}(a)|$ requires time $\mathcal{O}(\lg \lg \sigma \lg \lg \lg \sigma)$. Alternatively we can compute $f(i)$ and $|f^{-1}(a)|$ in time $\mathcal{O}(\lg \lg \sigma)$ and deliver any element of $f^{-1}(a)$ in $\mathcal{O}(1)$ time.*

We omit the other improvements of Theorem 1 and Corollary 1 for conciseness. We can also achieve interesting results with our theorems from Section 4, as runs arise naturally in many real-life functions. For example, suppose we decompose $f(1), \ldots, f(n)$ into ρ interleaved non-increasing or non-decreasing runs. Then we can store it as a combination of the permutation π that stably sorts the values $f(i)$, plus a compressed rank/select data structure storing a binary string $b[1..n + \sigma + 1]$ with $\sigma + 1$ bits set to 1: if f maps i values in $[1..n]$ to a value j in

---

[5] So that $[1..\sigma]$ is the effective alphabet size of string f. General functions with image of size $\sigma' < \sigma$ require $\mathcal{O}(\sigma' \lg(\sigma/\sigma')) + o(\sigma)$ extra bits, or we can handle them using $\mathcal{O}(\sigma \lg \lg n)$ bits with our structure M.

$[1..\sigma]$ then, in b, there are i bits set to 0 between the jth and $(j+1)$th bits set to 1. Therefore,

$$f(i) = b.\mathsf{rank}_1(b.\mathsf{select}_0(\pi(i)))$$

and the theorem below follows immediately from Theorem 3. Similarly, $f^{-1}(a)$ is obtained by applying $\pi^{-1}()$ to the area $b.\mathsf{rank}_0(b.\mathsf{select}_1(a))+1\ldots b.\mathsf{rank}_0(b.\mathsf{select}_1(a+1))$, and $|f^{-1}(a)|$ is computed in $\mathcal{O}(1)$ time. Notice that $H(\mathsf{runs}(\pi)) = H(\mathsf{runs}(f)) \le H_0(f)$, and that b can be stored in $\mathcal{O}\left(\sigma \lg \frac{n}{\sigma}\right) + o(n)$ bits [20].

Corollary 4. *Let $f : [1..n] \to [1..\sigma]$ be a surjective function[6] with $f(1),\ldots,f(n)$ consisting of ρ interleaved non-increasing or non-decreasing runs. Then we can store f in $2nH(\mathsf{runs}(f)) + o(n)(H(\mathsf{runs}(f))+1) + \mathcal{O}\left(\sigma \lg \frac{n}{\sigma}\right)$ bits and compute any $f(i)$, as well as retrieve any element in $f^{-1}(a)$, in $\mathcal{O}(\lg \lg \rho)$ time. The size $|f^{-1}(a)|$ can be computed in $\mathcal{O}(1)$ time.*

We can obtain a more competitive result if f is split into contiguous runs, but their entropy is no longer bounded by the zero-order entropy of string f.

Corollary 5. *Let $f : [1..n] \to [1..\sigma]$ be a surjective function with $f(1),\ldots,f(n)$ consisting of ρ contiguous non-increasing or non-decreasing runs. Then we can represent f in $nH(\mathsf{runs}(f))+o(n)(H(\mathsf{runs}(f))+1)+\mathcal{O}(\rho n^\epsilon)+\mathcal{O}\left(\sigma \lg \frac{n}{\sigma}\right)$ bits, for any constant $\epsilon > 0$, and compute any $f(i)$ in $\mathcal{O}(\lg \lg \sigma)$ time, as well as retrieve any element in $f^{-1}(a)$ in $\mathcal{O}(1/\epsilon)$ time. The size $|f^{-1}(a)|$ can be computed in $\mathcal{O}(1)$ time.*

Finally, we now give what is, to the best of our knowledge, the first result about storing a compressed collection of disjoint sets. The key point in the next theorem is that, as the sets in the collection C are merged, our space bound shrinks with the entropy of the distribution $\mathsf{sets}(C)$ of elements to sets.

Theorem 7. *Let C be a collection of disjoint sets whose union is $[1..n]$. For any constant $\epsilon > 0$, we can store C in $(1+\epsilon)nH(\mathsf{sets}(C)) + \mathcal{O}(|C|\lg n) + o(n)$ bits and perform any sequence of m union and find operations in a total of $\mathcal{O}((1/\epsilon)(m+n)\lg \lg n)$ time.*

Proof. We first use Theorem 1 to store the string $s[1..n]$ in which each $s[i]$ is the representative of the set containing i. We then store the representatives in a standard disjoint-set data structure D [22]. Together, our data structures take $nH(\mathsf{sets}(C)) + \mathcal{O}(|C|\lg n) + o(n)(H(\mathsf{sets}(C))+1)$ bits. We can perform a query find(i) on C by performing $D.\mathsf{find}(s[i])$, and perform a union(i,j) operation on C by performing $D.\mathsf{union}(D.\mathsf{find}(s[i]), D.\mathsf{find}(s[j]))$.

For our data structure to shrink as we union sets, we keep track of $H(\mathsf{sets}(C))$ and, whenever it shrinks by a factor of $1+\epsilon$, we rebuild our entire data structure on the updated values $s[i] \leftarrow \mathsf{find}(s[i])$. First, note that all those find operations take $\mathcal{O}(n)$ time because of path-compression [22]: Only the first time one accesses a node $v \in C$ it may occur that the representative is not directly v's parent.

---

[6] Otherwise we proceed as usual to map the domain to the effective one.

Reconstructing the structure of Theorem 1 takes also $\mathcal{O}(n)$ time: As we need just access on s, we need only rank and access on our multiary wavelet tree and access on the s_ℓ sequences. Thus the latter are implemented simply as arrays and the former are also easily built in linear time for these two queries [8].

Since $H(\text{sets}(C))$ is always less than $\lg n$, we rebuild only $\mathcal{O}\left(\log_{1+\epsilon} \lg n\right) = \mathcal{O}\left((1/\epsilon)\lg\lg n\right)$ times. Finally, the space term $o(n)H(\text{sets}(C))$ is absorbed by $\epsilon H(\text{sets}(C))$ by slightly adjusting ϵ.   □

6   Conclusions and Future Work

We have presented the first zero-order compressed representation of strings efficiently supporting queries access, rank, and select, so that the redundancy of the compressed representation is also compressed. That is, our space for string $s[1..n]$ over alphabet $[1..\sigma]$ is $nH_0(s) + o(n)(H_0(s) + 1)$ instead of the usual $nH_0(s) + o(n)\lg\sigma$ bits. This is very important in many practical applications where the data is highly compressible and the redundancy would otherwise dominate the overall space.

In the full paper we will work on several improvements and further applications. First, we can reduce the dependence on the alphabet size from $\mathcal{O}(\sigma\lg\lg n)$ to $\mathcal{O}(\sigma)$ by storing a length-restricted Shannon code in $\mathcal{O}(\sigma)$ bits [9] instead of the data structure M. To avoid the $\mathcal{O}(1)$ extra redundancy per character associated with using a length-restricted prefix code, we replace each character in s whose codeword length is at most $\lg\lg n$ by a distinct number in t. This increases the alphabet size of t by at most $\lg n$; calculation shows that our space bound increases by an $\mathcal{O}(1 + 1/\lg\lg n)$-factor and, thus, remains at most $nH_0(s) + o(n)(H_0(s)+1)$. Second, given any constant c, we can reduce the $\min(\sigma, n/|s|_a)$ in our time bounds by a factor of $(\lg n)^c$; to do this, we further partition each sub-alphabet into $(\lg n)^c$ sub-sub-alphabets. Third, our alphabet partitioning techniques yields a compressed representation of posting lists of sizes $(n_1, \ldots, n_\sigma)$ which supports access, rank and select on the rows in time $\mathcal{O}(\lg\lg\sigma)$, and uses total space for data and index proportional to the entropy $H(n_1, \ldots, n_\sigma)$ of the distribution of those sizes (if the posting lists refer to the words of a text, this is also the zero-order word-based entropy of the text). This is achieved by encoding the string of labels encountered during a row-first traversal, writing a special symbol (e.g. \$) at each change of row. This improves the space of previously known data structures [2], and improves the time complexity of previous compression results [5].

Naturally, the next challenge ahead is to obtain a data structure using space $nH_k(s)+o(n)(H_k(s)+1)$ bits rather than $nH_k(s)+o(n)\lg\sigma$, while still supporting the queries access, rank, and select, in reasonable time. Note that Barbay et al. [3] achieve $nH_k(s) + o(n)\lg\sigma$ for such a structure: we have reduced the redundancy to $o(n)(H_k(s) + 1)$ for the case $k = 0$ and for self-indexes, but not for the basic problem in the general case where $k = o(\log_\sigma n)$.

Acknowledgments. Many thanks to Djamal Belazzougui for helpful comments on a draft of this paper.

References

1. Andersson, A.: Sorting and searching revisited. In: Karlsson, R., Lingas, A. (eds.) SWAT 1996. LNCS, vol. 1097, pp. 185–197. Springer, Heidelberg (1996)
2. Barbay, J., Golynski, A., Munro, J.I., Rao, S.S.: Adaptive searching in succinctly encoded binary relations and tree-structured documents. Theoretical Computer Science 387(3), 284–297 (2007)
3. Barbay, J., He, M., Munro, J.I., Rao, S.S.: Succinct indexes for strings, binary relations and multi-labeled trees. In: Proc. 18th SODA, pp. 680–689 (2007)
4. Barbay, J., Navarro, G.: Compressed representations of permutations, and applications. In: Proc. 26th STACS, pp. 111–122 (2009)
5. Blandford, D., Blelloch, G.: Index compression through document reordering. In: Proc. DCC, pp. 342–351 (2002)
6. Burrows, M., Wheeler, D.: A block sorting lossless data compression algorithm. Technical Report 124, Digital Equipment Corporation (1994)
7. Claude, F., Navarro, G.: Practical rank/select queries over arbitrary sequences. In: Amir, A., Turpin, A., Moffat, A. (eds.) SPIRE 2008. LNCS, vol. 5280, pp. 176–187. Springer, Heidelberg (2008)
8. Ferragina, P., Manzini, G., Mäkinen, V., Navarro, G.: Compressed representations of sequences and full-text indexes. ACM Transactions on Algorithms 3(2)
9. Gagie, T., Navarro, G., Nekrich, Y.: Fast and compact prefix codes. In: van Leeuwen, J., Muscholl, A., Peleg, D., Pokorný, J., Rumpe, B. (eds.) SOFSEM 2010. LNCS, vol. 5901, pp. 419–427. Springer, Heidelberg (2010)
10. Golynski, A., Munro, J.I., Rao, S.S.: Rank/select operations on large alphabets: a tool for text indexing. In: Proc. 17th SODA, pp. 368–373 (2006)
11. Grossi, R., Gupta, A., Vitter, J.: High-order entropy-compressed text indexes. In: Proc. 14th SODA, pp. 841–850 (2003)
12. Hreinsson, J.B., Krøyer, M., Pagh, R.: Storing a compressed function with constant time access. In: Fiat, A., Sanders, P. (eds.) ESA 2009. LNCS, vol. 5757, pp. 730–741. Springer, Heidelberg (2009)
13. Levcopoulos, C., Petersson, O.: Sorting shuffled monotone sequences. Information and Computation 112(1), 37–50 (1994)
14. Mäkinen, V., Navarro, G.: Rank and select revisited and extended. Theoretical Computer Science 387(3), 332–347 (2007)
15. Manzini, G.: An analysis of the Burrows-Wheeler transform. Journal of the ACM 48(3), 407–430 (2001)
16. Mehlhorn, K.: Sorting presorted files. In: Weihrauch, K. (ed.) GI-TCS 1979. LNCS, vol. 67, pp. 199–212. Springer, Heidelberg (1979)
17. Munro, J.I., Raman, R., Raman, V., Rao, S.S.: Succinct representations of permutations. In: Baeten, J.C.M., Lenstra, J.K., Parrow, J., Woeginger, G.J. (eds.) ICALP 2003. LNCS, vol. 2719, pp. 345–356. Springer, Heidelberg (2003)
18. Navarro, G., Mäkinen, V.: Compressed full-text indexes. ACM Computing Surveys 39(1):article 2 (2007)
19. Rahman, N., Raman, R.: Rank and select operations on binary strings. In: Kao, M.-Y. (ed.) Encyclopedia of Algorithms. Springer, Heidelberg (2008)
20. Raman, R., Raman, V., Rao, S.: Succinct indexable dictionaries with applications to encoding k-ary trees and multisets. In: Proc. 13th SODA, pp. 233–242 (2002)
21. Sadakane, K.: New text indexing functionalities of the compressed suffix arrays. Journal of Algorithms 48(2), 294–313 (2003)
22. Tarjan, R.E., van Leeuwen, J.: Worst-case analysis of set union algorithms. Journal of the ACM 31(2), 245–281 (1984)

Entropy-Bounded Representation of Point Grids

Arash Farzan[1], Travis Gagie[2,*], and Gonzalo Navarro[2,*]

[1] Max-Planck-Institut für Informatik
afarzan@mpi-inf.mpg.de
[2] Department of Computer Science, University of Chile
{tgagie, gnavarro}@dcc.uchile.cl

Abstract. We give the first *fully compressed* representation of a set of m points on an $n \times n$ grid, taking $H + o(H)$ bits of space, where $H = \lg \binom{n^2}{m}$ is the entropy of the set. This representation supports range counting, range reporting, and point selection queries, with a performance that is comparable to that of uncompressed structures and that improves upon the only previous compressed structure. Operating within entropy-bounded space opens a new line of research on an otherwise well-studied area, and is becoming extremely important for handling large datasets.

1 Introduction

A point grid is an extremely basic structure underlying the representation of two-dimensional point sets, graphics, spatial databases, geographic data, binary relations, graphs, images, and so on. It has been intensively studied from a computational geometry viewpoint, where most of the focus has been on two basic primitives: (orthogonal) range counting (how many points are there in this rectangle?), and (orthogonal) range reporting (list the points falling within this rectangle). More sophisticated queries are possible if points have associated values, and also more general shapes than rectangles have been considered.

Consider a $n \times n$ grid containing m points. Currently the best results related to the focus of this paper are as follows. Range counting can be done in time $\mathcal{O}\left(\frac{\lg m}{\lg \lg m}\right)$ and linear space, that is, $\mathcal{O}(m)$ integers [12]. The preprocessing time is $\mathcal{O}(m\sqrt{\lg m})$ [5]. That counting time cannot be improved within space $\mathcal{O}(m \, \text{polylog}(m))$ [17]. Range reporting can be done in time $\mathcal{O}(\lg \lg m + k)$, where k is the number of points reported, using $\mathcal{O}(m \lg^\epsilon m)$ integers for any constant $\epsilon > 0$ [1]. The time raises to $\mathcal{O}(\lg \lg m (\lg \lg m + k))$ if the space is reduced to $\mathcal{O}(m \lg \lg m)$ integers [1], and it reaches $\mathcal{O}\left(\frac{\lg m}{\lg \lg m} + k \lg^\epsilon m\right)$ if the space is linear [15]. There are also some bounds that may be relevant when many points are to be reported: $\mathcal{O}(\lg m + k \lg \lg(4m/k))$ time using $\mathcal{O}(m \lg \lg m)$ integers, and $\mathcal{O}(\lg m + k \lg^\epsilon(2m/k))$ time using $\mathcal{O}(m)$ integers [6]. Some of these results have been matched even in the dynamic scenario [14].

Many of the application areas for this problem handle huge volumes of information, and thus even the "linear" space structures might be excessively large.

---

* Partially funded by Fondecyt Grant 1-080019, Chile.

O. Cheong, K.-Y. Chwa, and K. Park (Eds.): ISAAC 2010, Part II, LNCS 6507, pp. 327–338, 2010.
© Springer-Verlag Berlin Heidelberg 2010

Since each point requires storing two coordinates, the space complexities above should be multiplied by $2 \lg n$ bits. Unless m is very small, even storing these bare coordinates uses much more space than necessary, and moreover the constants hidden within the $\mathcal{O}(\ldots)$ notation are not negligible. Some succinct data structures have been designed using, for example, $m \lg m + o(m \lg m)$ bits when $m = n$, answering range counting queries in time $\mathcal{O}\left(\frac{\lg m}{\lg \lg m}\right)$ and reporting in time $\mathcal{O}\left((k+1)\frac{\lg m}{\lg \lg m}\right)$ [4]. That space is not the best possible when m is larger than n. A (worst-case) lower bound on the number of bits needed to represent a grid is the logarithm of the number of possible grids, called the "entropy" $H = \lg \binom{n^2}{m} = m \lg \frac{n^2}{m} + \mathcal{O}(m + \lg n)$.

To the best of our knowledge, the only previous work achieving this "compressed" space is by Barbay et al. [2]. They propose a data structure using $H + o(H) + \mathcal{O}(m) + o(n)$ bits (see our Thm. 2). Within this space they solve many interesting range counting, range reporting, and point selection queries (give the kth point in a rectangle, according to some order) in $\mathcal{O}(\lg n)$ time per delivered point. Contrarily to succinct indexes, they propose an integrated encoding, where the data and the index are stored together.

In this paper we push further in the direction of storing the grid data within its entropy bound, improving simultaneously the space redundancy and the time performance of queries. Most notably, we achieve a *fully compressed* representation taking $H + o(H)$ bits of space, while supporting the operations in constant time in some cases. Depending on m, we use different data structures to achieve this goal. The result is summarized in Thm. 1; see Section 2.2 for more details.

Theorem 1. *An $n \times n$ grid with m points can be represented within $H + o(H)$ bits of space, where $H = \lg \binom{n^2}{m}$, so that orthogonal range counting can be answered in $\mathcal{O}\left(\lg \frac{n^2}{m}\right)$ time, range reporting in time $\mathcal{O}\left(\lg^2 \frac{n^2}{m}\right)$ per delivered datum, and point selection queries in at most $\mathcal{O}(\lg^2 n)$ time. Depending on the density of the matrix the times are reduced down to $\mathcal{O}(1)$ per delivered datum.*

The paper is organized as follows. Section 2 gives basic concepts on bitmaps and point grids, defines the problems we address, proves some technical results needed later, and summarizes the results we achieve. Section 3 describes a "compressed" representation taking $H + o(n^2)$ bits of space and achieving constant time for range counting and reporting. Section 4 achieves the "fully compressed" space, in exchange for higher query times, and finishes with the proof of Thm. 1. Section 5 concludes and gives further research directions.

2   Basic Concepts

2.1   The One-Dimensional Case

The one-dimensional variant of the problem, i.e., on a bitmap $B[1, n]$, has been long studied. Let B have m bits set, then the entropy of the bitmap is $H = \lg \binom{n}{m}$.

All the range counting, range reporting, and point locating queries can be solved in terms of two primitives: $\mathsf{rank}_b(B, i)$ is the number of occurrences of bit b in $B[1, i]$, and $\mathsf{select}_b(B, j)$ is the position in B of the jth occurrence of bit b.

Clark [8] and Munro [13] showed that both rank and select can be solved in constant time using $n + o(n)$ bits of space, that is, B itself plus sublinear space. Golynski et al. [9] showed that the $o(n)$ term must be $\Omega(\frac{n \lg \lg n}{\lg n})$ if B is stored in plain form, and moreover achieved this bound. Raman et al. [19] provided a compressed representation retaining constant query times and taking $H + \mathcal{O}\left(m + \frac{n \lg \lg n}{\lg n}\right)$ bits. We prove now a technical lemma we will need later.

Lemma 1. *Let $0 < \alpha \leq 1$ be a constant and $b = \Theta(\lg^\alpha n)$. Let bitmap $B[1, n]$ be stored in a way such that we can only access pieces $B[(i - 1) \cdot b + 1, i \cdot b]$ at a time, for any i. Then we can perform rank and select in constant time using $\mathcal{O}\left(\frac{n \lg \lg n}{b}\right)$ bits of extra space, and this is optimal.*

Proof. Let us take any algorithm achieving constant time and $\mathcal{O}\left(\frac{n \lg \lg n}{\lg n}\right)$ extra space, say Golynski's [9], and adapt it to this restriction. The algorithm builds and uses several indexes and accesses B a constant number of times. Each such time, it reads a "word" of $w = \mathcal{O}(\lg n)$ consecutive bits of B, in order to either (a) count the number of 1s in a part of the word or (b) find the position of the kth 1 or 0 in a part of the word, using universal tables.

We introduce an indirection when accessing such universal tables. Each word is covered by $\frac{w}{b}$ pieces. For each piece, we store the *summary* number of 1s in the piece. This requires $\lg(b + 1)$ bits, so the total space is $\mathcal{O}\left(\frac{n \lg b}{b}\right) = \mathcal{O}\left(\frac{n \lg \lg n}{b}\right)$. Moreover, in a RAM machine with word size w we can read all the summary numbers of the pieces covering any word in $\mathcal{O}(1)$ accesses, as they add up to $\frac{w \lg b}{b} = o(\lg n)$ bits. With these summary numbers we can index a universal table of size $\mathcal{O}(2^{o(\lg n)}\mathrm{polylog}(n))$, telling (a) the number of bits set up to any given piece of the word, and (b) the piece where the kth 0/1 of the word occurs. A final access to one b-bit piece, with another universal table of size $\mathcal{O}(2^b \mathrm{polylog}(n))$, completes the query in constant time.

The lower bound comes directly from Golynski [9], who states that if one probes t bits and answers $\mathsf{rank}/\mathsf{select}$ in constant time, then the index must be of size $\Omega(\frac{m \lg t}{t})$. In the worst case $m = n$ and the algorithm can access at most $t = \mathcal{O}(b)$ bits in constant time. $\square$

2.2 Two Dimensions

We can identify a grid with a binary matrix. We will consider rectangular ranges of the form $(i_1, i_2) \times (j_1, j_2)$, where i_1 and i_2 are rows and j_1 and j_2 are columns. Over those ranges we define the queries

- $\mathsf{rank}(i_1, i_2, j_1, j_2)$ counts the number of points in the range; and
- $\mathsf{select}(i_1, i_2, j_1, j_2, k_1, k_2)$ gives the k_1th to the k_2th points in the range, in row-major or column-major order (this generalizes from range reporting and point selection queries).

The general case is called a 4-sided query. A particular case, a 3-sided query, arises when one of the coordinates is always 1 or n. A 2-sided query arises when two of the coordinates, one of row and one of column, is always 1 or n. A band-query has 1 and n for either the row or the column coordinates. Finally, a 1-sided query has only one coordinate different from 1 or n.

Since $\mathsf{rank}(i_1, i_2, j_1, j_2) = \mathsf{rank}(1, i_2, 1, j_2) - \mathsf{rank}(1, i_1 - 1, 1, j_2) - \mathsf{rank}(1, i_2, 1, j_1 - 1) + \mathsf{rank}(1, i_1 - 1, 1, j_1 - 1)$, we study only 2-sided queries for rank. For compliance with the existing literature, we prefer to study the queries in terms of selecting the kth point, $\mathsf{select}(i_1, i_2, j_1, j_2, k)$, and reporting (up to) k points in a range, $\mathsf{report}(i_1, i_2, j_1, j_2, k)$. Our solutions, however, can actually be combined to solve the general $\mathsf{select}(i_1, i_2, j_1, j_2, k_1, k_2)$ query within the time of selecting the k_1th point and then reporting the $k_2 - k_1 + 1$ points following it. Furthermore, our rank query is never slower than our select, and $\mathsf{select}(i_1, i_2, j_1, j_2, k) = \mathsf{select}(i_1, i_2, 1, n, k + x)$ with $x = \mathsf{rank}(i_1, i_2, 1, j_1 - 1)$ if select delivers in column-major order, and analogously in row-major order. Finally, our sublinear-sized indexes can be computed (or the algorithms trivially modified) for several rotations and reflections of the grid within the same asymptotic space. Therefore we can, without loss of generality, focus our study on the following queries:

- $\mathsf{rank}(i, j)$ is the number of points in $(1, i) \times (1, j)$;
- $\mathsf{select}(i_1, i_2, k)$ gives the kth point in the range $(i_1, i_2) \times (1, n)$, in column-major order (as explained this allows one to emulate any 4-sided query);
- $\mathsf{select}(i, k)$ gives the kth point in the range $(1, i) \times (1, n)$, in column-major order (this allows one to emulate any 3-sided query); and
- $\mathsf{report}(i_1, i_2, j_1, k)$ gives the first (up to) k points in the range $(i_1, i_2) \times (j_1, n)$, in column-major order.

Barbay et al. [2] propose a number of primitives on binary matrices, yet several can be reduced to others. Their maximal operations (in the sense that the others reduce to a constant number of applications of these) are `rel_rnk` (equivalent to our rank), `rel_sel_obj_maj` and `rel_sel_lab_maj` (equivalent to our select), and `lab_rnk` and `obj_rnk` (which count the number of nonempty rows/columns within a range and have no equivalent in this paper). By using *wavelet trees* [10], they achieve the following result (adapted and fixed here):

Theorem 2 ([2]). *A binary matrix of σ rows ("labels") by n columns ("objects") with t 1s can be represented within $H + o(H) + \mathcal{O}\left(t + \frac{n \lg \lg n}{\lg n}\right)$ bits, so that queries* `rel_rnk`(i_1, i_2, j_1, j_2) *(number of points in $(i_1, i_2) \times (j_1, j_2)$),* `rel_sel_lab_maj`$(i, k, j_1, j_2)$ *(kth point, in label-major order, in $(i, \sigma) \times (j_1, j_2)$), and* `rel_min_obj_maj`$(i_1, i_2, j)$ *and* `rel_acc_obj_maj`(i_1, i_2, j) *(first and successive points, in object-major order, in $(i_1, i_2) \times (j, n)$), can be answered in $\mathcal{O}(\lg \sigma)$ time per delivered datum.* `Rel_sel_obj_maj`(i_1, i_2, j, k) *(kth point, in object-major order, in $(i_1, i_2) \times (j, n)$), can be carried out in $\mathcal{O}(\lg \sigma \lg n)$ time.*

Table 1. Space and time complexities achieved by Barbay et al. [2] and in this paper. The "or" case depends on using row-major or column-major order. The times for Thm. 4 are simplified assuming $m = \mathcal{O}\left(\frac{n^2}{\lg^{1/4} n}\right)$; otherwise Thm. 3 takes over. The times to operate on 0s are the same for Thm. 3; for Thm. 4 we give them explicitly.

Source	Space	rank time	report time
Thm. 2 [2]	$H + o(H) + \mathcal{O}(m) + o(n)$	$\lg n$	$(k+1)\lg n$
Thm. 3	$H + \mathcal{O}\left(\frac{n^2 \lg \lg n}{\lg^{1/4} n}\right)$	1	$k+1$
Thm. 4	$H + o(H) + \mathcal{O}(m + \lg n)$	$\lg \frac{n^2}{m}$	$(k+1)\lg \frac{n^2}{m}$
Thm. 4 (0s)		$\lg \frac{n^2}{m}$	$(k+1)\lg^2 \frac{n^2}{m}$
Thm. 1	$H + o(H)$	$\lg \frac{n^2}{m}$	$(k+1)\lg^2 n$

Source	select time (4-sided)	select time (3-sided)
Thm. 2 [2]	$\lg n$ or $\lg^2 n$	$\lg n$
Thm. 3	$\lg n$	$\lg \lg n$
Thm. 4	$\lg n$ or $\lg n + \lg^2 \frac{n^2}{m}$	$\lg \frac{n^2}{m}$ or $\lg n + \lg^2 \frac{n^2}{m}$
Thm. 4 (0s)	$\lg n + \lg^2 \frac{n^2}{m}$	$\lg n + \lg^2 \frac{n^2}{m}$
Thm. 1	$\lg^2 n$	$\lg^2 n$

Proof. This is in their Thm. 2 [2]. Their space formula "$t \lg \sigma + o(t) \lg \sigma + \mathcal{O}(\min(t, n \lg \frac{t}{n}))$" should indeed be $t \lg \sigma + o(t) \lg \sigma + \lg \binom{n+t}{t} + \mathcal{O}(\min(n,t)) + \mathcal{O}\left(\frac{(n+t)\lg \lg(n+t)}{\lg(n+t)}\right)$. Since the last three terms are $t \lg \frac{n}{t} + \mathcal{O}\left(t + \frac{n \lg \lg n}{\lg n}\right)$, we have the total $t \lg \frac{n\sigma}{t} + o(t) \lg \frac{n\sigma}{t} + \mathcal{O}\left(t + \frac{n \lg \lg n}{\lg n}\right) = H + o(H) + \mathcal{O}\left(t + \frac{n \lg \lg n}{\lg n}\right)$. $\square$

Table 1 compares the previous and new complexities achieved for our operations. The previous compressed representation [2] achieves $H + o(H)$ bits only if $\omega\left(\frac{n \lg \lg n}{\lg^2 n}\right) = m = o(n^2)$ (note our m is their t). Also, it always supports rank in time $\mathcal{O}(\lg n)$. This time is $\mathcal{O}(1)$ in our "compressed" solution, and in our "fully compressed" solution it is $\mathcal{O}\left(\lg \frac{n^2}{m}\right)$ in the range $m = \mathcal{O}\left(\frac{n^2}{\lg^{1/4} n}\right)$. This is never worse than the previous result [2], and is strictly better if $m = n^{2-o(1)}$. For report and select we are faster or slower depending on the case.

3   A Compressed Representation

We first describe a solution using $n^2 + o(n^2)$ bits, and then convert it into one using $H + o(n^2)$ bits.

3.1   Constant-Time *Rank*

The matrix is first subdivided into *superblocks* of size $s \times s$, $s = \lg^2 n$. Each superblock is in turn subdivided into *blocks* of size $b \times b$, $b = \sqrt{\frac{\lg n}{2}}$. The n^2 bits of the matrix will be stored block-wise, that is, the $b^2 = \frac{\lg n}{2}$ bits of each block will be stored contiguously.

For each superblock in the matrix, we store the rank values at all the positions of the rightmost column and bottom row of the superblock. In other words, we store all $\mathsf{rank}(i, s \cdot j)$ and $\mathsf{rank}(s \cdot i, j)$ values. This requires $\mathcal{O}\left(\frac{n^2 \lg n}{\lg^2 n}\right) = o(n^2)$ bits. For each block within each superblock, we store the *local* (i.e. within its superblock) rank values at all the positions of the rightmost column and bottom row of the block. If we call rank_s those local rank values, what we store is all $\mathsf{rank}_s(i, b \cdot j)$ and $\mathsf{rank}_s(b \cdot i, j)$ values. This requires $\mathcal{O}\left(\frac{n^2 \lg \lg n}{\sqrt{\lg n}}\right) = o(n^2)$ bits.

This gives enough information to compute $\mathsf{rank}(i, j)$ in constant time. Let $i = s \cdot i_s + i_{rs}$ and $j = s \cdot j_s + j_{rs}$, so that $s \cdot i_s$ and $s \cdot j_s$ are the projections of i and j to the last superblock-aligned row and column, and $0 \le i_{rs}, j_{rs} < s$ are the local positions within their superblock. Similarly, let $i_{rs} = b \cdot i_b + i_{rb}$ and $j_{rs} = b \cdot j_b + j_{rb}$, with $0 \le i_{rb}, j_{rb} < b$ the projections into, and local coordinates within, the blocks. Then it is easy to verify that

$$
\begin{aligned}
\mathsf{rank}(i, j) = {} & \mathsf{rank}_b(i, j) \\
& + \mathsf{rank}_s(i, b \cdot j_b) + \mathsf{rank}_s(b \cdot i_b, j) - \mathsf{rank}_s(b \cdot i_b, b \cdot j_b) \\
& + \mathsf{rank}(i, s \cdot j_s) + \mathsf{rank}(s \cdot i_s, j) - \mathsf{rank}(s \cdot i_s, s \cdot j_s),
\end{aligned}
$$

where $\mathsf{rank}_b(i, j)$ is the local rank value within its block. All the rank and rank_s values in the formula are stored. As for $\mathsf{rank}_b(i, j)$, this is $\mathsf{rank}(i_{rb}, j_{rb})$ within its block. As there are only $2^{b^2} = \sqrt{n}$ different blocks, we can store all the answers to all possible (local) rank queries within $\mathcal{O}(\sqrt{n}\,\mathrm{polylog}(n)) = o(n)$ bits. Since we can read at once the $b^2 = \mathcal{O}(\lg n)$ bits of the block (stored contiguously as explained), we can look up a table entry in constant time.

3.2   Constant-Time *Report*

We first solve a subproblem that might have independent interest. Given a row range $[i_1, i_2]$ and a column j, $nextCol(i_1, i_2, j)$ is the smallest column number $j' > j$ that is nonempty (i.e., contains a 1) in the range $[i_1, i_2]$. We now show how to support this query in constant time and $o(n^2)$ extra space.

The key idea is to keep signature bit vectors which represent the bitwise-*or* of various contiguous ranges of matrix rows. First we divide the rows into *batches* of $s = \lg^2 n$ rows. Akin to the classical solution to range minimum queries (RMQs) [3], we explicitly store bit vectors of length n which are the *or* of batches i to $i + 2^k - 1$ for all $1 \le i \le n/s$, $0 \le k \le \lg(n/s)$. Furthermore, we enhance these bit vectors with one-dimensional constant-time rank and select structures. This requires $\mathcal{O}\left(\frac{n}{\lg^2 n} \cdot n \lg n\right) = o(n^2)$ bits and reduces the query $nextCol(s \cdot i_1, s \cdot i_2 - 1, j)$ to that of finding the next 1 after position j in either of two bit vectors (the one *or*-ing batches i_1 to $i_1 + 2^k - 1$ and the one *or*-ing batches $i_2 - 2^k$ to $i_2 - 1$, for $k = \lfloor \lg(i_2 - i_1) \rfloor$). Finding the next 1 in a bit vector is easily reduced to one-dimensional rank and select queries, $j' = \mathsf{select}_1(B, \mathsf{rank}_1(B, j) + 1)$.

A general range $[i_1, i_2]$ may contain several batches, plus possibly two within-batch areas at each extreme. Thus we have reduced the problem to within batches of size $\lg^2 n$. We now repeat the partition similarly within each batch. We divide the rows into *chunks* of $d = \lg^{1/4} n$ rows and again use the same machinery to isolate the problem to within chunks. The extra space for all the chunk-level bit vectors is $\mathcal{O}\left(\frac{n}{\lg^{1/4} n} \cdot n \lg(\lg^2 n)\right) = o(n^2)$.

Now, confined within a chunk of d rows, we consider bit vectors $B(i_1, i_2)$, $1 \leq i_1, i_2 < d$, such that $B(i_1, i_2)$ is the *or* of rows from i_1 to i_2. We cannot explicitly store all these vectors, as the space would be $\omega(n^2)$. However, we do explicitly store the rank and select *indexes* for each such bit vector. To simulate access to the virtual bit vector $B(i_1, i_2)$, we use our $b \times b$ matrix blocks stored contiguously, in order to provide in constant time any $\mathcal{O}(\sqrt{\lg n})$ bits of any horizontal strip of width $i_2 - i_1 + 1$. By Lemma 1, we can in this case achieve constant time for rank and select using extra indexes of size $\mathcal{O}\left(\frac{n \lg \lg n}{\sqrt{\lg n}}\right)$.

As there are $\mathcal{O}\left(\frac{n}{\lg^{1/4} n}\right)$ chunks, each storing $\mathcal{O}\left((\lg^{1/4} n)^2\right)$ indexes for $B(i_1, i_2)$, the total space is $\mathcal{O}\left(\frac{n}{\lg^{1/4} n} \cdot \sqrt{\lg n} \cdot \frac{n \lg \lg n}{\sqrt{\lg n}}\right) = o(n^2)$ bits. The $nextCol(i_1, i_2, j)$ query is thus solved by consulting at most 2 batch bitmaps, 4 chunk bitmaps, and 2 (virtual) $B(i_1, i_2)$ bitmaps. With rank and select on each, we easily find the next 1 after position j across the 8 bit vectors, in constant time.

Once $nextCol$ is solved, it is easy to address $report(i_1, i_2, j_1, j_2, k)$ queries. We store one-dimensional rank and select indexes for every column of the matrix. As already explained, their extra space adds up to $\mathcal{O}\left(\frac{n \lg \lg n}{\sqrt{\lg n}}\right) = o(n)$ per column as we can access only $\mathcal{O}(\sqrt{\lg n})$ contiguous bits of any column. The first points to report are at column $j = nextCol(i_1, i_2, j_1 - 1)$. With one-dimensional rank and select on column j, we can report the points at rows $[i_1, i_2]$ of that column, each in constant time. We go on with $j = nextCol(i_1, i_2, j)$, and so on, until either $j > j_2$ or we have reported k points. Thus the query takes time $\mathcal{O}(k + 1)$.

3.3   *Select* Queries

For $select(i_1, i_2, k)$ we binary search, using rank, the position of the kth point in $\mathcal{O}(\lg n)$ time. We can do better for the simpler $select(i, k)$ query. We have already stored the rank values at the rightmost columns of the superblocks. Assume these values are organized row-wise, and moreover in a y-fast trie data structure [20]. This sums to $\mathcal{O}\left(\frac{n \lg n}{\lg^2 n}\right) = o(n)$ bits per row. The trie for row i permits finding the superblock column containing the kth point in $(1, i) \times (1, n)$, in $\mathcal{O}(\lg \lg n)$ time (by finding the successor of k). Now a binary search over $\lg^2 n$ values gives, in another $\mathcal{O}(\lg \lg n)$ time, the precise column, and one-dimensional rank and select on the column give the position of the kth point. Thus the time is $\mathcal{O}(\lg \lg n)$.

3.4   Entropy-Bounded Space

We have assumed the $b \times b$ blocks are explicitly stored. Instead, we can replace them by a (c, o) pair, just as Raman et al. [19] do for one-dimensional bit vectors. Let a block contain m 1s. Then its *class* c is m and its *offset* o is the index of this particular $b \times b$ block among all the different blocks of class m. A table indexed by c and o storing the contents of all the possible bit vectors takes $\mathcal{O}(\sqrt{n} \lg n)$ bits, thus we can recover any block content in constant time.

Each c value is stored in $\lg(b^2+1) = \mathcal{O}(\lg \lg n)$ bits, adding up to $\mathcal{O}\left(\frac{n^2 \lg \lg n}{\lg n}\right) = o(n^2)$ bits in total. The number of bits required for all the o fields, assuming the rth block contains m_r bits set, is $\sum_r \lceil \lg \binom{b^2}{m_r} \rceil \leq \lg \binom{n^2}{m} + \mathcal{O}\left(\frac{n^2}{\lg n}\right)$ [19]. Finally, we also need pointers to find in constant time an o field, as these have variable-length representations. This can also be done within $\mathcal{O}\left(\frac{n^2 \lg \lg n}{\lg n}\right)$ bits [19] with techniques akin to one-dimensional rank.

Theorem 3. *A $n \times n$ matrix with m 1s and entropy $H = \lg \binom{n^2}{m}$ can be represented within $H + \mathcal{O}\left(\frac{n^2 \lg \lg n}{\lg^{1/4} n}\right)$ bits, so that operation* $\mathsf{rank}(i, j)$ *is computed in $\mathcal{O}(1)$ time,* $\mathsf{report}(i_1, i_2, j_1, j_2, k)$ *performs in time $\mathcal{O}(k + 1)$, $\mathsf{select}(i_1, i_2, k)$ is supported in $\mathcal{O}(\lg n)$ time, and* $\mathsf{select}(i, k)$ *is computed in $\mathcal{O}(\lg \lg n)$ time.*

Note that we can define the complementary queries, where 0s are considered instead of 1s. This is obvious for rank but not for report nor select. It is not hard to see that we can support in addition these complementary queries, by adding other similar $o(n^2)$ bits of space, that is, asymptotically for free. As explained, we can also support the select variants where rows and columns are exchanged, within $o(n^2)$ additional space.

4   A Fully-Compressed Representation

Our compressed representation achieves entropy-bounded space for the matrix itself, but the extra space is $o(n^2)$. This may dominate the entropy bound H. The key to achieving indexes sublinear in H is to adapt the partitioning into superblocks and blocks to the number of bits set in the matrix. The price will be superconstant time for all queries, due to our internal usage of Thm. 2.

4.1   *Rank* Query

We first divide the matrix into superblocks of size $s \times s$, where now $s = \frac{n^2 \lg m}{m}$ (assume for now $m = \Omega(n \lg m)$; we consider the other case later in Section 4.4). The superblocks are further divided into blocks of size $b \times b$, for $b = \frac{n^2 \lg \lg m}{m}$. Just as for Section 3.1, we store absolute ranks at the borders of superblocks and local ranks at the borders of blocks. As the former require $\lg m$ bits to be represented, they add up to $\mathcal{O}(m)$ bits. The latter require $\lg s^2$ bits per datum, adding up to $\mathcal{O}\left(\frac{n^2}{b} \cdot \lg s^2\right) = \mathcal{O}\left(\frac{m}{\lg \lg m} \cdot \lg \frac{n^2 \lg m}{m}\right) = o(H) + \mathcal{O}(m + \lg n)$.

As before, the problem is reduced to supporting local rank within a block of size b^2. We store each block using the wavelet tree of Thm. 2, which for the rth block with m_r bits set requires $\lg \binom{b^2}{m_r} + o(\lg \binom{b^2}{m_r}) + \mathcal{O}(m_r) + o(b)$.[1] It answers rank in time $\mathcal{O}(\lg b) = \mathcal{O}\left(\lg \frac{n^2}{m} + \lg \lg \lg m\right)$. Added over all the blocks, the space is $\lg \binom{n^2}{m} + o(\lg \binom{n^2}{m}) + \mathcal{O}(m) + o(m) = H + o(H) + \mathcal{O}(m)$.

4.2 *Select* Queries

As we have stored all the values $\mathsf{rank}(i, s \cdot j)$, $\mathsf{rank}(s \cdot i, j)$, and $\mathsf{rank}_s(i, b \cdot j)$, we can compute any $\mathsf{rank}(i_1, i_2, b \cdot j_1, b \cdot j_2)$ in constant time. Thus we can binary search for the column *of blocks* where the kth point of $(i_1, i_2) \times (1, n)$ lies. This takes time $\mathcal{O}\left(\lg \frac{n}{b}\right)$. For 3-sided queries we can arrange the superblock ranks of each row in a y-fast trie as before, so as to pay $\mathcal{O}(\lg \lg m)$ time to find the superblock, plus $\mathcal{O}\left(\lg \frac{s}{b}\right) = \mathcal{O}(\lg \lg m)$ to binary search for the block.

Let j_b be the column of blocks found, then the local rank of the (globally) kth point, within block-column j_b, is $k' = k - \mathsf{rank}(i_1, i_2, 1, b \cdot (j_b - 1))$. Now we refine the search to find the exact column where the answer lies. A general way to do this is to carry out a binary search within columns $[b \cdot (j_b - 1) + 1, b \cdot j_b]$ using rank. This rank is not constant-time because we are not in borders of blocks. Hence the time raises to $\mathcal{O}(\lg^2 b) = \mathcal{O}\left(\lg^2 \frac{n^2}{m} + (\lg \lg \lg m)^2\right)$. Once we know the precise column j, we must find the k''th point in it, within rows $[i_1, i_2]$, for $k'' = k - \mathsf{rank}(i_1, i_2, 1, j - 1)$. We first binary search the block vertically in time $\mathcal{O}\left(\lg \frac{n}{b}\right)$, since we can compute any $\mathsf{rank}(b \cdot i, j)$ value in constant time. Finally, confined within a block, we report the correct point in $\mathcal{O}(\lg^2 b)$ time using `rel_sel_obj_maj` on the wavelet tree of the block (or $\mathcal{O}(\lg b)$ using `rel_sel_lab_maj`, depending on the orientation).

A smarter way, but one which applies only to one direction (that is, we cannot have simultaneously the improved result for queries $\mathsf{select}(i_1, i_2, 1, n, k)$ and $\mathsf{select}(1, n, j_1, j_2, k)$), is to arrange the block contents in a different way. Instead of using one wavelet tree structure per block, we pack a whole block column per wavelet tree, taking the b columns as their σ "labels" and the n rows as their "objects" (recall Thm. 2). So the rank times are still $\mathcal{O}(\lg \sigma) = \mathcal{O}(\lg b)$, and the within-block rank used in Section 4.1 can still be carried out within this time. Furthermore, finding the k'th point, in label-major order, between objects i_1 and i_2 (operation `rel_sel_lab_maj`), also takes $\mathcal{O}(\lg b)$ time.

Therefore, if the band of our query is horizontal and we have stored block columns in wavelet trees (or vice versa), we find the kth point within time $\mathcal{O}(\lg n)$ (4-sided select) or $\mathcal{O}\left(\lg \lg m + \lg \frac{n^2}{m}\right)$ (3-sided select).

4.3 Range Reporting

Let us first assume that the query band is horizontal and we have stored rows of blocks in wavelet trees. To solve $\mathsf{report}(i_1, i_2, j_1, j_2, k)$, we first need to find

---

[1] The $o(\lg \binom{b^2}{m_r})$ term is asymptotic in b, which is $\omega(1)$ in terms of n, so it can be safely added up over many blocks later.

the next column after $j_1 - 1$ that is nonempty in the range $[i_1, i_2]$. We use the same RMQ-akin idea of Section 3.2. We form batches storing the *or* of ranges of rows of superblocks, for a total space of $\mathcal{O}(\frac{n}{s} \cdot n \lg \frac{n}{s}) = \mathcal{O}(m)$ bits, and chunks storing the *or* of ranges of rows of blocks within superblocks, for a total space of $\mathcal{O}(\frac{n}{b} \cdot n \lg \frac{s}{b}) = \mathcal{O}(m)$ bits. Instead of the virtual $B(i_1, i_2)$ bitmaps, to find the next 1 in the band $(i_1, i_2) \times (j + 1, n)$ we use the horizontally arranged wavelet trees. They find this point using a rel_min_obj_maj query, in $\mathcal{O}(\lg b)$ time.

Now we must be able to find the 1s in the current column j, before proceeding to the next one. Those 1s within the first block since global row i_1 are easily found with rel_acc_obj_maj in the wavelet tree of the block. In order to find the next block downwards containing points in column j, we store a signature bit vector $B_j[1, n/b]$ for each column j, so that $B_j[i] = 1$ iff there is a 1 in the range $(i \cdot (b-1) + 1, i \cdot b) \times (j, j)$ of the matrix. Using one-dimensional rank and select on the B_j vectors, we can easily find the next block downwards that has a 1 in the current column, in constant time. All the points in column j mod b of that block are then reported (unless they exceed the global row i_2). Bit vectors B_j require $\mathcal{O}\left(\frac{n^2}{b}\right) = o(m)$ bits of space in total.

If, instead, the band of the query is orthogonal to that of wavelet trees, we proceed as follows. The wavelet tree covering column j_1 can deliver all the points within rows $[i_1, i_2]$ since column j_1 onwards, in $\mathcal{O}(\lg b)$ time each, using rel_sel_lab_maj. We must now find the next nonempty block. We build a reduced matrix of $n \times (n/b)$, so that its cell (i', j') contains a 1 iff the original matrix contains a 1 in $(i', b \cdot (j' - 1) + 1) \times (i', b \cdot j')$. Then the block column that is nonempty in $[i_1, i_2]$ in the original matrix corresponds to the next column that is nonempty in $[i_1, i_2]$ in the reduced matrix. We can then apply the technique of Section 3.2, creating the batch and chunk bitmaps over the reduced matrix, taking overall space $\mathcal{O}\left(\frac{n^2}{b}\right) = o(m)$. Once we arrive at the next nonempty block column, its wavelet tree delivers its points in order, and so on.

4.4   The Final Result

A missing piece is to cover the case $m = o(n \lg m) = o(n \lg n)$. When the matrix is so sparse, $\lg \frac{n^2}{m} = \Theta(\lg n)$, and thus it is preferable to use the wavelet tree by itself. The only problem is the $o(n)$ extra space (see Thm. 2), which does not fit in $o(H)$ whenever $m = \mathcal{O}\left(\frac{n \lg \lg n}{\lg^2 n}\right)$. This is because Barbay et al. [2] chose Raman et al.'s representation [19] to achieve constant-time rank and select on a bit vector. By using instead a binary searchable representation [11], the $o(n)$ term becomes $o(H) + \mathcal{O}(m + \lg n)$ and the $\mathcal{O}(\lg \sigma)$ time becomes $\mathcal{O}(\lg \sigma + \lg m)$. This is $\mathcal{O}(\lg n)$ for us, which fits perfectly within our general result.

Theorem 4. *A $n \times n$ matrix with m 1s and entropy $H = \lg \binom{n^2}{m}$ can be stored in $H + o(H) + \mathcal{O}(m + \lg n)$ bits, so that operation rank(i, j) is computed in $\mathcal{O}(\tau)$ time and report(i_1, i_2, j_1, j_2, k) in time $\mathcal{O}((k + 1)\tau)$, where $\tau = \lg \frac{n^2}{m} + \lg \lg \lg m$. In one direction (that can be chosen), select(i_1, i_2, k) is computed in $\mathcal{O}(\lg n)$*

time, and select(i, k) *in time* $\mathcal{O}(\tau + \lg\lg m)$. *In the other direction, both* select *operations cost* $\mathcal{O}(\lg n + \tau^2)$ *time.*

Note that if $m = \mathcal{O}\left(\frac{n^2}{\lg^{1/4} n}\right)$, then $\tau + \lg\lg m = \mathcal{O}\left(\lg \frac{n^2}{m}\right)$; otherwise Thm. 3 is better in all aspects. Finally, if we want to report or select 0s, we can replicate all the extra structures considering the complemented matrix (still defining s and b in terms of 1s), within $o(H) + O(m + \lg n)$ bits . The wavelet trees, instead, must internally binary search on rank to simulate their local operations, all of which (except rank itself) would now cost $\mathcal{O}(\tau^2)$. Thus this complexity must be added to the times given for report (per delivered datum) and any select.

We now have all the necessary pieces to prove our main result.

Proof (of Thm. 1). Putting together our "compressed" (Thm. 3) and "fully compressed" (Thm. 4) solutions, the claimed time complexities are obtained. We achieve $H + o(H)$ bit space for all cases: (a) The $\mathcal{O}\left(\frac{n^2 \lg\lg n}{\lg^{1/4} n}\right)$ extra bits of our "compressed" solution is $o(H)$ as long as $\omega(\frac{n^2}{\lg^{1/4} n}) = m = n^2 - \omega(\frac{n^2}{\lg^{1/4} n})$. (b) The "fully compressed" solution uses $H + o(H) + \mathcal{O}(\lg n)$ space as long as $m = o(n^2)$. (c) These two cover the entire range for m except $m = n^2 - \mathcal{O}\left(\frac{n}{\lg^{1/4} n}\right)$; there, we complement the matrix and use the fully-compressed solution with 0-queries instead of 1-queries. (d) The final $\mathcal{O}(\lg n)$ is not $o(H)$ only if $m = \mathcal{O}(1)$, in which case we can encode the points in differential form (*e.g.* δ-encoding), and answer queries in constant time by scanning the points. □

5   Conclusions

Although the area of orthogonal range queries has received much attention, the extremely interesting case where the structure achieves entropy-bounded space is largely under-explored. This work completes a large portion of the picture, and hopefully opens the door to much further research.

A first line of future work is to find and reach the lower bounds on the time complexity under this new scenario. Existing lower bounds, such as $\Omega\left(\frac{\lg m}{\lg \frac{2M}{m}}\right)$ counting time when using M words of memory [7], or one-dimensional lower bounds [18] might be useful. Succinct-space results [4] suggest we can do better. Particularly intriguing is that select takes constant time on dense one-dimensional bitmaps, whereas we have not achieved that in two dimensions. Interestingly, schemes that achieve entropy-bounded extra space in one dimension [16], offer rank time analogous to ours, $\mathcal{O}(\lg \frac{n}{m})$, yet their select is constant-time.

As for space, $H = \lg \binom{n^2}{m}$ is a crude worst-case lower bound that does not account for regularities, such as clusters of points, that arise in real life. Our actual space is indeed much better: the sum of local entropies of small blocks. An interesting future work challenge is to improve the lower order term, $o(H)$.

Other natural directions for future work are to consider further operations [2], extending to d-dimensional spaces, achieving dynamic compressed structures, and secondary-memory variants.

Acknowledgements. We thank Jérémy Barbay for discussions and proofreading.

References

1. Alstrup, S., Brodal, G., Rauhe, T.: New data structures for orthogonal range searching. In: Proc. 41st FOCS, pp. 198–207 (2000)
2. Barbay, J., Claude, F., Navarro, G.: Compact rich-functional binary relation representations. In: López-Ortiz, A. (ed.) LATIN 2010. LNCS, vol. 6034, pp. 172–185. Springer, Heidelberg (2010)
3. Bender, M., Farach-Colton, M.: The level ancestor problem simplified. Theoretical Computer Science 321(1), 5–12 (2004)
4. Bose, P., He, M., Maheshwari, A., Morin, P.: Succinct orthogonal range search structures on a grid with applications to text indexing. In: Proc. 11th WADS, pp. 98–109 (2009)
5. Chan, T., Pătraşcu, M.: Counting inversions, offline orthogonal range counting, and related problems. In: Proc. 21st SODA, pp. 161–173 (2010)
6. Chazelle, B.: Filtering search: A new approach to query-answering. SIAM Journal of Computing 15, 703–724 (1986)
7. Chazelle, B.: Lower bounds for orthogonal range searching: II. The arithmetic model. Journal of the ACM 37(3), 430–463 (1990)
8. Clark, D.: Compact Pat Trees. PhD thesis, University of Waterloo, Canada (1996)
9. Golynski, A.: Optimal lower bounds for rank and select indexes. Theoretical Computer Science 387(3), 348–359 (2007)
10. Grossi, R., Gupta, A., Vitter, J.: High-order entropy-compressed text indexes. In: Proc. 14th SODA (2003)
11. Gupta, A., Hon, W.-K., Shah, R., Vitter, J.S.: Compressed data structures: Dictionaries and data-aware measures. In: Proc. 16th DCC, pp. 213–222 (2006)
12. Já Já, J., Mortensen, C.W., Shi, Q.: Space-efficient and fast algorithms for multidimensional dominance reporting and counting. In: Fleischer, R., Trippen, G. (eds.) ISAAC 2004. LNCS, vol. 3341, pp. 558–568. Springer, Heidelberg (2004)
13. Munro, I.: Tables. In: Chandru, V., Vinay, V. (eds.) FSTTCS 1996. LNCS, vol. 1180, pp. 37–42. Springer, Heidelberg (1996)
14. Nekrich, Y.: Space efficient dynamic orthogonal range reporting. In: Proc. 21st SCG, pp. 306–313 (2005)
15. Nekrich, Y.: Orthogonal range searching in linear and almost-linear space. Computational Geometry: Theory and Applications 42(4), 342–351 (2009)
16. Okanohara, D., Sadakane, K.: Practical entropy-compressed rank/select dictionary. In: Proc. 9th ALENEX (2007)
17. Pătraşcu, M.: Lower bounds for 2-dimensional range counting. In: Proc. 39th STOC, pp. 40–46 (2007)
18. Pătraşcu, M., Thorup, M.: Time-space trade-offs for predecessor search. In: Proc. 38th STOC, pp. 232–240 (2006)
19. Raman, R., Raman, V., Srinivasa Rao, S.: Succinct indexable dictionaries with applications to encoding k-ary trees and multisets. In: Proc. 13th SODA, pp. 233–242 (2002)
20. Willard, D.: Log-logarithmic worst-case range queries are possible in space $\theta(n)$. Information Processing Letters 17(2), 81–84 (1983)

Identifying Approximate Palindromes in Run-Length Encoded Strings*

Kuan-Yu Chen[1], Ping-Hui Hsu[1], and Kun-Mao Chao[1,2,3]

[1] Department of Computer Science and Information Engineering
[2] Graduate Institute of Biomedical Electronics and Bioinformatics
[3] Graduate Institute of Networking and Multimedia
National Taiwan University, Taipei, Taiwan 106

Abstract. We study the problem of identifying palindromes in compressed strings. The underlying compression scheme is called run-length encoding, which has been extensively studied and widely applied in diverse areas. Given a run-length encoded string $\mathrm{RLE}(T)$, we show how to preprocess $\mathrm{RLE}(T)$ to support efficient retrieval of the longest palindrome with a specified center position and a tolerated number of mismatches between its two arms. Let n be the number of runs of $\mathrm{RLE}(T)$ and k be the tolerated number of mismatches. We present two algorithms for the problem, both with preprocessing time polynomial in the number of runs. The first algorithm, devised for small k, identifies the desired palindrome in $O(\log n + \min\{k, n\})$ time with $O(n \log n)$ preprocessing time, while the second algorithm achieves $O(\log^2 n)$ query time, independent of k, after $O(n^2 \log n)$-time preprocessing.

1 Introduction

The explosion of digital data urges the need for effective compression methods. Representing data in compressed form reduces the consumption of disk space and the time needed for transmission. However, the drawback is that compressed data must be decompressed before existing text-mining algorithms can be applied. In 1992, Amir and Benson [1] addressed the two-dimensional matching problem with the text represented as run-length encoded format. Since then, it has become a challenging task to design algorithms directly searching or mining compressed data without any decompression. The most extensively studied problem under this paradigm is *compressed pattern matching*, for which several compression schemes have been considered, such as *run-length encoding*, *Lempel-Ziv encoding*, and *straight-line programs* (see [12,17] for more details). Another research direction is to identify featured patterns in compressed strings. For example, in [14] the authors studied the problem of finding all squares in a run-length encoded string. Finding palindromes and squares in LZ-compressed strings was discussed in [9]. In addition, the work of [16] demonstrated how

---

* Partially supported by NSC grants 97-2221-E-002-097-MY3 and 98-2221-E-002-081-MY3 from the National Science Council, Taiwan.

O. Cheong, K.-Y. Chwa, and K. Park (Eds.): ISAAC 2010, Part II, LNCS 6507, pp. 339–350, 2010.

to compute longest common substrings and palindromes in strings described in terms of straight-line programs.

The ultimate goal in this line of investigation is to design algorithms whose time depends solely on the compressed sizes. So far several positive results have been found for run-length encoding. Let m and n denote the number of runs of the two input run-length encoded strings. The exact string matching problem can be easily solved in $O(m+n)$ time. The k-mismatch with wildcards problem is solvable in $O(mn \log m)$ time [6]. Computing the longest common subsequence of two run-length encoded strings requires $O(mn \log mn)$ time [2], and computing their edit distance (Levenshtein distance) can be done in $O(mn^2)$ time [5]. Moreover, there exists an $O(n \log n)$-time algorithm for finding all squares in a run-length encoded string [14]. In this paper, we explore the problem of identifying approximate palindromes in a run-length encoded string. Two algorithms will be presented. Both of them have preprocessing time polynomial in the number of runs.

2   Problem Definition and Main Results

2.1   Run-Length Encoding

Run-length encoding (RLE) is a simple and effective coding scheme that performs lossless data compression. RLE compression represents consecutive and identical symbols with a run, usually denoted by σ^i, where σ is an alphabet symbol and i is its repetition times. For example, string $bbcccddaaaaa$ is optimally encoded as $b^2 c^3 d^2 a^5$. Transforming a non-optimal RLE format such as $b^2 c^1 c^2 d^2 a^3 a^2$ into an optimal one is easy. Thus, without loss of generality we assume that the input RLE string of our problem is optimally encoded.

2.2   Palindrome

A palindrome is a word that can be read the same way in either direction. Formally, a palindrome is a string of the form uu^r or $u\sigma u^r$, where u and u^r are two nonempty strings, one being the reverse string of the other, and σ is an alphabet symbol. A palindrome is called an *even palindrome* if it is of the form uu^r, and an *odd palindrome* otherwise. For example, strings $abbaabba$ and $abbabba$ are even and odd palindromes, respectively. Moreover, substrings u and u^r are referred to as the two *arms* of the palindrome. Identifying palindromes in a string has been an interesting topic from both theoretical and practical perspectives. In biological application, palindromic motifs are found in most genomes and have been especially investigated in bacterial chromosomes. In [7], the authors studied a new technique for the genome-wide analysis of palindrome formation. Due to the large scale of biological data, compression methods are often used to reduce the storage as well as to accelerate the computation.

2.3 Notation and Similarity Measure

Throughout the paper, we adopt the following notation. We let T denote the uncompressed string and RLE(T) denote its optimally encoded RLE format. Our problem will take RLE(T), instead of T, as input. We let capital letter N denote the string length of T and let small letter n denote the number of runs in RLE(T). We let $T[i \ldots j]$ denote the substring of T starting and ending at positions i and j. In particular, $T[i]$ is the i-th symbol of T. Alternatively, we use $T(c, \ell)$ to specify a substring of T, which is interpreted as the substring of T centered at position c with arm length ℓ. More specifically, we define $T(c, \ell)$, where $c \in \{1, \frac{3}{2}, 2, \frac{5}{2}, \ldots, N\}$ and $0 \leq \ell \leq \frac{N}{2}$, to be the substring $T[\lceil c \rceil - \ell \ldots \lfloor c \rfloor + \ell]$. For example, suppose that $T = abbacbcacabbacb$. The odd palindrome $T[4 \ldots 8] = acbca$ is substring $T(6, 2)$, and the even palindrome $T[1 \ldots 4] = abba$ is substring $T(\frac{5}{2}, 2)$. Analogously, $T[\lceil c \rceil - \ell \ldots \lceil c \rceil - 1]$ and $T[\lfloor c \rfloor + 1 \ldots \lfloor c \rfloor + \ell]$ are referred to as the *left* and *right arms* of $T(c, \ell)$.

A palindrome is said to be *exact* if its one arm exactly matches the other in the reverse order. In biological applications, genetic mutations occur during the evolutionary process; hence, it is more meaningful to identify *approximate palindromes* instead of exact ones [13]. We use *Hamming distance* as the similarity measure between the two arms of a palindrome.

Definition 1. *We define $d_H(T(c, \ell))$ to be the Hamming distance between the two arms of $T(c, \ell)$. Formally, $d_H(T(c, \ell)) = \sum_{i=1}^{\ell} \delta(T[\lceil c \rceil - i], T[\lfloor c \rfloor + i])$, where $\delta(a, b) = 0$ if symbol a matches symbol b and $\delta(a, b) = 1$ otherwise.*

2.4 Problem Definition

Since there are $2N - 1$ possible center positions for palindromes in T, it takes $\Omega(N)$ time to enumerate all approximate palindromes in T. An important feature of our approach is that instead of constructing all the palindromes explicitly, which easily leads to time complexity depending on the uncompressed size N, we aim at *preprocessing* the compressed string for later *retrieval* of palindromes at any given center. Formally, our problem is defined as a query-answering paradigm as follows. We preprocess RLE(T) in order to support on-line queries $Q(T, c, k) = \max\{\ell \mid d_H(T(c, \ell)) \leq k\}$, where $c \in \{1, \frac{3}{2}, 2, \frac{5}{2}, \ldots, N\}$ and $0 \leq k \leq \frac{N}{2}$. In other words, $Q(T, c, k)$ returns the length of the longest palindrome in T centered at position c and having no more than k mismatches between its arms. Note that once we obtain the value of $Q(T, c, k)$, the position of the palindrome is also located. We will present two algorithms whose preprocessing and query time depend only on the compressed size n.

2.5 Our Results

For notational convenience, we say that an algorithm has time complexity $\langle f(n), q(n) \rangle$, if it spends $O(f(n))$ time at the preprocessing stage and $O(q(n))$ time at the query stage. In this paper, we present two algorithms achieving the following time and space complexities.

1. The first algorithm takes $\langle n \log n, \log n + \min\{k, n\}\rangle$ time and $O(n)$ space.
2. The second algorithm takes $\langle n^2 \log n, \log^2 n\rangle$ time and $O(n^2 \log n)$ space.

In most applications, the number of mismatches is limited. Hence, the first algorithm is preferable because it has better preprocessing time and requires less extra space. From the theoretical point of view, however, since k ranges from 0 to $\frac{N}{2}$, our second algorithm achieves a considerable improvement on the query time when k is large. Moreover, the second algorithm reveals an interesting connection between the palindrome problem and the interval stabbing problem from computational geometry.

3   An $\langle n \log n, \log n + \min\{k, n\}\rangle$-Time Algorithm

Our first algorithm relies on *longest common extension* (LCE) queries. Let $\text{RLE}(T) = X_1 X_2 \ldots X_n$, where X_i denotes the i-th run of $\text{RLE}(T)$. Recall that each run X_i is composed of a run symbol, denoted by X_i^s, and a run length, denoted by X_i^l. Therefore, we can view a run, $X_i = (X_i^s, X_i^l)$, as a symbol drawn from $\Sigma \times \mathbb{Z}^+$. We write $X_i = X_j$ if $X_i^s = X_j^s$ and $X_i^l = X_j^l$.
At the preprocessing stage, we construct array $R[1 \ldots n]$ storing the start position of each run in the uncompressed string T. Formally, $R[i] = \sum_{j=1}^{i-1} X_j^l + 1$ for $1 \leq i \leq n$. Besides, we preprocess $\text{RLE}(T)$ to support online $\text{LCE}(\text{RLE}(T), i, j)$ queries, where $1 \leq i \leq n$ and $1 \leq j \leq n$, defined as follows.

$$\text{LCE}(\text{RLE}(T), i, j) = \max\{\ell \mid X_{i-1} = X_{j+1}, X_{i-2} = X_{j+2}, \ldots, X_{i-\ell} = X_{j+\ell}\}.$$

Lemma 1. *Answering an LCE query described above can be done in $\langle n \log n, 1\rangle$ time and $O(n)$ space.*

Proof. It is well known that the LCE problem can be solved by combining the techniques of *suffix trees* and *lowest common ancestor* (LCA) queries [10]. We construct a generalized suffix tree $\mathcal{T}$ of $\text{RLE}(T) = X_1 X_2 \ldots X_n$ and its reverse string $X_n X_{n-1} \ldots X_1$, and then preprocess $\mathcal{T}$ for constant-time LCA queries. Since $\text{RLE}(T)$ can be seen as a string over an unbounded alphabet, the first step takes $\Theta(n \log n)$ time and $O(n)$ space [8]. Preprocessing $\mathcal{T}$ for LCA queries takes $O(n)$ time and $O(n)$ space [11]. □

Figure 1 depicts the query procedure of our first algorithm. The procedure first performs a binary search on array R to locate which run the queried center lies in. Then, it simply extends the palindrome from the center position outwards as much as possible. The extension is based on either the left-side or the right-side extension, depending on which one encounters a run boundary first. The procedure is terminated when the promised k-mismatches budget is used up or the extension reaches one of the string ends.

Lemma 2. *The while-loop of QUERY is performed $O(\min\{k, n\})$ times.*

Procedure QUERY(c, k)
1 Perform binary search for the largest index i such that $R[i] \leq c$;
2 Initially, $i_1 \leftarrow i_2 \leftarrow i$, $\ell \leftarrow 0$, and $budget \leftarrow k$;
3 **while** $i_1 \geq 1$ and $i_2 \leq n$ **do**
4   **if** $X_{i_1}^s \neq X_{i_2}^s$ **then** $budget \leftarrow budget - (\min\{left(c, i_1), right(c, i_2)\} - \ell)$;
5   **if** $budget < 0$ **then**
6   Terminate the procedure and output $\min\{left(c, i_1), right(c, i_2)\} + budget$;
7   **else** // $budget \geq 0$
8   **if** $left(c, i_1) < right(c, i_2)$ **then**
9              $\ell \leftarrow left(c, i_1)$; $i_1 \leftarrow i_1 - 1$;
10   **else if** $right(c, i_2) < left(c, i_1)$ **then**
11             $\ell \leftarrow right(c, i_2)$; $i_2 \leftarrow i_2 + 1$;
12   **else** // $left(c, i_1) = right(c, i_2)$
13             $i_1 \leftarrow i_1 - \text{LCE}(\text{RLE}(T), i_1, i_2)$; $i_2 \leftarrow i_2 + \text{LCE}(\text{RLE}(T), i_1, i_2)$;
14             $\ell \leftarrow right(c, i_2)$; $i_1 \leftarrow i_1 - 1$; $i_2 \leftarrow i_2 + 1$;
15   **end if**
16   **end if**
17 **end while**
18 Output ℓ;

Fig. 1. The query procedure of the first algorithm. Let i_1 (resp., i_2) be a pointer tracking the run number of the current left (resp., right) extension. Function $left(c, i_1)$ (resp., $right(c, i_2)$) indicates the arm length if the extension reaches the left end (resp., right end) of run X_{i_1} (resp., X_{i_2}). Formally, $left(c, i_1) = \lfloor i \rfloor - R[i_1]$ and $right(c, i_2) = R[i_2 + 1] - \lceil i \rceil$. Parameter ℓ keeps track of the length of our last extension, and $budget$ records the remaining mismatches allowed. Note that when $left(c, i_1) = right(c, i_2)$, an additional LCE query is performed for a "free leap" over consecutive matched runs.

Proof. If the condition of line 4 is valid, we say that the while-loop enters a MISMATCH iteration; otherwise, it is in a MATCH iteration. Below, we consider MISMATCH and MATCH iterations separately. First, observe that (1) $budget$ is initialized as k and is decreased by a positive amount at MISMATCH iterations, and (2) the procedure is terminated when $budget < 0$. Therefore, the number of MISMATCH iterations is clearly bounded by $O(\min\{k, n\})$. Next, we prove the number of MATCH iterations satisfies the same bound by showing that a MATCH iteration must be immediately followed by a MISMATCH iteration, if any. The while-loop enters a MATCH iteration implies $X_{j_1}^s = X_{j_2}^s$. If lines 8–9 are executed, at the next iteration it cannot be the case that $X_{j_1-1}^s = X_{j_2}^s$, for runs X_{j_1} and X_{j_1-1} then encode the same symbol and should be encoded as a single run. It can be argued similarly for the case where lines 10–11 are executed. If lines 12–14 are executed, the definition of $\text{LCE}(\text{RLE}(T), j_1, j_2)$ directly implies that $X_{j_1}^s \neq X_{j_2}^s$ for the updated j_1 and j_2 of the next iteration.   □

Theorem 1. *The first algorithm runs in $\langle n \log n, \log n + \min\{k, n\} \rangle$ time and $O(n)$ space.*

Proof. Immediate from Lemmas 1 and 2.   □

4   An $\langle n^2 \log n, \log^2 n \rangle$-Time Algorithm

We note again that k ranges from 0 to $\frac{N}{2}$. Hence, the first algorithm may spend $O(n)$ time answering a query. In this section, we establish the connection between our problem and the *interval stabbing problem* [3]. Our second algorithm utilizes *segment trees*, initially proposed by Bentley [4], and achieves a query time of $O(\log^2 n)$, independent of k. The algorithm is devised in light of matrix D defined below.

Definition 2. *We define D to be an $N \times N$ matrix whose entry $D[i,j] = \delta(T[i], T[j])$, where $\delta(a,b) = 0$ if symbol a matches symbol b and $\delta(a,b) = 1$ otherwise.*

For those entries $D[i,j]$ where $i+j = d$, they are said to be on *anti-diagonal d* of D, denoted by D_d. The anti-diagonal number of D ranges from 2 to $2N$. Each run pair X_i and X_j corresponds to a *block* (sub-matrix) of D, denoted by $B_{i,j}$. Observe that all entries in $B_{i,j}$ are 0's if $X_i^s = X_j^s$, and are 1's otherwise. We call $B_{i,j}$ a *match block* if $X_i^s = X_j^s$, and a *mismatch block* if $X_i^s \neq X_j^s$. The blocks in D are further divided into different levels. For those blocks $B_{i,j}$ where $j-i = h$, they are said to be on *level h* of D. See Figure 2a for an illustration.

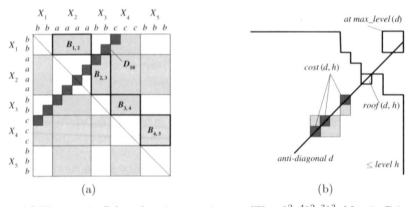

(a)   (b)

Fig. 2. (a) The matrix D based on input string $\text{RLE}(T) = b^2 a^4 b^2 c^3 b^3$. Matrix D is partitioned into mismatch blocks (light grey blocks) and match blocks (white blocks). The dark grey grids form anti-diagonal 10, and blocks $B_{1,2}$, $B_{2,3}$, $B_{3,4}$, and $B_{4,5}$ constitute level 1 of D. (b) An illustration of functions max_level, $roof$, and $cost$. The block with the maximum level, $max\_level(d)$, on anti-diagonal d is shown in the upper-right corner of the figure. The staircase depicts the exterior of blocks below level h, which intersects with anti-diagonal d at point $roof(d,h)$. The light-grey blocks depict the mismatch blocks on anti-diagonal d that are between level 0 and level h, and $cost(d,h)$ gives the number of entries passed through by anti-diagonal d.

Given a center position c and an arm length ℓ, observe that $d_H(T(c,\ell)) = \sum_{i=1}^{\ell} D[\lceil c \rceil - i, \lfloor c \rfloor + i]$. Therefore, we aim to find the longest extension on anti-diagonal $2c$, from entry $D[\lceil c \rceil - 1, \lfloor c \rfloor + 1]$ to entries in upper triangular matrix, with at most k entries that are 1's. With proper preprocessing, the second algorithm directly evaluates the number of entries that are 1's up to a certain level and a binary search on levels is then performed.

4.1   Query Decomposition

For presentation simplicity, each entry $D[i, j]$ is treated as a point (i, j) in the plane. That is, matrix $D = \{(i, j) \mid 1 \leq i \leq N \text{ and } 1 \leq j \leq N\}$, and anti-diagonals D_d and blocks $B_{i,j}$ are subsets of D. We will focus on blocks in the upper triangular matrix of D, i.e., blocks $B_{i,j}$ with $j - i \geq 0$.

Definition 3. *For $1 \leq i \leq n$ and $1 \leq j \leq n$, we define functions $corner_1(B_{i,j})$, $corner_2(B_{i,j})$, $corner_3(B_{i,j})$, and $corner_4(B_{i,j})$ to be the upper-left, upper-right, lower-right, and lower-left corner points of block $B_{i,j}$, respectively.*

With the help of array R, defined in the previous section, answering functions $corner_1$, $corner_2$, $corner_3$, and $corner_4$ can be easily done in constant time. Below, we define three additional functions max_level, $roof$, and $cost$. See Figure 2b for an illustration.

Definition 4. *For $2 \leq d \leq 2N$, we define $max\_level(d)$ to be the maximum block level on anti-diagonal d. Formally, $max\_level(d) = \max\{j - i \mid D_d \cap B_{i,j} \neq \phi\}$.*

Definition 5. *For $2 \leq d \leq 2N$ and $0 \leq h \leq n - 1$, we define $roof(d, h)$ to be the farthest point on anti-diagonal d that is below level h. Formally, $roof(d, h) = (x, y)$ such that $(x, y) \in D_d \cap B_{i,j}$, where $j - i \leq h$, and y is maximized.*

Definition 6. *For $2 \leq d \leq 2N$ and $0 \leq h \leq n - 1$, we define $cost(d, h)$ to be the total number of points on anti-diagonal d which belong to mismatch blocks between level 0 and level h. Formally, $cost(d, h) = |D_d \cap S|$, where $S = \{(x, y) \in B_{i,j} \mid B_{i,j} \text{ is a mismatch block and } 0 \leq j - i \leq h\}$.*

Procedure QUERY2(c, k)
1   **if** $cost(2c, max\_level(2c)) \leq k$ **then**
2   Output $j - \lfloor c \rfloor$, where $(i, j) = roof(2c, max\_level(2c))$;
3   **else**
4   Find the smallest $h \in [0, max\_level(2c)]$ such that $cost(2c, h) > k$;
5   Output $(j - \lfloor c \rfloor) - (cost(2c, h) - k)$, where $(i, j) = roof(2c, h)$;
6   **end if**

Fig. 3. The query procedure of the second algorithm

Sections 4.2, 4.3, and 4.4 are devoted to demonstrate that by a preprocessing of $O(n^2 \log n)$ time and space, functions max_level, $roof$, and $cost$ can be answered in $O(\log n)$ time. Our query algorithm based on these functions is presented in Figure 3. In QUERY2, the algorithm first evaluates the number of entries that are 1's of the longest extension that reaches the maximum level. If the value is below threshold k, we know the extension is able to reach the matrix border (see lines 1–2). Otherwise, a binary search is performed for the minimum level that exceeds threshold k (see line 4). If the claimed time for answering these functions is correct, the time required for QUERY2 is $O(\log^2 n)$.

4.2   Answering a *max_level* Query

We define $\mathcal{P}$ to be a list of points containing the upper-left corners of blocks in the first row of D and the upper-right corners of blocks in the last column of D. Formally, $\mathcal{P}(i) = corner_1(B_{1,i})$ for $1 \leq i \leq n$, and $\mathcal{P}(i) = corner_2(B_{i-n+1,n})$ for $n + 1 \leq i \leq 2n - 1$, where $\mathcal{P}(i)$ is the i-th point of list of $\mathcal{P}$.

Lemma 3. *Answering function max_level can be done in $\langle n, \log n \rangle$ time and $O(n)$ space.*

Proof. Preparing list $\mathcal{P}$ is easily done in $O(n)$ time and space. Observe that the anti-diagonal numbers of points in $\mathcal{P}$ are in sorted order. Hence, we can perform a binary search for the largest j such that the anti-diagonal number of $\mathcal{P}(j)$ is less than or equal to d. The value *max_level*(d) is then the block level of $\mathcal{P}(j)$. □

4.3   Answering a *roof* Query

Observe that for a fixed h, the points of $roof(d, h)$ for all $2 \leq d \leq 2N$ form a staircase in D. Let $\mathcal{Q}_h$ be the list of the upper-right corners of blocks on level h. Formally, $\mathcal{Q}_h(i) = corner_2(B_{i,i+h})$, where $\mathcal{Q}_h(i)$ is the i-th point in list $\mathcal{Q}_h$ for $1 \leq i \leq n - h$.

Lemma 4. *Answering function roof can be done in $\langle n^2, \log n \rangle$ time and $O(n^2)$ space.*

Proof. Preparing lists $\mathcal{Q}_h$, where $0 \leq h \leq n - 1$, is easily done in $O(n^2)$ time and space. Observe that the anti-diagonal numbers of points in $\mathcal{Q}_h$ are in sorted order. Hence, we perform a binary search for the largest j such that the anti-diagonal number of $\mathcal{Q}_h(j)$ is less than or equal to d. If no such j exists, then $roof(d, h)$ is the intersection of D_d and $\overline{uv}$, where $u = (1, 1)$ and $v = \mathcal{Q}_h(1)$. Similarly, if $j = n - h$, then $roof(d, h)$ is the intersection of D_d and $\overline{uv}$, where $u = \mathcal{Q}_h(n - h)$ and $v = (n, n)$. Otherwise, $roof(d, h)$ is the intersection of D_d with either $\overline{uv}$ or $\overline{vw}$, where $u = \mathcal{Q}_h(j)$, $w = \mathcal{Q}_h(j + 1)$, and $v = corner_2(B_{j+1,j+h})$. □

4.4   Answering a *cost* Query

We now show how to answer a *cost* query in $O(\log n)$ time after $O(n^2 \log n)$-time preprocessing.

Definition 7. *For each mismatch block $B_{i,j}$, where $0 \leq j - i \leq n - 1$, we define function $f_{i,j} : [2, 2N] \to \mathbb{N}$ to be the number of entries of $B_{i,j}$ that are on anti-diagonal d. Formally, $f_{i,j}(d) = |D_d \cap B_{i,j}|$ for $2 \leq d \leq 2N$.*

Let $(x, y) = corner_1(B_{i,j})$ and $(x', y') = corner_3(B_{i,j})$. If $x' + y \leq x + y'$, the value of $f_{i,j}(d)$ can be easily calculated by the following formula. (The values of $f_{i,j}$ in the other case where $x' + y > x + y'$ can be argued similarly.)

$$f_{i,j}(d) = \begin{cases} 0, & \text{for } 2 \leq d \leq x + y - 1; \\ d - x - y, & \text{for } x + y \leq d \leq x' + y - 1; \\ x' - x, & \text{for } x' + y \leq d \leq x + y'; \\ x' + y' - d, & \text{for } x + y' + 1 \leq d \leq x' + y'; \\ 0, & \text{for } x' + y' + 1 \leq d \leq 2N. \end{cases} \tag{1}$$

By Equation (1), the non-zero values of $f_{i,j}$ can be interpreted as three line segments in the plane (see Figure 4). Observe that to evaluate $f_{i,j}(d)$ we can compute the intersection of vertical line $x = d$ with the three line segments representing $f_{i,j}$. If there is no intersection, then $f_{i,j}(d) = 0$. Otherwise, the value of $f_{i,j}(d)$ is the y-coordinate of the intersection.

We let $\mathcal{F}$ denote the set of line segments obtained by interpreting all $f_{i,j}$ into segments. Note that $|\mathcal{F}| = O(n^2)$. Besides the two endpoints, each line segment of $f_{i,j}$ is also associated with value $j - i$, i.e., the level of block $B_{i,j}$. Given a line segment $L \in \mathcal{F}$, we let $(x_1(L), y_1(L))$ and $(x_2(L), y_2(L))$ denote its left and right endpoints, and let $z(L)$ denote its level.

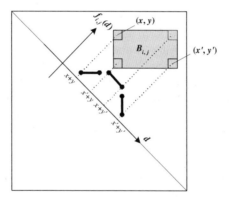

Fig. 4. The diagram of $f_{i,j}$ for the case where $x' + y \leq x + y'$. The values of $f_{i,j}$ are interpreted as three line segments in the plane (the bold lines). The x-axis represents anti-diagonal number, and the y-axis represents the value of $f_{i,j}$. If $x + y = x' + y - 1$ or $x' + y = x + y'$ or $x + y' + 1 = x' + y'$, the line segments become points. We note that this does not affect the correctness of our approach.

Now, answering a *cost* query can be interpreted as answering an interval stabbing query as follows. By definitions of *cost* and $f_{i,j}$, we have that $cost(d, h) = \sum_{j-i \leq h} f_{i,j}(d)$. Hence, to evaluate the value of $cost(d, h)$ we need to sum up the y-coordinates of intersections of line $x = d$ with $L \in \mathcal{F}$, where $z(L) \leq h$. That is, $cost(d, h) = \sum_{L \in \mathcal{G}} (y_1(L) + (d - x_1(L)) \times \frac{y_2(L) - y_1(L)}{x_2(L) - x_1(L)})$, where $\mathcal{G} = \{L \in \mathcal{F} \mid d \in [x_1(L), x_2(L)], z(L) \leq h\}$. Set $\mathcal{G}$ contains those segments below level h that are stabbed by vertical line $x = d$. Given set $\mathcal{F}$, reporting segments in $\mathcal{G}$ is in essence a two-dimensional interval stabbing query. However, our *cost* query is more like a *counting* query which asks for a value concerning the intersections. Below, we demonstrate how to utilize a segment tree to answer a *cost* query.

Let m denote the size of $\mathcal{F}$. The x-coordinates of endpoints of segments in $\mathcal{F}$ subdivide the x-axis into $2m + 1$ intervals, called *atomic intervals*.[1] The segment tree $\mathcal{T}$ built based on $\mathcal{F}$ is a balanced binary tree, whose leaves are in 1–1 correspondence with the atomic intervals, ordered from left to right. Each segment of $\mathcal{F}$ is split into several *pieces* and stored in at most $2 \log m$ nodes of $\mathcal{T}$ (see [3] for more details). An internal node u of $\mathcal{T}$ corresponds to the union of the atomic intervals of its descendent leaves, denoted by interval I_u. A segment is said to *span* I_u if its projection to the x-axis covers I_u. We store a piece of $L \in \mathcal{F}$ in node u if L spans I_u but does not span I_p, where p is u's parent. The stored piece will be the part of L whose x-coordinates lying in range I_u.

At answering a $cost(d, h)$, we traverse $\mathcal{T}$ from the root to a leaf node, where the path is determined by value d. All pieces found at the nodes along this path belong to segments stabbed by vertical line $x = d$. Recall that our *cost* query relates to intersections of the stabbed pieces that are below level h. Thus, we keep the stored pieces in a node ordered by their levels. Moreover, we replace each piece by an *accumulated segment* which sums up all pieces in the same node that are below its level. To sum up two pieces in a node we simply add up the corresponding y-coordinates of their two endpoints. Then, for each node in the query path we need only to perform a binary search to retrieve the accumulated segment with highest level below h and compute its intersection with vertical line $x = d$. A *cost* query can now be answered in $O(\log^2 m)$ time.

To reduce the time to $O(\log m)$, we apply the standard technique of *fractional cascading*, described in [15,18]. We add auxiliary lists of pointers to nodes of $\mathcal{T}$. With the help of auxiliary lists, only one binary search is needed at root node, and the binary searches in the rest of the query path are saved by using pointers (see Figure 5). More specifically, each accumulated segment is associated with a pointer, and the auxiliary lists are constructed bottom-up by merging the pointers in children's nodes. Besides, each pointer is linked to pointers in children's lists, if any, with level preceding it.

Lemma 5. *The segment tree $\mathcal{T}$ can be built in $O(n^2 \log n)$ time and space.*

Proof. Inserting a segment of $\mathcal{F}$ into $\mathcal{T}$ involves two search paths of length $O(\log m)$. All auxiliary pointers concerning the inserted segment are contained in nodes along these two paths. Hence, the total size of auxiliary lists is $O(m \log m)$. As long as we insert segments of $\mathcal{F}$ from the lowest level to higher levels in order, both replacing with accumulated segments and constructing links between pointers can be easily done in a dynamic programming fashion. Since $m = O(n^2)$, the lemma thus follows. □

---

[1] To simplify the discussion, we assume that the $2m$ endpoints of segments in $\mathcal{F}$ have distinct x-coordinates. Moreover, we do not construct a leaf node for each endpoint of the segments.

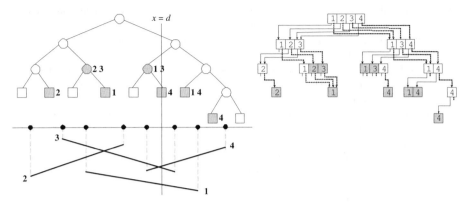

Fig. 5. The left side of the figure depicts the segment tree $\mathcal{T}$ which stores four two-dimensional segments as shown below the tree. The number next to each segment indicates the block level of the segment. Each segment is partitioned and stored as accumulated segments in (grey-shaded) nodes of $\mathcal{T}$. For example, the segment of level 1 is split into three pieces and stored as accumulated segments in two leaf nodes and one internal node. The nodes of $\mathcal{T}$ are associated with auxiliary lists of pointers, shown on the right side of the figure. To retrieve the accumulated segment stabbed by vertical line $x = d$ and having the highest level below level h, only one binary search is needed at the root's auxiliary list. The binary searches in the rest of nodes along the query path are saved by directly following the pointers.

Theorem 2. *The second algorithm runs in $\langle n^2 \log n, \log^2 n \rangle$ time and $O(n^2 \log n)$ space.*

Proof. Immediate from Lemmas 3, 4, 5 and procedure QUERY2.   □

5   Concluding Remarks

In this paper, we show how to preprocess an RLE string, with time polynomial in the number of runs, to support online queries of the longest approximate palindrome with a specified center position and a specified threshold of mismatches between the two arms. We would like to remark that both of our approaches can be easily extended to support queries of *gapped palindromes* whose two arms are separated by a nonempty substring called *loop* [13]. Moreover, in biological application a palindromic DNA sequence contains two paired strands of nucleotides complementary to each other. It is also not hard to adapt our approach to finding palindromes with complementary arms by redefining the notion of matching.

References

1. Amir, A., Benson, G.: Efficient Two-Dimensional Compressed Matching. In: Data Compression Conference, pp. 279–288 (1992)
2. Apostolico, A., Landau, G.M., Skiena, S.: Matching for Run-Length Encoded Strings. Journal of Complexity 15(1), 4–16 (1999)

3. de Berg, M., Cheong, O., van Kreveld, M., Overmars, M.: Computational Geometry: Algorithms and Applications. Springer, Heidelberg (2008)
4. Bentley, J.L.: Solutions to Klee's Rectangle Problems. Technical Report, Carnegie-Mellon University, Pittsburgh, PA (1977)
5. Chen, K.-Y., Chao, K.-M.: A Fully Compressed Algorithm for Computing the Edit Distance of Run-Length Encoded Strings. In: de Berg, M., Meyer, U. (eds.) ESA 2010. LNCS, vol. 6346, pp. 415–426. Springer, Heidelberg (2010)
6. Chen, K.-Y., Hsu, P.-H., Chao, K.-M.: Hardness of Comparing Two Run-Length Encoded Strings. Journal of Complexity (accepted) (A preliminary version appeared in CPM 2009)
7. Diede, S.J., Tanaka, H., Bergstrom, D.A., Yao, M.-C., Tapscott, S.J.: Tapscott: Genome-wide Analysis of Palindrome Formation. Nature Genetics 42(4), 279 (2010)
8. Farach-Colton, M., Ferragina, P., Muthukrishnan, S.: On the Sorting-Complexity of Suffix Tree Construction. Journal of ACM 47(6), 987–1011 (2000)
9. Gasieniec, L., Karpinski, M., Plandowski, W., Rytter, W.: Efficient Algorithms for Lempel-Ziv Encoding (Extended Abstract). In: Karlsson, R., Lingas, A. (eds.) SWAT 1996. LNCS, vol. 1097, pp. 392–403. Springer, Heidelberg (1996)
10. Gusfield, D.: Algorithms on Strings, Trees and Sequences: Computer Science and Computational Biology. Cambridge University Press, Cambridge (1997)
11. Harel, D., Tarjan, R.E.: Fast Algorithms for Finding Nearest Common Ancestors. SIAM Journal on Computing 13(2), 338–355 (1984)
12. Hermelin, D., Landau, G.M., Landau, S., Weimann, O.: A Unified Algorithm for Accelerating Edit-Distance Computation via Text-Compression. In: STACS, pp. 529–540 (2009)
13. Hsu, P.-H., Chen, K.-Y., Chao, K.-M.: Finding All Approximate Gapped Palindromes. In: ISAAC, pp. 1084–1093 (2009)
14. Liu, J.J., Huang, G.S., Wang, Y.L.: A Fast Algorithm for Finding the Positions of All Squares in a Run-Length Encoded String. Theoretical Computer Science 410(38-40), 3942–3948 (2009)
15. Lueker, G.S.: A Data Structure for Orthogonal Range Queries. In: FOCS, pp. 28–34 (1978)
16. Matsubara, W., Inenaga, S., Ishino, A., Shinohara, A., Nakamura, T., Hashimoto, K.: Efficient Algorithms to Compute Compressed Longest Common Substrings and Compressed Palindromes. Theoretical Computer Science 410(8-10), 900–913 (2009)
17. Rytter, W.: Algorithms on Compressed Strings and Arrays. In: Bartosek, M., Tel, G., Pavelka, J. (eds.) SOFSEM 1999. LNCS, vol. 1725, pp. 48–65. Springer, Heidelberg (1999)
18. Willard, D.E.: Predicate-Oriented Database Search Algorithms. Garland Publishing, New York (1978)

Minimum Cost Partitions
of Trees with Supply and Demand

Takehiro Ito[1], Takuya Hara[1], Xiao Zhou[1], and Takao Nishizeki[2]

[1] Graduate School of Information Sciences, Tohoku University,
Aoba-yama 6-6-05, Sendai, 980-8579, Japan
takehiro@ecei.tohoku.ac.jp,
hara@nishizeki.ecei.tohoku.ac.jp,
zhou@ecei.tohoku.ac.jp
[2] School of Science and Technology, Kwansei Gakuin University,
2-1 Gakuen, Sanda, 669-1337, Japan
nishi@kwansei.ac.jp

Abstract. Let T be a given tree. Each vertex of T is either a supply
vertex or a demand vertex, and is assigned a positive integer, called the
supply or the demand. Every demand vertex v of T must be supplied an
amount of "power," equal to the demand of v, from exactly one supply
vertex through edges in T. Each edge e of T has a direction, and is
assigned a positive integer which represents the cost required to delete e
from T or reverse the direction of e. Then one wishes to obtain subtrees
of T by deleting edges and reversing the directions of edges so that (a)
each subtree contains exactly one supply vertex whose supply is no less
than the sum of all demands in the subtree and (b) each subtree is rooted
at the supply vertex in a sense that every edge is directed away from the
root. We wish to minimize the total cost to obtain such rooted subtrees
from T. In the paper, we first show that this minimization problem is
NP-hard, and then give a pseudo-polynomial-time algorithm to solve the
problem. We finally give a fully polynomial-time approximation scheme
(FPTAS) for the problem.

1 Introduction

In this paper, we deal with a directed graph G in which each vertex is either
a supply vertex or a demand vertex. Each vertex v of G is assigned a positive
integer, which is called the *supply of v* if v is a supply vertex, otherwise, called
the *demand of v*. Each edge e of G is assigned a positive integer, called the
cost of e, which represents the cost required to delete e from G or reverse the
direction of e. Figure 1(a) illustrates a tree T, in which each edge has a direction,
each supply vertex is drawn as a rectangle and each demand vertex as a circle,
the supply or demand is written inside, and the cost is attached to each edge.
Assume that the "power" can flow only along the direction of an edge, and that
each demand vertex can receive power from exactly one supply vertex through
directed edges in G. One thus wishes to partition G into subtrees, that is, obtain

O. Cheong, K.-Y. Chwa, and K. Park (Eds.): ISAAC 2010, Part II, LNCS 6507, pp. 351–362, 2010.

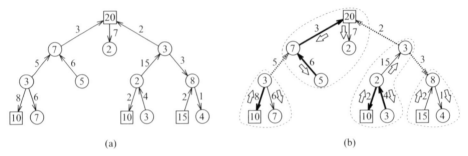

Fig. 1. (a) Tree T and (b) rooted subtrees of T

rooted subtrees from G by deleting some edges or reversing the directions of some edges, so that

(a) each subtree has exactly one supply vertex whose supply is no less than the sum of demands of all demand vertices in the subtree; and
(b) each subtree is rooted at the supply vertex in a sense that every edge is directed away from the root.

We wish to minimize the total cost to obtain such rooted subtrees from a given graph G; the total cost is the sum of costs of all edges that are deleted from G or are reversed the directions. The problem is called the *minimum partition problem* for G. Figure 1(b) illustrates an optimal solution for the tree T in Fig. 1(a), that is, a partition of T into rooted subtrees; each subtree is surrounded by a thin dotted curve; all the deleted edges are drawn by thick dotted lines, and all the edges whose directions are reversed are drawn by thick solid lines; the direction of each edge in rooted subtrees is indicated by a white arrow; the minimum total cost is $(5 + 2 + 3) + (8 + 6 + 3 + 2 + 4) = 33$.

The minimum partition problem has some applications to the power supply problem for power delivery networks [1,4,5,6,8,9]. Let G be a directed graph of a power delivery network. Each supply vertex represents a "feeder," which can supply electrical power. Each demand vertex represents a "load," which requires electrical power supplied from exactly one of the feeders through a network. Each directed edge of G represents a cable segment, which can be "turned off" by a switch in the segment; when the switch is "turned on," the power can flow through the cable segment only in the direction of the switch. The direction of the switch must be reversed if one wishes to flow power in the other direction. Thus, the minimum partition problem represents the "power supply switching problem" to minimize the sum of costs for "reconfiguring" the network by either turning off or reversing some switches in cable segments.

A simple problem to examine only whether a given graph has at least one partition (into subtrees satisfying (a) and (b) above) is called the *partition problem*. Ito *et al.* showed that the problem is NP-complete even for series-parallel graphs [6]. Therefore, it is very unlikely that the problem can be solved in polynomial time even for series-parallel graphs. However, they showed that the problem can be solved in linear time for trees [5]. One of their algorithms actually finds a partition of a given tree T if there is, but the cost to obtain a partition is not always minimum. Thus, it has been desired to obtain an efficient algorithm to find a partition of a tree with the minimum cost. (They also gave a fully polynomial-time approximation scheme (FPTAS) for the problem of maximizing the sum of demands supplied power when a given tree has no partition [5].)

In this paper, we give the following three results for the minimum partition problem for trees. The first is to show that the knapsack problem [2] can be reduced in linear time to the minimum partition problem for trees of height two, and hence the minimum partition problem is NP-hard even for trees. The second is to give a pseudo-polynomial-time algorithm to solve the problem for trees T. The algorithm is based on a sophisticated dynamic programming approach and takes time $O(n + W^2)$, where n is the number of vertices in T and W is the sum of costs of all edges in T. Thus, our algorithm takes polynomial time if W is bounded by a polynomial in n. As the third result, we give an FPTAS for the minimum partition problem for trees. The main idea of the FPTAS is to find an approximate solution by repeating the pseudo-polynomial-time algorithm with appropriately chosen scaling factors.

2 Pseudo-Polynomial-Time Algorithm

The minimum partition problem was defined informally in Introduction. In this section, we first formally define the minimum partition problem, and then give a pseudo-polynomial-time algorithm for the problem.

We deal with a directed graph $G = (V, E)$ with vertex set V and edge set E. V is often denoted by $V(G)$, and E by $E(G)$. Throughout the paper, we denote by (v, w) a directed edge in G joining vertices v and w whose direction is from v to w. We denote by $\text{sup}(v)$ the supply of a supply vertex v, by $\text{dem}(w)$ the demand of a demand vertex w, and by $c(e)$ the cost of an edge e. A *partition* $\mathcal{P}$ of G is to partition G into subtrees by deleting edges from G so that each subtree contains exactly one supply vertex whose supply is no less than the sum of all demands in the subtree. $\mathcal{P}$ often denotes the set of subtrees. For a partition $\mathcal{P}$ and an edge e, we denote by $c(\mathcal{P}, e)$ the cost paid by $\mathcal{P}$ for e; $c(\mathcal{P}, e) = c(e)$ if either e is deleted from G or the direction of e is toward the supply vertex in the subtree containing e; otherwise, $c(\mathcal{P}, e) = 0$. The *cost* $c(G, \mathcal{P})$ of a partition $\mathcal{P}$ is the sum of costs $c(\mathcal{P}, e)$ paid by $\mathcal{P}$ for all edges e in G, that is, $c(G, \mathcal{P}) = \sum_{e \in E} c(\mathcal{P}, e)$. We call $\mathcal{P}$ the *minimum partition* of G if $c(G, \mathcal{P})$ is minimum among all partitions of G. We denote by $\text{OPT}(G)$ the cost of a minimum partition of G; let $\text{OPT}(G) = +\infty$ if G has no partition. The *minimum partition problem* is to compute $\text{OPT}(G)$ for a given directed graph G, and to find a minimum partition unless $\text{OPT}(G) = +\infty$.

Our first result is the following theorem, whose proof is omitted from this extended abstract.

Theorem 1. *The minimum partition problem is NP-hard even for trees of height two.*

In this section, as the second result, we show that the minimum partition problem for a tree $T = (V, E)$ can be solved in pseudo-polynomial time. Indeed, we introduce a generalized problem, called the *minimum R-partition problem* for a subset R of E, and show that the problem can be solved in pseudo-polynomial time.

A partition $\mathcal{P}$ of T is called an *R-partition* if $c(\mathcal{P}, e) = 0$ for every edge $e \in R$, that is, e is neither deleted from T nor directed toward the supply vertex in the subtree containing e. We call $\mathcal{P}$ the *minimum R-partition* of T if $c(T, \mathcal{P})$ is minimum among all R-partitions of T. We denote by $\mathrm{OPT}(T, R)$ the cost of a minimum R-partition of T; let $\mathrm{OPT}(T, R) = +\infty$ if T has no R-partition. The *minimum R-partition problem* is to compute $\mathrm{OPT}(T, R)$, and to find a minimum R-partition $\mathcal{P}$ of T unless $\mathrm{OPT}(T, R) = +\infty$. The minimum partition problem is merely the minimum R-partition problem for $R = \emptyset$. (Our FPTAS in Section 3 finds an approximate solution as the best one among R-partitions of T for some subsets R of E.)

The main result of this section is the following theorem.

Theorem 2. *For a directed tree $T = (V, E)$ and a subset R of E, the minimum R-partition problem can be solved in time $O(n + W_R^2)$, where n is the number of vertices in T and $W_R = \sum_{e \in E \setminus R} c(e)$.*

In the remainder of this section, as a proof of Theorem 2, we give an algorithm to solve the minimum R-partition problem in time $O(n + W_R^2)$. We indeed show only how to compute the minimum cost $\mathrm{OPT}(T, R)$. It is easy to modify our algorithm so that it actually finds a minimum R-partition of T unless $\mathrm{OPT}(T, R) = +\infty$. Note that $c(T, \mathcal{P}) \le W_R$ for every R-partition $\mathcal{P}$ of T.

One may assume without loss of generality that T is a rooted tree with an arbitrarily chosen root r. For each vertex u of T, we denote by T_u the subtree of T which is rooted at u and is induced by all descendants of u in T. Every R-partition of T naturally induces a partition of the vertex set $V(T_u)$ of T_u. (See Figs. 2 and 3, where each subtree is indicated by a dotted closed curve.) However, the induced partition does not always form an R'-partition of T_u for $R' = R \cap E(T_u)$, because the subtree of T_u containing u may have no supply vertex. (See Figs. 3(a) and (b).) We will later define such a partition as an "extendable R'-partition" of T_u, and consider the power flow in the edge joining u and its parent in T. More precisely, we consider the following two types of extendable R'-partitions of T_u. The first is called a "*j-out partition*," which can deliver an amount j of marginal power outside T_u through u. The second is called a "*j-in partition*," which needs

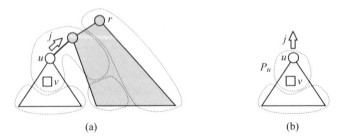

(a)   (b)

Fig. 2. (a) R-partition of tree T and (b) j-out partition of subtree T_u

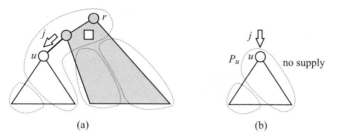

Fig. 3. (a) R-partition of tree T and (b) j-in partition of subtree T_u

an amount j of deficient power to be delivered inside T_u through u. For these extendable R'-partitions, we introduce two functions $f_{\text{out}}(T_u, x)$ and $f_{\text{in}}(T_u, x)$, $0 \leq x \leq W_R$; $f_{\text{out}}(T_u, x)$ is the maximum marginal power of a j-out partition $\mathcal{P}$ of T_u such that $c(T_u, \mathcal{P}) \leq x$; and $f_{\text{in}}(T_u, x)$ is the minimum deficient power of a j-in partition $\mathcal{P}$ of T_u such that $c(T_u, \mathcal{P}) \leq x$. We compute f_{out} and f_{in} from the leaves to the root r of T by means of dynamic programming.

We now formally define the notion of an extendable R'-partition of T_u for $R' \subseteq E(T_u)$. (Intuitively, $R' = R \cap E(T_u)$.) An *extendable R'-partition $\mathcal{P}$ of T_u* is a partition of T_u into subtrees such that

(a) the subtree $P_u \in \mathcal{P}$ containing u contains at most one supply vertex, and each subtree $P_i \in \mathcal{P} \setminus \{P_u\}$ contains exactly one supply vertex;
(b) if a subtree $P_i \in \mathcal{P}$ contains a supply vertex v, then $\sup(v) \geq \sum_{w \in V(P_i) \setminus \{v\}} \text{dem}(w)$; and
(c) $c(\mathcal{P}, e) = 0$ for each edge $e \in R'$.

Thus, an extendable R'-partition of T_u is an R'-partition of T_u if P_u contains a supply vertex.

Let m_s be the maximum supply, that is, $m_s = \max\{\sup(v) \mid v \text{ is a supply vertex in } T\}$. Let $\mathbb{Z}_s$ be the set of all integers j such that $0 \leq j \leq m_s$. We then classify the set of all extendable R'-partitions of T_u into several subclasses, called j-out partitions and j-in partitions for $j \in \mathbb{Z}_s$.

(a) An extendable R'-partition $\mathcal{P}$ of T_u is called a *j-out partition* if the subtree $P_u \in \mathcal{P}$ contains a supply vertex v and $\sup(v) \geq j + \sum_{w \in V(P_u) \setminus \{v\}} \text{dem}(w)$. (See Fig. 2(b).) A j-out partition of T_u corresponds to an R-partition of the whole tree T in which all demand vertices in P_u are supplied power from a supply vertex in T_u, as illustrated in Figs. 2(a) and (b); an amount j of power can be delivered outside T_u through u, and hence the "margin" of $\mathcal{P}$ is j. Note that any R'-partition of T_u is a j-out partition for some integer $j \in \mathbb{Z}_s$.

(b) An extendable R'-partition $\mathcal{P}$ of T_u is called a *j-in partition* if P_u contains no supply vertex and $\sum_{w \in V(P_u)} \text{dem}(w) \leq j$. (See Fig. 3(b).) Thus, if $\mathcal{P}$ is a j-in partition of T_u, then u is a demand vertex and $\text{dem}(u) \leq j$. A j-in partition of T_u corresponds to an R-partition of T in which all (demand) vertices in P_u including u are supplied power from a supply vertex outside T_u, as illustrated in Figs. 3(a) and (b); an amount j of power must be delivered inside T_u through

u, and hence the "deficiency" of $\mathcal{P}$ is j. Note that there is no 0-in partition of T_u, because $\mathrm{dem}(w) > 0$ for every demand vertex w in T.

We are now ready to give a formal definition of two functions f_{out} and f_{in}. Let $\mathbb{Z}_{W_R}$ be the set of all integers x such that $0 \le x \le W_R$. We first define $f_{\mathrm{out}}(T_u, x)$ for a subtree T_u and an integer $x \in \mathbb{Z}_{W_R}$, as follows:

$$f_{\mathrm{out}}(T_u, x) = \max\{j \in \mathbb{Z}_s \mid T_u \text{ has a } j\text{-out partition } \mathcal{P}$$
$$\text{such that } c(T_u, \mathcal{P}) \le x\}. \qquad (1)$$

Thus, $f_{\mathrm{out}}(T_u, x)$ is the maximum amount j of "marginal power" of a j-out partition $\mathcal{P}$ such that $c(T_u, \mathcal{P}) \le x$. If T_u has no j-out partition $\mathcal{P}$ with $c(T_u, \mathcal{P}) \le x$ for all integers $j \in \mathbb{Z}_s$, then let $f_{\mathrm{out}}(T_u, x) = -\infty$. We then define $f_{\mathrm{in}}(T_u, x)$ for a subtree T_u and an integer $x \in \mathbb{Z}_{W_R}$, as follows:

$$f_{\mathrm{in}}(T_u, x) = \min\{j \in \mathbb{Z}_s \mid T_u \text{ has a } j\text{-in partition } \mathcal{P} \text{ such that } c(T_u, \mathcal{P}) \le x\}.$$

Thus, $f_{\mathrm{in}}(T_u, x)$ is the minimum amount j of "deficient power" of a j-in partition $\mathcal{P}$ such that $c(T_u, \mathcal{P}) \le x$. If T_u has no j-in partition $\mathcal{P}$ with $c(T_u, \mathcal{P}) \le x$ for all integers $j \in \mathbb{Z}_s$, then let $f_{\mathrm{in}}(T_u, x) = +\infty$.

Our algorithm computes the two functions $f_{\mathrm{out}}(T_u, x)$ and $f_{\mathrm{in}}(T_u, x)$ for each vertex u of T from the leaves to the root r of T by means of dynamic programming. Since $T = T_r$, one can easily compute the minimum cost $\mathrm{OPT}(T, R)$ from $f_{\mathrm{out}}(T_r, x)$, as follows:

$$\mathrm{OPT}(T, R) = \min\{x \in \mathbb{Z}_{W_R} \mid f_{\mathrm{out}}(T_r, x) \ne -\infty\} \qquad (2)$$

if $f_{\mathrm{out}}(T_r, x) \ne -\infty$ for some integer $x \in \mathbb{Z}_{W_R}$; otherwise, $\mathrm{OPT}(T, R) = +\infty$.

Let u be a vertex of T, let $u_1, u_2, \cdots, u_l$ be the children of u ordered arbitrarily, and let e_i, $1 \le i \le l$, be the edge joining u and u_i. For each i, $1 \le i \le l$, let T_{u_i} be the subtree of T which is rooted at u_i and is induced by all descendants of u_i in T. We denote by T_u^i the subtree of T which consists of the vertex u, the edges $e_1, e_2, \cdots, e_i$ and the subtrees $T_{u_1}, T_{u_2}, \cdots, T_{u_i}$. Clearly $T_u = T_u^l$. For the sake of notational convenience, we denote by T_u^0 the tree consisting of a single vertex u.

One can easily compute $f_{\mathrm{out}}(T_u^0, x)$ and $f_{\mathrm{in}}(T_u^0, x)$ for each vertex u of T. (Due to the page limitation, we omit the details.)

We now compute the two functions $f_{\mathrm{out}}(T_u^i, x)$ and $f_{\mathrm{in}}(T_u^i, x)$, $1 \le i \le l$, for each internal vertex u of T from the counterparts of T_u^{i-1} and T_{u_i}, where l is the number of the children of u. However, due to the page limitation, we show only how to compute $f_{\mathrm{out}}(T_u^i, x)$; one can compute $f_{\mathrm{in}}(T_u^i, x)$ similarly. Remember that $T_u = T_u^l$, and that T_u^i is obtained from T_u^{i-1} and T_{u_i} by joining u and u_i as illustrated in Fig. 4, where T_u^{i-1} is indicated by a thin dotted closed curve. We denote by $\gamma(u, u_i)$ the cost paid for e_i by a partition in which e_i is not deleted and the power flows through e_i from u to u_i, that is,

$$\gamma(u, u_i) = \begin{cases} 0 & \text{if } e_i = (u, u_i); \\ c(e_i) & \text{if } e_i = (u_i, u). \end{cases}$$

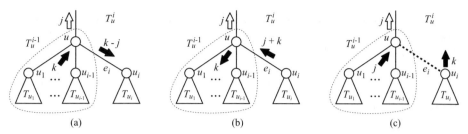

Fig. 4. Power flow in a j-out partition of T_u^i

We denote by $\gamma(u_i, u)$ the cost paid for e_i by a partition in which e_i is not deleted and the power flows through e_i from u_i to u.

We explain how to compute $f_{\mathrm{out}}(T_u^i, x)$. Let $\mathcal{P}$ be a j-out partition of T_u^i such that $c(T_u^i, \mathcal{P}) \leq x$ and $j = f_{\mathrm{out}}(T_u^i, x) \neq -\infty$. Then, there are the following three cases (a)–(c), as illustrated in Figs. 4(a)–(c), respectively:

Case (a): e_i is not deleted in $\mathcal{P}$ and the power flows through e_i from u to u_i;

Case (b): e_i is not deleted in $\mathcal{P}$ but the power flows through e_i from u_i to u; and

Case (c): e_i is deleted in $\mathcal{P}$.

Figures 4(a)–(c) illustrate the power flows in T_u^i for the three cases, where an arrow represents the direction of power flow and a thick dotted line represents a deleted edge. For two integers $x \in \mathbb{Z}_{W_R}$ and $y \in \mathbb{Z}_{W_R}$, we define $f_{\mathrm{out}}^{\mathrm{a}}(T_u^i, x, y)$, $f_{\mathrm{out}}^{\mathrm{b}}(T_u^i, x, y)$ and $f_{\mathrm{out}}^{\mathrm{c}}(T_u^i, x, y)$ for Cases (a), (b) and (c), respectively. Intuitively, x and y are the costs for T_u^i and T_u^{i-1}, respectively.

Case (a): *e_i is not deleted in $\mathcal{P}$ and the power flows through e_i from u to u_i.*

In this case, we have

$$c(\mathcal{P}, e_i) = \gamma(u, u_i) = \begin{cases} 0 & \text{if } e_i = (u, u_i); \\ c(e_i) & \text{if } e_i = (u_i, u). \end{cases}$$

(See Fig. 4(a).) Therefore, if $e_i \in R$ and $e_i = (u_i, u)$, then we define $f_{\mathrm{out}}^{\mathrm{a}}(T_u^i, x, y) = -\infty$ for every pair (x, y) of integers $x, y \in \mathbb{Z}_{W_R}$. One may thus assume that either $e_i \notin R$ or $e_i = (u, u_i)$ if $e_i \in R$. Then, u_i is supplied power from a vertex in T_u^{i-1}, and either u is a supply vertex or both u and u_i are supplied power from the same supply vertex in T_u^{i-1}. Therefore, for some integer $k \in \mathbb{Z}_s$ with $k \geq j$, the j-out partition $\mathcal{P}$ of T_u^i can be obtained by merging a k-out partition $\mathcal{P}_1$ of T_u^{i-1} and a $(k-j)$-in partition $\mathcal{P}_2$ of T_{u_i} such that

$$c(T_u^i, \mathcal{P}) = c(T_u^{i-1}, \mathcal{P}_1) + c(T_{u_i}, \mathcal{P}_2) + \gamma(u, u_i).$$

Since $c(T_u^i, \mathcal{P}) \leq x$, we have $c(T_u^{i-1}, \mathcal{P}_1) \leq y$ and $c(T_{u_i}, \mathcal{P}_2) \leq x - y - \gamma(u, u_i)$ for some integer $y \in \mathbb{Z}_{W_R}$. Since $\mathcal{P}_1$ is a k-out partition of T_u^{i-1} with $c(T_u^{i-1}, \mathcal{P}_1) \leq y$, one may assume by Eq. (1) that $k = f_{\mathrm{out}}(T_u^{i-1}, y)$. Similarly, one may assume that $k - j = f_{\mathrm{in}}(T_{u_i}, x - y - \gamma(u, u_i))$. Since $f_{\mathrm{out}}(T_u^i, x) = j = k - (k-j)$, we define

$$f_{\mathrm{out}}^{\mathrm{a}}(T_u^i, x, y) = f_{\mathrm{out}}(T_u^{i-1}, y) - f_{\mathrm{in}}(T_{u_i}, x - y - \gamma(u, u_i)). \tag{3}$$

Similarly as Case (a) above, we define $f_{\text{out}}^{\text{b}}(T_u^i, x, y)$ and $f_{\text{out}}^{\text{c}}(T_u^i, x, y)$ for two integers $x \in \mathbb{Z}_{W_R}$ and $y \in \mathbb{Z}_{W_R}$ for Cases (b) and (c), respectively. Then, we compute $f_{\text{out}}(T_u^i, x)$ as follows:

$$f_{\text{out}}(T_u^i, x) = \max\{f_{\text{out}}^{\text{a}}(T_u^i, x, y),\ f_{\text{out}}^{\text{b}}(T_u^i, x, y),\ f_{\text{out}}^{\text{c}}(T_u^i, x, y) \mid y \in \mathbb{Z}_{W_R}\}. \quad (4)$$

Proof of Theorem 2

For a vertex u of T and an index i, $0 \le i \le l$, we denote by $W_R(T_u^i)$ the sum of costs of all edges in $E(T_u^i) \setminus R$. Then, $c(T_u^i, \mathcal{P}) \le W_R(T_u^i)$ for every extendable R'-partition $\mathcal{P}$ of T_u^i, where $R' = R \cap E(T_u^i)$. Thus one can easily observe that it suffices to compute $f_{\text{out}}(T_u^i, x)$ and $f_{\text{in}}(T_u^i, x)$ only for integers x, $0 \le x \le W_R(T_u^i)$. We denote by $\text{Time}(T_u^i)$ the computation time of our algorithm for T_u^i.

Clearly, $W_R(T_u^0) = 0$ for each vertex u of T. Therefore, one can easily compute $f_{\text{out}}(T_u^0, 0)$ and $f_{\text{in}}(T_u^0, 0)$ for a vertex u of T in time $O(1)$. We thus have

$$\text{Time}(T_u^0) = O(1). \quad (5)$$

Therefore, $f_{\text{out}}(T_{u'}^0, 0)$ and $f_{\text{in}}(T_{u'}^0, 0)$ can be computed for all vertices u' of T_u^i in time $O(n_u^i)$, where n_u^i is the number of vertices in T_u^i.

We then show that $\text{Time}(T_u^i) = O(n_u^i + (W_R(T_u^i))^2)$. Since T_u^i is obtained from T_u^{i-1} and T_{u_i} by adding an edge e_i, we have

$$W_R(T_u^i) = \begin{cases} W_R(T_u^{i-1}) + W_R(T_{u_i}) + c(e_i) & \text{if } e_i \notin R; \\ W_R(T_u^{i-1}) + W_R(T_{u_i}) & \text{if } e_i \in R. \end{cases} \quad (6)$$

Using Eqs. (3) and (4), one can trivially compute $f_{\text{out}}(T_u^i, x)$ for all integers x, $0 \le x \le W_R(T_u^i)$, in time $O\big((W_R(T_u^i) + 1) \times (W_R(T_u^{i-1}) + 1)\big)$ because $0 \le x \le W_R(T_u^i)$ and $0 \le y \le W_R(T_u^{i-1})$. However, one can easily observe that it can be done in time $O\big(W_R(T_u^i) + 1 + (W_R(T_u^{i-1}) + 1) \times (W_R(T_{u_i}) + 1)\big)$ because Eq. (3), for example, can be rewritten as follows: for $0 \le y \le W_R(T_u^{i-1})$ and $0 \le z \le W_R(T_{u_i})$

$$f_{\text{out}}^{\text{a}}(T_u^i, y + z + \gamma(u, u_i), y) = f_{\text{out}}(T_u^{i-1}, y) - f_{\text{in}}(T_{u_i}, z)$$

if $0 \le y + z + \gamma(u, u_i) \le W_R(T_u^i)$. Similarly, one can compute $f_{\text{in}}(T_u^i, x)$ for all integers x, $0 \le x \le W_R(T_u^i)$, from the counterparts of T_u^{i-1} and T_{u_i} in time $O\big(W_R(T_u^i) + 1 + (W_R(T_u^{i-1}) + 1) \times (W_R(T_{u_i}) + 1)\big)$. Therefore, we have

$$\text{Time}(T_u^i) = \text{Time}(T_u^{i-1}) + \text{Time}(T_{u_i})$$
$$+ O\big(W_R(T_u^i) + 1 + (W_R(T_u^{i-1}) + 1) \times (W_R(T_{u_i}) + 1)\big).$$

Solving the recurrence equation above with Eqs. (5) and (6), we have $\text{Time}(T_u^i) = O(n_u^i + (W_R(T_u^i))^2)$.

Since T_r has n vertices and $W_R(T_r) = W_R$ for the root r of T, we have $\text{Time}(T_r) = O(n + W_R^2)$. One can thus compute $f_{\text{out}}(T_r, x)$ and $f_{\text{in}}(T_r, x)$ in time $O(n + W_R^2)$ for all integers x, $0 \leq x \leq W_R$. Then, by Eq. (2) one can compute $\mathrm{OPT}(T, R)$ from $f_{\mathrm{out}}(T_r, x)$ in time $O(W_R)$.

Thus, the minimum R-partition problem for a tree T can be solved in time $O(n + W_R^2)$.   □

3   FPTAS

The main result of this section is the following theorem.

Theorem 3. *There is a fully polynomial-time approximation scheme for the minimum partition problem for trees.*

In the remainder of this section, as a proof of Theorem 3, we give an algorithm to find a partition $\mathcal{P}$ of a tree $T = (V, E)$ with $c(T, \mathcal{P}) < (1+\varepsilon)\mathrm{OPT}(T)$ in time polynomial in both n and $1/\varepsilon$ for any real number $\varepsilon > 0$. Thus, our approximate value for T is $c(T, \mathcal{P})$, and the error must be bounded by $\varepsilon\mathrm{OPT}(T)$, that is,

$$c(T, \mathcal{P}) - \mathrm{OPT}(T) < \varepsilon\mathrm{OPT}(T). \tag{7}$$

We extend the ordinary "scaling and rounding" technique [2,3,4,5,7] to the minimum partition problem, and introduce new techniques. Trivially, the minimum partition problem can be solved in linear time if a given tree T has exactly one supply vertex; $\mathrm{OPT}(T)$ is the sum of costs of all edges that are directed toward the supply vertex unless $\mathrm{OPT}(T) = +\infty$. One may thus assume that T has two or more supply vertices. Then one or more edges must be deleted in every partition $\mathcal{P}$ of T and hence $c(\mathcal{P}, e) > 0$ for some edge e of T. Assume that the edges $e_1, e_2, \cdots, e_{n-1}$ in E are labeled in non-increasing order with respect to the cost c, that is, $c(e_1) \geq c(e_2) \geq \cdots \geq c(e_{n-1})$. We define subsets R_i, $0 \leq i \leq n-2$, of E as follows: $R_0 = \emptyset$, and $R_i = \{e_1, e_2, \cdots, e_i\}$, $1 \leq i \leq n-2$. Let $\overline{T}_i = (V, E)$, $0 \leq i \leq n-2$, be the tree with the same vertex set V and edge set E as T, but the cost $\bar{c}_i(e)$ of each edge e in $\overline{T}_i$ is defined as follows:

$$\bar{c}_i(e) = \left\lceil \frac{c(e)}{t_i} \right\rceil, \tag{8}$$

where t_i (> 0) is a scaling factor which will be decided later. Clearly, $\mathcal{P}$ is an R_i-partition of T if and only if $\mathcal{P}$ is an R_i-partition of $\overline{T}_i$.

[Algorithm]
Our algorithm finds an approximate solution $\mathcal{P}$ by repeating the algorithm in Section 2, as follows:

Step 1. Execute the following (1) and (2) for each i, $0 \leq i \leq n-2$:
  (1) find a minimum R_i-partition $\mathcal{P}_i$ of $\overline{T}_i$ with respect to the cost $\bar{c}_i$ by the algorithm in Section 2; and
  (2) regard $\mathcal{P}_i$ as an approximate solution for T, and compute its cost $c(T, \mathcal{P}_i)$ with respect to the original cost c;

Step 2. Choose a partition having the minimum cost among $\mathcal{P}_0, \mathcal{P}_1, \cdots, \mathcal{P}_{n-2}$ as our approximate solution $\mathcal{P}$ of the minimum partition problem for T, and hence

$$c(T, \mathcal{P}) = \min\{c(T, \mathcal{P}_i) \mid 0 \le i \le n - 2\}. \tag{9}$$

[Computation time]

We first show that our algorithm finds the partition $\mathcal{P}$ of T in time polynomial in both n and $1/\varepsilon$. It suffices to show that the minimum R_i-partition $\mathcal{P}_i$ of tree $\overline{T}_i$, $0 \le i \le n - 2$, can be found in time $O(n^4/\varepsilon^2)$. Then $\mathcal{P}$ can be found in time $O(n^5/\varepsilon^2)$.

Our idea is to choose appropriately the scaling factor t_i. Let $\overline{W}_{R_i} = \sum_{e \in E \setminus R_i} \bar{c}_i(e)$. Then, using the algorithm in Section 2, one can find the minimum R_i-partition $\mathcal{P}_i$ of $\overline{T}_i$, $0 \le i \le n - 2$, in time $O(n + \overline{W}^2_{R_i})$. Since $c(e_1) \ge c(e_2) \ge \cdots \ge c(e_{n-1})$ and $R_i = \{e_1, e_2, \cdots, e_i\}$, we have

$$\overline{W}_{R_i} = \sum_{e \in E \setminus R_i} \bar{c}_i(e) < n\bar{c}_i(e_{i+1}). \tag{10}$$

We choose the scaling factor t_i as follows:

$$t_i = \frac{\varepsilon c(e_{i+1})}{n}. \tag{11}$$

Then, by Eqs. (8), (10) and (11) we have

$$\overline{W}_{R_i} < n \left\lceil \frac{c(e_{i+1})}{t_i} \right\rceil < n \left(\frac{n}{\varepsilon} + 1 \right) = \frac{n^2}{\varepsilon} + n.$$

Thus $\mathcal{P}_i$ can be found in time $O(n + \overline{W}^2_{R_i}) = O(n^4/\varepsilon^2)$.

[Error]

We finally show that our approximate solution $\mathcal{P}$ satisfies Eq. (7).

Consider an arbitrary minimum partition $\mathcal{P}^*$ of T, whose cost is $\mathrm{OPT}(T)$. Clearly, $\mathcal{P}^*$ is an R_i-partition of T for some indices i, $0 \le i \le n-2$. Let k be the maximum one among these indices. Then $\mathcal{P}^*$ is indeed a minimum R_k-partition of T, and $c(\mathcal{P}^*, e_{k+1}) > 0$. We thus have

$$\mathrm{OPT}(T) = c(T, \mathcal{P}^*) = \mathrm{OPT}(T, R_k) \ge c(e_{k+1}). \tag{12}$$

The minimum R_k-partition $\mathcal{P}_k$ of $\overline{T}_k$ is of course a partition of T. It suffices to show that

$$c(T, \mathcal{P}_k) < (1 + \varepsilon)\mathrm{OPT}(T, R_k), \tag{13}$$

because Eqs. (9), (12) and (13) immediately imply $c(T, \mathcal{P}) < (1+\varepsilon)\mathrm{OPT}(T)$ and hence Eq. (7) holds.

We now verify Eq. (13). Let $E^* = \{e \in E \mid c(\mathcal{P}^*, e) > 0\}$, then by Eq. (12) we have

$$\mathrm{OPT}(T, R_k) = c(T, \mathcal{P}^*) = \sum_{e \in E^*} c(e). \tag{14}$$

Let $E_k = \{e \in E \mid c(\mathcal{P}_k, e) > 0\}$, then

$$c(T, \mathcal{P}_k) = \sum_{e \in E_k} c(e). \tag{15}$$

By Eq. (8) we have

$$t_k \bar{c}_k(e) \geq c(e) > t_k\big(\bar{c}_k(e) - 1\big) \tag{16}$$

for every edge $e \in E$. Therefore, by Eqs. (14) and (16) we have

$$\mathrm{OPT}(T, R_k) > \sum_{e \in E^*} t_k\big(\bar{c}_k(e) - 1\big) = \sum_{e \in E^*} t_k \bar{c}_k(e) - t_k |E^*|. \tag{17}$$

Clearly, for each edge $e \in E$, $\bar{c}_k(\mathcal{P}^*, e) > 0$ if and only if $c(\mathcal{P}^*, e) > 0$. We hence have $\bar{c}_k(\overline{T}_k, \mathcal{P}^*) = \sum_{e \in E^*} \bar{c}_k(e)$. $\mathcal{P}_k$ is a minimum R_k-partition of $\overline{T}_k$ with respect to the cost $\bar{c}_k$, while $\mathcal{P}^*$ is not necessarily a minimum R_k-partition of $\overline{T}_k$. Therefore, we have

$$\sum_{e \in E^*} \bar{c}_k(e) = \bar{c}_k(\overline{T}_k, \mathcal{P}^*) \geq \bar{c}_k(\overline{T}_k, \mathcal{P}_k) = \sum_{e \in E_k} \bar{c}_k(e)$$

and hence

$$\sum_{e \in E^*} t_k \bar{c}_k(e) \geq \sum_{e \in E_k} t_k \bar{c}_k(e). \tag{18}$$

Thus, by Eqs. (15)–(18) we have

$$\mathrm{OPT}(T, R_k) > \sum_{e \in E_k} t_k \bar{c}_k(e) - t_k |E^*| \geq \sum_{e \in E_k} c(e) - t_k |E^*|$$
$$= c(T, \mathcal{P}_k) - t_k |E^*|. \tag{19}$$

Since $|E^*| < n$, by Eqs. (11), (12) and (19) we have

$$c(T, \mathcal{P}_k) - \mathrm{OPT}(T, R_k) < t_k n = \varepsilon c(e_{k+1}) \leq \varepsilon \mathrm{OPT}(T, R_k).$$

We have thus verified Eq. (13).

This completes the proof of Theorem 3. $\qquad\qquad\square$

4 Conclusions

In this paper, we studied the minimum partition problem for trees. We first show that the problem is NP-hard even for trees of height two. We then give an algorithm to solve the minimum R-partition problem for a tree $T = (V, E)$ of n vertices in time $O(n + W_R^2)$, where $W_R = \sum_{e \in E \setminus R} c(e)$. Thus, the algorithm takes polynomial time if W_R is bounded by a polynomial in n. Based on the algorithm, we finally show that there is an FPTAS for the minimum partition problem for trees. It is easy to modify our algorithms so that they actually find a partition of a given tree.

In all partitions of a tree T, the same number $n_s - 1$ of edges must be deleted, where n_s is the number of supply vertices in T. Thus, if $c(e) = 1$ for all edges e, then the minimum partition problem is simply to obtain desired rooted subtrees from T by deleting $n_s - 1$ edges and reversing the directions of the minimum number of edges. Theorem 2 implies that such an instance can be solved in time $O(n^2)$.

Consider finally a slightly modified instance in which each edge e of a tree T is assigned two positive integers $c_{\mathrm{del}}(e)$ and $c_{\mathrm{rev}}(e)$; $c_{\mathrm{del}}(e)$ represents the cost required to delete e from T, while $c_{\mathrm{rev}}(e)$ represents the cost required to reverse the direction of e. Our results can be easily extended to such a case.

References

1. Boulaxis, N.G., Papadopoulos, M.P.: Optimal feeder routing in distribution system planning using dynamic programming technique and GIS facilities. IEEE Trans. on Power Delivery 17, 242–247 (2002)
2. Garey, M.R., Johnson, D.S.: Computers and Intractability: A Guide to the Theory of NP-Completeness. Freeman, San Francisco (1979)
3. Ibarra, O.H., Kim, C.E.: Fast approximation algorithms for the knapsack and sum of subset problems. J. ACM 22, 463–468 (1975)
4. Ito, T., Demaine, E.D., Zhou, X., Nishizeki, T.: Approximability of partitioning graphs with supply and demand. J. of Discrete Algorithms 6, 627–650 (2008)
5. Ito, T., Zhou, X., Nishizeki, T.: Partitioning trees of supply and demand. International J. of Foundations of Computer Science 16, 803–827 (2005)
6. Ito, T., Zhou, X., Nishizeki, T.: Partitioning graphs of supply and demand. Discrete Applied Mathematics 157, 2620–2633 (2009)
7. Kellerer, H., Pferschy, U., Pisinger, D.: Knapsack Problems. Springer, Heidelberg (2004)
8. Morton, A.B., Mareels, I.M.Y.: An efficient brute-force solution to the network reconfiguration problem. IEEE Trans. on Power Delivery 15, 996–1000 (2000)
9. Teng, J.-H., Lu, C.-N.: Feeder-switch relocation for customer interruption cost minimization. IEEE Trans. on Power Delivery 17, 254–259 (2002)

Computing the (t, k)-Diagnosability of Component-Composition Graphs and Its Application

Sun-Yuan Hsieh* and Chun-An Chen

Department of Computer Science and Information Engineering,
National Cheng Kung University,
No. 1, University Road, Tainan 701, Taiwan
{hsiehsy,p7696423}@mail.ncku.edu.tw

Abstract. (t, k)-Diagnosis, which is a generalization of sequential diagnosis, requires that at least k faulty processors should be identified and repaired in each iteration provided there are at most t faulty processors, where $t \geq k$. Let $\kappa(G)$ and $n(G)$ be the node connectivity and the number of nodes in G, respectively. We show that the class of m-dimensional component-composition graphs G for $m \geq 4$ is $(\Omega(h), \kappa(G))$-diagnosable, where $h = \frac{2^{m-2} \times \lg(m-1)}{m-1}$ if $2^{m-1} \leq n(G) < m!$; and $h = 2^{m-2}$ if $n(G) \geq m!$. Based on this result, the (t, k)-diagnosability of numerous multiprocessor systems can be computed efficiently.

Keywords: Diagnosability, component-composition graphs, applied graph theory, multiprocessor systems, the PMC model, (t, k)-diagnosis.

1 Introduction

Because of the rapid development of VLSI technology, a multiprocessor system may contain hundreds or even thousands of nodes, some of which may be faulty when the system is put into use. As the number of nodes in a multiprocessor system increases, node fault identification in such systems is crucial for reliable computing. The process of discriminating between faulty nodes and fault-free nodes in a system is called *fault diagnosis*. When a faulty node is identified, it is replaced by a fault-free node to maintain the system's reliability.

Preparata *et al.* [16] introduced a graph theoretical model, called the *PMC model* (Preparata, Metze, and Chien's model), for system level diagnosis. Under the PMC model, every node u is capable of testing whether another node v is faulty if there exists a communication link between them. Moreover, it is assumed that a test result is reliable (respectively, unreliable) if the node that initiates the test is fault-free (respectively, faulty). The PMC model has also been adopted in the approaches proposed [4,5,6].

Sequential diagnosis [16], also called diagnosis with repair, iteratively identifies subsets of faulty nodes; then, at the end of each iteration, all identified faulty nodes are replaced or repaired before the next iteration is initiated. This

---

* Corresponding author.

O. Cheong, K.-Y. Chwa, and K. Park (Eds.): ISAAC 2010, Part II, LNCS 6507, pp. 363–374, 2010.

process is repeated until all faulty nodes have been replaced or repaired. Araki and Shibata [3] further introduced a generalization of sequential diagnosis, called (t, k)-*diagnosis*, where $t \geq k$. If there are at most t faulty nodes, in each iteration, the method can identify at least k faulty nodes (or all faulty nodes if there are less than k faulty nodes remaining). A system is (t, k)-*diagnosable* if at least k faulty nodes can be identified in each iteration. Chang *et al.* [6,7] showed that n-dimensional *Matching Composition Networks* (MCNs) for $n > 5$ are $\left(\Omega \left(\frac{2^n \times \lg n}{n} \right), n \right)$-diagnosable. In [5], the authors showed d-dimensional grids and tori are, respectively, $\left(\Omega \left(N^{\frac{d}{d+1}} \right), \Omega(d) \right)$-diagnosable and $\left(\Omega \left(N^{\frac{d}{d+1}} \right), \Omega(2d) \right)$-diagnosable, where N is the number of nodes.

In this paper, we propose a unified approach to compute the (t, k)-diagnosability of numerous multiprocessor systems (graphs) under the PMC model, including hypercubes, crossed cubes, twisted cubes, locally twisted cubes, multiply-twisted cubes, generalized twisted cubes, recursive circulants, Möbius cubes, Mcubes, star graphs, bubble-sort graphs, pancake graphs, and burnt pancake graphs. Our approach first sketches the common properties of the above classes of graphs, and defines a super class of graphs, called m-dimensional component-composition graphs, to cover them. Let $\kappa(G)$ and $n(G)$ be the node connectivity and the number of nodes in G, respectively. We then show that the m-dimensional component-composition graph G for $m \geq 4$ is $(\Omega(h), \kappa(G))$-diagnosable, where $h = \frac{2^{m-2} \times \lg (m-1)}{m-1}$ if $2^{m-1} \leq n(G) < m!$; and $h = 2^{m-2}$ if $n(G) \geq m!$. Based on this result, the (t, k)-diagnosability of the above multiprocessor systems can be computed efficiently. Chang *et al.* [6] showed that n-dimensional hypercubes, n-dimensional twisted cubes, n-dimensional crossed cubes, and n-dimensional Möbius cubes, are $\left(\Omega \left(\frac{2^n \times \lg n}{n} \right), n \right)$-diagnosable, where $n > 5$. We improve on Chang's results because our t-value matches the asymptotic lower bound of the t-value derived in [6], but we extend the dimension from $n > 5$ to $n \geq 3$. To the best of our knowledge, except for Chang's results, the (t, k)-diagnosability of the other mentioned multiprocessor systems have not been resolved thus far.

2   Preliminaries

An *undirected graph* (*graph* for short) $G = (V, E)$ is comprised of a *node* (*vertex*) set V and an *edge set* E, where V is a finite set and E is a subset of $\{(u, v) | (u, v)$ is an unordered pair of $V \}$. We also use $V(G)$ and $E(G)$ to denote the node set and edge set of G, respectively. Let $n(G)$ be the number of nodes (*order*) in G and let $e(G)$ be the number of edges in G. For an edge (u, v), we call u and v the *end-nodes* of the edge. A *subgraph* of $G = (V, E)$ is a graph (V', E') such that $V' \subseteq V$ and $E' \subseteq E$. Given $U \subseteq V(G)$, the *subgraph of G induced by U* is defined as $G[U] = (U, \{(u, v) \in E(G) | u, v \in U\})$. For a node u in G, we let $N_G(u)$ denote the set of all its neighboring nodes, i.e., $N_G(u) = \{v \in V(G) : v$ is adjacent to $u\}$. For two non-empty sets $V_1, V_2 \subset V(G)$, the *neighborhood set* of V_1 in V_2 is defined as $N_G(V_1, V_2) = \{v \in V_2 : \exists$ a node $u \in V_1$ such that

$(u,v) \in E(G)\}$. A *multigraph* is a graph in which we allow multiple edges and self loops.

The *components* of a graph G are its maximal connected subgraphs. A component is *trivial* if it has no edge; otherwise, it is *non-trivial*. An *isolated node* is a node of degree 0; and a *cut-node* is a node whose deletion would increase the number of components. A *separating set* of a graph G is set $S \subset V(G)$ such that $G - S$ has more than one component. The *node connectivity* (*connectivity* for short) of a graph G, denoted by $\kappa(G)$, is the minimum size of a node set S such that $G - S$ is disconnected or contains only one node[1]. For two graphs $G_1 = (V_1, E_1)$ and $G_2 = (V_2, E_2)$, the *disjoint union of G_1 and G_2*, denoted by $G_1 \oplus G_2$, is the graph $(V_1 \bigcup V_2, E_1 \bigcup E_2)$. For a set W of unordered pairs of $V(G)$, we use $G \bigcup W$ to represent the graph $(V, E \bigcup W)$.

A *directed graph* $\vec{G} = (V, \vec{E})$ is comprised of a *node set* V and an *directed edge set* $\vec{E}$, where V is a finite set and $\vec{E}$ is a subset of $\{(u,v)|\ (u,v)$ is an ordered pair of $V\}$. For convenience, the notations defined for an undirected graph can also be adopted for directed graphs. For a directed edge (u,v), we call u and v the *tail* and *head* of the edge, respectively. Given $U \subseteq V(\vec{G})$, the *subgraph of $\vec{G} = (V, \vec{E})$ induced by U* is defined as $\vec{G}[U] = (U, \{(u,v) \in \vec{E}|\ u,v \in U\})$. A *spanning subgraph* of $\vec{G} = (V, \vec{E})$ is a subgraph of $\vec{G}$ with the node set V. A *directed path* in $\vec{G}$, denoted by $\langle u_1, u_2, \ldots, u_n \rangle$, is a sequence of nodes such that $(u_i, u_{i+1}) \in \vec{E}$ for $1 \le i \le n - 1$. For $v \in V(\vec{G})$, the *outdegree* $d^+(v)$ is the number of edges with tail v and the *indegree* $d^-(v)$ is the number of edges with head v.

In approaches based on the PMC model, a *self-diagnosable system* is often represented by a directed graph in which an arc directed from node u to node v means that u can test v. To diagnose faults, a number of tests are performed on nodes and the collection of all test results is referred to as a *syndrome*, which we define as follows.

Definition 1. [6] *Under the PMC model, a syndrome σ of a system $\vec{G}$ is defined as follows: For any directed edge (u,v), $\sigma(u,v) = 0$ if v is tested by u and found to be fault-free; and $\sigma(u,v) = 1$ if v is tested by u and found to be faulty.*

The test result initiated by a faulty node is unreliable; thus, more than one syndrome may be produced for G with faulty nodes. When G has faulty nodes, a syndrome σ is randomly generated for the purpose of fault diagnosis. We call the set of all faulty nodes, denoted as F, an *allowable faulty set* with σ under the PMC model if the following two conditions hold: (1) $\sigma(u,v) = 0$ for $u \in V - F$ and $v \in V - F$; (2) $\sigma(u,v) = 1$ for $u \in V - F$ and $v \in F$.

Definition 2. [3] *A system G is (t,k)-diagnosable, where $t \ge k$, if given any syndrome produced by the system under the presence of a faulty set F, the following conditions hold: (1) All faulty nodes can be identified when $|F| \le k$. (2) At least k faulty nodes can be identified when $k < |F| \le t$.*

---

[1] The notation $G - S$ represents the graph obtained by deleting all the nodes in S from G.

Let F be a faulty set in a system G. Under the sequential diagnosis approach, we need $|F|$ iterations to diagnose and repair all faulty nodes in the worst case. In (t, k)-diagnosable systems, it is guaranteed that the number of iterations required for diagnosis is at most $\left\lceil \frac{|F|}{k} \right\rceil$.

3   Component-Composition Graphs

A *matching* in a graph G is a set of non-loop edges with no shared end-nodes. The nodes incident to the edges of a matching M are *saturated* by M; the others are *unsaturated*. A *perfect matching* in a graph is a matching that saturates every node. The following theorem is useful to the definition of component-composition graphs.

Theorem 1. *Let $C_1, C_2, \ldots, C_l$ be l components of a graph G. There exists a perfect matching PM in $(V(G), \{(x, y) \mid x \in C_i \text{ and } y \in C_j \text{ for } 1 \leq i, j \leq l$ and $i \neq j\})$ such that $G \bigcup PM$ is connected if and only if the following three conditions hold: (a) $n(G)$ is even; (b) $2 \leq l \leq \frac{n(G)}{2} + 1$; (c) $n(C_i) \leq \sum_{\substack{1 \leq j \leq l \\ j \neq i}} n(C_j)$*

for all $1 \leq i \leq l$.

Proof. First, we prove the necessity part. Suppose, to the contrary, that G does not satisfy at least one condition. If Condition (a) does not hold (i.e., $n(G)$ is odd), then G obviously does not contain any perfect matching. If Condition (b) does not hold, then $0 \leq l \leq 1$ or $l \geq \frac{n(G)}{2} + 2$. If $0 \leq l \leq 1$, then $\{(x, y) \mid x \in C_i$ and $y \in C_j$ for $1 \leq i, j \leq l$ and $i \neq j\} = \emptyset$, which leads to $PM = \emptyset$; hence, $V(G)$ cannot be saturated by any matching edge in PM. If $l \geq \frac{n(G)}{2} + 2$ (i.e., $2l - n(G) \geq 4$), G will contain at least four trivial components (i.e., four isolated nodes) and each of the other non-trivial components is exactly a K_2 (complete graph with two nodes). This implies that $G \bigcup PM$ will be disconnected, which leads to a contradiction. If Condition (c) does not hold (i.e., $n(C_p) > \sum_{\substack{1 \leq j \leq l \\ j \neq p}} n(C_j)$

for some $1 \leq p \leq l$), then there exist at least $n(C_p) - \sum_{\substack{1 \leq j \leq l \\ j \neq i}} n(C_j) > 0$ nodes

that cannot be saturated by any matching edge in PM; hence, a contradiction occurs.

Next, we prove the sufficiency part by induction on $n(G)$ as follows. It is easily seen to hold for the base case of $n(G) = 2$. We consider now the case $n(G) \geq 4$. Without loss of generality, suppose that $n(C_1) \geq n(C_2) \geq \cdots \geq n(C_l)$. We start by pairing a non-cut-node $x \in V(C_1)$ with a non-cut-node $y \in V(C_l)$. Consider the reduced instance G' obtained by removing nodes x and y. Since Condition (b) is satisfied by G, it must be that $n(C_1) \geq 2$. Thus, it suffices to show that G' satisfies all conditions of the theorem; we can inductively find a perfect matching PM' such that $G' \bigcup PM'$ is connected, and add the edge (x, y) to it such that $PM = PM' \bigcup \{(x, y)\}$ is a perfect matching that saturates $V(G)$ and $G \bigcup PM$ is connected. Condition (a) continues to hold in G' since we remove an even number of nodes. If Condition (b) is violated, then it must be that initially $l = 2$ or $l = \frac{n(G)}{2} + 1$. In case of former, it must be that $n(C_1) = n(C_2)$ since G satisfies Condition (c). As $n(G) \geq 4$, the graph G' must also contain two components. In case of latter, we know that $n(C_l) = 1$ because there are $\frac{n(G)}{2} + 1$ components. Then, the number of components in G' is $l - 1 = \frac{n(G')}{2} + 1$ since removal of y eliminates the component C_l. Hence, Condition (b) is satisfied. Finally, for condition (c), if $l = 2$, then we are clearly done because it must be that $n(C_1) = n(C_2)$ in G. Otherwise, $l \geq 3$. If Condition (c) is violated in G', then it must be $n(C_2) > n(C_1) + \sum_{i=3}^{l} n(C_i) - 2$; and $n(C_1) = n(C_2)$ in G. But then $\sum_{i=3}^{l} n(C_i) < 2$, and hence $\sum_{i=3}^{l} n(C_i) = 1$ (since $l \geq 3$). This contradicts the fact that G satisfies Condition (a). □

Given that $G = C_1 \oplus C_2 \oplus \cdots \oplus C_l$ and PM both satisfy the conditions specified in Theorem 1, we use the notation $PM(C_1, C_2, \ldots, C_l)$ to represent the graph $G \bigcup PM$, where $V(PM(C_1, C_2, \ldots, C_l)) = V(C_1) \bigcup \cdots \bigcup V(C_l)$ and $E(PM(C_1, C_2, \ldots, C_l)) = E(C_1) \bigcup \cdots \bigcup E(C_l) \bigcup PM$. Based on Theorem 1, we now provide a recursive definition of a new class of graphs.

Definition 3. The class of *m-dimensional component-composition graphs*, denoted by CCG_m, is defined recursively as follows: (1) $CCG_1 = \{K_1\}$. (2) For $m \geq 2$, $CCG_m = \{PM(G_1, G_2, \ldots, G_l)\mid G_i \in CCG_{m-1}$ and $n(G_i) \leq \sum_{\substack{1 \leq j \leq l \\ j \neq i}} n(G_j)$ for

all $2 \leq l \leq \frac{\sum_{i=1}^{l} n(G_i)}{2} + 1\}$.

Based on Definition 3, it is not difficult to show the following result.

Lemma 1. *Every graph* $G = PM(G_1, G_2, \ldots, G_l) \in CCG_m$ *satisfies the following properties: (a) G is connected; (b) $n(G) \geq 2^{m-1}$; (c) G is $(m-1)$-regular; (d) $n(G)$ is even for $m \geq 2$; (e) $2 \leq l \leq \frac{n(G)}{2} + 1$ for $m \geq 2$; (f) $n(G_i) \leq \sum_{\substack{1 \leq j \leq l \\ j \neq i}} n(G_j)$*

for all $1 \leq i \leq l$.

4 (t,k)-Diagnosability of CCG_m

Under the PMC model, a self-diagnosable system G is represented by a directed graph $\vec{G} = (V, \vec{E})$. Suppose that $\langle u_1, u_2, \ldots, u_n \rangle$ is a directed path from node u_1 to node u_n in $\vec{G}$ and σ is a syndrome of $\vec{G}$. If $\sigma(u_i, u_{i+1}) = 0$ for all $1 \leq i \leq n-1$, then either u_1 is faulty or $u_1, u_2, \ldots, u_n$ are all fault-free. Let $\vec{G}'$ be a spanning subgraph of $\vec{G}$ that contains all arcs (x, y) with $\sigma(x, y) = 0$, i.e., $\vec{G}' = (V', \vec{E}')$, where $V' = V$ and $\vec{E}' = \{(x, y)\mid (x, y) \in E(\vec{G})$ and $\sigma(x, y) = 0\}$. In addition, let C' denote the set of all *strongly connected components* (also called component if there is no ambiguity) in $\vec{G}'$. Note that for every two distinct nodes u and v in the same strongly connected component, there is a directed path from u to v and vice versa. Clearly, in each component of C', all the nodes are either faulty or fault-free. Let a *faulty component* (respectively, *fault-free component*) be a component in which all the nodes are faulty (respectively, fault-free).

If C' has only one component, then all nodes in V are faulty or fault-free. Hence, we assume that C' contains at least two components. By regarding each component of C' as a *super node* and connecting two distinct components (super nodes) $X \in C'$ and $Y \in C'$ by an edge if there exists $(x, y) \in E(\vec{G})$ with $x \in X$ and $y \in Y$, we can construct an undirected graph $G'' = (V'', E'')$, where $V'' = C'$ and $E'' = \{(X, Y) \mid X \in V'', Y \in V'', \text{ and } (x, y) \in E(\vec{G}) \text{ with } x \in X \text{ and } y \in Y\}$. For $X \in V''$, let $N_{G''}(X)$ be the neighborhood of X in G'', i.e., $N_{G''}(X) = \{Y \mid Y \in V'' \text{ and } (X, Y) \in E(G'')\}$. For convenience, a super node X in G'' can be also used to represent the set of nodes of a strongly connected component (in $\vec{G'}$) corresponding to X.

Since all neighbors of a fault-free component are faulty components, faulty nodes can be found if some fault-free components are identified. Recall that in (t, k)-diagnosis, the number of faulty nodes is bounded from above by t. A node subset in G can be guaranteed to be a fault-free component, denoted by X, provided that $|X|$ is larger than or equal to $t + 1$. In other words, $t + 1$ is a threshold for guaranteeing a fault-free component.

The following function $\varPhi$ is to determine a lower bound on the degree of (t, k)-diagnosability of G.

Definition 4. Define $\varPhi(\chi_1, \chi_2)$ for G'' to be the largest integer p so that, for each nonempty subset S of V'' with $\chi_2 \leq \sum_{Z \in S} n(\vec{G}[Z]) \leq p$, $V'' - S$ is an independent set and there exists $X \in V'' - S$ that satisfies the following two conditions: (D1) $n(\vec{G}[X]) \geq \chi_1$; (D2) $\displaystyle\sum_{Y \in N_{G''}(X)} n(\vec{G}[Y]) \geq \chi_2$.

Recall that the process of (t, k)-diagnosis is iterative; in each iteration (except the last iteration), at least k faulty nodes can be identified provided there are at most t faulty nodes in G. In addition, we only need to find a component whose size is larger than or equal to the threshold $t + 1$. Let S described in Definition 4 be the faulty set F, and let $\chi_1 = t + 1$ and $\chi_2 = k$. Note that t is at most p provided that $\varPhi(t + 1, k) \geq t$.

4.1   A Feasible Value of k

Lemma 2. *For any graph $G = (V, E)$, if U be a subset of V and $W \subset V - U$ is a component of $V - U$, then $|N_G(W, U)| \geq \kappa(G)$.*

Proof. Since $W \subset V - U$ and W is maximal connected, there is no edge in G whose two end-nodes belong to W and $G - U - W$. Suppose, to the contrary, that $|N_G(W, U)| < \kappa(G)$. Then, by the assumption that $W \subset V - U$, $N_G(W, U)$ is a separating set whose size smaller than $\kappa(G)$, which leads to a contradiction. Hence, $|N_G(W, U)| \geq \kappa(G)$.   □

Due to the space limitation, the proof of the following lemma is omitted.

Lemma 3. $\varPhi(t + 1, \kappa(G)) = \varPhi(t + 1, q)$ *for all non-negative integers* $q < \kappa(G)$.

By Lemma 3, $\kappa(G)$ is a feasible value of k.

4.2   A Feasible Value of t

In Subsection 4.1, we have proved that $\Phi(t+1, \kappa(G)) = \Phi(t+1, 0) \geq t$. In order to evaluate $\Phi(t+1, 0)$, we define a function Ψ for G'' as follows: $\Psi(\chi_1)$ is the smallest integer p so that there exists a subset H of V'' satisfying $\sum_{Z \in H} n(\vec{G}[Z]) = p$. Moreover, $V'' - H$ is an independent set of G'' and there exists no $X \in V'' - H$ with $n(\vec{G}[X]) \geq \chi_1$. Let $\chi_1 = t + 1$ and $H = F$ be a faulty subset with $\sum_{Z \in F} n(\vec{G}[Z]) = \Psi(t + 1)$. Clearly, $\Phi(t + 1, 0) = \Psi(t + 1) - 1$. Therefore, we use $\Psi(t + 1) - 1 \geq t$ if a system $\vec{G}$ is (t, k)-diagnosable.

Let G be a graph in CCG_m. Given $\vec{W} \subseteq E(\vec{G})$, let $a(\vec{W})$ be the number of arcs in $\vec{W}$. Define $I(\beta) = \max\limits_{\substack{Z \subseteq V(G) \\ n(\vec{G}[Z]) = \beta}} a(E(\vec{G}[Z]))$, i.e., $I(\beta)$ is the maximal number of arcs in $\vec{G}$, whose two endpoints are contained in an β-node subset. Clearly, $I(1) = 0$.

Lemma 4. $0 \leq I(\beta) \leq \beta \times \lg \beta$, where $\beta \geq 1$.

Proof. Let $G = PM(G_1, G_2, \ldots, G_l)$ be a graph in CCG_m, where $G_i \in CCG_{m-1}$ for $1 \leq i \leq l$. For $U \subseteq V(G)$ with $\beta = |U| \geq 1$, let $U_i = U \cap V(G_i)$ for $1 \leq i \leq l$. We define $E_U = \{(x, y) \mid x \in U_i, \ y \in U_j, \ \text{and} \ (x, y) \in PM \ \text{for} \ i \leq i, j \leq l \ \text{and} \ i \neq j\}$, and let $\vec{E}_U$ be the set of directed edges obtained from E_U so that for each $(x, y) \in E_U$, there are two arcs (x, y) and (y, x) in $\vec{E}_U$.

Let $I_r(\beta)$ be the maximum number of arcs in $\vec{G}$ whose two endpoints belong to a β-node subset of an induced subgraph $G' \in CCG_r$ of G, where $1 \leq r \leq m$. Initially, $I_r(1) = 0$ for $1 \leq r \leq m$ and $I_r(\beta) \geq 0$. Clearly, $I(\beta) = I_m(\beta)$. Moreover, $I(|U|) = I_m(|U|) \leq I_{m-1}(|U_1|) + I_{m-1}(|U_2|) + \cdots + I_{m-1}(|U_l|) + a(\vec{E}_U)$, where $m \geq 2$. Let $U_{j_1}, U_{j_2}, \ldots, U_{j_s}$ be all the non-empty sets in $\{U_1, U_2, \ldots, U_l\}$. Then, $I(|U|) = I_m(|U|) \leq I_{m-1}(|U_{j_1}|) + I_{m-1}(|U_{j_2}|) + \cdots + I_{m-1}(|U_{j_s}|) + a(\vec{E}_U)$. Without loss of generality, we assume that $1 \leq |U_{j_1}| \leq |U_{j_2}| \leq \cdots \leq |U_{j_s}|$. Let $\beta_i = |U_{j_i}| \geq 1$ for all $1 \leq i \leq s$. Then, $\beta = \beta_1 + \beta_2 + \cdots + \beta_s$, where $1 \leq \beta_1 \leq \beta_2 \leq \cdots \leq \beta_s$. Moreover, $I(\beta) = I_m(\beta) \leq I_{m-1}(\beta_1) + I_{m-1}(\beta_2) + \cdots + I_{m-1}(\beta_s) + a(\vec{E}_U)$, where $s \geq 1$.

Next, we prove, by induction on m, that $I(\beta) \leq \beta \times \lg \beta$. First, it is straightforward to check that $I_1(\beta) \leq \beta \times \lg \beta$ for $\beta = 1$. Suppose that $I_p(\beta) \leq \beta \times \lg \beta$ holds when $m = p$. Now, we consider the case where $m = p + 1$. If $s = 1$, then $I(\beta) = I_{p+1}(\beta) \leq I_p(\beta) \leq \beta \times \lg \beta$. If $s \geq 2$, we have the following scenarios.

Case 1: $\beta_s \leq \beta_1 + \beta_2 + \cdots + \beta_{s-1}$. Note that $a(\vec{E}_U) \leq \lfloor \frac{\beta_1 + \beta_2 + \cdots + \beta_s}{2} \rfloor \times 2 \leq \beta_1 + \beta_2 + \cdots + \beta_s$. By the induction hypothesis, $I(\beta) = I_{p+1}(\beta) \leq I_p(\beta_1) + I_p(\beta_2) + \cdots + I_p(\beta_s) + (\beta_1 + \beta_2 + \cdots + \beta_s) \leq \beta_1 \times \lg \beta_1 + \beta_2 \times \lg \beta_2 + \cdots + \beta_s \times \lg \beta_s + (\beta_1 + \beta_2 + \cdots + \beta_s)$. We can prove this case based on the following claim. Due to the space limitation, the proof is omitted.

Claim 1. $\beta_1 \times \lg \beta_1 + \beta_2 \times \lg \beta_2 + \cdots + \beta_s \times \lg \beta_s + (\beta_1 + \beta_2 + \cdots + \beta_s) \leq \beta \times \lg \beta$.

Case 2: $\beta_s > \beta_1 + \beta_2 + \cdots + \beta_{s-1}$. In this case, $a(\vec{E}_U) \leq 2 \times (\beta_1 + \beta_2 + \cdots + \beta_{s-1})$. By the induction hypothesis, $I(\beta) = I_{p+1}(\beta) \leq I_p(\beta_1) + I_p(\beta_2) + \cdots + I_p(\beta_s) + 2 \times (\beta_1 + \beta_2 + \cdots + \beta_{s-1}) \leq \beta_1 \times \lg \beta_1 + \beta_2 \times \lg \beta_2 + \cdots + \beta_s \times \lg \beta_s + 2 \times (\beta_1 + \beta_2 + \cdots + \beta_{s-1})$. We prove this case based on the following claim. Due to the space limitation, the proof is omitted.

Claim 2. $\beta_1 \times \lg \beta_1 + \beta_2 \times \lg \beta_2 + \cdots + \beta_s \times \lg \beta_s + 2 \times (\beta_1 + \beta_2 + \cdots + \beta_{s-1}) \leq \beta \times \lg \beta$.

□

Lemma 5. $\Psi(t+1) = \left\lceil \frac{n(\vec{G}) \times ((m-1) - \lg t)}{2(m-1) - \lg t} \right\rceil$, where $G \in CCG_m$.

Proof. Let $G = PM(G_1, G_2, \ldots, G_l)$ be a graph in CCG_m, where $G_i \in CCG_{m-1}$ for $1 \leq i \leq l$. In addition, let $F = \{Y_1, Y_2, \ldots, Y_d\}$ be a set of faulty components and $V'' - F = \{X_1, X_2, \ldots, X_c\}$ be a set of fault-free components, where $c + d = n(G'')$. Recall that each node of $\vec{G}$ has indegree and outdegree $m - 1$. Clearly, $I(\sum_{j=1}^{d} n(\vec{G}[Y_j]))$ plus the number of arcs from F to $V'' - F$ does not exceed $(\sum_{j=1}^{d} n(\vec{G}[Y_j])) \times (m - 1)$; and the number of arcs from $V'' - F$ to F does not exceed $(\sum_{j=1}^{d} n(\vec{G}[Y_j])) \times (m - 1) - I(\sum_{j=1}^{d} n(\vec{G}[Y_j]))$. Hence, by Lemma 4, we have

$$a(E(\vec{G})) = \sum_{i=1}^{c} I(n(\vec{G}[X_i])) + I(\sum_{j=1}^{d} n(\vec{G}[Y_j])) + \text{(the number of arcs from } F \text{ to}$$

$$V'' - F) + \text{(the number of arcs from } V'' - F \text{ to } F)$$

$$\leq \sum_{i=1}^{c} (n(\vec{G}[X_i]) \times \lg(n(\vec{G}[X_i]))) + (\sum_{j=1}^{d} n(\vec{G}[Y_j])) \times (m-1)$$

$$+ (\sum_{j=1}^{d} n(\vec{G}[Y_j])) \times (m-1) - I(\sum_{j=1}^{d} n(\vec{G}[Y_j])).$$

$$(1)$$

Since G is $(m-1)$-regular, $a(E(\vec{G})) = n(\vec{G}) \times (m-1)$. Moreover, according to the definition of $\Psi(t+1)$, we need $n(\vec{G}[X_i]) \leq t$ for $1 \leq i \leq c$. Then, we have $\sum_{i=1}^{c} (n(\vec{G}[X_i]) \times \lg n(\vec{G}[X_i])) \leq \sum_{i=1}^{c} (n(\vec{G}[X_i])) \times \lg t = (n(\vec{G}) - \sum_{j=1}^{d} n(\vec{G}[Y_j])) \times \lg t$. Note that $I(\sum_{j=1}^{d} n(\vec{G}[Y_j])) \geq 0$. Thus, Formula (1) can be rewritten as follows: $n(\vec{G}) \times (m-1) \leq (n(\vec{G}) - \sum_{j=1}^{d} n(\vec{G}[Y_j])) \times \lg t + 2 \times (\sum_{j=1}^{d} n(\vec{G}[Y_j])) \times (m-1)$. From the above equation, we have $\sum_{j=1}^{d} n(\vec{G}[Y_j]) \geq \frac{n(\vec{G}) \times ((m-1) - \lg t)}{2(m-1) - \lg t}$. The definition of $\Psi(t+1)$ leads to $\Psi(t+1) = \left\lceil \frac{n(\vec{G}) \times ((m-1) - \lg t)}{2(m-1) - \lg t} \right\rceil$.

□

By Lemma 5, we have $\Phi(t+1, \kappa(G)) = \Phi(t+1, 0) = \Psi(t+1) - 1 = \left\lceil \frac{n(\vec{G}) \times ((m-1) - \lg t)}{2(m-1) - \lg t} \right\rceil - 1$. It is easy to see that each t satisfying $t \leq \left\lceil \frac{n(\vec{G}) \times ((m-1) - \lg t)}{2(m-1) - \lg t} \right\rceil - 1$ has $\Phi(t+1, \kappa(G)) \geq t$.

Lemma 6. *If $h(t) < 0$ for any $t \leq T = min\{a, 2^b\}$, then $(t-a) \times (\lg t - b) > h(t)$.*

Lemma 7. *$\Phi(t + 1, \kappa(G)) \geq t$ if $t \leq \frac{2^{m-2} \times \lg (m-1)}{m-1}$, where $G \in CCG_m$ and $2^{m-1} \leq n(\vec{G}) < m!$ for $m \geq 4$.*

Proof. For a positive number p and an integer q, $q - 1 < p$ if $\lceil p \rceil \geq q$. Since $\Phi(t + 1, \kappa(G)) = \Psi(t + 1) - 1 = \left\lceil \frac{n(\vec{G}) \times ((m-1) - \lg t)}{2(m-1) - \lg t} \right\rceil - 1 \geq t$ implies that $\frac{n(\vec{G}) \times ((m-1) - \lg t)}{2(m-1) - \lg t} > t$, $n(\vec{G}) \times (m-1) - n(\vec{G}) \times \lg t - 2(m-1) \times t + t \times \lg t > 0$. By factorizing the above inequality, we have $\left(t - \frac{2^{m-2} \times \lg (m-1)}{m-1}\right) \times (\lg t - 2(m-1)) > \lg t \times \left(n(\vec{G}) - \frac{2^{m-2} \times \lg (m-1)}{m-1}\right) + 2^{m-1} \times \lg (m - 1) - n(\vec{G}) \times (m - 1) = h(t)$. By Lemma 6, we can set $a = \frac{2^{m-2} \times \lg (m-1)}{m-1}$, $b = 2(m - 1)$, and $T = min\{a, 2^b\} = \frac{2^{m-2} \times \lg (m-1)}{m-1}$. Now, we prove that $h(t) < 0$ for any $t \leq T = \frac{2^{m-2} \times \lg (m-1)}{m-1}$ as follows. Since $2^{m-1} \leq n(\vec{G}) < m!$, we have

$$h(t) = \lg t \times \left(n(\vec{G}) - \frac{2^{m-2} \times \lg (m - 1)}{m - 1}\right) + 2^{m-1} \times \lg (m - 1) - n(\vec{G}) \times (m - 1)$$

$$\leq \lg \left(\frac{2^{m-2} \times \lg (m - 1)}{m - 1}\right) \times \left(n(\vec{G}) - \frac{2^{m-2} \times \lg (m - 1)}{m - 1}\right)$$
$$+ 2^{m-1} \times \lg (m - 1) - n(\vec{G}) \times (m - 1)$$

$$= \left(\lg \left(2^{m-2}\right) + \lg (\lg (m - 1)) - \lg (m - 1)\right) \times \left(n(\vec{G}) - \frac{2^{m-2} \times \lg (m - 1)}{m - 1}\right)$$
$$+ 2^{m-1} \times \lg (m - 1) - n(\vec{G}) \times (m - 1)$$

$$= n(\vec{G}) \times ((m - 2) + \lg (\lg (m - 1)) - \lg (m - 1))$$
$$- \left(\frac{2^{m-2} \times \lg (m - 1)}{m - 1}\right) \times ((m - 2) + \lg (\lg (m - 1)) - \lg (m - 1))$$
$$+ 2^{m-1} \times \lg (m - 1) - n(\vec{G}) \times (m - 1)$$

$$= n(\vec{G}) \times (-1 + \lg (\lg (m - 1)) - \lg (m - 1))$$
$$- \left(\frac{2^{m-2} \times \lg (m - 1)}{m - 1}\right) \times ((m - 2) + \lg (\lg (m - 1)) - \lg (m - 1))$$
$$+ 2^{m-1} \times \lg (m - 1)$$

$$\leq 2^{m-1} \times (-1 + \lg (\lg (m - 1)) - \lg (m - 1))$$
$$- \left(\frac{2^{m-2} \times \lg (m - 1)}{m - 1}\right) \times ((m - 2) + \lg (\lg (m - 1)) - \lg (m - 1))$$
$$+ 2^{m-1} \times \lg (m - 1)$$

$$= \left(\frac{2^{m-2}}{m - 1}\right) \times (-2(m - 1) + 2(m - 1) \times \lg (\lg (m - 1)) - (m - 2) \times \lg (m - 1)$$
$$- \lg (m - 1) \times \lg (\lg (m - 1)) + \lg (m - 1) \times \lg (m - 1)) = z(m).$$

It is not difficult to show that $h(t) \leq z(m) < 0$ as $m = 4, 5, \ldots, 16$.

When $m \geq 17$, $[\lg(m-1)]^2 \leq m - 1^2$. We have

$$h(t) \leq z(m) = \left(\frac{2^{m-2}}{m-1}\right) \times (2(m-1) \times \lg(\lg(m-1)) - (m-2) \times \lg(m-1)$$

$$- \lg(m-1) \times \lg(\lg(m-1)) + \lg(m-1) \times \lg(m-1) - 2(m-1))$$

$$= \left(\frac{2^{m-2}}{m-1}\right) \times \Big\{ \lg(\lg(m-1))^{2(m-1)} - \lg(m-1)^{(m-2)}$$

$$- \lg(\lg(m-1))^{\lg(m-1)} + \lg(m-1)^{\lg(m-1)} - \lg 2^{2(m-1)} \Big\}$$

$$= \left(\frac{2^{m-2}}{m-1}\right) \times \left\{ \lg\left(\frac{(\lg(m-1))^{2(m-1)-\lg(m-1)}}{(m-1)^{(m-2)}}\right) + \lg\left(\frac{(m-1)^{\lg(m-1)}}{2^{2(m-1)}}\right) \right\}$$

$$= \left(\frac{2^{m-2}}{m-1}\right) \times \left\{ \lg\left(\frac{[(\lg(m-1))^2]^{(m-1)-\frac{\lg(m-1)}{2}}}{(m-1)^{(m-2)}}\right) + \lg\left(\frac{(m-1)^{\lg(m-1)}}{(m-1)^{\frac{2(m-1)}{\lg(m-1)}}}\right) \right\}$$

$$< 0.$$

Hence, $\Phi(t+1, \kappa(G)) \geq t$ for any $t \leq \frac{2^{m-2} \times \lg(m-1)}{m-1}$.   □

Lemma 8. $\Phi(t+1, \kappa(G)) \geq t$ if $t \leq 2^{m-2}$, where $G \in CCG_m$ and $n(\vec{G}) \geq m!$ for $m \geq 4$.

Proof. Based on the argument used in the proof of Lemma 7, $\Phi(t+1, \kappa(G)) \geq t$ implies that $n(\vec{G}) \times (m-1) - n(\vec{G}) \times \lg t - 2(m-1) \times t + t \times \lg t > 0$. By factorizing the above inequality, we have $(t - n(\vec{G})) \times (\lg t - (m-2)) > m \times t - n(\vec{G}) = h(t)$. By Lemma 6, we can set $a = n(\vec{G})$, $b = m - 2$, and $T = min\{a, 2^b\} = 2^{m-2}$. Now, we prove $h(t) < 0$ for any $t \leq T = 2^{m-2}$ as follows. Since $n(\vec{G}) \geq m!$, we have $m \times t - n(\vec{G}) \leq m \times (2^{m-2}) - m! = m \times (2^{m-2} - (m-1)!) < 0$. Hence, $\Phi(t+1, \kappa(G)) \geq t$ for any $t \leq 2^{m-2}$.   □

Lemma 7 (respectively, Lemma 8) implies that $X \in V''$ with $n(\vec{G}[X]) \geq t+1$ provided that $t \leq T = \frac{2^{m-2} \times \lg(m-1)}{m-1}$ (respectively, $t \leq T = 2^{m-2}$) and $2^{m-1} \leq n(\vec{G}) < m!$ (respectively, $n(\vec{G}) \geq m!$). In other words, if a system contains at most T faulty nodes, all of these nodes can be identified by executing a (t, k)-diagnosis algorithm.

Since $n(\vec{G}) = n(G)$, by Lemma 7 and 8, we have the following theorem.

Theorem 2. If $G \in CCG_m$ for $m \geq 4$, then G is $(\Omega(h), \kappa(G))$-diagnosable, where $h = \frac{2^{m-2} \times \lg(m-1)}{m-1}$ if $2^{m-1} \leq n(G) < m!$; and $h = 2^{m-2}$ if $n(G) \geq m!$.

---

[2] Let $z(m) = [\lg(m-1)]^2 - (m-1)$. When $m \geq 17$, $\frac{\partial z(m)}{\partial m} = \frac{2 \times \lg(m-1)}{\ln 2 \times (m-1)} - 1 = \frac{\frac{2}{\ln 2} \times \lg(m-1) - (m-1)}{m-1} = \frac{\lg(m-1)^{\frac{2}{\ln 2}} - \lg 2^{(m-1)}}{m-1} < 0$. Since $z(m) \leq z(17) = 0 \leq 0$, we have $z(m) \leq 0$, which implies that $[\lg(m-1)]^2 \leq m-1$ for $m \geq 17$.

5 Application to Multiprocessor Systems

Definition 5. [18] A class of *n-dimensional hypercube-like graphs*, denoted by HL_n for $n \geq 0$, are simple, connected, undirected graphs that can be defined recursively as follows: (1) $HL_0 = \{K_1\}$. (2) For $n \geq 1$, a graph $G \in HL_n$ is constructed from two graphs $G_1 = (V_1, E_1)$ and $G_2 = (V_2, E_2)$, where each $G_i \in HL_{n-1}, V_1 = \{v_1, v_2, \ldots, v_n\}$, and $V_2 = \{w_1, w_2, \ldots, w_n\}$. Moreover, $V(G) = V_1 \bigcup V_2$ and $E(G) = E_1 \bigcup E_2 \bigcup E_3$, where $E_3 = \{(v_j, w_{i_j}) | 1 \leq j \leq n$ and $(i_1, i_2, \ldots, i_n)$ is a permutation of $1, 2, \ldots, n\}$.

The following lemma can be shown by induction on n.

Lemma 9. HL_n *is a subclass of* CCG_{n+1} *for* $n \geq 0$.

It has been shown that $\kappa(HL_n) = n$ [18]. Hence, by Theorem 2 and Lemma 9, we have the following result.

Corollary 1. *Every graph in* HL_n *is* $\left(\Omega(\frac{2^{n-1} \times \lg n}{n}), n \right)$*-diagnosable for* $n \geq 3$.

Numerous multiprocessor systems are in HL_n; for example, the n-dimensional hypercube Q_n, the n-dimensional crossed cube CQ_n, the n-dimensional twisted cube TQ_n, the n-dimensional locally twisted cube LTQ_n, the n-dimensional multiply-twisted cube MTQ_n, the n-dimensional generalized twisted cube GQ_n, recursive circulants $G(2^n, 4)$ for odd $n \geq 3$, the n-dimensional Möbius cube MQ_n, and the n-dimensional Mcubes M_n. Hence, by Corollary 1, the above multiprocessor systems are $\left(\Omega(\frac{2^{n-1} \times \lg n}{n}), n \right)$-diagnosable, where $n \geq 3$.

We can show that (1) the n-dimensional star graphs S_n defined in [1,2] belongs to CCG_n for $n \geq 1$; (2) The *n-dimensional bubble-sort graph* B_n $(n \geq 1)$ defined in [2,13] belongs to CCG_n for $n \geq 1$; (3) The *n-dimensional pancake graph* $\wp_n$ $(n \geq 1)$ defined in [2,10] belongs to CCG_n for $n \geq 1$; and (4) The *n-dimensional burnt pancake graph* BP_n $(n \geq 0)$ defined in [11,12] belongs to CCG_{n+1} for $n \geq 0$. Hence, based on Theorem 2, we have the following results.

Corollary 2. *(a)* S_n *for* $n \geq 4$ *is* $\left(\Omega(2^{n-2}), n - 1 \right)$*-diagnosable. (b)* B_n *for* $n \geq 4$ *is* $\left(\Omega(2^{n-2}), n - 1 \right)$*-diagnosable. (c)* $\wp_n$ *for* $n \geq 4$ *is* $\left(\Omega(2^{n-2}), n - 1 \right)$*-diagnosable. (d)* BP_n *for* $n \geq 3$ *is* $\left(\Omega(2^{n-1}), n \right)$*-diagnosable.*

6 Concluding Remarks

In this paper, we have investigated the (t, k)-diagnosability of component composition graphs. By applying our technical theorem, we have successfully demonstrated the (t, k)-diagnosability of several multiprocessor systems. Our results show that (t, k)-diagnosis can achieve high diagnosability, even if the degree of the graph is small. A future work is to apply our strategy to other classes of graphs.

References

1. Akers, S.B., Horel, D., Krishnamurthy, B.: The star graph: an attractive alternative to the n-cube. In: Proceedings of the International Conference on Parallel Processing, pp. 393–400 (1987)
2. Akers, S.B., Krishnamurthy, B.: A group-theoretic model for symmetric interconnection networks. IEEE Transactions on Computers 38(4), 555–566 (1989)
3. Araki, T., Shibata, Y.: (t, k)-diagnosable system: A generalization of the PMC models. IEEE Transactions on Computers 52(7), 971–975 (2003)
4. Armstrong, J.R., Gray, F.G.: Fault diagnosis in a boolean n-cube array of microprocessors. IEEE Transactions on Computers C-30(8), 587–590 (1981)
5. Chang, G.Y., Chen, G.H.: (t, k)-diagnosability of multiprocessor systems with applications to grids and tori. SIAM Journal on Computing 37(4), 1280–1298 (2007)
6. Chang, G.Y., Chen, G.H., Chang, G.J.: (t, k)-diagnosis for matching composition networks. IEEE Transactions on Computers 55(1), 88–92 (2006)
7. Chang, G.Y., Chen, G.H., Chang, G.J.: (t, k)-diagnosis for matching composition networks under the MM* model. IEEE Transactions on Computers 56(1), 73–79 (2007)
8. Esfahanian, A.H., Ni, L.M., Sagan, B.E.: The twisted n-cube with application to multiprocessing. IEEE Transactions on Computers 40(1), 88–93 (1991)
9. Fan, J.: Diagnosability of crossed cubes under the comparison diagnosis model. IEEE Transactions on Parallel and Distributed Systems 13(7), 687–692 (2002)
10. Jwo, J.S., Lakshmivarahan, S., Dhall, S.K.: A new class of interconnection networks based on the alternating group. Networks 23, 315–326 (1993)
11. Kaneko, K., Sawada, N.: An algorithm for node-to-set disjoint paths problem in burnt pancake graphs. IEICE Transactions on Information and Systems E86VD(12), 2588–2594 (2003)
12. Kaneko, K., Sawada, N.: An algorithm for node-to-node disjoint paths problem in burnt pancake graphs. IEICE Transactions on Information and Systems E90VD(1), 306–313 (2007)
13. Lakshmivarahan, S., Jwo, J.S., Dhall, S.K.: Symmetry in interconnection networks based on cayley graphs of permutation groups: a survey. Parallel Computing 19(4), 361–407 (1993)
14. Park, J.H., Kim, H.C., Lim, H.S.: Many-to-many disjoint path covers in hypercube-like interconnection networks with faulty elements. IEEE Transactions on Parallel and Distributed Systems 17(3), 227–240 (2006)
15. Park, J.H., Lim, H.S., Kim, H.C.: Panconnectivity and pancyclicity of hypercube-like interconnection networks with faulty elements. Theortical Computer Science 377, 170–180 (2007)
16. Preparata, F.P., Metze, G., Chien, R.T.: On the connection assignment problem of diagnosable systems. IEEE Transactions on Electronic Computers EC-16(6), 848–854 (1967)
17. Suzuki, Y., Kaneko, K.: An algorithm for node-disjoint paths in pancake graphs. IEICE Transactions on Information and Systems E86-D(3), 610–615 (2003)
18. Vaidya, A.S., Rao, P.S.N., Shankar, S.R.: A class of hypercube-like networks. In: Proc. 5th Symp. on Parallel and Distributed Processing, pp. 800–803 (1993)

Why Depth-First Search Efficiently Identifies Two and Three-Connected Graphs

Amr Elmasry[*]

Max-Planck Institut für Informatik and University of Copenhagen
elmasry@mpi-inf.mpg.de

Abstract. Given an undirected 3-connected graph G with n vertices and m edges, we modify depth-first search to produce a sparse spanning subgraph with at most $4n - 10$ edges that is still 3-connected. If G is 2-connected, to maintain 2-connectivity, the resulting graph will have at most $2n - 3$ edges. The way depth-first search discards irrelevant edges illustrates the reason behind its ability to verify and certify biconnectivity [1,2,3] and triconnectivity [4,5] in linear time. Dealing with a sparser graph, after the first depth-first-search calls, makes the algorithms in [2,5] more efficient. We also give a characterization of separation pairs of a 2-connected graph in terms of the resulting sparse graph.

1 Introduction

Depth-first search is a basic algorithm for graph traversal and for searching within a domain of outcomes with a uniform structure. The algorithm exists in most of the basic algorithmic textbooks [6,7]. Many graph problems have been efficiently solved in linear time using depth-first search. These include: finding strongly connected components [8], verifying biconnectivity [3], topological sorting [9], finding an *st* ordering [1,2], and finding triconnected components [5]. Using depth-first search, the classical path-addition method of Hopcroft and Tarjan was the first published linear-time planarity-testing algorithm [10].

Graph connectivity properties resemble a fundamental issue in the study of graph theory. Efficient algorithms for determining these properties are both theoretically and practically important. In the early seventies, several of the aforementioned papers were published illustrating the ability of depth-first search to solve graph problems and investigate connectivity properties. The question about the ability of depth-first search to efficiently handle two and three-connectivity, but not higher connectivity, is an intriguing question.

Around 50 years ago, Tutte [8] proved the fundamental result that every 3-connected graph on more than 4 vertices contains an edge whose contraction results yet in another 3-connected graph. Recently, Elmasry et al. [11] strengthened this result by showing that every depth-first-search tree of a 3-connected graph contains such a contractible edge. However, it is worth to mention that not every spanning tree of a 3-connected graph contains a contractible edge.

---

[*] On leave from Alexandria University of Egypt.

O. Cheong, K.-Y. Chwa, and K. Park (Eds.): ISAAC 2010, Part II, LNCS 6507, pp. 375–386, 2010.

The problem of finding a k-connected spanning subgraph of a k-connected graph is related, and interesting in its own sake. Finding a k-connected spanning subgraph with the minimum number of edges is known to be *NP-hard* for any $k \geq 2$ [12]. However, the problem is much easier if we want to produce a "sparse" k-connected spanning subgraph of a k-connected graph. This can be done in linear time using the algorithm of Nagamochi and Ibaraki [13], which produces a k-connected spanning subgraph with at most $kn - k(k+1)/2$ edges. The algorithm in [14] can also produce in linear time a 2-connected spanning subgraph with at most $2n - 2$ edges or a 3-connected spanning subgraph with at most $3n - 3$ edges. However, both algorithms do not use depth-first search.

In this paper, we give an algorithm that relies on depth-first search to produce a sparse spanning subgraph that is 2-connected if the input graph is 2-connected (Section 4), and 3-connected if the input graph is 3-connected (Section 5). The structure of the resulting graph is uniform in the sense that, in addition to the depth-first-search tree, a constant number of back edges are associated with each vertex. Once these constant pieces of information are computed, the succeeding depth-first-search calls efficiently extract the required information in linear time. These arguments seem not to apply for higher connectivity, and hence the job would not be done with the same efficiency. In Section 6, we characterize separation pairs in a 2-connected graph (more precisely than [5]) in terms of the information kept in the resulting sparse graph.

The st-ordering of a 2-connected graph is a certificate for biconnectivity. The algorithm in [2] performs two depth-first-search calls to produce such an st-ordering. Producing a sparse spanning subgraph by our algorithm, within the first depth-first-search call, makes the algorithm in [2] more efficient when performing the second call. A similar argument was used by Ebert [1] to efficiently produce an st-ordering. For finding the triconnected components of a 2-connected graph, the Hopcroft-Tarjan algorithm [5] works in two stages and performs several depth-first-search calls. Our method uses ideas related to the first stage of this algorithm. Producing a sparse spanning subgraph by our algorithm, within the depth-first-search calls of the first stage of the algorithm in [5], the second stage of the algorithm in [5] will be efficiently performed on the sparse graph and hence requiring $O(n)$ instead of $O(m)$ time.

2   Preliminaries

2.1   Definitions

Two vertices u and v are called connected if G contains a path from u to v. Otherwise, they are called disconnected.

A graph is called connected if every pair of distinct vertices in the graph can be connected through some path. A graph is called k-connected if every pair of distinct vertices in the graph can be connected through k disjoint paths.

A connected component is a maximal connected subgraph of G. Each vertex belongs to exactly one connected component, as does each edge. A separation class is the set of vertices of a connected component.

A *vertex cut (separation vertices)* is a set of vertices whose removal splits a connected graph into more than one connected component. A graph is k-connected if and only if the size of every vertex cut is at least k. In particular, a *separation pair* is a pair of vertices whose removal disconnects the graph.

A *palm tree P* is the directed graph produced by depth-first search when applied to G. There are two types of arcs in P: A *tree arc* joins a vertex u to its parent in the tree of P when u is first visited by depth-first search. The other arcs are called *cycle arcs* or *back edges*.

2.2   Notations

- $G = (V, E)$: an undirected graph on a set of vertices V and a set of edges E.
- $n = |V|$, and $m = |E|$.
- $\bar{u}v$: the edge between vertices u and v in G.
- P: the palm tree produced by the depth-first-search algorithm.
- P': the sparse palm tree produced by our algorithm.
- $u \rightarrow v$: the depth-first-search tree arc in P from vertex u to vertex v.
- $u \xrightarrow{*} v$: the simple directed path in the tree of P, constituting zero or more arcs, between vertices u and v.
- $u \hookrightarrow v$: the back edge in P from vertex u to vertex v.

3   Depth-First Search

In this section, we review the basic depth-first-search algorithm.

We use the following variables in Algorithm 1:

$num(u)$: is an integer that represents the order in which a vertex u is first visited by depth-first search.

$parent(u)$: is the parent of vertex u in the tree of P.

Algorithm 1. *DFS(u)*

1: $c = c + 1$
2: $num[u] = c$
3: **for all** w in the adjacency list of u **do**
4:      **if** $(num[w] == 0)$ **then**
5:          $parent[w] = u$
6:          *DFS(w)*
7:          add $u \rightarrow w$ as a tree arc in P
8:      **else**
9:          **if** $(num[w] < num[u]$ && $w \neq parent[u])$ **then**
10:             add $u \hookrightarrow w$ as a back edge in P
11:         **end if**
12:     **end if**
13: **end for**

Algorithm 2. $main()$

1: **for** $i = 1$ to n **do**
2:     $num[i] = parent[i] = 0$
3: **end for**
4: $c = 0$
5: $DFS(s)$ /* s is an arbitrary root of the DFS palm tree P

The following three lemmas are classical properties of depth-first search, which we will use later to prove our main lemmas.

Lemma 1. *If* $u \hookrightarrow v$ *is in* P, *then* $v \xrightarrow{*} u$ *is in* P.

Proof. Since $u \hookrightarrow v$ is in P, thus $num(v) < number(u)$ and $u \xrightarrow{*} v$ is not in P. Assume that $v \xrightarrow{*} u$ is also not in P. The way depth-fist search works implies that the adjacency list of v must have been fully scanned before the first time u is reached. Accordingly, $v \rightarrow u$ would have been in P and not $u \hookrightarrow v$. We conclude that $v \xrightarrow{*} u$ is in P.   □

Lemma 2. *If* $\bar{u}v \in E(G)$ *and* $num(v) < num(u)$, *then* $v \xrightarrow{*} u$ *is in* P.

Proof. Since $num(v) < num(u)$, then $u \xrightarrow{*} v$ is not in P. Assume that $v \xrightarrow{*} u$ is also not in P. Then, neither $u \rightarrow v$ nor $v \rightarrow u$ is in P. Also, by Lemma 1, neither $u \hookrightarrow v$ nor $v \hookrightarrow u$ is in P. But $\bar{u}v \in E(G)$, and hence it must exist in P either as a tree edge or as a back edge. We conclude that $v \xrightarrow{*} u$ is in P.   □

Lemma 3. *Consider any 3 vertices* v_1, v_2, v_3. *If* $num(v_1) < num(v_2) < num(v_3)$ *and* $v_1 \xrightarrow{*} v_3$ *is in* P, *then* $v_1 \xrightarrow{*} v_2$ *is in* P.

Proof. The way depth-first search works ensures that all the descendants of a vertex in the tree of P are fully scanned directly after and while this vertex is scanned. In other words, all the descendants of a vertex must be assigned consecutive numbers. It follows that $v_1 \xrightarrow{*} v_2$ is in P, for otherwise either $num(v_2) < num(v_1)$ or $num(v_2) > num(v_3)$ holds.   □

4   Biconnectivity

In this section, we show that at most one back edge is to be associated with every vertex to preserve biconnectivity. More precisely, for every vertex u, we only need to keep in P' the back edge $u \hookrightarrow w$ whose head w is the first vertex visited by depth-first search among all vertices v where $u \hookrightarrow v$ is in P. This property is implemented by modifying depth-first search; see Algorithm 3. The next lemma is a direct consequence of such property.

Lemma 4. *Consider the case when* $\bar{u}v \in E(G)$ *and* $num(v) < num(u)$, *but neither* $v \rightarrow u$ *nor* $u \hookrightarrow v$ *is in* P'. *There exists* $w \in V(G)$ *such that* $u \hookrightarrow w$ *is in* P' *and* $num(w) < num(v)$.

We use the following variable in Algorithm 3:

$low(u)$: is an integer that holds $num(w)$, where w is the vertex that has the smallest number among all vertices v with $u \hookrightarrow v$ in P.

Algorithm 3. *sparsify-2(u)*

1: $c = c + 1$
2: $low[u] = num[u] = c$
3: **for all** w in the adjacency list of u **do**
4:     **if** $(num[w] == 0)$ **then**
5:         $parent[w] = u$
6:         *sparsify-2(w)*
7:         add $u \to w$ as a tree arc in P'
8:     **else**
9:         **if** $(num[w] < num[u] \ \&\& \ w \neq parent[u])$ **then**
10:           **if** $(num[w] < low[u])$ **then**
11:             $low[u] = num[w]$
12:             $l = w$
13:           **end if**
14:         **end if**
15:     **end if**
16: **end for**
17: **if** $(low[u] \neq num[u])$ **then**
18:     add $u \hookrightarrow l$ as a back edge in P'
19: **end if**

Lemma 5. *P' is 2-connected if and only if G is 2-connected.*

Proof. As P' is a spanning subgraph of G, G is 2-connected if P' is 2-connected.

Assume that G is 2-connected, and suppose that P' is not 2-connected. Hence, there is a separation vertex a in P'. Let V_1 and V_2 be two separation classes in P' with respect to a. Since G is 2-connected, there exists an edge $\bar{u}v \in E(G)$ such that $u \in V_1$ and $v \in V_2$. Assume w.l.o.g. that $num(v) < num(u)$. Then, by Lemma 2, $v \xrightarrow{*} u$ is in P'. Hence $num(v) < num(a) < num(u)$.

Using Lemma 4, there is a back edge $u \hookrightarrow w$ in P' with $num(w) < num(v)$. Since $u \hookrightarrow w$ is in P', thus $w \in V_1$. Also, by Lemma 1, $w \xrightarrow{*} u$ is in P'. Since $num(w) < num(v) < num(u)$, Lemma 3 implies that $w \xrightarrow{*} v$ is in P'.

Because $num(w) < num(v) < num(a)$, $w \xrightarrow{*} v$ can not pass through a. Although $w \in V_1$ and $v \in V_2$, but still $w \xrightarrow{*} v$ does not pass through a. It follows that $w \xrightarrow{*} v$ is not in P'; this contradicts the implication of the previous paragraph. We conclude that P' is 2-connected. □

Lemma 6. *P' has at most $2n - 3$ edges.*

Proof. There are $n-1$ tree edges in P'. There is at most one back edge emanating from any vertex. In addition, there are no back edges emanating from vertices 1 and 2. The total number of back edges in P' is therefore at most $n - 2$. □

5   Triconnectivity

In this section, we show that at most three back edges are to be associated with every vertex of P' to preserve triconnectivity. Using the same notion as [5], we define $lowpt(u)$ as the lowest vertex reachable from u by traversing zero or more tree arcs in P followed by at most one back edge. More specifically,

$$lowpt(u) = \min(\{num(u)\} \cup \{num(w) \mid u \hookrightarrow w\} \cup \{lowpt(w) \mid u \to w\}).$$

Phase I computes such $lowpt$ values for all the vertices in a depth-first-search manner; see Algorithm 4.

Algorithm 4. *Phase I: compute-lowpt(u)*

1: $c = c + 1$
2: $lowpt[u] = num[u] = c$
3: **for all** w in the adjacency list of u **do**
4:     **if** $(num[w] == 0)$ **then**
5:       $parent[w] = u$
6:       compute-lowpt(w)
7:       **if** $(lowpt[w] < lowpt[u])$ **then**
8:         $lowpt[u] = lowpt[w]$
9:       **end if**
10:    **else**
11:       **if** $(num[w] < num[u] \ \&\& \ w \neq parent[u])$ **then**
12:         **if** $(num[w] < lowpt[u])$ **then**
13:           $lowpt[u] = num[w]$
14:         **end if**
15:       **end if**
16:    **end if**
17: **end for**

Accordingly, the following lemma relates the $lowpt$ value of a vertex and the low value of one of its descendants in the tree of P.

Lemma 7. *Given $r \in V(G)$, there exists $y \in V(G)$ such that $r \xrightarrow{*} y$ is in P and $lowpt(r) = low(y)$.*

Proof. We write $lowpt(r)$ as $lowpt(r) = \min(low(r), \{lowpt(w) \mid r \to w\})$. If $low(r) \le \min(\{lowpt(w) \mid r \to w\})$ or if r is a leaf in the tree of P, the lemma follows by noting that $lowpt(r) = low(r)$. Otherwise, there exists a vertex w' such that $r \to w'$ and $lowpt(r) = lowpt(w')$. By induction, the lemma applies for w', i.e., there exists a vertex y where $w' \xrightarrow{*} y$ is in P and $lowpt(w') = low(y)$. The lemma follows by noting that $r \to w' \xrightarrow{*} y$ is in P. □

Phase II sorts the adjacency lists of G in *descending* order according to the $lowpt$ values calculated in Phase I. To implement the sorting in linear time, we resort to random access and use *Radix Sort*; see Algorithm 5.

Algorithm 5. *Phase II: construct ordered adjacency lists*

1: **for** $i = 1$ to n **do**
2:     $bucket[i] = nill$
3: **end for**
4: **for all** $(u, w) \in E(G)$ **do**
5:     add (u, w) to $bucket[lowpt[w]]$
6: **end for**
7: **for** $i = 1$ to n **do**
8:     **for all** $(u, w) \in bucket[i]$ **do**
9:         add w to the *beginning* of $\mathcal{A}(u)$
10:     **end for**
11: **end for**

In the sequel, we assume that P is the palm tree produced by Phase II. We also assume that the vertices of P will be renumbered in Phase III by another depth-first search according to the new ordering of the adjacency lists.

The next lemma illustrates a nice property concerning such numbers assigned to the vertices by depth-first search when applied to the ordered adjacency lists.

Lemma 8. *Given $u, z \in V(G)$ where $num(u) < num(z)$, then either*

(i) $u \xrightarrow{*} z$ is in P, or
(ii) Let x be the lowest common ancestor of u and z, and let r be such that $x \rightarrow r \xrightarrow{*} z$ is in P (the vertex r may be z itself). Then, $num(u) < num(r)$ and $lowpt(r) \leq low(u)$.

Proof. Since $num(u) < num(z)$, thus $z \xrightarrow{*} u$ is not in P. If $u \xrightarrow{*} z$ is not in P, then there exist three vertices x, r, r' such that both $x \rightarrow r \xrightarrow{*} z$, $x \rightarrow r' \xrightarrow{*} u$ are in P and $r \neq r'$. Since $num(u) < num(z)$, depth-first search visits r' and all its descendants in the tree of P before r. It follows that $num(u) < num(r)$. The way we ordered the adjacency lists ensures that $lowpt(r) \leq lowpt(r')$. As $lowpt(r') \leq low(u)$, then $lowpt(r) \leq low(u)$. □

In Phase III, in addition to the tree of P, we only need to keep in P' the following three back edges associated with every vertex u:

- $u \hookrightarrow w_1$ and $u \hookrightarrow w_2$ whose heads w_1 and w_2 are the first and second vertices visited by depth-first search among all vertices v where $u \hookrightarrow v$ is in P.
- $z \hookrightarrow u$ whose tail z is the last vertex visited by depth-first search among all vertices v where $v \hookrightarrow u$ is in P.

To attain such a structure for P', Algorithm 6 once more applies depth-first search. We use the following variables in Algorithm 6:

$low2(u)$: is an integer that holds $num(w2)$, where $w2$ is the vertex that has the second-smallest number among all vertices v with $u \hookrightarrow v$ in P.
$high(u)$: is an integer that holds $num(z)$, where z is the vertex that has the largest number among all vertices v with $v \hookrightarrow u$ in P.
$vertex(c)$: is the vertex u that has $num(u) = c$.

Algorithm 6. *Phase III: sparsify-3(u)*

1: $c = c + 1$
2: $low[u] = low2[u] = high[u] = num[u] = c$
3: $vertex[c] = u$
4: **for all** $w \in \mathcal{A}(u)$ **do**
5:    **if** $(num[w] == 0)$ **then**
6:       $parent[w] = u$
7:       *sparsify-3(w)*
8:       add $u \rightarrow w$ as a tree arc in P'
9:    **else**
10:       **if** $(num[w] < num[u] \ \&\& \ w \neq parent[u])$ **then**
11:          **if** $(num[w] < low[u])$ **then**
12:             $low2[u] = low[u]$
13:             $low[u] = num[w]$
14:          **else**
15:             **if** $(num[w] < low2[u])$ **then**
16:                $low2[u] = num[w]$
17:             **end if**
18:          **end if**
19:          **if** $(num[u] > high[w])$ **then**
20:             $high[w] = num[u]$
21:          **end if**
22:       **end if**
23:    **end if**
24: **end for**
25: $l = vertex[low[u]]$
26: **if** $(l \neq u)$ **then**
27:    add $u \hookrightarrow l$ as a back edge in P'
28: **end if**
29: $l2 = vertex[low2[u]]$
30: **if** $(l2 \neq u)$ **then**
31:    add $u \hookrightarrow l2$ as a back edge in P'
32: **end if**
33: $h = vertex[high[u]]$
34: **if** $(h \neq u \ \&\& \ low[h] \neq num[u] \ \&\& \ low2[h] \neq num[u])$ **then**
35:    add $h \hookrightarrow u$ as a back edge in P'
36: **end if**

The next lemma directly follows and illustrates properties of P'.

Lemma 9. *Consider the case when $\bar{u}v \in E(G)$ and $num(v) < num(u)$, but neither $v \rightarrow u$ nor $u \hookrightarrow v$ is in P'. The following two facts must hold.*

(i) $\exists \ w_1, w_2 \in V(G)$ *such that both $u \hookrightarrow w_1$ and $u \hookrightarrow w_2$ are in P', and $num(w_1) < num(w_2) < num(v)$.*

(ii) $\exists \ z \in V(G)$ *such that $z \hookrightarrow v$ is in P', and $num(u) < num(z)$*

Algorithm 7. *main*()

1: **for** $i = 1$ to n **do**
2:     $num[i] = parent[i] = 0$
3: **end for**
4: $c = 0$
5: *compute-lowpt(s)*   /* s is an arbitrary root of the DFS palm tree P
6: *construct ordered adjacency lists*
7: **for** $i = 1$ to n **do**
8:     $num[i] = parent[i] = 0$
9: **end for**
10: $c = 0$
11: *sparsify-3(s)*

Lemma 10. *P' is 3-connected if and only if G is 3-connected.*

Proof. As P' is a spanning subgraph of G, G is 3-connected if P' is 3-connected.

Assume that G is 3-connected, and suppose that P' is not 3-connected. Following Lemma 5, P' is 2-connected. Hence, there is a separation pair a, b in P'. Let V_1 and V_2 be two separation classes in P' with respect to the cut $\{a, b\}$. Since G is 3-connected, there exists an edge $\bar{u}v \in E(G)$ such that $u \in V_1$ and $v \in V_2$. Assume w.l.o.g. that $num(v) < num(u)$. Then, by Lemma 2, $v \xrightarrow{} u$ is in P'. It follows that $v \xrightarrow{} u$ must pass through one of the separation vertices. Assume w.l.o.g. that $v \xrightarrow{} u$ passes through b, i.e., $v \xrightarrow{*} b \xrightarrow{*} u$ is in P'. Hence, $num(v) < num(b) < num(u)$.

From Lemma 9 part (i), there exist two back edges $u \hookrightarrow w1$ and $u \hookrightarrow w2$ in P' with $num(w_1) < num(w_2) < num(v)$. It follows that $w1, w2 \in V_1 \cup \{a, b\}$. Since $num(v) < num(b)$, thus $w1, w2 \neq b$. Accordingly, at least one of the two vertices $w1$ and $w2$ (namely $w1$) is $\notin \{a, b\}$. Call this vertex w, i.e., $w \in V_1$ and $num(w) < num(v)$. Also, by Lemma 1, $w \xrightarrow{*} u$ is in P'. Since $num(w) < num(v) < num(u)$, Lemma 3 implies that $w \xrightarrow{} v$ is in P'. It follows that $w \xrightarrow{*} a \xrightarrow{*} v$ is in P'. Hence, $num(w) < num(a) < num(v)$.

From Lemma 9 part (ii), there exists a back edge $z \hookrightarrow v$ in P' with $num(u) < num(z)$. As $v \in V_2$, it follows that $z \in V_2$. Since $num(u) < num(z)$, following Lemma 8 case (i), suppose that $u \xrightarrow{*} z$ is in P'. Then, either $u \xrightarrow{*} a$ or $u \xrightarrow{*} b$ is in P'. But, this is impossible because $num(a) < num(b) < num(u)$.

Using Lemma 8, we are only left with case (ii). Let x be the lowest common ancestor of u and z, and let r be such that $x \to r \xrightarrow{*} z$ is in P'. Lemma 8 implies $num(u) < num(r)$. As $num(u) < num(r) \leq num(z)$, then $r \xrightarrow{*} z$ can not pass through either a or b. Since $z \in V_2$, thus $r \in V_2$. By Lemma 7, there exists a vertex y such that $r \xrightarrow{*} y$ is in P (also in P') and $lowpt(r) = low(y)$. As $num(u) < num(r) \leq num(y)$, then $r \xrightarrow{*} y$ can not pass through either a or b. Since $r \in V_2$, thus $y \in V_2$. Also, by Lemma 8 case (ii), $lowpt(r) \leq low(u) = num(w)$. Then, $low(y) \leq num(w)$. Let q be the vertex with $num(q) = low(y)$. It follows that $num(q) \leq num(w)$, and $y \hookrightarrow q$ is in P'. Again, by Lemma 1,

$q \xrightarrow{*} y$ is in P'. Accordingly, q is in the same separation class as y, i.e., $q \in V_2$. Since $w \in V_1$, therefore $q \neq w$ ensuring that $num(q) < num(w)$.

As $num(q) < num(w) < num(y)$ and $q \xrightarrow{*} y$ is in P', it follows by lemma 3 that $q \xrightarrow{*} w$ is in P'. Although $q \in V_2$ and $w \in V_1$, still $q \xrightarrow{*} w$ can pass through neither a nor b because $num(w) < num(a) < num(b)$; a contradiction. We conclude that P' is 3-connected. $\square$

Lemma 11. *P' has at most $4n - 10$ edges.*

Proof. There are $n - 1$ tree edges in P'. There are at most two back edges emanating from any vertex, among the three associated with it. In addition, there are no back edges emanating from the two vertices numbered 1 and 2, and at most one back edge from the vertex numbered 3. This adds up to at most $2n - 5$ first-type back edges. There is at most one back edge entering any vertex, among the three associated with it. In addition, there are no back edges entering the two vertices numbered $n - 1$ and n. Also, a back edge entering one of the two vertices numbered 1 and 2 must have been counted among the first-type back edges. This accounts for at most $n - 4$ second-type back edges. The total number of edges in P' is then at most $(n-1) + (2n-5) + (n-4) = 4n - 10$. $\square$

6   Characterizing Separation Pairs

In this section, we show how to characterize any pair of separation vertices in a 2-connected graph. Most of the notion and ideas we describe here are similar to those in [5]; the difference is that we only use the information kept in the sparse palm tree P' instead of P.

Again, using the same notion as [5], we define $lowpt2(u)$ as the second lowest vertex reachable from u by traversing zero or more tree arcs in P followed by at most one back edge. The $lowpt2$ values solely depend on the low and $low2$ values, and can be as well computed in Phase I of the algorithm.

$$lowpt2(u) = \min(\{num(u)\} \cup (\{num(w) \mid u \hookrightarrow w\} \cup$$
$$\{lowpt(w) \mid u \to w\} - \{lowpt(u)\})).$$

Now, we give a characterization for any separation pair, which only depends on the low, $low2$ and $high$ values, in addition to the tree structure of P'.

Lemma 12. *Given a 2-connected graph G, and $a, b \in V(G)$ such that $num(a) < num(b)$. $\{a, b\}$ is a separation pair of G if and only if either (i) or (ii) holds.*

(i) Type 1 separation pairs:
  $\exists r$ *such that* $b \to r$ *and*
    1. $lowpt(r) = num(a)$.
    2. $lowpt2(r) \geq num(b)$.
    3. $\exists s \notin \{a, b\}$, *and* $r \xrightarrow{*} s$ *is not in* P'.

(ii) Type 2 separation pairs:
  - $num(a) \neq 1$.
  - $\exists\, r \neq b, a \rightarrow r \xrightarrow{\;*\;} b$ *such that* $\forall x$ *with* $r \xrightarrow{\;*\;} x$ *in* P' *and* $b \xrightarrow{\;*\;} x$ *not in* P' *we have:*
    1. $num(r) \leq num(x) < num(b)$.
    2. $low(x) \geq num(a)$.
    3. *Let* y *be the vertex having* $num(y) = high(x)$. *If* $b \rightarrow w \xrightarrow{\;*\;} y$, *then* $lowpt(w) \geq num(a)$.

Lemma 12 is pretty similar to [5, Lemma 13]. The only difference is in case $(ii)3$, where we save the checking of the back edges that are not in P'. As G is 2-connected, Lemma 5 implies that a separation pair in G is also a separation pair in P'. On the other hand, Lemma 10 implies that P' will not have a separation pair that is not a separation pair in G. Although it is possible to prove Lemma 12 from scratch, its correctness directly follows from the proof of [5, Lemma 13] and the arguments of Lemmas 5 and 10.

7   Conclusion

We have shown how to produce a sparse graph and maintain its two or three-connectivity using depth-first search. The nice feature of the resulting sparse graph is that, in addition to the tree arcs of the palm tree, we only need to maintain a constant number of back edges associated with every vertex (at most one back edge for biconnectivity and at most three back edges for triconnectivity). Being able to explore the graph in such a uniform way sheds the light on why it is possible to efficiently identify biconnectivity and triconnectivity in linear time using depth-first search. The ideas seem not to extend to higher connectivity, reasoning the inability to check higher connectivity in linear time. Our method can be directly used to improve the efficiency of the Hopcroft-Tarjan algorithm [5], because the underlying graph will have $O(n)$ (instead of m) edges after the first stage of the algorithm in [5], when replaced by our algorithm.

References

1. Ebert, J.: st-Ordering the vertices of biconnected graphs. Computing 30, 19–33 (1983)
2. Even, S., Tarjan, R.E.: Computing an st-numbering. Theoretical Computer Science 2, 339–344 (1976)
3. Tarjan, R.E.: Depth-first search and linear graph algorithms. SIAM Journal on Computing 1, 146–159 (1972)
4. Elmasry, A., Mehlhorn, K., Schmidt, J.M.: A linear-time certifying triconnectivity algorithm for Hamiltonian graphs. Available at the second author's home page (2010)
5. Hopcroft, J.E., Tarjan, R.E.: Dividing a graph into triconnected components. SIAM Journal on Computing 2(3), 135–158 (1973)
6. Cormen, T., Leiserson, C., Rivest, R., Stein, C.: Introduction to Algorithms, 2nd edn. MIT Press and McGraw-Hill (2001)

7. Knuth, D.: The Art of Computer Programming, 3rd edn., vol. 1. Addison-Wesley, Reading (1997)
8. Tutte, W.: A theory of 3-connected graphs. Indag. Math. 23, 441–455 (1961)
9. Tarjan, R.E.: Edge-disjoint spanning trees and depth-first search. Algorithmica 6(2), 171–185 (1976)
10. Hopcroft, J.E., Tarjan, R.E.: Efficient planarity testing. Journal of the Association for Computing Machinery 21(4), 549–568 (1974)
11. Elmasry, A., Mehlhorn, K., Schmidt, J.M.: Every DFS tree of a 3-connected graph contains a contractible edge. Available at the second author's home page (2010)
12. Garey, M.R., Johnson, D.S.: Computer and Intractability: A Guide to the Theory of NP-Completeness. W. Freeman, New York (1979)
13. Nagamochi, H., Ibaraki, T.: A linear-time algorithm for finding a sparse k-connected spanning subgraph of a k-connected graph. Algorithmica 7, 583–596 (1992)
14. Cheriyan, J., Kao, M.Y., Thurimella, R.: Scan-first search and sparse certificates: an improved parallel algorithm for k-vertex connectivity. SIAM Journal on Computing 22, 157–174 (1993)

Beyond Good Shapes: Diffusion-Based Graph Partitioning Is Relaxed Cut Optimization*

Henning Meyerhenke

University of Paderborn, Department of Computer Science
Fuerstenallee 11, D-33102 Paderborn, Germany
henningm@upb.de

Abstract. In this paper we study the prevalent problem of graph partitioning by analyzing the diffusion-based partitioning heuristic BUBBLE-FOS/C, a key component of the practically successful partitioner DIBAP [14]. Our analysis reveals that BUBBLE-FOS/C, which yields well-shaped partitions in experiments, computes a relaxed solution to an edge cut minimizing binary quadratic program (BQP). It therefore provides the first substantial theoretical insights (beyond intuition) why BUBBLE-FOS/C (and therefore indirectly DIBAP) yields good experimental results. Moreover, we show that in bisections computed by BUBBLE-FOS/C, at least one of the two parts is connected. Using arguments based on random walk techniques, we prove that in vertex-transitive graphs actually both parts must be connected components each. All these results may help to eventually bridge the gap between practical and theoretical graph partitioning.

Keywords: Diffusive graph partitioning, relaxed cut optimization, disturbed diffusion.

1 Introduction

Partitioning the vertices of a graph such that certain optimization criteria are met, occurs in many applications in computer science, engineering, and related fields. The most common formulation of the graph partitioning problem for an undirected (possibly edge-weighted) graph $G = (V, E)$ (or $G = (V, E, \omega)$) asks for a division Π of V into k pairwise disjoint subsets (*parts*) $\{\pi_1, \ldots, \pi_k\}$ of size at most $\lceil |V|/k \rceil$ each, such that the *edge cut* is minimized. The edge cut is defined as the total number (or total weight) of edges having their incident nodes in different subsets. Among many others, the applications of this $\mathcal{NP}$-hard problem include load balancing in numerical simulations [19] and image segmentation [6,20].

Despite recent approximation algorithms, simpler heuristics are preferred in practice, many of which can be found in the surveys [19] (graph partitioning) and [18] (graph clustering). Spectral algorithms have been widely used [8]; they are global optimizers based on graph eigenvectors. For computational efficiency or quality reasons, they have

---

* Partially supported by German Research Foundation (DFG) Priority Programme 1307 *Algorithm Engineering*.

O. Cheong, K.-Y. Chwa, and K. Park (Eds.): ISAAC 2010, Part II, LNCS 6507, pp. 387–398, 2010.

been mostly superseded by local improvement algorithms. Integrated into a multilevel framework, local optimizers such as Kernighan-Lin (KL) [10] can be found in several popular partitioning libraries [4,9]. Unfortunately, theoretical quality guarantees are not known for KL. Another class of improvement strategies comprises diffusion-based methods [14,17]. While they are slower than KL, diffusive methods often yield a better quality, also when repartitioning dynamic graphs [14,15].

Motivation. The hybrid algorithm DIBAP is a multilevel combination of the diffusive algorithms BUBBLE-FOS/C [15] and TRUNCCONS. Particularly on graphs arising in numerical simulations, DIBAP is very successful [14]. For example, it has computed for six of the eight largest graphs of a popular benchmark set [21] a large number (more than 80 out of 144 when DIBAP was published) of their best known partitions with respect to the edge cut. The algorithm BUBBLE-FOS/C, which is related to Lloyd's k-means method [11], is an integral part of DIBAP responsible for good solutions on smaller representations of the input graphs. In experiments with graphs from numerical simulations, BUBBLE-FOS/C computes partitions with well-shaped parts. This comes along with a small number of boundary nodes (i. e., nodes with at least one neighbor in a different part) and a small edge cut [15]. However, apart from intuition (see Section 2), there has been no satisfactory theoretical explanation why BUBBLE-FOS/C and ultimately DIBAP produce such good partitions.

Contribution. With this work we answer several open questions regarding diffusion-based partitioning with BUBBLE-FOS/C. In Section 3 we prove a major insight about the optimization criterion of BUBBLE-FOS/C. The heuristic computes a k-way ($k \geq 2$) balanced partition that is the relaxed solution of a binary quadratic program (BQP) for finding the partition with *minimum edge cut*. As a byproduct, computing new center nodes for each part is related to a very similar BQP by the contribution to the constraints. Note that, while the insight about relaxed cut optimization alone may not be sufficient to guarantee a heuristic's practical success, we pursue here the other way around and analyze a heuristic known to produce good partitions. These results may pave the way for finding bounds on BUBBLE-FOS/C's quality for certain graph classes.

The achievements in Section 4 concern the *connectedness* of the parts in a bipartition ($k = 2$) computed by BUBBLE-FOS/C. We prove a result known for spectral partitioning [5], which is new for BUBBLE-FOS/C: In any undirected connected graph G, at least one of the two parts is connected. For vertex-transitive graphs (such as torus or hypercube), we use the random walk measure *hitting times* and conditional expectations to show that *both* parts are always connected.

2   Notation and Related Work

2.1   Notation

We consider in this paper undirected edge-weighted graphs $G = (V, E, \omega)$, which are triples with the set of n *vertices* (or *nodes*) V, a set of m *edges* $E \subseteq V \times V$, and an *edge weight function* $\omega : E \to \mathbb{R}_{\geq 0}$. We also assume that the graphs are finite, connected, and simple, i. e., they do not contain self-loops (u, u) or multiple edges with the same endpoints. Connectedness can be enforced by focusing on the connected components.

Matrices $\mathbf{M}$ are written in bold font, a matrix entry at position (u, v) as $[\mathbf{M}]_{u,v}$. We use column vectors; the v-th entry of a vector w is denoted by $[w]_v$. In case we refer to the v-th entry of the i-th vector, we write $[w_i]_v$. The symmetric positive semidefinite *Laplace matrix* matrix $\mathbf{L}$ of G [3, p. 27ff.] has the entries $[\mathbf{L}]_{u,v} = -\omega(\{u, v\})$ for $\{u, v\} \in E$, $[\mathbf{L}]_{u,u} = \deg(u)$ (with $\deg(u) = -\sum_{v \neq u}[\mathbf{L}]_{u,v}$), and $[\mathbf{L}]_{u,v} = 0$ otherwise.

2.2   Diffusion-Based and Related Partitioning Techniques

Intuitively, a random walk [12] is likely to stay a long time in a dense graph region before leaving it via one of the few outgoing edges. There exist many graph clustering/partitioning techniques exploiting this notion (see [18]). Many diffusive processes are described by stochastic matrices and are therefore related to random walks [12]. Diffusion models many important transport phenomena such as heat flow; another application is localized load balancing in parallel computations. In such a discrete setting, diffusion is a local iterative process which exchanges splittable load entities between neighboring vertices, usually until all vertices have the same amount of load. For graph partitioning, diffusive algorithms and similarity measures are used to compute well-shaped partitions [15,17]. These works exploit the fact that diffusive processes send load faster into densely connected graph regions, which corresponds to the intuition that random walks stay longer in dense graph regions.

   Meila and Shi [13] connect random walks to spectral partitioning. Spectral methods such as [20] solve relaxations of IPs that minimize the edge cut or the related ratio cut. They build on Fiedler's seminal work on spectral partitioning [5] and use eigenvectors of Laplace or adjacency matrices for partitioning. A spectral relaxation to the geometric k-means clustering problem is given by Zha et al. [22].

   The diffusive partitioning algorithm BUBBLE-FOS/C, which we consider in this paper, is composed of the BUBBLE framework (described below) and the similarity measure FOS/C [15]. FOS/C (first order scheme with constant drain) introduces a drain-based disturbance into the first order diffusion scheme. With the disturbance, FOS/C reaches a steady state whose load vector w represents similarities of nodes. These similarities reflect whether nodes are connected by many paths of short length.

Definition 1. *(FOS/C) [15] Given a connected and undirected graph $G = (V, E, \omega)$ free of self-loops, a set of source nodes $\emptyset \neq S \subset V$, initial load vector $w^{(0)}$, and constants $0 < \alpha \leq (\deg(G) + 1)^{-1}$ and $\delta > 0$.[1] Let the drain vector d (which is responsible for the disturbance) be defined as $[d]_v = (\delta n / |S|) - \delta$ if $v \in S$ and $[d]_v = -\delta$ otherwise. Then, the FOS/C iteration in time step $t \geq 1$ is defined as $w^{(t)} = \mathbf{M}w^{(t-1)} + d$, where $\mathbf{M} = \mathbf{I} - \alpha\mathbf{L}$ is the doubly-stochastic diffusion matrix and $\mathbf{L}$ the Laplace matrix of G.*

Lemma 1. *[15] For any $d \perp (1, \ldots, 1)^T$ (i. e., $\langle d, (1, \ldots, 1)^T \rangle = 0$), the FOS/C iteration reaches a steady state, which can be computed by solving the linear system $\mathbf{L}w = d$.*

Definition 2. *If $|S| = 1$ ($|S| > 1$), we call the FOS/C iteration to the steady state or its corresponding linear system a single-source (multiple-source) FOS/C procedure. Also, let $[w^{(t)}]_v^u$ ($[w]_v^u$) denote the load on node v in time step t (in the steady state) of a single-source FOS/C procedure with node u as source.*

---

[1] Here, the maximum degree of G is defined as $\deg(G) := \max_{u \in V} \deg(u)$.

Remark 1. [16] $[w]_v^u = \lim_{t \to \infty}([\mathbf{M}^t w^{(0)}]_v^u + n\delta(\sum_{l=0}^{t-1}[\mathbf{M}^l]_{v,u}) - t\delta)$, where $[\mathbf{M}^l]_{v,u}$ is the probability of a random walk (defined by the stochastic diffusion matrix $\mathbf{M}$) starting at v to be on u after l steps. $[\mathbf{M}^t w^{(0)}]_v^u$ converges to the balanced load distribution. Thus, the important part of an FOS/C load in the steady state is $n\delta(\sum_{l=0}^{t-1}[\mathbf{M}^l]_{v,u})$ – the sum of transition probabilities of random walks with increasing lengths. Observe that load vectors w can be made comparable by normalizing them such that $\sum_{v \in V}[w]_v = n$.

Algorithm Bubble-FOS/C(G, k) → Π

01    $Z = \text{InitialCenters}(G, k)$ /* Arbitrary disjoint centers */

02   **for** $\tau = 1, 2, \ldots$ **until** convergence

       /* AssignPartition: */

03   **parallel for** each part π_p

04   Init d_p ($S_p = \{z_p\}$), solve and normalize $\mathbf{L}w_p = d_p$

05   **parallel for** each node $v \in V$

06            $\Pi(v) = \text{argmax}_{1 \le p \le k}[w_p]_v$

       /* ComputeCenters: */

07   **parallel for** each part π_p

08   Initialize d_p ($S_p = \pi_p$) and solve $\mathbf{L}w_p = d_p$

09            $z_p = \text{argmax}_{v \in \pi_p}[w_p]_v$

10   **return** Π

Fig. 1. Sketch of the main BUBBLE-FOS/C algorithm

The generic algorithmic framework behind the partitioning algorithm BUBBLE-FOS/C is the so-called BUBBLE framework. BUBBLE is related to Lloyd's k-means clustering algorithm [11] and transfers Lloyd's idea to graphs. The framework's first step chooses one initial representative (*center*) for each of the k parts. All remaining vertices are assigned to the closest (with respect to some measure) center vertex. After that, each part computes its new center for the next iteration. Then, the two latter operations are repeated alternately. BUBBLE-FOS/C is outlined in Figure 1, where $\Pi = \{\pi_1, \ldots, \pi_k\}$ denotes the set of parts, $\Pi(v)$ the part of node v, and $Z = \{z_1, \ldots, z_k\}$ the set of the corresponding center nodes. First, the algorithm determines possibly arbitrary, but pairwise disjoint initial centers (line 1). After that, with the new centers, the main loop is executed. It determines in alternating calls a new partition (AssignPartition, lines 3-6) and new centers (ComputeCenters, lines 7-9). BUBBLE-FOS/C implements these framework operations with k FOS/C procedures (more precisely with equivalent linear systems for efficiency) per major operation, single-source ones ($S_p = \{z_p\}$) for AssignPartition and multiple-source ($S_p = \pi_p$) ones for ComputeCenters. The loop is iterated until convergence is reached (convergence is guaranteed [14]) or, if time is important, a constant number of times.

Within the partitioner DIBAP one uses BUBBLE-FOS/C to compute solutions for smaller representations of the input graph with only a few thousand nodes and edges. This computation is reasonably fast and gives initial solutions that are usually better suited than KL-based ones. Initial solutions are refined by a faster local diffusion process (which yields initial solutions of lower quality, but refines well) within a multilevel process, see [14] for details. DIBAP is the combination of these two diffusive algorithms and yields very good experimental results in a reasonable amount of running time (as an example, for $k \leq 16$ and graphs with less than one million nodes and edges, DIBAP requires less than a minute on standard hardware). Except for the connection to random walks and other intuitive explanations mentioned earlier in this section, there has been no theoretical evidence until now why the important initial solutions produced by BUBBLE-FOS/C have a high quality.

3 Optimization Criterion of BUBBLE-FOS/C

It has been shown before [14, Thm. 10], that the iterative optimization performed by the graph partitioning heuristic BUBBLE-FOS/C can be described by a potential function. This function F sums up the diffusion load of each vertex $v \in V$ in a single-source FOS/C procedure in time step τ with v's closest center vertex z_p as source. In fact, the results computed by the operations AssignPartition and ComputeCenters each maximize F for their fixed input (centers or parts, respectively). Moreover, random walks (and also related diffusion processes) can identify dense vertex subsets because they do not escape these regions easily via one of the few external edges. However, it has been unclear so far how these facts relate to the good experimental results of BUBBLE-FOS/C with respect to metrics more specific to graph partitioning.

With the upcoming analysis of BUBBLE-FOS/C, we show that – under mild conditions – BUBBLE-FOS/C solves a relaxed edge cut minimization problem. This is slightly surprising: In previous experiments with numerical simulation graphs [15], BUBBLE-FOS/C was compared to the popular partitioning libraries KMETIS and JOS-TLE. The best improvements by BUBBLE-FOS/C could be seen regarding the number of boundary nodes and the shape of the parts. Yet, concerning the edge cut, the improvement over the other libraries was not as clear, probably because KMETIS and JOSTLE focus chiefly on the edge cut.

3.1 Edge Cut Minimization

Our plan is to express cut minimization by a binary quadratic programming problem (BQP) based on matrices and vectors equivalent or similar to those used in BUBBLE-FOS/C. For this purpose we introduce some notation first. Define a binary indicator vector $x_{(p)} \in \{0,1\}^n$ for part p, $1 \leq p \leq k$, with $[x_{(p)}]_v = 1 \Leftrightarrow v \in \pi_p$. Let $X \in \{0,1\}^{n \times k}$ be the matrix whose p-th column is $x_{(p)}$. Moreover, let $y_{(p,p')} := x_{(p)} - x_{(p')}$ and Y the matrix whose columns are the vectors $y_{(p,p')}$, $1 \leq p < p' \leq k$.

It is well-known [5] that $x^T L x = \sum_{\{u,v\} \in E} \omega(\{u,v\})([x]_u - [x]_v)^2$. Hence, finding a balanced partition with minimum edge cut can be written as:

$$\min_{X \in \{0,1\}^{n \times k}} \quad \sum_{1 \leq p \leq k} x_{(p)}^T L x_{(p)} \tag{1}$$

$$\text{subject to} \quad \|x_{(p)}\|_1 = \frac{n}{k} \quad \text{(balanced parts)}$$

$$\sum_{1 \leq p \leq k} [x_{(p)}]_v = 1 \ \forall v \in V \ \text{(exactly one part per node)}.$$

3.2 ASSIGNPARTITION **Computes Relaxed Minimum Cuts**

Assume we use BUBBLE-FOS/C to find a balanced ($|\pi_i| = |\pi_j| \ \forall 1 \leq i,j \leq k$) k-partition with minimum (or in practice at least small) edge cut of an undirected graph $G = (V, E, \omega)$ with n nodes, $n/k \in \mathbb{N}$. To find a good solution, BUBBLE-FOS/C alternates the operations AssignPartition and ComputeCenters. Eventually, it finds a local optimum of the potential function F described above [14]. In the original formulation of BUBBLE-FOS/C, we solve k linear systems $\mathbf{L}w_p = d_p$, $1 \leq p \leq k$, for each AssignPartition and ComputeCenters operation, respectively. Recall that d_p is the drain vector that changes according to the set of source nodes and w_p is the resulting load vector.

To ensure balanced parts, AssignPartition can be followed by an operation called ScaleBalance [15]. ScaleBalance searches iteratively for scalars β_p such that the assignment of vertices to parts according to $\mathrm{argmax}_{1 \leq p \leq k}[\beta_p w_p]_v$ (instead of $\mathrm{argmax}_{1 \leq p \leq k}[w_p]_v$) results in balanced parts. A simple iterative search for suitable β_p is not always successful in practice, but in many cases it is.

Remark 2. Let $1 \leq p \leq k$. If the β_p were known beforehand, they could be integrated into the drain vector. The resulting linear systems to solve would be $\mathbf{L}(\beta_p w_p) = (\beta_p d_p)$. Hence, it does not make a difference whether suitable β_p are searched such that $\mathrm{argmax}_{1 \leq p \leq k}[\beta_p w_p]_v$ results in balanced parts or if we solve $\mathbf{L}(\beta_p w_p) = (\beta_p d_p)$ from the very beginning.

That is why we assume the scalars β_p to be known for now with $0 < \beta_p \neq \beta_{p'} < 1$ for all $1 \leq p \neq p' \leq k$. For the BQP formulation this is feasible, as will become clear in the remainder of this section. It is essential that the drain vectors are adapted accordingly:

Definition 3. *The entry of vertex $v \in V$ in the drain vector $d_p^{(A)}$ (A for assign) for the FOS/C procedure of part π_p with center z_p in the operation* AssignPartition *with scale value β_p is defined as*

$$[d_p^{(A)}]_v = \delta \cdot \beta_p \cdot \begin{cases} (n-1) & \text{if } v = z_p \\ -1 & \text{otherwise.} \end{cases}$$

Remark 3. If $k = 2$, instead of solving $\mathbf{L}w = d_1^{(A)}$ and $\mathbf{L}w_2 = d_2^{(A)}$, it is sufficient to solve $\mathbf{L}(w_1 - w_2) = d_1^{(A)} - d_2^{(A)}$. Then, to assign vertices to parts, one does not search for argmax (the part with the highest load for the vertex) but makes a sign test. Such a new linear system $\mathbf{L}w_{(p,p')} = d_{(p,p')}^{(A)}$ with $w_{(p,p')} := w_p - w_{p'}$ and $d_{(p,p')}^{(A)} := d_p^{(A)} - d_{p'}^{(A)}$ (if $k = 2$, then $p = 1$ and $p' = 2$) is called *fused (linear) system*. We will see in the proofs of Lemmas 2 and 3 that this fusion technique can be extended easily to $k > 2$ parts.

Lemma 2. *Let $1 \leq p \leq k$. Given a graph $G = (V, E, \omega)$ with n vertices and $\frac{n}{k} \in \mathbb{N}$, its Laplace matrix $\mathbf{L}$, k center vertices $Z = \{z_1, \dots, z_k\}$, k pairwise different real scalars $0 < \beta_p < 1$ (with $\frac{1}{3} < \frac{\beta_p}{\beta_q} < 3$ for $1 \leq p \neq q \leq k$), the FOS/C drain constant δ, and the corresponding drain vectors $d_p^{(A)}$ (one for each part p).*

The BQP for finding a balanced k-partition $\Pi = \{\pi_1, \dots, \pi_k\}$ with minimum cut in G under the condition $z_p \in \pi_p$ can be reformulated as:

$$\min_{X \in \{0,1\}^{n \times k}} \sum_{1 \leq p \leq k} x_{(p)}^T \mathbf{L} x_{(p)} \tag{2}$$

$$\text{subject to } y_{(p,p')}^T d_{(p,p')}^{(A)} = n\delta(\beta_p + \beta_{p'}) \ \forall (p,p') \tag{3}$$

$$\text{with } y_{(p,p')} := x_{(p)} - x_{(p')} \text{ for all } 1 \leq p < p' \leq k.$$

In the proof (full paper version) it becomes clear that the $\binom{k}{2}$ constraints are chosen such that the center vertices do not change their parts and that the parts have equal size.

Corollary 1. *Let $\Pi = \{\pi_1, \dots, \pi_k\}$ be a balanced partition with minimum cut. If the set of center nodes $Z = \{z_1, \dots, z_k\}$ is chosen in Lemma 2 such that $z_p \in \pi_p$ $(1 \leq p \leq k)$, then the BQP (2), (3) computes Π or another balanced partition with minimum cut.*

Due to the $\mathcal{NP}$-hardness of BQP optimization, we aim at relaxed solutions. Instead of choosing only 0 or 1 in the indicator vectors, we now allow the entries of the relaxed indicator vectors $x_{(p)}$ to take on arbitrary real values. Moreover, we use $y_{(p,p')} := x_{(p)} - x_{(p')}$ in the objective function to use the fusion technique described in Remark 3 (in the integral problem, the use of $y_{(p,p')}$ instead of $x_{(p)}$ in the objective function would still model the edge cut, as the change is constant). These changes yield the new optimization problem

$$\min_{Y \in \mathbb{R}^{n \times \binom{k}{2}}} \sum_{1 \leq p < p' \leq k} y_{(p,p')}^T \mathbf{L} y_{(p,p')} \text{ with constraints as in (3).} \tag{4}$$

Lemma 3. *The global minimum $\overline{Y}$ of Problem (4) can be computed by solving and evaluating k linear equations of the form $\mathbf{L} z_p = -\frac{1}{2} d_p^{(A)}$ $(1 \leq p \leq k)$, where*

$$\overline{y}_{(p,p')} = \frac{n\delta(\beta_p + \beta_{p'})}{z_{(p,p')}^T \cdot d_{(p,p')}^{(A)}} \cdot z_{(p,p')} \text{ and } z_{(p,p')} := z_p - z_{p'}, \ 1 \leq p < p' \leq k.$$

Proof. Recall that `AssignPartition` solves k linear systems of the form $\mathbf{L} x_{(p)} = d_p^{(A)}$ and assigns each node to the part with the highest load for that node. It is essential to observe that this is equivalent to solving $\binom{k}{2}$ linear systems of the form $\mathbf{L} y_{(p,p')} = d_{(p,p')}^{(A)}$ and deciding a partial order with respect to the higher load for each node based on its sign in $y_{(p,p')} = x_{(p)} - x_{(p')}$. Note that, before performing scale balancing, all load vectors x are normalized by adding a proper multiple of $\mathbf{1} = (1, \dots, 1)^T$ such that $\sum_{v \in V} [x]_v = n$. This ensures a common basis for comparison and does not affect the equations, because $\mathbf{L1} = 0$ and $d_p \cdot \mathbf{1} = 0$. After $\binom{k}{2}$ comparisons for each node v, the "winning" part (i.e., the one with the highest load for v) has been determined. (Of course, for efficiency reasons, one would not perform such a large number of comparisons. Instead one solves k linear systems and makes $k - 1$ comparisons per node.)

Regarding Eq. (4), using standard multidimensional calculus, one can easily see that the function $f(Y) := \sum_{1 \leq p < p' \leq k} y_{(p,p')}^T \mathbf{L} y_{(p,p')}$ is differentiable over $\mathbb{R}^{n \times \binom{k}{2}}$, because it is a sum of differentiable functions. Furthermore, each constraint function

$h(y_{(p,p')}) := y^T_{(p,p')} d^{(A)}_{(p,p')} - n\delta(\beta_p + \beta_{p'})$ is continuously differentiable over $\mathbb{R}^n$. Hence, we can continue by using a Karush-Kuhn-Tucker argument (see [2, Ch. 4]) and let $\bar{Y} = (\bar{y}_{(1,2)}, \bar{y}_{(1,3)}, \ldots, \bar{y}_{(k-1,k)})$ be a feasible solution. For $\bar{Y}$ to be a global minimum, a vector $\Lambda = (\Lambda_{(1,2)}, \Lambda_{(1,3)}, \ldots, \Lambda_{(k-1,k)}) \in \mathbb{R}^{\binom{k}{2}}$ must exist with

$$\nabla f(\bar{Y}) + \Sigma_{1 \leq p < p' \leq k} \Lambda_{(p,p')} \nabla h(\bar{y}_{(p,p')}) = 0,$$

which yields $\qquad 2\mathbf{L}\Sigma_{1 \leq p < p' \leq k} \bar{y}_{(p,p')} = -\sum_{1 \leq p < p' \leq k} \Lambda_{(p,p')} d^{(A)}_{(p,p')}.$

Such a vector Λ indeed exists: We first solve the linear systems $\mathbf{L}z_p = -\frac{1}{2} d^{(A)}_p$ for all $1 \leq p \leq k$. With $z_{(p,p')} := z_p - z_{p'}$ we have for all $1 \leq p < p' \leq k : \mathbf{L}z_{(p,p')} = -\frac{1}{2} d^{(A)}_{(p,p')}$, so that $\mathbf{L}\Sigma_{1 \leq p < p' \leq k} z_{(p,p')} = -\frac{1}{2}\Sigma_{1 \leq p < p' \leq k} d^{(A)}_{(p,p')}$. Let $\bar{y}_{(p,p')} := \Lambda_{(p,p')} z_{(p,p')}$, so that we arrive at

$$\mathbf{L}\bar{y}_{(p,p')} = -\frac{1}{2}\Lambda_{(p,p')} d^{(A)}_{(p,p')} \; \forall 1 \leq p < p' \leq k$$

$$\Rightarrow \mathbf{L} \sum_{1 \leq p < p' \leq k} \bar{y}_{(p,p')} = -\frac{1}{2} \sum_{1 \leq p < p' \leq k} \Lambda_{(p,p')} d^{(A)}_{(p,p')}.$$

Hence, a suitable Λ exists. Following [2, Thm. 4.3.8], f and h are convex functions or the sum of convex functions (again, this is easy to check, for f by using that $\mathbf{L}$ is positive semidefinite ($x^T \mathbf{L}x \geq 0 \forall x$)), so that $\bar{Y}$ is a global optimum of Equation (4). Finally, some rearranging suffices to compute each $\bar{y}_{(p,p')}$ and $\Lambda_{(p,p')}$ from $z_{(p,p')}$. $\qquad \square$

Let us make clear now why the assumption of already known β_p is feasible. First, recall from Remark 2 that the result of the linear systems is the same regardless when the β_p are introduced into the equations. Lemma 2 tells us that the actual choice of the β_p is hardly relevant for the BQP to work – as long as they are not equal, lie between 0 and 1, and their quotient is neither too small nor too large. Hence, we choose suitable β_p such that the BQP works. In practice, however, we cannot make such a choice for BUBBLE-FOS/C a priori. Yet, given the mild conditions, we can assume that the scalars β_p computed by ScaleBalance will fulfill the constraints mentioned above in the vast majority of cases. Therefore, we can conclude this section with the following insight:

Theorem 1. *Let $k \geq 2$. Given a graph $G = (V, E, \omega)$ with n nodes ($n/k \in \mathbb{N}$) and a set Z with one center vertex for each of the k parts. Then, the two consecutive operations AssignPartition and ScaleBalance with suitable β_p ($1 \leq p \leq k$) together compute the global minimum of the Optimization Problem (4). Problem (4) is a relaxed version of BQP (2), (3). If $Z = \{z_1, ..., z_k\}$ is given such that $z_p \in \pi_p$ and $\Pi = \{\pi_1, \ldots, \pi_p\}$ is an (unknown) optimal (with respect to the edge cut) partition, then this BQP computes an optimal partition.*

Proof. We solve for each AssignPartition operation the linear systems $\mathbf{L}w_p = d_p$, where each d_p is the original drain vector without integration of β_p, $1 \leq p \leq k$. Performing ScaleBalance results in the load vector $\beta_p w_p$. With the Remarks 2 and 3 and the proof of Lemma 3, it follows that the assignment process can be regarded as making $\binom{k}{2}$ comparisons per vertex, i.e., vertices are assigned according to their sign in the $\binom{k}{2}$ fused load vectors $w_{(p,p')} = \beta_p w_p - \beta_{p'} w_{p'}$. As a direct consequence of the results above, for suitable β_p these load vectors $w_{(p,p')}$ correspond to a relaxed optimal solution of the BQP (2), (3). As shown before, this BQP would find the solution with minimum edge cut given an optimal placement of the center nodes.   □

3.3 ComputeCenters **Maximizes Constraint Contribution**

Recall that the iteration of BUBBLE-FOS/C with its alternating calls to AssignPartition and ComputeCenters maximizes the potential function F (see the beginning of Section 3). Insofar it is interesting to find out if a similar optimization property holds when ComputeCenters is described as the relaxation of a cut-minimizing BQP. Note that in the case of ComputeCenters we are given a fixed partition and need to return one center vertex for each part.

Compared to our derivation in Section 3.2, the drain vector for part π_p is not $d_p^{(A)}$ any more, but $d_p^{(C)}$ (C for centers). This change reflects that the total drain is not given to one center vertex, but shared among all vertices of the part under consideration. Moreover, the scale values β_p are not needed any more, i.e., they can be set to 1 here. Consequently, $[d_p^{(C)}]_v = \delta\beta_p(n/|\pi_p| - 1)$ if $v \in \pi_p$ and $[d_p^{(C)}]_v = -\beta_p\delta$ if $v \notin \pi_p$.

Remark 4. To establish a BQP for ComputeCenters given the input partition Π, we simply replace all occurrences of $d^{(A)}$ by $d^{(C)}$ in Equation (3), eliminate the unnecessary β_p, and use the indicator vectors $x_{(p)}$ here:

$$x_{(p)}^T d_p^{(C)} = \delta(n - |\pi_p|) \ \forall 1 \le p \le k. \tag{5}$$

As shown below, the modified constraints ensure that all vertices stay in their part. This is important because the operation ComputeCenters is not supposed to change the partition. In particular, the computed centers must come from different parts.

Lemma 4. *The constraints in Equation (5) ensure that the centers $Z = \{z_1, \ldots, z_k\}$ computed by the BQP (2), (5) are in pairwise different parts with respect to Π.*

Immediately the question arises how the computation of centers is supposed to minimize the edge cut. Indeed, the BQP formulation only computes an indicator vector that represents the input partition. Yet, the new centers do have an extremal property, the contribution to Constraint (5). Again, we relax the binary condition on $x_{(p)}$, i.e., let $x_{(p)} \in \mathbb{R}^n$. Since $d^{(C)}$ is constant for all vertices of the same part and $[x_{(p)}]_{z_p} = \text{argmax}_{1 \le v \le n}[x_{(p)}]_v$:

Corollary 2. *Given a partition $\Pi = \{\pi_1, \ldots, \pi_k\}$, let ComputeCenters compute the vertices $Z = \{z_1, \ldots, z_k\}$ as new centers. The respective entry $[x_{(p)}]_{z_p} d_{z_p}^{(C)}$ contributes the highest value of all vertices in π_p to $x_{(p)}^T \cdot d_p^{(C)}$, $1 \le p \le k$.*

4   Connectedness Properties of BUBBLE-FOS/C

For some applications that use partitioning as an intermediate step (e. g., tracking par-
ticles in parallel), it is advantageous that the parts are connected, i. e., that they have
exactly one connected component each. Experiments with graphs from finite element
discretizations reveal that the subdomains computed by BUBBLE-FOS/C are (nearly
always) connected if the algorithm is allowed to perform sufficiently many iterations.
Unfortunately, there has been no theoretical evidence for this observation until now.

In this section we make a step towards gaining more knowledge about the connectedness properties of BUBBLE-FOS/C. Similar to Fiedler's classical result [5] about
spectral bipartitioning (but by using a different proof approach), we state that at least
one part in a partition $\{\pi_1, \pi_2\}$ computed by BUBBLE-FOS/C is connected.

Theorem 2. *If the graph $G = (V, E, \omega)$ is connected, then at least one of $k = 2$ parts
computed by AssignPartition on G is connected.*

The proof relies on the fact that the diffusion loads increase monotonically on some path
from a vertex to a center. Note that the results of this section as well as some missing
auxiliary results are proved in the full version of this paper. Now we tighten the result
for all connected *vertex-transitive* graphs (a graph is vertex-transitive if for any pair of
distinct vertices there is an automorphism mapping one to the other [3]), where both
parts are shown to be connected. Two well-known vertex-transitive classes are torus
graphs and hypercubes, which are important network topologies.

Theorem 3. *Let $G = (V, E)$ be a connected vertex-transitive graph. Fix two arbitrary
different vertices $z_1, z_2 \in V$. Let the operation AssignPartition divide V into the two
subdomains $\pi_1 = \{u \in V \mid [w]_u^{z_1} \geq [w]_u^{z_2}\}$ and $\pi_2 = \{u \in V \mid [w]_u^{z_1} < [w]_u^{z_2}\}$. Then, π_1
and π_2 are each connected components in G.*

Proof. The random walk measure *hitting time* $H[u, v]$ between nodes u and v is the
expected timestep in which a random walk starting in u visits v for the first time. By
using [16, Thm. 1], we know that $\frac{1}{\alpha}([w]_u^v - [w]_v^v) = \delta(H[v, v] - H[u, v])$. First, we show
that hitting times are symmetric on vertex-transitive graphs. For this we use that $[w]_u^u =
[w]_v^v$, which follows from the fact that $[M^t]_{v,v} = [M^t]_{u,u}$ for all $u, v \in V$ and all $t \geq 0$ for
an unweighted vertex-transitive graph G [1, p. 151]. Also, the symmetry $[w]_v^u = [w]_u^v$
holds for all $u, v \in V$ [16] and $H[v, v]$ is zero (definition of hitting times). Thus:

$$([w]_v^v = [w]_u^u) \wedge ([w]_v^u = [w]_u^v) \Rightarrow [w]_u^v - [w]_v^v = [w]_v^u - [w]_u^u \Rightarrow H[u, v] = H[v, u].$$

Assume now for the sake of contradiction that π_2 is not connected. In this case there
exists a node-separator $T \subseteq \pi_1$ such that there are at least two components $A, B \subseteq \pi_2$
which are not connected by a path via nodes in π_2. Assume w. l. o. g. that $z_2 \in B$.
Then for each vertex $a \in A$ we obtain $[w]_a^{z_2} > [w]_a^{z_1} \Leftrightarrow [w]_a^{z_2} - [w]_{z_2}^{z_2} > [w]_a^{z_1} - [w]_{z_1}^{z_1} \Leftrightarrow
H[z_2, z_2] - H[a, z_2] > H[z_1, z_1] - H[a, z_1] \Leftrightarrow H[a, z_1] > H[a, z_2]$.

In the same manner we have for each vertex $x \in T$ that $H[x, z_1] \leq H[x, z_2]$. Let $X^{(t)}$
be the random variable representing the node visited in time step t by a random walk,
and let $\mathscr{F}_u(x)$ be the event that a fixed vertex x is the first vertex visited in T of a
random walk starting from $u \in V$. Furthermore, denote by $\tau_a(T) := \min_{t \in \mathbb{N}}\{X^{(t)} \in$

$T \mid X^{(0)} = a\}$ and let $\tau_{a,T}(z_1) := \min_{t \in \mathbb{N}}\{X^{(t)} = z_1 \mid X^{(0)} = a\} - \tau_a(T)$. By using conditional expectations ($\mathbb{E}[Y] = \sum_z \mathbf{Pr}[Z = z] \cdot \mathbb{E}[Y \mid Z = z]$) [7], we obtain $H[a, z_1] = \mathbb{E}[\tau_a(z_1)] = \mathbb{E}[\tau_a(T) + \tau_{a,T}(z_1)] = \sum_{x \in T} \mathbf{Pr}[\mathscr{F}_a(x)] \cdot (\mathbb{E}[\tau_a(T) + \tau_{a,T}(z_1) \mid \mathscr{F}_a(x)])$, which is transformed by using the linearity of conditional expectations into

$$H[a, z_1] = \sum_{x \in T} \mathbf{Pr}[\mathscr{F}_a(x)] \cdot \left(\mathbb{E}[\tau_a(T) \mid \mathscr{F}_a(x)] + \mathbb{E}[\tau_{a,T}(z_1) \mid \mathscr{F}_a(x)]\right)$$
$$= \sum_{x \in T} \mathbf{Pr}[\mathscr{F}_a(x)] \cdot \left(\mathbb{E}[\tau_a(x) \mid \mathscr{F}_a(x)] + \mathbb{E}[\tau_x(z_1) \mid \mathscr{F}_a(x)]\right)$$
$$= \sum_{x \in T} \mathbf{Pr}[\mathscr{F}_a(x)] \cdot \left(\mathbb{E}[\tau_a(x) \mid \mathscr{F}_a(x)] + H[x, z_1]\right).$$

Exactly the same arguments yield $H[a, z_2] = \sum_{x \in T} \mathbf{Pr}[\mathscr{F}_a(x)] \cdot (\mathbb{E}[\tau_a(x) \mid \mathscr{F}_a(x)] + H[x, z_2])$. Due to $H[x, z_1] \leq H[x, z_2]$ for each $x \in T$, we finally obtain

$$H[a, z_1] = \sum_{x \in T} \mathbf{Pr}[\mathscr{F}_a(x)] \cdot \left(\mathbb{E}[\tau_a(x) \mid \mathscr{F}_a(x)] + H[x, z_1]\right)$$
$$\leq \sum_{x \in T} \mathbf{Pr}[\mathscr{F}_a(x)] \cdot \left(\mathbb{E}[\tau_a(x) \mid \mathscr{F}_a(x)] + H[x, z_2]\right) = H[a, z_2],$$

which is a contradiction to our assumption $H[a, z_1] > H[a, z_2]$. Therefore, the subdomain π_2 has to be connected. The remainder of the proof is analogous (switch π_1 and π_2). □

Generalizing this result to other graph classes will probably require new techniques, as the FOS/C load property $[w]_v^v = [w]_u^u$ does not hold any more. Also, our hitting time argument in the proof cannot be generalized to $k > 2$ in a straightforward manner since the vertex separator may contain vertices from more than one part.

5 Conclusions and Future Work

As explained in the introduction, diffusion-based graph partitioning has proved to be very successful in practice. Here we have provided the first substantial theoretical evidence for this success by proving that the assignment of vertices to parts in the partitioning algorithm BUBBLE-FOS/C is relaxed cut optimization. In this sense BUBBLE-FOS/C is similar to spectral partitioning, but does not require the (possibly numerically problematic) computation of eigenvectors. Moreover, we have proved two results on the connectedness of parts, which is a property that is important for some applications.

With these new tools at hand, we would like to consider the iterative nature of BUBBLE-FOS/C and explore the faster partitioning algorithm DIBAP in future work. DIBAP uses BUBBLE-FOS/C as one of two key components. It will be interesting to learn more about the interaction of these components, whose combination is responsible for obtaining high quality at reasonable speed. Eventually, it might also be possible to derive an approximation guarantee on BUBBLE-FOS/C's and DIBAP's quality from our relaxed BQP results, at least for certain graph classes. Since there are no such guarantees known for the popular KL heuristic, such a result would be a major step towards uniting theoretical and practical graph partitioning.

Acknowledgments. The author thanks T. Sauerwald, who contributed to the proof of Thm. 3, and C. Buchheim, R. Feldmann, and B. Monien for helpful discussions.

References

1. Alon, N., Spencer, J.H.: The Probabilistic Method, 2nd edn. J. Wiley & Sons, Chichester (2000)
2. Bazaraa, M.S., Sherali, H.D., Shetty, C.M.: Nonlinear Programming. Theory and Algorithms, 2nd edn. John Wiley, Chichester (1993)
3. Biggs, N.: Algebraic Graph Theory. Cambridge University Press, Cambridge (1993)
4. Chevalier, C., Pellegrini, F.: Pt-scotch: A tool for efficient parallel graph ordering. Parallel Comput. 34(6-8), 318–331 (2008)
5. Fiedler, M.: A property of eigenvectors of nonnegative symmetric matrices and its application to graph theory. Czechoslovak Mathematical Journal 25, 619–633 (1975)
6. Grady, L., Schwartz, E.L.: Isoperimetric graph partitioning for image segmentation. IEEE Trans. Pattern Anal. Mach. Intell. 28(3), 469–475 (2006)
7. Grimmett, G.R., Stirzaker, D.R.: Probability and Random Processes, 3rd edn. Oxford University Press, Oxford (2001)
8. Hendrickson, B., Leland, R.: An improved spectral graph partitioning algorithm for mapping parallel computations. SIAM Journal on Scientific Computing 16(2), 452–469 (1995)
9. Karypis, G., Kumar, V.: Multilevel k-way partitioning scheme for irregular graphs. Journal of Parallel and Distributed Computing 48(1), 96–129 (1998)
10. Kernighan, B.W., Lin, S.: An efficient heuristic for partitioning graphs. Bell Systems Technical Journal 49, 291–308 (1970)
11. Lloyd, S.P.: Least squares quantization in PCM. IEEE Transactions on Information Theory 28(2), 129–136 (1982)
12. Lovász, L.: Random walks on graphs: A survey. Combinatorics, Paul Erdös is Eighty 2, 1–46 (1993)
13. Meila, M., Shi, J.: A random walks view of spectral segmentation. In: Eighth International Workshop on Artificial Intelligence and Statistics (AISTATS) (2001)
14. Meyerhenke, H., Monien, B., Sauerwald, T.: A new diffusion-based multilevel algorithm for computing graph partitions. Journal of Parallel and Distributed Computing 69(9), 750–761 (2009); Best Paper Awards and Panel Summary: IPDPS 2008
15. Meyerhenke, H., Monien, B., Schamberger, S.: Graph partitioning and disturbed diffusion. Parallel Computing 35(10-11), 544–569 (2009)
16. Meyerhenke, H., Sauerwald, T.: Analyzing disturbed diffusion on networks. In: Asano, T. (ed.) ISAAC 2006. LNCS, vol. 4288, pp. 429–438. Springer, Heidelberg (2006)
17. Pellegrini, F.: A parallelisable multi-level banded diffusion scheme for computing balanced partitions with smooth boundaries. In: Kermarrec, A.-M., Bougé, L., Priol, T. (eds.) Euro-Par 2007. LNCS, vol. 4641, pp. 195–204. Springer, Heidelberg (2007)
18. Schaeffer, S.E.: Graph clustering. Computer Science Review 1(1), 27–64 (2007)
19. Schloegel, K., Karypis, G., Kumar, V.: Graph partitioning for high performance scientific simulations. In: The Sourcebook of Parallel Computing, pp. 491–541. Morgan Kaufmann, San Francisco (2003)
20. Shi, J., Malik, J.: Normalized cuts and image segmentation. IEEE Transactions on Pattern Analysis and Machine Intelligence 22(8), 888–905 (2000)
21. Walshaw, C.: The graph partitioning archive (2010),
http://staffweb.cms.gre.ac.uk/~c.walshaw/partition/
(Last access: 1 March 2010)
22. Zha, H., He, X., Ding, C.H.Q., Gu, M., Simon, H.D.: Spectral relaxation for k-means clustering. In: Proceedings of Advances in Neural Information Processing Systems 14 (NIPS 2001), pp. 1057–1064. MIT Press, Cambridge (2001)

Induced Subgraph Isomorphism on Interval and Proper Interval Graphs*

Pinar Heggernes[1], Daniel Meister[2], and Yngve Villanger[1]

[1] Department of Informatics, University of Bergen, Norway
`pinar.heggernes@ii.uib.no, yngve.villanger@ii.uib.no`
[2] Theoretical Computer Science, RWTH Aachen University, Germany
`meister@cs.rwth-aachen.de`

Abstract. The INDUCED SUBGRAPH ISOMORPHISM problem on two input graphs G and H is to decide whether G has an induced subgraph isomorphic to H. Already for the restricted case where H is a complete graph the problem is NP-complete, as it is then equivalent to the CLIQUE problem. In a recent paper [7] Marx and Schlotter show that INDUCED SUBGRAPH ISOMORPHISM is NP-complete when G and H are restricted to be interval graphs. They also show that the problem is $W[1]$-hard with this restriction when parametrised by the number of vertices in H. In this paper we show that when G is an interval graph and H is a connected proper interval graph, the problem is solvable in polynomial time. As a more general result, we show that when G is an interval graph and H is an arbitrary proper interval graph, the problem is fixed parameter tractable when parametrised by the number of connected components of H. To complement our results, we prove that the problem remains NP-complete when G and H are both proper interval graphs and H is disconnected.

1 Introduction

Given two graphs G and H, where G has more vertices than H, the INDUCED SUBGRAPH ISOMORPHISM (ISI) problem is to decide whether G has an induced subgraph isomorphic to H. Equivalently, the question is whether we can delete vertices from G to obtain a graph isomorphic to H. ISI is a generalisation of several well known NP-complete problems like CLIQUE, INDEPENDENT SET, LONGEST INDUCED PATH, and GRAPH ISOMORPHISM, and it is thus NP-complete, as well as $W[1]$-hard when parametrised by the number of vertices in H.

As the problem is applicable in a variety of important practical areas [3], it has been studied with respect to polynomial-time solvability and fixed parameter tractability on restricted input graphs. ISI is solvable in polynomial time when G and H are both trees [8] but it remains NP-complete when G is a tree and

---

* This work is supported by the Deutsche Forschungsgemeinschaft and by the Research Council of Norway.

O. Cheong, K.-Y. Chwa, and K. Park (Eds.): ISAAC 2010, Part II, LNCS 6507, pp. 399–409, 2010.
© Springer-Verlag Berlin Heidelberg 2010

H is a forest [5] or when G is a cubic planar graph and H is a path [5]. When parametrised by the number of vertices in H, the problem is known to be fixed parameter tractable when G and H are planar [3] or have maximum degree bounded by a constant [2]. In a very recent paper by Marx and Schlotter, ISI is studied on interval graphs. When both G and H are interval graphs, the authors show that the problem is NP-complete and $W[1]$-hard when parametrised by the number of vertices in H [7].

Here, we show that when G is an interval graph and H is an arbitrary proper interval graph, ISI is fixed parameter tractable when parametrised by the number of connected components of H (and consequently also when parametrised by the number of vertices in H).

To indicate that these results are the best that we can hope for, we show that ISI remains NP-complete when G and H are both proper interval graphs and H is disconnected.

To achieve our polynomial-time algorithm, we give an intermediate algorithm for solving the following problem in polynomial time: Given two connected interval graphs G and H with a partial ordering of the vertices of each graph, is there an isomorphism between H and an induced subgraph of G that respects the given partial orderings? Our main result is obtained by showing that if H is isomorphic to an induced subgraph G' of G then the relative ordering of the vertices of G' is of a restricted type "fitting" the ordering of H, in any interval ordering for G.

Many NP-hard graph problems become solvable in polynomial time on interval graphs and even more on proper interval graphs. An example related to our problem is GRAPH ISOMORPHISM which can be solved in linear time on interval graphs [1]. From this point of view, the mentioned results of Marx and Schlotter and our hardness result are surprising.

Due to space limitations, all the proofs are excluded from this extended abstract.

2   Definitions and Notation

We consider simple finite undirected graphs. For a graph $G = (V, E)$, $V = V(G)$ is the *vertex set* of G and $E = E(G)$ is the *edge set* of G. For every edge $uv \in E$, vertices u and v are *adjacent* or *neighbours*. The *neighbourhood* of a vertex u in G is $N_G(u) =_{\text{def}} \{v \mid uv \in E\}$, and the *closed neighbourhood* of u is $N_G[u] =_{\text{def}} N_G(u) \cup \{u\}$. A set $X \subseteq V$ is called *clique* of G if the vertices in X are pairwise adjacent. A *maximal* clique is a clique that is not a proper subset of any other clique. For $U \subseteq V$, the *subgraph of G induced by U* is denoted by $G[U]$ and it is the graph with vertex set U and edge set equal to the set of edges $uv \in E$ with $u, v \in U$. For every $U \subseteq V$, $G' = G[U]$ is an *induced subgraph* of G. By $G \setminus X$ for $X \subseteq V$, we denote the graph $G[V \setminus X]$.

For two graphs G and H, G is *isomorphic* to H if there is a bijective mapping φ from $V(G)$ to $V(H)$ such that for every vertex pair u, v of G, $uv \in E(G)$ if and only if $\varphi(u)\varphi(v) \in E(H)$. Mapping φ is called an *isomorphism* from G to H. If

G has an induced subgraph G' such that G' is isomorphic to H then we say that G has an induced subgraph isomorphic to H or, equivalently, H is isomorphic to an induced subgraph of G. Let us formally define the problem we are working on.

INDUCED SUBGRAPH ISOMORPHISM (ISI)
Input: Two graphs G and H.
Question: Does G have an induced subgraph that is isomorphic to H?

For a graph G, vertices u, v of G and an integer $k \geq 0$, a u, v-*path of length k* is a sequence $(u_0, \ldots, u_k)$ of $k+1$ distinct vertices of G such that $u_i u_{i+1} \in E(G)$ for $0 \leq i < k$ and $u_0 = u$ and $u_k = v$. A path $(u_0, \ldots, u_k)$ is *chordless* if $u_i u_j \notin E(G)$ for $0 \leq i < i+1 < j \leq k$. A graph G is *connected* if there is a u, v-path in G for every vertex pair u, v of G. A *connected component* of G is a maximal connected induced subgraph of G. The *distance* between two vertices u and v in G is the smallest integer k such that G has a u, v-path of length k.

A graph is an *interval graph* if intervals of the real line can be assigned to its vertices such that two vertices are adjacent if and only if their assigned intervals overlap. A *clique path* of a graph G is an ordering $\langle A_1, \ldots, A_k \rangle$ of the maximal cliques of G that satisfies the following for every vertex x of G: if $1 \leq p < q < r \leq k$ and $x \in A_p \cap A_r$ then $x \in A_q$. A graph is an interval graph if and only if it has a clique path [4]. A clique path can be constructed in linear time [4]. Note that an interval graph can have many different clique paths. An *proper interval graph* is an interval graph whose vertices can be assigned intervals such that no interval is properly contained in any other interval. A *claw* is a graph that is isomorphic to $K_{1,3}$. A graph is *claw-free* if it does not have a claw as an induced subgraph. Proper interval graphs are exactly the claw-free interval graphs [10].

A *vertex ordering* for a graph G is a linear ordering $\sigma = \langle u_1, \ldots, u_n \rangle$ of the vertices of G. For two vertices u_i, u_j of G in σ, we write $u_i \preceq_\sigma u_j$ if $i \leq j$. If additionally $i \neq j$ then we write $u_i \prec_\sigma u_j$. A vertex ordering σ for $G = (V, E)$ is called *interval ordering* if for every vertex triple u, v, w of G, $u \prec_\sigma v \prec_\sigma w$ and $uw \in E$ imply $vw \in E$. A graph is an interval graph if and only if it admits an interval ordering [9]. A vertex ordering σ for G is called *proper interval ordering* if for every vertex triple u, v, w of G, $u \prec_\sigma v \prec_\sigma w$ and $uw \in E$ imply $uv, vw \in E$. A graph is a proper interval graph if and only if it admits a proper interval ordering [6]. Interval orderings and proper interval orderings can be computed in linear time, if they exist.

3   Polynomial-Time Solvable Cases of Induced Subgraph Isomorphism on Interval Graphs

We show that when G is an interval graph and H is a connected proper interval graph, ISI is solvable in polynomial time. From our intermediate results to reach this algorithm, it will follow that ISI is fixed-parameter tractable, parametrised by the number of connected components of H, when G is an interval graph and H is an arbitrary proper interval graph.

To obtain this result, we start by giving an intermediate result which is interesting on its own. In the first subsection we study the following problem: given two interval graphs G and H with clique paths $\langle A_1, \ldots, A_k \rangle$ and $\langle B_1, \ldots, B_l \rangle$ for G and H, respectively, decide whether there is an isomorphism from H to an induced subgraph of G that preserves the order of the maximal cliques given by the clique paths. We show that this problem is solvable in polynomial time.

3.1   Induced Subgraph Isomorphism on Ordered Interval Graphs

We start by showing that any isomorphism between an interval graph and an induced subgraph of another interval graph must map maximal cliques of the two graphs to each other.

Lemma 1. *Let G and H be interval graphs with clique paths $\langle A_1, \ldots, A_k \rangle$ and $\langle B_1, \ldots, B_l \rangle$, respectively. If there is an isomorphism φ from H to an induced subgraph of G then there is a mapping $\psi : \{1, \ldots, l\} \to \{1, \ldots, k\}$ such that $\varphi(B_l) \subseteq A_{\psi(l)}$ and $\varphi(B_i \setminus B_{i+1}) \subseteq A_{\psi(i)} \setminus A_{\psi(i+1)}$ for every $1 \le i < l$ and $\psi(i) \ne \psi(j)$ for every $1 \le i < j \le l$.*

This subsection considers isomorphisms that require $\psi(1) < \cdots < \psi(l)$ for function ψ of Lemma 1. We formalise this notion in the following way. Let G and H be graphs and let σ and τ be vertex orderings for respectively G and H. We say that H is (σ, τ)-*isomorphic* to an induced subgraph G' of G if there exists an isomorphism φ from H to G' such that $\varphi(u) \prec_\sigma \varphi(v)$ for every vertex pair u, v of H with $u \prec_\tau v$.

An interval graph $G = (V, E)$ may have many interval orderings. An interval ordering σ for G is a *preference interval ordering* if additionally the following condition is satisfied for every vertex triple u, v, w of G: if $u \prec_\sigma v \prec_\sigma w$ and $uw \in E$ and $uv \notin E$ then there is $x \in V$ such that $w \prec_\sigma x$ and $wx \in E$ and $vx \notin E$. (Informally, a preference interval ordering is a right endpoint ordering for an interval model where ties are broken by a left endpoint ordering.) Every interval graph has a preference interval ordering, and such an ordering can be computed in linear time. An interval graph can have many preference interval orderings. We need to relate preference interval orderings to clique paths and to (arbitrary) interval orderings. Let G be an interval graph with clique path $\langle A_1, \ldots, A_k \rangle$. A *preference interval ordering τ for G related to* $\langle A_1, \ldots, A_k \rangle$ satisfies for every vertex pair u, v of G and every $1 \le i < k$ that $u \in A_i \setminus A_{i+1}$ and $v \in A_{i+1} \cup \cdots \cup A_k$ implies $u \prec_\tau v$. Note that such an ordering always exists. Let σ be an interval ordering for G. A *preference interval ordering τ for (G, σ)* satisfies for every vertex pair u, v of G that $uv \notin E$ and $u \prec_\sigma v$ implies $u \prec_\tau v$. It is important to see that also such an ordering always exists and it is also obtainable from a given clique path.

The first algorithm that we consider is called LOCALORDERINGINDUCED-SUBGRAPH, LOIS for short, and presented in Figure 1. This algorithm solves a restricted version of ISI on interval graphs, namely it requires the same number of maximal cliques for the two input graphs and additionally checks for ordered isomorphisms only.

Algorithm LOCALORDERINGINDUCEDSUBGRAPH (LOIS)

Input   Graphs G and H with vertex orderings $\sigma = \langle x_1, \ldots, x_n \rangle$ and
$\tau = \langle y_1, \ldots, y_r \rangle$, respectively.

Output An isomorphism φ from an induced subgraph G' to H such that for all
vertex pairs u, v of G', $u \prec_\sigma v$ iff $u \prec_\tau v$, if such an isomorphism exists.

begin
　　for $i = r$ **downto** 1 **do** let $a_i = n - r + i$ **end for**;

　　while H is not (σ, τ)-isomorphic to $G[\{x_{a_1}, \ldots, x_{a_r}\}]$ **do**
　　　let y_i, y_j be a vertex pair of H where $i < j$ and j is largest possible
　　　　such that $y_i y_j \notin E(H) \Leftrightarrow x_{a_i} x_{a_j} \in E(G)$;

　　　if $y_i y_j \notin E(H)$ **then** PUSH(i) **else** PUSH(j) **end if**
　　end while;

　　return $a_1, \ldots, a_r$ and **accept**
end.

Subroutine PUSH(b)
begin
　　set $a_b = a_b - 1$;
　　while $a_b = a_{b-1}$ and $b \geq 2$ **do** set $b = b - 1$; set $a_b = a_b - 1$; **end while**;
　　if $a_1 = 0$ **then** reject **end if**
end.

Fig. 1. Algorithm LOIS

Lemma 2. *Let G and H be interval graphs with the same number of maximal cliques. Let σ and τ be preference interval orderings for respectively G and H. Algorithm LOIS on this input computes a (σ, τ)-isomorphism from H to an induced subgraph of G, if it exists.*

We extend the above problem to interval graphs with different numbers of maximal cliques. We apply a variant of Algorithm LOIS as a subroutine. The main difficulty is to determine the correct selection of maximal cliques of G. Our algorithm is called INDUCEDINTERVALSUBGRAPH, IIS for short, and it is given in Figure 2. It applies as a subroutine Algorithm LOIS*; this algorithm is described in the next paragraph.

Algorithm LOIS* mainly works as Algorithm LOIS given in Figure 1. As an additional input, there is a lower bound on the value of a_i for each vertex of H. The return value of LOIS* is a vertex or a special symbol instead of the values of $a_1, \ldots, a_r$. During the initialization or the execution of Subroutine PUSH, there may be an a_i that becomes smaller than the corresponding given lower bound. If such a lower bound violation occurs during the initialization step, let m be the largest integer such that a_m is smaller than its corresponding lower bound. The algorithm stops and returns vertex y_m. Otherwise, the initialization step is executed successfully, and a lower bound violation can occur only during the execution of Subroutine PUSH. Let y_i, y_j with $i < j$ be the (earliest) vertex pair of H for which a problem was encountered. If Subroutine PUSH was called with i as parameter then the algorithm returns y_j, if PUSH was called with j as

Algorithm INDUCEDINTERVALSUBGRAPH (IIS)
Input   An interval graph G with clique path $\langle A_1, \ldots, A_k \rangle$,
                an interval graph H with clique path $\langle B_1, \ldots, B_l \rangle$.
Output   **accept** if H is isomorphic to an induced subgraph of G under
                the extra conditions given in Lemma 3.

begin
    let τ be preference interval ordering for H related to $\langle B_1, \ldots, B_l \rangle$;
    let $\psi(0) = 0$;   **let** $\psi(1) = 0$;   **let** $m = 1$;   **let** $B_{l+1} = \emptyset$;
    loop
        INITIALIZE(m);
        let $G' = G[A_{\psi(1)} \cup \cdots \cup A_{\psi(l)}]$;
        let σ be preference interval ordering for G' related to $\langle A_{\psi(1)}, \ldots, A_{\psi(l)} \rangle$;
        for $i = 1$ **to** l **do**
            for each $x \in B_i \setminus B_{i+1}$ **do** $\lambda(x) = |A_{\psi(1)} \cup \cdots \cup A_{\psi(i-1)}| + 1$ **end for**
        end for;
        set $y = \text{LOIS}^*(G', H; \sigma, \tau; \lambda)$;
        if y is a vertex of H **then** **set** m such that $y \in B_m \setminus B_{m+1}$;
        else **set** $m = 0$; **end if**
    while $m > 0$;
    accept
end.

Subroutine INITIALIZE(s)
begin
    for $i = 1$ **to** s **do**
        let p be smallest such that $p > \psi(i-1)$ and $p > \psi(s)$;
        if p does not exist **then** **reject** **end if**;
        set $\psi(i) = p$
    end for
end.

Fig. 2. Algorithm IIS

parameter then the algorithm returns y_i. If no lower bound condition violation ever happens, which means that Algorithm LOIS* accepts, then the algorithm returns a special symbol. Note that LOIS already checks for a lower bound violation, namely it checks whether $a_1 < 1$ at the end of PUSH. So, it is clear that if Algorithm LOIS would reject then Algorithm LOIS* will not return the special symbol.

Lemma 3. *Let G and H be interval graphs with $\mathscr{A} = \langle A_1, \ldots, A_k \rangle$ and $\mathscr{B} = \langle B_1, \ldots, B_l \rangle$ clique paths for respectively G and H, where $l \leq k$. Algorithm* INDUCEDINTERVALSUBGRAPH *on this input accepts if and only if there are integers $s_1, \ldots, s_l$ satisfying $1 \leq s_1 < \cdots < s_l \leq k$ such that H is (σ, τ)-isomorphic to an induced subgraph of $G[A_{s_1} \cup \cdots \cup A_{s_l}]$ where σ is a preference interval ordering related to $\langle A_{s_1}, \ldots, A_{s_l} \rangle$ and τ is a preference interval ordering related to $\mathscr{B}$.*

We determine the running time of IIS. The running time is mainly determined by the number of executions of the main loop of IIS and the running time of a single execution of LOIS*. Let graph G have n vertices. The main loop is executed at most n^2 times. Each single loop execution, including the re-initialization, requires $\mathcal{O}(n^2)$ time plus the time for an execution of LOIS*. The running time of this procedure is of order the running time of LOIS. The main **while** loop is executed at most n^2 times. A single loop execution requires $\mathcal{O}(n^2)$ time for checking the isomorphism condition and finding a new vertex pair. This sums up to a total running time of $\mathcal{O}(n^6)$ for IIS.

3.2   Finding Induced Proper Interval Subgraphs of Interval Graphs

In the previous subsection, we gave an algorithm that, given interval graphs G and H, decides whether H is isomorphic to an induced subgraph of G where an additional ordering condition had to be satisfied. This additional ordering condition seems to be necessary to obtain a polynomial-time algorithm when both G and H are interval graphs, as without the ordering condition the problem is NP-complete by the results of Marx and Schlotter [7].

In this section, we show that INDUCED SUBGRAPH ISOMORPHISM is polynomial-time solvable if G is an interval graph and H is a connected proper interval graph. We will simply apply Algorithm IIS for deciding the question. Part of the input for this algorithm are clique paths. Our decision problem can be solved by trying all possible combinations of clique paths. Interval graphs can have many clique paths, which would result in a worst-case exponential-time algorithm. Connected proper interval graphs, however, have at most two clique paths [10]. For our algorithm, it will be of high importance that the clique path for G can be chosen arbitrarily.

Theorem 1. *Given an interval graph G and a connected proper interval graph H, it can be decided in $\mathcal{O}(n^6)$ time whether G has an induced subgraph isomorphic to H.*

For complementing the result of Theorem 1, we consider the case when input graph H disconnected. It can be shown that any isomorphism from H to an induced subgraph of given graph G maps the vertices of connected components consecutively with respect to any interval ordering for G. With this result and the algorithm of Theorem 1, the induced subgraph isomorphism problem can be solved in polynomial time when the order of the connected components of H is fixed with respect to an interval ordering for G. This implies the following result.

Theorem 2. *Given an interval graph G on n vertices and a proper interval graph H with r connected components, it can be decided in $\mathcal{O}(r! \cdot rn^6 \log n)$ time whether G has an induced subgraph isomorphic to H.*

Hence, when G is an interval graph and H is a proper interval graph, ISI is fixed-parameter tractable when parametrised by the number of connected components of H.

4   Induced Subgraph Isomorphism Is NP-Complete on Proper Interval Graphs

In this section, we will show that the algorithms obtained in the previous section can be considered optimal: If the order of the connected components of H is not fixed then the problem becomes NP-complete already when both G and H are proper interval graphs and G is connected. We will obtain the completeness result by a reduction from a variant of the HAMILTONIAN PATH problem.

Theorem 3 ([11]). *The* FIXED HAMILTONIAN PATH *problem, given a graph G and a vertex pair u, v of G, to decide whether G has a u, v-path that is Hamiltonian, is NP-complete.*

Let G be a graph and let u, v be a vertex pair of G. We will construct a graph pair (F, H) such that F and H are proper interval graphs and H is isomorphic to an induced subgraph of F if and only if there is a u, v-path in G that is Hamiltonian. Let $u_1, \ldots, u_n$ be the vertices of G. Without loss of generality, we will assume $u_1 = u$ and $u_n = v$. The main idea of the construction is that a u, v-path of G that is Hamiltonian is a sequence of $n - 1$ edges where consecutive pairs are adjacent. Our two graphs will have the following tasks:

- F provides a list of all edges of G and a means for checking whether $n - 1$ selected edges form a sequence of the desired type
- H provides a mechanism for selecting $n - 1$ edges of G.

We begin with the construction of graph F. The graph is composed of subgraphs as shown in Figure 3. The figure shows two graphs, where the upper one is a *graph type*. The graph type has a complete graph on six vertices on its left end and a complete graph on seven vertices on its right end. The two complete graphs are joined by a graph that is a sequence of n triangles, then a chordless path between the vertices c and d and then another sequence of n triangles. The graphs of the depicted type differ from each other just in the length of the path between c and d. For an integer l with $1 - n \le l \le n - 1$, let M_l be the graph of the depicted graph type where the path between c and d has length $(8n^3 + 2) + (n + l - 1)(2n + 5)$. By N_l, we denote the induced subgraph of M_l that is obtained by deleting two vertices of minimum degree from each of the two complete graphs. So, N_l has a complete graph on four vertices at its left end, then a sequence of n triangles, a chordless path, another sequence of n triangles and finally a complete graph on five vertices. Now, let i, j be an integer pair where $1 \le i, j \le n$ and $i \ne j$. We define $F_{i,j}$ and $H_{i,j}$ as the following graphs: $F_{i,j} =_{\text{def}} M_{j-i} \setminus \left(\{a_1, \ldots, a_n, b_1, \ldots, b_n\} \setminus \{a_i, b_j\} \right)$ and $H_{i,j} =_{\text{def}} N_{j-i} \setminus \left(\{a_1, \ldots, a_n, b_1, \ldots, b_n\} \setminus \{a_i, b_j\} \right)$.

This means that the two complete graphs of $F_{i,j}$ and $H_{i,j}$ are connected by a long path that contains only two triangles, namely the ones formed with vertex a_i and with vertex b_j.

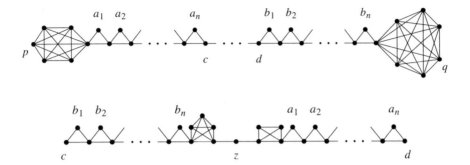

Fig. 3. Depicted on the upper part is a graph type, where vertices c and d are connected by a chordless path of arbitrary length. The graph type is used for the construction of graphs F and H. The lower part of the figure depicts a graph, that we call Q in the construction of F.

We consider the second (lower) graph in Figure 3; denote it by Q. We define an induced subgraph of Q for every vertex of G. Let $1 \le i \le n$: $Q_i =_{\text{def}}$ $Q \setminus \left(\{a_1, \ldots, a_n, b_1, \ldots, b_n\} \setminus \{a_i, b_i\} \right)$. We compound the graphs $Q_1, \ldots, Q_n$ to blocks, where we define three of them:

a) *middle block*
   join Q_i and Q_{i+1} by adding the edge between vertex d of Q_i and vertex c of Q_{i+1}, $1 \le i < n$
b) *start block*
   obtained from middle block by deleting all (remaining) vertices $a_2, \ldots, a_n$, $b_2, \ldots, b_n$ in $Q_2, \ldots, Q_n$
c) *end block*
   obtained from middle block by deleting all (remaining) vertices $a_1, \ldots, a_{n-1}$, $b_1, \ldots, b_{n-1}$ in $Q_1, \ldots, Q_{n-1}$.

Let B_1 and B_n be respectively a start and an end block, and let $B_2, \ldots, B_{n-1}$ be $n-2$ copies of the middle block. We denote by c_l^m, z_l^m, d_l^m the vertices c, z, d, respectively, of Q_m in block B_l. Let $P = P_{8n^3+1}$ be a chordless path on $8n^3 + 1$ vertices. One vertex of degree 1 of P is called *start vertex*, the other one is called *end vertex*. Obtain F^* from $B_1, \ldots, B_n$ and $n-1$ copies $P_1, \ldots, P_{n-1}$ of P first as the disjoint union of $B_1, \ldots, B_n$ and $P_1, \ldots, P_{n-1}$ and second adding the edge between vertex d_i^n and the start vertex of P_i and the edge between the end vertex of P_i and vertex c_{i+1}^1 for all $1 \le i < n$. For later arguments, it is important to observe that F^* is constructed from a long chordless path (with vertices c_1^1 and d_n^n as start and end vertex) by adding vertices, that are adjacent to exactly two adjacent vertices on the path.

We are ready for constructing graph F: F is the disjoint union of F^* and $F_{i,j}, F_{j,i}$ for every edge $u_i u_j \in E(G)$. Note that the number of connected components of F is $1+2|E(G)|$, since every edge of G is related to two connected components of F.

We continue with the construction of graph H. We define two new graphs, S and T:

s) S has a complete graph on six vertices that is connected to a sequence of n triangles; S is the induced subgraph of the upper graph type depicted in Figure 3 from the complete graph on the left hand side to vertex c.

t) T has a complete graph on seven vertices that is connected to a sequence of n triangles; T is the induced subgraph of the upper graph type depicted in Figure 3 from the complete graph on the right hand side to vertex d.

For each $1 \leq i \leq n$, we define S_i and T_i as follows: $S_i =_{\text{def}} S \setminus \left(\{a_1, \ldots, a_n\} \setminus \{a_i\} \right)$ and $T_i =_{\text{def}} T \setminus \left(\{b_1, \ldots, b_n\} \setminus \{b_i\} \right)$. With theses definitions, H is the disjoint union of $S_1, \ldots, S_{n-1}$ and $T_2, \ldots, T_n$ and $H_{i,j}, H_{j,i}$ for every $u_i u_j \in E(G)$. Note that the number of connected components of H is $2(n-1)+2|E(G)|$.

Lemma 4. *G has a u, v-path that is Hamiltonian if and only if H is isomorphic to an induced subgraph of F.*

Theorem 4. INDUCED SUBGRAPH ISOMORPHISM *is NP-complete when both input graphs are proper interval.*

As a final remark, we want to point out that we can make F connected without changing more of the construction of F and H. We would modify F by connecting the connected components of F by chordless paths of length $8n^3 + 2n(2n+5) = 8n^3 + 4n^2 + 10n$. Thus, ISI is NP-complete even when input graph G is connected proper interval.

5   Conclusion

Concluding from our results, we can summarise the knowledge on the tractability of INDUCED SUBGRAPH ISOMORPHISM when input graph G is an interval graph as follows:

- If H is interval, it is NP-complete and $W[1]$-hard ([7]).
- If H is a disconnected proper interval, it is NP-complete and fixed parameter tractable (this paper).
- If H is connected proper interval, it is polynomial-time solvable (this paper).

We would like to conclude with a couple of questions: (1) What is the computational complexity of ISI when G is a chordal graph and H is a connected proper interval graph? (2) For which subclasses C of proper interval graphs does ISI become polynomial-time solvable when G is interval and H is disconnected and belongs to C?

References

1. Booth, K.S., Lueker, G.S.: A linear time algorithm for deciding interval graph isomorphism. Journal of the ACM 26, 183–195 (1979)
2. Cai, L., Chan, S.M., Chan, S.O.: Random separation: A new method for solving fixed-cardinality optimization problems. In: Bodlaender, H.L., Langston, M.A. (eds.) IWPEC 2006. LNCS, vol. 4169, pp. 239–250. Springer, Heidelberg (2006)
3. Eppstein, D.: Subgraph isomorphism in planar graphs and related problems. J. Graph Algorithms Appl. 3(3), 1–27 (1999)
4. Fulkerson, D.R., Gross, O.A.: Incidence matrices and interval graphs. Pacific Journal of Mathematics 15, 835–855 (1965)
5. Garey, M.R., Johnson, D.S.: Computers and Intractability: A Guide to the Theory of NP-completeness. W. H. Freeman & Co., New York (1979)
6. Looges, P.J., Olariu, S.: Optimal greedy algorithms for indifference graphs. Computers & Mathematics with Applications 25, 15–25 (1993)
7. Marx, D., Schlotter, I.: Cleaning Interval Graphs. arXiv:1003.1260v1 (2010)
8. Matula, D.W.: Subtree isomorphism in $O(n^{5/2})$. Ann. Discrete Math. 2, 91–106 (1978)
9. Olariu, S.: An optimal greedy heuristic to color interval graphs. Information Processing Letters 37, 21–25 (1991)
10. Roberts, F.S.: Indifference Graphs. In: Proof Techniques in Graph Theory, pp. 139–146. Academic Press, New York (1969)
11. Sipser, M.: Introduction to the Theory of Computation. International Thomson Publishing (1996)

Testing Simultaneous Planarity When the Common Graph Is 2-Connected

Bernhard Haeupler[1], Krishnam Raju Jampani[2], and Anna Lubiw[2]

[1] CSAIL, Dept. of Computer Science,
Massachusetts Institute of Technology, Cambridge, MA 02139
haeupler@mit.edu
[2] David R. Cheriton School of Computer Science,
University of Waterloo, Waterloo, ON, Canada, N2L 3G1
{krjampan,alubiw}@uwaterloo.ca

Abstract. Two planar graphs G_1 and G_2 sharing some vertices and edges are *simultaneously planar* if they have planar drawings such that a shared vertex [edge] is represented by the same point [curve] in both drawings. It is an open problem whether simultaneous planarity can be tested efficiently. We give a linear-time algorithm to test simultaneous planarity when the two graphs share a 2-connected subgraph. Our algorithm extends to the case of k planar graphs where each vertex [edge] is either common to all graphs or belongs to exactly one of them.

Keywords: Simultaneous Embedding, Planar Graph, PQ Tree, Graph Drawing.

1 Introduction

Let $G_1 = (V_1, E_1)$ and $G_2 = (V_2, E_2)$ be two graphs sharing some vertices and edges. The simultaneous planar embedding problem asks whether there exist planar embeddings for G_1 and G_2 such that, in the two embeddings, each vertex $v \in V_1 \cap V_2$ is mapped to the same point and each edge $e \in E_1 \cap E_2$ is mapped to the same curve. We show that this problem can be solved efficiently when the common graph $(V_1 \cap V_2, E_1 \cap E_2)$ is 2-connected.

The study of planar graphs has a long history and has generated many deep results [24,23,25]. There is hope that some of the structure of planarity may carry over to simultaneous planarity. A possible analogy is with matroids, where optimization results carry over from one matroid to the intersection of two matroids [7]. On a more practical note, simultaneous planar embeddings are valuable for visualization purposes when two related graphs need to be displayed. For example, the two graphs may represent different relationships on the same node set, or they may be the "before" and "after" versions of a graph that has changed over time.

Over the last few years there has been a lot of work on simultaneous planar embeddings, which have been called "simultaneous embeddings with fixed

O. Cheong, K.-Y. Chwa, and K. Park (Eds.): ISAAC 2010, Part II, LNCS 6507, pp. 410–421, 2010.

edges" [8,9,12,13,14,15,21]. We mention a few results here and give a more detailed description in the Background section below. A major open question is whether simultaneous planarity of two graphs can be tested in polynomial time. The problem seems to be right on the feasibility boundary. The problem is NP-complete for three graphs [15] and the version where the planar drawings are required to be straight-line is already NP-hard for two graphs and only known to lie in PSPACE [10]. On the other hand several classes of (pairs of) graphs are known to always have simultaneous planar embeddings [9,8,14,13,21] and there are efficient algorithms to test simultaneous planarity for some very restricted graph-classes: biconnected outerplanar graphs [13], and the case where one graph has at most one cycle [12].

This paper shows how to efficiently test simultaneous planarity of any two graphs that share a 2-connected subgraph and thus greatly extends the classes of graph pairs for which a testing algorithm is known. Our algorithm builds on the planarity testing algorithm of Haeupler and Tarjan [17], which in turn unifies the planarity testing algorithms of Lempel-Even-Cederbaum [22], Shih-Hsu [27] and Boyer-Myrvold [5].

The paper is organized as follows: Section 1.1 gives more background and related work. In Section 2 we review and develop some techniques for PQ-trees, which are needed for our algorithm. We present our algorithm for simultaneous planarity in Section 3. We also show that our algorithm can be extended to solve a generalization of simultaneous planarity for k graphs, whose common graph is 2-connected.

1.1   Background

Versions of simultaneous planarity have received much attention in recent years. Brass et al. [6] introduced the concept of *simultaneous geometric embeddings* of a pair of graphs—these are planar straight-line drawings such that any common vertex is represented by the same point. Note that a common edge will necessarily be represented by the same line segment. It is NP-hard to test if two graphs have simultaneous geometric embeddings [10]. For other work on simultaneous geometric embeddings see [2] and its references.

The generalization to planar drawings where edges are not necessarily drawn as straight line segments, but any common edge must be represented by the same curve was introduced by Erten and Kobourov [9] and called *simultaneous embedding with consistent edges*. Most other papers follow the conference version of Erten and Kobourov's paper and use the term *simultaneous embedding with fixed edges (SEFE)*. In our paper we use the more self-explanatory term "simultaneous planar embeddings." A further justification for this nomenclature is that there are combinatorial conditions on a pair of planar embeddings that are equivalent to simultaneous planarity. Specifically, Jünger and Schultz give a characterization in terms of "compatible embeddings" [Theorem 4 in [21]]. Specialized to the case where the common graph is connected, their result says that two planar embeddings are simultaneously planar if and only if the cyclic orderings of common edges around common vertices are the same in both embeddings.

Several papers [9,8,14] show that pairs of graphs from certain restricted classes always have simultaneous planar embeddings, the most general result being that any planar graph has a simultaneous planar embedding with any tree [14]. On the other hand, there is an example of two outerplanar graphs that have no simultaneous planar embedding [14].

The graphs that have simultaneous planar embeddings when paired with any other planar graph have been characterized [13]. In addition, Jünger and Schultz [21] characterize the common graphs that permit simultaneous planar embeddings no matter what pairs of planar graphs they occur in. There are efficient algorithms to test simultaneous planarity for biconnected outerplanar graphs [13] and for a pair consisting of a planar graph and a graph with at most one cycle [12].

Simultaneously and independently of this work Angelini et al. [1] showed how to test simultaneous planarity of two graphs when the common graph is 2-connected. Their algorithm is based on SPQR-trees, takes $O(n^3)$ time and is restricted to the case where the two graphs have the same vertex set.

There is another, even weaker form of simultaneous planarity, where common vertices must be represented by common points, but the planar drawings are otherwise completely independent, with edges drawn as Jordan curves. Any set of planar graphs can be represented this way by virtue of the result that a planar graph can be drawn with any fixed vertex locations [26].

The idea of "simultaneous graph representations" has also been applied to intersection representations [19,18].

2   PQ-Trees

Many planarity testing algorithms in the literature use PQ-trees (or a variation) to obtain a linear-time implementation. PQ-trees were discovered by Booth and Lueker [4] and are used, not only for planarity testing, but for many other applications like recognizing interval graphs or testing matrices for the consecutive-ones property. We first review PQ-trees and then in Subsection 2.1 we show how to manipulate pairs of PQ-trees.

A PQ-tree represents the permutations of a set of elements satisfying a family of constraints. Each constraint specifies that a certain subset of elements must appear consecutively in any permutation. The leaves of a PQ-tree correspond to the elements of the set, and internal nodes are labeled 'P' or 'Q', and are drawn using a circle or a rectangle, respectively. PQ-trees are equivalent under arbitrary reorderings of the children of a P-node and reversals of the order of children of a Q-node. We consider a node with two children to be a Q-node. A leaf-order of a PQ-tree is the order in which its leaves are visited in an in-order traversal of the tree. The set of permutations represented by a PQ-tree is the set of leaf-orders of equivalent PQ-trees. Given a PQ-tree tree T on a set U of elements, adding a consecutivity constraint on a set $S \subseteq U$, *reduces* T to a PQ-tree T', such that the leaf-orders of T' are precisely the leaf-orders of T in which the elements of S appear consecutively. Booth and Lueker [4] gave an efficient implementation of PQ-trees that supports this operation in amortized $O(|S|)$ time.

Although PQ-trees were invented to represent linear orders, they can be reinterpreted to represent circular orders as well [17]: Given a PQ-tree we imagine that there is a new special leaf s attached as the "parent" of the root. A circular leaf order of the augmented tree is a circular order that begins at the special leaf, followed by a linear order of the remaining PQ-tree and ending at the special leaf. Again a PQ-tree represents all circular leaf-orders of equivalent PQ-trees. It is easy to see that a consecutivity constraint on such a set of circular orders directly corresponds to a consecutivity constraint on the original set of linear leaf-orders. Note that using PQ-trees for circular orders requires solely this different view on PQ-trees but does not need any change in their implementation.

2.1   Intersection and Projection of PQ-Trees

In this section we develop simple techniques to obtain consistent orders from two PQ-trees. More precisely when two PQ-trees share some but not (necessarily) all leaves, we want to find a permutation represented by each of them with a consistent ordering on the shared leaves. The idea for this is to first *project* both PQ-trees to the common elements, intersect the resulting PQ-trees, pick one remaining order and finally "lift" this order back. We now describe the individual steps of this process in more detail.

The *projection* of a PQ-tree on a subset of its leaves S is a PQ-tree obtained by deleting all elements not in S and simplifying the resulting tree. Simplifying a tree means that we (recursively) delete any internal node that has no children, and delete any node that has a single child by making the child's grandparent become its parent. This can easily be implemented in linear time.

Given two PQ-trees on the same set of leaves (elements) we define their *intersection* to be the PQ-tree T that represents exactly all orders that are leaf-orders in both trees. This intersection can be computed as follows.

1. Initialize T to be the first PQ-tree.
2. For each P-node in the second PQ-tree, reduce T by adding a consecutivity constraint on all its descendant leaves.
3. For each Q-node in the second tree, and for each pair of adjacent children of it, reduce T by adding a consecutivity constraint on all the descendant leaves of the two children.

Using the efficient PQ-tree implementation such an intersection can be computed in time linear in the size of the two PQ-trees (see Booth's thesis [3]).

These two operations are enough to achieve our goal. Given two PQ-trees T_1 and T_2 defined on different element (leaf) sets, we define S to be the set of common elements. Now we first construct the projections of both PQ-trees on S and then compute their intersection T as described above. Any permutation of S represented by T can now easily be "lifted" back to permutations of T_1 and of T_2 that respect the chosen ordering of S. Furthermore, any two permutations of T_1 and T_2 that are consistent on S can be obtained this way.

We note that techniques to "merge" PQ trees were also presented by Jünger and Leipert [20] in work on level planarity. Their merge is conceptually and technically different from ours in that the result of their merge is a single PQ tree whereas our structure of two orderings that are consistent on common elements cannot be captured by a single PQ tree.

3   Planarity

In this Section, we review the recent algorithm of Haeupler and Tarjan [17] for testing the planarity of a graph. Next we extend it to an algorithm for testing simultaneous planarity. We first begin with some basic definitions.

Let $G = (V, E)$ be a graph on vertex set $V = \{v_1, \cdots, v_n\}$ and let $\mathcal{O}$ be an ordering of the vertices of V. An edge $v_i v_j$ is an *in-edge* of v_i (in $\mathcal{O}$) if v_j appears before v_i in $\mathcal{O}$, and $v_i v_j$ is an *out-edge* of v_i if v_j appears after v_i in $\mathcal{O}$.

An st-ordering of G is an ordering $\mathcal{O}$ of the vertices of G, such that the first vertex of $\mathcal{O}$ is adjacent to the last vertex of $\mathcal{O}$ and every intermediate vertex has an in-edge and an out-edge. It is well-known that G has an st-ordering if and only if it is 2-connected. Further, an st-ordering can be computed in linear time [11].

A *combinatorial embedding* of G, denoted by $\mathcal{C}(G)$, is defined as a clockwise circular ordering of the incident edges of v_i, for each $i \in \{1, \cdots, n\}$, with respect to a planar drawing of G. If $\mathcal{C}$ is a combinatorial embedding of G, we use $\mathcal{C}(v_i)$ to denote the clockwise circular ordering of edges incident with v_i in $\mathcal{C}$.

3.1   Planarity Testing Using PQ-Trees

Let $G = (V, E)$ be a connected graph. The planarity testing algorithm of Haeupler and Tarjan embeds vertices (and their edges) one at a time and maintains all possible partial embeddings of the embedded subgraph at each stage. For the correctness of the algorithm it is crucial that the vertices are added in a leaf-to-root order of a spanning-tree. This guarantees that the remaining vertices induce a connected graph and hence lie in a single face of the partial embedding at any time. Using either a leaf-to-root order of a depth-first spanning tree or an st-order leads to particularly simple implementations that run in linear-time. Indeed these two orders are essentially the only two orders in which the algorithm runs in linear-time using the standard PQ-tree implementation. Our algorithm uses a mixture of the two orders: We first add the vertices that are contained in only one of the graphs using a depth-first search order and then add the common vertices using an st-ordering. We now give an overview of how the simple planarity test works for each of these orderings.

st-order:
Let $v_1, v_2, \cdots, v_n$ be an st-order of G. At any stage $i \in \{1, \cdots, n-1\}$ the vertices $\{v_1, \cdots, v_i\}$ form a connected component and the algorithm maintains all possible circular orderings of out-edges around this component using a PQ-tree T_i. Since $v_1 v_n$ is an out-edge at every stage, it can stay as the special leaf of T_i for all i. At stage 1, the tree T_1 consists of the special leaf $v_1 v_n$ and a P-node whose children are all other out-edges of v_1.

Now suppose we are at a stage $i \in \{1, \cdots, n-2\}$. We call the set of leaves of T_i that correspond to edges incident to v_{i+1}, the *black* leaves. To go to the next stage, we first reduce T_i so that all the black edges appear together. A non-leaf node in the reduced PQ-tree is said to be black if all its descendants are black edges. We next create a new P-node p_{i+1} and add all the out-edges of v_{i+1} as its children. Now T_{i+1} is constructed from T_i as follows:

Case 1: T_i *contains a black node x that is an ancestor of all the black leaves.* We obtain T_{i+1} from T_i by replacing x and all its descendants with p_{i+1}.

Case 2: T_i *contains a (non-black) Q-node containing a (consecutive) sequence of black children.* We obtain T_{i+1} from T_i by replacing these black children (and their descendants) with p_{i+1}.

Note that if the reduction step fails at any stage then the graph must be non-planar. Otherwise the algorithm concludes that the graph is planar.

Leaf-to-root order of a depth-first spanning tree:

Let $v_1, v_2, \cdots, v_n$ be a leaf-to-root order of a depth-first spanning tree of G. Note that at stage i, the vertices $\{v_1, \cdots, v_i\}$ may induce several components. We maintain a PQ-tree for each component representing the set of circular orderings of its out-edges. Using a depth-first spanning tree, in contrast to an arbitrary spanning tree, has the advantage that we can easily maintain the invariant that the edge to the smallest node greater than i will be the special leaf. Adding v_{i+1} can lead to merging several components into one.

To go to the next stage, we first reduce each PQ-tree corresponding to such a component by adding a consecutivity constraint that requires the set of out-edges that are incident to v_{i+1} to be consecutive and then deleting these edges. By the invariant stated above the special leaf is among these edges. Note that the resulting PQ-tree for a component now represents the set of linear-orders of the out-edges that are not incident to v_{i+1}. Now we construct the PQ-tree for the new merged component including v_{i+1} as follows:

Let v_l be the parent of v_{i+1} in the depth-first spanning tree. The PQ-tree for the new component consists of the edge $v_{i+1}v_l$ as the special leaf and a new P-node as a root and whose children are all the remaining out-edges of v_{i+1} and the roots of the PQ-trees of the reduced components (similar to the picture in Figure 1). Note that by choosing the edge $v_{i+1}v_l$ as the special leaf we again maintain the above mentioned invariant.

As before, if the reduction step fails for any component, then the graph is non-planar. Otherwise the algorithm concludes that the graph is planar.

3.2   Simultaneous Planarity

Let $G_1 = (V_1, E_1)$ and $G_2 = (V_2, E_2)$ be two planar connected graphs with $|V_1| = n_1$ and $|V_2| = n_2$. Let $G = (V_1 \cap V_2, E_1 \cap E_2)$ be 2-connected and $n = |V_1 \cap V_2|$. Let $v_1, v_2, \cdots, v_n$ be an st-ordering of $V_1 \cap V_2$. We call the edges and vertices of G *common* and all other vertices and edges *private*.

We say two linear or circular orderings of elements with some common elements are *compatible* if the common elements appear in the same relative order in both orderings. Similarly we say two combinatorial embeddings of G_1 and G_2 respectively are compatible if for each common vertex the two circular orderings of edges incident to it are compatible.

If G_1 and G_2 have simultaneous planar embeddings, they have combinatorial embeddings that are compatible with each other. If the common edges form a connected graph the converse is also true and is a special case of Theorem 4 of Jünger and Schultz [21]. Thus it is enough to compute a pair of compatible combinatorial embeddings.

We will find compatible combinatorial embeddings by adding vertices one by one, iteratively constructing two sets of PQ-trees, representing the partial planar embeddings of G_1 and of G_2 respectively. Each PQ-tree represents one connected component of G_1 or G_2. In the first phase we will add all private vertices of G_1 and G_2, and in the second phase we will add the common vertices in an st-ordering. When a common vertex is added, it will appear in two PQ-trees, one for G_1 and one for G_2 and we must take care to maintain compatibility.

Before describing the two phases, we give the main idea of maintaining compatibility between two PQ-trees. In Section 2.1 we found compatible orders for two PQ-trees using projection and intersection of PQ-trees, but we were unable to store a set of compatible orderings, which is what we really need, since planarity testing involves a sequence of PQ-trees.

To address this issue we introduce a Boolean "orientation" variable attached to each Q-node to encode whether it is ordered forward or backward. Compatibility is captured by equations relating orientation variables. At the conclusion of the algorithm, it is a simple matter to see if the resulting set of Boolean equations has a solution. If it does, we use the solution to create compatible orderings of the Q nodes of the two PQ-trees. Otherwise the graphs do not have simultaneous planar embeddings.

In more detail, we create a Boolean orientation variable $f(q)$ for each Q-node q, with the interpretation that $f(q) = \text{True}$ iff q has a "forward" ordering. We record the initial ordering of each Q-node in order to distinguish "forward" from "backward". During PQ-tree operations, Q-nodes may merge, and during planarity testing, parts of PQ-trees may be deleted. We handle these modifications to Q-nodes by the simple expedient of having an orientation variable for each Q-node, and equating the variables as needed. When Q-nodes q_1 and q_2 merge, we add the equation $f(q_1) = f(q_2)$ if q_1 and q_2 are merged in the same order (both forward or both backward), or $f(q_1) = \neg f(q_2)$ otherwise.

We now describe the two phases of our simultaneous planarity testing algorithm. To process the private vertices of G_1 and G_2 in the first phase we compute for each of them a reverse depth-first ordering by contracting G into a single vertex and then running a depth-first search from this vertex. With these orderings we can now run the algorithm of Haeupler and Tarjan for all private vertices as described in Section 3.1.

Now the processed vertices induce a collection of components, such that each component has an out-edge to a common vertex. Further, the planarity test provides us for each component with an associated PQ-tree representing all possible cyclic orderings of out-edges for that component. For each component we look at the out-edge that goes to the first common vertex in the st-order and re-root the PQ-tree for this component to have this edge represented by the special leaf. This completes the first phase.

For the second phase we insert the common vertices in an st-order. The algorithm is similar to that described in Section 3.1 for an st-order but in addition has to take care of merging in the private components as well. We first examine the procedure for a single graph. Adding the first common vertex v_1 is a special set-up phase; we will describe the general addition below. Adding v_1 joins some of the private components into a new component C_1 containing v_1.

For each of these private components we reduce the corresponding PQ-tree so that all the out-edges to v_1 appear together, and then delete those edges. Note that due to the re-rooting at the end of the first phase the special leaf is among those edges. Thus the resulting PQ-tree represents the linear orderings of the remaining edges. We now build a PQ-tree representing the circular orderings around the new component C_1 as follows: we take $v_1 v_n$ as the special leaf, create a new P-node as a root and add all the out-edges of v_1 and the roots of the PQ-trees of the merged private components as children of the root (see Fig. 1).

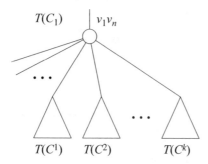

Fig. 1. Setting up $T(C_1)$. The P-node's children are the outgoing edges of v_1 and the PQ-trees for the components that are joined together by v_1.

Now consider the situation when we are about to add the common vertex v_i, $i \geq 2$. The graph so far may have many connected components but because of the choice of an st-ordering all common vertices embedded so far are in one component C_{i-1}, which we call the *main* component. When we add v_i, all components with out-edges to v_i join together to form the new main component C_i. This includes C_{i-1} and possibly some private components. The other private components do not change, nor do their associated PQ-trees.

We now describe how to update the PQ-tree T_{i-1} associated with C_{i-1} to form the PQ-tree T_i associated with C_i. This is similar to the approach described in

Section 3.1. We first reduce T_{i-1} so that all the *black* edges (the ones incident to v_i) appear together. As before, we call a non-leaf node in the reduced PQ-tree *black* if all its descendants are black leaves. For any private component with an out-edge to v_i, we reduce the corresponding PQ-tree so that all the out-going edges to v_i appear together, and then delete those edges. We make all the roots of the resulting PQ-trees into children of a new P-node p_i, and also add all the out-going edges of v_i as children of p_i. It remains to add p_i to T_{i-1} which we do as described below. In the process we also create a *black tree* J_i that represents the set of linear orderings of the black edges.

Case 1: T_{i-1} *contains a black node x such that all black edges are descendants of x.* Let J_i be the subtree rooted at x. We obtain T_i from T_{i-1} by replacing x and all its descendants with p_i.

Case 2: T_{i-1} *contains a non-black Q-node x that has a sequence of adjacent black children.* We group all the black children of x and add them as children (in the same order) of a new Q-node x'. Let J_i be defined as the subtree rooted at x'. We add an equation relating the orientation variables of x and x'. We obtain T_i from T_{i-1} by replacing the sequence of black children of x (and their descendants) with p_i (see Figure 2).

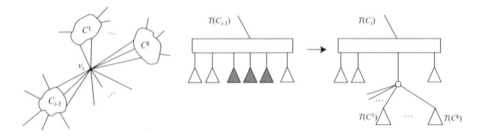

Fig. 2. (*left*) Adding vertex v_i which is connected to main component C_{i-1} and to private components $C^1 \ldots C^k$. (*right*) Creating T_i from T_{i-1} by replacing the black subtree by a P-node whose children are the outgoing edges of v_i and the PQ-trees for the newly joined private components.

Note that we use orientation variables above for a purpose other than compatibility. (We are only working with one graph so far). Standard planarity tests would simply keep track of the order of the deleted subtree J_i in relation to its parent. Since we have orientation variables anyway, we use them for this purpose.

We perform a similar procedure on graph G_2. We will distinguish the black trees of G_1 and G_2 using superscripts. Thus after adding v_i we have black trees J_i^1 and J_i^2. It remains to deal with compatibility. We claim that it suffices to enforce compatibility between each pair J_i^1 and J_i^2.

To do so, we perform a *unification step* in which we add equations between orientation variables for Q-nodes in the two trees.

Unification step for stage i

We first project J_i^1 and J_i^2 to the common edges, as described in Section 2.1, carrying over orientation variables from each original node to its copy in the projection. Next we create the PQ-tree R_i that is the intersection of these two projected trees as described in Section 2.1. Initially R_i is equal to the first tree. The step dealing with Q-nodes (Step 3) is enhanced as follows:

3. For each Q-node q of the second tree, and for each pair a_1, a_2 of adjacent children of q do the following: Reduce R_i by adding a consecutivity constraint on all the descendant leaves of a_1 and a_2. Find the Q-node that is the least common ancestor of the descendants of a_1 and a_2 in R_i. Add an equation relating the orientation variable of this ancestor with the orientation variable of q (using a negation if needed to match the orderings of the descendants).

Observe that any equations added during the unification step are necessary. Thus if the system of Boolean equations is inconsistent at the end of the algorithm, we conclude that G_1 and G_2 do not have a compatible combinatorial embedding. Finally, if the system of Boolean equations has a solution, then we obtain compatible leaf-orders for each pair J_i^1 and J_i^2 as follows: Pick an arbitrary solution to the system of Boolean equations. This fixes the truth values of all orientation variables and thus the orientations of all Q-nodes in all the trees. Subject to this, choose a leaf ordering I of R_i (by choosing the ordering of any P-nodes). I can then be lifted back to (compatible) leaf-orders of J_i^1 and J_i^2 that respect the ordering of I. The following Lemma shows that this is sufficient to obtain compatible combinatorial embeddings of G_1 and G_2.

Lemma 1. *If the system of Boolean equations has a solution then G_1 and G_2 have compatible combinatorial embeddings.*

Proof. The procedure described above produces compatible leaf orders for all pairs of black trees J_i^1 and J_i^2. Recall that the leaves of J_i^1 (resp. J_i^2) are the out-edges of the component C_{i-1} in G_1 (resp. G_2) and contain all the common in-edges of v_i. Focussing on G_1 individually, its planarity test has succeeded, and we have a combinatorial embedding such that the ordering of edges around v_i contains the leaf order of J_i^1. Also, we have a combinatorial embedding of G_2 such that the ordering of edges around v_i contains the leaf order of J_i^2.

The embedding of a graph imposes an ordering of the out-edges around every main component. We can show inductively, starting from $i = n$, that the ordering of the out-edges around the main component C_{i-1} in G_1 is compatible with the ordering of the out-edges in the corresponding main component in G_2. Moreover all the common edges incident to v_i, belong to either C_{i-1} or C_i. This implies that in both embeddings, the orderings of edges around any common vertex are compatible. Therefore G_1 and G_2 have compatible combinatorial embeddings.

A generalization of simultaneous planarity to k graphs. Consider a generalization of simultaneous planarity for k graphs, when each vertex [edge] is either present in all the graphs or present in exactly one of them. Our algorithm of section 3.2 can be readily extended to solve this generalized version, when the common graph is 2-connected (see the full paper [16] for more details).

3.3 Running Time

We show that our algorithm can be implemented to run in linear time. Computing the reverse depth-first ordering and the st-ordering are known to be feasible in linear-time [11]. The first phase of our algorithm uses PQ-tree based planarity testing with a reverse depth-first search order [17], which runs in linear time using the efficient PQ-tree implementation of Booth and Lueker [3,4]. The re-rooting between the two phases needs to be done only once and can easily be done in linear time. The second phase of our algorithm uses PQ-tree based planarity testing with an st-order, as discussed in Section 3.1. This avoids re-rooting of PQ-trees, and thus also runs in linear time [17,4,22]. The other part of the second phase is the unification step, which is only performed on the black trees, i.e. the edges connecting to the current vertex. Thus we can explicitly store the black trees and the intersection tree at every stage and allow the unification step to take time linear in the complete size of both black trees. The intersection algorithm needs to be implemented with a little bit more care but again, using the standard PQ-tree implementation and the intersection algorithm described in Booth's thesis [3], linear time is possible. The last thing that needs to be implemented efficiently is the handling of the orientation variables. Note that a Q-node is implemented as a doubly-linked list of its children [4]. By storing the variable and orientation information of a Q-node in one of the links of the doubly-linked list and generating equations lazily at the end of each unification step, we can generate all variable equations in linear time (see full paper for details). Once generated, the equations can clearly be solved in linear time.

Acknowledgment. We want to thank Bob Tarjan. Some of the ideas used here go back to joint research that was done for a complete version of [17].

References

1. Angelini, P., Di Battista, G., Frati, F., Patrignani, M., Rutter, I.: Testing the simultaneous embeddability of two graphs whose intersection is a biconnected graph or a tree. In: IWOCA. LNCS (2010)
2. Angelini, P., Geyer, M., Kaufmann, M., Neuwirth, D.: On a tree and a path with no geometric simultaneous embedding. In: CoRR, abs/1001.0555 (2010)
3. Booth, K.: PQ Tree Algorithms. PhD thesis, University of California, Berkeley (1975)
4. Booth, K., Lueker, G.: Testing for the consecutive ones property, interval graphs, and graph planarity using pq-tree algorithms. Journal of Computer and System Sciences 13, 335–379 (1976)
5. Boyer, J., Myrvold, W.: On the cutting edge: Simplified O(n) planarity by edge addition. Journal of Graph Algorithms and Applications 8(3), 241–273 (2004)
6. Brass, P., Cenek, E., Duncan, C., Efrat, A., Erten, C., Ismailescu, D., Kobourov, S.G., Lubiw, A., Mitchell, J.: On simultaneous planar graph embeddings. Computational Geometry: Theory and Applications 36(2), 117–130 (2007)
7. Cook, W.J., Cunningham, W.H., Pulleyblank, W.R., Schrijver, A.: Combinatorial Optimization. Wiley Interscience, Hoboken (1997)

8. DiGiacomo, G., Liotta, G.: Simultaneous embedding of outerplanar graphs, paths, and cycles. International Journal of Computational Geometry and Applications 17(2), 139–160 (2007)

9. Erten, C., Kobourov, S.G.: Simultaneous embedding of planar graphs with few bends. Journal of Graph Algorithms and Applications 9(3), 347–364 (2005)

10. Estrella-Balderrama, A., Gassner, E., Junger, M., Percan, M., Schaefer, M., Schulz, M.: Simultaneous geometric graph embeddings. In: Hong, S.-H., Nishizeki, T., Quan, W. (eds.) GD 2007. LNCS, vol. 4875, pp. 280–290. Springer, Heidelberg (2008)

11. Even, S., Tarjan, R.: Computing an st-Numbering. Theor. Comput. Sci. 2(3), 339–344 (1976)

12. Fowler, J., Gutwenger, C., Junger, M., Mutzel, P., Schulz, M.: An SPQR-tree approach to decide special cases of simultaneous embedding with fixed edges. In: Tollis, I.G., Patrignani, M. (eds.) GD 2008. LNCS, vol. 5417, pp. 1–12. Springer, Heidelberg (2009)

13. Fowler, J., Jünger, M., Kobourov, S.G., Schulz, M.: Characterizations of restricted pairs of planar graphs allowing simultaneous embedding with fixed edges. In: Broersma, H., Erlebach, T., Friedetzky, T., Paulusma, D. (eds.) WG 2008. LNCS, vol. 5344, pp. 146–158. Springer, Heidelberg (2008)

14. Frati, F.: Embedding graphs simultaneously with fixed edges. In: Kaufmann, M., Wagner, D. (eds.) GD 2006. LNCS, vol. 4372, pp. 108–113. Springer, Heidelberg (2007)

15. Gassner, E., Junger, M., Percan, M., Schaefer, M., Schulz, M.: Simultaneous graph embeddings with fixed edges. In: Fomin, F.V. (ed.) WG 2006. LNCS, vol. 4271, pp. 325–335. Springer, Heidelberg (2006)

16. Haeupler, B., Jampani, K.R., Lubiw, A.: Testing simultaneous planarity when the common graph is 2-connected (2010)

17. Haeupler, B., Tarjan, R.E.: Planarity algorithms via PQ-trees (extended abstract). Electronic Notes in Discrete Mathematics 31, 143–149 (2008)

18. Jampani, K.R., Lubiw, A.: The simultaneous representation problem for chordal, comparability and permutation graphs. In: WADS. LNCS, vol. 5664, pp. 387–398. Springer, Heidelberg (2009)

19. Jampani, K.R., Lubiw, A.: Simultaneous interval graphs (2010) (submitted)

20. Jünger, M., Leipert, S.: Level planar embedding in linear time. J. Graph Algorithms Appl. 6(1), 67–113 (2002)

21. Jünger, M., Schulz, M.: Intersection graphs in simultaneous embedding with fixed edges. Journal of Graph Algorithms and Applications 13(2), 205–218 (2009)

22. Lempel, A., Even, S., Cederbaum, I.: An algorithm for planarity testing of graphs. In: Rosenstiehl, P. (ed.) Theory of Graphs: International Symposium, pp. 215–232 (1967)

23. Mohar, B., Thomassen, C.: Graphs on Surfaces. Johns Hopkins University Press, Baltimore (2001)

24. Nishizeki, T., Chiba, N.: Planar graphs: theory and algorithms. Elsevier, Amsterdam (1988)

25. Nishizeki, T., Rahman, M.S.: Planar graph drawing. World Scientific, Singapore (2004)

26. Pach, J., Wenger, R.: Embedding planar graphs at fixed vertex locations. Graphs and Combinatorics 17(4), 717–728 (2001)

27. Shih, W.K., Hsu, W.-L.: A new planarity test. Theoretical Computer Science 223(1-2), 179–191 (1999)

Computing the Discrete Fréchet Distance with Imprecise Input

Hee-Kap Ahn[1], Christian Knauer[2], Marc Scherfenberg[2],
Lena Schlipf[3], and Antoine Vigneron[4]

[1] Department of Computer Science and Engineering, POSTECH, Pohang, Korea
heekap@postech.ac.kr
[2] Institute of Computer Science, Universität Bayreuth, 95440 Bayreuth, Germany
{Christian.knauer,marc.scherfenberg}@uni-bayreuth.de
[3] Institute of Computer Science, Freie Universität Berlin, Germany
schlipf@mi.fu-berlin.de
[4] INRA, UR 341 Mathématiques et Informatique Appliquées,
78352 Jouy-en-Josas, France
antoine.vigneron@jouy.inra.fr

Abstract. We consider the problem of computing the discrete Fréchet distance between two polygonal curves when their vertices are imprecise. An imprecise point is given by a region and this point could lie anywhere within this region. By modelling imprecise points as balls in dimension d, we present an algorithm for this problem that returns in time $2^{O(d^2)} m^2 n^2 \log^2(mn)$ the Fréchet distance lower bound between two imprecise polygonal curves with n and m vertices, respectively. We give an improved algorithm for the planar case with running time $O(mn \log^2(mn) + (m^2 + n^2) \log(mn))$. In the d-dimensional orthogonal case, where points are modelled as axis-parallel boxes, and we use the L_∞ distance, we give an $O(dmn \log(dmn))$-time algorithm.

We also give efficient $O(dmn)$-time algorithms to approximate the Fréchet distance upper bound, as well as the smallest possible Fréchet distance lower/upper bound that can be achieved between two imprecise point sequences when one is allowed to translate them. These algorithms achieve constant factor approximation ratios in "realistic" settings (such as when the radii of the balls modelling the imprecise points are roughly of the same size).

1 Introduction

Shape matching is an important ingredient in a wide range of computer applications such as computer vision, computer–aided design, robotics, medical imaging, and drug design. In shape matching, we are given two geometric objects and we compute their distance according to some geometric similarity measure. The Fréchet distance is a natural distance function for continuous shapes such as curves and surfaces, and is defined using reparameterizations of the shapes [3,4,5,16].

O. Cheong, K.-Y. Chwa, and K. Park (Eds.): ISAAC 2010, Part II, LNCS 6507, pp. 422–433, 2010.
© Springer-Verlag Berlin Heidelberg 2010

The discrete Fréchet distance is a variant of the Fréchet distance in which we only consider vertices of polygonal curves. In dimension d, given two polygonal curves with n and m vertices, respectively, there is a dynamic programming algorithm that computes the discrete Fréchet distance between them in $\Theta(dmn)$ time [9]. Later, Aronov et al. [6] presented efficient approximation algorithms for computing the discrete Fréchet distance of two natural classes of curves: κ-bounded curves and backbone curves. They also proposed a pseudo-output-sensitive algorithm for computing the discrete Fréchet distance exactly.

Most of previous works on the Fréchet distance assume that the input curves are given precisely. The input curve, however, could be only an approximation; In many cases, geometric data comes from measurements of continuous real-world phenomenons, and the measuring devices have finite precision. This impreciseness of geometric data has been studied lately, and quite a few algorithms that handle imprecise data have been given for fundamental geometric problems: for example, computing the Hausdorff distance [12], Voronoi diagrams [17], planar convex hulls [13], and Delaunay triangulations [11,14].

Imprecise data can be modelled in different ways. One possible model, for data that consists of points, is to assign each point to a region, typically a disk or a square. In this case, existing algorithms for computing the Fréchet distance could be too sensitive to the precision of the measurements, and they may return a solution without providing any guarantee on its correctness or preciseness. One solution to this problem is to take the impreciseness of the input into account in the design of algorithms, so that they return a solution with some additional information on its quality.

Our results. In this paper, we study the problem of computing the discrete Fréchet distance between two polygonal curves, where the vertices of a polygonal curve are imprecise. Each vertex belongs to a region, which is either a Euclidean ball or an axis-parallel box in $\mathbb{R}^d$. We consider two cases: the orthogonal case and the Euclidean case. In the orthogonal case, the regions are boxes, and we use the L_∞ distance. In the Euclidean case, the regions are balls and we use the Euclidean distance.

Typical applications of this problem include computing similarity of two spatio-temporal data sets such as polygonal trajectories of moving objects (e.g. cars, people, animals) whose vertex locations are obtained by some positioning services (e.g. the Global Positioning System), and therefore imprecise.

Given two imprecise sequences of n and m points, respectively, we give algorithms for computing the Fréchet distance lower bound between these two sequences. In the d-dimensional orthogonal case, our algorithm runs in time $O(dmn \log(dmn))$. In the Euclidean case, we give an $2^{O(d^2)} m^2 n^2 \log^2(mn)$-time algorithm for arbitrary dimension d, and we give an improved $O(mn \log^2(mn) + (m^2 + n^2) \log(mn))$-time algorithm in the plane.

We also give efficient $O(dmn)$-time algorithms to approximate the Fréchet distance upper bound, as well as the smallest possible Fréchet distance lower and upper bound that can be achieved between two imprecise point sequences when one is allowed to translate them. These algorithms achieve constant factor approximation ratios in realistic settings, such as when the radii of the balls modelling the imprecise points are roughly of the same size, or when any two consecutive imprecise points are well-separated (so that their imprecision regions do not overlap).

2   Notation and Preliminaries

We work in $\mathbb{R}^d$, and we use a metric $\text{dist}(\cdot, \cdot)$ which is either the Euclidean distance, or the L_∞ distance. Let $A = a_1, \ldots, a_n$ and $B = b_1, \ldots, b_m$ denote two sequences of points in $\mathbb{R}^d$. A *coupling* is a sequence of ordered pairs $(\alpha_1, \beta_1), \ldots, (\alpha_c, \beta_c)$ such that:

- $\alpha_1 = 1$, $\beta_1 = 1$, $\alpha_c = n$ and $\beta_c = m$.
- for each $1 \leqslant k < c$, one of the three statements below is true:
    - $\alpha_{k+1} = \alpha_k + 1$ and $\beta_{k+1} = \beta_k + 1$.
    - $\alpha_{k+1} = \alpha_k + 1$ and $\beta_{k+1} = \beta_k$.
    - $\beta_{k+1} = \beta_k + 1$ and $\alpha_{k+1} = \alpha_k$

The *discrete Fréchet distance* $\text{F}(A, B)$ is the minimum, over all couplings, of $\max_{1 \leqslant k \leqslant c} \text{dist}(a_{\alpha_k}, b_{\beta_k})$. (See Figure 1.)

In what follows, we consider the case where the two point-sequences A and B are *imprecise*. So, instead of knowing the position of each a_i, b_j, we are given two sequences of regions of $\mathbb{R}^d$ denoted by $H = h_1, \ldots, h_n$ and $V = v_1, \ldots, v_m$. These regions will be either Euclidean balls, or axis-aligned boxes. They specify where the points a_i, b_j may lie, and thus for each i, j, we have $a_i \in h_i$ and $b_j \in v_j$. For all $i \leqslant n$, we denote by H_i the subsequence $h_1, \ldots, h_i$, and for all $j \leqslant m$, we denote $V_j = v_1, \ldots, v_j$.

We will consider two different cases. In the *Euclidean case*, the regions are Euclidean balls in $\mathbb{R}^d$ and we use the Euclidean distance. In the *orthogonal case*, the regions are axis-aligned boxes and the distance we use is the L_∞ metric.

A *realization* of the region sequence H is a point sequence $A = a_1, \ldots, a_n$ such that $a_i \in h_i$ for all $1 \leqslant i \leqslant n$. Similarly, a realization of the region sequence V is a point sequence $B = b_1, \ldots, b_m$ such that $b_j \in v_j$ for all $1 \leqslant j \leqslant m$. We denote by $A \in_R H$ and $B \in_R V$ the fact that A is a realization of H, and B is a realization of V, respectively. When $A \in_R H$ and $B \in_R V$, we will say that (A, B) is a realization of (H, V). This will be denoted as $(A, B) \in_R (H, V)$.

Definition 1. *For two region sequences H and V, the* Fréchet distance lower bound $\text{F}^{\min}(H, V)$ *is the minimum, over all realizations (A, B) of (H, V), of the discrete Fréchet distance $\text{F}(A, B)$:*

$$\text{F}^{\min}(H, V) = \min_{(A,B) \in_R (H,V)} \text{F}(A, B).$$

The Fréchet distance upper bound $\text{F}^{\max}(H, V)$ *is the maximum, over all realizations (A, B) of (H, V), of the discrete Fréchet distance $\text{F}(A, B)$:*

$$\text{F}^{\max}(H, V) = \max_{(A,B) \in_R (H,V)} \text{F}(A, B).$$

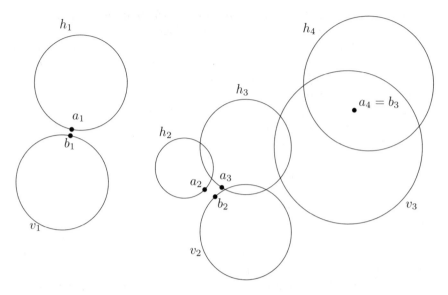

Fig. 1. The discrete Fréchet distance between the sequences $A = a_1, a_2, a_3, a_4$ and $B = b_1, b_2, b_3$ is achieved by the coupling $(1, 1), (2, 2), (3, 2), (4, 3)$, and we have $F(A, B) = \text{dist}(a_2, b_2) = \text{dist}(a_3, b_2)$. The sequences A and B are realizations of the sequences of regions $H = h_1, h_2, h_3, h_4$ and $V = v_1, v_2, v_3$, which is denoted by $(A, B) \in_R (H, V)$. The Fréchet distance lower bound $F^{\min}(H, V)$ is achieved by the realization (A, B), so we have $F^{\min}(H, V) = F(A, B)$.

3   Computing the Fréchet Distance Lower Bound $F^{\min}$

In this section, we give algorithms for computing $F^{\min}(H, V)$. We first give a decision algorithm that, given a real number $\delta \geqslant 0$, decides whether $F^{\min}(H, V) \leqslant \delta$. Then we give an improved decision algorithm for the Euclidean case. Based on these decision algorithms, we finally give optimization algorithms, which compute $F^{\min}(H, V)$ in the orthogonal case and in the Euclidean case.

We denote by h_i^δ (resp. v_j^δ) the set of points that are at distance at most δ from h_i (resp. v_j). In the Euclidean case, where h_i is a ball with radius r, the set h_i^δ is the concentric ball with radius $r + \delta$. In the orthogonal case, if $h_i = [x_1, y_1] \times \cdots \times [x_d, y_d]$, we have $h_i^\delta = [x_1 - \delta, y_1 + \delta] \times \cdots \times [x_d - \delta, y_d + \delta]$.

3.1   Decision Algorithm for the Orthogonal Case

Our decision algorithm is based on dynamic programming. In this sense, it is related to Eiter and Mannila's algorithm [9] for computing the discrete Fréchet distance, but we use additional invariants to address the impreciseness. These new invariants are carefully chosen *feasibility regions*, which indicate where the current points (a_i, b_j) may lie. Note that a straightforward discretization of the space of realizations of H, V would yield an exponential time bound, because one would have to consider the arrangement of nm surfaces in dimension $(m+n)d$ defined by the equation $\mathrm{dist}(a_i, b_j) \leqslant \delta$ for each pair i, j.

So in each cell of an array with n rows and m columns, we will store two feasibility regions $\mathrm{FH}_\delta(i,j) \subset \mathbb{R}^d$ and $\mathrm{FV}_\delta(i,j) \subset \mathbb{R}^d$. The ith row represents the region H_i, and the jth column represents V_j. We will compute these fields row by row, from $i=1$ to $i=n$.

Remember that A_i (resp. B_j) denotes the sequence $a_1, \ldots, a_i$ (resp. $b_1, \ldots, b_j$). As we shall see in Lemma 1, the feasibility region $\mathrm{FH}_\delta(i,j)$ represents the possible locations of a_i, where (A_i, B_j) is a realization of (H_i, V_j), and there exists a coupling that achieves $\mathrm{F}(A_i, B_j) \leqslant \delta$ whose last two pairs are not $(i-1,j), (i,j)$. The other feasibility region $\mathrm{FV}_\delta(i,j)$ represents the possible locations of b_j, when there is such a coupling whose last two pairs are not $(i, j-1), (i,j)$. Thus, the Fréchet distance lower bound $\mathrm{F}^{\min}(H_i, V_j)$ is more than δ if and only if both of these feasibility regions $\mathrm{FH}_\delta(i,j)$ and $\mathrm{FV}_\delta(i,j)$ are empty.

The pseudocode of our decision algorithm *DecideFréchetMin* is given below. Lines 1 to 8 initialize some of the fields of our array for the first row and column, as well as an extra zeroth column and row. It allows boundary cases when $i=1$ and $j=1$ to be handled correctly in the main loop. The main loop is from line 9 to 15. As we are in the orthogonal case, lines 12–15 consist in intersecting two axis-aligned boxes in dimension d. It can be done trivially in $O(d)$ time, so our algorithm runs in $O(dmn)$ time.

Algorithm *DecideFréchetMin*

Input: Two sequences of regions $H = h_1, \ldots, h_n$ and $V = v_1, \ldots, v_m$, and a value $\delta \geqslant 0$.

Output: TRUE when $\mathrm{F}^{\min}(H, V) \leqslant \delta$, and FALSE otherwise.

```
1.    for i ← 1 to n
2.           FH_δ(i,0) ← ∅
3.           FV_δ(i,0) ← ∅
4.    for j ← 1 to m
5.           FH_δ(0,j) ← ∅
6.           FV_δ(0,j) ← ∅
7.    FH_δ(0,0) ← ℝ^d
8.    FV_δ(0,0) ← ℝ^d
9.    for i ← 1 to n
10.          for j ← 1 to m
11.                 if FH_δ(i−1,j−1) = ∅ and FV_δ(i−1,j−1) = ∅
12.                    then FH_δ(i,j) ← FH_δ(i,j−1) ∩ v_j^δ
13.                         FV_δ(i,j) ← FV_δ(i−1,j) ∩ h_i^δ
14.                    else FH_δ(i,j) ← h_i ∩ v_j^δ
15.                         FV_δ(i,j) ← h_i^δ ∩ v_j
16.   if FH_δ(n,m) = ∅ and FV_δ(n,m) = ∅
17.      then return FALSE
18.      else  return TRUE
```

In order to prove that our decision algorithm *DecideFréchetMin* is correct, we need the following lemma.

Lemma 1. *For any* $2 \leqslant i \leqslant n$, $2 \leqslant j \leqslant m$, *we have* $\mathrm{F}^{\min}(H_i, V_j) \leqslant \delta$ *if and only if* $\mathrm{FH}_\delta(i,j) \neq \emptyset$ *or* $\mathrm{FV}_\delta(i,j) \neq \emptyset$. *More precisely, for any* $x, y \in \mathbb{R}^d$, *we have:*

(a) $x \in \mathrm{FH}_\delta(i,j)$ *if and only if there exists* $(A_i, B_j) \in_R (H_i, V_j)$ *such that* $a_i = x$, *and such that there exists a coupling achieving* $\mathrm{F}(A_i, B_j) \leqslant \delta$ *whose last two pairs are not* $(i-1, j), (i, j)$.

(b) $y \in \mathrm{FV}_\delta(i,j)$ *if and only if there exists* $(A_i, B_j) \in_R (H_i, V_j)$ *such that* $b_j = y$, *and such that there exists a coupling achieving* $\mathrm{F}(A_i, B_j) \leqslant \delta$ *whose last two pairs are not* $(i, j-1), (i, j)$.

We now prove Lemma 1 when $i, j \geqslant 3$. The boundary cases where $i = 2$ or $j = 2$ can be easily checked. We only prove Lemma 1(a); the proof of (b) is similar. Our proof is done by induction on (i, j), so we assume that Lemma 1 is true for all the cells that have been handled before cell (i, j) by our algorithm; in particular, it is true for all cells $(i', j') \neq (i, j)$ such that $i' \leqslant i$ and $j' \leqslant j$.

We first assume that $x \in \mathrm{FH}_\delta(i, j)$, and we want to prove that there exists $(A_i, B_j) \in_R (H_i, V_j)$ such that $a_i = x$, and such that there exists a coupling achieving $\mathrm{F}(A_i, B_j) \leqslant \delta$ whose last two pairs are not $(i-1, j), (i, j)$. We distinguish between two cases:

– First case: $\mathrm{FH}_\delta(i-1, j-1) \neq \emptyset$ or $\mathrm{FV}_\delta(i-1, j-1) \neq \emptyset$. Then, by induction, there exists $(A_{i-1}, B_{j-1}) \in_R (H_{i-1}, V_{j-1})$ such that $\mathrm{F}(A_{i-1}, B_{j-1}) \leqslant \delta$. We also know that $\mathrm{FH}_\delta(i, j)$ was set to $h_i \cap v_j^\delta$ at line 14. In other words, $x \in h_i$, and there exists $y' \in v_j$ such that $\mathrm{dist}(x, y') \leqslant \delta$. So we extend A_{i-1} and B_{j-1} by choosing $a_i = x$ and $b_j = y'$. We extend a coupling achieving $\mathrm{F}(A_{i-1}, B_{j-1}) \leqslant \delta$ with the pair (i, j), and obtain a coupling achieving $\mathrm{F}(A_i, B_j) \leqslant \delta$ whose last two pairs are $(i-1, j-1), (i, j)$.

– Second case: $\mathrm{FH}_\delta(i-1, j-1) = \emptyset$ and $\mathrm{FV}_\delta(i-1, j-1) = \emptyset$. Then $\mathrm{FH}_\delta(i, j)$ was set to $\mathrm{FH}_\delta(i, j-1) \cap v_j^\delta$ at line 12. Thus $x \in \mathrm{FH}_\delta(i, j-1)$, so by induction, there exists $(A_i, B_{j-1}) \in_R (H_i, V_{j-1})$ such that $a_i = x$ and $\mathrm{F}(A_i, B_{j-1}) \leqslant \delta$. Since $x \in v_j^\delta$, there exists $y' \in v_j$ such that $\mathrm{dist}(x, y') \leqslant \delta$. So we extend B_{j-1} by choosing $b_j = y'$. We extend a coupling achieving $\mathrm{F}(A_i, B_{j-1}) = \delta$ with the pair (i, j), and we obtain a coupling achieving $\mathrm{F}(A_i, B_j) \leqslant \delta$ whose last two pairs are $(i, j-1), (i, j)$.

Now we assume that there exists $(A_i, B_j) \in_R (H_i, V_j)$ such that there exists a coupling $\mathcal{C}$ achieving $\mathrm{F}(A_i, B_j) \leqslant \delta$ whose last two pairs are not $(i-1, j), (i, j)$. We want to prove that $a_i \in \mathrm{FH}_\delta(i, j)$. We distinguish between two cases:

– First case: $\mathrm{FH}_\delta(i-1, j-1) \neq \emptyset$ or $\mathrm{FV}_\delta(i-1, j-1) \neq \emptyset$. It implies that $\mathrm{FH}_\delta(i, j)$ was set to $h_i \cap v_j^\delta$ at line 14. Since $A_i \in_R H_i$, we have $a_i \in h_i$. Since $B_j \in_R V_j$ and $\mathrm{F}(A_i, B_j) \leqslant \delta$, it follows that $\mathrm{dist}(a_i, b_j) \leqslant \delta$, and thus $a_i \in v_j^\delta$. Thus, $a_i \in \mathrm{FH}_\delta(i, j)$.

– Second case: $\mathrm{FH}_\delta(i-1, j-1) = \emptyset$ and $\mathrm{FV}_\delta(i-1, j-1) = \emptyset$. Then, by induction, we have $\mathrm{F}^{\min}(H_{i-1}, V_{j-1}) > \delta$, which implies that $\mathrm{F}(A_{i-1}, B_{j-1}) > \delta$, so the pair $(i-1, j-1)$ cannot appear in $\mathcal{C}$. It follows that the last three pairs of $\mathcal{C}$ can only be $(i, j-2), (i, j-1), (i, j)$ or $(i-1, j-2), (i, j-1), (i, j)$.

So, by induction, we have $a_i \in \mathrm{FH}_\delta(i, j-1)$. Since $\mathrm{F}(A_i, B_j) \leqslant \delta$, we have $a_i \in v_j^\delta$. As $\mathrm{FH}_\delta(i-1, j-1) = \emptyset$ and $\mathrm{FV}_\delta(i-1, j-1) = \emptyset$, the value of $\mathrm{FH}_\delta(i, j)$ was set to $\mathrm{FH}_\delta(i, j-1) \cap v_j^\delta$ at line 14, so we have $a_i \in \mathrm{FH}_\delta(i, j)$.

This completes the proof of Lemma 1. It follows immediately from Lemma 1 that Algorithm *DecideFréchetMin* decides correctly whether $\mathrm{F}^{\min}(H, V) \leqslant \delta$. As we observed above, our algorithm runs in $O(dmn)$ time. Thus, we obtain the following result:

Theorem 1. *In the d-dimensional orthogonal case, given $\delta \geqslant 0$, and given two imprecise sequences H and V of n and m points, respectively, we can decide in $O(dmn)$ time whether $\mathrm{F}^{\min}(H, V) \leq \delta$.*

3.2   Decision Algorithm for the Euclidean Case

In this section, we give an efficient algorithm for the Euclidean case. A naive implementation of Algorithm *DecideFréchetMin* would require to construct the regions $\mathrm{FH}_\delta(i, j)$ and $\mathrm{FV}_\delta(i, j)$, which may be intersections of $\Omega(n)$ balls in $\mathbb{R}^d$. Even in $\mathbb{R}^2$, it would increase the running time of our algorithm by an order of magnitude. To improve the running time, we will show how to compute these intersections in amortized $2^{O(d^2)} \log(mn)$ time per step. We will need the following result:

Lemma 2. *We can decide in $2^{O(d^2)} k$ time whether k balls in d-dimensional Euclidean space have an empty intersection.*

Proof. We consider a collection of k balls in $\mathbb{R}^d$. We use the standard lifting-map [8, Section 1.2], which maps any point $x = (x_1, \ldots, x_d) \in \mathbb{R}^d$ to the point $\hat{x} = \left(x_1, \ldots, x_d, \sum_{i=1}^d x_i^2 \right) \in \mathbb{R}^{d+1}$. Then a ball $\mathcal{B} \subset \mathbb{R}^d$ can be mapped to an affine hyperplane $\mathcal{H} \subset \mathbb{R}^{d+1}$ such that $x \in \mathcal{B}$ if and only if $\hat{x}$ is below $\mathcal{H}$. Thus, deciding whether k balls have a non-empty intersection reduces to deciding whether there is a point x such that $\hat{x}$ is below all the corresponding hyperplanes. To do this, it suffices to decide whether there is a point $\hat{y} = (y_1, \ldots, y_{d+1})$ below all these hyperplanes and such that $\sum_{i=1}^d y_i^2 \leqslant y_{d+1}$. It can be done in $2^{O(d^2)} k$ time using an algorithm of Dyer [7] for some generalized linear programs in fixed dimension; in our case, the linear constraints for Dyer's algorithm are given by our set of hyperplanes, and the convex function we use is $(y_1, \ldots, y_{d+1}) \mapsto -y_{d+1} + \sum_{i=1}^d y_i^2$.

We now explain how we implement line 13 in amortized $2^{O(d^2)} \log n$ time. We fix the value of j, and we show how to build an incremental data structure that decides in amortized $2^{O(d^2)} \log n$ time whether $\mathrm{FV}_\delta(i, j) = \emptyset$. To achieve this, we do not maintain the region $\mathrm{FV}_\delta(i, j)$ explicitly: we only maintain an auxiliary data structure that allows us to decide quickly whether it is empty or not. During the course of Algorithm *DecideFréchetMin*, the region $\mathrm{FV}_\delta(i, j)$ can be reset to $h_i^\delta \cap v_j$ at line 15, and otherwise, it is the intersection of $\mathrm{FV}_\delta(i-1, j)$ with h_i^δ. So at any time, we have $\mathrm{FV}_\delta(i, j) = h_{i_0}^\delta \cap h_{i_0+1}^\delta \cdots \cap h_i^\delta \cap v_j$ for some $1 \leqslant i_0 \leqslant i$.

So our auxiliary data structure needs to perform three types of operations:

1. Set $\mathcal{S} = \emptyset$.
2. Insert the next ball into $\mathcal{S}$.
3. Decide whether the intersection of the balls in $\mathcal{S}$ is empty.

When we run Algorithm *DecideFréchetMin* on column j, the sequence of n balls $h_1^\delta, \ldots, h_n^\delta$ is known in advance, but not the sequence of operations. So this is the assumption we make for our auxiliary data structure: we know in advance the sequence of balls, but the sequence of operations is given online. A trivial implementation using Lemma 2 requires $2^{O(d^2)}n$ time per operation. Using exponential and binary search [15], we will show how to do it in amortized $2^{O(d^2)} \log n$ time per operation.

Operation 1 is trivial to implement. To implement operation 2, suppose that, before we perform this operation, the cardinality $|\mathcal{S}|$ of $\mathcal{S}$ is $s = 2^\ell$, for some integer ℓ. Then, using Lemma 2, we check whether the intersection of the balls in $\mathcal{S}$ and the next s balls is empty. If so, we find by binary search the first subsequence of balls, starting at the balls of $\mathcal{S}$, whose intersection is empty. By Lemma 2, it can be done in $2^{O(d^2)} s \log s$ time. Then we can perform in constant time each operation of type 2 or 3 until the next time operation 1 is performed. On the other hand, if the intersection of the balls in $\mathcal{S}$ and the next s balls is not empty, we record this fact. Then, until the cardinality of S reaches $2s = 2^{\ell+1}$, or we perform operation 1, we can perform each operation of type 2 or 3 in constant time.

This data structure needs only amortized $2^{O(d^2)} \log n$ time per operation. Keeping one such data structure for each value of j, we can perform line 13 of Algorithm *DecideFréchetMin* in amortized $2^{O(d^2)} \log n$ time. Similarly, we can implement line 12 in amortized $2^{O(d^2)} \log m$ time. Overall, we obtain the following result:

Theorem 2. *In the d-dimensional Euclidean case, given $\delta \geqslant 0$, and given two imprecise sequences H and V of n and m points, respectively, we can decide in $2^{O(d^2)} mn \log(mn)$ time whether $\mathrm{F}^{\min}(H, V) \leq \delta$.*

3.3   Optimization Algorithms

In this section, we give optimization algorithms for computing the Fréchet distance lower bound in the orthogonal case, and in the Euclidean case. They are based on the decision algorithms of sections 3.1 and 3.2.

We first consider the orthogonal case. The result of the decision algorithm may only change at some value of δ such that a box $\mathrm{FH}_\delta(i, j)$ or $\mathrm{FV}_\delta(i, j)$ degenerates to a box of dimension less than d. It may happen when the sides of two boxes of type h_i^δ, h_i, v_j^δ, or v_j have a common supporting hyperplane. Therefore, if we denote by $(x_1, \ldots, x_d, y_1, \ldots, y_d)$ the coordinates of the box $[x_1, y_1] \times \cdots \times [x_d, y_d]$, and if we denote by $(c_1, \ldots, c_k)$ the sequence of all these coordinates in increasing order, the optimal value $\mathrm{F}^{\min}(H, V)$ has to be of the form $c_j - c_i$ or $(c_j - c_i)/2$ for some $i \leqslant j$. The matrix with coefficients $c_{ij} = \max\{0, c_j - c_{k+1-i}\}$ is a k-by-k monotone matrix with $k \leqslant dmn$, so using the technique by Frederickson and Johnson [1,10] for searching in such a matrix, we can find $\mathrm{F}^{\min}(H,V)$ using $O(\log(dmn))$ calls to our decision algorithm. Thus, we obtained the following result:

Theorem 3. *In the d-dimensional orthogonal case, given two imprecise sequences H and V of n and m points, respectively, we can compute $\mathrm{F}^{\min}(H,V)$ in time $O(dmn\log(dmn))$.*

This approach does not work in the Euclidean case, so instead of using Frederickson and Johnson's technique, we use parametric search [1,2]. Using the algorithm from Theorem 2 both as the decision algorithm and the generic algorithm (without making it parallel), we obtain the following result:

Theorem 4. *In the d-dimensional Euclidean case, given two imprecise sequences H and V of n and m points, respectively, we can compute $\mathrm{F}^{\min}(H,V)$ in time $2^{O(d^2)}m^2n^2\log^2(mn)$.*

We can improve this result when $d=2$. To achieve this, we apply parametric search in a different way. Observe that the result of Algorithm *DecideFréchetMin* only changes when there is a change in the combinatorial structure of the arrangement of the circles bounding the disks $h_i, h_i^\delta, v_j, v_j^\delta$ for all i,j. So, as a generic algorithm, we use an algorithm that computes the arrangement of these $2m+2n$ circles. There exists such an algorithm with running time $O(\log(mn))$ using $O(m^2+n^2)$ processors [2]. The decision algorithm is just our algorithm *DecideFréchetMin*, which runs in $O(mn\log(mn))$ time. So we need a total of $O((m^2+n^2)\log(mn))$ time to run the generic algorithm, and a total of $O(mn\log^2(mn))$ time for the decision algorithm. Thus, we obtain the following result:

Theorem 5. *In the two-dimensional Euclidean case, given two imprecise sequences H and V of n and m points, respectively, we can compute $\mathrm{F}^{\min}(H,V)$ in $O(mn\log^2(mn)+(m^2+n^2)\log(mn))$ time.*

4   Approximation Algorithms

The running time of our algorithm for computing $\mathrm{F}^{\min}$ exactly in the Euclidean case, when the dimension is larger than 2, may be too large for some applications. The situation is worse for the problem of computing $\mathrm{F}^{\max}$ since we currently do not even have a polynomial time algorithm. The problem of matching imprecise shapes with respect to the discrete Fréchet distance under translations seems even more complicated; in particular, we currently do not know how to solve it in polynomial time.

Definition 2. *For two region sequences H and V, the* smallest Fréchet distance lower bound under translation *is the minimum over all translations t of the Fréchet distance lower bound [1] $\mathrm{F}^{\min}(H+t,V)$:*

$$\mathrm{F}^{\min}_{\mathrm{tr}}(H,V) = \min_t \mathrm{F}^{\min}(H+t,V).$$

---

[1] For a translation t and a region sequence $H = h_1,\ldots,h_n$ we denote by $H+t$ the translate of H by t. Formally $H+t = h_1\oplus t,\ldots,h_n\oplus t$ where $h_i\oplus t$ denotes the Minkowski sum of h_i and t, i.e., $h_i\oplus t = \{x+t\mid x\in h_i\}$.

The smallest Fréchet distance upper bound under translation *is the minimum over all translations t of the Fréchet distance upper bound* $\mathrm{F^{min}}(H + t, V)$:

$$\mathrm{F_{tr}^{max}}(H, V) = \min_t \ \mathrm{F^{max}}(H + t, V).$$

We obtained efficient algorithms to approximate $\mathrm{F^{min}}$, $\mathrm{F^{max}}$, $\mathrm{F_{tr}^{min}}$, and $\mathrm{F_{tr}^{max}}$ in arbitrary dimension d. Due to space limitation, we only state our results in this section, the proofs and the descriptions of the algorithms will be given in the full version of this paper.

As in the previous sections, we are given two input sequences H and V of n and m imprecise points, respectively, in d-dimensional space. In the Euclidean case, we use the Euclidean distance, and we assume that the imprecision regions h_i, v_j are Euclidean balls with centers a_i^0, b_j^0 and radius $0 < r_{min} \leq r(h_i), r(v_j) \leq r_{max}$. In the orthogonal case, we use the L_∞ distance, and the imprecision region h_i (resp. v_j) is an axis-parallel box that contains an L_∞ ball with radius r_{min} and center a_i^0 (resp. b_j^0), and is contained in a L_∞ ball with radius r_{max} and with the same center a_i^0 (resp. b_j^0). In both cases, we denote $A^0 = (a_1^0, \ldots, a_n^0)$ and $B^0 = (b_1^0, \ldots, b_m^0)$.

The approximation quality for $\mathrm{F_{tr}^{max}}$ and $\mathrm{F^{max}}$ depends on the error parameters r_{min}, r_{max}. In particular we get constant factor approximations for the case $r_{max} = \Theta(r_{min})$, which seems to be a reasonable assumption in practice. We obtain the following result for approximating the Fréchet distance upper bound.

Theorem 6. *In dimension d, given two imprecise sequences H and V of n and m points, respectively, we can compute in $O(dmn)$ time a value* $\mathrm{APP^{max}}(H, V)$ *such that*

$$\mathrm{F^{max}}(H, V) \leq \mathrm{APP^{max}}(H, V) \leq (1 + r_{max}/r_{min})\mathrm{F^{max}}(H, V).$$

The proof is omitted due to space limitation. The idea is to place each point at the center of its region, and take $\mathrm{APP^{max}}(H, V) = \mathrm{F}(A^0, B^0) + 2r_{max}$.

The approximation quality for $\mathrm{F_{tr}^{min}}$ and $\mathrm{F^{min}}$ depends on the error parameter r_{max} and an additional parameter measuring how well-separated any two *consecutive* points in an input sequence are:

Definition 3. *For a parameter $\Delta_{sep} > 0$, we say that a region sequence $H = h_1, \ldots, h_n$ is Δ_{sep}-separated if $\min_{x \in h_i, y \in h_{i+1}} \mathrm{dist}(x, y) \geq \Delta_{sep}$ for all $1 \leq i \leq n - 1$.*

We get constant factor approximations for the case $\Delta_{sep} = \Omega(r_{max})$, which again seems to be a realistic assumption. In particular, we obtain the following result for approximating the Fréchet distance lower bound. The proof is omitted due to space limitation.

Theorem 7. *In dimension d, given two Δ_{sep}-separated region sequences H and V of n and m points, respectively, we can compute in $O(dmn)$ time a value* $\mathrm{APP^{min}}(H, V)$ *such that*

$$\mathrm{F^{min}}(H, V) \leq \mathrm{APP^{min}}(H, V) \leq (1 + 8r_{max}/\Delta_{sep})\mathrm{F^{min}}(H, V).$$

Finally, we obtain the results below for approximating the Fréchet distance lower and upper bounds under translation. Our algorithms run in $O(dmn)$ time, and we currently do not know if these values can be computed exactly in polynomial time. The proof is omitted due to space limitation.

Theorem 8. *In dimension d, given two imprecise sequences H and V of n and m points, respectively, we can compute in $O(dmn)$ time two values $\mathrm{APP}_{\mathrm{tr}}^{\max}(H, V)$ and $\mathrm{APP}_{\mathrm{tr}}^{\min}(H, V)$ such that*

(i) $\mathrm{F}_{\mathrm{tr}}^{\max}(H, V) \leq \mathrm{APP}_{\mathrm{tr}}^{\max}(H, V) \leq (2 + 3r_{\max}/r_{\min})\mathrm{F}_{\mathrm{tr}}^{\max}(H, V)$, *and*
(ii) $\mathrm{F}_{\mathrm{tr}}^{\min}(H, V) \leq \mathrm{APP}_{\mathrm{tr}}^{\min}(H, V) \leq (2 + 20r_{\max}/\Delta_{\mathrm{sep}})\mathrm{F}_{\mathrm{tr}}^{\min}(H, V)$.

5   Conclusion

In this paper, we gave an efficient algorithm for computing the Fréchet distance lower bound between two imprecise point sequences. We also gave efficient approximation algorithms for the Fréchet distance upper bound, and for the Fréchet distance upper bound and lower bound under translations.

Unfortunately, our dynamic programming approach for the Fréchet distance lower bound does not seem to apply to the Fréchet distance upper bound. So we currently do not have a polynomial-time algorithm for computing the exact Fréchet distance upper bound. This problem may be hard, as it sometimes happens that a maximization problem for imprecise points is much harder than the corresponding minimization problem. For instance, Löffler and Van Kreveld [13] showed that computing the maximum area or perimeter of the convex hull of n imprecise points is NP-hard, even though the corresponding minimization problems can be solved in $O(n^2)$ and $O(n \log n)$ time respectively. Thus, it would be interesting to show that the exact Fréchet distance upper bound problem is NP-hard, or to find a polynomial-time algorithm.

Acknowledgements

Work by Ahn was supported by the Korea Research Foundation Grant funded by the Korean Government(KRF-2008-614-D00008). Work by Knauer and Scherfenberg was supported by the German Science Foundation (DFG) under grant Al 253/5-3. Work by Schlipf was supported by the Deutsche Forschungsgemeinschaft within the research training group 'Methods for Discrete Structures'(GRK 1408).

References

1. Agarwal, P.K., Sharir, M.: Efficient algorithms for geometric optimization. Computing Surveys 30(4), 412–458 (1998)
2. Agarwal, P.K., Sharir, M., Toledo, S.: Applications of parametric searching in geometric optimization. J. Algorithms 17(3), 292–318 (1994)
3. Alt, H., Godau, M.: Computing the Fréchet distance between two polygonal curves. International Journal of Computational Geometry and Applications 5, 75–91 (1995)

4. Alt, H., Knauer, C., Wenk, C.: Matching polygonal curves with respect to the Fréchet distance. In: Ferreira, A., Reichel, H. (eds.) STACS 2001. LNCS, vol. 2010, pp. 63–74. Springer, Heidelberg (2001)
5. Alt, H., Knauer, C., Wenk, C.: Comparison of distance measures for planar curves. Algorithmica 38(1), 45–58 (2003)
6. Aronov, B., Har-Peled, S., Knauer, C., Wang, Y., Wenk, C.: Fréchet distance for curves, revisited. In: Azar, Y., Erlebach, T. (eds.) ESA 2006. LNCS, vol. 4168, pp. 52–63. Springer, Heidelberg (2006)
7. Dyer, M.E.: A class of convex programs with applications to computational geometry. In: Proc. 8th Symposium on Computational Geometry, pp. 9–15. ACM, New York (1992)
8. Edelsbrunner, H.: Geometry and Topology for Mesh Generation. Cambridge University Press, Cambridge (2001)
9. Eiter, T., Mannila, H.: Computing discrete Fréchet distance. Tech. Rep. CD-TR 94/64, Christian Doppler Laboratory for Expert Systems, TU Vienna, Austria (1994)
10. Frederickson, G.N., Johnson, D.B.: Generalized selection and ranking: Sorted matrices. SIAM Journal on Computing 13(1), 14–30 (1984)
11. Khanban, A.A., Edalat, A.: Computing Delaunay triangulation with imprecise input data. In: Proc. 15th Canadian Conference on Computational Geometry, pp. 94–97 (2003)
12. Knauer, C., Löffler, M., Scherfenberg, M., Wolle, T.: The directed Hausdorff distance between imprecise point sets. In: ISAAC. LNCS, vol. 5878, pp. 720–729. Springer, Heidelberg (2009)
13. Löffler, M., van Kreveld, M.J.: Largest and smallest tours and convex hulls for imprecise points. In: Arge, L., Freivalds, R. (eds.) SWAT 2006. LNCS, vol. 4059, pp. 375–387. Springer, Heidelberg (2006)
14. Löffler, M., Snoeyink, J.: Delaunay triangulation of imprecise points in linear time after preprocessing. Computational Geometry: Theory and Applications 43(3), 234–242 (2010)
15. Moffat, A., Turpin, A.: Compression and Coding Algorithms. Kluwer, Dordrecht (2002)
16. Rote, G.: Computing the Fréchet distance between piecewise smooth curves. Computational Geometry: Theory and Applications 37(3), 162–174 (2007)
17. Sember, J., Evans, W.: Guaranteed Voronoi diagrams of uncertain sites. In: Proc. 20th Annual Canadian Conference on Computational Geometry (2008)

Connectivity Graphs of Uncertainty Regions[*]

Erin Chambers[1], Alejandro Erickson[2], Sándor Fekete[3], Jonathan Lenchner[4],
Jeff Sember[5], Srinivasan Venkatesh[2], Ulrike Stege[2], Svetlana Stolpner[6],
Christophe Weibel[7], and Sue Whitesides[2]

[1] Dept. of Mathematics and Computer Science, Saint Louis University
echambe5@slu.edu
[2] Dept. of Computer Science, University of Victoria
{ate,sue}@uvic.ca, {venkat,stege}@cs.uvic.ca
[3] Inst. of Operating and Computer Networks, TU Braunschweig
s.fekete@tu-bs.de
[4] IBM T.J. Watson Research Center
lenchner@us.ibm.com
[5] Dept. of Computer Science, University of British Columbia
jpsember@cs.ubc.ca
[6] School of Computer Science, McGill University
sveta@cim.mcgill.ca
[7] Computer Science Department, Dartmouth College
weibel@cs.dartmouth.edu

Abstract. We study a generalization of the well known bottleneck spanning tree problem called *Best Case Connectivity with Uncertainty*: Given
a family of geometric regions, choose one point per region, such that the
length of the longest edge in a spanning tree of a disc intersection graph
is minimized. We show that this problem is NP-hard even for very simple scenarios such as line segments and squares. We also give exact and
approximation algorithms for the case of line segments and unit discs
respectively.

1 Introduction

Finding an optimally connected substructure in a network is one of the fundamental combinatorial optimization problems in network design. The standard
problem of minimizing the total edge cost in the network amounts to a minimum spanning tree, which can be computed by straightforward greedy methods.
A closely related problem that has gained in importance in the context of wireless networking is to consider the "bottleneck" problem of minimizing the length
of the longest edge. This corresponds to choosing the necessary power and thus
range for the routers to be placed at nodes. Often, greedy methods still yield
optimal solutions. However, the situation changes when the location of devices
becomes part of the problem: How should each location be chosen from a given

---

[*] The authors are grateful for two Bellairs workshops supporting this research: the 8th
and 9th McGill—INRIA Workshop on Computational Geometry in 2009 and 2010.

O. Cheong, K.-Y. Chwa, and K. Park (Eds.): ISAAC 2010, Part II, LNCS 6507, pp. 434–445, 2010.

neighborhood, such that the solution to the resulting bottleneck connectivity problem is optimal? The neighborhoods can be the result of imprecise input data, or simply arise from a geometric range of possible locations; depending on the scenario, the choice of locations can be optimistic (i.e., best case) or adversarial (i.e., worst case).

Let U, $|U| = n$, denote a family of uncertainty regions, e.g., a family of disks, squares, line segments or pairs of points. For each uncertainty region $u_i \in U$, $1 \leq i \leq n$, one point p_i is to be chosen inside this region u_i. Let P be the set of points chosen. For some value $\alpha \in \mathbb{R}$, we define the *connectivity graph* $G_\alpha = (V, E)$ of P with respect to α as follows: $V = P$ and $E = \{(p_i \in P, p_j \in P), \|p_i - p_j\|_2 \leq 2\alpha\}$. Thus, the graph connects a pair of points with an edge whenever closed disks of radius α centered at these points intersect. We can now formally define the main problem, *Best Case Connectivity with Uncertainty* (BCU), that we study in this paper.

The BCU Problem. Given a set $U = \{u_1, \ldots, u_n\}$ of n uncertainty regions, find the minimum value α for which there exists a choice of point set $P = \{p_1, \ldots, p_n\}, p_i \in u_i$, such that the connectivity graph G_α of P is connected.

Related Work. If the n uncertainty regions are points (in other words, there is no uncertainty), then finding the minimum α for which the connectivity graph is connected amounts to finding a minimum Euclidean Bottleneck Spanning Tree (MBST) on the points. Since minimum spanning trees (MSTs) are also MBSTs, these can be found in time $O(n \log n)$.

The well-studied family of range assignment problems is closely related. In these problems the disks centered at each point can be of different radii, and the goal is to minimize the total power consumption under the constraint that the network satisfies certain structural properties like connectivity, strong connectivity, or a particular broadcast property. Most of the work on these problems has considered point sets rather than uncertainty regions (see [4,11,12,1,8]). Thus our work provides an early exploration of connectivity problems, arising in the context of wireless networks, for points lying in nontrivial uncertainty regions.

The minimum spanning tree problem has been studied in the setting of uncertainty regions. Yang *et al.* [16] showed that the problem of computing a spanning tree that minimizes the total edge length is NP-hard if the uncertainty regions are non-overlapping unit disks or rectangles. They also give a polynomial-time approximation scheme (PTAS) for the case where the uncertainty regions are unit disks; this is notably different from our problem, which does not admit a PTAS, unless P=NP. Other optimization problems with neighborhoods that have received attention include the Traveling Salesman Problem; e.g. see [2,10,7,5,13]. The bottleneck version of TSP is known to be NP-hard [9, p. 212]. A 2-approximation has been known since 1984 [14].

Our Main Results. After sketching that many variants of BCU are NP-hard (some even to approximate), we give exact and approximation algorithms for certain variants. Given the geometric nature of our problems, we use the Euclidean measure of distance. Our main results are as follows:

1. We show that BCU is NP-hard even in the simple case when the uncertainty regions are vertical line segments. Our proof technique also works when the regions are all squares or even pairs of points; We show that it is NP-hard to approximate BCU within a factor less than $\sqrt{5}/2$ when the uncertainty regions are pairs of points. See Section 2.
2. We present an exact algorithm for BCU when the instance consists of n fixed points and k line segments. The algorithm is polynomial in n for constant k. See Section 3.
3. For uncertainty regions that are all unit disks, we give a simple constant additive approximation algorithm for this problem. A slight modification of this algorithm gives a constant multiplicative approximation in case the disks are *non-overlapping*. See Section 4.

2   Hardness Results

We prove hardness results for two variants of the BCU problem. Our first theorem shows NP-hardness when the uncertainty regions are line segments or point pairs. Interestingly, this result also implies a hardness of approximation result for the case of point pairs. Our second theorem proves NP-hardness for the case of non-overlapping square uncertainty regions. In the remainder of this section, we sketch the main ideas behind the first result. Detailed proofs are available in the full version of this article [3].

NP-hardness of BCU for line segments or point pairs. We consider the BCU problem for non-overlapping uncertainty regions of vertically aligned pairs of points, unit distance apart with integer coordinates. We study the decision version of BCU problem for $\alpha = 1$, *i.e.*, we want to decide if $G_\alpha = G_1$ is connected for some choice of points, one for each uncertainty pair. By using a reduction from Planar 3-SAT, we will show that this problem is NP-hard.

Overview of the Reduction. We use a reduction from Planar 3-SAT, 3-SAT with the added condition that the input formula can be represented as a planar graph. We make use of the fact that, given a planar 3-SAT instance Φ with formula graph $H(\Phi)$, this graph has a planar layout on an $O(n) \times O(n)$ grid [6,15]. Further, in this layout, the vertices (variables and clauses) can be drawn as horizontal line segments and edges as vertical line segments.

   To reduce from Planar 3-SAT to an instance of BCU, where the uncertainty regions are pairs of points, we design various gadgets. Specifically, given a layout of a Planar 3-SAT instance using line segments as described above, we replace each horizontal line segment corresponding to a variable by a variable gadget, each horizontal line segment corresponding to a clause by a clause gadget, and each edge by an appropriate vertical sequence of uncertainty pairs. We will argue that there exists a choice of point in each of these uncertainty pairs such that the connectivity graph for $\alpha = 1$, G_1, is connected if and only if the corresponding Planar 3-SAT instance is satisfiable.

Overview of the Gadgets. We give the main ideas behind the clause gadget, variable gadget and connector gadgets linking others.

A clause gadget is designed so that it contains three "gates", one for each of the literals in the clause. The gate for each literal will be either on the top or the bottom of the clause gadget depending on whether the literal appears below or above the clause in the planar grid layout with horizontal and vertical segments. For the connectivity subgraph corresponding to the clause to be connected to the rest of the graph in G_1, at least one of these three gates must be open. This will correspond to setting the literal to "true" in the clause. This, in turn, ensures that the clause is satisfied.

The role of a variable gadget is to choose and propagate a truth value for the variable to all the clauses containing it in a consistent manner. The variable gadget contains three types of constructs. Type I and type II constructs will help link the variable to all the clauses that contain it and are either above or below it. We have one such type I-type II pair for every occurrence of the variable in a clause. A construct of type III is used to ensure that the truth assignment to the variable in all the copies of type I-type II pairs are the same. If not, the subgraph of G_1 corresponding to this variable gadget is not connected. Furthermore, G_1 itself cannot be connected as it cannot join subgraphs arising from parts of a variable gadget.

The variable and clause gadgets are linked to each other using two types of connectors. The connector linking a clause to a variable ensures a consistent assignment of truth value to a variable and a clause that contains this variable or its negation. Inconsistent assignments will result in G_1 being disconnected. The connector linking a variable to another variable can be more flexible. It should help connect one variable gadget to another variable gadget in G_1 irrespective of the choice of truth values for each of them.

We observe that there is no approximation algorithm for the maximum edge length, polynomial in the size of the input, with approximation ratio less than $\sqrt{5}/2$, unless $P = NP$. Indeed, our instances have a Bottleneck Spanning Tree of maximum edge length 2 ($\alpha = 1$) if the underlying 3-SAT instance has a valid assignment and at least $\sqrt{5}$ otherwise. Hence, if we had a polynomial time approximation to the solution with a ratio less than $\sqrt{5}/2$, we could use the approximation to find a Bottleneck Spanning Tree with maximum edge length not greater than 2, a contradiction (unless $P = NP$).

We finish this section by noting that if we replace the point pairs by a unit length vertical line segment joining them, we can also prove that the BCU problem for line segments is NP-hard. However, we cannot prove that it is NP-hard to approximate.

3   An Exact Algorithm Solving BCU for n Fixed Points and k Segments

We present an exact algorithm that solves our problem when the input consist of n fixed points and k line segments of any length and orientation (as uncertainty regions). For the ease of presentation, we assume the line segments to be in general position. Our algorithm determines, in a time that is polynomial in n for any fixed k, a set of point positions on the segments, such that there is a spanning tree connecting all the points on the segments as well as all fixed points, with no tree edge longer than L, where $L = 2\alpha$ and α is minimized. In other words, we seek to find a spanning tree connecting exactly one point of each segment and all fixed points, with its longest edges being of minimum length.

Algorithm Overview. A key feature of our recursive algorithm is that it restricts an exhaustive search for an optimal solution to a search of candidate solutions in the set of what we call *minimum solution trees*, optimal solutions that satisfy additional properties. Our algorithm computes the combinatorial structure of candidates for a minimum solution tree by exhaustive search for possible "support sequences" $(E_1, \ldots, E_m)$ for candidate *critical paths*. Their support sequences implicitly determine the locations of the tree vertices on the segments of the support sequences. The coordinates are specified up to user-defined precision δ. Once a critical path is found, the original problem is updated with new fixed points, specified to precision δ, and fewer segments. At the end the combinatorial structure of a minimum solution tree candidate is known, together with support sequences for critical paths through tree vertices on segments. The exact locations of these vertices are given implicitly by the support sequences. Whenever a fixed point is determined for each segment, we compute a MBST on these points to obtain the α for this minimum solution tree candidate. The candidate tree with the minimum α gives an optimum solution to our input.

Minimum Solution Trees. To prepare for the description of our algorithm, we start with describing properties of the positions of points on segments in an optimal solution. We remark that in an optimal solution for $k > 1$ we can have considerable freedom on the placement of points on segments not incident to longest edges, and therefore even an infinite number of optimal solutions may exist. To reduce the search space, we constrain the solution to be determined in defining a way to compare different (spanning tree) solutions. For this, consider the set of all spanning trees taken over all fixed points and all point choices on the k segments. We define a linear ordering on this set as follows.

For any two selections of points on the k segments, and for two of their corresponding spanning trees, let $\mathcal{L}$ and $\mathcal{L}'$ be ordered lists of lengths of all edges in the two trees, sorted from longest to shortest. That is, $\mathcal{L} = (l_1, l_2, \ldots, l_{n+k-1})$ and $\mathcal{L}' = (l'_1, l'_2, \ldots, l'_{n+k-1})$, with $l_i \geq l_{i+1}$ and $l'_i \geq l'_{i+1}$ for all i. We say that $\mathcal{L}$ *is preferred over* $\mathcal{L}'$ if for a certain i, $l_i < l'_i$, and $l_j = l'_j$ for all $j < i$. This defines a general ordering on lists. Our algorithm seeks to choose points on segments and a spanning tree such that the corresponding list of edge lengths is preferred over *all other possibilities* of point selections on the segments. We call such a tree a *minimum solution tree* $\mathcal{T}$.[1] In a minimum solution tree $\mathcal{T}$, not only are longest edges as short as possible, but also the number of longest edges is minimum. In other words, the tree with a smallest number of shortest longest edges is preferred over the ones with more edges of the same length. Further, for

---

[1] We remark that for lists $\mathcal{L}$ and $\mathcal{L}'$ for two different spanning tress with two different sets of points on the segments, it is possible that $\mathcal{L} = \mathcal{L}'$, that is we can have a tie.

all i the ith longest edge is as small as possible, and the number of edges of that length is minimum. A choice of points on segments that results in a minimum solution tree $\mathcal{T}$ is an *optimal point set* for $\mathcal{T}$.

The above conditions imply convenient properties on the optimal point set w.r.t. a minimum solution tree. Note that, for any point p of an optimal point set, it is impossible to *improve* the solution by slightly moving p on its segment; in fact, any perturbation of a point must lengthen at least one of the edges that is longest among all edges incident to p. We distinguish the possibilities for a point p on a segment in an optimal point set (see Figure 1). Given a point p on a segment, we call an edge of a minimum solution tree incident to p *locally longest* if no other tree edge incident to p is longer.

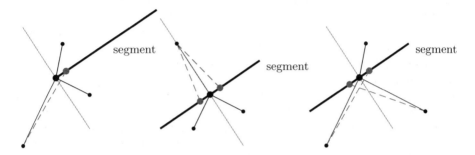

Fig. 1. Points (in blue) of type 1, 2, and 3 respectively, with longest incident edges in blue. In each case, moving the point along the segment results in a longer longest incident edge.

1. Point p lies at an extremity of the segment. Then, one of the locally longest edges incident to p lies on the half plane that is delimited by a line perpendicular to the segment and that does not contain the segment. Moving p would lengthen that edge.
2. Point p is on the relative interior of the segment and one of the locally longest edges incident to p is perpendicular to the segment. Moving p in any direction would lengthen that edge.
3. Point p is on the relative interior of the segment and not of type 2. Then there are two locally longest edges incident to p lying on different half-planes delimited by a line perpendicular to the segment passing through p. Moving p in any direction would increase the length of one of these two edges.

Notably, if we know for any point p on a segment of type 1 or 2 that it has a locally longest incident edge that would become longer if p were moved or if we know for a point p of type 3 that it has a pair of locally longest incident edges where one of them would become longer when moving p, then we can deduce p's position without any knowledge of other incident edges of p.

Critical Paths. To define this concept, let $(E_1, \ldots, E_m)$ be a given sequence of fixed points and segments, where E_1 and E_m are fixed points or segments, and $E_2, \ldots, E_{m-1}$ are segments. A *critical path* consists of points p_i located at selected positions on the segments E_i and are connected by edges e_i such that (1) the edges all have the same length, and (2) no other selection of point locations on these segments results in a sequence where edges are not longer but some are strictly shorter.

Since it is impossible to shorten an edge of a critical path by moving a single point p_i without causing another edge to be longer, all the p_i on the critical path must be of one of the three types described above.

Before we can describe how to compute a critical path and argue that a sequence $(E_1, \ldots, E_M)$ supports at most one critical path, we need a few more definitions. Given $(E_1, \ldots, E_m)$ and a positive number Λ, let $U_1(\Lambda)$ be the set of points in the plane reachable from E_1 by an edge of length Λ or less, and let $U_i(\Lambda)$ for $i > 1$ be the set of points on the plane reachable from $U_{i-1}(\Lambda) \cap E_i$ by an edge of length Λ or less. Let $S_1(\Lambda)$ be the set of points on the plane reachable from E_1 by an edge e_1 of length *exactly* Λ, that is in the case that E_1 is a segment, no point in $S_1(\Lambda)$ can be reached from E_1 by an edge shorter than Λ. Similarly, let $S_i(\Lambda)$ for $i > 1$ be the set of points on the plane reachable from $S_{i-1}(\Lambda) \cap E_i$ by an edge e_i of length exactly Λ, such that $e_1, \ldots, e_i$ form a critical path (assuming the endpoint of e_i that is not on E_i is a fixed point).

We study the properties of $U_i(\Lambda)$ and $S_i(\Lambda)$. First, we can deduce inductively that if $U_{i-1}(\Lambda) \cap E_i \neq \emptyset$, then it consists of a single point or a subsegment (a connected subset) of E_i: If the set $U_i(\Lambda) \neq \emptyset$, then $U_i(\Lambda)$ is either a ball of radius Λ, or the Minkowski sum of a ball of radius Λ and a segment where $U_i(\Lambda) = E_1$ if $i = 1$, and $U_i(\Lambda) = U_{i-1}(\Lambda) \cap E_i$ otherwise (Figure 2).

Fig. 2. Examples of $U_1(\Lambda)$ and $S_1(\Lambda)$ for the cases that E_1 is a fixed point or a segment

Lemma 1. $S_i(\Lambda) \subseteq U_i(\Lambda)$.

Proof. Sketch: By definition, $S_1(\Lambda)$ is a subset of $U_1(\Lambda)$. We prove by induction that $S_i(\Lambda)$ is a subset of $U_i(\Lambda)$ for all i. We remark that it is sufficient to prove that any open set containing a point of $S_i(\Lambda)$ intersects with the boundary of $U_i(\Lambda)$, which can be done with some care by contradiction.

It follows that if $U_{i-1}(\Lambda) \cap E_i = \emptyset$, then $S_{i-1}(\Lambda) \cap E_i = \emptyset$. If $U_{i-1}(\Lambda) \cap E_i$ consists of a single point p, then either $S_{i-1}(\Lambda) \cap E_i = \emptyset$ or $S_{i-1}(\Lambda) \cap E_i = \{p\}$. If $U_{i-1}(\Lambda) \cap E_i$ is a subsegment of E_i, then $S_{i-1}(\Lambda) \cap E_i$ can be empty, one extremity of the subsegment, both extremities of the subsegment, or the complete subsegment. In fact, we can prove the following lemma.

Lemma 2. *(a) The set $S_1(\Lambda)$ is the boundary of $U_1(\Lambda)$. (b) For all $i > 1$, the set $S_i(\Lambda)$ is the intersection of the boundary of $U_i(\Lambda)$ with the Minkowski sum of a circle of radius Λ and $S_{i-1}(\Lambda) \cap E_i$.*

Proof. (a) This follows from the definition of $S_1(\Lambda)$. (b) By definition, $S_i(\Lambda)$ contains only points at distance Λ from $S_{i-1} \cap E_i$. From Lemma 1, we know that $S_i(\Lambda) \subseteq U_i(\Lambda)$. It is therefore sufficient to prove that every point in the intersection is in $S_i(\Lambda)$. We prove this by induction. Let p be any point in the intersection. Since p is in the Minkowski sum of a circle of radius Λ and $S_{i-1} \cap E_i$, there exists $q \in S_{i-1} \cap E_i$ at distance exactly Λ from p, and by the induction hypothesis, there is a path of edges of length Λ from q, which we can extend to p. We need to prove that there is no other path to p that uses edges no longer than Λ, and some shorter. Suppose there is a choice of $p_1, \ldots, p_i$ such that $p_1, \ldots, p_i, p$ is such a path. Suppose first that the last edge is shorter than Λ by some $\varepsilon > 0$. Then, by changing the length of the last edge by less than ε, we can find paths to any point in some open set around p. This contradicts the assumption that p is part of the boundary of $U_i(\Lambda)$. Therefore, the last edge of the path is of length Λ exactly. But that means that p is at distance Λ from both q and p_i, which are both in $U_i(\Lambda) \cap E_i$. This means that the midpoint of the segment from q to p_i is in $U_i(\Lambda) \cap E_i$ and at distance less than Λ from p, yielding a contradiction. $\square$

The following cases are possible (Figure 3).

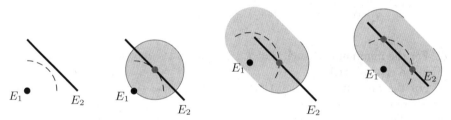

Fig. 3. Shapes of $S_i(\Lambda)$ of type a, b, c and d respectively. $S_1(\Lambda)$ is indicated with a dashed line, $S_1(\Lambda) \cap E_2$ and $S_2(\Lambda)$ are in red.

a. $S_{i-1}(\Lambda) \cap E_i = \emptyset$ and therefore $S_i(\Lambda) = \emptyset$.
b. $U_{i-1}(\Lambda) \cap E_i$ consists of a single point p and $S_{i-1}(\Lambda) \cap E_i = \{p\}$. Then $S_i(\Lambda)$ is a circle of radius Λ centered around p.
c. $U_{i-1}(\Lambda) \cap E_i$ is a subsegment of E_i and $S_{i-1}(\Lambda) \cap E_i$ is a single extremity of the subsegment. In this case, $U_i(\Lambda)$ is the Minkowski sum of the subsegment and a ball of radius L, and $S_i(\Lambda)$ is the half circle of radius Λ centered on $S_{i-1}(\Lambda) \cap E_i$ at one extremity of $U_i(\Lambda)$.
d. $U_{i-1}(\Lambda) \cap E_i$ is a subsegment of E_i and $S_{i-1}(\Lambda) \cap E_i$ consists of both extremities of the subsegment. In this case, $U_i(\Lambda)$ is the Minkowski sum of the subsegment and a ball of radius Λ, and $S_i(\Lambda)$ consists of both half circles of radius Λ each centered on a point of $S_{i-1}(\Lambda) \cap E_i$ at each extremity of $U_i(\Lambda)$.

For a given $(E_1, \ldots, E_m)$ and length Λ it is therefore possible to compute successively the $U_i(\Lambda)$'s and $S_i(\Lambda)$'s. In order to obtain a critical path we require $U_{m-1}(\Lambda) \cap E_m = S_{m-1}(\Lambda) \cap E_m$. Notably, this can only happen for the smallest Λ such that $U_{m-1}(\Lambda) \cap E_m \neq \emptyset$.

In our exact algorithm critical paths contribute to the set of candidate solutions. Given $(E_1, \ldots, E_m)$ and Λ, the following algorithm determines an existing critical path with a precision, say δ.

(1) Determine $U_{m-1}(\Lambda)$. (2) If $U_{m-1}(\Lambda) \cap E_m = \emptyset$, then increase Λ and go to (1). (3) If $U_{m-1}(\Lambda) \cap E_m \neq \emptyset$, then decrease Λ and go to (1). Once the approximate minimum $L^* \in [\Lambda - \delta, \Lambda + \delta]$ is found, check whether $U_{m-1}(L^*) \cap E_m = S_{m-1}(L^*) \cap E_m$.

Possible outcomes for $U_{m-1}(\Lambda) \cap E_m \neq \emptyset$ (Figure 4) are:

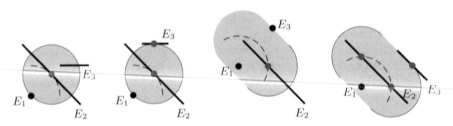

Fig. 4. Sequence (E_1, E_2, E_3) with outcomes of type α, β, γ, and δ respectively for the case that Λ is minimum such that $U_2(\Lambda) \cap E_3 \neq \emptyset$. $S_1(\Lambda)$ is depicted by dashed lines. $S_1(\Lambda) \cap E_2$, $S_2(\Lambda)$ and $S_2(\Lambda) \cap E_3$ are shown in red.

α. For the minimum L^* such that E_m intersects the interior of $U_{m-1}(L^*)$, there is no critical path.

β. For the minimum L^* such that $U_{m-1}(L^*) \cap E_m = S_{m-1}(L^*) \cap E_m = \{p\}$, there is a critical path.

γ. For the minimum L^* such that $U_{m-1}(L^*) \cap E_m$ is a single point and $S_{m-1}(L^*) \cap E_m = \emptyset$, there is no critical path.

δ. For the minimum L^* such that $U_{m-1}(L^*) \cap E_m$ is a segment and $S_{m-1}(L^*) \cap E_m$ consists only of extremities of the segment, there is no critical path.

Corollary 1. *If $(E_1, \ldots, E_m)$ supports a critical path of locally longest edges in a minimum solution tree then there is a unique choice of point locations that defines the critical path.*

Description of the Algorithm. We show how to compute an optimal solution by examining possible critical paths, using the following corollary.

Corollary 2. *If $(E_1, \ldots, E_m)$ supports a critical path of locally longest edges in a minimum solution tree then this choice is part of an optimal solution.*

Note that if we had an oracle giving us a $(E_1, \ldots, E_m)$ that supports a critical path of locally longest edges in the optimal solution, then we could determine the choice of points on these elements in the optimal solution. We could then replace the segments in the sequence by fixed points and solve the rest of the problem separately. This eventually would allow us to replace all segments by fixed points, and then to solve the problem by finding an associated MBST.

Lacking an oracle we determine these sequences *by complete enumeration* of *all possible sequences* $(E_1, \ldots, E_m)$, where E_1 and E_m are fixed points or segments, and $E_2, \ldots, E_{m-1}$ are segments. There are $O((n + k)^2 \cdot k! \cdot k)$ such sequences. This enumeration accounts for most of the complexity of our algorithm. For each sequence in the enumeration, we check whether it supports a critical path and whether the edge-length for this path is best so far. If not, we discard this path. Otherwise we recurse on the updated set of sequences and points. That is, we prune these sequences in the enumeration as we find them. Once we have gone through the complete list of possible sequences, we will have found the critical path with shortest locally longest edges. Then we replace the segments of the sequence with fixed points defined by the critical path. We then execute the algorithm recursively on the thus reduced instance, using edges of length no greater than the ones in the critical path just found. Once the instance does not contain any more segments, we connect all remaining connected components with a greedy algorithm, in polynomial time.

We remark that it is crucial to determine critical paths with progressively decreasing edge lengths since positions of points on segments should only be determined using the locally longest incident edges.

The enumeration in our algorithm described above is superexponential in the number k of segments, which is not surprising since we have shown the problem with no fixed points is NP-hard. For constant k the problem is, however, polynomial in the number n of points, as our running time analysis will show.

The (multiple recursive) enumeration results in a search tree of size $O(((n + k)^2 \cdot k! \cdot k^2)^k)$ with a $O(k)$ running time for each node in the search tree. Thus the total time complexity is $O(((n + k)^2 \cdot k! \cdot k^2)^k \cdot k)$.

4   Constant Factor and Additive Approximations

Our approximations are based on computing minimum bottleneck spanning trees, MBSTs, which are spanning trees that minimize the maximum length edge in the tree.

Lemma 3. *Given a set of uncertainty regions that are unit disks $D_1, \ldots, D_n$ with centers $p_1, \ldots, p_n$, let L be the largest edge of a bottleneck spanning tree on $\{p_i\}$. Then choosing broadcasting locations $\ell_i = p_i$ and $\alpha = L/2$ is at worst an OPT+1 approximation to the BCU Problem. In other words, if OPT denotes the smallest radius α for any choice of $\ell_i \in D_i$, then $L/2 \leq OPT+1$. This approximation can be computed in polynomial time.*

Proof. Let L be the maximum length edge of a MBST on $\{p_i\}$. Consider the best choice of the $\ell_i \in D_i$ and an associated MBST on these ℓ_i. The edges of this MBST are each at most 2 shorter than the corresponding edges of a spanning tree, S, on the corresponding p_i. Thus the maximum length of any edge in S is at most 2 greater than the maximum length edge in the MBST on the ℓ_i, and, similarly, the maximum length, L, of any edge of a MBST on the $\{p_i\}$ must be at most 2 greater than the maximum length edge in the MBST on the ℓ_i. The result follows.   □

Our approximation for the BCU Problem is not a constant factor approximation, since if one takes n unit disks with non-empty intersection, then the ℓ_i can all be taken to equal one of the intersection points so that OPT= 0 while $L/2$ can be non-zero (and as big as 1). However, it is a constant factor approximation for *non-overlapping* unit disks.

5   Other Related Results and Conclusions

In this work, we also studied a closely related problem, *Worst Case Connectivity with Uncertainty* (WCU):

The WCU Problem. Find the minimum value α such that for *any* choice of points P, the connectivity graph G_α of P is connected.

We were able to show a simple approximation algorithm for WCU that is within an additive factor of 1 and a muliplicative factor of 2 when the uncertainty regions are unit disks (result omitted due to space restriction). Although we have been able to obtain several NP-hardness results for BCU, we do not have any complexity lower bounds for WCU which, a priori, seems harder. It is an interesting open question to improve our approximation algorithms for both these problems.

It would be interesting to show NP-hardness results for the BCU problem for other uncertainty regions such as disks. It is also possible that techniques from convex optimization could be used to design approximation algorithms for BCU for, say, line segments or squares. From the perspective of fixed parameter tractability, we observed that BCU is in FPT when the instance consists of n fixed points and k pairs of points; the parameter is k. However, we conjecture that BCU for the case of line segments is W[1]-hard and hence our exact algorithm is unlikely to be improved upon significantly.

In conclusion, our work on connectivity problems for uncertainty regions motivated by wireless network scenarios suggests that this area provides a rich collection of problems for further investigation.

Note. *The details of the reductions, as well as full proofs of our results, can be found in the full version of this article [3].*

References

1. Alt, H., Arkin, E., Brönnimann, H., Erickson, J., Fekete, S., Knauer, C., Lenchner, J., Mitchell, J., Whittlesey, K.: Minimum-cost coverage of point sets by disks. In: Proc. 22nd ACM Symp. Comp. Geom. (SoCG), pp. 449–458 (2006)

2. Arkin, E.M., Hassin, R.: Approximation algorithms for the geometric covering salesman problem. Disc. Appl. Mathematics 55(3), 197–218 (1994)
3. Chambers, E., Erickson, A., Fekete, S., Lenchner, J., Sember, J., Venkatesh, S., Stege, U., Stolpner, S., Weibel, C., Whitesides, S.: Connectivity graphs of uncertainty regions. arXiV:1009.3469 (2010)
4. Clementi, A.E.F., Penna, P., Silvestri, R.: On the power assignment problem in radio networks. Technical Report TR00-054, Electronic Colloquium on Computational Complexity (2000)
5. de Berg, M., Gudmundsson, J., Katz, M.J., Levcopoulos, C., Overmars, M.H., van der Stappen, A.F.: TSP with neighborhoods of varying size. J. Algorithms 57(1), 22–36 (2005)
6. Duchet, P., Hamidoune, Y.O., Vergnas, M.L., Meyniel, H.: Representing a planar graph by vertical lines joining different levels. Disc. Mathematics 46(3), 319–321 (1983)
7. Dumitrescu, A., Mitchell, J.S.B.: Approximation algorithms for TSP with neighborhoods in the plane. In: Proc. 12th ACM-SIAM Symp. on Disc. Algorithms (SODA), pp. 38–46 (2001)
8. Fuchs, B.: On the hardness of range assignment problems. In: Calamoneri, T., Finocchi, I., Italiano, G.F. (eds.) CIAC 2006. LNCS, vol. 3998, pp. 127–138. Springer, Heidelberg (2006)
9. Garey, M.R., Johnson, D.S.: Computers and Intractability: A Guide to the Theory of NP-Completeness. W.H. Freeman, New York (1979)
10. Gudmundsson, J., Levcopoulos, C.: A fast approximation algorithm for TSP with neighborhoods. Nord. J. Comput. 6(4), 469 (1999)
11. Lev-Tov, N., Peleg, D.: Exact algorithms and approximation schemes for base station placement problems. In: Penttonen, M., Schmidt, E.M. (eds.) SWAT 2002. LNCS, vol. 2368, pp. 90–99. Springer, Heidelberg (2002)
12. Lev-Tov, N., Peleg, D.: Polynomial time approximation schemes for base station coverage with minimum total radii. Computer Networks 47(4), 489–501 (2005)
13. Mitchell, J.S.B.: A PTAS for TSP with neighborhoods among fat regions in the plane. In: Proc. 18th ACM-SIAM Symp. on Disc. Algorithms (SODA), pp. 11–18 (2007)
14. Parker, G., Rardin, R.L.: Guaranteed performance heuristics for the bottleneck traveling salesman problem. Operations Research Letters 2(6), 269–272 (1984)
15. Rosenstiehl, P., Tarjan, R.E.: Rectilinear planar layouts and bipolar orientations of planar graphs. Disc. Comp. Geom. 1, 343–353 (1986)
16. Yang, Y., Lin, M., Xu, J., Xie, Y.: Minimum spanning tree with neighborhoods. In: Kao, M.-Y., Li, X.-Y. (eds.) AAIM 2007. LNCS, vol. 4508, pp. 306–316. Springer, Heidelberg (2007)

π/2-Angle Yao Graphs Are Spanners

Prosenjit Bose[1,*], Mirela Damian[2,**], Karim Douïeb[3,*], Joseph O'Rourke[4],
Ben Seamone[5], Michiel Smid[6,*], and Stefanie Wuhrer[7]

[1] School of Computer Science, Carleton University, Ottawa, Canada
jit@scs.carleton.ca
[2] Department of Computer Science, Villanova University, Villanova, USA
mirela.damian@villanova.edu
[3] School of Computer Science, Carleton University, Ottawa, Canada
kdouieb@ulb.ac.be
[4] Department of Computer Science, Smith College, Northampton, USA
orourke@cs.smith.edu
[5] School of Mathematics and Statistics, Carleton University, Ottawa, Canada
bseamone@connect.carleton.ca
[6] School of Computer Science, Carleton University, Ottawa, Canada
michiel@scs.carleton.ca
[7] Institute for Information Technology, National Research Council, Ottawa, Canada
stefanie.wuhrer@nrc-cnrc.gc.ca

Abstract. We show that the Yao graph Y_4 in the L_2 metric is a spanner
with stretch factor $8\sqrt{2}(29 + 23\sqrt{2})$.

1   Introduction

Let V be a finite set of points in the plane and let $G = (V, E)$ be the complete
Euclidean graph on V. We will refer to the points in V as *nodes*, to distinguish
them from other points in the plane. The *Yao graph* [7] with an integer parameter
$k > 0$, denoted Y_k, is defined as follows. Any k equally-separated rays starting
at the origin define k cones. Pick a set of arbitrary, but fixed cones. We can now
translate the cones to each node $u \in V$. In each cone, pick a shortest edge uv,
if there is one, and add to Y_k the directed edge $\overrightarrow{uv}$. Ties are broken arbitrarily.
Note that the Yao graph differs from the Θ-graph in how the shortest edge is
chosen. While the Yao graph chooses the shortest edge in terms of the Euclidean
distance, the Θ-graph chooses the shortest edge as the one that has the shortest
distance to u after being projected to the bisector of the cone. Most of the time
we ignore the direction of an edge uv; we refer to the directed version $\overrightarrow{uv}$ of
uv only when its origin (u) is important and unclear from the context. We will
distinguish between Y_k, the Yao graph in the Euclidean L_2 metric, and Y_k^∞, the
Yao graph in the L_∞ metric. Unlike Y_k however, in constructing Y_k^∞ ties are
broken by always selecting the most counterclockwise edge; the reason for this
choice will become clear in Section 2.

---

* Supported by NSERC.
** Supported in part by NSF grant CCF-0728909 and by Villanova's CEET.

O. Cheong, K.-Y. Chwa, and K. Park (Eds.): ISAAC 2010, Part II, LNCS 6507, pp. 446–457, 2010.

For a given subgraph $H \subseteq G$ and a fixed $t \geq 1$, H is called a *t-spanner* for G if, for any two nodes $u, v \in V$, the shortest path in H from u to v is no longer than t times the length of uv. The value t is called the *dilation* or the *stretch factor* of H. If t is constant, then H is called a *length spanner*, or simply a *spanner*.

The class of graphs Y_k has been much studied. Bose et al. [2] showed that, for $k \geq 9$, Y_k is a spanner with stretch factor $\frac{1}{\cos \frac{2\pi}{k} - \sin \frac{2\pi}{k}}$. In [1] we improve the stretch factor and show that, in fact, Y_k is a spanner for any $k \geq 7$. Recently, Molla [5] showed that Y_2 and Y_3 are not spanners, and that Y_4 is a spanner with stretch factor $4(2 + \sqrt{2})$, for the special case when the nodes in V are in convex position (see also [3]). The authors conjectured that Y_4 is a spanner for arbitrary point sets. In this paper, we settle their conjecture and prove that Y_4 is a spanner with stretch factor $8\sqrt{2}(29 + 23\sqrt{2})$.

The paper is organized as follows. In Section 2, we prove that the graph Y_4^∞ is a spanner with stretch factor 8. In Section 3 we establish several properties for the graph Y_4. Finally, in Section 4, we use the properties of Section 3 to prove that, for every edge ab in Y_4^∞, there exists a path between a and b in Y_4 not much longer than the Euclidean distance between a and b. By combining this with the result of Section 2, it follows that Y_4 is a spanner.

2 Y_4^∞ in the L_∞ Metric

In this section we focus on Y_4^∞, which has a nicer structure compared to Y_4. First we prove that Y_4^∞ is a plane graph. Then we use this property to show that Y_4^∞ is an 8-spanner. To be precise, we prove that for any two nodes a and b, the graph Y_4^∞ contains a path between a and b whose length (in the L_∞-metric) is at most $8|ab|_\infty$.

We need a few definitions. We say that two edges ab and cd *properly cross* (or *cross*, for short) if they share a point other than an endpoint (a, b, c or d); we say that ab and cd *intersect* if they share a point (either an interior point or an endpoint). Let $Q_1(a), Q_2(a), Q_3(a)$ and $Q_4(a)$ be the four quadrants at a, as in

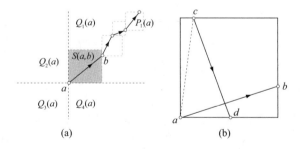

(a)   (b)

Fig. 1. (a) Definitions: $Q_i(a)$, $P_i(a)$ and $S(a, b)$. (b) Lemma 1: ab and cd cannot cross.

Figure 1a. Let $P_i(a)$ be the path that starts at point a and follows the directed Yao edges in quadrant Q_i. Let $P_i(a, b)$ be the subpath of $P_i(a)$ that starts at a and ends at b. Let $|ab|_\infty$ be the L_∞ distance between a and b. Let $sp(a, b)$ denote a shortest path in Y_4^∞ between a and b. Let $S(a, b)$ denote the open square with corner a whose boundary contains b, and let $\partial S(a, b)$ denote the boundary of $S(a, b)$. These definitions are illustrated in Figure 1a. For a node $a \in V$, let $x(a)$ denote the x-coordinate of a and $y(a)$ denote the y-coordinate of a.

Lemma 1. Y_4^∞ is a plane graph.

Proof. The proof is by contradiction. Assume the opposite. Then there are two edges $\overrightarrow{ab}, \overrightarrow{cd} \in Y_4^\infty$ that cross each other. Since $\overrightarrow{ab} \in Y_4^\infty$, $S(a, b)$ must be empty of nodes in V, and similarly for $S(c, d)$. Let j be the intersection point between ab and cd. Then $j \in S(a, b) \cap S(c, d)$, meaning that $S(a, b)$ and $S(c, d)$ must overlap. However, neither square may contain a, b, c or d. It follows that $S(a, b)$ and $S(c, d)$ coincide, meaning that c and d lie on $\partial S(a, b)$ (see Figure 1b). Since cd intersects ab, c and d must lie on opposite sides of ab. Thus either ac or ad lies counterclockwise from ab. Assume without loss of generality that ac lies counterclockwise from ab; the other case is identical. Because $S(a, c)$ coincides with $S(a, b)$, we have that $|ac|_\infty = |ab|_\infty$. In this case however, Y_4^∞ would break the tie between ac and ab by selecting the most counterclockwise edge, which is $\overrightarrow{ac}$. This contradicts that $\overrightarrow{ab} \in Y_4^\infty$.   □

Theorem 1. Y_4^∞ is an 8-spanner in the L_∞ metric space.

Proof. We show that, for any pair of points $a, b \in V$, $|sp(a, b)|_\infty < 8|ab|_\infty$. The proof is by induction on the pairwise distance between the points in V. Assume without loss of generality that $b \in Q_1(a)$, and $|ab|_\infty = |x(b) - x(a)|$. Consider the case in which ab is a closest pair of points in V (the base case for our induction). If $ab \in Y_4^\infty$, then $|sp(a, b)|_\infty = |ab|_\infty$. Otherwise, there must be $ac \in Y_4^\infty$, with $|ac|_\infty = |ab|_\infty$. But then $|bc|_\infty < |ab|_\infty$ (see Figure 2a), a contradiction.

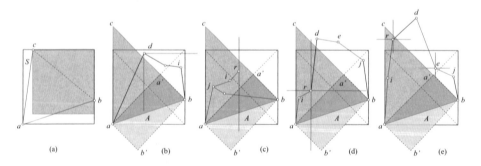

Fig. 2. (a) Base case. (b) $\triangle abc$ empty (c) $\triangle abc$ non-empty, $P_{ar} \cap P_2(b) = \{j\}$ (d) $\triangle abc$ non-empty, $P_{ar} \cap P_2(b) = \emptyset$, e above r (e) $\triangle abc$ non-empty, $P_{ar} \cap P_2(b) = \emptyset$, e below r.

Assume now that the inductive hypothesis holds for all pairs of points closer than $|ab|_\infty$. If $ab \in Y_4^\infty$, then $|sp(a,b)|_\infty = |ab|_\infty$ and the proof is finished. If $ab \notin Y_4^\infty$, then the square $S(a,b)$ must be nonempty.

Let A be the rectangle $ab'ba'$ as in Figure 2b, where ba' and bb' are parallel to the diagonals of S. If A is nonempty, then we can use induction to prove that $|sp(a,b)|_\infty <= 8|ab|_\infty$ as follows. Pick $c \in A$ arbitrary. Then $|ac|_\infty + |cb|_\infty = |x(c) - x(a)| + |x(b) - x(c)| = |ab|_\infty$, and by the inductive hypothesis $sp(a,c) \oplus sp(c,b)$ is a path in Y_4^∞ no longer than $8|ac|_\infty + 8|cb|_\infty = 8|ab|_\infty$; here $\oplus$ represents the concatenation operator. Assume now that A is empty. Let c be at the intersection between the line supporting ba' and the vertical line through a (see Figure 2b). We discuss two cases, depending on whether $\triangle abc$ is empty of points or not.

Case 1: $\triangle abc$ is empty of points. Let $ad \in P_1(a)$. We show that $P_4(d)$ cannot contain an edge crossing ab. Assume the opposite, and let $st \in P_4(d)$ cross ab. Since $\triangle abc$ is empty, s must lie above bc and t below ab, therefore $|st|_\infty \geq |y(s) - y(t)| > |y(s) - y(b)| = |sb|_\infty$, contradicting the fact that $st \in Y_4^\infty$. It follows that $P_4(d)$ and $P_2(b)$ must meet in a point $i \in P_4(d) \cap P_2(b)$ (see Figure 2b). Now note that $|P_4(d,i) \oplus P_2(b,i)|_\infty \leq |x(d) - x(b)| + |y(d) - y(b)| < 2|ab|_\infty$. Thus we have that $|sp(a,b)|_\infty \leq |ad \oplus P_4(d,i) \oplus P_2(b,i)|_\infty < |ab|_\infty + 2|ab|_\infty = 3|ab|_\infty$.

Case 2: $\triangle abc$ is nonempty. In this case, we seek a short path from a to b that does not cross to the underside of ab, to avoid oscillating paths that cross ab arbitrarily many times. Let r be the rightmost point that lies inside $\triangle abc$. Arguments similar to the ones used in Case 1 show that $P_3(r)$ cannot cross ab and therefore it must meet $P_1(a)$ in a point i. Then $P_{ar} = P_1(a,i) \oplus P_3(r,i)$ is a path in Y_4^∞ of length

$$|P_{ar}|_\infty < |x(a) - x(r)| + |y(a) - y(r)| < |ab|_\infty + 2|ab|_\infty = 3|ab|_\infty. \quad (1)$$

The term $2|ab|_\infty$ in the inequality above represents the fact that $|y(a) - y(r)| \leq |y(a) - y(c)| \leq 2|ab|_\infty$. Consider first the simpler situation in which $P_2(b)$ meets P_{ar} in a point $j \in P_2(b) \cap P_{ar}$ (see Figure 2c). Let $P_{ar}(a,j)$ be the subpath of P_{ar} extending between a and j. Then $P_{ar}(a,j) \oplus P_2(b,j)$ is a path in Y_4^∞ from a to b, therefore $|sp(a,b)|_\infty \leq |P_{ar}(a,j) \oplus P_2(b,j)|_\infty < 2|y(j) - y(a)| + |ab|_\infty \leq 5|ab|_\infty$.

Consider now the case when $P_2(b)$ does not intersect P_{ar}. We argue that, in this case, $Q_1(r)$ may not be empty. Assume the opposite. Then no edge $st \in P_2(b)$ may cross $Q_1(r)$. This is because, for any such edge, $|sr|_\infty < |st|_\infty$, contradicting $st \in Y_4^\infty$. This implies that $P_2(b)$ intersects P_{ar}, again a contradiction to our assumption. This establishes that $Q_1(r)$ is nonempty. Let $rd \in P_1(r)$. The fact that $P_2(b)$ does not intersect P_{ar} implies that d lies to the left of b. The fact that r is the rightmost point in $\triangle abc$ implies that d lies outside $\triangle abc$ (see Figure 2d). It also implies that $P_4(d)$ shares no points with $\triangle abc$. This along with arguments similar to the ones used in case 1 show that $P_4(d)$ and $P_2(b)$ meet in a point $j \in P_4(d) \cap P_2(b)$. Thus we have found a path

$$P_{ab} = P_1(a,i) \oplus P_3(r,i) \oplus rd \oplus P_4(d,j) \oplus P_2(b,j) \quad (2)$$

extending from a to b in Y_4^∞. If $|rd|_\infty = |x(d) - x(r)|$, then $|rd|_\infty < |x(b) - x(a)| = |ab|_\infty$, and the path P_{ab} has length

$$|P_{ab}|_\infty \le 2|y(d) - y(a)| + |ab|_\infty < 7|ab|_\infty. \tag{3}$$

In the above, we used the fact that $|y(d) - y(a)| = |y(d) - y(r)| + |y(r) - y(a)| < |ab|_\infty + 2|ab|_\infty$. Suppose now that

$$|rd|_\infty = |y(d) - y(r)|. \tag{4}$$

In this case, it is unclear whether the path P_{ab} defined by (2) is short, since rd can be arbitrarily long compared to ab. Let e be the clockwise neighbor of d along the path P_{ab} (e and b may coincide). Then e lies below d, and either $de \in P_4(d)$, or $ed \in P_2(e)$ (or both). If e lies above r, or at the same level as r (i.e., $e \in Q_1(r)$, as in Figure 2d), then

$$|y(e) - y(r)| < |y(d) - y(r)| \tag{5}$$

Since $rd \in P_1(r)$ and e is in the same quadrant of r as d, we have $|rd|_\infty \le |re|_\infty$. This along with inequalities (4) and (5) implies $|re|_\infty > |y(e) - y(r)|$, which in turn implies $|re|_\infty = |x(e) - x(r)| \le |ab|_\infty$, and so $|rd|_\infty \le |ab|_\infty$. Then inequality (3) applies here as well, showing that $|P_{ab}|_\infty < 7|ab|_\infty$.

If e lies below r (as in Figure 2e), then

$$|ed|_\infty \ge |y(d) - y(e)| \ge |y(d) - y(r)| = |rd|_\infty. \tag{6}$$

Assume first that $ed \in P_2(e)$, or $|ed|_\infty = |x(e) - x(d)|$. In either case, $|ed|_\infty \le |er|_\infty < 2|ab|_\infty$. This along with inequality (6) shows that $|rd|_\infty < 2|ab|_\infty$. Substituting this upper bound in (2), we get $|P_{ab}|_\infty \le 2|y(d) - y(a)| + 2|ab|_\infty < 8|ab|_\infty$. Assume now that $ed \notin P_2(e)$, and $|ed|_\infty = |y(e) - y(d)|$. Then $ee' \in P_2(e)$ cannot go above d (otherwise $|ed|_\infty < |ee'|_\infty$, contradicting $ee' \in P_2(e)$). This along with the fact $de \in P_4(d)$ implies that $P_2(e)$ intersects P_{ar} in a point k. Redefine $P_{ab} = P_{ar}(a, k) \oplus P_2(e, k) \oplus P_4(e, j) \oplus P_2(b, j)$. Then P_{ab} is a path in Y_4^∞ from a to b of length $|P_{ab}| \le 2|y(r) - y(a)| + |ab|_\infty \le 5|ab|_\infty$.   □

This theorem will be employed in Section 4.

3 Y_4 in the L_2 Metric

In this section we establish basic properties of Y_4. Due to space restrictions, some of these properties are stated without proofs. The proofs can be found in [1]. The ultimate goal of this section is to show that, if two edges in Y_4 cross, there is a short path between their endpoints (Lemma 8). We begin with a few definitions.

Let $Q(a, b)$ denote the infinite quadrant with origin at a that contains b. For a pair of nodes $a, b \in V$, define recursively a directed path $\mathcal{P}(a \to b)$ from a to b in Y_4 as follows. If $a = b$, then $\mathcal{P}(a \to b) = null$. If $a \ne b$, there must exist $\vec{ac} \in Y_4$ that lies in $Q(a, b)$. In this case, define

$$\mathcal{P}(a \to b) = \vec{ac} \oplus \mathcal{P}(c \to b).$$

Recall that $\oplus$ represents the concatenation operator. This definition is illustrated in Figure 3a. Fischer et al. [4] show that $\mathcal{P}(a \to b)$ is well defined and lies entirely inside the square centered at b whose boundary contains a.

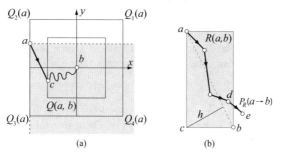

Fig. 3. Definitions: (a) $Q(a, b)$ and $\mathcal{P}(a \to b)$. (b) $\mathcal{P}_R(a \to b)$.

For any node $a \in V$, let $D(a, r)$ denote the open disk centered at a of radius r, and let $\partial D(a, r)$ denote the boundary of $D(a, r)$. Let $D[a, r] = D(a, r) \cup \partial D(a, r)$. For any path P and any pair of nodes $a, b \in P$, let $P[a, b]$ be the subpath of P from a to b. Let $R(a, b)$ be the closed rectangle with diagonal ab.

For a fixed pair of nodes $a, b \in V$, define a path $\mathcal{P}_R(a \to b)$ as follows. Let $e \in V$ be the first node along $\mathcal{P}(a \to b)$ that is not strictly interior to $R(a, b)$. Then $\mathcal{P}_R(a \to b)$ is the subpath of $\mathcal{P}(a \to b)$ that extends between a and e. In other words, $\mathcal{P}_R(a \to b)$ is the path that follows the Y_4 edges pointing towards b, truncated as soon as it reaches b or leaves $R(a, b)$. Formally, $\mathcal{P}_R(a \to b) = \mathcal{P}(a \to b)[a, e]$. This definition is illustrated in Figure 3b. Our proofs will make use of the following two propositions.

Proposition 1. *The sum of the lengths of crossing diagonals of a non-degenerate (necessarily convex) quadrilateral abcd is strictly greater than the sum of the lengths of either pair of opposite sides:*

$$|ac| + |bd| > |ab| + |cd|$$
$$|ac| + |bd| > |bc| + |da|$$

Proposition 2. *For any triangle $\triangle abc$, the following inequalities hold:*

$$|ac|^2 \begin{cases} < |ab|^2 + |bc|^2, & \text{if } \angle abc < \pi/2 \\ = |ab|^2 + |bc|^2, & \text{if } \angle abc = \pi/2 \\ > |ab|^2 + |bc|^2, & \text{if } \angle abc > \pi/2 \end{cases}$$

Lemma 2. *For each pair of nodes $a, b \in V$,*

$$|\mathcal{P}_R(a \to b)| \leq |ab|\sqrt{2} \tag{7}$$

Furthermore, each edge of $\mathcal{P}_R(a \to b)$ is no longer than $|ab|$.

Proof. Let c be one of the two corners of $R(a, b)$, other than a and b. Let $\overrightarrow{de} \in \mathcal{P}_R(a \to b)$ be the last edge on $\mathcal{P}_R(a \to b)$, which necessarily intersects $\partial R(a, b)$ (note that it is possible that $e = b$). Refer to Figure 3b. Then $|de| \leq |db|$, otherwise $\overrightarrow{de}$ could not be in Y_4. Since db lies in the rectangle with diagonal ab, we have that $|db| \leq |ab|$, and similarly for each edge on $\mathcal{P}_R(a \to b)$. This establishes the latter claim of the lemma. For the first claim of the lemma, let $p = \mathcal{P}_R(a \to b)[a, d] \oplus db$. Since $|de| \leq |db|$, we have that $|\mathcal{P}_R(a \to b)| \leq |p|$. Since p lies entirely inside $R(a, b)$ and consists of edges pointing towards b, we have that p is an xy-monotone path. It follows that $|p| \leq |ac| + |cb|$, which is bounded above by $|ab|\sqrt{2}$. □

Lemma 3. *Let $a, b, c, d \in V$ be four disjoint nodes such that $\overrightarrow{ab}, \overrightarrow{cd} \in Y_4$, $b \in Q_i(a)$ and $d \in Q_i(c)$, for some $i \in \{1, 2, 3, 4\}$. Then ab and cd cannot cross.*

The next four lemmas (4–8) each concern a pair of crossing Y_4 edges, culminating (in Lemma 8) in the conclusion that there is a short path in Y_4 between a pair of endpoints of those edges.

Lemma 4. *Let a, b, c and d be four disjoint nodes in V such that $\overrightarrow{ab}, \overrightarrow{cd} \in Y_4$, and ab crosses cd. Then (i) the ratio between the shortest side and the longer diagonal of the quadrilateral acbd is no greater than $1/\sqrt{2}$, and (ii) the shortest side of the quadrilateral acbd is strictly shorter than either diagonal.*

Lemma 5. *Let a, b, c, d be four distinct nodes in V, with $c \in Q_1(a)$, such that (i) $\overrightarrow{ab} \in Q_1(a)$ and $\overrightarrow{cd} \in Q_2(c)$ are in Y_4 and cross each other, and (ii) ad is a shortest side of quadrilateral acbd. Then $\mathcal{P}_R(a \to d)$ and $\mathcal{P}_R(d \to a)$ have a nonempty intersection.*

Lemma 6. *Let a, b, c, d be four distinct nodes in V, with $c \in Q_1(a)$, such that (i) $\overrightarrow{ab} \in Q_1(a)$ and $\overrightarrow{cd} \in Q_3(c)$ are in Y_4 and cross each other, and (ii) ad is a shortest side of quadrilateral acbd. Then $\mathcal{P}_R(d \to a)$ does not cross ab.*

The next lemma relies on all of Lemmas 2–6.

Lemma 7. *Let $a, b, c, d \in V$ be four distinct nodes such that $\overrightarrow{ab} \in Y_4$ crosses $\overrightarrow{cd} \in Y_4$, and let xy be a shortest side of the quadrilateral abcd. Then there exist two paths $\mathcal{P}_x$ and $\mathcal{P}_y$ in Y_4, where $\mathcal{P}_x$ has x as an endpoint and $\mathcal{P}_y$ has y as an endpoint, with the following properties:*

(i) $\mathcal{P}_x$ and $\mathcal{P}_y$ have a nonempty intersection.
(ii) $|\mathcal{P}_x| + |\mathcal{P}_y| \leq 3\sqrt{2}|xy|$.
(iii) Each edge on $\mathcal{P}_x \cup \mathcal{P}_y$ is no longer than $|xy|$.

Proof. Assume without loss of generality that $b \in Q_1(a)$. We discuss the following exhaustive cases:

1. $c \in Q_1(a)$, and $d \in Q_1(c)$. In this case, ab and cd cannot cross each other (by Lemma 3), so this case is finished.

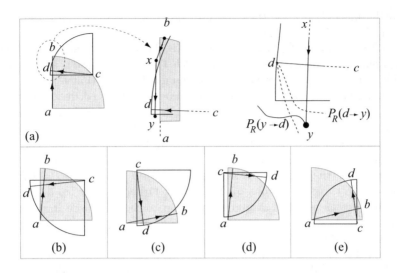

Fig. 4. Lemma 7: (a, b) $c \in Q_1(a)$ (c) $c \in Q_2(a)$ (d) $c \in Q_4(a)$.

2. $c \in Q_1(a)$, and $d \in Q_2(c)$, as in Figure 4a. Since ab crosses cd, $b \in Q_2(c)$. Since $\overrightarrow{ab} \in Y_4$, $|ab| \leq |ac|$. Since $\overrightarrow{cd} \in Y_4$, $|cd| \leq |cb|$. These along with Lemma 4 imply that ad and db are the only candidates for a shortest edge of $acbd$. Assume first that ad is a shortest edge of $acbd$. By Lemma 3, $\mathcal{P}_a = \mathcal{P}_R(a \rightarrow d)$ does not cross cd. It follows from Lemma 5 that $\mathcal{P}_a$ and $\mathcal{P}_d = \mathcal{P}_R(d \rightarrow a)$ have a nonempty intersection. Furthermore, by Lemma 2, $|\mathcal{P}_a| \leq |ad|\sqrt{2}$ and $|\mathcal{P}_d| \leq |ad|\sqrt{2}$, and no edge on these paths is longer than $|ad|$, proving the lemma true for this case. Consider now the case when db is a shortest edge of $acbd$ (see Figure 4a). Note that d is below b (otherwise, $d \in Q_2(c)$ and $|cd| > |cb|$) and, therefore, $b \in Q_1(d)$). By Lemma 3, $\mathcal{P}_d = \mathcal{P}_R(d \rightarrow b)$ does not cross ab. If $\mathcal{P}_b = \mathcal{P}_R(b \rightarrow d)$ does not cross cd, then $\mathcal{P}_b$ and $\mathcal{P}_d$ have a nonempty intersection, proving the lemma true for this case. Otherwise, there exists $\overrightarrow{xy} \in \mathcal{P}_R(b \rightarrow d)$ that crosses cd (see Figure 4a). Define

$$\mathcal{P}_b = \mathcal{P}_R(b \rightarrow d) \oplus \mathcal{P}_R(y \rightarrow d)$$
$$\mathcal{P}_d = \mathcal{P}_R(d \rightarrow y)$$

By Lemma 3, $\mathcal{P}_R(y \rightarrow d)$ does not cross cd. Then $\mathcal{P}_b$ and $\mathcal{P}_d$ must have a nonempty intersection. We now show that $\mathcal{P}_b$ and $\mathcal{P}_d$ satisfy conditions (i) and (iii) of the lemma. Proposition 1 applied on the quadrilateral $xdyc$ tells us that $|xc| + |yd| < |xy| + |cd|$. We also have that $|cx| \geq |cd|$, since $\overrightarrow{cd} \in Y_4$ and x is in the same quadrant of c as d. This along with the inequality above implies $|yd| < |xy|$. Because $xy \in \mathcal{P}_R(b \rightarrow d)$, by Lemma 2 we have that $|xy| \leq |bd|$, which along with the previous inequality shows that $|yd| < |bd|$. This along with Lemma 2 shows that condition (iii) of the lemma is satisfied.

Furthermore, $|\mathcal{P}_R(y \to d)| \le |yd|\sqrt{2}$ and $|\mathcal{P}_R(d \to y)| \le |yd|\sqrt{2}$. It follows that $|\mathcal{P}_b| + |\mathcal{P}_d| \le 3\sqrt{2}|bd|$.

3. $c \in Q_1(a)$, and $d \in Q_3(c)$, as in Figure 4b. Then $|ac| \ge \max\{ab, cd\}$, and by Lemma 4 ac is not a shortest edge of $acbd$. The case when bd is a shortest edge of $acbd$ is settled by Lemmas 3 and 2: Lemma 3 tells us that $\mathcal{P}_d = \mathcal{P}_R(d \to b)$ does not cross ab, and $\mathcal{P}_b = \mathcal{P}_R(b \to d)$ does not cross cd. It follows that $\mathcal{P}_d$ and $\mathcal{P}_b$ have a nonempty intersection. Furthermore, Lemma 2 guarantees that $\mathcal{P}_d$ and $\mathcal{P}_b$ satisfy conditions (ii) and (iii) of the lemma. Consider now the case when ad is a shortest edge of $acbd$; the case when bc is shortest is symmetric. By Lemma 6, $\mathcal{P}_R(d \to a)$ does not cross ab. If $\mathcal{P}_R(a \to d)$ does not cross cd, then this case is settled: $\mathcal{P}_d = \mathcal{P}_R(d \to a)$ and $\mathcal{P}_a = \mathcal{P}_R(a \to d)$ satisfy the three conditions of the lemma. Otherwise, let $\overrightarrow{xy} \in \mathcal{P}_R(a \to d)$ be the edge crossing cd. Arguments similar to the ones used in case 1 above show that $\mathcal{P}_a = \mathcal{P}_R(a \to d) \oplus \mathcal{P}_R(y \to d)$ and $\mathcal{P}_d = \mathcal{P}_R(d \to y)$ are two paths that satisfy the conditions of the lemma.

4. $c \in Q_1(a)$, and $d \in Q_4(c)$, as in Figure 4c. Note that a horizontal reflection of Figure 4c, followed by a rotation of $\pi/2$, depicts a case identical to case 1, which has already been settled.

5. $c \in Q_2(a)$, as in Figure 4d. Note that Figure 4d rotated by $\pi/2$ depicts a case identical to case 1, which has already been settled.

6. $c \in Q_3(a)$. Then it must be that $d \in Q_1(c)$, otherwise cd cannot cross ab. By Lemma 3 however, ab and cd may not cross, unless one of them is not in Y_4.

7. $c \in Q_4(a)$, as in Figure 4e. Note that a vertical reflection of Figure 4e depicts a case identical to case 1, so this case is settled as well. □

We are now ready to establish the main lemma of this section, showing that there is a short path between the endpoints of two intersecting edges in Y_4.

Lemma 8. *Let $a, b, c, d \in V$ be four distinct nodes such that $\overrightarrow{ab} \in Y_4$ crosses $\overrightarrow{cd} \in Y_4$, and let xy be a shortest side of the quadrilateral $abcd$. Then Y_4 contains a path $p(x, y)$ connecting x and y, of length $|p(x, y)| \le \frac{6}{\sqrt{2}-1} \cdot |xy|$. Furthermore, no edge on $p(x, y)$ is longer than $|xy|$.*

Proof. Let $\mathcal{P}_x$ and $\mathcal{P}_y$ be the two paths whose existence in Y_4 is guaranteed by Lemma 7. By condition (iii) of Lemma 7, no edge on $\mathcal{P}_x$ and $\mathcal{P}_y$ is longer than $|xy|$. By condition (i) of Lemma 7, $\mathcal{P}_x$ and $\mathcal{P}_y$ have a nonempty intersection. If $\mathcal{P}_x$ and $\mathcal{P}_y$ share a node $u \in V$, then the path $p(x, y) = \mathcal{P}_x[x, u] \oplus \mathcal{P}_y[y, u]$ is a path from x to y in Y_4 no longer than $3\sqrt{2}|xy|$; the length restriction follows from guarantee (ii) of Lemma 7. Otherwise, let $\overrightarrow{a'b'} \in \mathcal{P}_x$ and $\overrightarrow{c'd'} \in \mathcal{P}_y$ be two edges crossing each other. Let $x'y'$ be a shortest side of the quadrilateral $a'c'b'd'$, with $x' \in \mathcal{P}_x$ and $y' \in \mathcal{P}_y$. Lemma 7 tells us that $|a'b'| \le |xy|$ and $|c'd'| \le |xy|$. These along with Lemma 4 imply that $|x'y'| \le |xy|/\sqrt{2}$. This enables us to derive a recursive formula for computing a path $p(x, y) \in Y_4$ as follows:

$$p(x, y) = \begin{cases} x, & \text{if } x = y \\ \mathcal{P}_x[x, x'] \oplus \mathcal{P}_y[y, y'] \oplus p(x', y'), & \text{if } x \ne y \end{cases}$$

Simple induction on the length of xy establishes the claim of the lemma. □

4 Y_4^∞ and Y_4

We prove that every individual edge of Y_4^∞ is spanned by a short path in Y_4. This, along with the result of Theorem 1, establishes that Y_4 is a spanner. Fix an edge $\overrightarrow{xy} \in Y_4^\infty$. Define an edge or a path as *t-short* (with respect to $|xy|$) if its length is within a constant factor t of $|xy|$. In our proof that ab is spanned by a t-short path with respect to $|ab|$ in Y_4, we will make use of the following three statements.

S1 If ab is t-short, then $\mathcal{P}_R(a \to b)$, and therefore its reverse, $\mathcal{P}_R^{-1}(a \to b)$, are $t\sqrt{2}$-short by Lemma 2.

S2 If $ab \in Y_4$ is t_1-short and $cd \in Y_4$ is t_2-short, and if ab intersects cd, Lemmas 4 and 8 show that there is a t_3-short path between any two of the endpoints of these edges with $t_3 = t_1 + t_2 + 3(2 + \sqrt{2}) \max(t_1, t_2)$.

S3 If $p(a, b)$ is a t_1-short path and $p(c, d)$ is a t_2-short path and the two paths intersect, then there is a t_3-short path P between any two of the endpoints of these paths with $t_3 = t_1 + t_2 + 3(2 + \sqrt{2}) \max(t_1, t_2)$, by **S2**.

Lemma 9. *For any edge $ab \in Y_4^\infty$, there is a path $p(a, b) \in Y_4$ between a and b, of length $|p(a, b)| \leq t|ab|$, for $t = 29 + 23\sqrt{2}$.*

Proof. For the sake of clarity, we only prove here that there is a short path $p(a, b)$ between a and b, and skip the calculations of the actual stretch factor t (which are detailed in the appendix of [1]). We refer to an edge or a path as *short* if its length is within a constant factor of $|ab|$. Assume without loss of generality that $\overrightarrow{ab} \in Y_4^\infty$, and $\overrightarrow{ab} \in Q_1(a)$. If $\overrightarrow{ab} \in Y_4$, then $p(a, b) = ab$ and the proof is finished. So assume the opposite, and let $\overrightarrow{ac} \in Q_1(a)$ be the edge in Y_4; since $Q_1(a)$ is nonempty, $\overrightarrow{ac}$ exists. Because $\overrightarrow{ac} \in Y_4$ and b is in the same quadrant of a as c, we have that

$$|ac| \leq |ab| \qquad \text{(i)}$$
$$|bc| \leq |ac|\sqrt{2} \qquad \text{(ii)} \qquad\qquad (8)$$

Thus both ac and bc are short. And this in turn implies that $\mathcal{P}_R(b \to c)$ is short by **S1**. We next focus on $\mathcal{P}_R(b \to c)$. Let $b' \notin R(b, c)$ be the other endpoint of $\mathcal{P}_R(b \to c)$. We distinguish three cases.

Case 1: $\mathcal{P}_R(b \to c)$ and ac intersect. Then by **S3** there is a short path $p(a, b)$ between a and b.

Case 2: $\mathcal{P}_R(b \to c)$ and ac do not intersect, and $\mathcal{P}_R(b' \to a)$ and ab do not intersect (see Figure 5b). Note that because b' is the endpoint of the short path $\mathcal{P}_R(b \to c)$, the triangle inequality on $\triangle abb'$ implies that ab' is short, and therefore $\mathcal{P}_R(b' \to a)$ is short. We consider two cases:

(i) $\mathcal{P}_R(b' \to a)$ intersects ac. Then by **S3** there is a short path $p(a, b')$. So

$$p(a, b) = p(a, b') \oplus \mathcal{P}_R^{-1}(b \to c)$$

is short.

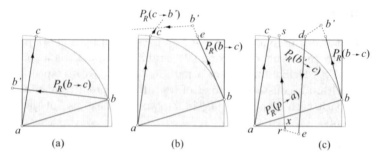

Fig. 5. Lemma 9: (a) Case 1: $\mathcal{P}_R(b \to c)$ and ac have a nonempty intersection. (b) Case 2: $\mathcal{P}_R(b' \to a)$ and ab have an empty intersection. (c) Case 3: $\mathcal{P}_R(b' \to a)$ and ab have a non-empty intersection.

(ii) $\mathcal{P}_R(b' \to a)$ does not intersect ac. Then $\mathcal{P}_R(c \to b')$ must intersect $\mathcal{P}_R(b \to c) \oplus \mathcal{P}_R(b' \to a)$. Next we establish that $b'c$ is short. Let $\overrightarrow{eb'}$ be the last edge of $\mathcal{P}_R(b \to c)$, and so incident to b' (note that e and b may coincide). Because $\mathcal{P}_R(b \to c)$ does not intersect ac, b' and c are in the same quadrant for e. It follows that $|eb'| \le |ec|$ and $\angle b'ec < \pi/2$. These along with Proposition 2 for $\triangle b'ec$ imply that $|b'c|^2 < |b'e|^2 + |ec|^2 \le 2|ec|^2 < 2|bc|^2$ (this latter inequality uses the fact that $\angle bec > \pi/2$, which implies that $|ec| < |bc|$). It follows that

$$|b'c| \le |bc|\sqrt{2} \le 2|ac| \qquad \text{(by (8)ii)} \qquad (9)$$

Thus $b'c$ is short, and by **S1** we have that $\mathcal{P}_R(c \to b')$ is short. Since $\mathcal{P}_R(c \to b')$ intersects the short path $\mathcal{P}_R(b \to c) \oplus \mathcal{P}_R(b' \to a)$, there is by **S3** a short path $p(c, b)$, and so

$$p(a, b) = ac \oplus p(c, b)$$

is short.

Case 3: $\mathcal{P}_R(b \to c)$ and ac do not intersect, and $\mathcal{P}_R(b' \to a)$ intersects ab (see Figure 5c). If $\mathcal{P}_R(b' \to a)$ intersects ab at a, then $p(a, b) = \mathcal{P}_R(b \to c) \oplus \mathcal{P}_R(b' \to a)$ is short. So assume otherwise, in which case there is an edge $\overrightarrow{de} \in \mathcal{P}_R(b' \to a)$ that crosses ab. Then $d \in Q_1(a)$, $e \in Q_3(a) \cup Q_4(a)$, and e and a are in the same quadrant for d. Note however that e cannot lie in $Q_3(a)$, since in that case $\angle dae > \pi/2$, which would imply $|de| > |da|$, which in turn would imply $\overrightarrow{de} \notin Y_4$. So it must be that $e \in Q_4(a)$.

Next we show that $\mathcal{P}_R(e \to a)$ does not cross ab. Assume the opposite, and let $\overrightarrow{rs} \in \mathcal{P}_R(e \to a)$ cross ab. Then $r \in Q_4(a)$, $s \in Q_1(a) \cup Q_2(a)$, and s and a are in the same quadrant for r. Arguments similar to the ones above show that $s \notin Q_2(a)$, so s must lie in $Q_1(a)$. Let d be the L_∞ distance from a to b. Let x be the projection of r on the horizontal line through a. Then

$$|rs| \ge |rx| + d \ge |rx| + |xa| > |ra| \qquad \text{(by the triangle inequality)}$$

Because a and s are in the same quadrant for r, the inequality above contradicts $\vec{rs} \in Y_4$.

We have established that $\mathcal{P}_R(e \to a)$ does not cross ab. Then $\mathcal{P}_R(a \to e)$ must intersect $\mathcal{P}_R(e \to a) \oplus de$. Note that de is short because it is in the short path $\mathcal{P}_R(b' \to a)$. Thus ae is short, and so $\mathcal{P}_R(a \to e)$ and $\mathcal{P}_R(e \to a)$ are short. Thus we have two intersecting short paths, and so by **S3** there is a short path $p(a, e)$. Then

$$p(a, b) = p(a, e) \oplus \mathcal{P}_R^{-1}(b' \to a) \oplus \mathcal{P}_R^{-1}(b \to c)$$

is short. Straightforward calculations show that, in each of these cases, the stretch factor for $p(a, b)$ does not exceed $29 + 23\sqrt{2}$.   □

Our main result follows immediately from Theorem 1 and Lemma 9:

Theorem 2. Y_4 is a t-spanner, for $t \geq 8\sqrt{2}(29 + 23\sqrt{2})$.

5   Conclusion

Our results settle a long-standing open problem, asking whether Y_4 is a spanner or not. We answer this question positively, and establish a loose stretch factor of $8\sqrt{2}(29 + 23\sqrt{2})$. Experimental results, however, indicate a stretch factor of the order $1 + \sqrt{2}$, a factor of 200 smaller. Finding tighter stretch factors for both Y_4^∞ and Y_4 remain interesting open problems. Establishing whether Y_5 and Y_6 are spanners or not is also open.

References

1. Bose, P., Damian, M., Douïeb, K., O'Rourke, J., Seamone, B., Smid, M., Wuhrer, S.: $\pi/2$-Angle Yao Graphs are Spanners. Technical Report, arXiv:1001.2913v1 (2010)
2. Bose, P., Maheshwari, A., Narasimhan, G., Smid, M., Zeh, N.: Approximating geometric bottleneck shortest paths. Computational Geometry: Theory and Applications 29, 233–249 (2004)
3. Damian, M., Molla, N., Pinciu, V.: Spanner properties of $\pi/2$-angle Yao graphs. In: Proc. of the 25th European Workshop on Computational Geometry, pp. 21–24 (March 2009)
4. Fischer, M., Lukovszki, T., Ziegler, M.: Geometric searching in walkthrough animations with weak spanners in real time. In: Bilardi, G., Pietracaprina, A., Italiano, G.F., Pucci, G. (eds.) ESA 1998. LNCS, vol. 1461, pp. 163–174. Springer, Heidelberg (1998)
5. Molla, N.: Yao spanners for wireless ad hoc networks. M.S. Thesis, Department of Computer Science, Villanova University (December 2009)
6. Green, J.W.: A note on the chords of a convex curve. Portugaliae Mathematica 10(3), 121–123 (1951)
7. Yao, A.C.-C.: On constructing minimum spanning trees in k-dimensional spaces and related problems. SIAM Journal on Computing 11(4), 721–736 (1982)

Identifying Shapes Using Self-assembly
(Extended Abstract)

Matthew J. Patitz[1] and Scott M. Summers[2]

[1] Department of Computer Science, University of Texas–Pan American,
Edinburg, TX, 78539, USA
mpatitz@cs.panam.edu
[2] Department of Computer Science and Software Engineering,
University of Wisconsin–Platteville, Platteville, WI 53818, USA
summerss@uwplatt.edu

Abstract. In this paper, we introduce the following problem in the theory of algorithmic self-assembly: given an input shape as the seed of a tile-based self-assembly system, design a finite tile set that can, in some sense, uniquely identify whether or not the given input shape–drawn from a very general class of shapes–matches a particular target shape. We first study the complexity of correctly identifying squares. Then we investigate the complexity associated with the identification of a considerably more general class of non-square, hole-free shapes.

1 Introduction

As amazingly complex as biological organisms are, at the nanoscale they are composed of "simple" pieces that spontaneously self-assemble–a bottom-up process by which a relatively small group of fundamental components combine according to local rules in order to form a complex structure. This very basic process is responsible for the vast diversity and complexity of life–from the most simple single-cell organisms to human beings.

Inspired by nature, scientists have developed and studied a wide variety of artificial self-assembling systems in order to produce structures as varied as nanowires [25], crystals [11], nanofiber scaffoldings [10], landscapes for nanoscale robots [9, 13] and dozens of other novel supramolecules (see [4, 15, 19] for more examples). In addition to experimental work, there has also been a plethora of theoretical work in the design and analysis of the complexities and limitations of self-assembling systems, with notable examples including [6, 8, 14, 16].

Much of the research in algorithmic self-assembly (both theoretical and experimental) can be loosely categorized into four "genres:" the self-assembly of shapes [17, 18], evaluating computable functions to direct nanoscale self-assembly [24] replicating input shapes [1], and creating novel materials that have various chemical properties [26]. In this paper, we introduce a novel (theoretical) self-assembly problem that is motivated by not only the behavior of biological systems but also the practical need to verify artificial laboratory-based self-assembly systems. We call this new problem the *shape identification problem*, and define it as the task

O. Cheong, K.-Y. Chwa, and K. Park (Eds.): ISAAC 2010, Part II, LNCS 6507, pp. 458–469, 2010.
© Springer-Verlag Berlin Heidelberg 2010

of designing a tile-based self-assembly system that positively identifies a target structure that has a pre-specified shape (and size) from among possible "junk" structures drawn from a very general pool of objects.

Motivation: Shape identification is a fundamental process of nature and is explicitly used by biological systems in a variety of ways. First and foremost, the immune system generates complexes whose express purpose is to selectively identify–and ultimately bind to–precisely-shaped locations on the surface of foreign objects in order to mark them for destruction (by, for example, killer T cells). Also, cellular transport systems, such as those which transport amino acids or sugars, work by moving specifically-shaped molecules from one side of a membrane to the other. Furthermore, the power of a self-assembling system (natural or artificial) ultimately arises from the information encoded in its constituent components. In the notable case of proteins, it is the information embedded in their precise three-dimensional geometry that allow them to match and combine with the necessary specificity to build the fundamental building blocks of life.

The ability to correctly identify only the completely formed products of an artificial self-assembly system is also of extreme importance to practitioners. Unfortunately, accomplishing this task is difficult because the self-assembly environment is often variable and chaotic, where mistakes are likely to be made and partially-formed products common. Current methods of imaging the results of nanoscale self-assembling systems provide insufficient resolution for automated visual inspection of assemblies and require error-prone manual inspection (for instance, by pouring over atomic force microscope images). Methods such as gel electrophoresis allow for the separation of products based loosely on their mass and shape, but unfortunately with far less shape specificity than desired. With accurate nanoscale shape identification schemes, however, the accuracy of the techniques that experimenters use to identify the products of self-assembling systems could be improved dramatically.

In this paper, we formulate the shape identification problem in algorithmic self-assembly (defined formally in Section 2.1) and exhibit a variety of solutions thereof while working in the *RNAse enzyme model*–a discrete mathematical model of two-handed tile-based self-assembly (based on Winfree's abstract Tile Assembly Model [16, 21]) that distinguishes DNA tiles from RNA tiles and permits the usage of an RNAse enzyme that dissolves all of the RNA tiles in a given assembly. This model was initially suggested by Rothemund and Winfree in the final section of [16] and formally defined by Abel, Benbernou, Damian, Demaine, Demaine, Flatland, Kominers and Schweller [1]. We focus our attention on the design of "small" tile sets that identify certain types of target shapes by tagging them with a border of DNA tiles. Note that the borders, which signify positive identification, could also be "functionalized" with bindings sites that facilitate the easy extraction of only the correct assemblies. It is worthy of note that, while the results presented in this paper are based on tile-based self-assembly systems identifying tile-based assemblies, the underlying principles of this paper are applicable to the identification of any type of precisely shaped shaped object (e.g., a DNA origami complex [15]) so long as its perimeter advertises the necessary binding domains, which in the case of this paper, are single-stranded DNA sequences.

Statement of Results: In Section 3, we exhibit a *planar* tile assembly system capable of identifying any $n \times n$ square using $O\left(\frac{\log n}{\log \log n}\right)$. We subsequently prove a matching lower bound on the minimum number of unique tile types required to identify an $n \times n$ square for *almost all* n. We conclude Section 3 with a $O(1)$ size planar tile assembly system that "universally" identifies whether or not any hole-free input shape is an $n \times n$ square. In Section 4, we develop a non-planar tile assembly system that identifies a wide variety of "hole-free" shapes that have a kind of "perimeter-rectangle decomposition" that uses an optimal number of unique tile types in the sense of Kolmogorov complexity. We then mildly extend the aforementioned result to identify a more general class of shapes–assuming the use of two different types of RNAse enzymes is permitted.

2   Preliminaries and Notation

In this paper, we work in a variant of Erik Winfree's abstract Tile Assembly Model [21, 22] modified to model unseeded growth, known as the *two-handed* aTAM, which has been studied previously under various names [1–3, 6, 7, 12, 23]. In the two-handed aTAM, any two assemblies can attach to each other, rather than enforcing that tiles can only accrete one at a time to an existing seed assembly. A *tile type* is a unit square with four sides, each having a *glue* consisting of a *label* (a finite string) and *strength* (a natural number). We represent tiles as squares. Notches on the sides of tile types represent the glue strength of that side. The thick notches represent strength 4, and otherwise each single notch contributes a strength of one to the glue strength of that side.

We assume a finite set T of tile types, but an infinite number of copies of each tile type, each copy referred to as a *tile*. A *supertile* (a.k.a., *assembly*) is a positioning of tiles on the integer lattice $\mathbb{Z}^2$. Two adjacent tiles in a supertile *interact* if the glues on their abutting sides are equal and have positive strength. Each supertile induces a *binding graph*, a grid graph whose vertices are tiles, with an edge between two tiles if they interact. The supertile is τ-*stable* if every cut of its binding graph has strength at least τ, where the weight of an edge is the strength of the glue it represents. That is, the supertile is stable if at least energy τ is required to separate the supertile into two parts. A *tile assembly system* (TAS) is a pair $\mathcal{T} = (T, \sigma, \tau)$, where T is a finite tile set, σ is an initial seed configuration and $\tau \in \mathbb{N}$ is the *temperature*. Given a TAS $\mathcal{T} = (T, \sigma, \tau)$, a supertile is *producible* if either it is a single tile from T, σ, or it is the τ-stable result of translating two producible assemblies. A supertile α is *terminal* if for every producible supertile β, α and β cannot be τ-stably attached. A TAS is *directed* (a.k.a. *deterministic* or *confluent*) if it has only one terminal, producible supertile and all producible assemblies are finite. Given a connected shape $X \subseteq \mathbb{Z}^2$, a TAS $\mathcal{T}$ produces X *uniquely* if every producible, terminal supertile places tiles only on positions in X (appropriately translated if necessary).

We also assume that each tile type is defined as being either composed of DNA or of RNA. The utility of RNA-based tile types comes from that fact that, at prescribed points during the assembly process, we assume that the experimenter can add an RNAse enzyme to the solution which causes all tiles composed of RNA to dissolve. We assume that when this occurs, all portions of all RNA tiles are completely dissolved, including glue portions that may be bound to DNA tiles, returning the previously bound edges of those DNA tiles to unbound states.

In other words, for a given supertile α that is stable at temperature τ, when the RNAse enzyme is added to the solution, all positions in α which are occupied by RNA tiles become undefined (locations at which no tiles exist). The resultant supertile may not be τ-stable and thus defines a multiset of subsupertiles consisting of the maximal stable supertiles of α at temperature τ, which we denote as $BREAK^\tau(\alpha)$.

Unless explicitly stated, in this paper we subscribe to the restriction that the RNAse enzyme *must* be added exactly once–and *only* after an initial phase of two-handed self-assembly (involving both DNA and RNA tiles at temperature τ) reaches some (intermediate) terminal state. Of course, after the RNAse enzyme has completely dissolved all of the RNA tiles, two-handed self-assembly of only DNA tiles is allowed to proceed until a final terminal state is reached. We also assume that tile types *cannot* be added at any point of the self-assembly process, whence all of our constructions in this paper have $O(1)$ *stage complexity*.

The two handed RNAse enzyme assembly model was initially suggested in [16] and was recently used to study the problem of replicating shapes using self-assembly [1].

2.1   The Shape Identification Problem in the RNAse Enzyme Model

For every *shape* (a.k.a., a finite, connected subset of $\mathbb{Z}^2$) X, define the assembly σ_X as the placement of specially designated seed (DNA) tiles at every point in X subject to the restrictions that σ_X must be τ-stable, the strengths of all of the "external" glues must be 1 and be labeled with the empty string. [1]

We now define the *shape identification problem* in self-assembly. For any set of shapes $\mathcal{C}$ and any target shape $X \in \mathcal{C}$, the ordered pair $(\mathcal{C}, X)$ is an *instance* of the shape identification problem. A *solution* to the shape identification problem instance $(\mathcal{C}, X)$ is an ordered pair (T, τ), where T is a finite set of tile types, $\tau \in \mathbb{N}$ is the temperature and for any $Y \in \mathcal{C}$, the tile assembly system $\mathcal{T} = (T, \sigma_Y, \tau)$ satisfies the following conditions. If $X = Y$ then $\mathcal{T}$ uniquely produces a fully-connected final assembly α consisting of σ_Y with a fully connected ring of easily-distinguishable "border" tiles along the perimeter of σ_Y. However, if $X \neq Y$, then σ_Y is terminal (no tiles remain attached to it) and the size of (number of tiles in) any terminal assembly $\alpha \neq \sigma_Y$ is at most 5. Here, (T, τ) *identifies the shape X with respect to $\mathcal{C}$ at temperature τ*, a.k.a., (T, τ) *solves* $(\mathcal{C}, X)$.

---

[1] We speculate that one possible molecular implementation of this might be achieved using Rothemund's DNA origami as a seed structure [5, 15] to which DNA and RNA tiles can subsequently attach.

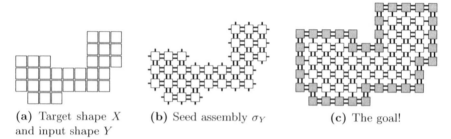

(a) Target shape X (b) Seed assembly σ_Y (c) The goal!
and input shape Y

Fig. 1. The desired outcome for a "yes" instance of the shape identification problem

An example of an instance of–and solution to–the shape identification problem (for some shape with respect to some class of shapes) is depicted in Figure 1. We say that the *identification complexity of* (C, X) is the number of unique tile types in the smallest tile set of any solution that solves (C, X) (this definition should be somewhat reminiscent of the *tile complexity* [16] of a shape X being defined as the minimum number of tile types necessary to uniquely produce X).

3   Identification of $n \times n$ Squares with Planarity

We begin our investigation into the identification of shapes using self-assembly by studying the planar identification of $n \times n$ squares. A tile assembly system T is *planar* if, for every possible assembly sequence in T, all attaching supertiles have obstacle-free paths to their mates and therefore require use of only two spacial dimensions. The interested reader is encouraged to consult [7] for more discussion of planarity in self-assembly. We say that a solution (T, τ) to (C, X) is *planar* if, for every shape $Y \subseteq \mathbb{Z}^2$, the tile assembly system $T = (T, \sigma_Y, \tau)$ is planar.

In this section, we exhibit two planar solutions to the shape identification problem that efficiently identify $n \times n$ squares with respect to the set of all *hole-free* shapes: shapes whose complements are infinite, connected subsets of $\mathbb{Z}^2$. We also construct a universal tile set that is capable of identifying whether a given input shape is in fact a square of *any* dimension.

For each $n \in \mathbb{N}$, let $S_n = \{0, 1, \ldots, n-1\}^2$ be the $n \times n$ square whose lower-left corner is positioned at the origin. Throughout this paper, C denotes the class of all hole-free shapes. The motivating factor behind defining C this way is because we want our constructions to be able to distinguish a target shape from among many different possible "junk" (i.e., non-square) input shapes.

3.1   Planar Identification of $n \times n$ Squares with $O(\log n)$ Unique Tile Types

Our first main result of this section is the following theorem, which essentially states that there is an efficient planar identification scheme for $n \times n$ squares, i.e., the identification complexity of S_n with respect to C is $O(\log n)$.

Theorem 1. *For every $6 < n \in \mathbb{N}$, there exists a tile set T_n with $|T_n| = O(\log n)$ such that $(T_n, \tau = 4)$ is planar and solves $(\mathcal{C}, S_n)$.*

The proof idea of Theorem 1 is as follows. Suppose we are trying to identify S_n for some $6 < n \in \mathbb{N}$. Given an input shape $Y \in \mathcal{C}$, our construction first attaches "verification modules" to north- south- and west-facing sides of Y (if Y is an $n \times n$ square, then there will be exactly one of each of these types of sides). These modules are side-by-side pairs of unary counters and binary counters that do not interact with each other as they count. The unary counters count (in unary) the length of the side to which they are attached and the binary counters essentially count (in binary) up to n (the dimension of the target square). Each verification module compares the length of the side to which it is attached with n. If all three verification modules report success *and* agree with each other, then the input shape is in some sense "almost" a square. The three verification modules then cooperate to allow DNA border tiles to start attaching to the east-facing side of the input shape. If border tiles can attach to all but the two bottom rightmost points along the east-facing side, then the input shape is in fact S_n and our construction reaches an intermediate terminal state. At this point, we add the RNAse enzyme leaving only the input shape to which the east-facing border tiles are attached. The remaining border tiles attach in a clockwise fashion until a complete and fully connected border is assembled. However, if not all east-facing border tiles can attach (i.e., the input shape is not S_n), then after the RNAse enzyme is added all previously-attached border tiles will disassociate one at a time until no tiles are attached to the input shape.

3.2 $\Omega\left(\frac{\log n}{\log \log n}\right)$ Unique Tile Types Are Necessary to Identify an $n \times n$ Square

In [16], Rothemund and Winfree established an $\Omega\left(\frac{\log n}{\log \log n}\right)$ lower bound on the number of tile types required to uniquely assemble an $n \times n$ full square for almost all n. In this section, we adapt their information-theoretic proof technique to the shape identification problem for $n \times n$ squares under the RNAse enzyme model. Formally, we say that $P(n)$ is true for *almost all n* if and only if $\lim_{n \to \infty} \frac{|\{1 \leq m \leq n | P(m) \text{ is true }\}|}{n} = 1$.

Theorem 2. *For almost all $n \in \mathbb{N}$, if (T, τ) solves $(\mathcal{C}, S_n)$, then $|T| = \Omega\left(\frac{\log n}{\log \log n}\right)$.*

Proof. For each $n \in \mathbb{N}$, define the *Kolmogorov complexity of n* as $K(n) = \min\{|\pi| \mid U(\pi) = n\}$ where U is some fixed universal Turing machine. The reader is encouraged to consult [20] for a more detailed discussion of Kolmogorov complexity. An easy application of the pigeonhole principle tells us that for almost all $n \in \mathbb{N}$, $K(n) = \Omega(\log n)$.

For $n \in \mathbb{N}$ satisfying $K(n) \geq \log n$ and temperature $\tau \in \mathbb{N}$, there exists a constant size Turing machine M that takes as input a tile set T_n that uniquely identifies S_n, a seed assembly representing the input shape σ (as discussed in

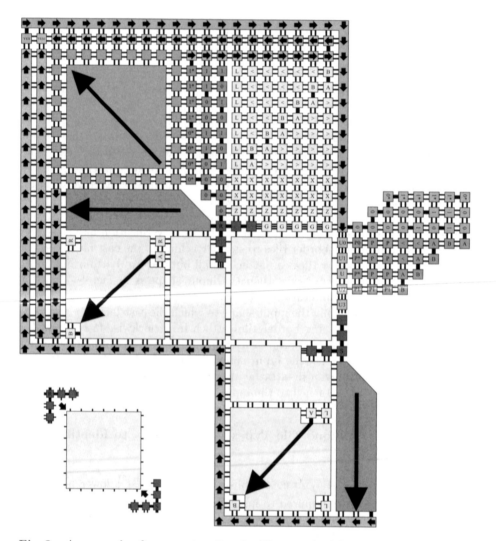

Fig. 2. An example of our construction for Theorem 1 with $n = 7$. Our tile set is partitioned into several logical groups–each given a different color in this figure to represent the relative order in which they assemble (i.e., Red, Orange, Yellow, Green, Blue, Indigo, Violet). First, the red supertiles assemble and attach to the corners of the input shape. The orange group essentially encodes the length of the to-be-identified square via a binary counter and requires $O(\log n)$ unique tile types. The "U" border tiles attach along the east-facing side of the input shape. All tiles are RNA tiles except for the "U" tiles and, of course, the tiles that make up the initial seed square.

Section 2.1) and outputs the maximum extent (height or width) of the uniquely produced terminal assembly. We can then use M as a subroutine in another constant size Turing machine N that takes as input T_n and sequentially simulates M on T_n with the seed assembly σ_{S_i} for $i \geq 0$ *in order* while checking if the maximum extent (height or width) of the ith uniquely produced terminal assembly is $i + 2$. Since T_n uniquely identifies S_n, we are guaranteed that this search will eventually terminate, at which point N halts and outputs $i = n$. This implies that the size of (number of bits in) the encoding for T_n must be $\Omega(\log n)$. Since we can encode an arbitrary tile set T with $O(|T| \log |T|)$ bits (assuming T has a diagonal strength function) and τ with $O(1)$ bits, we have that $|T_n| = \Omega \left(\frac{\log n}{\log \log n} \right)$.

3.3 Planar Identification of $n \times n$ Squares with $O \left(\frac{\log n}{\log \log n} \right)$ Unique Tile Types

The construction for Theorem 1 can be modified to prove the following asymptotically optimal result for the identification of $n \times n$ squares.

Theorem 3. *For every $6 < n \in \mathbb{N}$, there exists a tile set T_n with $|T_n| = O \left(\frac{\log n}{\log \log n} \right)$ such that $(T_n, \tau = 4)$ is planar and solves $(\mathcal{C}, S_n)$.*

3.4 Universal Planar Identification of Squares with $O(1)$ Unique Tile Types

In the previous subsections, we focused our attention on the problem of identifying $n \times n$ squares for particular values of n from among any input shape drawn from the set of all hole-free shapes. We now study the related problem of universally identifying whether or not a given input shape is an $n \times n$ square for some $n \in \mathbb{N}$. Here, we are given an arbitrary hole-free input shape and we wish to correctly identify it (in the sense of tagging its border with special tiles) if and only if it is in fact a square.

Theorem 4. *There exists a finite tile set T with $|T| = O(1)$ such that, for every $6 < n \in \mathbb{N}$, $(T, \tau = 4)$ is planar and solves $(\mathcal{C}, S_n)$.*

Intuitively, we prove Theorem 4 by constructing a constant size tile set that (1) grows unary counters off of the north, west and south sides of the input shape and then (2) allows a border of DNA tiles to assemble if and only if all of the counters agree on the same value (in addition to the right side of the input shape being consistent with that of a square).

4 Non-planar Identification of More Shapes

We now exhibit a non-planar self-assembly system that efficiently identifies a wide variety of shapes with respect to the set of all hole-free shapes but at the expense of sacrificing planarity. We first define some notation.

Let $(x, y) = \boldsymbol{a} \in \mathbb{Z}^2$ and $(w, z) = \boldsymbol{b} \in \mathbb{Z}^2$ and define $d_\infty(\boldsymbol{a}, \boldsymbol{b}) = \max\{|x - w|, |y - z|\}$. If X is a shape, then we say that the *feature size* of X is the minimum $d_\infty(\boldsymbol{a}, \boldsymbol{b})$ such that $\boldsymbol{a}$ and $\boldsymbol{b}$ are on two non-adjacent edges of X.

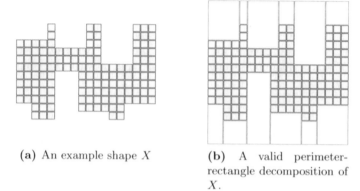

(a) An example shape X

(b) A valid perimeter-rectangle decomposition of X.

Fig. 3. An example of a perimeter-rectangle decomposition of a particular shape

We say that a shape X is *x-monotone* if its intersection with any vertical line is a connected line. If X is a shape, then let $\widetilde{R}(X)$ be the smallest rectangle that contains $X + \{(0,3),(0,-3)\}$, where, for any set $A \subseteq \mathbb{Z}^2$, $X + A = \{x + a \mid x \in X \text{ and } a \in A\}$. We say that X has a *perimeter-rectangle decomposition*, denoted as $\{R_i\}_{i=0}^{n-1}$ for some $n \in \mathbb{N}$, if: for each $0 \le i < n$, R_i is a rectangle, for all $0 \le j < n$, $i \ne j \Rightarrow R_i \cap R_j = \emptyset$, $height(R_i) \le 2^{width(R_i)} + 3$, $\widetilde{R}(X) - X = \bigcup_{i=0}^{n-1} R_i$, and for each $0 \le i < n$, the perimeter of R_i intersects the perimeter of $\widetilde{R}(X)$. For any rectangle R, we write $h(R) = height(R)$, and $w(R) = width(R)$. See Figure 3 for an example of a shape and a valid perimeter-rectangle decomposition thereof. Recall that $\mathcal{C}$ is the set of all hole-free shapes.

Theorem 5. *Fix a universal Turing machine U. Let X be a shape and π_X be any program such that $U(\pi_X) = \langle X \rangle$, where $\langle \cdot \rangle$ is a standard binary encoding of a finite object. If X is x-monotone, has feature size 5 and has perimeter-rectangle decomposition $\{R_i\}_{i=0}^{n-1}$, then the identification complexity of X with respect to $\mathcal{C}$ is $O\left(\frac{|\pi_X|}{\log|\pi_X|}\right)$.*

Note that by choosing π_X to be the shortest program such that $U(\pi_X) = \langle X \rangle$, then $|\pi_X| = K(X)$. The proof idea of Theorem 5 is as follows. Given a shape X that satisfies the hypothesis, our construction first converts X into a string xyz such that x encodes all of the "north-facing" features of X, y encodes $h\left(\widetilde{R}(X)\right)$ and z encodes all of the "south-facing" features of X. Our construction then uses this string as a seed in order to assemble a frame to which an input shape can attach. Once the input shape attaches to the frame, a single-tile-wide border assembles around the perimeter of the input shape and fills in completely if and only if the input shape matches the target shape. Once the RNAse enzyme is added, the frame dissolves and if the input shape has a full border of DNA tiles, then we are done and the target shape has been correctly identified. However, if the input shape does not match the target shape, then our construction ensures that a full border around the input shape is not allowed to assemble. Moreover, if the border is not fully formed after the RNAse enzyme is added, then the partially formed border will disassemble in a counter-clockwise fashion one tile at a time eventually leaving the input shape completely free of DNA border tiles. We encode X as a program π_X using the optimal encoding scheme of Soloveichik and Winfree [18], whence the identification complexity of X with respect to $\mathcal{C}$ is $O\left(\frac{|\pi_X|}{\log|\pi_X|}\right)$.

5 Non-planar Identification of *Even More* Shapes

Throughout this paper, we have assumed that the RNAse enzyme (the agent responsible for removing all of the RNA tile types) is added once–and only–after the initial stage of two-handed self-assembly is allowed to reach a terminal state. Under this assumption, the RNAse enzyme universally dissolves *every* RNA tile in all of the produced assemblies. In this section, we relax this restriction and allow for the use of *two* different types of RNAse enzymes in two separate dissolve stages that each affect a different group of RNA tiles. Doing so leads to a mild refinement of Theorem 4, stated precisely as the following theorem.

Theorem 6. *Fix a universal Turing machine U and let X be a shape and π_X be any program such that $U(\pi_X) = \langle X \rangle$. If X is x-monotone, has feature size 6 and if the use of two different types of RNAse enzymes in two separate dissolve stages is permitted, then the identification complexity of X with respect to $\mathcal{C}$ is* $O\left(\frac{|\pi_X|}{\log|\pi_X|}\right)$.

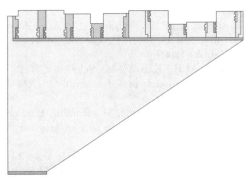

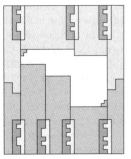

(a) Very high-level overview of the construction for Theorem 6. The grey wedge represents a self-assembly simulation of a Turing machine that unpacks a compact description of all of the rectangles that eventually assemble into a frame that accepts the input shape.

(b) After the first type of RNAse is added, all of the supertiles are free to assemble into a frame that accepts the input shape.

Fig. 4. Overview of the construction for Theorem 6

Notice that, unlike Theorem 4, in Theorem 6, we no longer require that X have a perimeter-rectangle decomposition, which implies that now X may potentially have a "jagged" border–perhaps even drastically so!

The proof idea of Theorem 6 is similar to that of Theorem 4 in that we assemble a frame that accepts an input shape Y and allows a border of DNA tiles to assemble if and only if the $Y = X$. In order to overcome the assumption that the Y must have a perimeter-rectangle decomposition, we use two dissolve stages.

Acknowledgment. The authors are indebted to Paul Rothemund and John Mayfield for helpful discussions as well as to an anonymous reviewer for providing detailed comments regarding several technical aspects of this paper.

References

1. Abel, Z., Benbernou, N., Damian, M., Demaine, E., Demaine, M., Flatland, R., Kominers, S., Schweller, R.: Shape replication through self-assembly and RNAse enzymes. In: SODA 2010: Proceedings of the Twentyfirst Annual ACM-SIAM Symposium on Discrete Algorithms, pp. 1045–1064 (2010)
2. Adleman, L.: Toward a mathematical theory of self-assembly (extended abstract), Tech. Report 00-722, University of Southern California (2000)
3. Adleman, L., Cheng, Q., Goel, A., Huang, M.-D., Wasserman, H.: Linear self-assemblies: Equilibria, entropy and convergence rates. In: Sixth International Conference on Difference Equations and Applications. Taylor and Francis, Abington (2001)
4. Andersen, E.S., Dong, M., Nielsen, M.M., Jahn, K., Subramani, R., Mamdouh, W., Golas, M.M., Sander, B., Stark, H., Oliveira, C.L.P., Pedersen, J.S., Birkedal, V., Besenbacher, F., Gothelf, K.V., Kjems, J.: Self-assembly of a nanoscale dna box with a controllable lid. Nature 459(7243), 73–76 (2009)
5. Barish, R.D., Schulman, R., Rothemund, P.W., Winfree, E.: An information-bearing seed for nucleating algorithmic self-assembly. Proceedings of the National Academy of Sciences 106(15), 6054–6059 (2009)
6. Cheng, Q., Aggarwal, G., Goldwasser, M.H., Kao, M.-Y., Schweller, R.T., de Espanés, P.M.: Complexities for generalized models of self-assembly. SIAM Journal on Computing 34, 1493–1515 (2005)
7. Demaine, E.D., Demaine, M.L., Fekete, S.P., Ishaque, M., Rafalin, E., Schweller, R.T., Souvaine, D.L.: Staged self-assembly: nanomanufacture of arbitrary shapes with $O(1)$ glues. Natural Computing 7(3), 347–370 (2008)
8. Fu, Y., Schweller, R.: Temperature 1 self-assembly: Deterministic assembly in 3d and probabilistic assembly in 2d. In: Proceedings of the ACM-SIAM Symposium on Discrete Algorithms (SODA 2011) (to appear, 2011)
9. Gu, H., Chao, J., Xiao, S.-J., Seeman, N.C.: A proximity-based programmable dna nanoscale assembly line. Nature 465(7295), 202–205 (2010)
10. Hartgerink, J.D., Beniash, E., Stupp, S.I.: Self-Assembly and Mineralization of Peptide-Amphiphile Nanofibers. Science 294(5547), 1684–1688 (2001)
11. Kalsin, A.M., Fialkowski, M., Paszewski, M., Smoukov, S.K., Bishop, K.J.M., Grzybowski, B.A.: Electrostatic Self-Assembly of Binary Nanoparticle Crystals with a Diamond-Like Lattice. Science 312(5772), 420–424 (2006)

12. Luhrs, C.: Polyomino-safe DNA self-assembly via block replacement. In: Goel, A., Simmel, F.C., Sosík, P. (eds.) DNA14. LNCS, vol. 5347, pp. 112–126. Springer, Heidelberg (2008)
13. Lund, K., Manzo, A.J., Dabby, N., Michelotti, N., Johnson-Buck, A., Nangreave, J., Taylor, S., Pei, R., Stojanovic, M.N., Walter, N.G., Winfree, E., Yan, H.: Molecular robots guided by prescriptive landscapes. Nature 465(7295), 206–210 (2010)
14. Majumder, U., LaBean, T.H., Reif, J.H.: Activatable tiles for compact error-resilient directional assembly. In: 13th International Meeting on DNA Computing (DNA 13), Memphis, Tennessee, June 4-8 (2007)
15. Rothemund, P.W.K.: Folding DNA to create nanoscale shapes and patterns. Nature 440(7082), 297–302 (2006)
16. Rothemund, P.W.K., Winfree, E.: The program-size complexity of self-assembled squares (extended abstract). In: STOC 2000: Proceedings of the Thirty-second Annual ACM Symposium on Theory of Computing, pp. 459–468. ACM, New York (2000)
17. Rothemund, P.W.K., Papadakis, N., Winfree, E.: Algorithmic self-assembly of DNA Sierpinski triangles. PLoS Biology 2(12), 2041–2053 (2004)
18. Soloveichik, D., Winfree, E.: Complexity of self-assembled shapes. SIAM Journal on Computing 36(6), 1544–1569 (2007)
19. Tang, Z., Zhang, Z., Wang, Y., Glotzer, S.C., Kotov, N.A.: Self-Assembly of CdTe Nanocrystals into Free-Floating Sheets. Science 314(5797), 274–278 (2006)
20. Vitányi, P., Li, M.: An introduction to kolmogorov complexity and its applications. Springer, Heidelberg (1997)
21. Winfree, E.: Algorithmic self-assembly of DNA, Ph.D. thesis, California Institute of Technology (June 1998)
22. Winfree, E.: Simulations of computing by self-assembly, Tech. Report CaltechC-STR:1998.22, California Institute of Technology (1998)
23. Winfree, E.: Self-healing tile sets. In: Chen, J., Jonoska, N., Rozenberg, G. (eds.) Nanotechnology: Science and Computation. Natural Computing Series, pp. 55–78. Springer, Heidelberg (2006)
24. Winfree, E., Yang, X., Seeman, N.C.: Universal computation via self-assembly of dna: Some theory and experiments. In: DNA Based Computers II. DIMACS, vol. 44, pp. 191–213. American Mathematical Society, Providence (1996)
25. Yan, H., Park, S.H., Finkelstein, G., Reif, J.H., LaBean, T.H.: DNA-Templated Self-Assembly of Protein Arrays and Highly Conductive Nanowires. Science 301(5641), 1882–1884 (2003)
26. Zeng, H., Li, J., Liu, J.P., Wang, Z.L., Sun, S.: Exchange-coupled nanocomposite magnets by nanoparticle self-assembly. Nature 420(6914), 395–398 (2002)

Author Index

Printing: Mercedes-Druck, Berlin
Binding: Stein+Lehmann, Berlin